ELECTRONICS
A SYSTEMS APPROACH

PEARSON

At Pearson, we take learning personally. Our courses and resources are available as books, online and via multi-lingual packages, helping people learn whatever, wherever and however they choose.

We work with leading authors to develop the strongest learning experiences, bringing cutting-edge thinking and best learning practice to a global market. We craft our print and digital resources to do more to help learners not only understand their content, but to see it in action and apply what they learn, whether studying or at work.

Pearson is the world's leading learning company. Our portfolio includes Penguin, Dorling Kindersley, the Financial Times and our educational business, Pearson International. We are also a leading provider of electronic learning programmes and of test development, processing and scoring services to educational institutions, corporations and professional bodies around the world.

Every day our work helps learning flourish, and wherever learning flourishes, so do people.

To learn more please visit us at: **www.pearson.com/uk**

ELECTRONICS
A SYSTEMS APPROACH

Fifth Edition

Neil Storey
University of Warwick

PEARSON

Harlow, England • London • New York • Boston • San Francisco • Toronto • Sydney
Auckland • Singapore • Hong Kong • Tokyo • Seoul • Taipei • New Delhi
Cape Town • São Paulo • Mexico City • Madrid • Amsterdam • Munich • Paris • Milan

PEARSON EDUCATION LIMITED
Edinburgh Gate
Harlow CM20 2JE
United Kingdom
Tel: +44 (0)1279 623623
Web: www.pearson.com/uk

———————————————

First published 1992 (print)
Fifth edition published 2013 (print and electronic)

ISBN: 978-0-273-77327-6 (print)
     978-0-273-77864-6 (eText)

British Library Cataloguing-in-Publication Data
A catalogue record for the print edition is available from the British Library

Library of Congress Cataloging-in-Publication Data
A catalog record for the print edition is available from the Library of Congress

10 9 8 7 6 5 4 3 2 1
16 15 14 13

Print edition typeset in 10/11.5pt by 35
Print edition printed and bound in Malaysia

NOTE THAT ANY PAGE CROSS REFERENCES REFER TO THE PRINT EDITION

Brief Contents

Contents

Companion Website

For open-access **student resources** specifically written to complement this textbook and support your learning, please visit **www.pearsoned.co.uk/storey-elec**

ON THE WEBSITE

Preface

Electronics represents one of the most important, and rapidly changing, areas of engineering. It is used at the heart of a vast range of products that extends from mobile phones to computers, and from cars to nuclear power stations. For this reason, *all* engineers, scientists and technologists need a basic understanding of such systems, while many will require a far more detailed knowledge of this area.

When the first edition of this text was published it represented a very novel approach to the teaching of electronics. At that time most texts adopted a decidedly 'bottom-up' approach to the subject, starting by looking at semiconductor materials and working their way through diodes and transistors before eventually, several chapters later, looking at the uses of the circuits being considered. *Electronics: A Systems Approach* pioneered a new, 'top-down' approach to the teaching of electronics by explaining the uses and required characteristics of circuits, before embarking on detailed analysis. This aids comprehension and makes the process of learning much more interesting.

One of the great misconceptions concerning this approach is that it is in some way less rigorous in its treatment of the subject. A top-down approach does *not* define the depth to which a subject is studied but only the order and manner in which the material is presented. Many students *will* need to look in detail at the operation of electronic components and understand the physics of its materials; however, this will be more easily absorbed if the characteristics and uses of the components are understood first.

A great benefit of a top-down approach is that it makes the text more accessible for *all* its potential readers. For those who intend to specialise in electronic engineering the material is presented in a way that makes it easy to absorb, providing an excellent grounding for further study. For those intending to specialise in other areas of engineering or science, the order of presentation allows them to gain a good grounding in the *basics*, and to progress into the *detail* only as far as is appropriate for their needs.

While a top-down approach offers a very accessible route to understanding electronics, it is much more effective if one starts with a thorough understanding of the basic components used in such circuits. Some readers of this text will already be familiar with this material from previous study, while others will have little or no knowledge of such topics. For this reason, the book is divided into two parts. Part 1 provides an introduction to **Electrical Circuits and Components** and makes very few assumptions about prior knowledge. This section gives a gentle and well-structured introduction to this area, and readers can select from the various topics depending on their needs and interests. Part 2 then provides a thorough introduction to **Electronic Systems**, adopting the well-tried, top-down approach for which this text is renowned. The text therefore provides a comprehensive introduction to both Electrical and Electronic Engineering, making it appropriate for a wide range of first-level courses in areas such as **Electronic Engineering**, **Electrical Engineering** and **Electrical and Electronic Engineering**.

New in this edition

This fifth edition represents a significant expansion and an opportunity to update the text to take account of developments in a rapidly changing field. The most significant of these changes are as follows:

- A major new chapter on **Communications**, including both analogue and digital techniques.
- A new section on the **Design of Sequential Logic Circuits** within the chapter on Sequential Logic.
- A new section on **Implementing Complex Gates in CMOS** in the chapter on Digital Devices.
- A new chapter on **Electric Motors and Generators**.
- A new section on **Power Dissipation in Digital Systems** in the chapter on Digital Devices.
- Revised and extended treatment of topics such as **Timers, Microcomputer Programming** and **System on a Chip (SOC) Devices**.
- The inclusion of topics for **further study** at the end of each chapter, many based on more challenging, real-world, problems.

Video Tutorials

Also new within this edition is the provision of over a hundred supporting videos. These provide tutorial support for topics throughout the text and also aim to provide guidance and encourage creativity within the various 'further study' exercises. These videos are *not* hour-long 'lectures' covering broad-ranging themes, but are short, succinct tutorials, lasting only a few minutes, that describe in detail various aspects of design or analysis.

Video 0

Accessing the tutorials couldn't be easier. Icons of the form shown on the left are placed throughout the text to indicate the topics covered. If you are using a paper copy of the text, simply scan the QR code in the icon with your phone or laptop to go straight to the video. If you are using an e-book it's even easier, as the icon contains a direct link to the video. Alternatively, you can go to the companion website (see below) which gives a full list of all the videos available and provides direct access.

Who should read this text

This text is intended for undergraduate students in all fields of engineering and science. For students of electronics or electrical engineering it provides a first-level introduction to electronics that will give a sound basis for further study. For students of other disciplines it includes most of the electronics material that they will need throughout their course.

Assumed knowledge

The book assumes very little by way of prior knowledge, except for an understanding of the basic principles of physics and mathematics.

Companion website

The text is supported by a comprehensive companion website that will greatly increase both your understanding and your enjoyment of it. The site contains a range of support material, including computer-marked self-assessment exercises for each chapter. These exercises not only give you instant feedback on your understanding of the material, but also give useful guidance on areas of difficulty. The website also gives easy access to the numerous video tutorials mentioned above. To visit the site, go to www.pearsoned.co.uk/storey-elec.

Circuit simulation

Circuit simulation offers a powerful and simple means of gaining an insight into the operation of electronic circuits. Throughout the text there are numerous **Computer Simulation Exercises** that support the material in the text. These are marked by icons in the margin as shown on the left. The exercises may be performed using any circuit simulation package, although perhaps the most widely used are those produced by National Instruments and by Cadence. Both these packages are widely used within industry and within Universities and Colleges. Each comes as part of a suite of programs that provides schematic capture of circuits and the graphical display of simulation results.

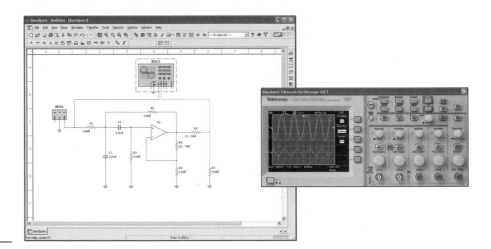

Circuit simulation using the National Instruments Multisim package

Many students will have access to simulation tools within their University or College, but you may also obtain the software for use on your own computer if you wish. For some packages demonstration versions may be downloaded free of charge from the manufacturer's website or obtained on a free CD. In other cases, low-cost student versions are available. Details of how to obtain your free or reduced-cost simulation software are given on the text's companion website.

To simplify the use of simulation as an aid to understanding the material within the text, a series of demonstration files for the National Instruments Multisim simulation packages can also be downloaded from the website. The name of the relevant demonstration file is given in the margin under the computer icon for the associated computer simulation exercise. Computer icons are also found next to some circuit

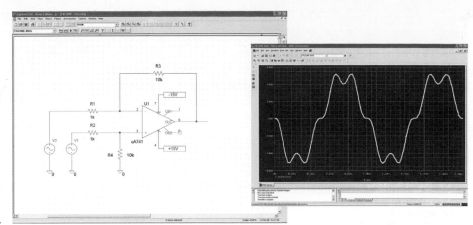

Circuit simulation using the Cadence OrCAD Capture CIS package

diagrams where simulation files have been provided to aid understanding of the operation of the circuit. The demonstration files come with full details of how to carry out the various simulation exercises.

Problems using simulation have also been included within the exercises at the end of the various chapters. These exercises do *not* have demonstration files and are set to develop and test the reader's understanding of the use of simulation as well as the circuits concerned.

To the instructor

A comprehensive set of online support material is available for instructors using this text as course reading. This includes a set of editable **PowerPoint slides** to aid in the preparation of lectures, plus an **Instructor's Manual** that gives **fully worked solutions** to all the numerical problems and **sample answers** for the various non-numerical exercises. Guidance is also given on **course preparation** and on the selection of topics to meet the needs of students with different backgrounds and interests. This material, together with the various online study aids, simulation exercises and self-assessment tests, should greatly assist both the instructor and the student to gain maximum benefit from courses based on this text. Instructors adopting this text should visit the companion website at www.pearsoned.co.uk/storey-elec for details of how to gain access to the secure website that holds the instructor's support material.

Guider Tour of the Book

Chapter 11 Electronic Systems

Objectives

When you have studied the material in this chapter, you should be able

- discuss the need for electronic systems in a wide range of applicatio
- describe the characteristics and advantages of a 'systems approach
  engineering
- identify the inputs and outputs of an engineering system and underst
  significance of the choice of system boundary
- explain the varied characteristics of physical quantities and the need
  represent these quantities by electrical signals
- use block diagrams to represent complex engineering systems.

11.1 Introduction

Having looked at a range of basic electrical components and circuits i
the text, we are now in a position to turn our attention to more comple
systems.

In recent years electronic systems have found their way into almost al
our lives. Such systems wake us in the morning; control the operation of
we drive to work; maintain a comfortable working environment in our
homes; allow us to communicate worldwide; provide access to informa

Objectives at the start of each chapter set out clearly what the student should have achieved by the end of the chapter.

Introduction at the start of each chapter sets the chapter in context, describes the content and indicates how to get the most out of the content.

Example 6.8

Calculate the output voltage v_o in the circuit of Figure 6.14(a) if $C = 20($
$R_1 = 5\,\Omega$, $L = 50$ mH, $R_2 = 50\,\Omega$ and the input voltage $v = 10\sin 500t$.

Since $v = 10\sin 500t$, $\omega = 500$ rad/s. We will take the input voltage as the refe
phase, so v corresponds to $10\angle 0$.

Our first task is to calculate the impedances corresponding to Z_1 and $Z
Figure 6.14(c):

$$Z_1 = R_1 - \mathrm{j}X_C$$
$$= R_1 - \mathrm{j}\frac{1}{\omega C}$$
$$= 5 - \mathrm{j}\frac{1}{500 \times 200 \times 10^{-6}}$$
$$= 5 - \mathrm{j}10\,\Omega$$

Examples occur frequently to show step by step how theory is applied.

Video Tutorials provide short expositions of an issue in the design or analysis of an electronic system. They are indicated by the 'play' icon and can be accessed by scanning the QR code with an appropriate device, or via the companion website.

Video 1A

1.8 Kirchhoff's laws

1.8.1 Current law

At any instant, the algebraic sum of all the currents flowing into any ju
circuit is zero:

$$\Sigma I = 0$$

A **junction** is any point where electrical paths meet. The law comes about
sideration of conservation of charge – the charge flowing into a point mus
flowing out.

1.8.2 Voltage law

At any instant, the algebraic sum of all the voltages around any loop in a cir

Further Study sections at the end of chapters provide a video tutorial, often based on a more-challenging real-world problem.

Further study

In this chapter we have looked at various aspects of electromagnetic compatibility. While EMC is of great importance in a wide range of applications, it is of particular significance in situations where failure of a system could have safety implications. Electronic systems within vehicles invariably come within this category.

Identify some of the safety-related systems within a modern, high-performance car and consider the EMC related factors that could affect their operation. What design measures could be taken to maximise the dependability of these systems in relation to

---

Computer simulation exercise 16.1

File 16A

Simulate the circuit in Example 16.1 using one of the operational amplifiers supported by your simulation package. Apply a 100 mV DC input to the circuit and measure the output voltage. Then, deduce the voltage gain of the circuit and confirm that this is as expected. Experiment with different values for the two resistors and see how this affects the voltage gain. Experiment with different values for the input voltage (including both positive and negative values) and confirm that the circuit behaves as you expect.

16.3.2 An inverting amplifier

The second of our standard circuits is that of an **inverting amplifier**. This is shown in Figure 16.6. As in the previous circuit, because the gain of the op-amp is infinite

Computer Simulation Exercises provide the opportunity to run exercises that simulate the operation of electrical circuits.

Key Points summarise the chapter contents and are useful in review and revision of learning.

Key points

- Electrical machines can be broadly divided into generators, which mechanical energy into electrical energy, and motors, which convert energy into mechanical energy.
- In most cases, generators can also function as motors, and vice versa
- Electrical machines can be divided into DC machines and AC machines
- All electrical machines operate through the interaction between a field and a set of windings.
- The rotation of a coil in a uniform magnetic field produces a sinusoi This principle is at the heart of an AC generator or alternator.
- A commutator can be used to convert the above sinusoidal e.m.f. into a form. This is the basis of a DC generator or dynamo.
- While the magnetic field in an electrical machine can be produ permanent magnet, it is more common to produce this electrically u

---

Exercises

16.1 What is meant by the term 'integrated circuit'?
16.2 Explain the acronyms DIL and SMT as applied to IC packages.
16.3 What are typical values for the positive and negative supply voltages of an operational amplifier?
16.4 Outline the characteristics of an 'ideal' op-amp.
16.5 Sketch an equivalent circuit of an ideal operational amplifier.
16.6 Determine the gain of the following circuit.

Exercises at the end of each chapter test the student's ability to apply the learning in the chapter to the solution of problems. Answers to numerical-type exercises are provided at the back of the book.

List of Videos

Acknowledgements

I would like to express my gratitude to the various people who have assisted in the preparation of this text. In particular I would like to thank my colleagues at the University of Warwick who have provided useful feedback and ideas. My thanks also go to Cliff Armstrong for providing a number of helpful suggestions.

I would also like to thank the companies who have given permission to reproduce their material: RS Components for the photographs in Figures 2.14, 2.15, 10.15, 12.2(a) and (b), 12.3(a) and (b), 12.4, 12.5(a) and (b), 12.6, 12.7, 12.9(a) and (b), 12.12, 13.1, 13.2, 13.3 and 13.4; Farnell Electronic Components Limited for the photographs in Figures 2.12, and 2.13; Texas Instruments for Figures 21.4(a) and (b), 21.6, 21.7(a) and (b), 26.22, 26.34; Lattice Semiconductor Corporation for Figures 27.11, 27.12, 27.13; Anachip Corporation for Figure 27.14; Microchip Technology for Figures 27.41 and 27.42. Certain materials herein are reprinted with the permission of Microchip Technology Incorporated. No further reprints or reproductions may be made of said materials without Microchip Technology Inc.'s prior written consent. Cadence Design Systems Inc. for permission to use the OrCAD Capture CIS images in the Preface; and National Instruments for permission to use the Multisim images in the Preface. Photos on pages 17, 38, 63, 80, 100, 121, 134, 165, 182, 198, 209, 229, 240, 259, 276, 304, 334, 382, 446, 474, 488, 512, 526, 581, 625, 683, 733, 743, 760 and 784 courtesy of Shutterstock; photo on page 797 courtesy of Fotolia.

Finally, I would like to give special thanks to my family for their help and support during the writing of this text. In particular I wish to thank my wife Jillian for her never-failing encouragement and understanding.

Trademarks

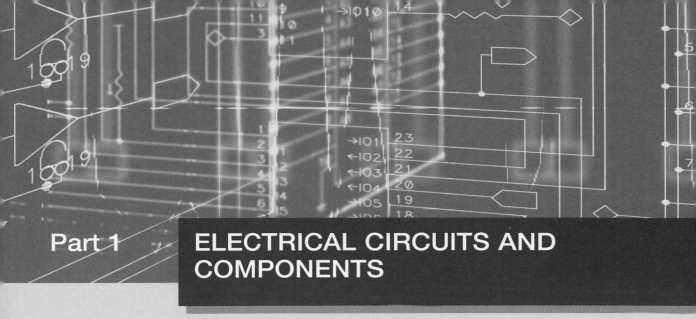

Part 1   ELECTRICAL CIRCUITS AND COMPONENTS

Chapter 1 Basic Electrical Circuits and Components

Objectives

When you have studied the material in this chapter, you should be able to:

- give the Système International (SI) units for a range of electrical quantities
- use a range of common prefixes to represent multiples of these units
- describe the basic characteristics of resistors, capacitors and inductors
- apply Ohm's law, and Kirchhoff's voltage and current laws, to simple electrical circuits
- calculate the effective resistance of resistors in series or in parallel, and analyse simple resistive potential divider circuits
- define the terms 'frequency' and 'period' as they apply to sinusoidal quantities
- draw the circuit symbols for a range of common electrical components.

1.1 Introduction

While the title of this text refers to 'electronic systems', this first part refers to 'electrical circuits and components' and it is perhaps appropriate to start by explaining what is meant by the terms 'electronic' and 'electrical' in this context. Both terms relate to the use of electrical energy, but *electrical* is often used to refer to circuits that use only simple passive components such as resistors, capacitors and inductors, while the term *electronic* implies circuits that also use more sophisticated components such as transistors or integrated circuits. Therefore, before looking in detail at the operation of electronic systems, we need to have a basic understanding of the world of electrical engineering, since the components and circuits of this domain also form the basis of more sophisticated electronic applications.

Unfortunately, while this use of the words electrical and electronic is common, it is not universal. Engineers sometimes use the term *electrical* when describing applications associated with the generation, transmission or use of large amounts of electrical energy, and use *electronic* when describing applications associated with smaller amounts of power, where the electrical energy is used to convey information rather than as a source of power. For this reason, within this text we will be fairly liberal with our use of the two terms, since much of the material covered is relevant to all forms of electrical and electronic systems.

Most readers will have met the basic concepts of electrical circuits long before embarking on study at this level, and later chapters will assume that the reader is familiar with this elementary material. In the chapters that follow we will look at these basic concepts in more detail and extend them to give a greater understanding of the behaviour of the circuits and systems that we will be studying. However, it is essential that you are familiar with some elementary material before continuing.

The list below gives an indication of the topics that you should be familiar with before reading the following chapters:

- The Système International (SI) units for quantities such as energy, power, temperature, frequency, charge, potential, resistance, capacitance and inductance. You should also know the symbols used for these units.
- The prefixes used to represent common multiples of these units and their symbols (for example, 1 kilometre = 1 km = 1000 metres).
- Electrical circuits and quantities such as charge, e.m.f. and potential difference.
- Direct and alternating currents.
- The basic characteristics of resistors, capacitors and inductors.
- Ohm's law, Kirchhoff's laws and power dissipation in resistors.
- The effective resistance of resistors in series and parallel.
- The operation of resistive potential dividers.
- The terms used to describe sinusoidal quantities.
- The circuit symbols used for resistors, capacitors, inductors, voltage sources and other common components.

If, having read through the list above, you are confident that you are familiar with all these topics you can move on immediately to Chapter 2. However, just in case there are a few areas that might need some reinforcement, the remainder of this chapter provides what might be seen as a *revision* section on this material. This does not aim to give a detailed treatment of these topics (where appropriate this will be given in later chapters) but simply explains them in sufficient detail to allow an understanding of the early parts of the text.

In this chapter, worked examples are used to illustrate several of the concepts involved. One way of assessing your understanding of the various topics is to look quickly through these examples to see if you can perform the calculations involved, before looking at the worked solutions. Most readers will find the early examples trivial, but experience shows that many will feel less confident about some of the later topics, such as those related to **potential dividers**. These are very important topics, and a clear understanding of these circuits will make it much easier to understand the remainder of the book.

The exercises at the end of this chapter are included to allow you to test your understanding of the 'assumed knowledge' listed above. If you can perform these exercises easily you should have no problems with the technical content of the next few chapters. If not, you would be well advised to invest a little time in looking at the relevant sections of this chapter before continuing.

1.2 Système International units

The Système International (SI) d'Unités (International System of Units) defines units for a large number of physical quantities but, fortunately for our current studies, we need very few of them. These are shown in Table 1.1. In later chapters, we will introduce additional units as necessary, and Appendix B gives a more comprehensive list of units relevant to electrical and electronic engineering.

Table 1.1 Some important units.

Quantity	Quantity symbol	Unit	Unit symbol
Capacitance	C	farad	F
Charge	Q	coulomb	C
Current	I	ampere	A
Electromotive force	E	volt	V
Frequency	f	hertz	Hz
Inductance (self)	L	henry	H
Period	T	second	s
Potential difference	V	volt	V
Power	P	watt	W
Resistance	R	ohm	Ω
Temperature	T	kelvin	K
Time	t	second	s

1.3 Common prefixes

Table 1.2 lists the most commonly used unit prefixes. These will suffice for most purposes although a more extensive list is given in Appendix B.

Table 1.2 Common unit prefixes.

Prefix	Name	Meaning (multiply by)
T	tera	10^{12}
G	giga	10^{9}
M	mega	10^{6}
k	kilo	10^{3}
m	milli	10^{-3}
μ	micro	10^{-6}
n	nano	10^{-9}
p	pico	10^{-12}

1.4 Electrical circuits

1.4.1 Electric charge

Charge is an amount of electrical energy and can be either positive or negative. In atoms, protons have a positive charge and electrons have an equal negative charge. While protons are fixed within the atomic nucleus, electrons are often weakly bound and may therefore be able to move. If a body or region develops an excess of electrons it will have an overall negative charge, while a region with a deficit of electrons will have a positive charge.

1.4.2 Electric current

An electric current is a flow of electric charge, which in most cases is a flow of electrons. Conventional current is defined as a flow of electricity from a positive to a negative region. This conventional current is in the opposite direction to the flow of the negatively charged electrons. The unit of current is the **ampere** or **amp** (A).

1.4.3 Current flow in a circuit

A sustained electric current requires a complete circuit for the recirculation of electrons. It also requires some stimulus to cause the electrons to flow around this circuit.

1.4.4 Electromotive force and potential difference

The stimulus that causes an electric current to flow around a circuit is termed an electromotive force or e.m.f. The e.m.f. represents the energy introduced into the circuit by a source such as a battery or a generator. The circuit or component in which the current is induced is sometimes called a load.

The energy transferred from the source to the load results in a change in the electrical potential at each point in the load. Between any two points in the load there will exist a certain potential difference, which represents the energy associated with the passage of a unit of charge from one point to the other.

Both e.m.f. and potential difference are expressed in units of **volts**, and clearly these two quantities are related. Figure 1.1 illustrates the relationship between them: an e.m.f. is a quantity that produces an electric current, while a potential difference is the effect on the circuit of this passage of energy.

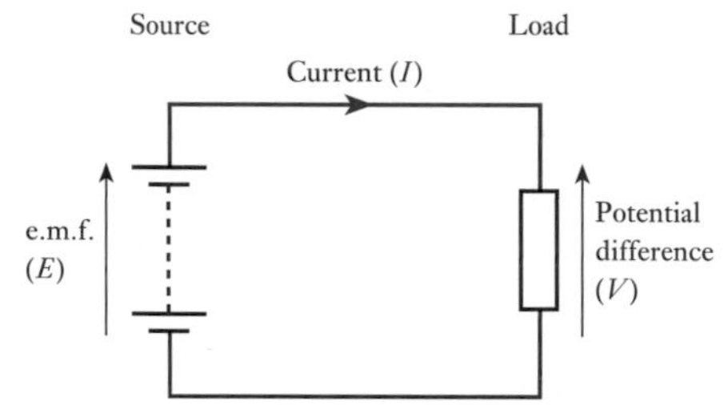

Figure 1.1
Electromotive force
and potential
difference.

If you have difficulty visualising an e.m.f., a potential difference, a resistance or a current, you may find it helpful to use an analogy. Consider, for example, the arrangement shown in Figure 1.2. Here a water pump forces water to flow around a series of pipes and through some form of restriction. While no analogy is perfect, this model illustrates the basic properties of the circuit of Figure 1.1. In the water-based diagram, the *water pump* forces water around the arrangement and is equivalent to the *voltage source* (or battery), which pushes electric charge around the corresponding electrical circuit. The flow of water through the pipe corresponds to the flow of charge around the circuit and therefore the *flow rate* represents the *current* in the circuit. The *restriction* within the pipe opposes the flow of water and is equivalent to the *resistance* of the electrical circuit. As water flows through the restriction the pressure will fall,

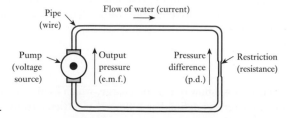

Figure 1.2
A water-based analogy
of an electrical circuit.

creating a *pressure difference* across it. This is equivalent to the *potential difference* across the resistance within the electrical circuit. The flow rate of the water will increase with the output pressure of the pump and decrease with the level of restriction present. This is analogous to the behaviour of the electrical circuit, where the current increases with the e.m.f. of the voltage source and decreases with the magnitude of the resistance.

1.4.5 Voltage reference points

Electromotive forces and potential differences in circuits produce different potentials (or voltages) at different points in the circuit. It is normal to describe the voltages throughout a circuit by giving the potential at particular points with respect to a single reference point. This reference is often called the **ground** or **earth** of the circuit. Since voltages at points in the circuit are measured with respect to ground, it follows that the voltage on the ground itself is zero. Therefore, the ground is also called the **zero–volts line** of the circuit.

In a circuit, a particular point or junction may be taken as the zero–volt reference and this may then be labelled as 0 V, as shown in Figure 1.3(a). Alternatively, the ground point of the circuit may be indicated using the ground symbol, as shown in Figure 1.3(b).

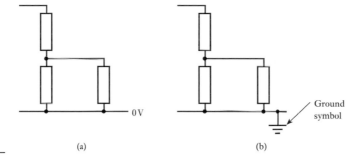

Figure 1.3
Indicating voltage
reference points.

1.4.6 Representing voltages in circuit diagrams

Conventions for representing voltages in circuit diagrams vary considerably between countries. In the UK, and in this text, it is common to indicate a potential difference by an arrow, which is taken to represent the voltage at the head of the arrow with respect to that at the tail. This is illustrated in Figure 1.4(a). In many cases, the tail of the arrow will correspond to the zero–volt line of the circuit (as shown in V_A in the figure). However, it can indicate a voltage difference between any two points in the circuit (as shown by V_B).

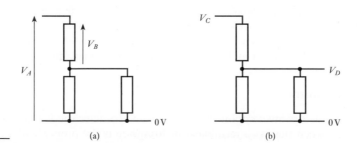

Figure 1.4
Indicating voltages in
circuit diagrams.

In some cases, it is inconvenient to use arrows to indicate voltages in circuits and simple labels are used instead, as shown in Figure 1.4(b). Here the labels V_C and V_D represent the voltage at the corresponding points *with respect to ground* (that is, with respect to the zero–volt reference).

1.4.7 Representing currents in circuit diagrams

Currents in circuit diagrams are conventionally indicated by an arrow in the direction of the *conventional* current flow (that is, in the opposite direction to the flow of electrons). This was illustrated in Figure 1.1. This figure also shows that for positive voltages and currents the arrow for the current flowing out of a voltage source is in the *same direction* as the arrow representing its e.m.f. However, the arrow representing the current in a resistor is in the *opposite direction* to the arrow representing the potential difference across it.

1.5   Direct current and alternating current

The currents associated with electrical circuits may be constant or may vary with time. Where currents vary with time they may also be unidirectional or alternating.

When the current in a conductor always flows in the same direction this is described as a direct current (DC). Such currents will often be associated with voltages of a single polarity. Where the direction of the current periodically changes, this is referred to as alternating current (AC), and such currents will often be associated with alternating voltages. One of the most common forms of alternating waveform is the sine wave, as discussed in Section 1.13.

1.6   Resistors, capacitors and inductors

1.6.1 Resistors

Resistors are components whose main characteristic is that they provide resistance between their two electrical terminals. The **resistance** of a circuit represents its opposition to the flow of electric current. The unit of resistance is the **ohm (Ω)**. One may also define the **conductance** of a circuit as its ability to *allow* the flow of electricity. The conductance of a circuit is equal to the reciprocal of its resistance and has the units of **siemens (S)**. We will look at resistance in some detail later in the text (see Chapter 3).

1.6.2 Capacitors

Capacitors are components whose main characteristic is that they exhibit capacitance between their two terminals. **Capacitance** is a property of two conductors that are electrically insulated from each other, whereby electrical energy is stored when a potential difference exists between them. This energy is stored in an electric field that is created between the two conductors. Capacitance is measured in **farads (F)**, and we will return to look at capacitance in more detail later in the text (see Chapter 4).

1.6.3 Inductors

Inductors are components whose main characteristic is that they exhibit inductance between their two terminals. **Inductance** is the property of a coil that results in an e.m.f. being induced in the coil as a result of a change in the current in the coil. Like capacitors, inductors can store electrical energy and in this case it is stored in a magnetic field. The unit of inductance is the **henry (H)**, and we will look at inductance later in the text (see Chapter 5).

1.7 Ohm's law

Ohm's law states that the current I flowing in a conductor is directly proportional to the applied voltage V and inversely proportional to its resistance R. This determines the relationship between the units for current, voltage and resistance, and the **ohm** is defined as the resistance of a circuit in which a current of 1 **amp** produces a potential difference of 1 **volt**.

The relationship between voltage, current and resistance can be represented in a number of ways, including:

$$V = IR \tag{1.1}$$

$$I = \frac{V}{R} \tag{1.2}$$

$$R = \frac{V}{I} \tag{1.3}$$

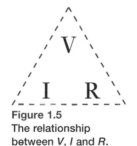

Figure 1.5
The relationship between V, I and R.

A simple way of remembering these three equations is to use the 'virtual triangle' of Figure 1.5. The triangle is referred to as 'virtual' simply as a way of remembering the order of the letters. Taking the first three letters of VIRtual and writing them in a triangle (starting at the top) gives the arrangement shown in the figure. If you place your finger on one of the letters, the remaining two show the expression for the selected quantity. For example, to find the expression for 'V' put your finger on the V and you see I next to R, so $V = IR$. Alternatively, to find the expression for 'I' put your finger on the I and you are left with V above R, so $I = V/R$. Similarly, covering 'R' leaves V over I, so $R = V/I$.

Example 1.1

Voltage measurements (with respect to ground) on part of an electrical circuit give the values shown in the diagram below. If the resistance of R_2 is 220 Ω, what is the current I flowing through this resistor?

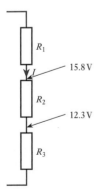

From the two voltage measurements, it is clear that the voltage difference across the resistor is $15.8 - 12.3 = 3.5\,\text{V}$. Therefore using the relationship

$$I = \frac{V}{R}$$

we have

$$I = \frac{3.5\,\text{V}}{220\,\Omega} = 15.9\,\text{mA}$$

Video 1A

1.8 Kirchhoff's laws

1.8.1 Current law

At any instant, the algebraic sum of all the currents flowing into any junction in a circuit is zero:

$$\sum I = 0 \tag{1.4}$$

A **junction** is any point where electrical paths meet. The law comes about from consideration of conservation of charge – the charge flowing into a point must equal that flowing out.

1.8.2 Voltage law

At any instant, the algebraic sum of all the voltages around any loop in a circuit is zero:

$$\sum V = 0 \tag{1.5}$$

The term **loop** refers to any continuous path around the circuit, and the law comes about from consideration of conservation of energy.

With both laws, it is important that the various quantities are assigned the correct sign. When summing currents, those flowing *into* a junction are given the opposite polarity to those flowing *out* from it. Similarly, when summing the voltages around a loop, *clockwise* voltages will be assigned the opposite polarity to *anticlockwise* ones.

Example 1.2 Use Kirchhoff's current law to determine the current I_2 in the following circuit.

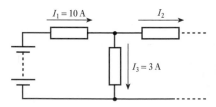

From Kirchhoff's current law

$$I_2 = I_1 - I_3$$
$$= 10 - 3$$
$$= 7\,\text{A}$$

Example 1.3 | Use Kirchhoff's voltage law to determine the magnitude of V_1 in the following circuit.

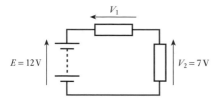

From Kirchhoff's voltage law (summing the voltages clockwise around the loop)

$$E - V_1 - V_2 = 0$$

or, rearranging,

$$V_1 = E - V_2$$
$$= 12 - 7$$
$$= 5\,\text{V}$$

1.9 Power dissipation in resistors

The instantaneous power dissipation P of a resistor is given by the product of the voltage across the resistor and the current passing through it. Combining this result with Ohm's law gives a range of expressions for P. These are

$$P = VI \tag{1.6}$$

$$P = I^2 R \tag{1.7}$$

$$P = \frac{V^2}{R} \tag{1.8}$$

Example 1.4 | Determine the power dissipation in the resistor R_3 in the following circuit.

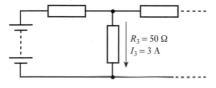

From Equation 1.7

$$P = I^2 R$$
$$= 3^2 \times 50$$
$$= 450\,\text{W}$$

1.10	**Resistors in series**

The effective resistance of a number of resistors in series is equal to the sum of their individual resistances:

$$R = R_1 + R_2 + R_3 + \cdots + R_n \tag{1.9}$$

For example, for the three resistors shown in Figure 1.6 the total resistance R is given by

$$R = R_1 + R_2 + R_3$$

Figure 1.6
Three resistors in series.

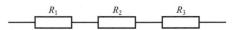

Example 1.5 Determine the equivalent resistance of the following combination.

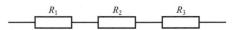

From above

$$R = R_1 + R_2 + R_3 + R_4$$
$$= 10 + 20 + 15 + 25$$
$$= 70 \ \Omega$$

1.11	**Resistors in parallel**

The effective resistance of a number of resistors in parallel is given by the following expression:

$$\frac{1}{R} = \frac{1}{R_1} + \frac{1}{R_2} + \frac{1}{R_3} + \cdots + \frac{1}{R_n} \tag{1.10}$$

For example, for the three resistors shown in Figure 1.7 the total resistance R is given by

$$\frac{1}{R} = \frac{1}{R_1} + \frac{1}{R_2} + \frac{1}{R_3}$$

Figure 1.7
Three resistors in parallel.

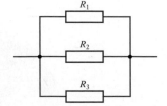

Example 1.6 | Determine the equivalent resistance of the following combination.

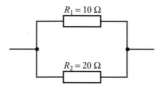

From above

$$\frac{1}{R} = \frac{1}{R_1} + \frac{1}{R_2}$$

$$= \frac{1}{10} + \frac{1}{20}$$

$$= \frac{3}{20}$$

$$\therefore R = \frac{20}{3} = 6.67\ \Omega$$

Note that the effective resistance of a number of resistors in parallel will always be less than that of the lowest-value resistor.

Video 1B

1.12 ## Resistive potential dividers

When several resistors are connected in series the current flowing through each resistor is the same. The magnitude of this current is given by the voltage divided by the total resistance. For example, if we connect three resistors in series, as shown in Figure 1.8, the current is given by

$$I = \frac{V}{R_1 + R_2 + R_3}$$

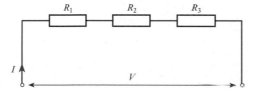

Figure 1.8
A resistive potential divider.

The voltage across each resistor is then given by this current multiplied by its resistance. For example, the voltage V_1 across resistor R_1 will be given by

$$V_1 = IR_1 = \left(\frac{V}{R_1 + R_2 + R_3}\right)R_1 = V\left(\frac{R_1}{R_1 + R_2 + R_3}\right)$$

Therefore, the *fraction* of the total voltage across each resistor is equal to its *fraction* of the total resistance, as shown in Figure 1.9, where

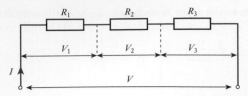

Figure 1.9
The division of voltages in a potential divider.

$$\frac{V_1}{V} = \frac{R_1}{R_1 + R_2 + R_3} \quad \frac{V_2}{V} = \frac{R_2}{R_1 + R_2 + R_3} \quad \frac{V_3}{V} = \frac{R_3}{R_1 + R_2 + R_3}$$

or, rearranging,

$$V_1 = V\frac{R_1}{R_1 + R_2 + R_3} \quad V_2 = V\frac{R_2}{R_1 + R_2 + R_3} \quad V_3 = V\frac{R_3}{R_1 + R_2 + R_3}$$

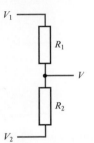

Figure 1.10
A simple potential divider.

To calculate the voltage at a point in a chain of resistors, one must determine the voltage across the complete chain, calculate the voltage across those resistors between that point and one end of the chain and add this to the voltage at that end of the chain. For example, in Figure 1.10

$$V = V_2 + (V_1 - V_2)\frac{R_2}{R_1 + R_2} \tag{1.11}$$

Example 1.7 | Determine the voltage V in the following circuit.

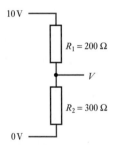

As described above, we first determine the voltage across the chain (by subtracting the voltages at either end of the chain). Then we calculate the voltage across the relevant resistor and add this to the voltage at the appropriate end of the chain.

In this case one end of the chain of resistors is at zero volts, so the calculation is very straightforward. The voltage across the chain is 10 V, and V is simply the voltage across R_2, which is given by

$$V = 10\frac{R_2}{R_1 + R_2}$$

$$= 10\frac{300}{200 + 300}$$

$$= 6\,V$$

Note that a common mistake in such calculations is to calculate $R_1/(R_1 + R_2)$, rather than $R_2/(R_1 + R_2)$. The value used as the numerator in this expression represents the resistor across which the voltage is to be calculated.

Potentiometer calculations are slightly more complicated where neither end of the chain of resistors is at zero volts.

Example 1.8 Determine the voltage V in the following circuit.

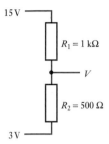

Again, we first determine the voltage across the chain (by subtracting the voltages at either end of the chain). Then we calculate the voltage across the relevant resistor and add this to the voltage at the appropriate end of the chain. Therefore

$$V = 3 + (15 - 3)\frac{R_2}{R_1 + R_2}$$
$$= 3 + 12\,\frac{500}{1000 + 500}$$
$$= 3 + 4$$
$$= 7\,\text{V}$$

In this case we pick one end of the chain of resistors as our reference point (we picked the lower end) and calculate the voltage on the output with respect to this point. We then add to this calculated value the voltage at the reference point.

1.13 Sinusoidal quantities

Sinusoidal quantities have a magnitude that varies with time in a manner described by the **sine** function. The variation of any quantity with time can be described by drawing its **waveform**. The waveform of a sinusoidal quantity is shown in Figure 1.11. The length of time between corresponding points in successive cycles of the waveform is termed its **period**, which is given the symbol T. The number of cycles of the waveform within 1 second is termed its **frequency**, which is usually given the symbol f.

The frequency of a waveform is related to its period by the expression

$$f = \frac{1}{T}$$

(1.12)

Magnitude

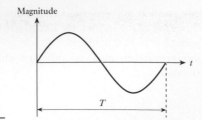

Figure 1.11
A sine wave.

Example 1.9 What is the period of a sinusoidal quantity with a frequency of 50 Hz?
From above we know that

$$f = \frac{1}{T}$$

and therefore the period is given by

$$T = \frac{1}{f} = \frac{1}{50} = 0.02\,\text{s} = 20\,\text{ms}$$

1.14 Circuit symbols

The following are circuit symbols for a few basic electrical components.

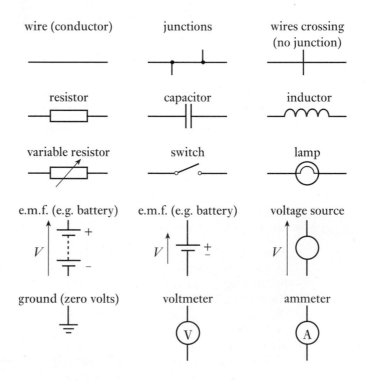

In later chapters we will meet a number of additional component symbols, but these
are sufficient for our current needs.

Video 1C

Further study

So far in this chapter we have concentrated on analysing the behaviour of given circuits. While this is a useful skill, often our task is to *design* a circuit to provide a particular functionality. To test your comprehension of the material in this chapter, attempt to design an arrangement to meet the following requirement.

The circuit below shows a component used within an automotive application. The module is connected directly across the car's 12 V battery and has two functions. The first is to act as a light source and R_1 represents the resistance of a built-in light bulb. The second is to produce a definable output voltage that is determined by the ratio of R_2 and R_3. For the purposes of this exercise you may assume that negligible current is drawn by whatever circuitry is connected to this output. The resistor R_3 is sealed within the unit and cannot be altered. However, R_2 is external to the unit and can be chosen to produce the required output voltage. What value of R_2 is required to produce an output voltage of 8 V?

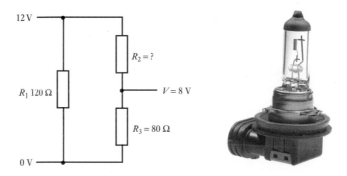

It is now required to use the same element in another application where the supply voltage is 18 V, but where the output voltage is still required to be 8 V. Since the sealed unit (and in particular the light bulb) is designed for use within cars, it has a maximum operating voltage of 12 V. Devise an arrangement to allow the sealed unit to be used in this arrangement and determine the total power consumption of your circuit – including the sealed unit.

Key points

Since this chapter introduces no new material there are very few key points. However, the importance of a good understanding of this 'assumed knowledge' encourages me to emphasise the following:

- Understanding the next few chapters relies on understanding the various topics covered in this chapter.

- A clear concept of voltage and current is essential for all readers.

- Ohm's law and Kirchhoff's voltage and current laws are used extensively in later chapters.

- Experience shows that students have most problems with potential dividers – a topic that is widely used in these early chapters. You are therefore advised to make very sure that you are happy with this material before continuing.

Exercises

1.1 Give the prefixes used to denote the following powers: 10^{-12}; 10^{-9}; 10^{-6}; 10^{-3}; 10^{3}; 10^{6}; 10^{9}; 10^{12}.

1.2 Explain the difference between 1 ms, 1 m/s and 1 mS.

1.3 Explain the difference between 1 mΩ and 1 MΩ.

1.4 If a resistor of 1 kΩ has a voltage of 5 V across it, what is the current flowing through it?

1.5 A resistor has 9 V across it and a current of 1.5 mA flowing through it. What is its resistance?

1.6 A resistor of 25 Ω has a voltage of 25 V across it. What power is being dissipated by the resistor?

1.7 If a 400 Ω resistor has a current of 5 μA flowing through it, what power is being dissipated by the resistor?

1.8 What is the effective resistance of a 20 Ω resistor in series with a 30 Ω resistor?

1.9 What is the effective resistance of a 20 Ω resistor in parallel with a 30 Ω resistor?

1.10 What is the effective resistance of a series combination of a 1 kΩ resistor, a 2.2 kΩ resistor and a 4.7 kΩ resistor?

1.11 What is the effective resistance of a parallel combination of a 1 kΩ resistor, a 2.2 kΩ resistor and a 4.7 kΩ resistor?

1.12 Calculate the effective resistance between the terminals A and B in the following arrangements.

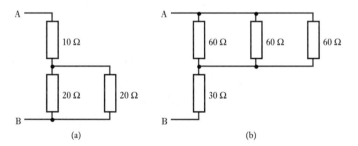

1.13 Calculate the effective resistance between the terminals A and B in the following arrangements.

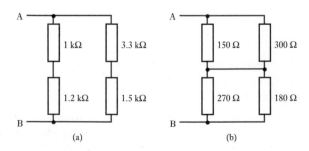

1.14 Calculate the voltages V_1, V_2 and V_3 in the following arrangements.

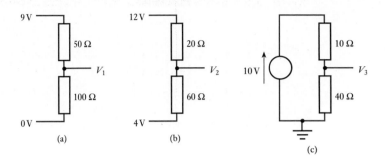

1.15 Calculate the voltages V_1, V_2 and V_3 in the following arrangements.

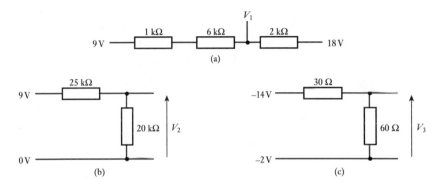

1.16 A sinusoidal quantity has a frequency of 1 kHz. What is its period?

1.17 A sinusoidal quantity has a period of 20 μs. What is its frequency?

Chapter 2   Measurement of Voltages and Currents

Objectives

When you have studied the material in this chapter, you should be able to:

- describe several forms of alternating waveform, such as sine waves, square waves and triangular waves
- define terms such as peak value, peak-to-peak value, average value and r.m.s. value as they apply to alternating waveforms
- convert between these various values for both sine waves and square waves
- write equations for sine waves to represent their amplitude, frequency and phase angle
- configure moving-coil meters to measure currents or voltages within a given range
- describe the problems associated with measuring non-sinusoidal alternating quantities using analogue meters and explain how to overcome these problems
- explain the operation of digital multimeters and describe their basic characteristics
- discuss the use of oscilloscopes in displaying waveforms and measuring parameters such as phase shift.

2.1   Introduction

In the previous chapter we looked at a range of electrical components and noted their properties and characteristics. An understanding of the operation of these components will assist you in later chapters as we move on to analyse the behaviour of electronic circuits in more detail. In order to do this, we need first to look at the measurement of voltages and currents in electrical circuits, and in particular at the measurement of alternating quantities.

Alternating currents and voltages vary with time and periodically change their direction. Figure 2.1 shows examples of some alternating waveforms. Of these, by far the most

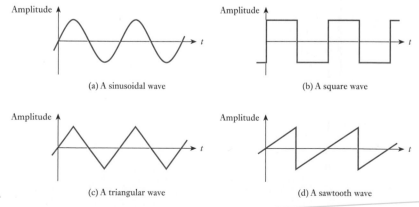

Figure 2.1
Examples of alternating waveforms.

(a) A sinusoidal wave

(b) A square wave

(c) A triangular wave

(d) A sawtooth wave

important is the **sinusoidal** waveform or **sine wave**. Indeed, in many cases, when engineers use the terms 'alternating current' or 'alternating voltage' they are referring to a sinusoidal quantity. Since sine waves are so widely used, it is important that we understand the nature of these waveforms and the ways in which their properties are defined.

Video 2A

2.2 Sine waves

In Chapter 1, we noted that the length of time between corresponding points in successive cycles of a sinusoidal waveform is termed its **period** T and that the number of cycles of the waveform within 1 s is termed its **frequency** f. The frequency of a waveform is related to its period by the expression

$$f = \frac{1}{T}$$

The maximum amplitude of the waveform is termed its **peak** value, and the difference between the maximum positive and maximum negative values is termed its **peak-to-peak** value. Because of the waveform's symmetrical nature, the peak-to-peak value is twice the peak value.

Figure 2.2 shows an example of a sinusoidal voltage signal. This illustrates that the period T can be measured between any convenient corresponding points in successive cycles of the waveform. It also shows the peak voltage V_p and the peak-to-peak voltage V_{pk-pk}. A similar waveform could be plotted for a sinusoidal *current* waveform indicating its peak current I_p and peak-to-peak current I_{pk-pk}.

Figure 2.2
A sinusoidal voltage signal.

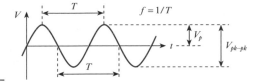

Example 2.1 | Determine the period, frequency, peak voltage and peak-to-peak voltage of the following waveform.

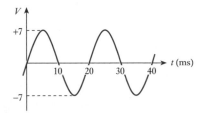

From the diagram the period is 20 ms or 0.02 s, so the frequency is 1/.02 = 50 Hz. The peak voltage is 7 V and the peak-to-peak voltage is therefore 14 V.

2.2.1 Instantaneous value

The shape of a sine wave is defined by the sine mathematical function. Thus we can describe such a waveform by the expression

$$y = A \sin \theta$$

where y is the value of the waveform at a particular point on the curve, A is the peak value of the waveform and θ is the angle corresponding to that point. It is conventional to use lower-case letters for time-varying quantities (such as y in the above equation) and upper-case letters for fixed quantities (such as A).

In the voltage waveform of Figure 2.2, the peak value of the waveform is V_p, so this waveform could be represented by the expression

$$v = V_p \sin \theta$$

One complete cycle of the waveform corresponds to the angle θ going through one complete cycle. This corresponds to θ changing by 360°, or 2π radians. Figure 2.3 illustrates the relationship between angle and magnitude for a sine wave.

Figure 2.3
Relationship between instantaneous value and angle for a sine wave.

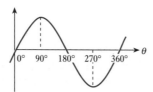
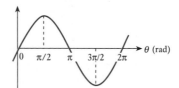

2.2.2 Angular frequency

The frequency f of a waveform (in hertz) is a measure of the number of cycles of that waveform that pass within 1 s. Each cycle corresponds to 2π radians, and it follows that there will be $2\pi f$ radians per second. The number of radians per second is termed the **angular frequency** of the waveform and is given the symbol ω. Therefore

$$\omega = 2\pi f \text{ rad/s} \tag{2.1}$$

2.2.3 Equation of a sine wave

The angular frequency can be thought of as the rate at which the angle of the sine wave changes. Therefore, the phase angle at a particular point in the waveform, θ, is given by

$$\theta = \omega t \text{ rad}$$

Thus our earlier expression for a sine wave becomes

$$y = A \sin \theta$$
$$= A \sin \omega t$$

and the equation of a sinusoidal voltage waveform becomes

$$v = V_p \sin \omega t \tag{2.2}$$

or

$$v = V_p \sin 2\pi f t \tag{2.3}$$

A sinusoidal current waveform might be described by the equation

$$i = I_p \sin \omega t \tag{2.4}$$

or

$$i = I_p \sin 2\pi ft \qquad (2.5)$$

Example 2.2 | Determine the equation of the following voltage signal.

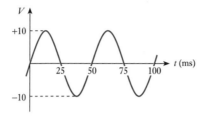

From the diagram the period is 50 ms or 0.05 s, so the frequency is 1/0.05 – 20 Hz. The peak voltage is 10 V. Therefore, from Equation 2.3

$$v = V_p \sin 2\pi ft$$
$$= 10 \sin 2\pi 20 t$$
$$= 10 \sin 126 t$$

2.2.4 Phase angles

The expressions of Equations 2.2 to 2.5 assume that the angle of the sine wave is zero at the origin of the time measurement ($t = 0$) as in the waveform of Figure 2.2. If this is not the case, then the equation is modified by adding the angle at $t = 0$. This gives an equation of the form

$$y = A \sin(\omega t + \phi) \qquad (2.6)$$

where ϕ is the phase angle of the waveform at $t = 0$. It should be noted that at $t = 0$ the term ωt is zero, so $y = A \sin \phi$. This is illustrated in Figure 2.4.

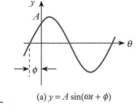 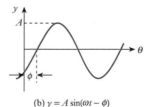

Figure 2.4
The effects of phase angles.

(a) $y = A \sin(\omega t + \phi)$ (b) $y = A \sin(\omega t - \phi)$

Example 2.3 | Determine the equation of the following voltage signal.

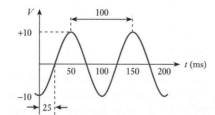

In this example, the period is 100 ms or 0.1 s, so the frequency is $1/0.1 = 10$ Hz. The peak voltage is 10 V. Here the point corresponding to zero degrees of the sine wave occurs at $t = 25$ ms, so at $t = 0$ the phase angle (ϕ) is given by $-25/100 \times 360° = -90°$ (or $\pi/2$ rad). Therefore

$$v = V_p \sin(2\pi ft + \phi)$$
$$= 10 \sin(2\pi 10t + \phi)$$
$$= 10 \sin(63t - \pi/2)$$

2.2.5 Phase differences

Two waveforms of the same frequency may have a constant **phase difference** between them, as shown in Figure 2.5. In this case, we will often say that one waveform is **phase shifted** with respect to the other. To describe the phase relationship between the two, we often take one of the waveforms as our reference and describe the way in which the other *leads* or *lags* this waveform. In Figure 2.5, waveform A has been taken as the reference in each case. In Figure 2.5(a), waveform B reaches its maximum value some time *after* waveform A. We therefore say that B lags A. In this example B lags A by 90°. In Figure 2.5(b), waveform B reaches its maximum value *before* waveform A. Here B leads A by 90°. In the figure the phase angles are shown in degrees, but they could equally well be expressed in radians.

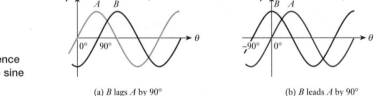

Figure 2.5
Phase difference between two sine waves.

(a) B lags A by 90°   (b) B leads A by 90°

It should be noted that the way in which the phase relationship is expressed is a matter of choice. For example, if A leads B by 90°, then clearly B lags A by 90°. These two statements are equivalent, and the one used will depend on the situation and personal preference.

Figure 2.5 illustrates phase difference using two waveforms of the same magnitude, but this is not a requirement. Phase difference can be measured between any two waveforms of the same frequency, regardless of their relative size. We will consider methods of measuring phase difference later in this chapter.

2.2.6 Average value of a sine wave

Clearly, if one measures the average value of a sine wave over one (or more) complete cycles, this average will be zero. However, in some situations we are interested in the average magnitude of the waveform independent of its polarity (we will see an example of this later in this chapter). For a symmetrical waveform such as a sine wave, we can visualise this calculation as taking the average of just the positive half-cycle of the waveform. In this case, the average is the area within this half-cycle divided by half the period. This process is illustrated in Figure 2.6(a). Alternatively, one can view the calculation as taking the average of a **rectified sine wave** (that is, a sine wave where the polarity of the negative half-cycles has been reversed). This is shown in Figure 2.6(b).

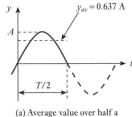

(a) Average value over half a
cycle of a sine wave

(b) Average value of a
rectified sine wave

Figure 2.6
Calculation of the
average value of a sine
wave.

We can calculate this average value by integrating a sinusoidal quantity over half a cycle and dividing by half the period. For example, if we consider a sinusoidal voltage $v = V_p \sin \theta$, the period is equal to 2π, so

$$V_{av} = \frac{1}{\pi} \int_0^{\pi} V_p \sin \theta \, \mathrm{d}\theta$$

$$= \frac{V_p}{\pi} [-\cos \theta]_0^{\pi}$$

$$= \frac{2V_p}{\pi}$$

Therefore

$$V_{av} = \frac{2}{\pi} \times V_p = 0.637 \times V_p \tag{2.7}$$

and similarly, for a sinusoidal current waveform,

$$I_{av} = \frac{2}{\pi} \times I_p = 0.637 \times I_p \tag{2.8}$$

2.2.7 The r.m.s. value of a sine wave

Often of more interest than the average value is the **root mean square** or **r.m.s.** value of the waveform. This is true not only for sine waves but also for other alternating waveforms.

In Chapter 1, we noted that when a voltage V is applied across a resistor R this will produce a current I (determined by Ohm's law), and that the power dissipated in the resistor will be given by three equivalent expressions:

$$P = VI \quad P = I^2 R \quad P = \frac{V^2}{R}$$

If the voltage has a varying magnitude, then the *instantaneous* power will be related to the instantaneous voltage and instantaneous current in a similar manner. As before, we use lower-case characters to represent varying quantities, so the instantaneous power p is related to the instantaneous voltage v and instantaneous current i by the expressions

$$p = vi \quad p = i^2 R \quad p = \frac{v^2}{R}$$

The *average* power will be given by the average (or *mean*) values of these expressions. Since the resistance is constant, we could say that the average power is given by

$$P_{av} = \frac{[\text{average (or mean) of } v^2]}{R} = \frac{\overline{v^2}}{R}$$

or

$$P_{av} = [\text{average (or mean) of } i^2]R = \overline{i^2}R$$

Placing a line (a *bar*) above an expression is a common notation for the mean of that expression. The term $\overline{v^2}$ is referred to as the **mean-square voltage** and $\overline{i^2}$ as the **mean-square current**.

While the mean-square voltage and current are useful quantities, we more often use the square root of each quantity. These are termed the root-mean-square voltage (V_{rms}) and the root-mean-square current (I_{rms}) where

$$V_{rms} = \sqrt{\overline{v^2}}$$

and

$$I_{rms} = \sqrt{\overline{i^2}}$$

We can evaluate each of these expressions by integrating a corresponding sinusoidal quantity over a complete cycle and dividing by the period. For example, if we consider a sinusoidal voltage $v = V_p \sin \omega t$, we can see that

$$V_{rms} = \left(\frac{1}{T} \int_0^T V_p^2 \sin^2 \omega t \; dt \right)^{1/2}$$

$$= \left(\frac{V_p^2}{T} \int_0^T \frac{1}{2}(1 - \cos 2\omega t) \; dt \right)^{1/2} = \frac{V_p}{\sqrt{2}}$$

Therefore

$$V_{rms} = \frac{1}{\sqrt{2}} \times V_p = 0.707 \times V_p \tag{2.9}$$

and similarly

$$I_{rms} = \frac{1}{\sqrt{2}} \times I_p = 0.707 \times I_p \tag{2.10}$$

Combining these results with the earlier expressions gives

$$P_{av} = \frac{\overline{v^2}}{R} = \frac{V_{rms}^2}{R}$$

and

$$P_{av} = \overline{i^2}R = I_{rms}^2 R$$

If we compare these expressions with those for the power produced by a constant voltage or current, we can see that the r.m.s. value of an alternating quantity produces the same power as a constant quantity of the same magnitude. Thus for alternating quantities

$$P_{av} = V_{rms} I_{rms} \tag{2.11}$$

$$P_{av} = \frac{V_{rms}^2}{R} \tag{2.12}$$

$$P_{av} = I_{rms}^2 R \tag{2.13}$$

This is illustrated in the following example.

Example 2.4 Calculate the power dissipated in a 10 Ω resistor if the applied voltage is:

(a) a constant 5 V;
(b) a sine wave of 5 V r.m.s.;
(c) a sine wave of 5 V peak.

(a) $P = \dfrac{V^2}{R} = \dfrac{5^2}{10} = 2.5 \text{ W}$

(b) $P_{av} = \dfrac{V_{rms}^2}{R} = \dfrac{5^2}{10} = 2.5 \text{ W}$

(c) $P_{av} = \dfrac{V_{rms}^2}{R} = \dfrac{(V_p/\sqrt{2})^2}{R} = \dfrac{V_p^2/2}{R} = \dfrac{5^2/2}{10} = 1.25 \text{ W}$

2.2.8 Form factor and peak factor

The **form factor** of any waveform is defined as

$$\text{form factor} = \frac{\text{r.m.s. value}}{\text{average value}} \tag{2.14}$$

For a sine wave

$$\text{form factor} = \frac{0.707 V_p}{0.637 V_p} = 1.11 \tag{2.15}$$

The significance of the form factor will become apparent in Section 2.5.

The **peak factor** (also known as the **crest factor**) for a waveform is defined as

$$\text{peak factor} = \frac{\text{peak value}}{\text{r.m.s. value}} \tag{2.16}$$

For a sine wave

$$\text{peak factor} = \frac{V_p}{0.707 V_p} = 1.414 \tag{2.17}$$

Although we have introduced the concepts of average value, r.m.s. value, form factor and peak factor for sinusoidal waveforms, it is important to remember that these measures may be applied to any repetitive waveform. In each case, the meanings of the terms are unchanged, although the numerical relationships between these values will vary. To illustrate this, we will now turn our attention to square waves.

2.3 Square waves

2.3.1 Period, frequency and magnitude

Frequency and period have the same meaning for all repetitive waveforms, as do the peak and peak-to-peak values. Figure 2.7 shows an example of a square-wave voltage signal and illustrates these various parameters.

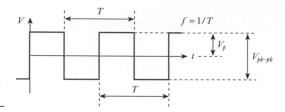

Figure 2.7
A square wave voltage signal.

2.3.2 Phase angle

We can if we wish divide the period of a square wave into 360° or 2π radians, as in a sine wave. This might be useful if we were discussing the phase difference between two square waveforms, as shown in Figure 2.8. Here two square waves have the same frequency but have a phase difference of 90° (or $\pi/2$ radians). In this case B lags A by 90°. An alternative way of describing the relationship between the two waveforms is to give the time delay of one with respect to the other.

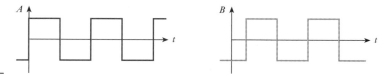

Figure 2.8
Phase-shifted square waves.

2.3.3 Average and r.m.s. values

Since the average value of a symmetrical alternating waveform is its average value over the positive half-cycle, the average value of a symmetrical square wave (as in Figure 2.8) is equal to its peak value. Thus for a voltage waveform the average value is V_p and for a current waveform it is I_p.

Since the instantaneous value of a symmetrical square wave is always equal to either its positive or its negative peak value, the square of this value is constant. For example, for a voltage waveform the instantaneous value will always be either $+V_p$ or $-V_p$ and in either case the square of this value will be constant at V_p^2. Thus the mean of the voltage squared will be V_p^2, and the square root of this will be V_p. Therefore, the r.m.s. value of a square wave is simply equal to its peak value.

2.3.4 Form factor and peak factor

Using the definitions given in Section 2.2.8, we can now determine the form factor and peak factor for a square wave. Since the average and r.m.s. values of a square wave are both equal to the peak value, it follows that

$$\text{form factor} = \frac{\text{r.m.s. value}}{\text{average value}} = 1.0$$

$$\text{peak factor} = \frac{\text{peak value}}{\text{r.m.s. value}} = 1.0$$

Video 2B

The relationship between the peak, average and r.m.s. values depends on the shape of a waveform. We have seen that this relationship is very different for a square wave and a sine wave, and further analysis would show similar differences for other waveforms, such as triangular waves.

2.4 Measuring voltages and currents

A wide range of instruments is available for measuring voltages and currents in electrical circuits. These include analogue ammeters and voltmeters, digital multimeters, and oscilloscopes. While each of these devices has its own characteristics, there are some issues that are common to the use of all these instruments.

2.4.1 Measuring voltage in a circuit

To measure the voltage between two points in a circuit, we place a voltmeter (or other measuring instrument) between the two points. For example, to measure the voltage drop across a component we connect the voltmeter *across* the part as shown in Figure 2.9(a).

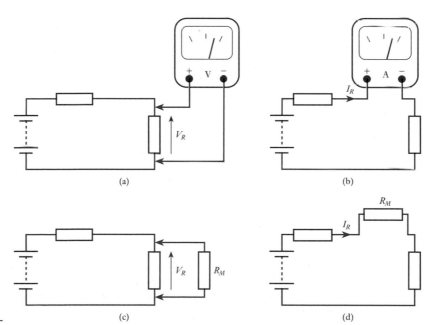

Figure 2.9
Measuring voltage and current.

2.4.2 Measuring current in a circuit

To measure the current flowing through a conductor or a component, we connect an ammeter *in series* with the element, as shown in Figure 2.9(b). Note that the ammeter is connected so that conventional current flows from the positive to the negative terminal.

2.4.3 Loading effects

Unfortunately, connecting additional components to a circuit can change the behaviour of that circuit. These **loading** effects can also occur when a voltmeter or an ammeter is connected to a circuit. The result is that the process of measurement actually changes the quantity being measured.

These loading effects are illustrated in Figures 2.9(c) and 2.9(d), which show equivalent circuits for the measurement processes of Figures 2.9(a) and 2.9(b). In each case, the measuring instrument is replaced by its equivalent resistance R_M, and it is clear that the presence of these additional resistances will affect the operation of the circuits. When measuring voltages (as in Figure 2.9(c)), the presence of the voltmeter reduces the effective resistance of the circuit and therefore tends to lower the voltage between these two points in the circuit. To minimise this effect, the resistance of the voltmeter should be as *high* as possible to reduce the current that it passes. When measuring currents (as in Figure 2.9(d)), the ammeter tends to increase the resistance in the circuit and therefore tends to reduce the current flowing. To minimise this effect, the ammeter should have as *low* a resistance as possible to reduce the voltage drop across it.

When using analogue voltmeters and ammeters (as described in the next section), loading effects should always be considered. Instruments will normally indicate their effective resistance (which will usually be different for each range of the instrument), and this information can be used to quantify any loading errors. If these are appreciable it may be necessary to make corrections to the measured values. When using digital voltmeters or oscilloscopes, loading effects are usually less of a problem but should still be considered.

2.5 Analogue ammeters and voltmeters

Most modern analogue ammeters and voltmeters are based on moving-coil meters (we will look at the characteristics of such meters in Chapter 13). These produce an output in the form of movement of a pointer, where the displacement is directly proportional to the current through the meter. Meters are characterised by the current required to produce **full-scale deflection (f.s.d.)** of the meter and their effective resistance R_M. Typical meters produce a full-scale deflection for a current of between 50 μA and 1 mA and have a resistance of between a few ohms and a few kilohms.

2.5.1 Measuring direct currents

Since the deflection of the meter's pointer is directly proportional to the current through the meter, currents up to the f.s.d. value can be measured directly. For larger currents, **shunt resistors** are used to scale the meter's effective sensitivity. This is illustrated in Figure 2.10, where a meter with an f.s.d. current of 1 mA is used to measure a range of currents.

In Figure 2.10(a), the meter is being used to measure currents in the range 0–1 mA, that is currents up to its f.s.d. value. In Figure 2.10(b), a shunt resistor R_{SH} of $R_M/9$ has been placed in parallel with the meter. Since the same voltage is applied across the meter and the resistor, the current through the resistor will be nine times greater than that through the meter. To put this another way, only one-tenth of the input current I will pass through the meter. Therefore, this arrangement has one-tenth the sensitivity of the meter alone and will produce an f.s.d. for a current of 10 mA. Figure 2.10(c) shows a similar arrangement for measuring currents up to 100 mA, and

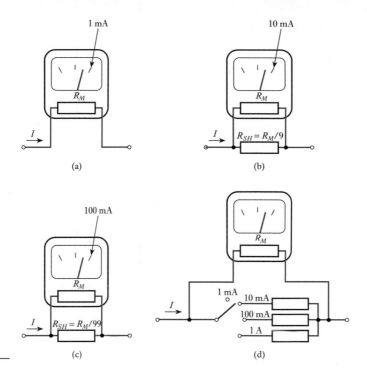

Figure 2.10
Use of a meter as an
ammeter

clearly this technique can be extended to measure very large currents. Figure 2.10(d) shows a **switched-range ammeter** arrangement, which can be used to measure a wide range of currents. It can be seen that the effective resistance of the meter is different for each range.

Example 2.5

A moving-coil meter produces an f.s.d. for a current of 1 mA and has a resistance of 25 Ω. Select a shunt resistor to turn this device into an ammeter with an f.s.d. of 50 mA.

We need to reduce the sensitivity of the meter by a factor of

$$\frac{50 \text{ mA}}{1 \text{ mA}} = 50$$

Therefore, we want 1/50 of the current to pass through the meter. Therefore, R_{SH} must be equal to $R_M \div 49 = 510$ mΩ.

2.5.2 Measuring direct voltages

To measure direct voltages, we place a resistor in series with the meter and measure the resultant current, as shown in Figure 2.11. In Figure 2.11(a), the meter has an f.s.d. current of 1 mA and the series resistor R_{SE} has been chosen such that $R_{SE} + R_M$ = 1 kΩ. The voltage V required to produce a current of 1 mA is given by Ohm's law and is 1 mA × 1 kΩ = 1 V. Therefore, an f.s.d. of the meter corresponds to an input voltage of 1 V.

In Figure 2.11(b), the series resistor has been chosen such that the total resistance is 10 kΩ, and this will give an f.s.d. for an input voltage of 10 V. In this way, we can tailor

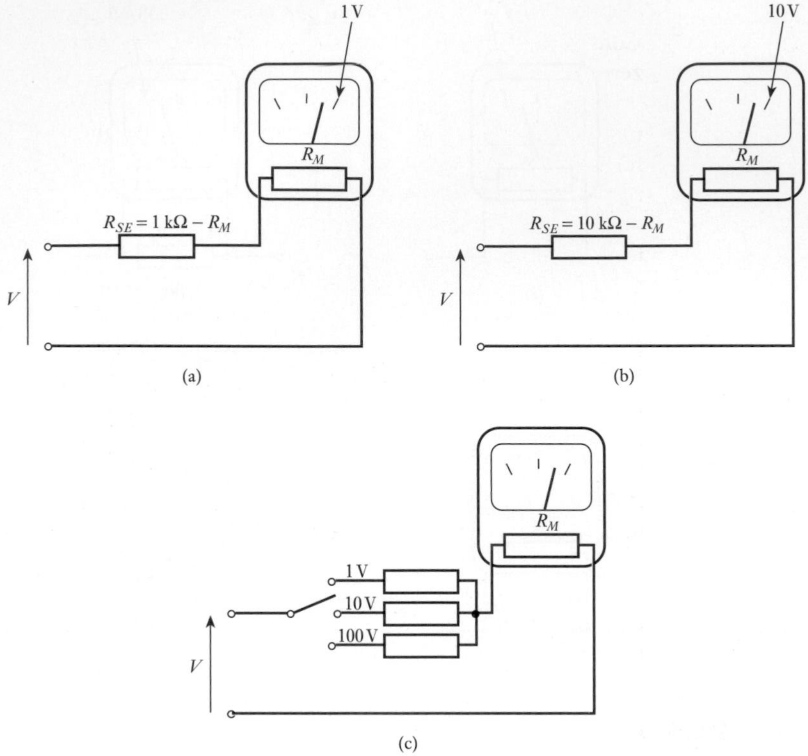

Figure 2.11
Use of a meter as a
voltmeter.

the sensitivity of the arrangement to suit our needs. Figure 2.11(c) shows a switched–range voltmeter that can be used to measure a wide range of voltages. As with the ammeter, the effective resistance of the meter changes as the range is switched.

Example 2.6 | A moving-coil meter produces an f.s.d. for a current of 1 mA and has a resistance of 25 Ω. Select a series resistor to turn this device into a voltmeter with an f.s.d. of 50 V.

The required total resistance of the arrangement is given by the f.s.d. current divided by the full-scale input voltage. Hence

$$R_{SE} + R_M = \frac{50 \text{ V}}{1 \text{ mA}} = 50 \text{ k}\Omega$$

Therefore

$$R_{SE} = 50 \text{ k}\Omega - R_M$$

$$= 49.975 \text{ k}\Omega$$

$$\approx 50 \text{ k}\Omega$$

2.5.3 Measuring alternating quantities

Moving-coil meters respond to currents of either polarity, each producing deflections in opposite directions. Because of the mechanical inertia of the meter, it cannot respond to rapid changes in current and so will average the readings over time. Consequently, a symmetrical alternating waveform will cause the meter to display zero.

In order to measure an alternating current, we can use a **rectifier** to convert it into a unidirectional current that *can* be measured by the meter. This process was illustrated for a sine wave in Figure 2.6(b). The meter responds by producing a deflection corresponding to the average value of the rectified waveform.

We noted in Section 2.2 that when measuring sinusoidal quantities we are normally more interested in the r.m.s. value than in the average value. Therefore, it is common to calibrate AC meters so that they effectively multiply their readings by 1.11, this being the form factor of a sine wave. The result is that the meter (which responds to the average value of the waveform) gives a direct reading of the r.m.s. value of a sine wave. However, a problem with this arrangement is that it gives an incorrect reading for non-sinusoidal waveforms. For example, we noted in Section 2.3 that the form factor for a square wave is 1.0. Consequently, if we measure the r.m.s. value of a square wave using a meter designed for use with sine waves, it will produce a reading that is about 11 per cent too high. This problem can be overcome by adjusting our readings to take account of the form factor of the waveform we are measuring.

Like all measuring devices, meters are only accurate over a certain range of frequencies determined by their frequency response. Most devices will work well at the frequencies used for AC power distribution (50 or 60 Hz), but all will have a maximum frequency at which they can be used.

2.5.4 Analogue multimeters

General-purpose instruments use a combination of switches and resistors to achieve a large number of voltage and current ranges within a single unit. Such units are often referred to as analogue **multimeters**. A rectifier is also used to permit both unidirectional and alternating quantities to be measured, and additional circuitry is used to allow resistance measurement. While such devices are very versatile, they often have a relatively low input resistance and therefore can have considerable loading effects on the circuits to which they are connected. A typical analogue multimeter is shown in Figure 2.12.

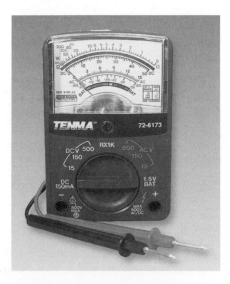

Figure 2.12
An analogue
multimeter.

2.6 Digital multimeters

A standard measuring instrument in any electronics laboratory is a **digital multimeter** (**DMM**). This combines high accuracy and stability in a device that is very easy to use. It also normally has a very high input resistance when used as a voltmeter and a very low input resistance when measuring currents, so minimising loading effects. While these instruments are capable of measuring voltage, current and resistance, they are often (inaccurately) referred to as **digital voltmeters** or simply **DVMs**. At the heart of the meter is an **analogue-to-digital converter** (**ADC**), which takes as its input a voltage signal and produces as its output a digital measurement that is used to drive a numeric display. We will look at the operation of such ADCs in Chapter 28.

Measurements of voltage, current and resistance are achieved by using appropriate circuits to generate a voltage proportional to the quantity to be measured. When measuring voltages, the input signal is connected to an attenuator, which can be switched to vary the input range. When measuring currents, the input signal is connected across an appropriate shunt resistor, which generates a voltage proportional to the input current. The value of the shunt resistance is switched to select different input ranges. In order to measure resistance the inputs are connected to an **ohms converter**, which passes a small current between the two input connections. The resultant voltage is a measure of the resistance between these terminals.

In simple DMMs, an alternating voltage is rectified, as in an analogue multimeter, to give its average value. This is then multiplied by 1.11 (the form factor of a sine wave) to display the corresponding r.m.s. value. As discussed earlier, this approach gives inaccurate readings when the alternating input signal is not sinusoidal. For this reason, more sophisticated DMMs use a **true r.m.s. converter**, which accurately produces a voltage proportional to the r.m.s. value of an input waveform. Such instruments can be used to make measurements of alternating quantities even when they are not sinusoidal. However, all DMMs are accurate over only a limited range of frequencies.

Figure 2.13(a) shows a typical hand-held DMM and Figure 2.13(b) is a simplified block diagram of a such a device.

(a) A typical digital multimeter

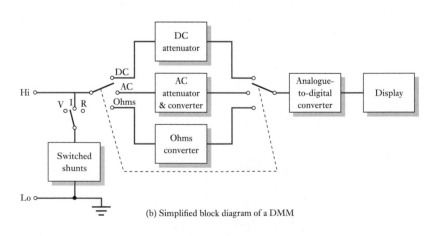

(b) Simplified block diagram of a DMM

Figure 2.13
A digital multimeter (DMM).

Oscilloscopes

An **oscilloscope** is an instrument that allows voltages to be measured by displaying the corresponding voltage waveform on a **cathode ray tube (CRT)** or **LCD** display. The oscilloscope effectively acts as an automated graph plotter that plots the input voltage against time.

2.7.1 Analogue oscilloscopes

In an **analogue CRT oscilloscope** a **timebase** circuit is used to scan a spot repeatedly from left to right across the screen at a constant speed by applying a 'sawtooth' waveform to the horizontal deflection circuitry. An input signal is then used to generate a vertical deflection proportional to the magnitude of the input voltage. Most oscilloscopes can display two input quantities by switching the vertical deflection circuitry between two input signals. This can be done by displaying one complete trace of one waveform, then displaying one trace of the other (**ALT** mode) or by rapidly switching between the two waveforms during each trace (**CHOP** mode). The choice between these modes is governed by the timebase frequency, but in either case the goal is to switch between the two waveforms so quickly that both are displayed steadily and with no noticeable flicker or distortion. In order to produce a stable trace, the timebase circuitry includes a **trigger** circuit that attempts to synchronise the beginning of the timebase sweep so that it always starts at the same point in a repetitive waveform – thus producing a stationary trace. Figure 2.14(a) shows a typical analogue laboratory oscilloscope and Figure 2.14(b) shows a simplified block diagram of such an instrument.

(a) A typical analogue oscilloscope

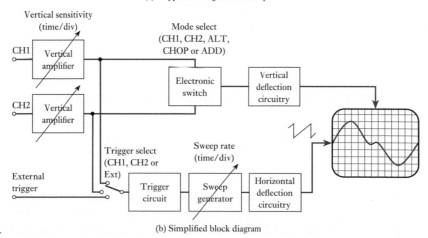

Figure 2.14
An analogue laboratory oscilloscope.

(b) Simplified block diagram

2.7.2 Digital oscilloscopes

In recent years, analogue oscilloscopes have largely been replaced by more advanced digital designs. These duplicate many of the basic features of analogue devices but use an ADC to change the input waveforms into a form than can be more easily stored and manipulated than is possible within analogue instruments. Digital oscilloscopes are particularly useful when looking at very slow waveforms or short transients as their ability to store information enables them to display a steady trace. Many instuments also provide measurement facilities to enable fast and accurate measurements to be made, as well as mathematical functions for displaying information such as the frequency content of a signal. Figure 2.15(a) shows a typical digital laboratory oscilloscope and Figure 2.15(b) shows a simplified block diagram of such an instrument.

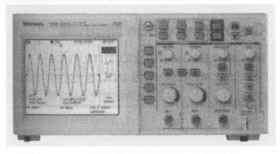

(a) A typical digital oscilloscope

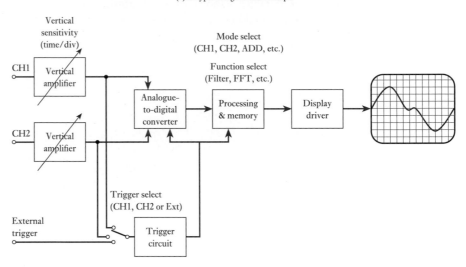

Figure 2.15
A digital laboratory oscilloscope.

(b) Simplified block diagram

2.7.3 Making measurements with an oscilloscope

One of the advantages of oscilloscopes is that they permit measurements to be made over a much wider frequency range than is possible with analogue or digital multimeters. This often extends to some hundreds of megahertz or perhaps several gigahertz. Another great advantage is that an oscilloscope allows the user to see the shape of a waveform. This is invaluable in determining whether circuits are functioning correctly and may also permit distortion or other problems to be detected. Oscilloscopes are also useful when signals have both DC and AC components. This last point is illustrated in Figure 2.16. The figure shows a voltage signal that has a large DC component with

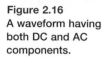

Figure 2.16
A waveform having
both DC and AC
components.

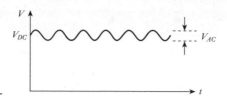

a small AC component superimposed on it. If this signal were applied to an analogue or digital multimeter, the reading would simply reflect the magnitude of the DC component. However, using an oscilloscope the presence of the alternating component is visible, and the true nature of the signal becomes apparent. Oscilloscopes also allow the DC component to be 'blocked' by selecting an **AC-coupled input**, permitting the AC component to be seen and measured easily. We will see how capacitors can be used to block DC signals in Chapter 8.

When used with AC signals, it is normal to measure the peak-to-peak voltage of a waveform when using an oscilloscope, since this is the quantity that is most readily observed. Care must be taken when comparing such readings with those taken using a multimeter, since the latter will normally give r.m.s. values.

Oscilloscopes also allow the direct comparison of waveforms and permit the temporal relationship between them to be investigated. For example, we might use the two traces to display the input and output signals of a module, and hence determine the phase difference between the input and the output.

The measurement of phase difference is illustrated in Figure 2.17. The horizontal scale (which corresponds to time) is used to measure the period of the waveforms (T) and also the time difference between corresponding points in the two waveforms (t). The term t/T now represents the fraction of a complete cycle by which the waveforms are phase shifted. Since one cycle represents $360°$ or 2π radians, the phase difference ϕ is given by

$$\text{phase difference } \phi = \frac{t}{T} \times 360° = \frac{t}{T} \times 2\pi \text{ radians} \qquad (2.18)$$

In Figure 2.17, waveform B lags waveform A by approximately one-eighth of a complete cycle, or about $45°$ ($\pi/4$ rad).

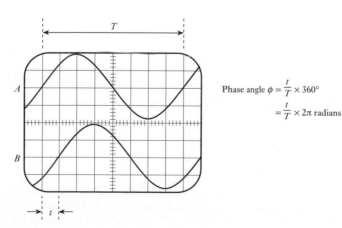

$$\text{Phase angle } \phi = \frac{t}{T} \times 360°$$
$$= \frac{t}{T} \times 2\pi \text{ radians}$$

Figure 2.17
Measurement of phase
difference using an
oscilloscope.

Video 2C

Further study

The voltage V in the following circuit represents the output of a controllable voltage source. This source can produce an output of a range of forms, including the four waveforms shown. A control on the front of the unit allows the magnitude of the output signal to be increased or decreased, although this control is not calibrated, so it is not possible to set the magnitude to a predefined value.

Our goal is to adjust the output signal so that the power dissipated in the 10 Ω resistor is exactly 10 W. Since the voltage source does not indicate the magnitude of its output, we have connected an analogue voltmeter across this resistor. This meter is calibrated to indicate the r.m.s. value of a sinusoidal input voltage, which simplifies our task when using the sinusoidal waveform. Your task is to determine the readings required on the voltmeter for each of the waveforms shown, in order to produce a power dissipation of 10 W in the resistor.

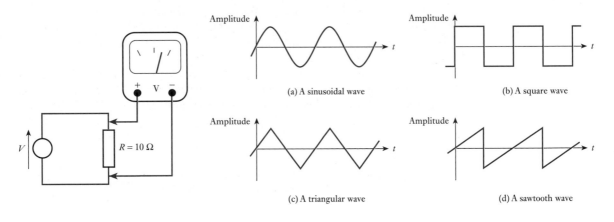

(a) A sinusoidal wave

(b) A square wave

(c) A triangular wave

(d) A sawtooth wave

Key points

- Alternating waveforms vary with time and periodically change their direction. By far the most important form of alternating waveform is the sine wave.

- The frequency of a periodic waveform f is equal to the reciprocal of its period T.

- The magnitude of an alternating waveform can be described by its *peak* value, its *peak-to-peak* value, its *average* value or its *r.m.s.* value.

- A sinusoidal voltage signal can be described by the expressions

$$v = V_p \sin(2\pi f t + \phi)$$

or

$$v = V_p \sin(\omega t + \phi)$$

where V_p is the peak voltage, f is the frequency (in hertz), ω is the angular frequency (in radians/second) and ϕ is the angle of the waveform at $t = 0$.

- A sinusoidal current signal can be described by the expressions

$$i = I_p \sin(2\pi f t + \phi)$$

or

$$i = I_p \sin(\omega t + \phi)$$

where I_p is the peak current and the other terms are as before.

- Two waveforms of the same frequency may have a constant phase difference between them. One waveform is said to *lead* or *lag* the other.

- The average value of a repetitive alternating waveform is defined as the average over the positive half-cycle.

- The root-mean-square (r.m.s.) value of an alternating waveform is the value that will produce the same power as an equivalent direct quantity.

- For a sinusoidal signal, the *average* voltage or current is $2/\pi$ (or 0.637) times the corresponding *peak* value, and the *r.m.s.* voltage or current is $1/\sqrt{2}$ (or 0.707) times the corresponding *peak* value.

- For square waves, the average and r.m.s. values of voltage and current are equal to the corresponding peak values.

- Simple analogue ammeters and voltmeters are often based on moving-coil meters. These can be configured to measure currents or voltages over a range of magnitudes through the use of series or shunt resistors.

- Meters respond to the average value of a rectified alternating waveform and are normally calibrated to read the r.m.s. value of a sine wave. These will give inappropriate readings when used with non-sinusoidal signals.

- Digital multimeters are easy to use and offer high accuracy. Some have a true r.m.s. converter, allowing them to be used with non-sinusoidal alternating signals.

- Oscilloscopes display the form of a signal and allow distortion to be detected and measured. They also allow comparison between signals and the measurement of parameters such as phase shift.

Exercises

2.1   Sketch three common forms of alternating waveform.

2.2   A sine wave has a period of 10 s. What is its frequency (in hertz)?

2.3   A square wave has a frequency of 25 Hz. What is its period?

2.4   A triangular wave (of the form shown in Figure 2.1) has a peak amplitude of 2.5 V. What is its peak-to-peak amplitude?

2.5   What is the peak-to-peak current of the waveform described by the following equation?

$$i = 10 \sin \theta$$

2.6   A signal has a frequency of 10 Hz. What is its angular frequency?

2.7   A signal has an angular frequency of 157 rad/s. What is its frequency in hertz?

2.8 Determine the peak voltage, the peak-to-peak voltage, the frequency (in hertz) and the angular frequency (in rad/s) of the following waveform.

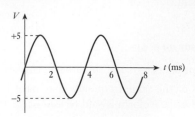

2.9 Write an equation to describe a voltage waveform with an amplitude of 5 V peak and a frequency of 50 Hz.

2.10 Write an equation to describe a current waveform with an amplitude of 16 A peak-to-peak and an angular frequency of 150 rad/s.

2.11 What are the frequency and peak amplitude of the waveform described by the following equation?

$$v = 25 \sin 471t$$

2.12 Determine the equation of the following voltage signal.

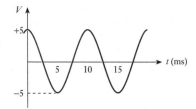

2.13 A sine wave has a peak value of 10. What is its average value?

2.14 A sinusoidal current signal has an average value of 5 A. What is its peak value?

2.15 Explain what is meant by the mean-square value of an alternating waveform. How is this related to the r.m.s. value?

2.16 Why is the r.m.s. value a more useful quantity than the average value?

2.17 A sinusoidal voltage signal of 10 V peak is applied across a resistor of 25 Ω. What power is dissipated in the resistor?

2.18 A sinusoidal voltage signal of 10 V r.m.s. is applied across a resistor of 25 Ω. What power is dissipated in the resistor?

2.19 A sinusoidal waveform with an average voltage of 6 V is measured by an analogue multimeter. What voltage will be displayed?

2.20 A square-wave voltage signal has a peak amplitude of 5 V. What is its average value?

2.21 A square wave of 5 V peak is applied across a 25 Ω resistor. What will be the power dissipated in the resistor?

2.22 A moving-coil meter produces a full-scale deflection for a current of 50 μA and has a resistance of 10 Ω. Select a shunt resistor to turn this device into an ammeter with an f.s.d. of 250 mA.

2.23 A moving-coil meter produces a full-scale deflection for a current of 50 μA and has a resistance of 10 Ω. Select a series resistor to turn this device into a voltmeter with an f.s.d. of 10 V.

2.24 What percentage error is produced if we measure the voltage of a square wave using an analogue multimeter that has been calibrated to display the r.m.s. value of a sine wave?

2.25 A square wave of 10 V peak is connected to an analogue multimeter that is set to measure alternating voltages. What voltage reading will this show?

2.26 Describe the basic operation of a digital multimeter.

2.27 How do some digital multimeters overcome the problem associated with different alternating waveforms having different form factors?

2.28 Explain briefly how an analogue oscilloscope displays the amplitude of a time-varying signal.

2.29 How is an analogue oscilloscope able to display two waveforms simultaneously?

2.30 What is the difference between the ALT and CHOP modes on an analogue oscilloscope?

2.31 What is the function of the trigger circuitry in an oscilloscope?

2.32 A sinusoidal waveform is displayed on an oscilloscope and has a peak-to-peak amplitude of 15 V. At the same time, the signal is measured on an analogue multimeter that is set to measure alternating voltages. What value would you expect to be displayed on the multimeter?

2.33 Comment on the relative accuracies of the two measurement methods outlined in the last exercise.

2.34 What is the phase difference between waveforms A and B in the following oscilloscope display? Which waveform is leading and which lagging?

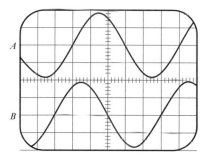

Chapter 3 — Resistance and DC Circuits

Objectives

When you have studied the material in this chapter, you should be able to:

- define terms such as current, charge, electromotive force, potential difference, resistance and power, and write equations relating them
- apply Ohm's law and Kirchhoff's voltage and current laws in a range of situations
- derive simple equivalent circuits for electrical networks to aid in their analysis
- explain the principle of superposition and its use in circuit analysis
- describe the processes of nodal analysis and mesh analysis, and explain their uses and importance
- use nodal analysis and mesh analysis to determine currents and voltages in electrical networks
- discuss the selection of analytical techniques for use with electrical circuits.

3.1 Introduction

Many electrical and electronic circuits can be analysed, and in some cases designed, using little more than Ohm's law. However, in some cases additional techniques are required and in this chapter we will start to look in more detail at the analysis of electrical circuits. We will begin by reviewing some of the basic elements that are used in such circuits and provide a more detailed understanding of their characteristics. We will then look at a range of techniques for modelling and analysing electrical and electronic circuits.

3.2 Current and charge

An electric **current** represents a flow of electric **charge** and therefore

$$I = \frac{dQ}{dt} \tag{3.1}$$

where I is the current in amperes, Q is the charge in coulombs and dQ/dt represents the rate of flow of charge (with units of coulombs per second). Conventionally, current is assumed to represent a flow of positive charge.

At an atomic level, a current represents a flow of **electrons**, each of which carries a minute *negative* charge of about 1.6×10^{-19} coulombs. For this reason, the flow of a conventional current in one direction actually represents the passage of electrons in the opposite direction. However, unless we are looking at the physical operation of devices, this distinction is unimportant.

Rearranging the expression of Equation 3.1, we can obtain an expression for the charge passed as a result of the flow of a current.

$$Q = \int I \mathrm{d}t \qquad\qquad (3.2)$$

If the current is constant, this results in the simple relationship that charge is equal to the product of the current and time.

$$Q = I \times t$$

3.3 Voltage sources

A voltage source produces an electromotive force (e.m.f.), which causes a current to flow within a circuit. Despite its name, an e.m.f. is not a force in the conventional sense but represents the energy associated with the passage of charge through the source. The unit of e.m.f. is the **volt**, which is defined as the potential difference between two points when 1 joule of energy is used to move 1 coulomb of charge from one point to the other.

Real voltage sources, such as batteries, have resistance associated with them, which limits the current that they can supply. When analysing circuits, we often use the concept of an **ideal voltage source** that has no resistance. Such sources can represent constant or alternating voltages, and they can also represent voltages that vary in response to some other physical quantity (**controlled** or **dependent voltage sources**). Figure 3.1 shows examples of the symbols used to represent various forms of voltage source.

Unfortunately, a range of notations is used to represent voltages in electrical circuits. Most textbooks published in North America adopt a notation where the polarity of a voltage is indicated using a '+' symbol. In the UK, and many other countries, it is more common to use the notation shown in Figure 3.1, where an arrow is used to indicate polarity. Here the label associated with the arrow represents the voltage at the head of the arrow with respect to the voltage at its tail. An advantage of this notation is that the label can unambiguously represent a positive, negative or alternating quantity.

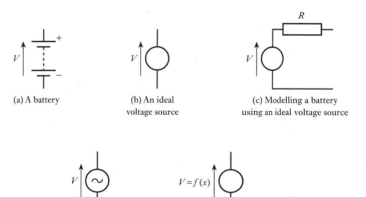

(a) A battery

(b) An ideal voltage source

(c) Modelling a battery using an ideal voltage source

(d) An alternating voltage source

(e) A controlled voltage source

Figure 3.1
Voltage sources.

3.4 Current sources

In addition to the concept of an ideal voltage source, it is sometimes convenient to model an **ideal current source**. As with its voltage counterpart, such a component is not physically realisable, but the use of such a conceptual model can greatly simplify some forms of circuit analysis. Just as an ideal voltage source produces a certain voltage no matter what is connected to it, so an ideal current source will always pass a particular current. This current could be constant or alternating (depending on the nature of the current source), or it might be determined by some physical quantity within a circuit (a **controlled** or **dependent current source**). The circuit symbol for a current source is shown in Figure 3.2.

It is interesting to note that, while an ideal voltage source has *zero* output resistance, an ideal current source has *infinite* output resistance. This is evident if we consider loading effects and the situation required for the output current to remain constant regardless of variations in load resistance.

Figure 3.2
An ideal current source.

3.5 Resistance and Ohm's law

Readers are already familiar with one of the best-known relationships in electrical engineering, which is that the voltage across a conductor is directly proportional to the current flowing in it (Ohm's law):

$$V \propto I$$

The constant of proportionality of this relationship is termed the **resistance** of the conductor (R), which gives rise to the well-known expressions

$$V = IR \quad I = \frac{V}{R} \quad R = \frac{V}{I}$$

The unit of resistance is the ohm (Ω), which can be defined as the resistance of a circuit in which a current of 1 A produces a **potential difference** of 1 V (see Chapter 1).

When current flows through a resistance, power is dissipated in it. This power is dissipated in the form of heat. The power (P) is related to V, I and R by the expressions

$$P = IV \quad P = \frac{V^2}{R} \quad P = I^2 R$$

Components designed to provide resistance in electrical circuits are termed **resistors**. The resistance of a given sample of material is determined by its dimensions and by the electrical characteristics of the material used in its construction. The latter is described by the **resistivity** of the material ρ (Greek letter *rho*) or sometimes by its **conductivity** σ (Greek letter *sigma*), which is the reciprocal of the resistivity. Figure 3.3 shows a piece of resistive material with electrical contacts on each end. If the body of the component is uniform, its resistance will be directly related to its length (l) and

Figure 3.3
The effects of component dimensions on resistance.

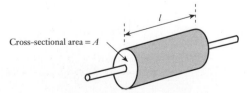

Cross-sectional area = A

l

inversely related to its cross-sectional area (A). Under these circumstances, the resistance of the device will be given by

$$R = \frac{\rho l}{A} \qquad (3.3)$$

The units of resistivity are ohm metres ($\Omega \cdot m$). Copper has a resistivity of about 1.6×10^{-8} $\Omega \cdot m$ at 0 °C, while carbon has a resistivity of 6500×10^{-8} $\Omega \cdot m$ at 0 °C.

Since the flow of current through a resistor produces heat, this will cause the temperature of the resistor to rise. Unfortunately, the resistance of most materials changes with temperature, this variation being determined by its **temperature coefficient of resistance** α. Pure metals have positive temperature coefficients, meaning that their resistance increases with temperature. Many other materials (including most insulators) have negative coefficients. The materials used in resistors are chosen to minimise these temperature-related effects. In addition to altering its resistance, an excessive increase in temperature would inevitably result in damage to a resistor. Consequently, any particular component has a maximum **power rating**, which should not be exceeded. Larger components have a greater surface area and can therefore dissipate heat more effectively. Consequently, power ratings tend to increase with the physical size of resistors (although it is also affected by other factors). A small general-purpose resistor might have a power rating of an eighth or a quarter of a watt, while larger components might handle several watts.

3.6 Resistors in series and parallel

In Chapter 1, we noted the effective resistance produced by connecting a number of resistors in series or in parallel. Before moving on, it is perhaps appropriate to ensure that the reasons for these relationships are clear.

Figure 3.4(a) shows an arrangement in which a voltage V is applied across a series arrangement of resistors $R_1, R_2, \ldots, R_N$. The voltage across each individual resistor is given by the product of the current (I) and its resistance. The applied voltage V must be equal to the sum of the voltages across the resistors, and therefore

$$V = IR_1 + IR_2 + \cdots + IR_N$$
$$= I(R_1 + R_2 + \cdots + R_N)$$
$$= IR$$

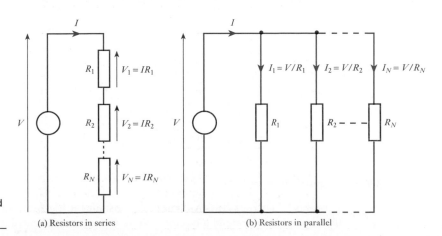

Figure 3.4
Resistors in series and parallel.

(a) Resistors in series

(b) Resistors in parallel

where $R = (R_1 + R_2 + \cdots + R_N)$. Therefore, the circuit behaves as if the series of resistors were replaced by a single resistor with a value equal to their sum.

Figure 3.4(b) shows an arrangement where several resistors are connected in parallel. The voltage across each resistor is equal to the applied voltage V, so the current in each resistor is given by this voltage divided by its resistance. The total current I is equal to the sum of the currents in the individual resistors, and therefore

$$I = \frac{V}{R_1} + \frac{V}{R_2} + \cdots + \frac{V}{R_N}$$

$$= V \left(\frac{1}{R_1} + \frac{1}{R_2} + \cdots + \frac{1}{R_N} \right)$$

$$= V \left(\frac{1}{R} \right)$$

where $1/R = 1/R_1 + 1/R_2 + \cdots + 1/R_N$. Therefore, the circuit behaves as if the combination of resistors were replaced by a single resistor whose value is given by the reciprocal of the sum of the reciprocals of their values.

3.6.1 Notation

Parallel combinations of resistors are very common in electrical circuits, so there is a notation to represent the effective resistance of resistors in parallel. This consists of the resistor names or values separated by '//'. Therefore, $R_1//R_2$ would be read as 'the effective resistance of R_1 in parallel with R_2'. Similarly, 10 kΩ//10 kΩ simply means the resistance of two 10 kΩ resistors connected in parallel (this is 5 kΩ).

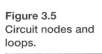

3.7 Kirchhoff's laws

A point in a circuit where two or more circuit components are joined together is termed a **node**, while any closed path in a circuit that passes through no node more than once is termed a **loop**. A loop that contains no other loop is called a **mesh**. These definitions are illustrated in Figure 3.5. Here points A, B, C, D, E and F are nodes in the circuit, while the paths ABEFA, BCDEB and ABCDEFA represent loops. It can be seen that the first and second of these loops are also meshes, while the last is not (since it contains smaller loops).

We can apply Kirchhoff's current law to the various nodes in a circuit and Kirchhoff's voltage law to the various loops and meshes.

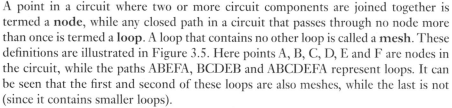

Figure 3.5
Circuit nodes and loops.

3.7.1 Current law

Kirchhoff's current law says that at any instant the algebraic sum of all the currents flowing *into* any node in a circuit is zero. If we consider currents flowing *into* a node

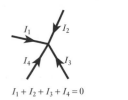
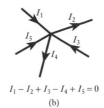

Figure 3.6
Application of
Kirchhoff's current law.

$I_1 + I_2 + I_3 + I_4 = 0$

(a)

$I_1 - I_2 + I_3 - I_4 + I_5 = 0$

(b)

to be positive and currents flowing *out of* that node to be negative, then all the various currents must sum to zero. That is

$$\sum I = 0$$

This is illustrated in Figure 3.6. In Figure 3.6(a), the currents are each defined as flowing *into* the node and therefore their magnitudes simply sum to zero. It is evident that one or more of these currents must be negative in order for this to be possible (unless they are all zero). A current of $-I$ flowing into a node is clearly equivalent to a current of I flowing out from it. In Figure 3.6(b), some currents are defined as flowing *into* the node while others flow *out*. This results in the equation shown.

Example 3.1 | Determine the magnitude of I_4 in the following circuit.

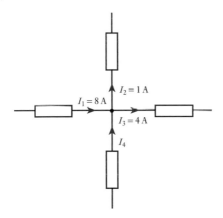

Summing the currents flowing *into* the node, we have

$$I_1 - I_2 - I_3 + I_4 = 0$$
$$8 - 1 - 4 + I_4 = 0$$
$$I_4 = -3 \text{ A}$$

Therefore, I_4 is equal to -3 A: that is, a current of 3 A flowing in the opposite direction to the arrow in the diagram.

3.7.2 Voltage law

Kirchhoff's voltage law says that at any instant the algebraic sum of all the voltages around any loop in a circuit is zero

$$\sum V = 0$$

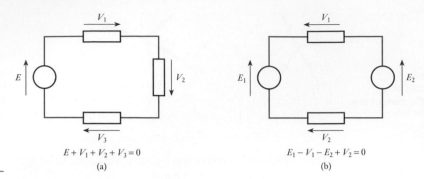

Figure 3.7
Applying Kirchhoff's voltage law.

$$E + V_1 + V_2 + V_3 = 0$$

(a)

$$E_1 - V_1 - E_2 + V_2 = 0$$

(b)

Our only difficulty in applying this law is in ensuring that we get the polarities of the voltages correct in our arithmetic. A simple way of ensuring this is to use arrows within our circuit diagrams to represent the polarity of each e.m.f. or potential difference (as in earlier circuits). We then move around the loop in a clockwise direction and any arrow in this direction represents a positive voltage while any arrow in the opposite direction represents a negative voltage. This is illustrated in Figure 3.7.

In Figure 3.7(a), all the e.m.f.s and potential differences are defined in a clockwise direction. Therefore, their magnitudes simply sum to zero. Note that the directions of the arrows show how the voltages are defined (or measured) and do *not* show the polarity of the voltages. If E has a positive value, the top of the voltage source in Figure 3.7(a) will be positive with respect to the bottom. If E has a negative value, the polarity will be reversed. Similarly, E could represent a varying or alternating voltage, but the relationship shown in the equation would still hold. In Figure 3.7(b), some of the e.m.f.s and potential differences are defined in a clockwise direction and some are defined in the opposite direction. This results in the equation shown, where clockwise terms are added and anticlockwise terms are subtracted.

Example 3.2   **Determine the magnitude of V_2 in the following circuit.**

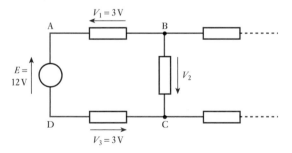

Summing the voltages clockwise around the loop ABCDA, we have

$$E - V_1 + V_2 - V_3 = 0$$

$$12 - 3 + V_2 - 3 = 0$$

$$V_2 = -6\,\text{V}$$

Therefore, V_2 has a value of $-6\,$V: that is, a potential difference of $6\,$V with node B being more positive than node C. Had we chosen to define V_2 by an arrow pointing in the opposite direction, our calculations would have found it to have a value of $+6\,$V, which would again have represented a potential difference of $6\,$V with node B being more positive than node C.

Thévenin's theorem and Norton's theorem

It is often convenient to represent electrical circuits by simpler **equivalent circuits** that model their behaviour. For example, we could represent a real voltage source (such as a battery) by an ideal voltage source and a series resistor. This representation is an example of what is termed a **Thévenin equivalent circuit**. This is an arrangement based on **Thévenin's theorem**, which may be paraphrased as:

> *As far as its appearance from outside is concerned, any two terminal networks of resistors and energy sources can be replaced by a series combination of an ideal voltage source V and a resistor R, where V is the open-circuit voltage of the network and R is the resistance that would be measured between the output terminals if the energy sources were removed and replaced by their internal resistance.*

It can be seen that this simple equivalent circuit can be used to represent not only the outputs of batteries but also any arrangement of resistors, voltage sources and current sources that have two output terminals. However, it is important to note that the equivalence is valid only 'as far as its appearance from outside' is concerned. The equivalent circuit does not represent the internal characteristics of the network, such as its power consumption.

While Thévenin equivalent circuits are useful in a wide range of applications, there are situations where it would be more convenient to have an equivalent circuit that uses a current source rather than a voltage source. Fortunately, such circuits are described by **Norton's theorem**, which may be paraphrased as:

> *As far as its appearance from outside is concerned, any two terminal networks of resistors and energy sources can be replaced by a parallel combination of an ideal current source I and a resistor R, where I is the short-circuit current of the network and R is the resistance that would be measured between the output terminals if the energy sources were removed and replaced by their internal resistance.*

The implications of these two theorems are summarised in Figure 3.8. This shows that any such circuit can be represented by *either* form of equivalent circuit. The form that is used depends on the application. When modelling simple batteries we might find it convenient to use a Thévenin form. However, we will see in later chapters that when we consider devices such as transistors it is often more convenient to use a Norton form.

Since the three arrangements of Figure 3.8 are equivalent, it follows that they should each produce the same output in all circumstances. If we connect nothing to the terminals of each circuit their outputs should be the same: this is the **open-circuit voltage** V_{OC}. Similarly, if we connect the output terminals together, each circuit should produce the same current: the **short-circuit current** I_{SC}. These equivalences allow us to deduce relationships between the various values used in the equivalent circuits.

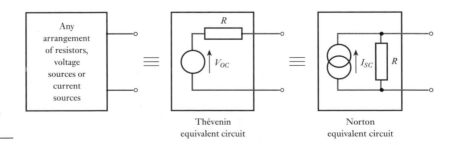

Figure 3.8
Thévenin and Norton equivalent circuits.

From the theorems above, it is clear that the same resistance R is used in each equivalent circuit. If we look at the Thévenin circuit and consider the effect of joining the output terminals together it is clear, from Ohm's law, that the resultant current I_{SC} would be given by

$$I_{SC} = \frac{V_{OC}}{R}$$

Similarly, looking at the Norton circuit we can see, again from Ohm's law, that the open-circuit voltage V_{OC} is given by

$$V_{OC} = I_{SC}R$$

Rearranging either of these relationships gives us the same result, which is that the resistance in the two equivalent circuits is given by

$$R = \frac{V_{OC}}{I_{SC}} \tag{3.4}$$

Therefore, the resistance can be determined from the open-circuit voltage and the short-circuit current, or the resistance seen 'looking into' the output of the circuit with the effects of the voltage and current sources removed.

Values for the components required to model a particular circuit can be found by analysis of its circuit diagram or by taking measurements on a physical circuit. These approaches are illustrated in the following examples.

Example 3.3 Determine the Thévenin and Norton equivalent circuits of the following arrangement.

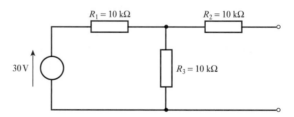

If nothing is connected across the output, no current can pass through R_2 and there will be no voltage drop across it. Therefore, the output voltage will be determined simply by the voltage source and the potential divider formed by R_1 and R_2. Since the two resistors are equal, the output voltage will be half the voltage of the source, so

$$V_{OC} = \frac{30}{2} = 15 \text{ V}$$

If the output terminals are shorted together, R_2 is effectively in parallel with R_3, so their combined resistance is $R_2//R_3 = 10 \text{ k}\Omega//10 \text{ k}\Omega = 5 \text{ k}\Omega$. The total resistance connected across the voltage source is therefore $R_1 + 5 \text{ k}\Omega = 15 \text{ k}\Omega$, and the current taken from the source will be $30 \text{ V}/15 \text{ k}\Omega = 2 \text{ mA}$. Since R_2 and R_3 are the same size and are in parallel, the current through each will be the same. Therefore, the current through each resistor will be $2 \text{ mA}/2 = 1 \text{ mA}$. Here the current through R_2 is the output current (in this case the short-circuit output current), so

$$I_{SC} = 1 \text{ mA}$$

From Equation 3.4, we know that the resistance in Thévenin and Norton equivalent circuits is given by the ratio of V_{OC} to I_{SC}, therefore

$$R = V_{OC}/I_{SC} = 15\,\text{V}/1\,\text{mA} = 15\,\text{k}\Omega$$

Alternatively, R could be obtained by noting the effective resistance seen looking into the output of the circuit with the voltage source replaced by its internal resistance. The internal resistance of an ideal voltage source is zero, so this produces a circuit where R_1 is effectively in parallel with R_3. The resistance seen at the output is therefore $R_2 + (R_1//R_3) = 10\,\text{k}\Omega + (10\,\text{k}\Omega//10\,\text{k}\Omega) = 15\,\text{k}\Omega$, as before.

Therefore our equivalent circuits are

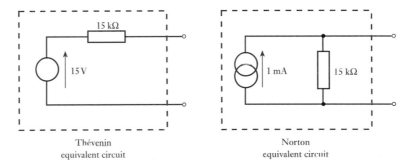

Thévenin
equivalent circuit

Norton
equivalent circuit

To obtain the equivalent circuit for an actual circuit (rather than from its circuit diagram), we make appropriate measurements of its characteristics. Since we can obtain the necessary values from a knowledge of the open-circuit voltage V_{OC} and the short-circuit current I_{SC}, one approach is simply to measure these quantities. If a high-resistance voltmeter is placed across the output of the circuit this will give a reasonable estimate of its open-circuit output voltage, provided that the input resistance of the meter is high compared with the output resistance of the circuit. However, direct measurement of the short-circuit current is often more difficult, since shorting out the circuit may cause damage. An alternative approach is to take other measurements and to use these to deduce V_{OC} and I_{SC}.

Example 3.4 A two-terminal network, which has an unknown internal circuit, is investigated by measuring its output voltage when connected to different loads. It is found that when a resistance of 25 Ω is connected the output voltage is 2 V, and when a load of 400 Ω is connected the output voltage is 8 V. Determine the Thévenin and Norton equivalent circuits of the unknown circuit.

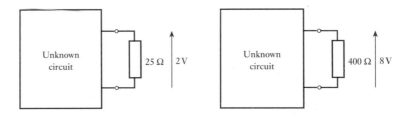

Method 1

One approach to this problem is to plot a graph of the output current against the output voltage. When the output voltage is 2 V the output current is $2\,\text{V}/25\,\Omega = 80\,\text{mA}$, and when the output voltage is 8 V the output current is 8 V/400 Ω = 20 mA. This gives the following graph.

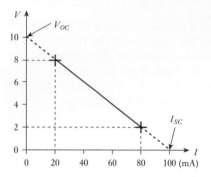

Extrapolating this line shows that when the current is zero the voltage is 10 V (the open-circuit voltage), and when the voltage is zero the current is 100 mA (the short-circuit current). From Equation 3.4

$$R = \frac{V_{OC}}{I_{SC}} = \frac{10 \text{ V}}{100 \text{ mA}}$$

$$= 100 \, \Omega$$

Therefore the equivalent circuits are

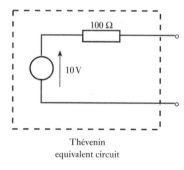

Thévenin
equivalent circuit

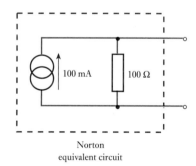

Norton
equivalent circuit

Method 2

Non-graphical methods are also available. For example, if we assume that the circuit is replaced by a Thévenin equivalent circuit consisting of a voltage source V_{OC} and a resistor R, we have the following arrangements.

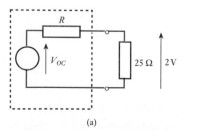

(a)

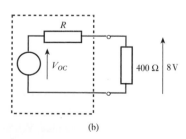

(b)

Applying the equation of a potential divider to (a) and (b) gives

$$V_{OC}\frac{25}{R+25}=2 \quad \text{and} \quad V_{OC}\frac{400}{R+400}=8$$

which can be rearranged to give the simultaneous equations

$$25V_{OC}=2R+50$$

$$400V_{OC}=8R+3200$$

which can be solved in the normal way to give $V_{OC}=10$ and $R=100$. The value of I_{SC} can now be found using Equation 3.4, and we have the same values as before.

3.9 Superposition

When a circuit contains more than one energy source, we can often simplify the analysis by applying the **principle of superposition**. This allows the effects of each voltage and current source to be calculated separately and then added to give their combined effect. More precisely, the principle states:

> *In any linear network of resistors, voltage sources and current sources, each voltage and current in the circuit is equal to the algebraic sum of the voltages or currents that would be present if each source were to be considered separately. When determining the effects of a single source, the remaining sources are replaced by their internal resistance.*

The use of superposition is most easily appreciated through the use of examples.

Example 3.5 Calculate the output voltage V of the following circuit.

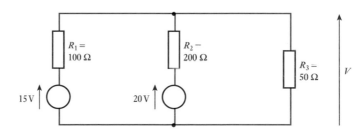

First we consider the effect of the 15 V source. When doing this we replace the other voltage source by its internal resistance, which for an ideal voltage source is zero (that is, a short circuit). This gives us the following circuit.

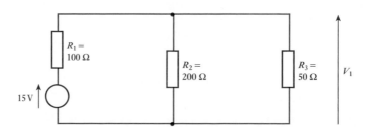

This is a potential divider circuit formed by R_1 and the parallel combination of R_2 and R_3. Using the equations for resistors in parallel and for a potential divider, this gives

$$V_1 = 15 \frac{200//50}{100 + 200//50}$$

$$= 15 \frac{40}{100 + 40}$$

$$= 4.29 \text{ V}$$

Next we consider the effect of the 20 V source, replacing the 15 V source with a short circuit. This gives

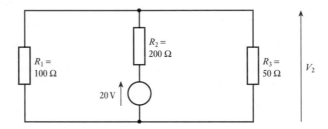

This is a potential divider circuit formed by R_2 and the parallel combination of R_1 and R_3, which gives

$$V_2 = 20 \frac{100//50}{200 + 100//50}$$

$$= 20 \frac{33.3}{200 + 33.3}$$

$$= 2.86 \text{ V}$$

Note that R_1 and R_3 are in parallel – they could equally well be drawn side by side. The output of the original circuit is now found by summing these two voltages

$$V = V_1 + V_2 = 4.29 + 2.86 = 7.15 \text{ V}$$

File 3A

Computer simulation exercise 3.1

Use circuit simulation to investigate the circuit of Example 3.5. Determine the magnitude of the voltage V and confirm that this is as expected.

The effective internal resistance of a current source is infinite. Therefore, when removing the effects of a current generator, we replace it by an open circuit. This is illustrated in the following example.

Example 3.6 | Calculate the output current *I* in the following circuit.

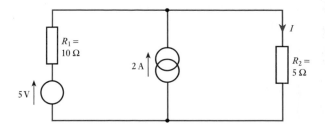

First we consider the voltage source. When doing this we replace the current source by its internal resistance, which for an ideal current source is infinite (that is, an open circuit). This gives the following circuit.

Therefore

$$I_1 = \frac{5\,\text{V}}{10\,\Omega + 5\,\Omega} = 0.33\,\text{A}$$

Next we consider the effect of the current source, replacing the voltage source with a short circuit. This gives

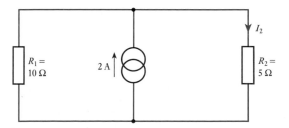

The two resistors are effectively in parallel across the current source. Since their combined resistance is $10\,\Omega//5\,\Omega = 3.33\,\Omega$, the voltage across the two resistors will be $2\,\text{A} \times 3.33\,\Omega = 6.66\,\text{V}$. Therefore, the current I_2 is given by

$$I_2 = \frac{6.66\,\text{V}}{5\,\Omega} = 1.33\,\text{A}$$

The output current of the original circuit is now found by summing these two currents

$$I = I_1 + I_2 = 0.33 + 1.33 = 1.66\,\text{A}$$

File 3B

Computer simulation exercise 3.2
Use circuit simulation to investigate the circuit of Example 3.6. Determine the magnitude of the current *I* and confirm that this is as expected.

Video 3B

3.10 Nodal analysis

In Section 3.7, we saw that we can apply Kirchhoff's current law to any node in a circuit and Kirchhoff's voltage law to any loop. The analysis of real circuits often requires us to apply these laws to a group of nodes or loops, and this produces a series of simultaneous equations that must be solved to find the various voltages and currents in the circuit. Unfortunately, as the complexity of the circuit increases, the number of nodes and loops increases and the analysis becomes more involved. In order to simplify this process, we often use one of two systematic approaches to the production of these simultaneous equations, namely **nodal analysis** and **mesh analysis**. We will look at nodal analysis in this section and at mesh analysis in the next.

Nodal analysis is a systematic method of applying Kirchhoff's current law to nodes in a circuit in order to produce an appropriate set of simultaneous equations. The technique involves six distinct steps:

1   One of the nodes in the circuit is chosen as a reference node. This selection is arbitrary, but it is normal to select the ground or earth node as the reference point, and all voltages will then be measured with respect to this node.
2   The voltages on the remaining nodes in the circuit are then labelled V_1, V_2, V_3, etc. Again, the numbering of these node voltages is arbitrary.
3   If the voltages on any of the nodes are known (due to the presence of fixed-voltage sources), then these values are added to the diagram.
4   Kirchhoff's current law is then applied to each node for which the voltage is not known. This produces a set of simultaneous equations.
5   These simultaneous equations are solved to determine each unknown node voltage.
6   If necessary, the node voltages are then used to calculate appropriate currents in the circuit.

To illustrate the use of this technique consider Figure 3.9(a), which shows a relatively simple circuit. Although none of the nodes is specifically labelled as the earth, we will choose the lower point in the circuit as our reference node. We then label the three remaining node voltages as V_1, V_2 and V_3 as shown in Figure 3.9(b). It is clear that V_1 is equal to E, since this is set by the voltage source, and this is marked on the diagram. The next step is to apply Kirchhoff's current law to each node for which the voltage is unknown. In this case only V_2 and V_3 are unknown, so we need only consider these two nodes.

Let us initially consider the node associated with V_2. Figure 3.9(c) labels the currents flowing into this node as I_A, I_B and I_C, and by applying Kirchhoff's current law we see that

$$I_A + I_B + I_C = 0$$

These currents can easily be determined from the circuit diagram. Each current is given by the voltages across the associated resistor, and in each case this voltage is simply the difference between two node voltages. Therefore

$$I_A = \frac{V_1 - V_2}{R_A} = \frac{E - V_2}{R_A} \quad I_B = \frac{V_3 - V_2}{R_B} \quad I_C = \frac{0 - V_2}{R_C}$$

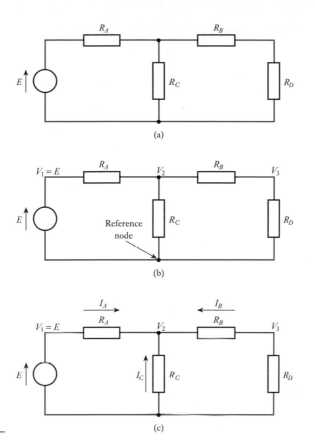

Figure 3.9
The application of
nodal analysis.

Note that in each case we want the current flowing *into* the node associated with V_2, and therefore we subtract V_2 from the other node voltage in each case. Summing these currents then gives

$$\frac{E - V_2}{R_A} + \frac{V_3 - V_2}{R_B} + \frac{0 - V_2}{R_C} = 0$$

A similar treatment of the node associated with V_3 gives

$$\frac{V_2 - V_3}{R_B} + \frac{0 - V_3}{R_D} = 0$$

We therefore have two equations that we can solve to find V_2 and V_3. From these values we can then calculate the various currents if necessary.

The circuit of Figure 3.9 contains a single voltage source, but nodal analysis can be applied to circuits containing multiple voltage sources or current sources. In a voltage source, the voltage across the device is known but the current flowing through it is not. In a current source, the current flowing through it is known but the voltage across it is not.

Example 3.7 | Calculate the current I_1 in the following circuit.

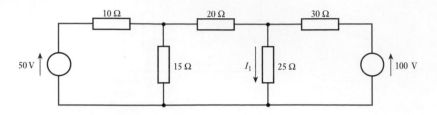

First we pick our reference node and label the various node voltages, assigning values where these are known.

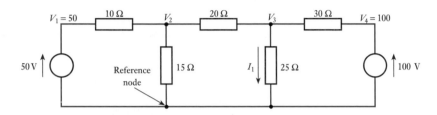

Next we sum the currents flowing into the nodes for which the node voltages are unknown. This gives

$$\frac{50-V_2}{10}+\frac{V_3-V_2}{20}+\frac{0-V_2}{15}=0$$

and

$$\frac{V_2-V_3}{20}+\frac{100-V_2}{30}+\frac{0-V_3}{25}=0$$

Solving these two equations (which is left as an exercise for the reader) gives

$$V_2 = 32.34\,\text{V}$$

$$V_3 = 40.14\,\text{V}$$

and the required current I_1 is given by

$$I_1 = \frac{V_3}{25\,\Omega} = \frac{40.14\,\text{V}}{25\,\Omega} = 1.6\,\text{A}$$

File 3C

Computer simulation exercise 3.3

Use circuit simulation to investigate the circuit of Example 3.7. Determine the voltages V_2 and V_3 and the current I_1 and confirm that these are as expected.

Video 3C

3.11 Mesh analysis

Mesh analysis, like nodal analysis, is a systematic means of obtaining a series of simultaneous equations describing the behaviour of a circuit. In this case, Kirchhoff's voltage law is applied to each mesh in the circuit. Again a series of steps are involved:

1 Identify the meshes in the circuit and assign a clockwise-flowing current to each. Label these currents I_1, I_2, I_3, etc.
2 Apply Kirchhoff's law to each mesh by summing the voltages clockwise around each mesh, equating the sum to zero. This produces a number of simultaneous equations (one for each loop).
3 Solve these simultaneous equations to determine the currents I_1, I_2, I_3, etc.
4 Use the values obtained for the various currents to compute voltages in the circuit as required.

This process is illustrated in Figure 3.10. The circuit of Figure 3.10(a) contains two meshes, which are labelled in Figure 3.10(b). Next we need to define the polarities of the various voltages in the circuit. This assignment is arbitrary at this stage, but it will enable us to interpret correctly the polarity of the voltages that we calculate. Figure 3.10(c) shows the way in which the various voltages have been defined. Note that a positive current in one direction in a resistor produces a voltage drop in the other direction. Thus, in Figure 3.10(c), if I_1 is positive, V_A will also be positive.

Having defined our various voltages and currents, we are now in a position to write down our equations. We do this by summing the voltages clockwise around each loop and equating this sum to zero. For the first mesh, this gives

$$E - V_A - V_C = 0$$
$$E - I_1 R_A - (I_1 - I_2)R_C = 0$$

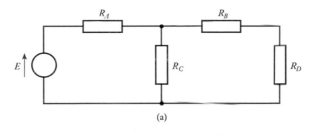

(a)

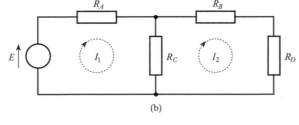

(b)

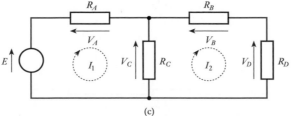

(c)

Figure 3.10
Mesh analysis.

Note that only I_1 flows through R_A, so V_A is simply I_1R_A. However, in R_C the current I_1 flows in one direction, while I_2 flows in the other direction. Consequently, the voltage across this resistor is $(I_1 - I_2)R_C$. Applying the same approach to the second loop gives

$$V_C - V_B - V_D = 0$$

$$(I_1 - I_2)R_C - I_2R_B - I_2R_D = 0$$

Hence we have two equations relating I_1 and I_2 to the circuit values. These simultaneous equations can be solved to obtain their values, which can then be used to calculate the various voltages.

As with nodal analysis, mesh analysis can be applied to circuits with any number of voltage or current sources.

Example 3.8 **Calculate the voltage across the 10 Ω resistor in the following circuit.**

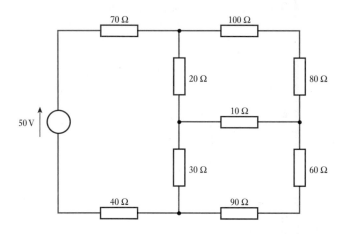

The circuit contains three meshes. To these we assign loop currents I_1, I_2 and I_3, as shown below. The diagram also defines the various voltages and, to aid explanation, assigns symbolic names to each resistor.

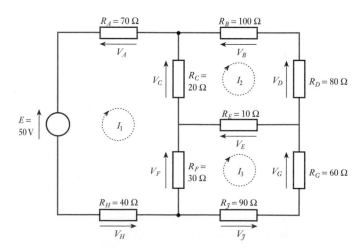

The next stage is to apply Kirchhoff's voltage law to each mesh. Normally, we would write down these equations directly using the component values and currents. However, to aid comprehension of the process, we will first write these symbolically. Considering the three loops in turn, this gives

$$E - V_A - V_C - V_F - V_H = 0$$

$$V_C - V_B - V_D + V_E = 0$$

$$V_F - V_E - V_G - V_J = 0$$

which gives the following simultaneous equations:

$$50 - 70I_1 - 20(I_1 - I_2) - 30(I_1 - I_3) - 40I_1 = 0$$

$$20(I_1 - I_2) - 100I_2 - 80I_2 + 10(I_3 - I_2) = 0$$

$$30(I_1 - I_3) - 10(I_3 - I_2) - 60I_3 - 90I_3 = 0$$

These can be rearranged to give

$$50 - 160I_1 + 20I_2 + 30I_3 = 0$$

$$20I_1 - 210I_2 + 10I_3 = 0$$

$$30I_1 + 10I_2 - 190I_3 = 0$$

and these three simultaneous equations may be solved to give

$$I_1 = 326 \text{ mA}$$

$$I_2 = 34 \text{ mA}$$

$$I_3 = 53 \text{ mA}$$

The voltage across the 10 Ω resistor is the product of its resistance and the current through it, so

$$V_E = R_E(I_3 - I_2)$$

$$= 10(0.053 - 0.034)$$

$$= 0.19 \text{ V}$$

Since the calculated voltage is positive, the polarity is as shown by the arrow, with the left-hand end of the resistor more positive than the right-hand end.

File 3D

Computer simulation exercise 3.4

Use circuit simulation to investigate the circuit of Example 3.8. Determine the three currents I_1, I_2 and I_3 and the voltage V_E and confirm that these are as expected.

The direction in which the currents are defined and the direction in which the voltages are summed are arbitrary. However, if you are consistent in the direction that you choose you are less likely to make mistakes. This is why the list of steps above specifies a clockwise direction in each case.

3.12 | Solving simultaneous circuit equations

We have seen that both nodal analysis and mesh analysis produce a series of simultaneous equations that must be solved in order to determine the required voltages and currents. When considering simple circuits with few nodes or meshes, the number of equations generated may be small enough to be solved 'by hand' as illustrated above. However, with more complex circuits this approach becomes impractical.

A more attractive method of solving simultaneous equations is to express them in matrix form and to use the various tools of matrix algebra to obtain the solutions. For example, in Example 3.8 we obtained the following set of equations:

$$50 - 160I_1 + 20I_2 + 30I_3 = 0$$

$$20I_1 - 210I_2 + 10I_3 = 0$$

$$30I_1 + 10I_2 - 190I_3 = 0$$

These can be rearranged as

$$160I_1 - 20I_2 - 30I_3 = 50$$

$$20I_1 - 210I_2 + 10I_3 = 0$$

$$30I_1 + 10I_2 - 190I_3 = 0$$

and expressed in matrix form as

$$\begin{bmatrix} 160 & -20 & -30 \\ 20 & -210 & 10 \\ 30 & 10 & -190 \end{bmatrix} \begin{bmatrix} I_1 \\ I_2 \\ I_3 \end{bmatrix} = \begin{bmatrix} 50 \\ 0 \\ 0 \end{bmatrix}$$

This can be solved by hand using Cramer's rule or some other matrix algebra technique. Alternatively, automated tools can be used. Many scientific calculators can solve such problems when they involve a handful of equations, while computer-based packages such as MATLAB or Mathcad can solve large numbers of simultaneous equations if expressed in this form.

3.13 | Choice of techniques

In this chapter, we have looked at a number of techniques that can be used to analyse electrical circuits. This raises the question of how we know which technique to use in a given situation. Unfortunately, there are no simple rules to aid this choice, and it often comes down to the general form of the circuit and which technique seems most appropriate. Techniques such as nodal and mesh analysis will work in a wide range of situations, but these are not always the simplest methods. To investigate the choice of technique in a given situation, take a look at the further study section below.

For any given circuit some approaches will be easier than others, and inevitably, your skill in picking the simplest method will increase with practice. While the analysis of simple circuits is generally straightforward, more complex circuits can be very time consuming. In these cases, we normally make use of computer-based network analysis tools. These often use nodal analysis and, if necessary, can handle circuits of great complexity. However, in many cases the manual techniques described in this chapter are completely adequate.

Further study

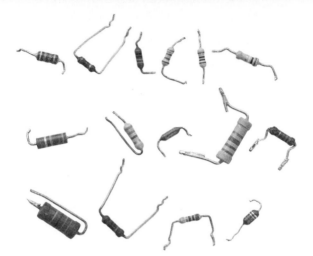

To see how a range of techniques can be used to analyse a given circuit, consider the circuit below.

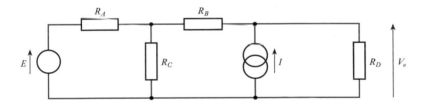

This circuit can be analysed by nodal or mesh analysis as described earlier. Alternatively, the effects of each source could be investigated separately (using Ohm's law) and combined using the principle of superposition. Another approach is to use Thévenin's or Norton's theorems to simplify the circuit. Investigate these approaches and decide which is the easiest.

Key points

- An electric current is a flow of charge.
- A voltage source produces an e.m.f., which causes a current to flow in a circuit. An ideal voltage source has zero output resistance, but all real voltage sources have associated resistance.
- An ideal current source produces a constant current no matter what is connected to it. Such a source has an infinite output resistance.
- The current in a conductor is directly proportional to the voltage across it (this is Ohm's law). This voltage divided by the current gives the resistance of the conductor.
- The resistance of several resistors in series is given by the sum of their resistances.

- The resistance of several resistors in parallel is equal to the reciprocal of the sum of the reciprocals of their resistances.
- At any instant, the currents flowing into any node in a circuit sum to zero (Kirchhoff's current law).
- At any instant, the voltages around any loop in a circuit sum to zero (Kirchhoff's voltage law).
- Any two-terminal network of resistors and energy sources can be replaced by a series combination of a voltage source and a resistor (Thévenin's theorem).
- Any two-terminal network of resistors and energy sources can be replaced by a parallel combination of a current source and a resistor (Norton's theorem).
- In any linear network containing more than one energy source, the currents and voltages are equal to the sum of the voltages or currents that would be present if each source were considered separately (principle of superposition).
- Nodal analysis and mesh analysis each offer systematic methods of obtaining a set of simultaneous equations that can be solved to determine the voltages and currents in a circuit.
- When considering any particular circuit, a range of circuit analysis techniques can be used. The choice of technique should be based on an assessment of the nature of the circuit.

Exercises

3.1   Write down an equation relating current and charge.

3.2   What quantity of charge is transferred if a current of 5 A flows for 10 seconds?

3.3   What is the internal resistance of an ideal voltage source?

3.4   What is meant by a *controlled* voltage source?

3.5   What is the internal resistance of an ideal current source?

3.6   Determine the voltage V in each of the following circuits, being careful to note its polarity in each case.

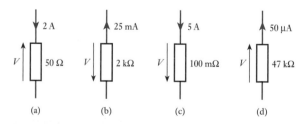

3.7   For each circuit in Exercise 3.6, determine the power dissipated in the resistor.

3.8   Estimate the resistance of a copper wire with a cross-sectional area of 1 mm² and a length of 1 m at 0 °C.

3.9 Determine the resistance of each of the following combinations.

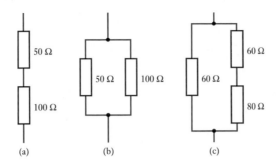

3.10 What resistance corresponds to 10 kΩ//10 kΩ?

3.11 Define the terms 'node', 'loop' and 'mesh'.

3.12 Derive Thévenin and Norton equivalent circuits for the following arrangements.

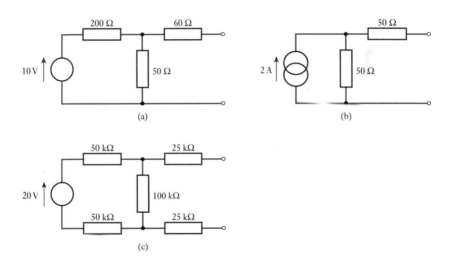

3.13 A two-terminal network is investigated by measuring the output voltage when connected to different loads. When a resistance of 12 Ω is connected across the output the output voltage is 16 V, and when a load of 48 Ω is connected the output voltage is 32 V. Use a graphical method to determine the Thévenin and Norton equivalent circuits of this arrangement.

3.14 Repeat Exercise 3.13 using a non-graphical approach.

3.15 Use the principle of superposition to determine the voltage V in each of the following circuits.

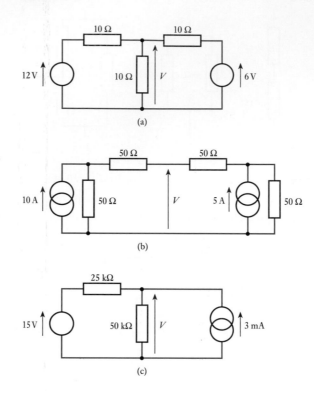

(a)

(b)

(c)

3.16 Use nodal analysis to determine the voltage V in the following circuit.

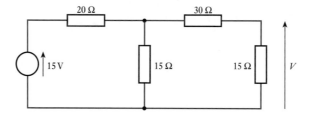

3.17 Simulate the circuit of Exercise 3.16 and hence confirm your answer to this exercise.

3.18 Use nodal analysis to determine the current I_1 in the following circuit.

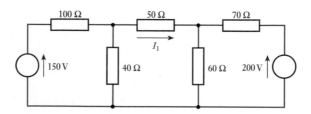

3.19 Simulate the circuit of Exercise 3.18 and hence confirm your answer to this exercise.

3.20 Use nodal analysis to determine the current I_1 in the following circuit.

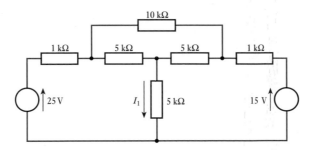

3.21 Simulate the circuit of Exercise 3.20 and hence confirm your answer to this exercise.

3.22 Use mesh analysis to determine the voltage V in the following circuit.

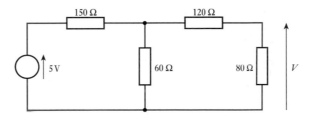

3.23 Simulate the circuit of Exercise 3.22 and hence confirm your answer to this exercise.

3.24 Use mesh analysis to determine the voltage V in the following circuit.

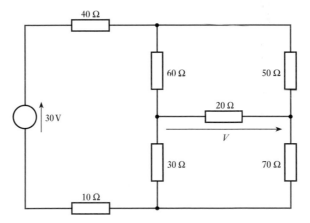

3.25 Simulate the circuit of Exercise 3.24 and hence confirm your answer to this exercise.

3.26 Use mesh analysis to determine the current I in the following circuit.

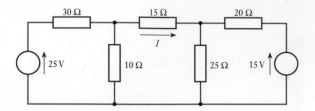

3.27 Simulate the circuit of Exercise 3.26 and hence confirm your answer to this exercise.

3.28 Use an appropriate form of analysis to determine the voltage V_o in the following circuit.

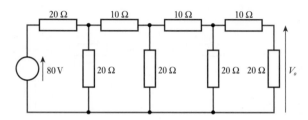

3.29 Simulate the circuit of Exercise 3.28 and hence confirm your answer to this exercise.

Chapter 4

Capacitance and Electric Fields

Objectives

When you have studied the material in this chapter, you should be able to:

- describe the construction and form of a range of capacitors
- explain the mechanism whereby charge is stored in a capacitor and the relationship between quantities such as charge, voltage, current and capacitance
- define terms such as absolute permittivity and relative permittivity, and use these quantities to calculate the capacitance of a component from its dimensions
- explain concepts such as electric field strength and electric flux density, and calculate the magnitudes of these quantities
- determine the effective capacitance of a number of capacitors when connected in series or parallel
- describe the relationship between the voltage and the current in a capacitor for both DC and AC signals
- calculate the energy stored in a charged capacitor.

4.1 Introduction

An electric current represents a flow of electric charge (see Chapter 3). A capacitor is a component that can store electric charge and can therefore store energy. Capacitors are often used in association with alternating currents and voltages, and they are a key component in almost all electronic circuits.

4.2 Capacitors and capacitance

Capacitors consist of two conducting surfaces separated by an insulating layer called a **dielectric**. Figure 4.1(a) shows a simple capacitor consisting of two rectangular metal sheets separated by a uniform layer of dielectric. The gap between the conducting layers could be filled with air (since this is a good insulator) or with another dielectric material. In some cases the conductors are made of metal foil and the dielectric is flexible, allowing the arrangement to be rolled into a cylindrical shape as in Figure 4.1(b). In integrated circuits, a capacitor might be formed by depositing a layer of metal onto an insulating layer above a conducting semiconductor layer as shown in Figure 4.1(c). A wide range of construction methods are used to form capacitors, but in each case their basic operation is the same.

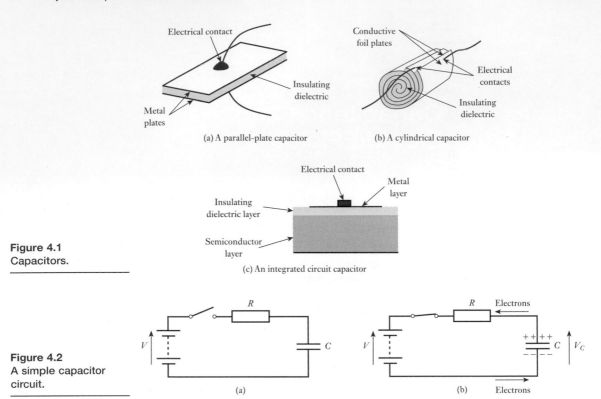

Figure 4.1
Capacitors.

Figure 4.2
A simple capacitor
circuit.

To understand the operation of a capacitor, consider the circuit of Figure 4.2(a). Here a battery, a resistor, a switch and a capacitor are connected in series. While the switch is open no current will flow around the circuit. However, when the switch is closed (as in Figure 4.2(b)), the e.m.f. produced by the battery will attempt to drive a current around the network. Electrons flowing from the negative terminal of the battery will flow onto the lower plate of the capacitor, where they will repel electrons from the upper plate. As electrons are repelled from this plate, they leave a residual positive charge (in the form of a deficit of negatively charged electrons). The combined effect of these two processes is that electrons flow into the lower plate of the capacitor and flow out from the upper plate. Note that this represents a flow of conventional current in the opposite direction.

Since electrons flow *into* one side of the capacitor and flow *out of* the other, it might appear that current is flowing *through* it. However, this is an illusion. The dielectric between the plates of the capacitor is an insulator, and electrons do not actually cross this barrier. It should also be noted that this flow of electrons cannot last indefinitely. As electrons flow around the circuit they produce an increasing positive charge on one side of the capacitor and an increasing negative charge on the other. The result is an increasing **electric field** between the two plates. This produces a potential difference V_C across the capacitor, which opposes the e.m.f. of the battery. Eventually, the voltage across the capacitor is equal to that of the battery and the current falls to zero.

In this state, the capacitor is storing electric charge and is therefore storing electrical energy. If the switch is opened at this point there is no path by which this charge can flow, and the capacitor will remain charged with a voltage of V_C across it. If now a resistor is connected across the charged capacitor, the stored energy will drive a current through the resistor, discharging the capacitor and releasing the stored energy. The capacitor therefore acts a little like a 'rechargeable battery', although the mechanism used to store the electrical energy is very different and the amount of energy stored is normally very small.

For a given capacitor, the charge stored Q is directly proportional to the voltage across it V. The relationship between these two quantities is determined by the capacitance C of the capacitor, such that

$$C = \frac{Q}{V}$$

(4.1)

If the charge is measured in *coulombs* and the voltage in *volts*, then the capacitance has the units of **farads**.

Example 4.1

A 10 µF capacitor has 10 V across it. What quantity of charge is stored in it?

From Equation 4.1 we have

$$C = \frac{Q}{V}$$
$$Q = CV$$
$$= 10^{-5} \times 10$$
$$= 100 \ \mu C$$

Remember that C is the unit symbol for coulombs, while C is the symbol used for capacitance.

4.3   Capacitors and alternating voltages and currents

It is clear from the discussion above that a constant current (that is, a direct current) cannot flow through a capacitor. However, since the voltage across a capacitor is proportional to the charge on it, it follows that a changing voltage must correspond to a changing charge. Consequently, a changing voltage must correspond to a current flowing into or out of the capacitor. This is illustrated in Figure 4.3.

The voltage produced by the alternating voltage source in Figure 4.3 is constantly changing. When this voltage becomes more positive, this will produce a positive current *into* the top plate of the capacitor (a flow of electrons out of the top plate), making this plate more positive. When the source voltage becomes more negative, this will cause current to flow *out of* the top plate (a flow of electrons into this plate), making this plate more negative. Thus the alternating voltage produces an alternating current around the complete circuit.

It is important to remember at this point that the observed current does not represent electrons flowing from one plate to the other – these plates are separated by an insulator. However, the circuit behaves as if an alternating current is flowing through

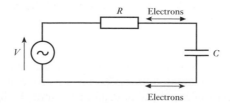

Figure 4.3
A capacitor and an alternating voltage.

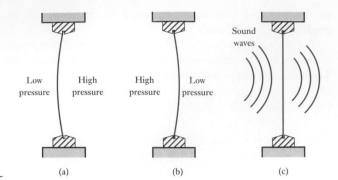

Figure 4.4
A mechanical analogy
of a capacitor – a
window.

the capacitor. Perhaps one way of understanding this apparent paradox is to consider a mechanical analogy, as shown in Figure 4.4.

Figure 4.4(a) shows a section through a glass window that is fitted within a window frame. In this figure, the pressure on one side of the glass is higher than on the other, causing a deflection of the glass. In deflecting the glass in this way the air has done work on it, and energy is now stored in the glass (as it would be in a spring under tension). Figure 4.4(b) shows a similar position where the imbalance in air pressure is reversed. Again the glass is deflected and is storing mechanical energy. In Figure 4.4(c), the average pressure on each side of the glass is equal, but sound waves are striking the glass on one side. Sound waves represent fluctuations in air pressure, and these will cause the window to vibrate backwards and forwards. This in turn will cause sound waves to be generated on the other side of the glass. We can see that a constant pressure difference across the window does not result in the passage of air through it. However, an alternating pressure difference (sound) is transmitted from one side to the other, even though no air passes through the window.

While a window is not a perfect analogy for a capacitor, it does illustrate quite well the way that a capacitor *blocks* direct currents but appears to *pass* alternating currents. However, it should be noted that, just as a window attenuates sounds that pass through it, so a capacitor impedes the flow of the current that appears to pass through it. In the case of a window, the attenuation will depend on the size of the window and the nature of the sound (its frequency range). Similarly, the effect of the capacitor will depend on its size (its capacitance) and the frequency of the signals present. We will return to look at this phenomenon in more detail when we look at alternating voltages and currents (see Chapter 6).

4.4   The effect of a capacitor's dimensions on its capacitance

The capacitance of a capacitor is directly proportional to the area A of the conducting plates and inversely proportional to the distance d between them. Therefore $C \propto A/d$. The constant of proportionality of this relationship is the **permittivity** ε of the dielectric. The permittivity is normally expressed as the product of two terms: the **absolute permittivity** ε_0 and the **relative permittivity** ε_r of the dielectric used,

$$C = \frac{\varepsilon A}{d} = \frac{\varepsilon_0 \varepsilon_r A}{d} \tag{4.2}$$

ε_0 is also referred to as the **permittivity of free space** and is the permittivity of a vacuum. It has a value of about 8.85 picofarads per metre (pF/m). ε_r represents the ratio of the permittivity of a material to that of a vacuum. Air has an ε_r very close to 1, while insulators have values from about 2 to a thousand or more. While capacitors can be produced using air as the dielectric material, much smaller components can be produced by using a material with a much higher relative permittivity.

Example 4.2

The conducting plates of a capacitor are 10×25 mm and have a separation of 7 μm. If the dielectric has a relative permittivity of 100, what will be the capacitance of the device?

From Equation 4.2 we have

$$C = \frac{\varepsilon_0 \varepsilon_r A}{d}$$

$$= \frac{8.85 \times 10^{-12} \times 100 \times 10 \times 10^{-3} \times 25 \times 10^{-3}}{7 \times 10^{-6}} = 31.6 \text{ nF}$$

Note that to achieve the same capacitance using air as the dielectric we would need a device of 100×250 mm (assuming the same plate separation).

Capacitance is present not only between the plates of a capacitor but also between any two conductors that are separated by an insulator. Therefore, a small amount of capacitance exists between each of the conductors in electrical circuits (for example, between each wire) and between the various elements in electrical components. These small, unintended capacitances are called stray capacitances and can have a very marked effect on circuit behaviour. The need to charge and discharge these stray capacitances limits the speed of operation of circuits and is a particular problem in high-speed circuits and those that use small signal currents.

4.5   Electric field strength and electric flux density

Electric charges of the same polarity repel each other, while those of opposite polarities attract. When charged particles experience a force as a result of their charge, we say that an electric field exists in that region. The magnitude of the force exerted on a charged particle is determined by the **electric field strength**, E, at that point in space. This quantity is defined as the force exerted on a unit charge at that point. When a voltage V exists between two points a distance d apart, the electric field strength is given by

$$E = \frac{V}{d} \tag{4.3}$$

and has units of volts per metre (V/m).

The charge stored in a capacitor produces a potential across the capacitor and an electric field across the dielectric material. This is illustrated in Figure 4.5, which shows a capacitor that has two plates each of area A, separated by a dielectric of thickness d. The capacitor holds a charge of Q and the potential across the capacitor is V. From Equation 4.3, we know that the electric field strength in the dielectric material is V/d volts per metre.

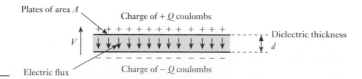

Figure 4.5
A charged capacitor.

Example 4.3

The conducting plates of a capacitor have a separation of 10 μm. If the potential across the capacitor is 100 V, what is the electric field strength in the dielectric?

From Equation 4.3 we have

$$E = \frac{V}{d}$$
$$= \frac{100}{10^{-5}}$$
$$= 10^7 \text{ V/m}$$

All insulating materials have a maximum value for the field strength that they can withstand before they break down. This is termed their **dielectric strength** E_m. For this reason, all capacitors have a maximum operating voltage, which is related to the material used for the dielectric and its thickness. From Equation 4.2, it is clear that, to obtain the maximum capacitance for a device of a given plate size, we want to make the dielectric as thin as possible. Equation 4.3 shows that in doing this we increase the electrical stress on the insulating material, and in practice we need to compromise between physical size and breakdown voltage.

The force between positive and negative charges is often described in terms of an **electric flux** linking them. This is measured using the same units as electric charge (coulombs) and thus a charge of Q coulombs will produce a total electric flux of Q coulombs. We also define what is termed the **electric flux density**, D, as the amount of flux passing through a defined area perpendicular to the flux. In a capacitor, the size of the plates is always much greater than their separation, so 'edge effects' can be ignored and we can assume that all the flux produced by the stored charge passes through the area of the dielectric. Returning to Figure 4.5, we see that here a total flux of Q passes through an area of A, so the electric flux density is given by

$$D = \frac{Q}{A} \qquad (4.4)$$

Example 4.4

The conducting plates of a capacitor have an area of 200 mm². If the charge on the capacitor is 15 μC, what is the electric flux density within the dielectric?

From Equation 4.4 we have

$$D = \frac{Q}{A}$$
$$= \frac{15 \times 10^{-6}}{200 \times 10^{-6}}$$
$$= 75 \text{ mC/m}^2$$

By combining the results of Equations 4.1–4.4, it is relatively easy to show that

$$\varepsilon = \frac{D}{E} \qquad (4.5)$$

and thus the permittivity of the dielectric in a capacitor is equal to the ratio of the electric flux density to the electric field strength.

Video 4A

4.6 Capacitors in series and in parallel

We have already considered the combined effect of putting several resistors in series and in parallel (see Chapter 3), and it is appropriate to look at combinations of capacitors in the same way. We will begin by looking at a parallel arrangement.

4.6.1 Capacitors in parallel

Consider a voltage V applied across two capacitors C_1 and C_2 connected in parallel, as shown in Figure 4.6(a). If we call the charge on these two capacitors Q_1 and Q_2, then

$$Q_1 = VC_1 \quad \text{and} \quad Q_2 = VC_2$$

If the two capacitors are now replaced by a single component with a capacitance of C such that this has the same capacitance as the parallel combination, then clearly the charge stored on C must be equal to $Q_1 + Q_2$. Therefore

$$\text{charge stored on } C = Q_1 + Q_2$$
$$VC = VC_1 + VC_2$$
$$C = C_1 + C_2$$

Therefore, the effective capacitance of two capacitors in parallel is equal to the sum of their capacitances. This result can clearly be extended to any number of components, and in general, for N capacitors in parallel,

$$C = C_1 + C_2 + \cdots + C_N \tag{4.6}$$

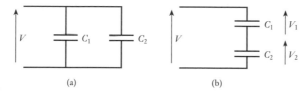

(a)　　　(b)

Figure 4.6
Capacitors in parallel
and in series.

Example 4.5　What is the effective capacitance of this arrangement?

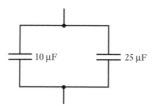

From Equation 4.6 we have

$$C = C_1 + C_2$$
$$= 10 + 25$$
$$= 35 \ \mu\text{F}$$

4.6.2 Capacitors in series

Consider a voltage V applied across two capacitors C_1 and C_2 connected in series, as shown in Figure 4.6(b). In this series arrangement, the only charge that can be delivered to the lower plate of C_1 is the charge supplied from the upper plate of C_2 and therefore the charge on each capacitor must be identical. We will call the charge on each capacitor Q.

If the two capacitors are now replaced by a single component with a capacitance of C such that this has the same capacitance as the series combination, then clearly the charge stored on C must also be equal to Q. From the diagram, it is clear that the applied voltage V is equal to $V_1 + V_2$ and therefore

$$V = V_1 + V_2$$

$$\frac{Q}{C} = \frac{Q}{C_1} + \frac{Q}{C_2}$$

$$\frac{1}{C} = \frac{1}{C_1} + \frac{1}{C_2}$$

Therefore, the effective capacitance of two capacitors in series is equal to the reciprocal of the sum of the reciprocals of their capacitances. This result can clearly be extended to any number of components, and in general, for N capacitors in series,

$$\frac{1}{C} = \frac{1}{C_1} + \frac{1}{C_2} + \cdots + \frac{1}{C_N} \tag{4.7}$$

Example 4.6 | What is the effective capacitance of this arrangement?

From Equation 4.7 we have

$$\frac{1}{C} = \frac{1}{C_1} + \frac{1}{C_2}$$

$$= \frac{1}{10} + \frac{1}{25} = \frac{35}{250}$$

$$C = 7.14 \ \mu\text{F}$$

4.7 Relationship between voltage and current in a capacitor

While the voltage across a resistor is directly proportional to the current flowing though it, this is *not* the case in a capacitor. From Equation 4.1, we know that the voltage across a capacitor is directly related to the *charge* on the capacitor, and we also know (Equation 3.2) that charge is given by the integral of the current with respect to time. Therefore, the voltage across a capacitor V is given by

$$V = \frac{Q}{C} = \frac{1}{C}\int I \, dt \tag{4.8}$$

One implication of this relationship is that the voltage across a capacitor cannot change instantaneously, since this would require an infinite current. The rate at which the voltage changes is controlled by the magnitude of the current. For this reason, capacitors tend to stabilise the voltage across them.

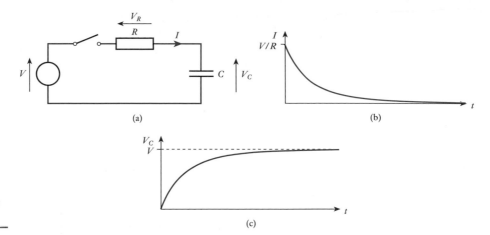

Figure 4.7
Relationship between
voltage and current in
a DC circuit.

Another way of looking at this relationship is to consider the current into the capacitor as a function of voltage, rather than the other way around. By differentiating our basic equation $Q = CV$, we obtain

$$\frac{dQ}{dt} = C\frac{dV}{dt}$$

and since dQ/dt is equal to current, this gives

$$I = C\frac{dV}{dt} \tag{4.9}$$

To investigate these relationships further, consider the circuit of Figure 4.7(a). If the capacitor is initially discharged, then the voltage across it will be zero (since $V = Q/C$). If now the switch is closed (at $t = 0$) then the voltage across the capacitor cannot change instantly, so initially $V_C = 0$. By applying Kirchhoff's voltage law around the circuit, it is clear that $V = V_R + V_C$, and if initially $V_C = 0$, then the entire supply voltage V will appear across the resistor. Therefore $V_R = V$, and the initial current flowing around the circuit, and thus into the capacitor, is given by $I = V/R$.

As the current flows into the capacitor its charge increases and V_C grows. This reduces the voltage across the resistor and hence the current. Therefore, as the voltage on the capacitor increases, the charging current decreases. The result is that the charging current is initially V/R but falls exponentially with time, and the voltage across the capacitor is initially zero but rises with time. Eventually, the voltage on the capacitor is virtually equal to the supply voltage and the charging current becomes negligible. This behaviour is shown in Figures 4.7(b) and 4.7(c).

File 4A

Computer simulation exercise 4.1

Simulate the circuit of Figure 4.7(a) with $V = 1$ V, $R = 1$ kΩ and $C = 200$ μF. Include in your circuit a switch that closes at $t = 0$ and another switch (not shown in Figure 4.7(a)) that opens at $t = 0$. This second switch should be connected directly across C to ensure that the capacitor is initially discharged. Use transient simulation to investigate the behaviour of the circuit during the first second after the switches change. Plot V_C and I against time on separate graphs and confirm that the circuit behaves as expected. Experiment with different values of the circuit components and note the effects on the voltage and current graphs.

It is clear from the above that the charging current is determined by R and the voltage across it. Therefore, increasing R will increase the time needed to charge up the capacitor, while reducing R will speed up the process. It is also clear that increasing the capacitance C will also increase the time taken for it to charge to a given voltage (since the voltage on a capacitor is inversely proportional to its capacitance for a given charge). Thus the time taken for the capacitor to charge to a particular voltage increases with both C and R. This leads to the concept of the **time constant** for the circuit, which is equal to the product CR. The time constant is given the symbol T (upper-case Greek letter *tau*). It can be shown that the charging rate of the circuit of Figure 4.7(a) is determined by the value of the time constant of the circuit rather than the actual values of C and R. We will return to look at the effects of the time constant in later chapters.

File 4A

Computer simulation exercise 4.2

Repeat Computer simulation exercise 4.1 noting the effect of different component values. Begin with the same values as in the previous exercise and then change the values of C and R while keeping their product constant. Again plot V_C and I against time on separate graphs and confirm that the characteristics are unchanged. Hence confirm that the characteristics are determined by the time constant CR rather than the actual values of C and R.

4.8 Sinusoidal voltages and currents

So far we have considered the relationship between voltage and current in a DC circuit containing a capacitor. It is also important to look at the situation when using sinusoidal quantities.

Consider the arrangement of Figure 4.8(a), where an alternating voltage is applied across a capacitor. Figure 4.8(b) shows the sinusoidal voltage waveform across the capacitor, which in turn dictates the charge on the capacitor and the current into it. From Equation 4.9, we know that the current into a capacitor is given by $C\,dV/dt$, so the current is directly proportional to the *time differential* of the voltage. Since the differential of a sine wave is a cosine wave, we obtain a current waveform as shown in Figure 4.8(c). The current waveform is phase shifted with respect to the voltage wave-

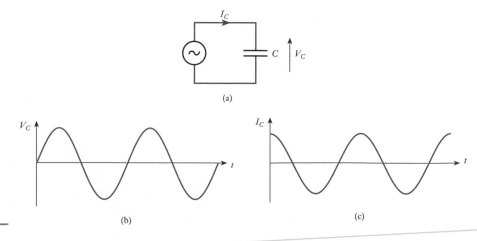

Figure 4.8
Capacitors and
alternating quantities.

form by 90° (or $\pi/2$ radians). It is also clear that the current waveform *leads* the voltage waveform. This is a very important property of capacitors, and we will return to look at the mathematics of this relationship later (see Chapter 6).

File 4B

> ### Computer simulation exercise 4.3
>
> Simulate the circuit of Figure 4.8(a) using any value of capacitor. Use a sinusoidal voltage source to apply a voltage of 1 V peak at 1 Hz and use transient analysis to display the voltage across the capacitor, and the current through it, over a period of several seconds. Note the phase relationship between the two waveforms and hence confirm that the current leads the voltages by 90° (or $\pi/2$ radians). Note the effect of varying the capacitor value, and the frequency used.

Another implication of the fact that the current into a capacitor is equal to $C\,dV/dt$ is that the magnitude of the current is determined by the rate at which the voltage changes. In a sinusoidal waveform, this rate of change is related to its frequency. This **frequency dependence** will be investigated in more detail later (see Chapter 6).

4.9 Energy stored in a charged capacitor

We noted earlier that capacitors store energy, and it is now time to quantify this effect. To move a charge Q through a potential difference V requires an amount of energy QV. As we progressively charge up a capacitor, we can consider that we are repeatedly adding small amounts of charge ΔQ by moving them through the voltage on the capacitor V. The energy needed to do this is clearly $V \times \Delta Q$. Since $Q = CV$, it follows that ΔQ is equivalent to $C\Delta V$ (since C is constant) and hence the energy needed to add an amount of charge ΔQ is $V \times C\Delta V$ or, rearranging, $CV\Delta V$. If we now take an uncharged capacitor and calculate the energy needed to add successive charges to raise the voltage to V, this is

$$E = \int_0^V CV\,dV = \frac{1}{2}CV^2 \tag{4.10}$$

Alternatively, since $V = Q/C$, the energy can be written as

$$E = \frac{1}{2}CV^2 = \frac{1}{2}C\left(\frac{Q}{C}\right)^2 = \frac{1}{2}\frac{Q^2}{C}$$

Thus the energy stored within a charged capacitor is $^1/_2CV^2$ or $^1/_2Q^2/C$. The unit of energy is the joule (J).

Example 4.7 | Calculate the energy stored in a 10 µF capacitor when it is charged to 100 V.

From Equation 4.10

$$E = \frac{1}{2}CV^2$$

$$= \frac{1}{2} \times 10^{-5} \times 100^2$$

$$= 50 \text{ mJ}$$

4.10	Circuit symbols

Although we have used a single circuit symbol to represent capacitors, other symbols are sometimes used to differentiate between different types of device. Figure 4.9 shows a range of symbols that are commonly used. Figure 4.9(a) shows the standard symbol for a fixed capacitor and Figure 4.9(b) the symbol for a variable capacitor. Some devices, such as **electrolytic capacitors**, use construction methods that mean that they can only be used with voltages of a single polarity. In this case, the symbol is modified to indicate how the component should be connected. This can be done by adding a '+' sign, as in Figure 4.9(c), or by using a modified symbol as in Figure 4.9(d). In this last figure, the upper terminal of the component is the positive terminal.

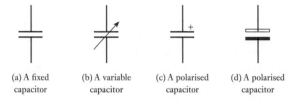

Figure 4.9
Circuit symbols for capacitors.

(a) A fixed capacitor   (b) A variable capacitor   (c) A polarised capacitor   (d) A polarised capacitor

Video 4B

Further study

The circuit below consists of two charged capacitors and a switch. One of the capacitors has a capacitance of $100\,\mu F$ and is initially charged to $50\,V$. The other has a capacitance of $50\,\mu F$ and is charged to $100\,V$. Calculate the charge on each capacitor and the energy stored in each component.

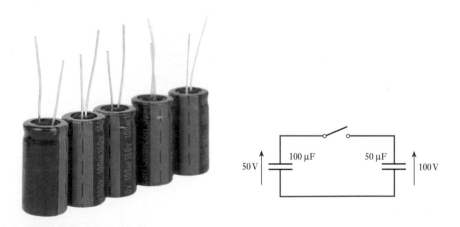

The switch in the circuit is now closed, placing the two capacitors in parallel. Determine the charge on the parallel combination and the voltage across each capacitor. What is the total energy now stored in the combination?

Key points

■ A capacitor consists of two conducting plates separated by an insulating dielectric.

■ Electrons flowing around a circuit can produce a positive charge on one side of the capacitor and an equal negative charge on the other. This results in an electric field between the two plates.

■ The charge stored on a capacitor is directly proportional to the voltage across it.

$$Q = CV \quad \text{or} \quad C = \frac{Q}{V}$$

■ A capacitor *blocks* direct currents but appears to *pass* alternating currents.

■ The capacitance of a parallel-plate capacitor is proportional to the surface area of the plates and inversely proportional to their separation. The constant of proportionality is the permittivity of the dielectric:

$$C = \frac{\varepsilon A}{d} = \frac{\varepsilon_0 \varepsilon_r A}{d}$$

■ The charge on the plates of the capacitor produces an electric field with a strength $E = V/d$.

■ The stored charge produces an electric flux in the dielectric. The flux density $D = Q/A$.

■ The capacitance of several capacitors in parallel is given by the sum of their individual capacitances.

■ The capacitance of several capacitors in series is given by the reciprocal of the sum of the reciprocals of the individual capacitances.

■ The voltage on a capacitor is given by

$$V = \frac{1}{C} \int I \, dt$$

■ The current into a capacitor is given by

$$I = C \frac{dV}{dt}$$

■ When a capacitor is charged through a resistor, the charging rate is determined by the time constant CR.

■ When capacitors are used with sinusoidal signals, the current leads the voltage by $90°$ ($\pi/2$ radians).

■ The energy stored in a charged capacitor is $^1/_2 CV^2$ or $^1/_2 Q^2/C$.

Exercises

4.1 Explain what is meant by a dielectric.

4.2 If electrons represent negative charge in a capacitor, what constitutes positive charge?

4.3 If the two plates of a capacitor are insulated from each other, why does it appear that under some circumstances a current flows between them?

4.4 Why does the presence of charge on the plates of a capacitor represent the storage of energy?

4.5 How is the voltage across a capacitor related to the stored charge?

4.6 A 22 μF capacitor holds 1 mC of stored charge. What voltage is seen across its terminals?

4.7 A capacitor has a voltage of 25 V across it when it holds 500 μC of charge. What is its capacitance?

4.8 Why does a capacitor appear to pass AC signals while blocking DC signals?

4.9 How is the capacitance of a parallel–plate capacitor related to its dimensions?

4.10 The conducting plates of a capacitor are 5×15 mm and have a separation of 10 μm. What would be the capacitance of such a device if the space between the plates were filled with air?

4.11 What would be the capacitance of the device described in Exercise 4.10 if the space between the plates were filled with a dielectric with a relative permittivity of 200?

4.12 What is meant by stray capacitance, and why is this sometimes a problem?

4.13 Explain what is meant by an electric field and by electric field strength.

4.14 The plates of a capacitor have 250 V across them and have a separation of 15 μm. What is the electric field strength in the dielectric?

4.15 What is meant by dielectric strength?

4.16 Explain what is meant by electric flux and by electric flux density.

4.17 The plates of a capacitor are 15×35 mm and store a charge of 35 μC. Calculate the electric flux density in the dielectric.

4.18 Determine the effective capacitance of each of the following arrangements.

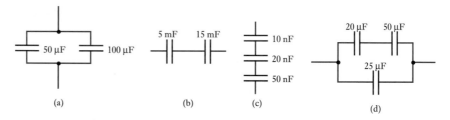

(a)   (b)   (c)   (d)

4.19 How is voltage related to current in a capacitor?

4.20 Repeat Computer simulation exercise 4.1 with $V = 5$ V, $R = 100$ kΩ and $C = 1$ μF. Plot the voltage across the capacitor as a function of time and hence estimate the time taken for the capacitor voltage to reach 2.5 V.

4.21 Explain what is meant by a time constant. What is the time constant of the circuit in Exercise 4.20?

4.22 The circuit of Exercise 4.20 is modified by changing R to 10 kΩ. What value should be chosen for C so that the time taken for the capacitor to charge to 2.5 V is unchanged?

4.23 Confirm your answer to Exercise 4.22 using computer simulation.

4.24 Describe the relationship between the voltage across a capacitor and the current if the voltage is sinusoidal.

4.25 Give an expression for the energy stored in a charged capacitor.

4.26 A 5 mF capacitor is charged to 15 V. What is the energy stored in the capacitor?

4.27 A 50 μF capacitor contains 1.25 mC of charge. What energy is stored in the capacitor?

Chapter 5 | Inductance and Magnetic Fields

Chapter 5

Objectives

When you have studied the material in this chapter, you should be able to:

- explain the meaning and significance of terms such as magnetic field strength, magnetic flux, permeability, reluctance and inductance
- outline the basic principles of electromagnetism and apply these to simple calculations of magnetic circuits
- describe the mechanisms of self-induction and mutual induction
- estimate the inductance of simple inductors from a knowledge of their physical construction
- describe the relationship between the current and voltage in an inductor for both DC and AC signals
- calculate the energy stored in an inductor in terms of its inductance and its current
- describe the operation and characteristics of transformers
- explain the operation of a range of inductive sensors.

5.1 Introduction

Capacitors store energy by producing an electric field within a piece of dielectric material (see Chapter 4). Inductors also store energy, but in this case it is stored within a *magnetic* field. In order to understand the operation and characteristics of inductors, and related components such as transformers, we need first to look at *electromagnetism*.

5.2 Electromagnetism

A wire carrying an electrical current causes a **magnetomotive force** (m.m.f.), F, which produces a **magnetic field** about it, as shown in Figure 5.1(a). One can think of an m.m.f. as being similar in some ways to an e.m.f. in an electrical circuit. The presence of an e.m.f. results in an electric field and in the production of an electric current. Similarly, in magnetic circuits, the presence of an m.m.f. results in a magnetic field and the production of magnetic flux. The m.m.f. has units of amperes and for a single wire F is simply equal to the current I.

The magnitude of the field is defined by the **magnetic field strength**, H, which in this arrangement is given by

$$H = \frac{I}{l} \tag{5.1}$$

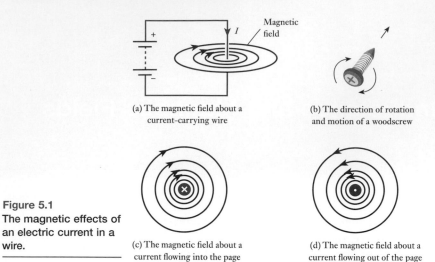

Figure 5.1
The magnetic effects of an electric current in a wire.

(a) The magnetic field about a current-carrying wire

(b) The direction of rotation and motion of a woodscrew

(c) The magnetic field about a current flowing into the page

(d) The magnetic field about a current flowing out of the page

where I is the current flowing in the wire and l is the length of the magnetic circuit. The units of H are amperes per metre. The length of the circuit increases as the circumference of the circles increases, and hence the field gets weaker as we move further from the wire. Since the circumference of a circle is linearly related to its radius (being equal to $2\pi r$), the field strength is directly proportional to the current I and inversely proportional to the distance from the wire.

Example 5.1

A straight wire carries a current of 5 A. What is the magnetic field strength, H, at a distance of 100 mm from the wire?

Since the field about a straight wire is symmetrical, the length of the magnetic path at a distance r from the wire is given by the circumference of a circle of this radius. When $r = 100$ mm, the circumference is equal to $2\pi r = 0.628$ m. Therefore, from Equation 5.1,

$$\text{magnetic field strength}, H = \frac{I}{l}$$

$$= \frac{5}{0.628}$$

$$= 7.96 \text{ A/m}$$

The direction of the magnetic field is determined by the direction of the current in the wire. For a long straight wire the magnetic field is circular about its axis, and one way of remembering the direction of the magnetic field is to visualise a woodscrew lying along the axis of the wire. In this arrangement, the rotation of the screw bears the same relationship to the direction of motion of the screw as the direction of the magnetic field has to the flow of current in the wire. This is shown in Figure 5.1(b). If we imagine a wire running perpendicular through this page, then a current flowing into the page would produce a clockwise magnetic field, while one flowing out of the page would result in an anticlockwise field, as shown in Figures 5.1(c) and 5.1(d). The direction of current flow in these figures is indicated by a cross to show current into the page and a dot to show current coming out of the page. To remember this notation, you may find it useful to visualise the head or the point of the screw of Figure 5.1(b).

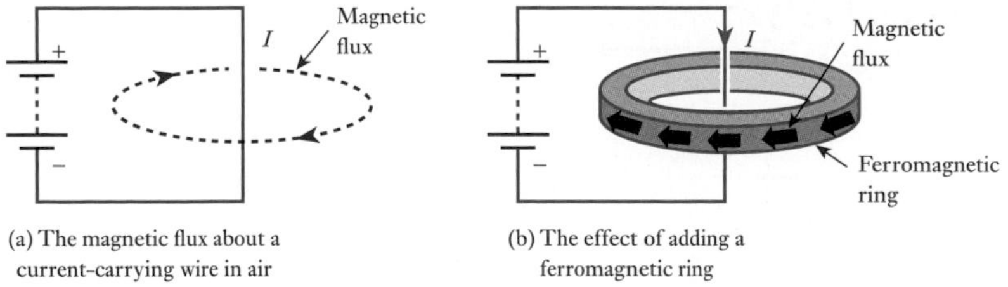

Figure 5.2
Magnetic flux
associated with a
current-carrying wire.

(a) The magnetic flux about a
current-carrying wire in air

(b) The effect of adding a
ferromagnetic ring

The magnetic field produces a **magnetic flux** that flows in the same direction as the field. Magnetic flux is given the symbol Φ, and the unit of flux is the **weber** (Wb).

The strength of the flux at a particular location is measured in terms of the **magnetic flux density**, B, which is the flux per unit area of cross-section. Therefore

$$B = \frac{\Phi}{A} \tag{5.2}$$

The unit of flux density is the tesla (T), which is equal to 1 Wb/m^2.

The flux density at a point depends on the strength of the field at that point, but it is also greatly affected by the material present. If a current-carrying wire is surrounded by air, this will result in a relatively small amount of magnetic flux as shown in Figure 5.2(a). However, if the wire is surrounded by a ferromagnetic ring, the flux within the ring will be orders of magnitude greater, as illustrated in Figure 5.2(b).

Magnetic flux density is related to the field strength by the expression

$$B = \mu H \tag{5.3}$$

where μ is the **permeability** of the material through which the field passes. One can think of the permeability of a material as a measure of the ease with which a magnetic flux can pass through it. This expression is often rewritten as

$$B = \mu_0 \mu_r H \tag{5.4}$$

where μ_0 is the permeability of free space, and μ_r is the relative permeability of the material present. μ_0 is a constant with a value of $4\pi \times 10^{-7} \text{ H/m}$. μ_r is the ratio of the flux density produced in a material to that produced in a vacuum. For air and most non-magnetic materials, $\mu_r = 1$ and $B = \mu_0 H$. For ferromagnetic materials, μ_r may have a value of 1000 or more. Unfortunately, for ferromagnetic materials μ_r varies considerably with the magnetic field strength.

When a current-carrying wire is formed into a coil, as shown in Figure 5.3, the magnetic field is concentrated within the coil, and it increases as more and more turns are added. The m.m.f. is now given by the product of the current I and the number of turns of the coil N, so that

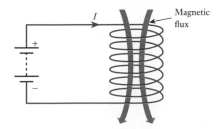

Figure 5.3
The magnetic field in a
coil.

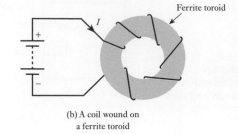

Figure 5.4
The use of
ferromagnetic materials
in coils.

(a) A coil wound on an iron rod
to increase the flux density

(b) A coil wound on
a ferrite toroid

$$F = IN \qquad (5.5)$$

For this reason, the m.m.f. is often expressed in *ampere-turns*, although formally its units are amperes, since the number of turns is dimensionless.

In a long coil with many turns, most of the magnetic flux passes through the centre of the coil. Therefore, it follows from Equations 5.1 and 5.5 that the magnetic field strength produced by such a coil is given by

$$H = \frac{IN}{l} \qquad (5.6)$$

where l is the length of the flux path as before.

As discussed earlier, the flux density produced as a result of a magnetic field is determined by the permeability of the material present. Therefore, the introduction of a ferromagnetic material in a coil will dramatically increase the flux density. Figure 5.4 shows examples of arrangements that use such materials in coils. The first shows an iron bar placed within a linear coil to increase its flux density. The second shows a coil wound on a ferrite toroid (a ring with a circular cross-section).

Example 5.2

A coil is formed by winding 500 turns of wire onto a non-magnetic toroid that has a mean circumference of 400 mm and a cross-sectional area of 300 mm². If the current in the coil is 6 A, calculate:

(a) the magnetomotive force;
(b) the magnetic field strength within the coil;
(c) the flux density in the coil;
(d) the total flux.

How would these quantities be affected if the toroid were replaced by one of similar dimensions but constructed of a magnetic material with $\mu_r = 100$?

(a) The magnetomotive force is given by the 'ampere-turns' of the coil and is therefore

$$F = IN$$

$$= 6 \times 500$$

$$= 3000 \text{ ampere-turns}$$

(b) The magnetic field strength is given by the m.m.f. divided by the length of the magnetic path. In this case, the length of the magnetic path is the mean circumference of the coil, so

$$H = \frac{IN}{l}$$

$$= \frac{3000}{0.4}$$

$$= 7500 \text{ A/m}$$

(c) For a non-magnetic material $B = \mu_0 H$, so

$$B = \mu_0 H$$

$$= 4\pi \times 10^{-7} \times 7500$$

$$= 9.42 \text{ mT}$$

(d) The total flux can be deduced from Equation 5.2, from which it is clear that $\Phi = BA$. Hence

$$\Phi = BA$$

$$= 9.42 \times 10^{-3} \times 300 \times 10^{-6}$$

$$= 2.83 \text{ }\mu\text{Wb}$$

If the toroid were replaced by a material with $\mu_r = 100$, this would have no effect on (a) and (b) but would increase (c) and (d) by a factor of 100.

5.3　Reluctance

As we know, in electrical circuits, when an e.m.f. is applied across a resistive component a current is produced. The ratio of the voltage to the resultant current is termed the **resistance** of the component and is a measure of how the component opposes the flow of electricity.

A directly equivalent concept exists in magnetic circuits. Here an m.m.f. produces a magnetic flux, and the ratio of one to the other is termed the **reluctance**, S, of the magnetic circuit. In this case, the reluctance is a measure of how the circuit opposes the flow of *magnetic flux*. Just as resistance is equal to V/I, so the reluctance is given by the m.m.f. (F) divided by the flux (Φ) and hence

$$S = \frac{F}{\Phi} \tag{5.7}$$

The units of reluctance are amperes per weber (A/Wb).

5.4　Inductance

A changing magnetic flux induces an electrical voltage (an e.m.f.) in any conductor within the field. The magnitude of the effect is given by **Faraday's law**, which states that:

The magnitude of the e.m.f. induced in a circuit is proportional to the rate of change of the magnetic flux linking the circuit.

Also of importance is **Lenz's law**, which states that:

The direction of the e.m.f. is such that it tends to produce a current that opposes the change of flux responsible for inducing that e.m.f.

When a circuit forms a single loop, the e.m.f. induced by changes in the magnetic flux associated with that circuit is simply given by the rate of change of the flux. When a circuit contains many loops, then the resulting e.m.f. is the sum of the e.m.f.s produced by each loop. Therefore, if a coil of N turns experiences a change in magnetic flux, then the induced voltage V is given by

$$V = N \frac{\mathrm{d}\Phi}{\mathrm{d}t} \tag{5.8}$$

where $\mathrm{d}\Phi/\mathrm{d}t$ is the rate of change of flux in Wb/s.

This property, whereby an e.m.f. is induced into a wire as a result of a change in magnetic flux, is referred to as **inductance**.

5.5   Self-inductance

We have seen that a current flowing in a coil (or in a single wire) produces a magnetic flux about it, and that changes in the current will cause changes in the magnetic flux. We have also seen that, when the magnetic flux associated with a circuit changes, this induces an e.m.f. in that circuit which opposes the changing flux. It follows therefore that when the current in a coil changes, an e.m.f. is induced in that coil which tends to oppose the change in the current. This process is known as **self-inductance**.

The voltage produced across the inductor as a result of changes in the current is given by the expression

$$V = L \frac{\mathrm{d}I}{\mathrm{d}t} \tag{5.9}$$

where L is the inductance of the coil. The unit of inductance is the **henry** (symbol H), which can be defined as the inductance of a circuit when an e.m.f. of $1\,\mathrm{V}$ is induced by a change in the current of $1\,\mathrm{A/s}$.

5.5.1   Notation

It should be noted that some textbooks assign a negative polarity to the voltages of Equations 5.8 and 5.9 to reflect the fact that the induced voltage *opposes* the change in flux or current. This notation reflects the implications of Lenz's law. However, either polarity can be used provided that the calculated quantity is applied appropriately, and in this text we will use the *positive* notation since this is consistent with the treatment of voltages across resistors and capacitors.

Example 5.3   The current in a 10 mH inductor changes at a constant rate of 3 A/s. What voltage is induced across this coil?

From Equation 5.9

$$V = L \frac{\mathrm{d}I}{\mathrm{d}t}$$

$$= 10 \times 10^{-3} \times 3$$

$$= 30\,\mathrm{mV}$$

5.6 # Inductors

Circuit elements that are designed to provide inductance are called **inductors**. Typical components for use in electronic circuits will have an inductance of the order of microhenries or millihenries, although large components may have an inductance of the order of henries.

Small-value inductors can be produced using air-filled coils, but larger devices normally use ferromagnetic materials. As we noted earlier, the presence of a ferromagnetic material dramatically increases the flux density in a coil and consequently also increases the rate of change of flux. Therefore, adding a ferromagnetic core to a coil greatly increases its inductance. Inductor cores may take many forms, including rods, as in Figure 5.4(a), or rings, as in Figure 5.4(b). Small inductor cores are often made from iron oxides called **ferrites**, which have very high permeability. Larger components are often based on laminated steel cores.

Unfortunately, the permeability of ferromagnetic materials decreases with increasing magnetic field strength, making inductors non-linear. Air does not suffer from this problem, so air-filled inductors are linear. For this reason, air-filled devices may be used in certain applications even though they may be physically larger than components using ferromagnetic cores.

5.6.1 Calculating the inductance of a coil

The inductance of a coil is determined by its dimensions and by the material around which it is formed. Although it is fairly straightforward to calculate the inductance of simple forms from first principles, designers often use standard formulae. Here we will look at a couple of examples, as shown in Figure 5.5.

Figure 5.5(a) shows a simple, helical, air-filled coil of length l and cross-sectional area A. The characteristics of this arrangement vary with the dimensions, but provided that the length is much greater than the diameter, the inductance of this coil is given by the expression

$$L = \frac{\mu_0 A N^2}{l} \tag{5.10}$$

Figure 5.5(b) shows a coil wound around a toroid that has a mean circumference of l and a cross-sectional area of A. The inductance of this arrangement is given by

$$L = \frac{\mu_0 \mu_r A N^2}{l} \tag{5.11}$$

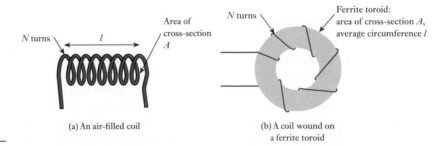

Figure 5.5
Examples of standard inductor formats.

(a) An air-filled coil

(b) A coil wound on a ferrite toroid

where μ_r is the relative permeability of the material used for the toroid. If this is a non-magnetic material then μ_r will be equal to 1, and the inductance becomes

$$L = \frac{\mu_0 A N^2}{l} \tag{5.12}$$

which is the same as for the long air-filled coil described earlier (although the meaning of l is slightly different). Although these two examples have very similar equations, other coil arrangements will have different characteristics.

In these two examples, and in many other inductors, the inductance increases as the square of the number of turns.

Example 5.4 | **Calculate the inductance of a helical, air-filled coil 200 mm in length, with a cross-sectional area of 30 mm² and having 400 turns.**

From Equation 5.10

$$L = \frac{\mu_0 A N^2}{l}$$

$$= \frac{4\pi \times 10^{-7} \times 30 \times 10^{-6} \times 400^2}{200 \times 10^{-3}}$$

$$= 30\ \mu\text{H}$$

5.6.2 Equivalent circuit of an inductor

So far we have considered inductors as idealised components. In practice, all inductors are made from wires (or other conductors) and therefore all real components will have resistance. We can model a real component as an **ideal inductor** (that is, one that has inductance but no resistance) in series with a resistor that represents its internal resistance. This is shown in Figure 5.6.

Figure 5.6
An equivalent circuit of a real inductor.

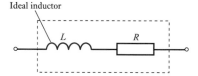

Ideal inductor

L R

5.6.3 Stray inductance

While circuit designers will often use inductors to introduce inductance into circuits, the various conductors in *all* circuits introduce **stray inductance** that is often unwanted. We have seen that even a straight wire exhibits inductance, and though this is usually small (perhaps 1 nH per mm length of wire) the combined effects of these small amounts of inductance can dramatically affect circuit operation – particularly in high–speed circuits. In such cases, great care must be taken to reduce both stray inductance and stray capacitance (as discussed in Chapter 4).

5.7 Inductors in series and in parallel

When several inductors are connected together, their effective inductance is computed in the same way as when resistors are combined, *provided that they are not linked magnetically*. Therefore, when inductors are connected in series their inductances add. Similarly, when inductors are connected in parallel their combined inductance is given by the reciprocal of the sum of the reciprocals of the individual inductances. This is shown in Figure 5.7.

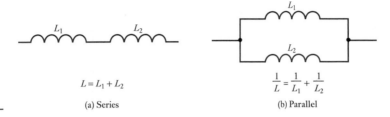

Figure 5.7
Inductors in series and parallel.

$L = L_1 + L_2$

(a) Series

$\dfrac{1}{L} = \dfrac{1}{L_1} + \dfrac{1}{L_2}$

(b) Parallel

Example 5.5 Calculate the inductance of:

(a) a 10 H and a 20 H inductor in series;
(b) a 10 H and a 20 H inductor in parallel.

(a) Inductances in series add

$$L = L_1 + L_2$$

$$= 10 + 20$$

$$= 30 \text{ H}$$

(b) Inductances in parallel sum as their reciprocals

$$\frac{1}{L} = \frac{1}{L_1} + \frac{1}{L_2}$$

$$= \frac{1}{10} + \frac{1}{20}$$

$$= \frac{30}{200}$$

$$L = 6.67 \text{ H}$$

5.8 Relationship between voltage and current in an inductor

From Equation 5.9, we know that the relationship between the voltage across an inductor and the current through it is given by

$$V = L \frac{dI}{dt}$$

This implies that when a constant current is passed through an inductor ($dI/dt = 0$) the voltage across it is zero. However, when the current changes, a voltage is produced that tends to oppose this change in current. Another implication of the equation is

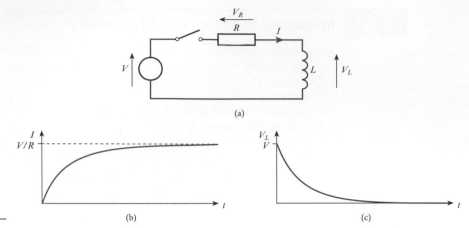

(a)

Figure 5.8
Relationship between
voltage and current in
an inductor.

(b)

(c)

that the current through an inductor cannot change instantaneously, since this would correspond to $dI/dt = \infty$ and would produce an infinite induced voltage opposing the change in current. Thus inductors tend to stabilise the *current* flowing through them. (You may recall that in capacitors the voltage cannot change instantaneously, so capacitors tend to stabilise the *voltage* across them.)

The relationship between the voltage and the current in an inductor is illustrated in Figure 5.8. In the circuit of Figure 5.8(a), the switch is initially open and no current flows in the circuit. If now the switch is closed (at $t = 0$), then the current through the inductor cannot change instantly, so initially $I = 0$, and consequently $V_R = 0$. By applying Kirchhoff's voltage law around the circuit, it is clear that $V = V_R + V_L$, and if initially $V_R = 0$, then the entire supply voltage V will appear across the inductor, and $V_L = V$.

The voltage across the inductor dictates the initial rate of change of the current (since $V_L = L\, dI/dt$) and hence the current steadily increases. As I grows, the voltage across the resistor grows and V_L falls, reducing dI/dt. Therefore, the rate of increase of the current decreases with time. Gradually, the voltage across the inductor tends to zero and all the applied voltage appears across the resistor. This produces a steady-state current of V/R. The result is that the current is initially zero but increases with time, and the voltage across the inductor is initially V but falls exponentially with time. This behaviour is shown in Figures 5.8(b) and 5.8(c). You might like to compare these curves with the corresponding results produced for a capacitor in Figure 4.7.

File 5A

Computer simulation exercise 5.1
Simulate the circuit of Figure 5.8(a) with $V = 1$ V, $R = 1\ \Omega$ and $L = 1$ H. Include in your circuit a switch that closes at $t = 0$. Use transient simulation to investigate the behaviour of the circuit during the first 5 s after the switch changes. Plot V_L against time and confirm that the circuit behaves as expected. Experiment with different values of the circuit components and note the effects on the voltage graph.

The time taken for a capacitor to charge increases with both the capacitance C and the series resistance R, and a term called the time constant, equal to the product CR, determines the charging time (see Chapter 4). In the inductor circuit discussed above, the rate at which the circuit approaches its steady-state condition increases with the inductance L but *decreases* with the value of R. Therefore, such circuits have a time constant (T) equal to L/R.

File 5A

> ### Computer simulation exercise 5.2
>
> Repeat Computer simulation exercise 5.1 noting the effect of different component values. Begin with the same values as in the previous simulation exercise and then change the values of L and R while keeping the ratio L/R constant. Again plot V_L against time and confirm that the characteristics are unchanged. Hence confirm that the characteristics are determined by the time constant L/R rather than the actual values of L and R.

It is interesting to consider what happens in the circuit of Figure 5.8(a) if the switch is opened some time after being closed. From Figure 5.8(b), we know that the current stabilises at a value of V/R. If the switch is now opened, this would suggest that the current would instantly go to zero. This would imply that dI/dt would be infinite and that an infinite voltage would be produced across the coil. In practice, the very high induced voltage appears across the switch and causes 'arcing' at the switch contacts. This maintains the current for a short time after the switch is operated and reduces the rate of change of current. This phenomenon is used to advantage in some situations such as in automotive ignition coils. However, arcing across switches can cause severe damage to the contacts and also generates electrical interference. For this reason, when it is necessary to switch inductive loads, we normally add circuitry to reduce the rate of change of the current. This circuitry may be as simple as a capacitor placed across the switch.

So far in this section we have assumed the use of an ideal inductor and have ignored the effects of any internal resistance. In Section 5.6, we noted that an inductor with resistance can be modelled as an ideal inductor in series with a resistor. In Chapter 6, we will look at the characteristics of circuits containing elements of various types (resistive, inductive and capacitive), so we will leave the effects of internal resistance until then.

5.9 Sinusoidal voltages and currents

Having looked at the relationship between voltage and current in a DC circuit containing an inductor, it is now time to turn our attention to circuits using sinusoidal quantities.

Consider the arrangement of Figure 5.9(a), where an alternating current is passed through an inductor. Figure 5.9(c) shows the sinusoidal current waveform in the inductor, which in turn dictates the voltage across the inductor. From Equation 5.9 we know that the voltage across an inductor is given by $L\, dI/dt$, so the voltage is directly proportional to the *time differential* of the current. Since the differential of a sine wave is a cosine wave, we obtain a voltage waveform as shown in Figure 5.9(b). The current waveform is phase shifted with respect to the voltage waveform by $90°$ (or $\pi/2$ radians). It is also clear that the current waveform *lags* the voltage waveform. You might like to compare this result with that shown in Figure 4.8 for a capacitor. You will note that in a capacitor the current *leads* the voltage, while in an inductor the current *lags* the voltage. We will return to the analysis of sinusoidal waveforms in Chapter 6.

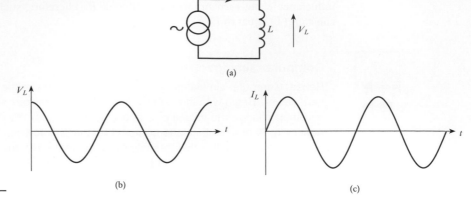

Figure 5.9
Inductors and
alternating quantities.

File 5B

Computer simulation exercise 5.3

Simulate the circuit of Figure 5.9(a) using any value of inductor. Use a sinusoidal current source to produce a current of 1 A peak at 1 Hz and use transient analysis to display the voltage across the inductor, and the current through it, over a period of several seconds. Note the phase relationship between the two waveforms and hence confirm that the current lags the voltages by 90° (or $\pi/2$ radians). Note the effect of varying the inductor value, and the frequency used.

5.10   Energy storage in an inductor

Inductors store energy within a magnetic field. The amount of energy stored in this way can be determined by considering an initially unenergised inductor of inductance L, in which a current is gradually increased from zero to I amperes. If the rate of change of the current at a given time is di/dt, then the instantaneous voltage across the inductor (v) will be given by

$$v = L\frac{di}{dt}$$

In a small amount of time dt, the amount of energy added to the magnetic field is equal to the product of the instantaneous voltage (v), the instantaneous current (i) and the time interval (dt):

$$\text{energy added} = vi\,dt$$

$$= L\frac{di}{dt}i\,dt$$

$$= Li\,di$$

Therefore, the energy added to the magnetic field as the current increases from zero to I is given by

$$\text{stored energy} = L \int_0^I i \, \mathrm{d}t$$

$$\boxed{\text{stored energy} = \frac{1}{2} L I^2} \tag{5.13}$$

Example 5.6 | What energy is stored in an inductor of 10 mH when a current of 5 A is passing through it?

From Equation 5.13

$$\text{stored energy} = \frac{1}{2} L I^2$$

$$= \frac{1}{2} \times 10^{-2} \times 5^2$$

$$= 125 \, \text{mJ}$$

5.11 Mutual inductance

If two conductors are linked magnetically, then a changing current in one of these will produce a changing magnetic flux associated with the other and will result in an induced voltage in this second conductor. This is the principle of **mutual inductance**.

Mutual inductance is quantified in a similar way to self-inductance, such that, if a current I_1 flows in one circuit, the voltage induced in a second circuit V_2 is given by

$$\boxed{V_2 = M \frac{\mathrm{d}I_1}{\mathrm{d}t}} \tag{5.14}$$

where M is the mutual inductance between the two circuits. The unit of mutual inductance is the henry, as for self-inductance. Here, a henry would be defined as the mutual inductance between two circuits when an e.m.f. of 1 V is induced in one by a change in the current of 1 A/s in the other. The mutual inductance between two circuits is determined by their individual inductances and the magnetic linkage between them.

Often our interest is in the interaction of coils, as in a **transformer**. Here a changing current in one coil (the primary) is used to induce a changing current in a second coil (the secondary). Figure 5.10 shows arrangements of two coils that are linked magnetically. In Figure 5.10(a), the two coils are loosely coupled with a relatively small part of the flux of the first coil linking with the second. Such an arrangement would have a relatively low mutual inductance. The degree of coupling between circuits is described by their **coupling coefficient**, which defines the fraction of the flux of one coil that links with the other. A value of 1 represents total flux linkage, while a value of 0 represents no linkage. The coupling between the two coils can be increased in a number of ways, such as by moving the coils closer together, by wrapping one coil around the other, or by adding a **ferromagnetic core** as in Figure 5.10(b). Excellent coupling is achieved by wrapping coils around a continuous ferromagnetic loop as in Figures 5.10(c) and 5.10(d). In these examples, the cores increase the inductance of the coils and increase the flux linkage between them.

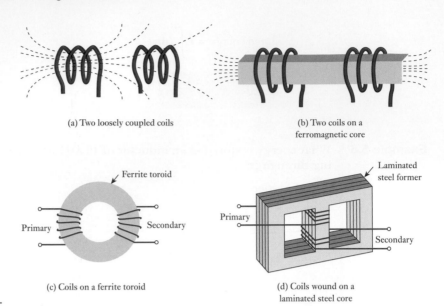

(a) Two loosely coupled coils

(b) Two coils on a ferromagnetic core

(c) Coils on a ferrite toroid

(d) Coils wound on a laminated steel core

Figure 5.10
Mutual inductance between two coils.

5.12 Transformers

Video 5A

The basic form of a transformer is illustrated in Figure 5.11(a). Two coils, a primary and a secondary, are wound onto a ferromagnetic core or former in an attempt to get a coupling coefficient as close as possible to unity. In practice, many transformers are very efficient and for the benefit of this discussion we will assume that all the flux from the primary coil links with the secondary. That is, we will assume an ideal transformer with a coupling coefficient of 1.

If an alternating voltage V_1 is applied to the primary, this will produce an alternating current I_1, which in turn will produce an alternating magnetic field. Since the variation in the magnetic flux associated with the primary coil is the same as that associated with the secondary, the voltage induced *in each turn* of the primary and the secondary will be the same. Let us call this V_T. Now, if the number of turns in the primary is N_1, then the voltage induced across the primary will be $N_1 V_T$. Similarly, if the number of turns in the secondary is N_2, then the voltage across the secondary will be $N_2 V_T$. Therefore, the ratio of the output voltage V_2 to the input voltage V_1 is given by

$$\frac{V_2}{V_1} = \frac{N_2 V_T}{N_1 V_T}$$

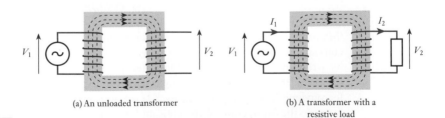

Figure 5.11
A transformer.

(a) An unloaded transformer

(b) A transformer with a resistive load

and thus

$$\frac{V_2}{V_1} = \frac{N_2}{N_1} \qquad (5.15)$$

Thus the transformer works as a voltage amplifier with a gain determined by the ratio of the number of turns in the secondary to that in the primary. N_2/N_1 is often called the **turns ratio** of the transformer.

However, there are several points to note about this arrangement. The first is that this voltage amplification clearly applies only to alternating voltages – a constant voltage applied to the primary will not produce a changing magnetic flux and consequently no output voltage will be induced. Second, it must be remembered that this 'amplifier' has no energy source other than the input signal (that is, it is a passive amplifier) and consequently the power delivered at the output cannot be greater than that absorbed at the input. This second point is illustrated in Figure 5.11(b), where a resistive load has been added to our transformer. The addition of a load means that a current will now flow in the secondary circuit. This current will itself produce magnetic flux, and the nature of induction means that this flux will oppose that generated by the primary circuit. Consequently, the current flowing in the secondary coil tends to reduce the voltage in that coil. The overall effect of this mechanism is that when the secondary is open circuit, or when the output current is very small, the output voltage is as predicted by Equation 5.15, but as the output current increases the output voltage falls.

The efficiency of modern transformers is very high and therefore the power delivered at the output is almost the same as that absorbed at the input. For an ideal transformer

$$V_1 I_1 = V_2 I_2 \qquad (5.16)$$

If the secondary of a transformer has many more turns than the primary we have a **step-up transformer**, which provides an output voltage that is much higher than the input voltage, but it can deliver a smaller output current. If the secondary has fewer turns than the primary we have a **step-down transformer**, which provides a smaller output voltage but can supply a greater current. Step-down transformers are often used in power supplies for low-voltage electronic equipment, where they produce an output voltage of a few volts from the supply voltage. An additional advantage of this arrangement is that the transformer provides **electrical isolation** from the supply lines, since there is no electrical connection between the primary and the secondary circuits.

5.13   Circuit symbols

We have looked at several forms of inductor and transformer, and some of these may be indicated through the use of different circuit symbols. Figure 5.12 shows various symbols and identifies their distinguishing characteristics. Figure 5.12(f) shows a transformer with two secondary coils. This figure also illustrates what is termed the **dot notation** for indicating the polarity of coil windings. Current flowing *into* each winding at the connection indicated by the dot will produce m.m.f.s in the same direction within the core. Reversing the connections to a coil will invert the corresponding voltage waveform. The dot notation allows the required connections to be indicated on the circuit diagram.

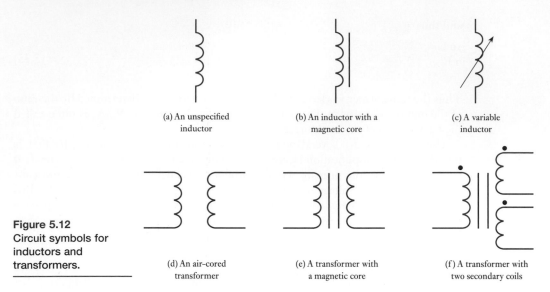

Figure 5.12
Circuit symbols for inductors and transformers.

(a) An unspecified inductor

(b) An inductor with a magnetic core

(c) A variable inductor

(d) An air-cored transformer

(e) A transformer with a magnetic core

(f) A transformer with two secondary coils

5.14   The use of inductance in sensors

Inductors and transformers are used in a wide range of electrical and electronic systems, and we will be meeting several such applications in later chapters. However, at this point, it might be useful to look at a couple of situations where inductance is used as a means of measuring physical quantities.

5.14.1   Inductive proximity sensors

The essential elements of an inductive proximity sensor are shown in Figure 5.13. The device is basically a coil wrapped around a ferromagnetic rod. The arrangement is used as a sensor by combining it with a ferromagnetic plate (attached to the object to be sensed) and a circuit to measure the self-inductance of the coil. When the plate is close to the coil it increases its self-inductance, allowing its presence to be detected. The sensor can be used to measure the separation between the coil and the plate but is more often used in a binary mode to sense its presence or absence. A typical application might mount the coil arrangement on the frame of a door and the plate on the door itself. The sensor would then be used to detect whether the door is open or closed.

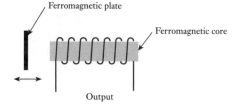

Ferromagnetic plate

Ferromagnetic core

Figure 5.13
An inductive proximity sensor.

Output

5.14.2 Linear variable differential transformers (LVDTs)

An LVDT consists of three coils wound around a hollow, non-magnetic tube, as shown in Figure 5.14. The centre coil forms the primary of the transformer and is exited by an alternating voltage. The remaining coils form identical secondaries, positioned symmetrically either side of the primary. The two secondary coils are connected in series in such a way that their output voltages are out of phase (note the position of the dots in Figure 5.14) and therefore cancel. If a sinusoidal signal is applied to the primary coil, the symmetry of the arrangement means that the two secondary coils produce identical signals that cancel each other, and the output is zero. This assembly is turned into a useful sensor by the addition of a movable 'slug' of ferromagnetic material inside the tube. The material increases the mutual inductance between the primary and the secondary coils and thus increases the magnitude of the voltages induced in the secondary coils. If the slug is positioned centrally with respect to the coils, it will affect both coils equally and the output voltages will still cancel. However, if the slug is moved slightly to one side or the other, it will increase the coupling to one and decrease the coupling to the other. The arrangement will now be out of balance, and an output voltage will be produced. The greater the displacement of the slug from its central position, the greater the resulting output signal. The output is in the form of an alternating voltage where the magnitude represents the offset from the central position and the phase represents the direction in which the slug is displaced. A simple circuit can be used to convert this alternating signal into a more convenient DC signal if required.

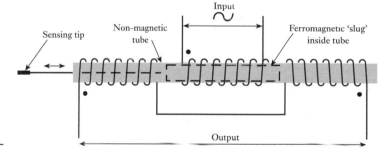

Figure 5.14
A linear variable differential transformer (LVDT).

LVDTs can be constructed with ranges from a few metres down to a fraction of a millimetre. They typically have a resolution of about 0.1 per cent of their full range and have good linearity. Unlike resistive potentiometers, they do not require a frictional contact and so can have a very low operating force and long life.

Video 5B

Further study

Inductive sensors are used in a wide range of applications and in this chapter we have looked at just two of their many forms. Consider how inductive sensors might be used in a range of industrial applications and evaluate their possible advantages and disadvantages in each situation.

Amongst other applications, you might want to consider the use of inductive sensors in:

- a computer keyboard;
- various forms of metal detector;
- an automatic barrier at the entrance to a car park; and
- limit switches in machine tools to prevent moving parts from crashing into end stops.

Key points

- Inductors store energy within a magnetic field.
- A wire carrying an electric current causes a magnetomotive force (m.m.f.), which produces a magnetic field about it.
- The magnetic field strength, H, is proportional to the current and inversely proportional to the length of the magnetic circuit.
- The magnetic field produces a magnetic flux, Φ, which flows in the same direction.
- The flux density is determined by the field strength and the permeability of the material present.
- When a current-carrying wire is formed into a coil, the magnetic field is concentrated. The m.m.f. increases with the number of turns of the coil.
- A changing magnetic flux induces an electrical voltage in any conductors within the field.
- The direction of the induced e.m.f. is such that it opposes the change of flux.
- When the current in a coil changes, an e.m.f. is induced in that coil which tends to oppose the change in the current. This is self-inductance.
- The induced voltage is proportional to the rate of change of the current in the coil.
- Inductors can be made by coiling wire in air, but much greater inductance is produced if the coil is wound around a ferromagnetic core.
- All real inductors have some resistance.
- When inductors are connected in series their inductances add. When inductors are connected in parallel the resultant inductance is the reciprocal of the sum of the reciprocals of the individual inductances.
- The current in an inductor cannot change instantly.

■ When using sinusoidal signals the current lags the voltage by 90° (or $\pi/2$ radians).

■ The energy stored in an inductor is equal to $^1/_2 LI^2$.

■ When two conductors are linked magnetically, a changing current in one will induce a voltage in the other. This is mutual induction.

■ When a transformer is used with alternating signals, the voltage gain is determined by the turns ratio.

■ Several forms of sensor make use of variations in inductance.

Exercises

5.1 Explain what is meant by a magnetomotive force (m.m.f.).

5.2 Describe the field produced by a current flowing in a straight wire.

5.3 A straight wire carries a current of 3 A. What is the magnetic field at a distance of 1 m from the wire? What is the direction of this field?

5.4 What factors determine the flux density at a particular point in space adjacent to a current-carrying wire?

5.5 Explain what is meant by the permeability of free space. What are its value and units?

5.6 Give an expression for the magnetomotive force produced by a coil of N turns that is passing a current of I amperes.

5.7 A coil is formed by wrapping wire around a wooden toroid. The cross-sectional area of the coil is 400 mm^2, the number of turns is 600, and the mean circumference of the toroid is 900 mm. If the current in the coil is 5 A, calculate the magnetomotive force, the magnetic field strength in the coil, the flux density in the coil and the total flux.

5.8 If the toroid in Exercise 5.7 were to be replaced by a ferromagnetic toroid of the same size with a relative permeability of 500, what effect would this have on the values calculated?

5.9 If an m.m.f. of 15 ampere-turns produces a total flux of 5 mWb, what is the reluctance of the magnetic circuit?

5.10 State Faraday's law and Lenz's law.

5.11 Explain what is meant by inductance.

5.12 Explain what is meant by self-inductance.

5.13 How is the voltage induced in a conductor related to the rate of change of the current within it?

5.14 Define the henry as it applies to the measurement of self-inductance.

5.15 The current in an inductor changes at a constant rate of 50 mA/s, and there is a voltage across it of 150 μV. What is its inductance?

5.16 Why does the presence of a ferromagnetic core increase the inductance of an inductor?

5.17 Calculate the inductance of a helical, air-filled coil 500 mm in length, with a cross-sectional area of 40 mm^2 and having 600 turns.

5.18 Calculate the inductance of a coil wound on a ferromagnetic toroid of 300 mm mean circumference and 100 mm^2 cross-sectional area, if there are 250 turns on the coil and the relative permeability of the toroid is 800.

5.19 Calculate the effective inductance of the following arrangements.

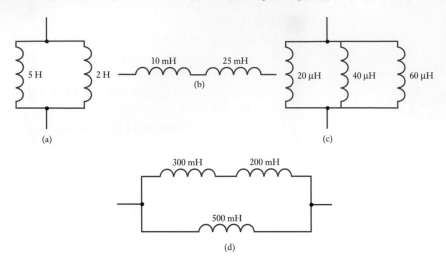

(a)

(b)

(c)

(d)

5.20 Describe the relationship between voltage and current in an inductor.

5.21 Why is it not possible for the current in an inductor to change instantaneously?

5.22 Repeat Computer simulation exercise 5.1 with $V = 15\,\text{V}$, $R = 5\,\Omega$ and $L = 10\,\text{H}$. Plot the voltage across the inductor as a function of time and hence estimate the time taken for the inductor voltage to fall to $5\,\text{V}$.

5.23 Explain what is meant by a time constant. What is the time constant of the circuit in Exercise 5.22?

5.24 The circuit of Exercise 5.22 is modified by changing R to $10\,\Omega$. What value should be chosen for L so that the time taken for the inductor voltage to fall to $5\,\text{V}$ is unchanged?

5.25 Confirm your answer to Exercise 5.24 using computer simulation.

5.26 Discuss the implications of induced voltages when switching inductive circuits.

5.27 How do real inductors differ from ideal inductors?

5.28 What is the relationship between the sinusoidal current in an inductor and the voltage across it?

5.29 What is the energy stored in an inductor of 2 mH when a current of 7 A is passing through it?

5.30 Explain what is meant by mutual inductance.

5.31 Define the henry as it applies to the measurement of mutual inductance.

5.32 What is meant by a coupling coefficient?

5.33 What is meant by the turns ratio of a transformer?

5.34 A transformer has a turns ratio of 10. A sinusoidal voltage of 5 V peak is applied to the primary coil, with the secondary coil open circuit. What voltage would you expect to appear across the secondary coil?

5.35 Explain the dot notation used when representing transformers in circuit diagrams.

5.36 Describe the operation of an inductive proximity sensor.

5.37 Describe the construction and operation of an LVDT.

Chapter 6

Alternating Voltages and Currents

Objectives

When you have studied the material in this chapter, you should be able to:

■ describe the relationship between sinusoidal voltages and currents in resistors, capacitors and inductors

■ explain the meaning of terms such as reactance and impedance, and calculate values for these quantities for individual components and simple circuits

■ use phasor diagrams to determine the relationship between voltages and currents in a circuit

■ analyse circuits containing resistors, capacitors and inductors to determine the associated voltages and currents

■ explain the use of complex numbers in the description and analysis of circuit behaviour

■ use complex notation to calculate the voltages and currents in AC circuits.

6.1 Introduction

We know that a sinusoidal voltage waveform can be described by the equation

$$v = V_p \sin(\omega t + \phi)$$

where V_P is the **peak voltage** of the waveform, ω is the **angular frequency** and ϕ represents its **phase angle**.

The angular frequency of the waveform is related to its natural frequency by the expression

$$\omega = 2\pi f$$

It follows that the period of the waveform, T, is given by

$$T = \frac{1}{f} = \frac{2\pi}{\omega}$$

If the phase angle ϕ is expressed in radians, then the corresponding time delay t is given by

$$t = \frac{\phi}{\omega}$$

These relationships are illustrated in Figure 6.1, which shows two voltage waveforms of similar magnitude and frequency but with different phase angles.

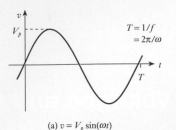

Figure 6.1
Sinusoidal voltage
waveforms.

(a) $v = V_p \sin(\omega t)$ (b) $v = V_p \sin(\omega t + \phi)$

Video 6A

6.2 Relationship between voltage and current

In earlier chapters, we looked at the relationship between voltage and current in a range of components. It is interesting to compare the voltages across a resistor, an inductor and a capacitor when a similar sinusoidal current is passed through each. In this case, we will use a current given by

$$i = I_P \sin(\omega t)$$

6.2.1 Resistors

In a resistor, the relationship between the voltage and the current is given by Ohm's law, and we know that

$$v_R = iR$$

If $i = I_P \sin(\omega t)$, then

$$v_R = I_P R \sin(\omega t) \tag{6.1}$$

6.2.2 Inductors

In an inductor, the voltage is related to the current by the expression

$$v_L = L\frac{\mathrm{d}i}{\mathrm{d}t}$$

If $i = I_P \sin(\omega t)$, then

$$v_L = L\frac{\mathrm{d}(I_P \sin(\omega t))}{\mathrm{d}t} = \omega L I_P \cos(\omega t) \tag{6.2}$$

6.2.3 Capacitors

In a capacitor, the voltage is related to the current by the expression

$$v_C = \frac{1}{C}\int i\,\mathrm{d}t$$

If $i = I_P \sin(\omega t)$, then

$$v_C = \frac{1}{C}\int I_P \sin(\omega t)\mathrm{d}t = -\frac{I_P}{\omega C}\cos(\omega t) \tag{6.3}$$

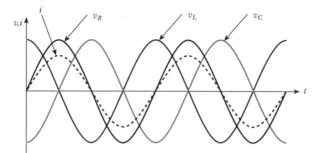

Figure 6.2
Relationship between
the voltage across a
resistor, a capacitor
and an inductor.

Figure 6.2 shows the corresponding voltages across a resistor, an inductor and a capacitor when the same sinusoidal current is passed through each. In this figure, the magnitudes of the various traces are unimportant, since these will depend on component values. The dashed line shows the current waveform and acts as a reference for the other traces. The voltage across the resistor v_R is in phase with the applied current, as indicated by Equation 6.1. The voltage across the inductor v_L is given as a cosine wave in Equation 6.2 and is therefore phase shifted by 90° with respect to the current waveform (and therefore with respect to v_R). The voltage across the capacitor v_C is also given as a cosine wave in Equation 6.3, but here the magnitude is negative, inverting the waveform as shown in Figure 6.2. Inverting a sinusoidal waveform has the same effect as shifting its phase by 180°. Thus the voltage across the resistor is *in phase* with the current; the voltage across the inductor *leads* the current by 90°, and the voltage across the capacitor *lags* the current by 90°. This is consistent with the discussion of resistors, inductors and capacitors in earlier chapters.

While it is relatively easy to remember that there is a phase difference of 90° between the current and the voltage in an inductor and a capacitor, it is important to remember which leads which in each case. One way of remembering this is to use the simple mnemonic

C I V I L

In C, I leads V; V leads I in L.

In other words, in a capacitor the current leads the voltage, while the voltage leads the current in an inductor.

6.3   Reactance of inductors and capacitors

Equations 6.1 to 6.3 show the relationship between current and voltage for a resistor, an inductor and a capacitor. Let us ignore, for the moment, the phase relationship between the current and the voltage and consider instead the relationship between the magnitudes of these quantities. In each case, we will consider the relationship between the peak voltage and the peak current of each component.

6.3.1   Resistance

From Equation 6.1, the ratio of the peak magnitude of the voltage to the peak magnitude of the current is given by

$$\frac{\text{peak value of voltage}}{\text{peak value of current}} = \frac{\text{peak value of } I_p R \sin(\omega t)}{\text{peak value of } I_p \sin(\omega t)} = \frac{I_p R}{I_p} = R$$

6.3.2 Inductance

From Equation 6.2, the ratio of the peak magnitude of the voltage to the peak magnitude of the current is given by

$$\frac{\text{peak value of voltage}}{\text{peak value of current}} = \frac{\text{peak value of } \omega L I_p \cos(\omega t)}{\text{peak value of } I_p \sin(\omega t)}$$

$$= \frac{\omega L I_p}{I_p} = \omega L$$

6.3.3 Capacitance

From Equation 6.3, the ratio of the peak magnitude of the voltage to the peak magnitude of the current is given by

$$\frac{\text{peak value of voltage}}{\text{peak value of current}} = \frac{\text{peak value of } -\dfrac{I_p}{\omega C} \cos(\omega t)}{\text{peak value of } I_p \sin(\omega t)}$$

$$= \frac{\dfrac{I_p}{\omega C}}{I_p} = \frac{1}{\omega C}$$

Note that the three expressions derived above would have been identical had we chosen to compare the r.m.s. value of the voltage with the r.m.s. value of the current in each case, since this would have simply multiplied the upper and lower term in each expression by $1/\sqrt{2}$ (0.707).

This ratio of the voltage to the current, ignoring any phase shift, is a measure of how the component opposes the flow of electricity. In the case of a resistor, we already know that this is termed its resistance. In the case of inductors and capacitors, this quantity is termed its **reactance**, which is given the symbol X. Therefore

$$\text{reactance of an inductor, } X_L = \omega L \qquad (6.4)$$

$$\text{reactance of a capacitor, } X_C = \frac{1}{\omega C} \qquad (6.5)$$

Since reactance represents the ratio of voltage to current it has the units of ohms.

Example 6.1   Calculate the reactance of an inductor of 1 mH at an angular frequency of 1000 rad/s.

$$\text{reactance, } X_L = \omega L$$

$$= 1000 \times 10^{-3}$$

$$= 1\ \Omega$$

Example 6.2   Calculate the reactance of a capacitor of 2 μF at a frequency of 50 Hz.

At a frequency of 50 Hz, the angular frequency is given by

$$\omega = 2\pi f$$
$$= 2 \times \pi \times 50$$
$$= 314.2 \text{ rad/s}$$

$$\text{reactance, } X_C = \frac{1}{\omega C}$$

$$= \frac{1}{314.2 \times 2 \times 10^{-6}}$$

$$= 1.59 \text{ k}\Omega$$

The reactance of a component can be used to calculate the voltage across it from a knowledge of the current through it, and vice versa, in much the same way as we use resistance for a resistor. Therefore, for an inductor

$$V = IX_L$$

and for a capacitor

$$V = IX_C$$

Note that these relationships are true whether V and I represent r.m.s., peak or peak-to-peak values, provided that the same measure is used for both.

Example 6.3
> **A sinusoidal voltage of 5 V peak and 100 Hz is applied across an inductor of 25 mH. What will be the peak current in the inductor?**
>
> At this frequency, the reactance of the inductor is given by
>
> $$X_L = \omega L$$
> $$= 2\pi f L$$
> $$= 2 \times \pi \times 100 \times 25 \times 10^{-3} = 15.7 \ \Omega$$
>
> Therefore
>
> $$I_L = \frac{V_L}{X_L}$$
> $$= \frac{5}{15.7}$$
> $$= 318 \text{ mA}$$

Example 6.4
> **A sinusoidal current of 2 A r.m.s. at 25 rad/s flows through a capacitor of 10 mF. What voltage will appear across the capacitor?**
>
> At this frequency, the reactance of the capacitor is given by
>
> $$X_C = \frac{1}{\omega C}$$
> $$= \frac{1}{25 \times 10 \times 10^{-3}}$$
> $$= 4 \ \Omega$$

Therefore

$$V_C = I_C X_C$$

$$= 2 \times 4$$

$$= 8 \, \text{V r.m.s.}$$

When describing sinusoidal quantities, we very often use r.m.s. values (for reasons discussed in Chapter 2). However, since the calculation of currents and voltages is essentially the same whether we are using r.m.s., peak or peak-to-peak quantities, in the remaining examples in this chapter we will simply give magnitudes in volts and amps, ignoring the form of the measurement.

6.4 Phasor diagrams

We have seen that sinusoidal signals are characterised by their *magnitude*, their *frequency* and their *phase*. In many cases, the voltages and currents at different points in a system are driven by a single source (such as the AC supply voltage) such that they all have a common frequency. However, the magnitudes of the signals at different points will be different and, as we have seen above, the phase relationship between these signals may also be different. We therefore are often faced with the problem of combining or comparing signals of the same frequency but of differing magnitude and phase. A useful tool in this area is the **phasor diagram**, which allows us to represent both the magnitude and the phase of a signal in a single diagram.

Figure 6.3(a) shows a single phasor representing a sinusoidal voltage. The length of the phasor, L, represents the magnitude of the voltage, while the angle ϕ represents its phase angle with respect to some reference waveform. The end of the phasor is marked with an arrowhead to indicate its direction and also to make the end visible should it coincide with one of the axes or another phasor. Phase angles are traditionally measured anticlockwise from the right-pointing horizontal axis.

The use of phasors clearly indicates the phase relationship between signals of the same frequency. For example, Figure 6.3(b) shows the three voltage waveforms v_R, v_L and v_C of Figure 6.2 represented by phasors $\mathbf{V_R}, \mathbf{V_L}$ and $\mathbf{V_C}$. Here the current waveform is taken as the reference phase, and therefore $\mathbf{V_R}$ has a phase angle of zero, $\mathbf{V_L}$ has a phase angle of +90° (+π/2 rad), and $\mathbf{V_C}$ is shown with a phase angle of −90° (−π/2 rad). Conventionally, the name of a phasor is written in bold (for example, $\mathbf{V_R}$), the magnitude of the phasor is written in italics (for example, V_R) and the instantaneous value of the sinusoidal quantity it represents is written in lower-case italics (for example, v_R).

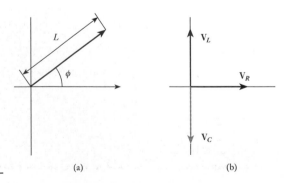

Figure 6.3
Phasor diagrams.

(a) (b)

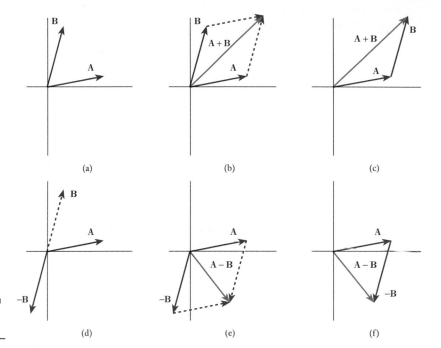

Figure 6.4
Representing the addition or subtraction of waveforms using phasors.

Phasor diagrams can be used to represent the addition or subtraction of signals of the same frequency but different phase. Consider for example Figure 6.4(a), which shows two phasors, **A** and **B**. Since the two sinusoids have different phase angles, if we add them together the magnitude of the resulting signal will *not* be equal to the arithmetic sum of the magnitudes of **A** and **B**. A phasor diagram can be used to determine the effect of adding the two signals, using techniques similar to those used in **vector analysis** to add vectors. Figure 6.4(b) shows the technique of 'completing the parallelogram' used to compute the effect of adding **A** and **B**. It can be seen that the diagram gives the magnitude and the phase of the resulting waveform. Another way of combining phasors is illustrated in Figure 6.4(c), where the phasors **A** and **B** are added by drawing **B** from the end of **A**. This process can be seen as adding **B** to **A** and can be repeated to combine any number of phasors. The *subtraction* of one sinusoid from another is equivalent to *adding* a sinusoid of the opposite polarity. This in turn is equivalent to adding a signal that is phase shifted by 180°. Figure 6.4(d) shows the phasors **A** and **B** from Figure 6.4(a) and also shows the phasor −**B**. It can be seen that −**B** is represented by a line of equal magnitude to **B** but pointing in the opposite direction. These two vectors can again be combined by completing the parallelogram (as in Figure 6.4(e)) or by drawing one on the end of the other (as in Figure 6.4(f)).

Fortunately, the nature of circuit components means that we are normally concerned with phasors that are at right angles to each other (as illustrated in Figure 6.3(b)). This makes the use of phasor diagrams much simpler.

6.4.1 Phasor analysis of an *RL* circuit

Consider the circuit of Figure 6.5, which shows a circuit containing a resistor and an inductor in series with a sinusoidal voltage source. The reference phase for the diagram is the current signal i and consequently the voltage across the resistor (represented by V_R in the phasor diagram) has zero phase (since this is in phase with the current),

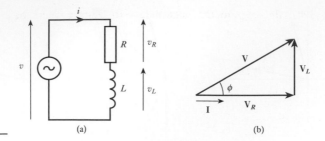

Figure 6.5
Phasor analysis of an
RL circuit.

while the voltage across the inductor (represented by $\mathbf{V}_L$) has a phase angle of 90° (or $\pi/2$ radians). The voltage across the voltage source v is represented by the phasor $\mathbf{V}$, and from the circuit diagram it is clear that $v = v_L + v_R$. Thus in the phasor diagram $\mathbf{V} = \mathbf{V}_R + \mathbf{V}_L$; the diagram shows that the voltage across the circuit is not in phase with the current and that v leads i by a phase angle of ϕ.

Phasor diagrams can be used in connection with values for resistance and reactance to determine the voltages and currents within a circuit, and the phase relationships between them.

Example 6.5 A sinusoidal current of 5 A at 50 Hz flows through a series combination of a resistor of 10 Ω and an inductor of 25 mH. Determine:

(a) the voltage across the combination;
(b) the phase angle between this voltage and the current.

(a) The voltage across the resistor is given by

$$V_R = IR$$
$$= 5 \times 10$$
$$= 50\,\text{V}$$

At a frequency of 50 Hz, the reactance of the inductor is given by

$$X_L = 2\pi fL$$
$$= 2 \times \pi \times 50 \times 0.025$$
$$= 7.85\,\Omega$$

Therefore, the magnitude of the voltage across the inductor is given by

$$V_L = IX_L$$
$$= 5 \times 7.85$$
$$= 39.3\,\text{V}$$

These can be combined using a phasor diagram as follows:

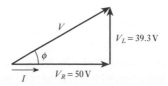

From the diagram, the magnitude of the voltage across the combination is given by

$$V = \sqrt{(V_R^2 + V_L^2)}$$
$$= \sqrt{(50^2 + 39.3^2)}$$
$$= 63.6 \text{ V}$$

(b) The phase angle is given by

$$\phi = \tan^{-1} \frac{V_L}{V_R}$$
$$= \tan^{-1} \frac{39.3}{50}$$
$$= 38.2°$$

Therefore, the voltage leads the current by 38.2°.

File 6A

Computer simulation exercise 6.1

Simulate the circuit of Example 6.5 using a sinusoidal current source. Use transient analysis to investigate the magnitude and the phase of the voltage across the resistor/inductor combination and compare these with the values calculated above.

6.4.2 Phasor analysis of an *RC* circuit

A similar approach can be used with circuits involving resistors and capacitors, as shown in Figure 6.6. Again the reference phase is that of the current *i*. Since in a capacitor the voltage lags the current by 90°, the voltage across the capacitor has a phase angle of −90°, and the phasor is drawn vertically downwards. The resultant phase angle ϕ is now negative, showing that the voltage across the combination lags the current through it.

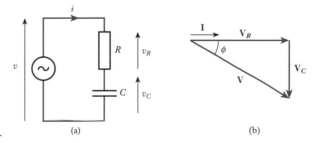

Figure 6.6
Phasor analysis of an
RC circuit.

As before, we can use the phasor diagram to determine the relationships between currents and voltages in the circuit.

Example 6.6 A sinusoidal voltage of 10 V at 1 kHz is applied across a series combination of a resistor of 10 kΩ and a capacitor of 30 nF. Determine:

(a) the current flowing through the combination;
(b) the phase angle between this current and the applied voltage.

(a) In this example, the current in the circuit is unknown – we will call this current I. The magnitude of the voltage across the resistor is given by

$$V_R = IR = I \times 10^4 \, \text{V}$$

At a frequency of 1 kHz, the reactance of the capacitor is given by

$$X_C = \frac{1}{2\pi f C}$$

$$= \frac{1}{2 \times \pi \times 10^3 \times 3 \times 10^{-8}} = 5.3 \, \text{k}\Omega$$

Therefore, the magnitude of the voltage across the capacitor is given by

$$V_C = IX_C = I \times 5.3 \times 10^3 \, \text{V}$$

These can be combined using a phasor diagram as follows:

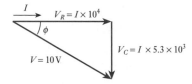

From the diagram, we see that

$$V^2 = V_R^2 + V_C^2$$

$$10^2 = (I \times 10^4)^2 + (I \times 5.3 \times 10^3)^2$$

$$= I^2 \times 1.28 \times 10^8$$

which can be solved to give

$$I = 884 \, \mu\text{A}$$

(b) The phase angle is given by

$$\phi = \tan^{-1} \frac{V_C}{V_R}$$

$$= \tan^{-1} \frac{I \times 5.3 \times 10^3}{I \times 10^4}$$

$$= -27.9°$$

The phase angle is negative by inspection of the diagram. This means that the voltage lags the current by nearly 28° or, alternatively, the current leads the voltage by this amount.

File 6B

Computer simulation exercise 6.2

Simulate the circuit of Example 6.6 using a sinusoidal voltage source. Use transient analysis to investigate the magnitude and the phase of the current flowing in the resistor/capacitor combination and compare these with the values calculated above.

6.4.3 **Phasor analysis of *RLC* circuits**

The techniques described above can be used with circuits containing any combination of resistors, inductors and capacitors. For example, the circuit of Figure 6.7(a) contains two resistors, one inductor and one capacitor, and Figure 6.7(b) shows the corresponding phasor diagram. It is interesting to note that summing the voltages across the four components in a different order *should* give us the same overall voltage. Figure 6.7(c) illustrates this and shows that this does indeed give us the same voltage (both in magnitude and phase angle).

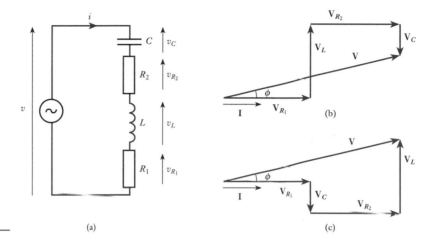

Figure 6.7
Phasor analysis of an
RLC circuit.

6.4.4 **Phasor analysis of parallel circuits**

The phasor diagrams shown above relate to series combinations of components. In such circuits, the current is the same throughout the network and we are normally interested in the voltages across each component. Our phasor diagram therefore shows the relationship between the various voltages in the circuit. In a parallel network, the voltage across each component is the same and it is the currents that are of interest. We can also use phasor diagrams to represent currents as shown in Figure 6.8. In

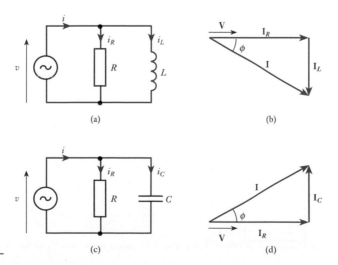

Figure 6.8
Phasor diagrams of
parallel networks.

these circuits, the applied voltage v is taken as the reference phase, and therefore the current though the resistor i_R has a phase angle of zero. Since the current through an inductor *lags* the applied voltage, the phasor $\mathbf{I}_L$ is shown vertically downwards. Similarly, since the current in a capacitor *leads* the applied voltage, $\mathbf{I}_C$ is shown vertically upwards. These directions are clearly opposite to those in voltage phasor diagrams.

As with series arrangements, phasor diagrams can be used with parallel circuits containing any number of components, and calculations can be performed in a similar manner.

6.5 Impedance

In circuits containing only resistive elements, the current is related to the applied voltage by the *resistance* of the arrangement. In circuits containing reactive components, the relationship between the current and the applied voltage can be described by the **impedance Z** of the arrangement, which represents the effect of the circuit not only on the *magnitude* of the current but also on its *phase*. Impedance can be used in reactive circuits in a manner similar to the way that resistance is used in resistive circuits, but it should be remembered that impedance changes with frequency.

Consider Figure 6.9, which shows a series RL network as in Figure 6.5. From the phasor diagram of Figure 6.9(b), it is clear that the magnitude of the voltage across the RL combination, V, is given by

$$
\begin{aligned}
V &= \sqrt{(V_R^2 + V_L^2)} \\
&= \sqrt{(IR)^2 + (IX_L)^2} \\
&= I\sqrt{R^2 + X_L^2} \\
&= IZ
\end{aligned}
$$

where $Z = \sqrt{(R^2 + X_L^2)}$. As before, we use letters in italics to represent the magnitude of a quantity that has both magnitude and phase, so $Z = |\mathbf{Z}|$ is the magnitude of the impedance of the RL combination. Similar consideration of the RC circuit of Figure 6.6 shows that the magnitude of the impedance of this arrangement is given by $Z = \sqrt{(R^2 + X_C^2)}$. Thus in each case the magnitude of the impedance is given by the expression

$$
Z = \sqrt{R^2 + X^2} \tag{6.6}
$$

The impedance of a circuit has not only a magnitude (Z) but also a phase angle (ϕ), which represents the phase between the voltage and the current. From Figure 6.9, we can see that the phase angle ϕ is given by

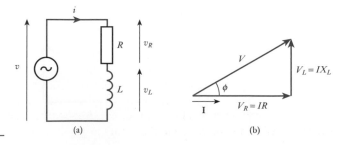

Figure 6.9
A series *RL* network.

$$\phi = \tan^{-1}\frac{V_L}{V_R} = \tan^{-1}\frac{IX_L}{IR} = \tan^{-1}\frac{X_L}{R}$$

and a similar analysis can be performed for a series RC circuit, which produces the result that $\phi = \tan^{-1}X_C/R$. This leads to the general observation that

$$\phi = \tan^{-1}\frac{X}{R} \tag{6.7}$$

where the sign of the resulting phase angle can be found by inspection of the phasor diagram.

Figure 6.9(b) shows a phasor representation of circuit *voltages*. However, a similar diagram could be used to combine *resistive* and *reactive* components to determine their combined impedance. This is shown in Figure 6.10, where Figure 6.10(a) shows the impedance of a series RL combination, and Figure 6.10(b) shows the impedance of a series RC arrangement. The impedance (**Z**) can be expressed in a **rectangular form** as a resistive plus a reactive component, or in a **polar form** by giving the magnitude Z and the angle ϕ.

Figure 6.10
A graphical representation of impedance.

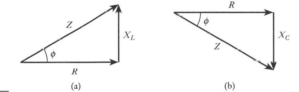

(a)   (b)

The technique demonstrated in Figure 6.10 can be extended to compute the impedance of other combinations of resistors, inductors and capacitors by adding the various circuit elements in turn. However, it is often more convenient to use what is termed **complex notation** to represent impedances, so we will turn our attention to this approach.

6.6   Complex notation

Readers who are familiar with complex mathematics will have noticed several similarities between the phasor diagrams discussed above and **Argand diagrams** used to represent complex quantities. If you are not familiar with this topic, you are advised to read Appendix D before continuing with this section, since this gives a brief introduction to complex numbers.

The distinction between real and imaginary numbers in complex arithmetic is in many ways similar to the difference between the resistive and the reactive components of impedance. When a sinusoidal current is passed through a resistor, the resulting voltage is in phase with the current and for this reason we consider the impedance of a resistor to be *real*. When a similar current is passed though an inductor or a capacitor, the voltage produced leads or lags the current with a phase angle of 90°. For this reason, we consider reactive elements to represent an *imaginary* impedance. Because the phase shift produced by inductors is opposite to that produced by capacitors, we assign different polarities to the imaginary impedances associated with these elements. By convention, inductors are taken to have *positive* impedance and capacitors are assumed to have *negative* impedance. The *magnitude* of the impedance of capacitors and inductors is determined by the reactance of the component.

Therefore, the impedance of resistors, inductors and capacitors is given by

resistors: $\quad Z_R = R$

inductors: $\quad Z_L = jX_L \ = j\omega L$

capacitors: $\quad Z_C = -jX_C = -j\dfrac{1}{\omega C} = \dfrac{1}{j\omega C}$

Figure 6.11 shows the complex impedances of the arrangements of Figure 6.10.

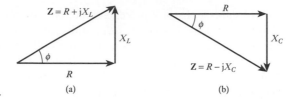

Figure 6.11
Complex impedances.

An attractive characteristic of complex impedances is that they can be used for sinusoidal signals in a similar manner to the way resistances are used for DC circuits.

6.6.1 Series combinations

For a series combination of impedances $Z_1, Z_2, \ldots , Z_N$, the effective total impedance is equal to the sum of the individual impedances.

$$Z = Z_1 + Z_2 + \cdots + Z_N \tag{6.8}$$

This is illustrated in Figure 6.12(a).

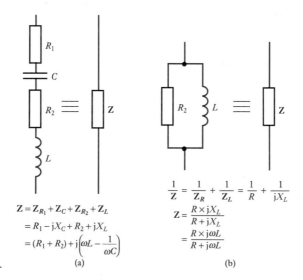

Figure 6.12
The impedance of series and parallel arrangements.

6.6.2 Parallel combinations

For a parallel combination of impedances $Z_1, Z_2, \ldots , Z_N$, the effective total impedance is equal to the reciprocal of the sum of the reciprocals of the individual impedances.

$$\frac{1}{Z} = \frac{1}{Z_1} + \frac{1}{Z_2} + \cdots + \frac{1}{Z_N}$$ (6.9)

This is illustrated in Figure 6.12(b).

Example 6.7 Determine the complex impedance of the following series arrangement at a frequency of 50 Hz.

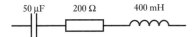

A frequency of 50 Hz corresponds to an angular frequency of

$$\omega = 2\pi f$$
$$= 2 \times \pi \times 50$$
$$= 314 \text{ rad/s}$$

Therefore

$$Z = Z_C + Z_R + Z_L$$
$$= R + j(X_L - X_C)$$
$$= R + j\left(\omega L - \frac{1}{\omega C}\right)$$
$$= 200 + j\left(314 \times 400 \times 10^{-3} - \frac{1}{314 \times 50 \times 10^{-6}}\right)$$
$$= 200 + j62 \ \Omega$$

Note that *at this frequency* the impedance of the arrangement is equivalent to a resistor of 200 Ω in series with an inductor of $X_L = 62$ Ω. Since $X_L = \omega L$, this equivalent inductance $L = X_L/\omega = 62/314 = 197$ mH. Therefore, at this single frequency, the circuit above is equivalent to

6.6.3 Expressing complex quantities

We can express complex quantities in a number of ways, the most common forms being the rectangular form ($a + jb$), the polar form ($r\angle\theta$) and the **exponential form** ($re^{j\theta}$). If you are unfamiliar with these forms, and with arithmetic based on them, you are advised to read Appendix D before continuing with this section.

If we wish to add (or subtract) complex quantities (for example, two impedances), this is easiest to achieve if they are both expressed in a rectangular form, since

$$(a + jb) + (c + jd) = (a + c) + j(b + d)$$

However, if we wish to multiply (or divide) complex quantities (for example, multiplying a sinusoidal current by a complex impedance), this is easier using either polar or exponential forms. For example, using a polar form

$$\frac{A\angle\alpha}{B\angle\beta} = \frac{A}{B}\angle(\alpha - \beta)$$

Fortunately, conversion between the various forms is straightforward (see Appendix D), and we will often change the form of a quantity in order to simplify the arithmetic.

Sinusoidal voltages and currents of the same frequency are often described in polar form by their magnitude and their phase angle. Since phase angles are relative, we need to define a reference phase and this will often be the input voltage or the current in a circuit. For example, if a circuit has a sinusoidal input voltage with a magnitude of 20 V, and we choose this signal to define our reference phase, then we can describe this signal in polar form as $20\angle 0$ V. If we determine that another signal in the circuit is $5\angle 30°$, then this will have a magnitude of 5 V and a phase angle of 30° with respect to the reference (input) waveform. Positive values for this phase angle mean that the signal *leads* the reference waveform, while negative values indicate a phase *lag*.

6.6.4 Using complex impedance

Complex impedances can be used with sinusoidal signals in a similar manner to the way resistances are used in DC circuits. Consider, for example, the circuit of Figure 6.13, where we wish to determine the current i.

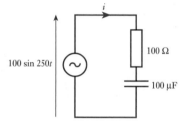

Figure 6.13
Use of impedance in a simple *RC* circuit.

100 sin 250*t*

100 Ω

100 μF

If this were a purely resistive circuit, then the current would be given by $i = v/R$. So, in this case the current is given by $i = v/\mathbf{Z}$, where $\mathbf{Z}$ is the complex impedance of the circuit. In this circuit, the driving voltage is equal to 100 sin 250*t*, so the angular frequency ω is equal to 250. Thus the impedance $\mathbf{Z}$ of the circuit is given by

$$\mathbf{Z} = R - jX_C$$
$$= R - j\frac{1}{\omega C}$$
$$= 100 - j\frac{1}{250 \times 10^{-4}}$$
$$= 100 - j40$$

The current is given by $v/\mathbf{Z}$, and this is much easier to compute if we express each quantity in its polar form. If we define the reference phase to be that of the input voltage, then $v = 100\angle 0$. The polar form of $\mathbf{Z}$ is given by

$$\mathbf{Z} = 100 - j40$$

$$|\mathbf{Z}| = \sqrt{100^2 + 40^2} = 107.7$$

$$\angle\mathbf{Z} = \tan^{-1}\frac{-40}{100} = -21.8°$$

$$\mathbf{Z} = 107.7\angle{-21.8°}$$

Therefore

$$i = \frac{v}{Z}$$

$$= \frac{100\angle 0}{107.7\angle -21.8}$$

$$= 0.93\angle 21.8°$$

Another example of the use of complex impedance is demonstrated by the circuit of Figure 6.14(a), which contains two resistors, a capacitor and an inductor. To analyse this circuit, we first replace each component with its impedance as shown in Figure 6.14(b). $\mathbf{Z}_C$ and $\mathbf{Z}_{R_1}$ are in series and may be combined to form a single impedance $\mathbf{Z}_1$ as discussed in the last section. Similarly, $\mathbf{Z}_L$ and $\mathbf{Z}_{R_2}$ are in parallel and may be combined to form a single impedance $\mathbf{Z}_2$. This reduces our circuit to that shown in Figure 6.14(c), from which the output voltage can be determined using our standard formula for a potential divider (with the resistances replaced by impedances).

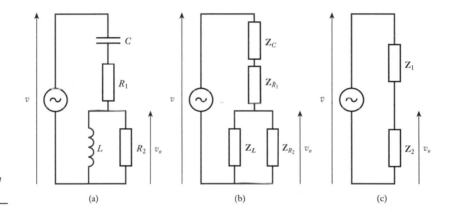

Figure 6.14
The use of impedance in circuit analysis.

(a) (b) (c)

Example 6.8

Calculate the output voltage v_o in the circuit of Figure 6.14(a) if $C = 200$ μF, $R_1 = 5$ Ω, $L = 50$ mH, $R_2 = 50$ Ω and the input voltage $v = 10 \sin 500t$.

Since $v = 10 \sin 500t$, $\omega = 500$ rad/s. We will take the input voltage as the reference phase, so v corresponds to $10\angle 0$.

Our first task is to calculate the impedances corresponding to $\mathbf{Z}_1$ and $\mathbf{Z}_2$ in Figure 6.14(c):

$$\mathbf{Z}_1 = R_1 - jX_C$$

$$= R_1 - j\frac{1}{\omega C}$$

$$= 5 - j\frac{1}{500 \times 200 \times 10^{-6}}$$

$$= 5 - j10 \ \Omega$$

$$\frac{1}{\mathbf{Z}_2} = \frac{1}{R_2} + \frac{1}{jX_L}$$

$$Z_2 = \frac{jX_L R_2}{R_2 + jX_L}$$

$$= \frac{jX_L R_2(R_2 - jX_L)}{(R_2 + jX_L)(R_2 - jX_L)}$$

$$= \frac{R_2 X_L^2 + jR_2^2 X_L}{R_2^2 + X_L^2}$$

$$= \frac{R_2 \omega^2 L^2 + jR_2^2 \omega L}{R_2^2 + \omega^2 L^2}$$

$$= \frac{(50 \times 500^2 \times 0.05^2) + (j \times 50^2 \times 500 \times 0.05)}{50^2 + (500^2 \times 0.05^2)}$$

$$= \frac{31,250 + j62,500}{3125}$$

$$= 10 + j20 \ \Omega$$

From Figure 6.14(c), it is clear that

$$v_o = v \times \frac{Z_2}{Z_1 + Z_2}$$

$$= v \times \frac{10 + j20}{(5 - j10) + (10 + j20)}$$

$$= v \times \frac{10 + j20}{15 + j10}$$

Division and multiplication are often easier using the polar form, which gives

$$v_o = v \times \frac{22.4\angle 63.4°}{18.0\angle 33.7°}$$

$$= v \times 1.24\angle 29.7°$$

$$= 10\angle 0 \times 1.24\angle 29.7°$$

$$= 12.4\angle 29.7°$$

Therefore, since the frequency of v_o is equal to that of v,

$$v_o = 12.4 \sin(500t + 29.7°)$$

and the output voltage leads the input voltage by 29.7°.

File 6C

Computer simulation exercise 6.3

Simulate the circuit of Example 6.8 using a sinusoidal voltage source for *v*. Use transient analysis to investigate the magnitude and the phase of the output voltage v_o and compare these with the values calculated above.

Impedances can be used in place of resistances in calculations involving Ohm's law, and Kirchhoff's voltage and current laws, allowing the various circuit analysis techniques described in Chapter 3 to be applied to alternating voltages and currents.

Video 6B

Further study

Figure (a) below shows a circuit consisting of four components. If it is known that this circuit will be used at a frequency of 100 Hz, design a simpler arrangement to reproduce this circuit's behaviour.

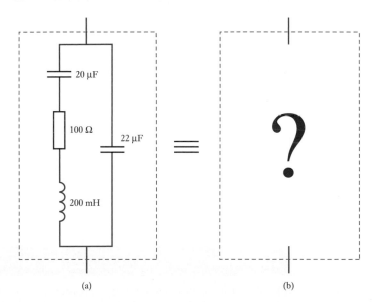

(a)   (b)

Key points

- A sinusoidal voltage waveform can be described by the equation

$$v = V_P \sin(\omega t + \phi)$$

where V_P is the peak voltage of the waveform, ω is the angular frequency and ϕ represents its phase angle.

- When a sinusoidal current flows through a resistor, an inductor and a capacitor:
  - the voltage across the resistor is *in phase* with the current;
  - the voltage across the inductor *leads* the current by 90°;
  - the voltage across the capacitor *lags* the current by 90°.
- The magnitude of the voltage across an inductor or a capacitor is determined by its reactance, where
  - the reactance of an inductor, $X_L = \omega L$
  - the reactance of a capacitor, $X_C = \dfrac{1}{\omega C}$
- The *magnitudes* of the voltages across resistors, inductors and capacitors are given by
  - resistor: $V = IR$
  - inductor: $V = IX_L$
  - capacitor: $V = IX_C$
- Phasor diagrams can be used to represent both the magnitude and the phase of quantities of the same frequency.
- Representing alternating voltages by phasors allows signals of different phase to be added or subtracted easily.
- The relationship between the current and the voltage in a circuit containing reactive components is described by its impedance **Z**.
- Complex notation simplifies calculations of currents and voltages in circuits that have reactive elements. Using this notation, the impedance of resistors, inductors and capacitors is given by
  - resistors: $\mathbf{Z}_R = R$
  - inductors: $\mathbf{Z}_L = \mathrm{j}\omega L$
  - capacitors: $\mathbf{Z}_C = -\mathrm{j}\dfrac{1}{\omega C} = \dfrac{1}{\mathrm{j}\omega C}$
- Complex quantities may be expressed in a number of ways, the most common forms being the *rectangular form* $(a + \mathrm{j}b)$, the *polar form* $(r\angle\theta)$ and the *exponential form* $(r\mathrm{e}^{\mathrm{j}\theta})$.
- Complex impedances can be used with sinusoidal signals in a similar manner to the way resistances are used in DC circuits.

Exercises

6.1 A signal v is described by the expression $v = 15 \sin 100t$. What is the angular frequency of this signal, and what is its peak magnitude?

6.2 A signal v is described by the expression $v = 25 \sin 250t$. What is the frequency of this signal (in Hz), and what is its r.m.s. magnitude?

6.3 Give an expression for a sinusoidal signal with a peak voltage of 20 V and an angular frequency of 300 rad/s.

6.4 Give an expression for a sinusoidal signal with an r.m.s. voltage of 14.14 V and a frequency of 50 Hz.

6.5 Give an expression relating the voltage across an inductor to the current through it.

6.6 Give an expression relating the voltage across a capacitor to the current through it.

6.7 If a sinusoidal current is passed through a resistor, what is the phase relationship between this current and the voltage across the component?

6.8 If a sinusoidal current is passed through a capacitor, what is the phase relationship between this current and the voltage across the component?

6.9 If a sinusoidal current is passed through an inductor, what is the phase relationship between this current and the voltage across the component?

6.10 Explain what is meant by the term 'reactance'.

6.11 What is the reactance of a resistor?

6.12 What is the reactance of an inductor?

6.13 What is the reactance of a capacitor?

6.14 Calculate the reactance of an inductor of 20 mH at a frequency of 100 Hz, being sure to include the units in your answer.

6.15 Calculate the reactance of a capacitor of 10 nF at an angular frequency of 500 rad/s, being sure to include the units in your answer.

6.16 A sinusoidal voltage of 15 V r.m.s. at 250 Hz is applied across a 50 μF capacitor. What will be the current in the capacitor?

6.17 A sinusoidal current of 2 mA peak at 100 rad/s flows through an inductor of 25 mH. What voltage will appear across the inductor?

6.18 Explain briefly the use of a phasor diagram.

6.19 What is the significance of the length and direction of a phasor?

6.20 Estimate the magnitude and phase of (**A** + **B**) and (**A** − **B**) in the following phasor diagram.

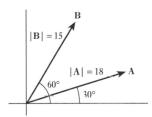

6.21 A voltage is formed by summing two sinusoidal waveforms of the same frequency. The first has a magnitude of 20 V and is taken as the reference phase (that is, its phase angle is taken as 0°). The second has a magnitude of 10 V and leads the first waveform by 45°. Draw a phasor diagram of this arrangement and hence estimate the magnitude and phase of the resultant signal.

6.22 A sinusoidal current of 3 A at 100 Hz flows through a series combination of a resistor of 25 Ω and an inductor of 75 mH. Use a phasor diagram to determine the voltage across the combination and the phase angle between this voltage and the current.

6.23 Use circuit simulation to confirm your results for Exercise 6.22. In doing this, you might find it useful to start with your circuit for Computer simulation exercise 6.1.

6.24 A sinusoidal voltage of 12 V at 500 Hz is applied across a series combination of a resistor of 5 kΩ and a capacitor of 100 nF. Use a phasor diagram to determine the current through the combination and the phase angle between this current and the applied voltage.

6.25 Use circuit simulation to confirm your results for Exercise 6.24. In doing this, you might find it useful to start with your circuit for Computer simulation exercise 6.2.

6.26 Use a phasor diagram to determine the magnitude and phase angle of the impedance formed by the series combination of a resistance of 25 Ω and a capacitance of 10 μF, at a frequency of 300 Hz.

6.27 If $x = 5 + j7$ and $y = 8 - j10$, evaluate $(x + y)$, $(x - y)$, $(x \times y)$ and $(x \div y)$.

6.28 What is the complex impedance of a resistor of 1 kΩ at a frequency of 1 kHz?

6.29 What is the complex impedance of a capacitor of 1 μF at a frequency of 1 kHz?

6.30 What is the complex impedance of an inductor of 1 mH at a frequency of 1 kHz?

6.31 Determine the complex impedance of the following arrangements at a frequency of 200 Hz.

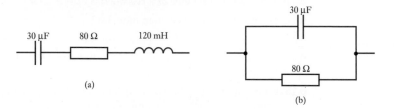

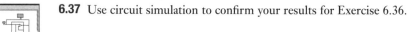

(a)

(b)

6.32 Express $x = 20 + j30$ in polar form and in exponential form.

6.33 Express $y = 25\angle{-40°}$ in rectangular form and in exponential form.

6.34 A voltage $v = 60 \sin 314t$ is applied across a series combination of a 10 Ω resistor and an inductance of 50 mH. Determine the magnitude and phase of the resulting current.

6.35 Use circuit simulation to confirm your results for Exercise 6.34.

6.36 A current of $i = 0.5 \sin 377t$ is passed through a parallel combination of a resistance of 1 kΩ and a capacitance of 5 μF. Determine the magnitude and phase of the resulting voltage across the combination.

6.37 Use circuit simulation to confirm your results for Exercise 6.36.

Power in AC Circuits

Objectives

When you have studied the material in this chapter, you should be able to:

- explain concepts such as apparent power, active power, reactive power and power factor
- calculate the power dissipated in circuits containing resistors, capacitors and inductors when these are used with AC signals
- discuss the importance of the power factor in determining the efficiency of power utilisation and distribution
- determine the power factor of a given circuit arrangement and propose appropriate additional components to achieve power factor correction if necessary
- describe the measurement of power in both single-phase and three-phase arrangements.

7.1 Introduction

The instantaneous power dissipated in a resistor is given by the product of the instantaneous voltage and the instantaneous current. In a DC circuit, this produces the well-known relationship

$$P = VI$$

while in an AC circuit

$$p = vi$$

where v and i are the instantaneous values of an alternating waveform, and p is the instantaneous power dissipation.

In a purely resistive circuit v and i are in phase, and calculation of p is straight-forward. However, in circuits containing reactive elements, there will normally be a phase difference between v and i, and calculating the power is slightly more complicated. Here we will begin by looking at the power dissipated in resistive loads before going on to consider inductive, capacitive and mixed loads.

7.2 Power dissipation in resistive components

If a sinusoidal voltage $v = V_p \sin \omega t$ is applied across a resistance R, then the current i through the resistance will be

$$i = \frac{v}{R}$$
$$= \frac{V_p \sin \omega t}{R}$$
$$= I_p \sin \omega t$$

where $I_P = V_P/R$.

The resultant power p is given by

$$p = vi$$
$$= V_p \sin \omega t \times I_p \sin \omega t$$
$$= V_p I_p (\sin^2 \omega t)$$
$$= V_p I_P \left(\frac{1 - \cos 2\omega t}{2} \right)$$

We see that p varies at a frequency *double* that of v and i. The relationship between v, i and p is shown in Figure 7.1. Since the average value of a cosine function is zero, the average value of $(1 - \cos 2\omega t)$ is 1, and the average value of p is $^1/_2 V_P I_P$. Therefore, the average power P is equal to

$$P = \frac{1}{2} V_P I_P$$
$$= \frac{V_P}{\sqrt{2}} \times \frac{I_P}{\sqrt{2}}$$
$$= VI$$

where V and I are the r.m.s. voltage and current. This is consistent with our discussion of r.m.s. values in Chapter 2.

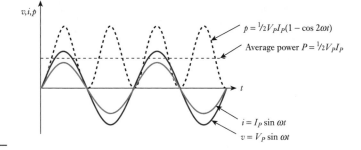

Figure 7.1
Relationship between voltage, current and power in a resistor.

7.3 Power in capacitors

We know that the current in a capacitor leads the voltage by 90° (see Chapter 4). Therefore, if $v = V_p \sin \omega t$, then the resultant current will be given by $i = I_p \cos \omega t$, and the power p will be

$$p = vi$$
$$= V_p \sin \omega t \times I_p \cos \omega t$$
$$= V_p I_p (\sin \omega t \times \cos \omega t)$$
$$= V_p I_P \left(\frac{\sin 2\omega t}{2} \right)$$

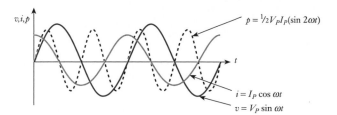

Figure 7.2
Relationship between
voltage, current and
power in a capacitor.

The relationship between v, i and p is shown in Figure 7.2. Again p varies at a frequency *double* that of v and i, but since the average value of a sine function is zero, the average value of p is also zero. Therefore, power flows *into* the capacitor during part of the cycle and then flows *out* of the capacitor again. Thus the capacitor stores energy for part of the cycle and returns it to the circuit again, with the average power dissipated P in the capacitor being zero.

7.4   Power in inductors

We know that the current in an inductor lags the voltage by 90° (see Chapter 5). Therefore, if $v = V_P \sin \omega t$, the resultant current will be given by $i = -I_P \cos \omega t$. Therefore, the power p is given by

$$p = vi$$
$$= V_P \sin \omega t \times -I_P \cos \omega t$$
$$= -V_P I_P (\sin \omega t \times \cos \omega t)$$
$$= -V_P I_P \left(\frac{\sin 2\omega t}{2} \right)$$

The situation is very similar to that of the capacitor, and again the power dissipated in the inductor is zero. The relationship between v, i and p is shown in Figure 7.3.

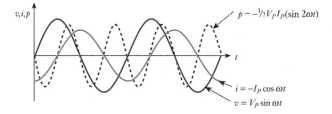

Figure 7.3
Relationship between
voltage, current and
power in an inductor.

Video 7A

7.5   Power in circuits with resistance and reactance

When a sinusoidal voltage $v = V_P \sin \omega t$ is applied across an impedance that contains both resistive and inductive elements, the resulting current will be of the general form $i = I_P \sin(\omega t - \phi)$. Therefore, the instantaneous power p is given by

$$p = vi$$
$$= V_P \sin \omega t \times I_P (\sin \omega t - \phi)$$
$$= \frac{1}{2} V_P I_P [\cos \phi - \cos(2\omega t - \phi)]$$

$$p = \frac{1}{2}V_P I_P \cos\phi - \frac{1}{2}V_P I_P \cos(2\omega t - \phi) \tag{7.1}$$

It can be seen that the expression for p has two components. The second is a function of $2\omega t$ and consequently this component oscillates at twice the frequency of v. The average value of this term over a complete cycle, or over a long period of time, is zero. This represents the energy that is stored in the reactive elements and is then returned to the circuit within each voltage cycle. The first component is independent of time and therefore has a constant value. This represents the power dissipated in the resistive elements in the circuit. Consequently, the average power dissipated is given by

$$P = \frac{1}{2}V_P I_P (\cos\phi)$$

$$= \frac{V_P}{\sqrt{2}} \times \frac{I_P}{\sqrt{2}} \times (\cos\phi)$$

$$P = VI \cos\phi \tag{7.2}$$

where V and I are r.m.s. values of the voltage and current. This average power dissipation P is called the **active power** and is measured in watts.

If one were to measure the r.m.s. voltage V across a load and the r.m.s. current through the load I, and multiply these two quantities together, one would obtain a quantity equal to VI, which is termed the **apparent power**. One could imagine that an inexperienced engineer, unfamiliar with the effects of phase angle, might take such measurements and perform such a calculation in order to calculate the power dissipated in the load. From the discussion above, we know that this product does *not* give the dissipated power, but it is still a quantity of interest and the apparent power is given the symbol S. To avoid confusion with dissipated power, S is given the units of volt amperes (VA).

From Equation 7.2, we know that $P = VI \cos\phi$, and therefore

$$P = S \cos\phi$$

In other words, the active power is equal to the apparent power times the cosine of the phase angle. This cosine term is referred to as the **power factor**, which is defined as the ratio of the active to the apparent power. Thus

$$\frac{\text{active power (in watts)}}{\text{apparent power (in volt amperes)}} = \text{power factor} \tag{7.3}$$

From the above

$$\text{power factor} = \frac{P}{S} = \cos\phi \tag{7.4}$$

Example 7.1   The voltage across a component is measured as 50 V r.m.s. and the current through it is 5 A r.m.s. If the current leads the voltage by 30°, calculate:

(a) the apparent power;
(b) the power factor;
(c) the active power.

(a) The apparent power is

$$\text{apparent power } S = VI$$

$$= 50 \times 5$$

$$= 250 \, \text{VA}$$

(b) The power factor is

$$\text{power factor} = \cos \phi$$

$$= \cos 30°$$

$$= 0.866$$

(c) The active power is

$$\text{active power } P = S \cos \phi$$

$$= 250 \times 0.866$$

$$= 216.5 \, \text{W}$$

File 7A

Computer simulation exercise 7.1

Simulate a circuit consisting of a sinusoidal voltage source and a load resistor. Use transient analysis to look at a few cycles of the waveform and plot the voltage, the current and the power dissipated ($v_R \times i_R$). Observe the relationship between these waveforms and confirm that the power waveform has a frequency double that of the voltage source. Also confirm that the average value of the power is half its peak value.

Now repeat this experiment with the resistor replaced first with a capacitor, then with an inductor and compare the results. Confirm that the average power is now zero in each case.

Finally, repeat the experiment using a series combination of a resistor and a capacitor and note the effect. When doing this last part, you should be careful to measure the voltage across the series combination of components.

7.6 Active and reactive power

From Equation 7.1, it is clear that, when a load has both resistive and reactive elements, the resultant power will have two components. The first is that which is *dissipated* in the resistive element of the load, which we have described as the active power. The second element is not dissipated but is *stored* and *returned* by the reactive elements in the circuit. We refer to this as the **reactive power** in the circuit, which is given the symbol Q.

While reactive power is not dissipated in the load, its presence does have an effect on the rest of the system. The need to supply power to the reactive elements during part of the supply cycle, and to accept power from them during other parts of the cycle, increases the current that must be supplied by the power source and also increases the losses due to resistance in power cables.

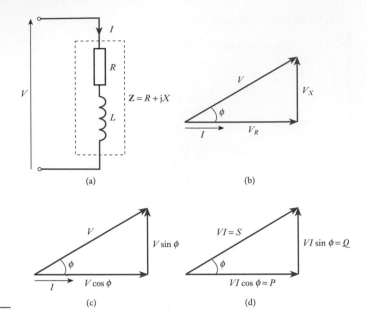

Figure 7.4
Active and reactive power.

In order to quantify this effect, let us consider the situation where a sinusoidal voltage V is applied to a complex load given by $\mathbf{Z} = R + jX$ as shown in Figure 7.4(a). A voltage phasor diagram of this arrangement is shown in Figure 7.4(b), and this is redrawn in Figure 7.4(c) expressing V_R and V_X in terms of the applied voltage V and the phase angle ϕ. If we take the *magnitudes* of the various elements of this phasor diagram and multiply each by I, we obtain a triangle of a similar shape, as shown in Figure 7.4(d). If V and I represent the r.m.s. voltage and current, then the hypotenuse of this triangle is of length VI, which we identified earlier as being the apparent power S; the base of the triangle is of length $VI \cos \phi$, which we identified as being the active power P; and the vertical line is of length $VI \sin \phi$, which represents the reactive power Q in the circuit. Reactive power is given the units of **volt amperes reactive** or **var** (to differentiate it from active power, with which it is dimensionally equivalent). For obvious reasons, Figure 7.4(d) is referred to as a **power triangle**. Therefore

$$\text{active power } P = VI \cos \phi \text{ W} \tag{7.5}$$

$$\text{reactive power } Q = VI \sin \phi \text{ var} \tag{7.6}$$

$$\text{apparent power } S = VI \text{ VA} \tag{7.7}$$

$$S^2 = P^2 + Q^2 \tag{7.8}$$

Example 7.2

A 2 kVA motor operates from a 240 V supply at 50 Hz and has a power factor of 0.75. Determine the apparent power, the active power, the reactive power and the current in the motor.

The apparent power S of the motor is 2000 VA, since this is the rating of the motor. The power factor ($\cos \phi$) is 0.75. Therefore, the active power in the motor is active power $P = S \cos \phi$

$$= 2000 \times 0.75$$

$$= 1500 \text{ W}$$

Since $\cos \phi = 0.75$, it follows that $\sin \phi = \sqrt{(1 - \cos^2\phi)} = 0.661$. Therefore

reactive power $Q = S \sin \phi$

$$= 2000 \times 0.6614$$

$$= 1323 \text{ var}$$

The current is given by the apparent power divided by the voltage

current $I = \dfrac{S}{V}$

$$= \dfrac{2000}{240}$$

$$= 8.33 \text{ A}$$

<h2>7.7   Power factor correction</h2>

The power factor is of particular importance in high-power applications. A given power supply or generator normally has a maximum output voltage and a maximum output current, and in general we wish to maximise the useful power that we can obtain from such a source. If the load connected to the supply has a power factor of unity, then the available power will be maximised, and all the current supplied by the generator will be used to produce active power. If the power factor is less than 1, the power available at the load will be less than this maximum value, and reactive currents will result in increased losses in the associated cables.

Inductive loads are said to have *lagging* power factors, since the current lags the applied voltage, while capacitive loads are said to have *leading* power factors. Many high-power devices, such as motors, are inductive, and a typical AC motor will have a lagging power factor of 0.9 or less. Consequently, the total load applied to national power distribution grids tends to have a power factor of 0.8–0.9 lagging. This leads to major inefficiencies in the power generation and distribution process, and power companies therefore charge fees that penalise industrial users who introduce a poor power factor.

The problems associated with inductive loads can be tackled by adding additional components to bring the power factor closer to unity. A capacitor of an appropriate size in parallel with a *lagging* load can improve the load factor by 'cancelling out' the inductive element in the load's impedance. The effect is that the reactive currents associated with the inductance now flow in and out of the capacitor, rather than to and from the power supply. This not only reduces the current flowing into and out of the power supply but also reduces the resistive losses in the cables. A similar effect can also be obtained by adding a capacitor in series with the load, although a parallel arrangement is more common because it does not alter the voltage on the load.

Example 7.3 | **A capacitor is to be added in parallel with the motor of Example 7.2 to increase its power factor to 1.0. Calculate the value of the required capacitor, and calculate the active power, the apparent power and the current after power factor correction.**

In Example 7.2, we determined that for this motor

$$\text{apparent power } S = 2000 \text{ VA}$$

$$\text{active power } P = 1500 \text{ W}$$

$$\text{current } I = 8.33 \text{ A}$$

$$\text{reactive power } Q = 1323 \text{ var}$$

The capacitor is required to cancel the *lagging* reactive power. We therefore need to add a capacitive element with a *leading* reactive power Q_C of −1323 var.

Now, just as $P = V^2/R$, so $Q = V^2/X$. Since capacitive reactive power is negative

$$Q_C = -\frac{240^2}{X_C} = -1323 \text{ var}$$

$$X_C = \frac{240^2}{1323} = 43.54 \ \Omega$$

$X_C = 1/\omega C$ which is equal to $1/2\pi f C$. Therefore

$$\frac{1}{2\pi f C} = 43.54$$

$$C = \frac{1}{43.54 \times 2 \times \pi \times f}$$

$$= \frac{1}{43.54 \times 2 \times 3.142 \times 50}$$

$$= 73 \ \mu\text{F}$$

The power factor correction does not affect the active power in the motor, and P is therefore unchanged at 1500 W. However, since the power factor is now 1, the apparent power is now $S = P = 1500$. The current is now given by

$$\text{current } I = \frac{S}{V} = \frac{1500}{240} = 6.25 \text{ A}$$

Thus the apparent power is reduced from 2000 VA to 1500 VA as a result of the addition of the capacitor, while the current drops from 8.33 A to 6.25 A. The active power dissipated by the motor remains unchanged at 1500 W.

In Example 7.3, we chose a capacitor to increase the power factor to unity, but this is not always appropriate. High-voltage capacitors suitable for this purpose are expensive, and it may be more cost effective to increase the power factor by a more modest amount, perhaps up to about 0.9.

7.8 Three-phase systems

So far, our description of AC signals has been restricted to **single-phase** arrangements. This is the most common arrangement and is that used in conventional domestic supplies, where power is delivered using a single pair of cables (often with the addition of an earth conductor). While this arrangement works well for heating and lighting applications, there are certain situations (in particular when using large electric motors) when single phase arrangements are unsatisfactory. In such situations,

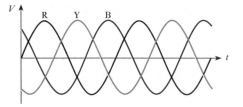

Figure 7.5
Voltage waveforms
of a three-phase
arrangement.

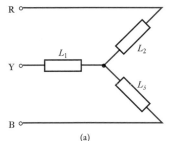

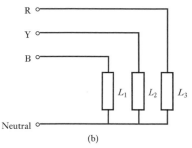

Figure 7.6
Three-phase
connections.

(a)   (b)

it is common to use a three-phase supply, which provides power using three alternating waveforms, each differing in phase by 120°. The three **phases** are given the labels red, yellow and blue, which are normally abbreviated to R, Y and B. The relationship between these phases is shown in Figure 7.5.

Three-phase power can be supplied using three or four conductors. Where three are used each provides one of the phases, and loads are connected between the conductors, as shown in Figure 7.6(a). In a four-line system, the additional wire is a neutral conductor. Loads may then be connected between each phase and neutral, as shown in Figure 7.6(b).

Three-phase arrangements are common in high-power industrial applications, particularly where electrical machines are involved.

7.9 Power measurement

When using sinusoidal signals, the power dissipated in a load is determined not only by the r.m.s. values of the voltage and current but also by the phase angle between the voltage and current waveforms (which determines the power factor). Consequently, it is not possible to calculate the power simply by taking independent measurements of the voltage and the current.

In single-phase AC circuits, power is normally measured using an **electro-dynamic wattmeter**. This device passes the load current through a series of low-resistance field coils and places the load voltage across a high-resistance armature coil. The resulting deflection is directly related to the product of the instantaneous current and voltage and hence to the instantaneous power in the load. The inertia of the coil smooths out the line frequencies and produces a reading that is proportional to the average value of the power. The device can therefore be directly calibrated in watts.

In three-phase circuits, it is necessary to sum the power taken from each phase in order to measure the total power consumption. In a three-wire system, because of the interaction between the phases, the readings from two wattmeters can be used to measure the total power. Summing the readings from the two meters gives the total power, whether or not the same power is taken from each phase. In a four-line arrangement, it may be necessary to use three wattmeters to measure the power taken from each of the phases. However, if it is known that the system is balanced (that is, that equal power is taken from each phase), then a single wattmeter can be used. This is connected to any one of the phases and its reading multiplied by three to get the total power.

Video 7B

Further study

We have seen that power factor correction plays an important role in the design of high-power systems. In a particular industrial application, a 3 kVA electric motor is connected to a single-phase electrical supply of 240 V at 50 Hz. Preliminary measurements of this arrangement show that the power factor is 0.7.

In order to decide on the appropriate form of power factor correction it is necessary to compare the cost of performing the correction (in terms of the cost of the necessary capacitor) with the potential energy savings. To facilitate this analysis, calculate the size of the capacitor that must be connected in parallel with the motor in order to produce a power factor of 1.0, and also the size required to produce a power factor of 0.9. Also estimate the percentage energy savings in each case.

Key points

- In both DC and AC circuits, the instantaneous power in a resistor is given by the product of the instantaneous voltage and the instantaneous current.
- In purely resistive circuits, the current is in phase with the voltage and therefore calculating power is straightforward: the average power is equal to VI, where V and I are r.m.s. values.
- In a capacitor, the current leads the voltage by 90° and the average power dissipation is zero.
- In an inductor, the current lags the voltage by 90° and again the average power dissipation is zero.
- In circuits that have both resistive and inductive elements, the average power $P = VI \cos \phi$, where ϕ is the phase angle between the current and the voltage waveforms. P is termed the active power, which has units of watts (W).
- The product of the r.m.s. voltage V and the r.m.s. current I is termed the apparent power S. This has units of volt amperes (VA).
- The ratio of the active power to the apparent power is the power factor:

$$\text{power factor} = \frac{P}{S} = \cos \phi$$

- The power stored and returned to the system by reactive elements is termed the reactive power Q. This has units of volt amperes reactive (var).
- The efficiency of power utilisation and distribution can be increased by increasing the power factor – a process called power factor correction. Since most high-power loads have an inductive element, correction is normally achieved by adding a capacitor in parallel with the load.
- High-power applications often make use of three-phase supplies.
- Power can be measured directly using a wattmeter, which multiplies instantaneous values of voltage and current to measure power directly.

Exercises

7.1 A sinusoidal voltage $v = 10 \sin 377t$ is applied to a resistor of 50 Ω. Calculate the average power dissipated in it.

7.2 The voltage of Exercise 7.1 is applied across a capacitor of 1 μF. Calculate the average power dissipated in the capacitor in this arrangement.

7.3 The voltage of Exercise 7.1 is applied across an inductor of 1 mH. Calculate the average power dissipated in the inductor in this arrangement.

7.4 The voltage across a component is 100 V r.m.s. and the current is 7 A r.m.s. If the current lags the voltage by 60°, calculate the apparent power, the power factor and the active power.

7.5 Explain the difference between the units of watts, VA and var.

7.6 A sinusoidal voltage of 100 V r.m.s. at 50 Hz is applied across a series combination of a 40 Ω resistor and an inductor of 100 mH. Determine the r.m.s. current, the apparent power, the power factor, the active power and the reactive power.

7.7 A machine operates on a 250 V supply at 60 Hz; it is rated at 500 VA and has a power factor of 0.8. Determine the apparent power, the active power, the reactive power and the current in the machine.

7.8 Explain what is meant by power factor correction and explain why this is of importance in high-power systems.

7.9 Calculate the value of capacitor required to be added in parallel with the machine of Exercise 7.7 to achieve a power factor of 1.0.

7.10 Calculate the value of capacitor required to be added in parallel with the machine of Exercise 7.7 to achieve a power factor of 0.9.

7.11 A sinusoidal signal of 20 V peak at 50 Hz is applied to a load consisting of a 10 Ω resistor and a 16 mH inductor connected in series. Calculate the power factor of this arrangement and the active power dissipated in the load.

7.12 Simulate the arrangement of Exercise 7.11 and plot the voltage and the current. Estimate the phase difference between these two waveforms and hence confirm the value you calculated for the power factor. Plot the product of the voltage and the current and estimate its average value. Hence confirm your calculated value for the active power of the circuit.

7.13 Determine the value of capacitor needed to be added in series with the circuit of Exercise 7.11 to produce a power factor of 1.0. Calculate the active power that would be dissipated in the circuit with the addition of such a capacitor.

7.14 Simulate the arrangement of Exercise 7.13 and confirm your predictions for its behaviour.

7.15 Explain the difference between three- and four-conductor arrangements of three-phase power supplies.

7.16 Explain why it is not possible to calculate the power dissipated in an AC network by multiplying the readings of a voltmeter and an ammeter.

7.17 Explain how it is possible to measure power directly in a single-phase system.

Frequency Characteristics of AC Circuits

8.1 Introduction

Having studied the AC behaviour of some basic circuit components, we are now in a position to consider their effects on the frequency characteristics of simple circuits.

While the properties of a pure resistance are not affected by the frequency of the signal concerned, this is not true of reactive components. The reactance of both inductors and capacitors is dependent on frequency, and therefore the characteristics of any circuit that includes capacitors or inductors will change with frequency. However, the situation is more complex than this because (as noted in Chapters 4 and 5) all real circuits have both stray capacitance and stray inductance. Inevitably, therefore, the characteristics of all circuits will change with frequency.

In order to understand the nature of these frequency-related effects we will look at simple combinations of resistors, capacitors and inductors and see how their characteristics change with frequency. However, before looking at these circuits it is useful to introduce a couple of new concepts and techniques.

8.2 Two-port networks

A two-port network is, as its name suggests, simply a circuit configuration that has two 'ports', namely the input port and the output port. Such an arrangement is shown in Figure 8.1(a).

When connected to some input circuitry (perhaps a voltage source) and to some output circuitry (such as a load resistor) we can then identify the voltages at the input

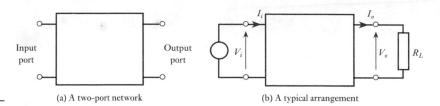

Figure 8.1
A two-port network.

(a) A two-port network   (b) A typical arrangement

and the output (V_i and V_o) and the currents flowing into and out of the network (I_i and I_o). Such an arrangement is illustrated in Figure 8.1(b). Clearly the relationship between the output voltage and the output current is determined by the value of the *load resistance* R_L. Similarly, the relationship between the input voltage and the input current determines the effective resistance looking into the input port of the arrangement. This is termed the **input resistance** of the network, which is given the symbol R_i and clearly is equal to V_i/I_i. We can then use these various voltages and currents to describe the characteristics of the two-port network.

The ratio of the output voltage to the input voltage is termed the **voltage gain** of the circuit, while the ratio of the output current to the input current is termed the **current gain**. The **power gain** of the network is the ratio of the power supplied to a load to the power absorbed from the source. The input power can be calculated from the input voltage and the input current, and the output power can be determined from the output voltage and output current. Thus

$$\text{voltage gain } (A_v) = \frac{V_o}{V_i} \tag{8.1}$$

$$\text{current gain } (A_i) = \frac{I_o}{I_i} \tag{8.2}$$

$$\text{power gain } (A_p) = \frac{P_o}{P_i} \tag{8.3}$$

Example 8.1

Calculate the voltage, current gain and power gain of the following two-port network.

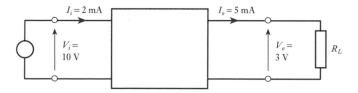

From the diagram, and from Equations 8.1 to 8.3 above, we see that

$$\text{voltage gain } (A_v) = \frac{V_o}{V_i} = \frac{3\text{ V}}{10\text{ V}} = 0.3$$

$$\text{current gain } (A_i) = \frac{I_o}{I_i} = \frac{5\text{ mA}}{2\text{ mA}} = 2.5$$

$$\text{power gain } (A_p) = \frac{P_o}{P_i} = \frac{V_o \times I_o}{V_i \times I_i} = \frac{3\text{ V} \times 5\text{ mA}}{10\text{ V} \times 2\text{ mA}} = 0.75$$

Note that the various gains can each be greater or less than unity. A gain of greater than unity represents *amplification* while a gain of less than unity represents *attenuation*. A power gain of greater than unity implies that the circuit is delivering more power to the load than it is accepting at its input. Such an arrangement requires some form of external power source. *Passive* circuits, such as combinations of resistors, capacitors and inductors, will always have a power gain that is no greater than unity. *Active* circuits, which use an external power supply, can have a power gain that is very much greater than unity.

The power gain of a modern electronic amplifier may be very high, gains of 10^6 or 10^7 being common. With these large numbers it is often convenient to use a logarithmic expression of gain rather than a simple ratio. This is often done using **decibels**.

8.3   The decibel (dB)

The decibel (dB) is a dimensionless figure for **power gain** and is defined by

$$\text{power gain (dB)} = 10\log_{10}\frac{P_2}{P_1} \tag{8.4}$$

where P_2 is the output power and P_1 is the input power of the amplifier or other circuit.

Example 8.2   **Express a power gain of 2500 in decibels.**

$$\text{power gain (dB)} = 10\log_{10}\frac{P_2}{P_1}$$

$$= 10\log_{10}2500$$

$$= 10 \times 3.40$$

$$= 34.0 \text{ dB}$$

Decibels may be used to represent both amplification and attenuation, and, in addition to making large numbers more manageable, the use of decibels has several other advantages. For example, when several stages of amplification or attenuation are connected in series (this is often referred to as **cascading** circuits), the overall gain of the combination can be found simply by adding the individual gains of each stage when these are expressed in decibels. This is illustrated in Figure 8.2. The use of decibels also simplifies the description of the frequency response of circuits, as we will see later in this chapter.

For certain values of gain, the decibel equivalents are easy to remember or to calculate using mental arithmetic. Since $\log_{10} n$ is simply the power to which 10 must be raised to equal n, for powers of 10 it is easy to calculate. For example, $\log_{10} 10 = 1$, $\log_{10} 100 = 2$, $\log_{10} 1000 = 3$, and so on. Similarly, $\log_{10} 1/10 = -1$, $\log_{10} 1/100 = -2$ and $\log_{10} 1/1000 = -3$. Therefore, gains of 10, 100 and 1000 are simply 10 dB, 20 dB and 30 dB respectively, and attenuations of $1/10$, $1/100$ and $1/1000$ are simply -10 dB, -20 dB and -30 dB. A circuit that doubles the power has a gain of $+3$ dB,

Figure 8.2
Calculating the gain of several stages in series.

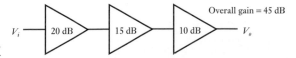

Table 8.1 Expressing power amplification and attenuation in decibels.

Power gain (ratio)	Decibels (dB)
1000	30
100	20
10	10
2	3
1	0
0.5	−3
0.1	−10
0.01	−20
0.001	−30

while a circuit that halves the power has a gain of −3 dB. A circuit that leaves the power unchanged (a power gain of 1) has a gain of 0 dB. These results are summarised in Table 8.1.

In many cases, our knowledge of a circuit relates to its voltage gain rather than to its power gain. Clearly, these two measures are related, and we know that the power dissipated in a resistance R is related to the applied voltage V by the expression V^2/R. Therefore, the gain of an amplifier expressed in decibels can be written as

$$\text{power gain (dB)} = 10 \log_{10} \frac{P_2}{P_1} = 10 \log_{10} \frac{V_2^2/R_2}{V_1^2/R_1}$$

where V_1 and V_2 are the input and output voltages, respectively, and R_1 and R_2 are the input and load resistances, respectively.

If, and only if, R_1 and R_2 are equal, the power gain of the amplifier is given by

$$\text{power gain (dB)} = 10 \log_{10} \frac{V_2^2}{V_1^2}$$

$$= 20 \log_{10} \frac{V_2}{V_1}$$

$$= 20 \log_{10} (\text{voltage gain})$$

Some networks do have equal input and load resistance, and in these cases it is often useful to express the gain in decibels rather than as a simple ratio. Note that it is not strictly correct to say, for example, that a circuit has a voltage gain of 10 dB, even though you will often hear such statements. Decibels represent power gain, and what is meant is that the circuit has a voltage gain that corresponds to a power gain of 10 dB. However, it is very common to describe the voltage gain of a circuit in dB as

$$\text{voltage gain (dB)} = 20 \log_{10} \frac{V_2}{V_1} \qquad (8.5)$$

even when R_1 and R_2 are not equal.

Example 8.3 Calculate the gain in decibels of circuits that have power gains of 5, 50 and 500 and voltage gains of 5, 50 and 500.

Power gain of 5	Gain (dB) = $10 \log_{10}(5)$	7.0 dB
Power gain of 50	Gain (dB) = $10 \log_{10}(50)$	17.0 dB
Power gain of 500	Gain (dB) = $10 \log_{10}(500)$	27.0 dB
Voltage gain of 5	Gain (dB) = $20 \log_{10}(5)$	14.0 dB
Voltage gain of 50	Gain (dB) = $20 \log_{10}(50)$	34.0 dB
Voltage gain of 500	Gain (dB) = $20 \log_{10}(500)$	54.0 dB

Converting from gains expressed in decibels to simple power or voltage ratios requires the reversal of the operations used above. For example, since

$$\text{power gain (dB)} = 10 \log_{10}(\text{power gain})$$

it follows that

$$10 \log_{10}(\text{power gain}) = \text{power gain (dB)}$$

$$\log_{10}(\text{power gain}) = \frac{\text{power gain (dB)}}{10}$$

$$\text{power gain} = 10^{(\text{power gain (dB)}/10)} \tag{8.6}$$

Similarly,

$$\text{voltage gain} = 10^{(\text{power gain (dB)}/20)} \tag{8.7}$$

Example 8.4 Express gains of 20 dB, 30 dB and 40 dB as both power gains and voltage gains.

20 dB	$20 = 10 \log_{10}(\text{power gain})$ power gain = 10^2	power gain = 100
	$20 = 20 \log_{10}(\text{voltage gain})$ voltage gain = 10	voltage gain = 10
30 dB	$30 = 10 \log_{10}(\text{power gain})$ power gain = 10^3	power gain = 1000
	$30 = 20 \log_{10}(\text{voltage gain})$ voltage gain = $10^{1.5}$	voltage gain = 31.6
40 dB	$40 = 10 \log_{10}(\text{power gain})$ power gain = 10^4	power gain = 10,000
	$40 = 20 \log_{10}(\text{voltage gain})$ voltage gain = 10^2	voltage gain = 100

8.4 Frequency response

Since the characteristics of reactive components change with frequency, the behaviour of circuits using these components will also change. The way in which the gain of a circuit changes with frequency is termed its **frequency response**. These changes

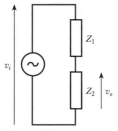

Figure 8.3
A potential divider circuit.

take the form of variations in the magnitude of the gain and in its phase angle, leading to two aspects of the response, namely the *amplitude response* and the *phase response*. In some situations both aspects are of importance, while in others only the amplitude response is needed. For this reason, the term *frequency response* is often used to refer simply to the amplitude response of a system.

In order to understand the nature of these frequency–related effects, we will start by looking at very simple circuits containing resistors and capacitors, or resistors and inductors. In Chapter 6, we looked at circuits involving impedances, including the potential divider arrangement shown in Figure 8.3. From our earlier consideration of the circuit, we know that the output voltage of this circuit is given by

$$v_o = v_i \times \frac{\mathbf{Z}_2}{\mathbf{Z}_1 + \mathbf{Z}_2}$$

Another way of describing the behaviour of this circuit is to give an expression for the output voltage divided by the input voltage. In this case, this gives

$$\frac{v_o}{v_i} = \frac{\mathbf{Z}_2}{\mathbf{Z}_1 + \mathbf{Z}_2} \tag{8.8}$$

This ratio is the **voltage gain** of the circuit, but it is also referred to as its **transfer function**. We will now use this expression to analyse the behaviour of simple *RC* and *RL* circuits.

8.5 ## A high-pass *RC* network

Consider the circuit of Figure 8.4(a), which shows a potential divider, formed from a capacitor and a resistor. This circuit is shown redrawn in Figure 8.4(b), which is electrically identical. Applying Equation 8.8, we see that

$$\frac{v_o}{v_i} = \frac{\mathbf{Z}_R}{\mathbf{Z}_R + \mathbf{Z}_C} = \frac{R}{R - j\dfrac{1}{\omega C}} = \frac{1}{1 - j\dfrac{1}{\omega CR}} \tag{8.9}$$

At high frequencies, ω is large and the value of $1/j\omega CR$ is small compared with 1. Therefore, the denominator of the expression is close to unity and the voltage gain is approximately 1.

However, at lower frequencies the magnitude of $1/\omega CR$ becomes more significant and the gain of the network decreases. Since the denominator of the expression for the gain has both real and imaginary parts, the magnitude of the voltage gain is given by

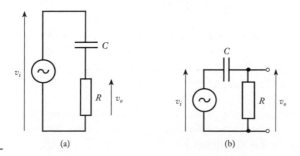

Figure 8.4
A simple *RC* network.

$$|\text{voltage gain}| = \frac{1}{\sqrt{1^2 + \left(\dfrac{1}{\omega CR}\right)^2}}$$

When the value of $1/\omega CR$ is equal to 1, this gives

$$|\text{voltage gain}| = \frac{1}{\sqrt{1+1}} = \frac{1}{\sqrt{2}} = 0.707$$

Since power gain is proportional to the square of the voltage gain, this is a halving of the power gain (or a fall of 3 dB) compared with the gain at high frequencies. This is termed the **cut-off frequency** of the circuit. If the angular frequency corresponding to this cut-off frequency is given the symbol ω_c, then $1/\omega_c CR$ is equal to 1, and

$$\omega_c = \frac{1}{CR} = \frac{1}{T} \text{ rad/s} \qquad (8.10)$$

where $T = CR$ is the time constant of the capacitor–resistor combination that produces the cut-off frequency.

Since it is often more convenient to deal with *cyclic* frequencies (which are measured in hertz) rather than *angular* frequencies (which are measured in radians per second) we can use the relationship $\omega = 2\pi f$ to calculate the corresponding cyclic cut-off frequency f_c:

$$f_c = \frac{\omega_c}{2\pi} = \frac{1}{2\pi CR} \text{ Hz} \qquad (8.11)$$

Example 8.5

Calculate the time constant T, the angular cut-off frequency ω_c and the cyclic cut-off frequency f_c of the following arrangement.

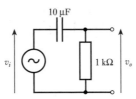

From above

$$T = CR = 10 \times 10^{-6} \times 1 \times 10^3 = 0.01 \text{ s}$$

$$\omega_c = \frac{1}{T} = \frac{1}{0.01} = 100 \text{ rad/s}$$

$$f_c = \frac{\omega_c}{2\pi} = \frac{100}{2\pi} = 15.9 \text{ Hz}$$

If we substitute for ω (where $\omega = 2\pi f$) and CR (where $CR = 1/2\pi f_c$) in Equation 8.9, we obtain an expression for the gain of the circuit in terms of the signal frequency f and the cut-off frequency f_c:

$$\frac{v_0}{v_i} = \frac{1}{1 - j\dfrac{1}{\omega CR}} = \frac{1}{1 - j\dfrac{1}{(2\pi f)(1/2\pi f_c)}} = \frac{1}{1 - j\dfrac{f_c}{f}} \qquad (8.12)$$

This is a general expression for the voltage gain of this form of *CR* network.

From Equation 8.12, it is clear that the voltage gain is a function of the signal frequency *f* and that the magnitude of the gain varies with frequency. Since the gain has an imaginary component, it is also clear that the circuit produces a **phase shift** that changes with frequency. To investigate how these two quantities change with frequency, let us consider the gain of the circuit in different frequency ranges.

8.5.1 When $f \gg f_c$

When the signal frequency *f* is much greater than the cut-off frequency f_c, then in Equation 8.12 f_c/f is much less than unity, and the voltage gain is approximately equal to 1. Here the imaginary part of the gain is negligible and the gain of the circuit is effectively real. Hence the phase shift produced is negligible. This situation is shown in the phasor diagram of Figure 8.5(a).

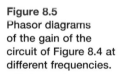

Figure 8.5
Phasor diagrams
of the gain of the
circuit of Figure 8.4 at
different frequencies.

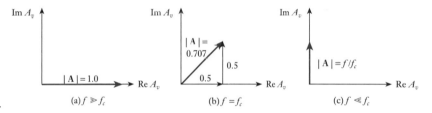

8.5.2 When $f = f_c$

When the signal frequency *f* is equal to the cut-off frequency f_c, then Equation 8.12 becomes

$$\frac{v_o}{v_i} = \frac{1}{1 - j\dfrac{f_c}{f}} = \frac{1}{1 - j}$$

Multiplying the numerator and the denominator by $(1 + j)$ gives

$$\frac{v_o}{v_i} = \frac{(1 + j)}{(1 - j)(1 + j)} = \frac{(1 + j)}{2} = 0.5 + 0.5j$$

This is illustrated in the phasor diagram of Figure 8.5(b), which shows that the magnitude of the gain at the cut-off frequency is 0.707. This is consistent with our earlier analysis, which predicted that the gain at the cut-off frequency should be $1/\sqrt{2}$ (or 0.707) times the mid-band gain. In this case, the mid-band gain is the gain some way above the cut-off frequency, which we have just shown to be 1. The phasor diagram also shows that at this frequency the phase angle of the gain is $+45°$. This shows that the output voltage *leads* the input voltage by 45°. The gain is therefore $0.707\angle 45°$.

8.5.3   When $f \ll f_c$

The third region of interest is where the signal frequency is well below the cut-off frequency. Here f_c/f is much greater than 1, and Equation 8.12 becomes

$$\frac{v_o}{v_i} = \frac{1}{1-j\dfrac{f_c}{f}} \approx \frac{1}{-j\dfrac{f_c}{f}} = j\frac{f}{f_c}$$

The 'j' signifies that the gain is imaginary, as shown in the phasor diagram of Figure 8.5(c). The magnitude of the gain is simply f/f_c and the phase shift is +90°, the '+' sign meaning that the output voltage *leads* the input voltage by 90°.

Since f_c is a constant for a given circuit, in this region the voltage gain is linearly related to frequency. If the frequency is halved the voltage gain will be halved. Therefore, the gain falls by a factor of 0.5 for every octave drop in frequency (an **octave** is a doubling or halving of frequency and is equivalent to an octave jump on a piano or other musical instrument). A fall in voltage gain by a factor of 0.5 is equivalent to a change in gain of −6 dB. Therefore, the rate of change of gain can be expressed as 6 dB per octave. An alternative way of expressing the rate of change of gain is to specify the change of gain for a decade change in frequency (a **decade**, as its name suggests, is a change in frequency of a factor of 10). If the frequency falls to 0.1 of its previous value, the voltage gain will also drop to 0.1 of its previous value. This represents a change in gain of −20 dB. Thus the rate of change of gain is 20 dB per decade.

Example 8.6	Determine the frequencies corresponding to:

(a) an octave above 1 kHz;
(b) three octaves above 10 Hz;
(c) an octave below 100 Hz;
(d) a decade above 20 Hz;
(e) three decades below 1 MHz;
(f) two decades above 50 Hz.

(a) an octave above 1 kHz = 1000 × 2 = 2 kHz
(b) three octaves above 10 Hz = 10 × 2 × 2 × 2 = 80 Hz
(c) an octave below 100 Hz = 100 ÷ 2 = 50 Hz
(d) a decade above 20 Hz = 20 × 10 = 200 Hz
(e) three decades below 1 MHz = 1,000,000 ÷ 10 ÷ 10 ÷ 10 = 1 kHz
(f) two decades above 50 Hz = 50 × 10 × 10 = 5 kHz

8.5.4   Frequency response of the high-pass *RC* network

Figure 8.6 shows the gain and phase response of the circuit of Figure 8.4 for frequencies above and below the cut-off frequency. It can be seen that, at frequencies much greater than the cut-off frequency, the magnitude of the gain tends to a straight line corresponding to a gain of 0 dB (that is, a gain of 1). Therefore, this line (shown dashed in Figure 8.6) forms an **asymptote** to the response. At frequencies much less than the cut-off frequency, the response tends to a straight line drawn at a slope of 6 dB per octave (20 dB per decade) change in frequency. This line forms a second asymptote to the response and is also shown dashed in Figure 8.6. The two asymptotes intersect at the cut-off frequency. At frequencies considerably above or below the cut-off

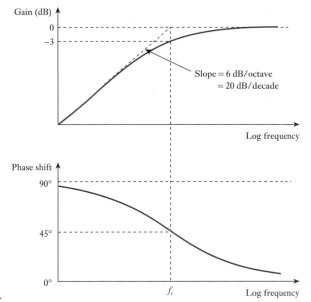

Figure 8.6
Gain and phase responses (or Bode diagram) for the high-pass *RC* network.

frequency, the gain response tends towards these two asymptotes. Near the cut-off frequency, the gain deviates from the two straight lines and is 3 dB below their inter-section at the cut-off frequency.

Figure 8.6 also shows the variation of phase with frequency of the *RC* network. At frequencies well above the cut-off frequency, the network produces very little phase shift and its effects may generally be ignored. However, as the frequency decreases the phase shift produced by the arrangement increases, reaching 45° at the cut-off frequency and increasing to 90° at very low frequencies.

Asymptotic diagrams of gain and phase of the form shown in Figure 8.6 are referred to as **Bode diagrams** (or sometimes **Bode plots**). These plot logarithmic gain (usually in dB) and phase against logarithmic frequency. Such diagrams are easy to plot and give a useful picture of the characteristic of the circuit. We will look at the Bode diagrams for a range of other circuits in this chapter and then consider how they may be easily drawn and used.

It can be seen that the *RC* network passes signals of some frequencies with little effect but that signals of other frequencies are attenuated and are subjected to a phase shift. The network therefore has the characteristics of a **high-pass filter**, since it allows high-frequency signals to pass but filters out low-frequency signals. We will look at filters in more detail later in this chapter.

File 8A

Computer simulation exercise 8.1

Calculate the cut-off frequency of the circuit of Figure 8.4 if $R = 1\ \text{k}\Omega$ and $C = 1\ \mu\text{F}$. Simulate the circuit using these component values and perform an AC sweep to measure the response over a range from 1 Hz to 1 MHz. Plot the gain (in dB) and the phase of the output over this frequency range, estimate the cut-off frequency from these plots and compare this with the predicted value. Measure the phase shift at the estimated cut-off frequency and compare this with the value predicted above. Repeat this exercise for different values of R and C.

8.6 A low-pass *RC* network

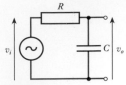

Figure 8.7
A low-pass *RC*
network.

The circuit of Figure 8.7 shows an *RC* arrangement similar to the earlier circuit but with the positions of the resistor and the capacitor reversed. Applying Equation 8.8 produces

$$\frac{v_o}{v_i} = \frac{\mathbf{Z}_C}{\mathbf{Z}_R + \mathbf{Z}_C} = \frac{-j\dfrac{1}{\omega C}}{R - j\dfrac{1}{\omega C}} = \frac{1}{1 + j\omega CR} \tag{8.13}$$

Comparing this expression with that of Equation 8.9 shows that it has a very different frequency characteristic. At low frequencies, ω is small and the value of $j\omega CR$ is small compared with 1. Therefore, the denominator of the expression is close to unity and the voltage gain is approximately 1. At high frequencies, the magnitude of ωCR becomes more significant and the gain of the network decreases. We therefore have a **low-pass filter** arrangement.

A similar analysis to that in the last section will show that the magnitude of the voltage gain is now given by

$$|\,\text{voltage gain}\,| = \frac{1}{\sqrt{1 + (\omega CR)^2}}$$

When the value of ωCR is equal to 1, this gives

$$|\,\text{voltage gain}\,| = \frac{1}{\sqrt{1+1}} = \frac{1}{\sqrt{2}} = 0.707$$

and again this corresponds to a cut-off frequency. The angular frequency of the cut-off ω_c corresponds to the condition that $\omega CR = 1$, therefore

$$\omega_c = \frac{1}{CR} = \frac{1}{\mathsf{T}} \ \text{rad/s} \tag{8.14}$$

as before. Therefore the expression for the cut-off frequency is identical to that in the previous circuit.

Example 8.7

Calculate the time constant T, the angular cut-off frequency ω_c and the cyclic cut-off frequency f_c of the following arrangement.

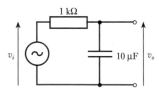

From above

$$\mathsf{T} = CR = 10 \times 10^{-6} \times 1 \times 10^{3} = 0.01\,\text{s}$$

$$\omega_c = \frac{1}{\mathsf{T}} = \frac{1}{0.01} = 100\,\text{rad/s}$$

$$f_c = \frac{\omega_c}{2\pi} = \frac{100}{2\pi} = 15.9\,\text{Hz}$$

While the cut-off frequency of this circuit is identical to that of the previous arrangement, you should note that in the circuit of Figure 8.4 the cut-off attenuates low-frequency signals and is therefore a **low-frequency cut-off**. However, in the circuit of Figure 8.7 high frequencies are attenuated, so this circuit has a **high-frequency cut-off**.

Substituting into Equation 8.13 gives

$$\frac{v_o}{v_i} = \frac{1}{1 + j\omega CR} = \frac{1}{1 + j\dfrac{\omega}{\omega_c}} = \frac{1}{1 + j\dfrac{f}{f_c}} \tag{8.15}$$

You might like to compare this with the expression for a high-pass network in Equation 8.12. As before, we can investigate the behaviour of this arrangement in different frequency ranges.

8.6.1 When $f \ll f_c$

When the signal frequency f is much lower than the cut-off frequency f_c, then in Equation 8.15 f/f_c is much less than unity, and the voltage gain is approximately equal to 1. The imaginary part of the gain is negligible and the gain of the circuit is effectively real. This situation is shown in the phasor diagram of Figure 8.8(a).

Figure 8.8
Phasor diagrams of the gain of the low-pass network at different frequencies.

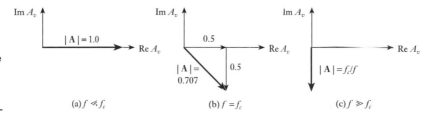

(a) $f \ll f_c$ (b) $f = f_c$ (c) $f \gg f_c$

8.6.2 When $f = f_c$

When the signal frequency f is equal to the cut-off frequency f_c, then Equation 8.15 becomes

$$\frac{v_o}{v_i} = \frac{1}{1 + j\dfrac{f}{f_c}} = \frac{1}{1 + j}$$

Multiplying the numerator and the denominator by $(1 - j)$ gives

$$\frac{v_o}{v_i} = \frac{(1 - j)}{(1 + j)(1 - j)} = \frac{(1 - j)}{2} = 0.5 - 0.5j$$

This is illustrated in the phasor diagram of Figure 8.8(b), which shows that the magnitude of the gain at the cut-off frequency is 0.707 and the phase angle of the gain is $-45°$. This shows that the output voltage *lags* the input voltage by 45°. The gain is therefore $0.707\angle-45°$.

8.6.3 When $f \gg f_c$

At high frequencies f/f_c is much greater than 1, and Equation 8.15 becomes

$$\frac{v_o}{v_i} = \frac{1}{1 + j\dfrac{f}{f_c}} \approx \frac{1}{j\dfrac{f}{f_c}} = -j\frac{f_c}{f}$$

The 'j' signifies that the gain is imaginary, and the minus sign indicates that the output lags the input. This is shown in the phasor diagram of Figure 8.8(c). The magnitude of the gain is simply f_c/f and, since f_c is a constant, the voltage gain is inversely proportional to frequency. If the frequency is halved, the voltage gain will be doubled. Therefore, the rate of change of gain can be expressed as −6 dB/octave or −20 dB/decade.

8.6.4 Frequency response of the low-pass *RC* network

Figure 8.9 shows the gain and phase response (or Bode diagram) of the low-pass network for frequencies above and below the cut-off frequency. The magnitude response is very similar in form to that of the high-pass network shown in Figure 8.6, with the frequency scale reversed. The phase response is a similar shape to that in Figure 8.6, but here the phase goes from 0° to −90° as the frequency is increased, rather than from +90° to 0° as in the previous arrangement. From the figure it is clear that this is a low-pass filter arrangement.

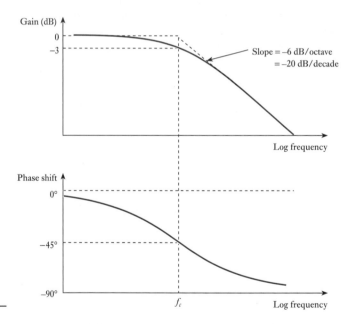

Figure 8.9
Gain and phase responses (or Bode diagram) for the low-pass *RC* network.

File 8B

Computer simulation exercise 8.2

Calculate the cut-off frequency of the circuit of Figure 8.7 if $R = 1\,k\Omega$ and $C = 1\,\mu F$. Simulate the circuit using these component values and perform an AC sweep to measure the response over a range from 1 Hz to 1 MHz. Plot the gain (in dB) and the phase of the output over this frequency range, estimate the cut-off frequency from these plots and compare this with the predicted value. Measure the phase shift at the estimated cut-off frequency and compare this with the value predicted above. Repeat this exercise for different values of R and C.

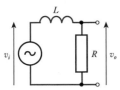

Figure 8.10
A low-pass *RL*
network.

8.7 A low-pass *RL* network

High-pass and low-pass arrangements may also be formed using combinations of resistors and inductors. Consider for example the circuit of Figure 8.10. This shows a circuit similar to that of Figure 8.4, but with the capacitor replaced by an inductor. If we apply a similar analysis to that used above, we obtain

$$\frac{v_o}{v_i} = \frac{\mathbf{Z}_R}{\mathbf{Z}_R + \mathbf{Z}_L} = \frac{R}{R + j\omega L} = \frac{1}{1 + j\omega\dfrac{L}{R}} \tag{8.16}$$

A similar analysis to that in the last section will show that the magnitude of the voltage gain is now given by

$$|\,\text{voltage gain}\,| = \frac{1}{\sqrt{1 + \left(\omega\dfrac{L}{R}\right)^2}}$$

When the value of $\omega L/R$ is equal to 1, this gives

$$|\,\text{voltage gain}\,| = \frac{1}{\sqrt{1+1}} = \frac{1}{\sqrt{2}} = 0.707$$

and this corresponds to a cut-off frequency. The angular frequency of the cut-off ω_c corresponds to the condition that $\omega L/R = 1$, therefore

$$\omega_c = \frac{R}{L} = \frac{1}{\mathsf{T}} \ \text{rad/s} \tag{8.17}$$

As before, T is the time constant of the circuit, and in this case T is equal to L/R.

Example 8.8 Calculate the time constant T, the angular cut-off frequency ω_c and the cyclic cut-off frequency f_c of the following arrangement.

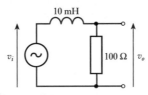

From above

$$\mathsf{T} = \frac{L}{R} = \frac{10 \times 10^{-3}}{100} = 10^{-4} \ \text{s}$$

$$\omega_c = \frac{1}{\mathsf{T}} = \frac{1}{1 \times 10^{-4}} = 10^4 \ \text{rad/s}$$

$$f_c = \frac{\omega_c}{2\pi} = \frac{1 \times 10^4}{2\pi} = 1.59 \ \text{kHz}$$

Substituting into Equation 8.16, we have

$$\frac{v_o}{v_i} = \frac{1}{1 + j\omega\dfrac{L}{R}} = \frac{1}{1 + j\dfrac{\omega}{\omega_c}} = \frac{1}{1 + j\dfrac{f}{f_c}} \tag{8.18}$$

This expression is identical to that of Equation 8.15, and thus the frequency behaviour of this circuit is identical to that of the circuit of Figure 8.7.

File 8C

Computer simulation exercise 8.3

Calculate the cut-off frequency of the circuit of Figure 8.10 if $R = 10\ \Omega$ and $L = 5$ mH. Simulate the circuit using these component values and perform an AC sweep to measure the response over a range from 1 Hz to 1 MHz. Plot the gain (in dB) and the phase of the output over this frequency range, estimate the cut-off frequency from these plots and compare this with the predicted value. Measure the phase shift at the estimated cut-off frequency and compare this with the value predicted above. Repeat this exercise for different values of R and L.

| 8.8 | **A high-pass *RL* network** |

Interchanging the components of Figure 8.10 gives the circuit of Figure 8.11. Analysing this as before, we obtain

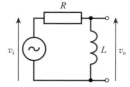

Figure 8.11
A high-pass *RL* network.

$$\frac{v_o}{v_i} = \frac{\mathbf{Z}_L}{\mathbf{Z}_R + \mathbf{Z}_L} = \frac{j\omega L}{R + j\omega L} = \frac{1}{1 + \dfrac{R}{j\omega L}} = \frac{1}{1 - j\dfrac{R}{\omega L}} \tag{8.19}$$

If we substitute $\omega_c = R/L$ as before, this gives

$$\frac{v_o}{v_i} = \frac{1}{1 - j\dfrac{R}{\omega L}} = \frac{1}{1 - j\dfrac{\omega_c}{\omega}} = \frac{1}{1 - j\dfrac{f_c}{f}} \tag{8.20}$$

This expression is identical to that of Equation 8.12, and thus the frequency behaviour of this circuit is identical to that of the circuit of Figure 8.4.

File 8D

Computer simulation exercise 8.4

Calculate the cut-off frequency of the circuit of Figure 8.11 if $R = 10\ \Omega$ and $L = 5$ mH. Simulate the circuit using these component values and perform an AC sweep to measure the response over a range from 1 Hz to 1 MHz. Plot the gain (in dB) and the phase of the output over this frequency range, estimate the cut-off frequency from these plots and compare this with the predicted value. Measure the phase shift at the estimated cut-off frequency and compare this with the value predicted above. Repeat this exercise for different values of R and L.

A comparison of *RC* and *RL* networks

From the above it is clear that *RC* and *RL* circuits have many similarities. The behaviour of the circuits we have considered is summarised in Figure 8.12. Each of the circuits has a cut-off frequency, and in each case this frequency is determined by the time constant T of the circuit. In the *RC* circuits $\mathsf{T} = CR$, and in the *RL* circuits $\mathsf{T} = L/R$. In each case the angular cut-off frequency is then given by $\omega_c = 1/\mathsf{T}$ and the cyclic cut-off frequency by $f_c = \omega_c/2\pi$.

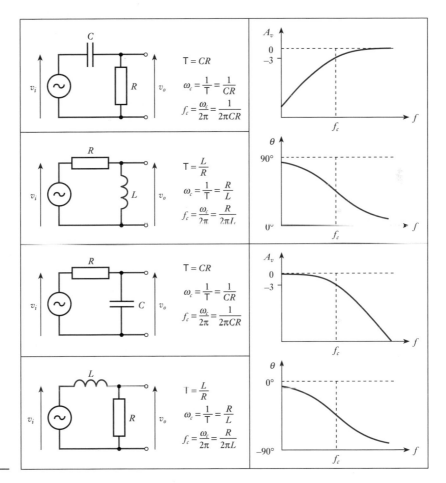

Figure 8.12
A comparison of
RC and *RL* networks.

Two of the circuits of Figure 8.12 have high-frequency cut-offs (low–pass circuits), and two have low-frequency cut-offs (high-pass circuits). Transposing the components in a particular circuit will change it from a high-pass to a low-pass circuit, and vice versa. Replacing a capacitor by an inductor, or replacing an inductor by a capacitor, will also change it from a high-pass to a low-pass circuit, and vice versa.

File 8E

Computer simulation exercise 8.5

Calculate the time constants of the circuits of Figure 8.12 if $R = 1\ \text{k}\Omega$, $C = 1\ \text{nF}$ and $L = 1\ \text{mH}$. Simulate the first of these circuits (using these component values) and use an AC sweep to plot the gain and phase responses of the circuit (as in the earlier simulation exercises in this chapter). Make a note of the cut-off frequency and confirm that this is a low-frequency cut-off. Now interchange the capacitor and resistor and again plot the circuit's characteristics. Note the effect on the cut-off frequency and the nature of the cut-off (that is, whether it is now a high- or a low-frequency cut-off).

Replace the capacitor by an inductor of 1 mH and again note the effect on the cut-off frequency and the nature of the cut-off. Finally, interchange the inductor and resistor and repeat the analysis. Hence confirm the form of the characteristics given in Figure 8.12.

Video 8A

8.10   Bode diagrams

Earlier we looked at Bode diagrams (also called Bode plots) as a means of describing the gain and phase response of a circuit (as in Figures 8.6 and 8.9). In the circuits we have considered, the gain at high and low frequencies has an asymptotic form, greatly simplifying the drawing of the diagram. The phase response is also straightforward, changing progressively between defined limits.

It is often sufficient to use a 'straight-line approximation' to the Bode diagram, simplifying its construction. For the circuits shown in Figure 8.12, we can construct the gain section of these diagrams simply by drawing the two asymptotes. One of these will be horizontal, representing the frequency range in which the gain is approximately constant. The other has a slope of +6 dB/octave (+20 dB/decade) or −6 dB/octave (−20 dB/decade), depending on whether this is a high-pass or low-pass circuit. These two asymptotes cross at the cut-off frequency of the circuit. The phase section of the response is often adequately represented by a straight-line transition between the two limiting values. The position of this line is defined by the phase shift at the cut-off frequency, which in these examples is 45°. A reasonable approximation to the response can be gained by drawing a straight line with a slope of −45°/decade through this point. Using this approach, the line starts one decade below the cut-off frequency and ends one decade above, making it very easy to construct. Straight-line approximations to the Bode diagrams for the circuits shown in Figure 8.12 are shown in Figure 8.13.

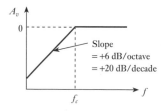

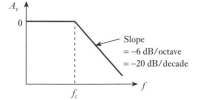

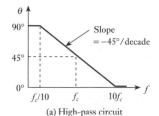

(a) High-pass circuit

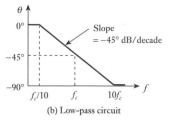

(b) Low-pass circuit

Figure 8.13
Simple straight-line
Bode diagrams.

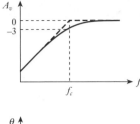

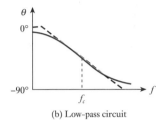

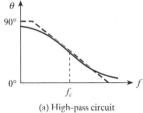

Figure 8.14
Drawing Bode
diagrams from their
straight-line
approximations.

(a) High-pass circuit   (b) Low-pass circuit

Once the straight-line Bode plots have been constructed, it is simple to convert these to a more accurate curved-line form if required. This can usually be done by eye by noting that the gain at the cut-off frequency is −3 dB, and that the phase response is slightly steeper at the cut-off frequency and slightly less steep near each end than the straight-line approximation. This is illustrated in Figure 8.14.

8.11 Combining the effects of several stages

While simple circuits may produce a single cut-off frequency, more complex circuits often possess a number of elements that each have some form of frequency dependence. Thus a circuit might have both high-pass and low-pass characteristics, or might have several high- or low-pass elements.

One of the advantages of the use of Bode diagrams is that they make it very easy to see the effects of combining several different elements. We noted in Section 8.3 that, when several stages of amplification are connected in series, the overall gain is equal to the product of their individual gains, or the *sum* of their gains when these are expressed in decibels. Similarly, the phase shift produced by several amplifiers in series is equal to the sum of the phase shifts produced by each amplifier separately. Therefore, the combined effects of a series of stages can be predicted by 'adding' the Bode diagrams of each stage. This is illustrated in Figure 8.15, which shows the effects of combining a high-pass and a low-pass element. In this case, the cut-off frequency of the high-pass element is lower than that of the low-pass element, resulting in a **band-pass filter** characteristic as shown in Figure 8.15(c). Such a circuit passes a given range of frequencies while rejecting low- and high-frequency components.

Bode diagrams can also be used to investigate the effects of combining more than one high-pass or low-pass element. This is illustrated in Figure 8.16, which shows the effects of combining two elements that each contain a single high- and a single low-pass element. In this case, the cut-off frequencies of each element are different, resulting in four transitions in the characteristic. For obvious reasons, these frequencies are known as **break** or **corner frequencies**.

In Figure 8.16, the first element is a band-pass amplifier that has a gain of A dB within its pass band, a low-frequency cut-off of f_1 and a high-frequency cut-off of f_3. The second element is also a band-pass amplifier, this time with a gain of B dB, a low-frequency cut-off of f_2 and a high-frequency cut-off of f_4. Within the frequency range from f_2 to f_3, the gains of both amplifiers are approximately constant, so the gain

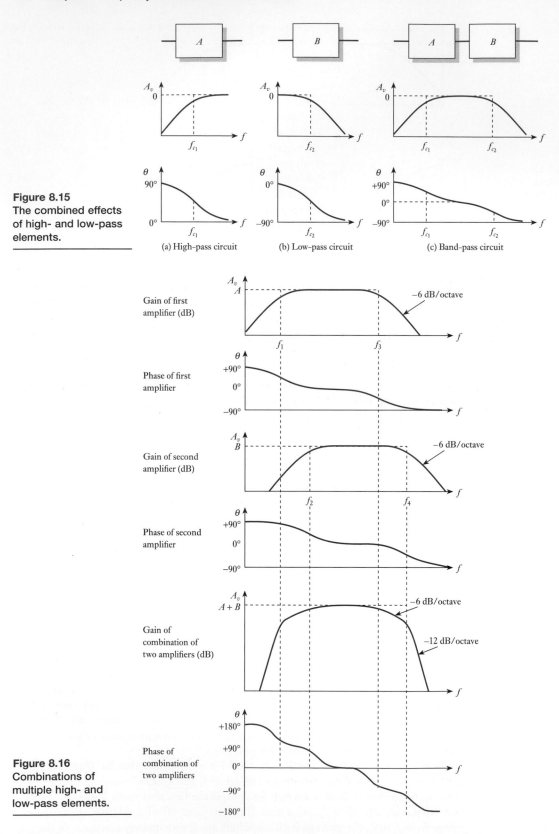

Figure 8.15
The combined effects
of high- and low-pass
elements.

(a) High-pass circuit (b) Low-pass circuit (c) Band-pass circuit

Figure 8.16
Combinations of
multiple high- and
low-pass elements.

of the combination is also approximately constant, with a value of $(A + B)$ dB. In the range f_3 to f_4, the gain of the second amplifier is approximately constant, but the gain of the first falls at a rate of 6 dB/octave. Therefore, in this range the gain of the combination also falls at 6 dB/octave. At frequencies above f_4, the gain of both amplifiers is falling at a rate of 6 dB/octave, so the gain of the combination falls at 12 dB/octave. A similar combination of effects causes the gain to fall at first by 6 dB/octave, and then by 12 dB/octave as the frequency decreases below f_2. The result is a band–pass filter with a gain of $(A + B)$.

Within the pass band both amplifiers produce relatively little phase shift. However, as we move to frequencies above f_3 the first amplifier produces a phase shift that increases to $-90°$, and as we move above f_4 the second amplifier produces an additional shift, taking the total phase shift to $-180°$. This effect is mirrored at low frequencies, with the two amplifiers producing a total phase shift of $+180°$ at very low frequencies.

While the arrangement represented in Figure 8.16 includes a total of two low-frequency and two high-frequency cut-offs, clearly more complex arrangements can include any number of cut-offs. As more are added, each introduces an additional 6 dB/octave to the maximum rate of increase or decrease of gain with frequency, and also increases the phase shift introduced at high or low frequencies by $90°$.

8.12 *RLC* circuits and resonance

8.12.1 Series *RLC* circuit

Having looked at *RC* and *RL* circuits, it is now time to look at circuits containing resistance, inductance and capacitance. Consider for example the series arrangement of Figure 8.17. This can be analysed in a similar manner to the circuits discussed above, by considering it as a potential divider. The voltages across each component can be found by dividing its complex impedance by the total impedance of the circuit and multiplying this by the applied voltage. For example, the voltage across the resistor is given by

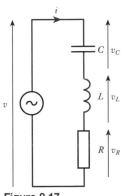

Figure 8.17
A series *RLC*
arrangement.

$$v_R = v \times \frac{\mathbf{Z}_R}{\mathbf{Z}_R + \mathbf{Z}_L + \mathbf{Z}_C} = v \times \frac{R}{R + j\omega L + \dfrac{1}{j\omega C}} \tag{8.21}$$

It is also interesting to consider the impedance of this arrangement, which is given by

$$\mathbf{Z} = R + j\omega L + \frac{1}{j\omega C} = R + j\left(\omega L - \frac{1}{\omega C}\right) \tag{8.22}$$

It can be seen that, if the magnitude of the reactance of the inductor and the capacitor are equal (that is, if $\omega L = 1/\omega C$), the imaginary part of the impedance is zero. Under these circumstances, the impedance of the arrangement is simply equal to R. This condition occurs when

$$\omega L = \frac{1}{\omega C} \qquad \omega^2 = \frac{1}{LC} \qquad \omega = \frac{1}{\sqrt{LC}}$$

This situation is referred to as **resonance**, and the frequency at which it occurs is called the **resonant frequency** of the circuit. An arrangement that exhibits such behaviour is known as a **resonant circuit**. The angular frequency at which resonance occurs is given the symbol ω_0, and the corresponding cyclic frequency is given the symbol f_0. Therefore

Figure 8.18
Variation of current with frequency for a series *RLC* arrangement.

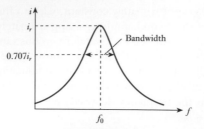

$$\omega_0 = \frac{1}{\sqrt{LC}} \tag{8.23}$$

$$f_0 = \frac{1}{2\pi\sqrt{LC}} \tag{8.24}$$

From Equation 8.22 it is clear that in the circuit of Figure 8.17 the impedance is at a *minimum* at resonance, and therefore the current will be at a *maximum* under these conditions. Figure 8.18 shows the current in the circuit as the frequency varies above and below resonance. Since the current is at a maximum at resonance, it follows that the voltages across the capacitor and the inductor are also large. Indeed, at resonance the voltages across these two components can be many times greater than the applied voltage. However, these two voltages are out of phase with each other and therefore cancel out, leaving only the voltage across the resistor.

Power is not dissipated in capacitors or inductors (as noted in Chapter 7) as these components simply store energy before returning it to the circuit. Therefore, the current flowing into and out of the inductor and capacitor at resonance results in energy being repeatedly stored and returned. This allows the resonant effect to be quantified by measuring the ratio of the energy stored to the energy dissipated during each cycle. This ratio is termed the **quality factor** or *Q* of the circuit. Since the energies stored in the inductor and the capacitor are equal, we can choose either of them to calculate *Q*. If we choose the inductor, we have

$$\text{quality factor } Q = \frac{I^2 X_L}{I^2 R} = \frac{X_L}{R} \tag{8.25}$$

and if we choose the capacitor, we have

$$\text{quality factor } Q = \frac{I^2 X_C}{I^2 R} = \frac{X_C}{R} \tag{8.26}$$

If we take either of these expressions and multiply top and bottom by *I*, we get the corresponding voltages across the associated component. Therefore, *Q* may also be defined as

$$\text{quality factor } Q = \frac{V_L}{V_R} = \frac{V_C}{V_R} \tag{8.27}$$

Since at resonance V_R is equal to the supply voltage, it follows that

$$\text{quality factor } Q = \frac{\text{voltage across } L \text{ or } C \text{ at resonance}}{\text{supply voltage}} \tag{8.28}$$

and thus *Q* represents the voltage magnification at resonance.

Combining Equations 8.23 and 8.28 gives us an expression for the Q of a series *RLC* circuit, which is

$$Q = \frac{1}{R}\sqrt{\left(\frac{L}{C}\right)} \qquad (8.29)$$

The series *RLC* circuit is often referred to as an **acceptor circuit**, since it passes signals at frequencies close to its resonant frequency but rejects signals at other frequencies. We can define the bandwidth B of a resonant circuit as the frequency range between the points where the gain (or in this case the current) falls to $1/\sqrt{2}$ (or 0.707) times its mid-band value. This is illustrated in Figure 8.18. An example of an application of an acceptor circuit is in a radio, where we wish to accept the frequencies associated with a particular station while rejecting others. In such situations, we need a resonant circuit with an appropriate bandwidth to accept the wanted signal while rejecting unwanted signals and interference. The 'narrowness' of the bandwidth is determined by the Q of the circuit, and it can be shown that the resonant frequency and the bandwidth are related by the expression

$$\text{quality factor } Q = \frac{\text{resonant frequency}}{\text{bandwidth}} = \frac{f_0}{B} \qquad (8.30)$$

Combining Equations 8.24, 8.29 and 8.30, we can obtain an expression for the bandwidth of the circuit in terms of its component values. This is

$$B = \frac{R}{2\pi L} \text{ Hz} \qquad (8.31)$$

It can be seen that reducing the value of R increases the Q of the circuit and reduces its bandwidth. In some situations it is desirable to have very high values of Q, and Equation 8.29 would suggest that if the resistor were omitted (effectively making $R = 0$) this would produce a resonant circuit with infinite Q. However, in practice all real components exhibit resistance (and inductors are particularly 'non-ideal' in this context), so the Q of such circuits is limited to a few hundred.

Example 8.9 | For the following arrangement, calculate the resonant frequency f_0, the impedance of the circuit at this frequency, the quality factor Q of the circuit and its bandwidth B.

5 Ω   30 μF   15 mH

From Equation 8.24

$$f_0 = \frac{1}{2\pi\sqrt{LC}}$$

$$= \frac{1}{2\pi\sqrt{15\times10^{-3}\times30\times10^{-6}}}$$

$$= 237 \text{ Hz}$$

At the resonant frequency the impedance is equal to R, so $\mathbf{Z} = 5$ Ω.

From Equation 8.29

$$Q = \frac{1}{R}\sqrt{\left(\frac{L}{C}\right)} = \frac{1}{5}\sqrt{\left(\frac{15 \times 10^{-3}}{30 \times 10^{-6}}\right)} = 4.47$$

and from Equation 8.31

$$B = \frac{R}{2\pi L} = \frac{5}{2\pi \times 15 \times 10^{-3}} = 53 \text{ Hz}$$

File 8F

Computer simulation exercise 8.6

Simulate a circuit that applies a sinusoidal voltage to the arrangement of Example 8.9 and use an AC sweep to plot the variation of current with frequency. Measure the resonant frequency of the arrangement and its bandwidth, and hence calculate its Q. Measure the peak current in the circuit and, from a knowledge of the excitation voltage used, estimate the impedance of the circuit at resonance. Hence confirm the findings of Example 8.9 above.

8.12.2 Parallel *RLC* circuit

Consider now the parallel circuit of Figure 8.19. The impedance of this circuit is given by

$$Z = \frac{1}{\dfrac{1}{R} + j\omega C + \dfrac{1}{j\omega L}} = \frac{1}{\dfrac{1}{R} + j\left(\omega C - \dfrac{1}{\omega L}\right)} \tag{8.32}$$

and it is clear that this circuit also has a resonant characteristic. When $\omega C = 1/\omega L$, the term within the brackets is equal to zero and the imaginary part of the impedance disappears. Under these circumstances, the impedance is purely resistive, and $Z = R$. The frequency at which this occurs is the resonant frequency, which is given by

$$\omega C = \frac{1}{\omega L}$$

$$\omega^2 = \frac{1}{LC}$$

$$\omega = \frac{1}{\sqrt{LC}}$$

which is the same as for the series circuit. Therefore, as before, the resonant angular and cyclic frequencies are given by

$$\omega_0 = \frac{1}{\sqrt{LC}} \tag{8.33}$$

Figure 8.19
A parallel *RLC*
arrangement.

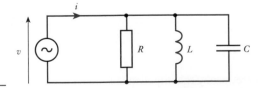

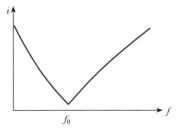

Figure 8.20
Variation of current
with frequency for a
parallel *RLC*
arrangement.

$$f_0 = \frac{1}{2\pi\sqrt{LC}} \tag{8.34}$$

From Equation 8.32, it is clear that the impedance of the parallel resonant circuit is a *maximum* at resonance and that it decreases at higher and lower frequencies. This arrangement is therefore a **rejector** circuit, and Figure 8.20 shows how the current varies with frequency.

As for the series resonant circuit, we can define both the bandwidth B and the quality factor Q for the parallel arrangement (although the definitions of these terms are a little different). The corresponding expressions for these quantities are

$$Q = R\sqrt{\left(\frac{C}{L}\right)} \tag{8.35}$$

and

$$B = \frac{1}{2\pi RC}\,\text{Hz} \tag{8.36}$$

A comparison between series and parallel resonant circuits is shown in Table 8.2. It should be noted that in a series resonant circuit Q is increased by *reducing* the value

Table 8.2 Series and parallel resonant circuits.

	Series resonant circuit	Parallel resonant circuit
Circuit		
Impedance, **Z**	$Z = R + j\left(\omega L - \dfrac{1}{\omega C}\right)$	$Z = \dfrac{1}{\dfrac{1}{R} + j\left(\omega C - \dfrac{1}{\omega L}\right)}$
Resonant frequency, f_0	$f_0 = \dfrac{1}{2\pi\sqrt{LC}}$	$f_0 = \dfrac{1}{2\pi\sqrt{LC}}$
Quality factor, Q	$Q = \dfrac{1}{R}\sqrt{\left(\dfrac{L}{C}\right)}$	$Q = R\sqrt{\left(\dfrac{C}{L}\right)}$
Bandwidth, B	$B = \dfrac{R}{2\pi L}\,\text{Hz}$	$B = \dfrac{1}{2\pi RC}\,\text{Hz}$

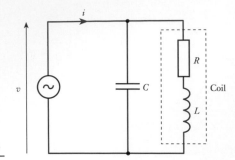

Figure 8.21
An *LC* resonant circuit.

of R, while in a parallel resonant circuit Q is increased by *increasing* the value of R. In each case, Q is increased when the losses are reduced.

While the circuit of Figure 8.19 represents a generalised parallel RLC circuit, it is not the most common form. In practice, the objective is normally to maximise the Q of the arrangement, and this is achieved by removing the resistive element. However, in practice all inductors have appreciable resistance, so it is common to model this in the circuit as shown in Figure 8.21. Capacitors also exhibit resistance, but this is generally quite small and can often be ignored.

The resonant frequency of the circuit of Figure 8.21 is given by

$$f_0 = \frac{1}{2\pi} \sqrt{\frac{1}{LC} - \frac{R^2}{L^2}} \qquad (8.37)$$

As the resistance of the coil tends to zero, this expression becomes equal to that of Equation 8.34. This circuit has similar characteristics to the earlier parallel arrangement and has a Q given by

$$Q = \sqrt{\frac{L}{R^2 C} - 1} \qquad (8.38)$$

8.13   Filters

8.13.1   *RC* filters

Earlier in this chapter we looked at *RC* high-pass and low-pass networks and noted that these have the characteristics of filters since they pass signals of certain frequencies while attenuating others. These simple circuits, which contain only a single time constant, are called **first-order** or **single-pole filters**. Circuits of this type are often used in systems to select or remove components of a signal. However, for many applications the relatively slow roll-off of the gain (6 dB/octave) is inadequate to remove unwanted signals effectively. In such cases, filters with more than one time constant are used to provide a more rapid roll-off of gain. Combining two high-pass time constants produces a second-order (two-pole) high-pass filter in which the gain will roll off at 12 dB/octave (as seen in Section 8.11). Similarly, the addition of three or four stages can produce a roll-off rate of 18 or 24 dB/octave.

In principle, any number of stages can be combined in this way to produce an *n*th-order (*n*-pole) filter. This will have a cut-off slope of 6*n* dB/octave and produce

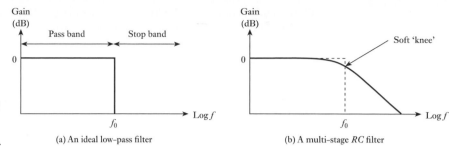

Figure 8.22
Gain responses of ideal and real low-pass filters.

(a) An ideal low-pass filter

(b) A multi-stage *RC* filter

up to $n \times 90°$ of phase shift. It is also possible to combine high-pass and low-pass characteristics into a single band-pass filter if required.

For many applications, an **ideal filter** would have a constant gain and zero phase shift within one range of frequencies (its **pass band**) and zero gain outside this range (its **stop band**). The transition from the pass band to the stop band occurs at the **corner frequency** f_0. This is illustrated for a low-pass filter in Figure 8.22(a).

Unfortunately, although adding more stages to the *RC* filter increases the *ultimate* rate of fall of gain within the stop band, the sharpness of the 'knee' of the response is not improved (see Figure 8.22(b)). To produce a circuit that more closely approximates an ideal filter, different techniques are required.

8.13.2 *LC* filters

The combination of inductors and capacitors allows the production of filters with a very sharp cut-off. Simple *LC* filters can be produced using the series and parallel resonant circuits discussed in the last section. These are also known as **tuned circuits** and are illustrated in Figure 8.23.

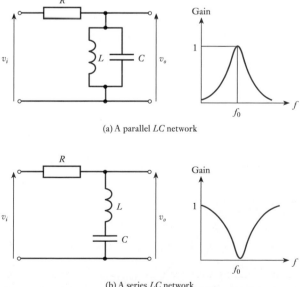

(a) A parallel *LC* network

Figure 8.23
LC filters.

(b) A series *LC* network

These combinations of inductors and capacitors produce narrow-band filters with centre frequencies corresponding to the resonant frequency of the tuned circuit, so

$$f_0 = \frac{1}{2\pi\sqrt{LC}}$$

(8.39)

The bandwidth of the filters is determined by the **quality factor** Q as discussed in the last section.

Other configurations of inductors, capacitors and resistors can be used to form high-pass, low-pass, band-pass and band-stop filters and can achieve very high cut-off rates.

8.13.3 Active filters

Although combinations of inductors and capacitors can produce very high-performance filters, the use of inductors is inconvenient since they are expensive, bulky and suffer from greater losses than other passive components. Fortunately, a range of very effective filters can be constructed using an operational amplifier and suitable arrangements of resistors and capacitors. Such filters are called **active filters**, since they include an active component (the operational amplifier) in contrast to the other filters we have discussed, which are purely passive (ignoring any buffering). A detailed study of the operation and analysis of active filters is beyond the scope of this text, but it is worth looking at the characteristics of these circuits and comparing them with those of the RC filters discussed earlier.

To construct multiple-pole filters, it is often necessary to cascade many stages. If the time constants and the gains of each stage are varied in a defined manner, it is possible to create filters with a wide range of characteristics. Using these techniques, it is possible to construct filters of a number of different types to suit particular applications.

In simple RC filters, the gain starts to fall towards the edge of the pass band and so is not constant throughout the band. This is also true of active filters, but here the gain may actually rise towards the edge of the pass band before it begins to fall. In some circuits the gain fluctuates by small amounts right across the band. These characteristics are illustrated in Figure 8.24.

The ultimate rate of fall of gain with frequency for any form of active filter is $6n$ dB/octave, where n is the number of poles in the filter, which is often equal to the number of capacitors in the circuit. Thus the performance of the filter in this respect is related directly to circuit complexity.

Although the ultimate rate of fall of gain of a filter is defined by the number of poles, the sharpness of the 'knee' of the filter varies from one design to another.

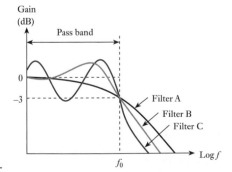

Figure 8.24
Variations of gain with frequency for various filters.

Filters with a very sharp knee tend to produce more variation in the gain of the filter within the pass band. This is illustrated in Figure 8.24, where it is apparent that filters B and C have a more rapid roll-off of gain than filter A but also have greater variation in their gain within the pass band.

Of great importance in some applications is the **phase response** of the filter: that is, the variation of phase lag or lead with frequency as a signal passes through the filter. We have seen that *RC* filters produce considerable amounts of phase shift within the pass band. All filters produce a phase shift that varies with frequency, but the way in which it is related to frequency varies from one type of filter to another. The phase response of a filter is of particular importance where pulses are to be used.

A wide range of filter designs are available, enabling one to be selected to favour any of the above characteristics. Unfortunately, the requirements of each are often mutually exclusive, so there is no universal optimum design, and an appropriate circuit must be chosen for a given application. From the myriad of filter designs, three basic types are discussed here, first because they are widely used and second because they are each optimised for a particular characteristic.

The **Butterworth filter** is optimised to produce a flat response within its pass band, which it does at the expense of a less sharp 'knee' and a less than ideal phase performance. This filter is sometimes called a **maximally flat filter** as it produces the flattest response of any filter type.

The **Chebyshev filter** produces a sharp transition from the pass band to the stop band but does this by allowing variations in gain throughout the pass band. The gain ripples within specified limits, which can be selected according to the application. The phase response of the Chebyshev filter is poor, and it creates serious distortion of pulse waveforms.

The **Bessel filter** is optimised for a linear phase response and is sometimes called a **linear phase filter**. The 'knee' is much less sharp than for the Chebyshev or the Butterworth types (though slightly better than a simple *RC* filter), but its superior phase characteristics make it preferable in many applications, particularly where pulse waveforms are being used. The phase shift produced by the filter is approximately linearly related to the input frequency. The resultant phase shift therefore has the appearance of a fixed time delay, with all frequencies being delayed by the same time interval. The result is that complex waveforms that consist of many frequency components (such as pulse waveforms) are filtered without distorting the phase relationships between the various components of the signal. Each component is simply delayed by an equal time interval.

Figure 8.25 compares the characteristics of these three types of filter. Parts (a), (b) and (c) show the frequency responses for Butterworth, Chebyshev and Bessel filters, each with six poles (the Chebyshev is designed for 0.5 dB ripple), while (d), (e) and (f) show the responses of the same filters to a step input.

Over the years a number of designs have emerged to implement various forms of filter. The designs have different characteristics, and each has advantages and disadvantages. We will look at examples of active filter circuits in Chapter 16, when we look at operational amplifiers.

While active filters have several advantages over other forms of filter, it should be noted that they rely on the operational amplifier having sufficient gain at the frequencies being used. Active filters are widely used with audio signals (which are limited to a few tens of kilohertz) but are seldom used at very high frequencies. In contrast, *LC* filters can be used very successfully at frequencies up to several hundred megahertz. At very high frequencies, a range of other filter elements are available including SAW, ceramic and transmission line filters.

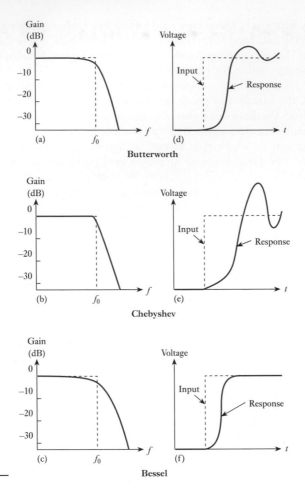

Figure 8.25
A comparison
of Butterworth,
Chebyshev and
Bessel filters.

8.14 Stray capacitance and inductance

While many circuits will include a number of capacitors and inductors that have been intentionally introduced by the circuit designer, *all* circuits also include additional 'unintended' stray capacitances and stray inductances (as discussed in Chapters 4 and 5). Stray capacitance tends to introduce unintended low-pass filters in circuits, as illustrated in Figure 8.26(a). It also produces unwanted coupling of signals between circuits, resulting in a number of undesirable effects such as cross-talk. Stray inductance can also produce undesirable effects. For example, in Figure 8.26(b) a stray inductance L_s appears in series with a load resistor, producing an unintended low-pass effect. Stray effects also have a dramatic effect on the stability of circuits. This is illustrated in Figure 8.26(c), where stray capacitance C_s across an inductor L results in an unintended resonant circuit. We will return to look at stability in more detail in Chapter 23.

Stray capacitances and inductances are generally relatively small and therefore tend to be insignificant at low frequencies. However, at high frequencies they can have dramatic effects on the operation of circuits. In general, it is the presence of these unwanted circuit elements that limits the high-frequency performance of circuits.

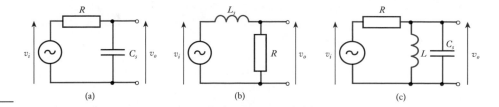

Figure 8.26
The effects of stray capacitance and inductance.

Further study

In most parts of the world the band from 520 kHz to 1,610 kHz of the radio frequency spectrum is used primarily for AM broadcasting (AM radio). This band is divided up, giving individual stations a particular range of frequencies. Within most of the world the spacing between stations is set at 9 kHz, while in the Americas it is 10 kHz.

In later chapters we will look at the techniques used to transmit and receive amplitude modulated (AM) signals, but here we will concern ourselves simply with the task of selecting a single station from the multitude of signals being transmitted.

Without considering in detail the circuit to be used, but assuming that the circuit has the broad characteristics of the series RLC filter described in Section 8.12.1, design a filter to select a single station that is being transmitted with a centre frequency of 909 kHz and a bandwidth of 9 kHz. You may assume that the design is to make use of an inductor of 250 µH.

In practice, radios are not designed to receive a single, predetermined station, but can be adjusted to select any station within a particular frequency range. How might your filter be modified to allow it to select any station in the frequency band given above?

Key points

- The reactance of capacitors and inductors is dependent on frequency. Therefore, the behaviour of any circuit that contains these components will change with frequency.
- Since all real circuits include stray capacitance and stray inductance, all real circuits have characteristics that change with frequency.
- Combinations of a single resistor and a single capacitor, or a single resistor and a single inductor, can produce circuits with a single high- or low-frequency cut-off. In each case, the angular cut-off frequency ω_c is given by the reciprocal of the time constant T of the circuit.
- For an RC circuit $T = CR$, while in an RL circuit $T = L/R$.
- These single time constant circuits have certain similar characteristics:
  - their cut-off frequency $f_c = \omega_c/2\pi = 1/2\pi T$;
  - at frequencies well away from their cut-off frequency within their pass band, they have a gain of 0 dB and zero phase shift;

- at their cut-off frequency, they have a gain of −3 dB and ±45° phase shift;
  - at frequencies well away from their cut-off frequency within their stop band, their gain changes by ±6 dB/octave (±20 dB/decade) and they have a phase shift of ±90°.
- Gain and phase responses are often given in the form of a Bode diagram, which plots gain (in dB) and phase against log frequency.
- When several stages are used in series, the gain of the combination at a given frequency is found by multiplying their individual gains, while the phase shift is found by adding their individual phase shifts.
- Combinations of resistors, inductors and capacitors can be analysed using the tools covered in earlier chapters. Of particular interest is the condition of resonance, when the reactance of the capacitive and inductive elements cancels. Under these conditions, the impedance of the circuit is simply resistive.
- The 'sharpness' of the resonance is measured by the quality factor Q.
- Simple RC and RL circuits represent first-order, or single-pole, filters. Although these are useful in certain applications, they have a limited 'roll-off' rate and a soft 'knee'.
- Combining several stages of RC filters increases the roll-off rate but does not improve the sharpness of the knee. Higher performance can be achieved using LC filters, but inductors are large, heavy and have high losses.
- Active filters produce high performance without using inductors. Several forms are available to suit a range of applications.
- Stray capacitance and stray inductance limit the performance of all high-frequency circuits.

Exercises

8.1 Calculate the reactance of a 1 µF capacitor at a frequency of 10 kHz, and the reactance of a 20 mH inductor at a frequency of 100 rad/s. In each case include the units in your answer.

8.2 Express an angular frequency of 250 rad/s as a cyclic frequency (in Hz).

8.3 Express a cyclic frequency of 250 Hz as an angular frequency (in rad/s).

8.4 Determine the transfer function of the following circuit.

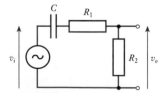

8.5 A series RC circuit is formed from a resistor of 33 kΩ and a capacitor of 15 nF. What is the time constant of this circuit?

8.6 Calculate the time constant T, the angular cut-off frequency ω_c and the cyclic cut-off frequency f_c of the following arrangement. Is this a high- or a low-frequency cut-off?

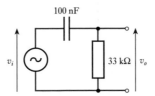

8.7 Simulate the arrangement of Exercise 8.6 and use an AC sweep to display the gain response. Measure the cut-off frequency of the circuit and hence confirm your results for the previous exercise.

8.8 Determine the frequencies that correspond to:

(a) an octave below 30 Hz;

(b) two octaves above 25 kHz;

(c) three octaves above 1 kHz;

(d) a decade above 1 MHz;

(e) two decades below 300 Hz;

(f) three decades above 50 Hz.

8.9 Calculate the time constant T, the angular cut-off frequency ω_c and the cyclic cut off frequency f_c of the following arrangement. Is this a high- or a low-frequency cut-off?

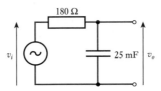

8.10 Simulate the arrangement of Exercise 8.9 and use an AC sweep to display the gain response. Measure the cut-off frequency of the circuit and hence confirm your results for the previous exercise.

8.11 A parallel RL circuit is formed from a resistor of 150 Ω and an inductor of 30 mH. What is the time constant of this circuit?

8.12 Calculate the time constant T, the angular cut-off frequency ω_c and the cyclic cut-off frequency f_c of the following arrangement. Is this a high- or a low-frequency cut-off?

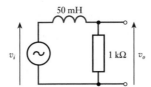

8.13 Simulate the arrangement of Exercise 8.12 and use an AC sweep to display the gain response. Measure the cut-off frequency of the circuit and hence confirm your results for the previous exercise.

8.14 Calculate the time constant T, the angular cut-off frequency ω_c and the cyclic cut-off frequency f_c of the following arrangement. Is this a high- or a low-frequency cut-off?

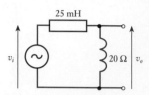

8.15 Simulate the arrangement of Exercise 8.14 and use an AC sweep to display the gain response. Measure the cut-off frequency of the circuit and hence confirm your results for the previous exercise.

8.16 Sketch a straight-line approximation to the Bode diagram of the circuit of Exercise 8.14. Use this approximation to produce a more realistic plot of the gain and phase responses of the circuit.

8.17 A circuit contains three high-frequency cut-offs and two low-frequency cut-offs. What are the rates of change of gain of this circuit at very high and very low frequencies?

8.18 Explain what is meant by the term 'resonance'.

8.19 Calculate the resonant frequency f_0, the quality factor Q and the bandwidth B of the following circuit.

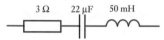

8.20 Simulate a circuit that applies a sinusoidal voltage to the arrangement of Exercise 8.19 and use an AC sweep to plot the variation of current with frequency. Measure the resonant frequency of the arrangement and its bandwidth, and hence calculate its Q. Hence confirm your results for the previous exercise.

8.21 Explain the difference between a passive and an active filter.

8.22 Why are inductors often avoided in the construction of filters?

8.23 What form of active filter is optimised to produce a flat response within its pass band?

8.24 What form of active filter is optimised to produce a sharp transition from the pass band to the stop band?

8.25 What form of filter is optimised for a linear phase response?

8.26 Explain why stray capacitance and stray inductance affect the frequency response of electronic circuits.

Chapter 9   Transient Behaviour

Objectives

When you have studied the material in this chapter, you should be able to:

- explain concepts such as steady-state response, transient response and total response as they apply to electronic circuits
- describe the transient behaviour of simple RC and RL circuits
- predict the transient response of a generalised first-order system from a knowledge of its initial and final values
- sketch increasing or decreasing waveforms and identify their key characteristics
- describe the output of simple RC and RL circuits in response to a square-wave input
- outline the transient behaviour of various forms of second-order systems.

9.1   Introduction

In earlier chapters, we looked at the behaviour of circuits in response to either fixed DC signals or constant AC signals. Such behaviour is often referred to as the **steady-state response** of the system. Now it is time to turn our attention to the performance of circuits before they reach this steady-state condition: for example, how the circuits react when a voltage or current source is initially turned on or off. This is referred to as the **transient response** of the circuit.

We will begin by looking at simple RC and RL circuits and then progress to more complex arrangements.

9.2   Charging of capacitors and energising of inductors

9.2.1   Capacitor charging

Figure 9.1(a) shows a circuit that charges a capacitor C from a voltage source V through a resistor R. The capacitor is assumed to be initially uncharged, and the switch in the circuit is closed at time $t = 0$.

When the switch is first closed the charge on the capacitor is zero, and therefore the voltage across it is also zero. Thus all the applied voltage is across the resistor, and the initial current is given by V/R. As this current flows into the capacitor the charge on it builds and the voltage across it increases. As the voltage across the capacitor increases, the voltage across the resistor decreases, causing the current in the circuit to fall. Gradually, the voltage across the capacitor increases until it is equal to the

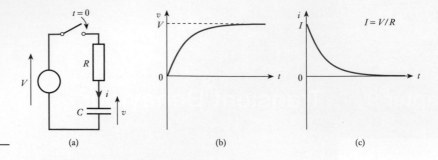

Figure 9.1
Capacitor charging.

applied voltage, and the current goes to zero. We can understand this process more fully by deriving expressions for the voltage across the capacitor v and the current flowing into the capacitor i.

Applying Kirchhoff's voltage law to the circuit of Figure 9.1(a), we see that

$$iR + v = V$$

From Chapter 4, we know that the current in a capacitor is related to the voltage across it by the expression

$$i = C\frac{dv}{dt}$$

therefore, substituting,

$$CR\frac{dv}{dt} + v = V$$

This is a first-order differential equation with constant coefficients and is relatively easy to solve. First we rearrange the expression to give

$$\frac{dv}{dt} = \frac{V - v}{CR}$$

and then again to give

$$\frac{dt}{CR} = \frac{dv}{V - v}$$

Integrating both sides then gives

$$\frac{t}{CR} = -\ln(V - v) + A$$

where A is the constant of integration.

In this case we know (from our assumption that the capacitor is initially uncharged) that when $t = 0$, $v = 0$. Substituting this into the previous equation gives

$$\frac{0}{CR} = -\ln(V - 0) + A$$
$$A = \ln V$$

Therefore

$$\frac{t}{CR} = -\ln(V - v) + \ln V = \ln\frac{V}{V - v}$$

and

$$e^{t/CR} = \frac{V}{V - v}$$

Finally, rearranging we have

$$v = V(1 - e^{-t/CR}) \tag{9.1}$$

From this expression, we can also derive an expression for the current I, since

$$i = C\frac{dv}{dt} = CV\frac{d}{dt}(1 - e^{-t/CR}) = \frac{V}{R}e^{-t/CR}$$

We noted earlier that at $t = 0$ the voltage across the capacitor is zero and the current is given by V/R. If we call this initial current I, then our expression for the current becomes

$$i = Ie^{-t/CR} \tag{9.2}$$

In Equations 9.1 and 9.2, you will note that the exponential component contains the term t/CR. You will recognise CR as the time constant T of the circuit, and thus t/CR is equal to t/T and represents time as a fraction of the time constant. For this reason, it is common to give these two equations in a more general form, replacing CR by T:

$$v = V(1 - e^{-t/\mathsf{T}}) \tag{9.3}$$

$$i = Ie^{-t/\mathsf{T}} \tag{9.4}$$

From Equations 9.3 and 9.4, it is clear that in the circuit of Figure 9.1(a) the voltage rises with time, while the current falls exponentially. These two waveforms are shown in Figures 9.1(b) and 9.1(c).

Example 9.1 The switch in the following circuit is closed at $t = 0$. **Derive an expression for the output voltage v after this time and hence calculate the voltage on the capacitor at $t = 25$ s.**

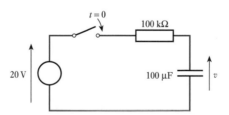

The time constant of the circuit $\mathsf{T} = CR = 100 \times 10^3 \times 100 \times 10^{-6} = 10$ s. From Equations 9.3

$$v = V(1 - e^{-t/\mathsf{T}})$$

$$= 20(1 - e^{-t/10})$$

At $t = 25$ s

$$v = 20(1 - e^{-25/10})$$

$$= 18.36\,\mathrm{V}$$

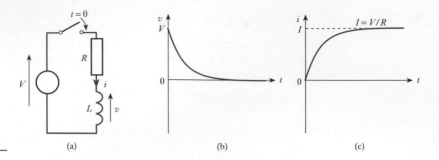

Figure 9.2
Inductor energising.

9.2.2 Inductor energising

Figure 9.2(a) shows a circuit that energises an inductor L using a voltage source V and a resistor R. The circuit is closed at time $t = 0$, and before that time no current flows in the inductor.

When the switch is first closed the current in the circuit is zero, since the nature of the inductor prevents the current from changing instantly. If the current is zero there is no voltage across the resistor, so all the applied voltage appears across the inductor. The applied voltage causes the current to increase, producing a voltage drop across the resistor and reducing the voltage across the inductor. Eventually, the voltage across the inductor falls to zero and all the applied voltage appears across the resistor, producing a steady current of V/R. As before, it is interesting to look at expressions for v and i.

Applying Kirchhoff's voltage law to the circuit of Figure 9.2(a), we see that

$$iR + v = V$$

From Chapter 5, we know that the voltage across an inductor is related to the current through it by the expression

$$v = L\frac{\mathrm{d}i}{\mathrm{d}t}$$

therefore, substituting,

$$iR + L\frac{\mathrm{d}i}{\mathrm{d}t} = V$$

This first-order differential equation can be solved in a similar manner to that derived for capacitors above. This produces the equations

$$v = V\mathrm{e}^{-Rt/L} \tag{9.5}$$

$$i = I(1 - \mathrm{e}^{-Rt/L}) \tag{9.6}$$

where I represents the final (maximum) current in the circuit and is equal to V/R. In Equations 9.5 and 9.6, you will note that the exponential component contains the term Rt/L. Now L/R is the time constant T of the circuit, thus Rt/L is equal to t/T. We can therefore rewrite these two equations as

$$v = V\mathrm{e}^{-t/\mathsf{T}} \tag{9.7}$$

$$i = I(1 - \mathrm{e}^{-t/\mathsf{T}}) \tag{9.8}$$

The forms of v and i are shown in Figures 9.2(b) and 9.2(c). You might like to compare these figures with Figures 9.1(b) and 9.1(c) for a charging capacitor. You might also like to compare Equations 9.7 and 9.8, which describe the energising of an inductor, with Equations 9.3 and 9.4, which we derived earlier to describe the charging of a capacitor.

Example 9.2 An inductor is connected to a 15 V supply as shown below. How long after the switch is closed will the current in the coil reach 300 mA?

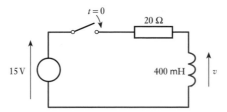

The time constant of the circuit $T = L/R = 0.4 \div 20 = 0.02$ s. The final current I is given by $V/R = 15/20 = 750$ mA.

From Equation 9.8

$$i = I(1 - e^{-t/T})$$

$$300 - 750(1 - e^{-t/0.02})$$

which can be evaluated to give

$$t = 10.2 \text{ ms}$$

9.3 **Discharging of capacitors and de-energising of inductors**

The charging of a capacitor or the energising of an inductor stores energy in that component that can be used at a later time to produce a current in a circuit. In this section, we look at the voltages and currents associated with this process.

9.3.1 Capacitor discharging

In order to look at the discharging of a capacitor, first we need to charge it up. Figure 9.3(a) shows a circuit in which a capacitor C is initially connected to a voltage source V and

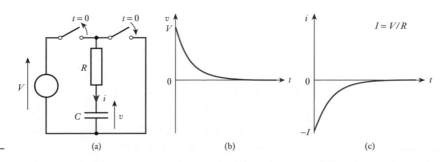

Figure 9.3
Capacitor discharging.

is then discharged through a resistor R. The discharge is initiated at $t = 0$ by opening one switch and closing another. In this diagram, the defining direction of the current i is *into* the capacitor, as in Figure 9.1(a), but clearly during the discharge process charge flows *out* of the capacitor, so i is negative.

The charged capacitor produces an electromotive force that drives a current around the circuit. Initially, the voltage across the capacitor is equal to the voltage of the source used to charge it (V), so the initial current is equal to V/R. However, as charge flows out of the capacitor its voltage decreases and the current falls. v and i can be determined in a similar manner to that used above for the charging arrangement. Applying Kirchhoff's voltage law to the circuit gives

$$iR + v = 0$$

giving

$$CR\frac{dv}{dt} + v = 0$$

Solving this as before leads to the expressions

$$v = Ve^{-t/CR} = Ve^{-t/\mathsf{T}} \tag{9.9}$$

$$i = -Ie^{-t/CR} = -Ie^{-t/\mathsf{T}} \tag{9.10}$$

As before, the voltage and current have an exponential form, and these are shown in Figures 9.3(b) and 9.3(c). Note that if i were defined in the opposite direction (as the current flowing out of the capacitor) then the polarity of the current in Figure 9.3(c) would be reversed. In this case, both the voltage and current would be represented by similar decaying exponential waveforms.

9.3.2 Inductor de-energising

In the circuit of Figure 9.4(a), a voltage source is used to energise an inductor by pass-ing a constant current through it. At time $t = 0$ one switch is closed and the other is opened, so the energy stored in the inductor is now dissipated in the resistor. Since the current in an inductor cannot change instantly, initially the current flowing in the coil is maintained. To do this the inductor produces an electromotive force that is in the opposite direction to the potential created across it by the voltage source. With time, the energy stored in the inductor is dissipated and the e.m.f. decreases and the current falls.

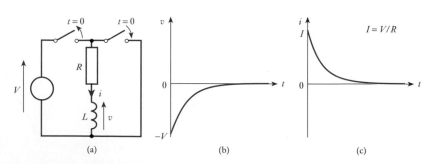

Figure 9.4
Inductor de-energising.

As before, v and i can be determined by applying Kirchhoff's voltage law to the circuit. This gives

$$iR + v = 0$$

and thus

$$iR + L\frac{\mathrm{d}i}{\mathrm{d}t} = 0$$

Solving this as before leads to the expressions

$$v = -Ve^{-Rt/L} = -Ve^{-t/\mathsf{T}} \tag{9.11}$$

$$i = Ie^{-Rt/L} = Ie^{-t/\mathsf{T}} \tag{9.12}$$

As before, the voltage and current have an exponential form, and these are shown in Figures 9.4(b) and 9.4(c).

9.4   Generalised response of first-order systems

We have seen in Sections 9.2 and 9.3 that circuits containing resistance and either capacitance *or* inductance can be described by first-order differential equations. For this reason, such circuits are described as **first-order systems**. We have also seen that the transient behaviour of these circuits produces voltages and currents that change exponentially with time. However, although the various waveforms are often similar in form, they are not identical for different circuits. Fortunately, there is a simple method of determining the response of such systems to sudden changes in their environment.

Video 9A

9.4.1   Initial and final value theorems

Increasing and decreasing exponential waveforms (for either voltage or current) can be found from the expressions

$$v = V_f + (V_i - V_f)e^{-t/\mathsf{T}} \tag{9.13}$$

$$i = I_f + (I_i - I_f)e^{-t/\mathsf{T}} \tag{9.14}$$

where V_i and I_i are the *initial* values of the voltage and current, and V_f and I_f are the *final* values. The first element in these two expressions represents the steady-state response of the circuit, which lasts indefinitely. The second element represents the transient response of the circuit. This has a magnitude determined by the step change applied to the circuit, and it decays at a rate determined by the time constant of the arrangement. The combination of the steady-state and the transient response gives the **total response** of the circuit. To see how these formulae can be used, Table 9.1 shows them applied to the circuits discussed in Sections 9.2 and 9.3.

These **initial and final value theorems** are not restricted to situations where a voltage or current changes to, or from, zero. They can be used wherever there is a step change in the voltage or current applied to a first-order network. This is illustrated in Example 9.3.

Table 9.1 Transient response of first-order systems.

	$V_i = 0 \quad V_f = V$ $I_i = V/R = I \quad I_f = 0$ $T = CR$	$v = V_f + (V_i - V_f)e^{-t/T}$ $= V + (0 - V)e^{-t/T}$ $= V(1 - e^{-t/T})$	
		$i = I_f + (I_i - I_f)e^{-t/T}$ $= 0 + (I - 0)e^{-t/T}$ $= Ie^{-t/T}$	
	$V_i = V \quad V_f = 0$ $I_i = 0 \quad I_f = V/R = I$ $T = L/R$	$v = V_f + (V_i - V_f)e^{-t/T}$ $= 0 + (V - 0)e^{-t/T}$ $= Ve^{-t/T}$	
		$i = I_f + (I_i - I_f)e^{-t/T}$ $= I + (0 - I)e^{-t/T}$ $= I(1 - e^{-t/T})$	
	$V_i = V \quad V_f = 0$ $I_i = -V/R = -I \quad I_f = 0$ $T = CR$	$v = V_f + (V_i - V_f)e^{-t/T}$ $= 0 + (V - 0)e^{-t/T}$ $= Ve^{-t/T}$	
		$i = I_f + (I_i - I_f)e^{-t/T}$ $= 0 + (-I - 0)e^{-t/T}$ $= -Ie^{-t/T}$	
	$V_i = -V \quad V_f = 0$ $I_i = V/R = I \quad I_f = 0$ $T = L/R$	$v = V_f + (V_i - V_f)e^{-t/T}$ $= 0 + (-V - 0)e^{-t/T}$ $= -Ve^{-t/T}$	
		$i = I_f + (I_i - I_f)e^{-t/T}$ $= 0 + (I - 0)e^{-t/T}$ $= Ie^{-t/T}$	

Example 9.3 | The input voltage to the following CR network undergoes a step change from 5 V to 10 V at time $t = 0$. Derive an expression for the resulting output voltage.

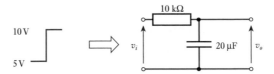

In this example the initial value of the output voltage is 5 V and the final value is 10 V. The time constant of the circuit is equal to $CR = 10 \times 10^3 \times 20 \times 10^{-6} = 0.2$ s.

Therefore, from Equation 9.13, for $t \geq 0$

$$v_o = V_f + (V_i - V_f)e^{-t/T}$$

$$= 10 + (5 - 10)e^{-t/0.2}$$

$$= 10 - 5e^{-t/0.2} \, \text{V}$$

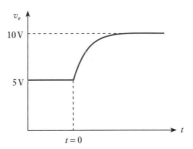

9.4.2 The nature of exponential curves

We have seen that the transients associated with first-order systems contain terms of the form $A(1 - e^{-t/T})$ or $Ae^{-t/T}$. The first of these represents a **saturating exponential** waveform and the second a **decaying exponential** waveform. The characteristics of these expressions are shown in Figure 9.5.

In general, one does not need to produce exact plots of such waveforms, but it is useful to know some of their basic properties. Perhaps the most important properties of exponential curves of this form are:

Figure 9.5
Exponential
waveforms.

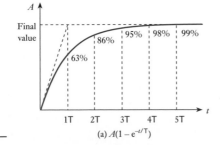

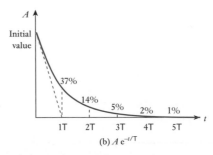

1 The initial slope of the curve crosses the final value of the waveform at a time $t = T$.
2 At a time $t = T$, the waveform has achieved approximately 63 per cent of its total transition.
3 The transition is 99 per cent complete after a period of time equal to 5T.

9.4.3 Response of first-order systems to pulse and square waveforms

Having looked at the transient response of first-order systems, we are now in a position to consider their response to pulse and square waveforms. Such signals can be viewed as combinations of positive-going and negative-going transitions and can therefore be treated in the same way as the transients discussed above. This is illustrated in Figure 9.6, which shows how a square waveform of fixed frequency is affected by RC and RL networks with different time constants.

Figure 9.6(a) shows the action of an RC network. We looked at the transient response of such an arrangement in Sections 9.2 and 9.3 and at typical waveforms in Figures 9.1 and 9.3. We noted that the response is exponential, with a rate of change that is determined by the time constant of the circuit. Figure 9.6(a) shows the effect of passing a square wave with a frequency of 1 kHz through RC networks with time constants of 0.01 ms, 0.1 ms and 1 ms. The first of these passes the signal with little

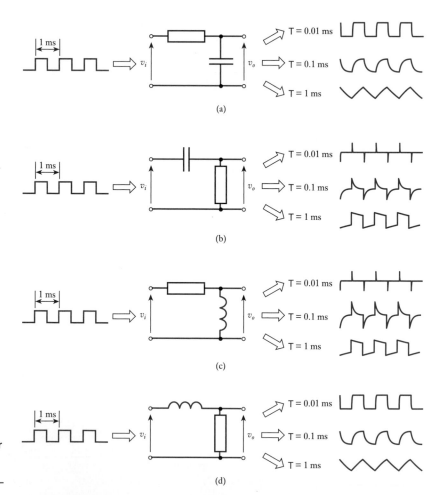

Figure 9.6
Response of first-order systems to a square wave.

distortion, since the wavelength of the signal is relatively long compared with the time constant of the circuit. As the time constant is increased to 0.1 ms and then to 1 ms, the distortion becomes more apparent as the network responds more slowly. When the time constant of the RC network is large compared with the period of the input waveform, the operation of the circuit resembles that of an **integrator**, and the output represents the integral of the input signal.

Transposing the positions of the resistor and the capacitor in the circuit of Figure 9.6(a) produces the arrangement shown in Figure 9.6(b). The output voltage is now the voltage across the resistor and is therefore proportional to the *current* in the circuit (and hence to the current in the capacitor). We would therefore expect the transients to be similar in shape to the current waveforms shown in Figures 9.1 and 9.3. The steady-state value of the output is zero in this circuit and, when a signal of 1 kHz is passed through a network with a time constant of 0.01 ms, the signal is reduced to a series of spikes. The circuit responds rapidly to the transient change in the input, and the output then decays quickly to its steady-state output value of zero. Here the time constant of the RL network is small compared with the period of the input waveform, and the operation of the circuit resembles that of a **differentiator**. As the time constant is increased, the output decays more slowly and the output signal is closer to the input.

Figures 9.6(c) and 9.6(d) show first-order RL networks and again illustrate the effects of the time constant on the characteristics of the circuits. The pair of circuits produces similar signals to the RC circuits (when the configurations are reversed), and again one circuit approximates to an integrator while the other approximates to a differentiator.

File 9A

Computer simulation exercise 9.1

Simulate the circuit of Figure 9.6(a) choosing appropriate component values to produce a time constant of 0.01 ms. Use a digital clock generator to produce a square-wave input signal to this circuit, setting the frequency of the clock to 1 kHz. Observe the output of this circuit and compare this with that predicted in Figure 9.6. Change one of the component values to alter the time constant to 0.1 ms, and then to 1 ms, observing the output in each case. Hence confirm the form of the waveforms shown in the figure. Experiment with both longer and shorter time constants and note the effect on the output.

Repeat this exercise for the remaining three circuits of Figure 9.6.

The shapes of the waveforms in Figure 9.6 are determined by the *relative* values of the time constant of the network and the period of the input waveform. Another way of visualising this relationship is to look at the effect of passing signals of different frequencies through the same network. This is shown in Figure 9.7. Note that the horizontal (time) axis is different in the various waveform plots.

The RC network of Figure 9.7(a) is a low-pass filter and therefore low-frequency signals are transmitted with little distortion. However, as the frequency increases the circuit has insufficient time to respond to changes in the input and becomes distorted. At high frequencies, the output resembles the **integral** of the input.

The RC network of Figure 9.7(b) is a high-pass filter and therefore high-frequency signals are transmitted with little distortion. At low frequencies, the circuit has plenty of time to respond to changes in the input signal and the output resembles that of a differentiator. As the frequency of the input increases, the network has progressively less time to respond and the output becomes more like the input waveform.

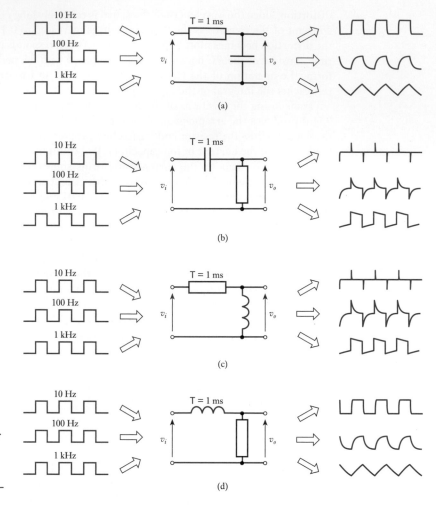

Figure 9.7
Response of first-order systems to square waves of different frequencies.

The *RL* network of Figure 9.7(c) represents a high-pass filter and therefore has similar characteristics to those of the *RC* network of Figure 9.7(b). Similarly, the circuit of Figure 9.7(d) is a low-pass filter and behaves in a similar manner to the circuit of Figure 9.7(a).

File 9B

Computer simulation exercise 9.2

Simulate the circuit of Figure 9.7(a) choosing appropriate component values to produce a time constant of 1 ms. Use a digital clock generator to produce a square-wave input signal to this circuit, setting the frequency of the clock to 10 Hz. Observe the output of this circuit and compare this with that predicted in Figure 9.7. Change the frequency of the clock generator to 100 Hz and then to 1 kHz, observing the output in each case. Hence confirm the form of the waveforms shown in the figure. Experiment with both higher and lower frequencies and note the effect on the output.

Repeat this exercise for the remaining three circuits of Figure 9.7.

Circuits that contain both capacitance and inductance are normally described by **second-order differential equations** (which may also describe some other circuit configurations). Arrangements described by these equations are termed **second-order systems**. Consider for example the *RLC* circuit of Figure 9.8. Applying Kirchhoff's voltage law to this circuit gives

$$L\frac{di}{dt} + Ri + v_C = V$$

Since i is equal to the current in the capacitor, this is equal to $C\, dv_C/dt$. Differentiating this with respect to t gives $di/dt = C\, d^2v_C/dt^2$, and therefore

$$LC\frac{d^2v_C}{dt^2} + RC\frac{dv_C}{dt} + v_C = V$$

which is a second-order differential equation with constant coefficients.

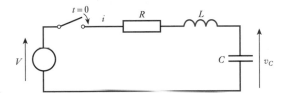

Figure 9.8
A series *RLC*
arrangement.

When a step input is applied to a second-order system, the form of the resultant transient depends on the relative magnitudes of the coefficients of its differential equation. The general form of the differential equation is

$$\frac{1}{\omega_n^2}\frac{d^2y}{dt^2} + \frac{2\zeta}{\omega_n}\frac{dy}{dt} + y = x$$

where ω_n is the **undamped natural frequency** in rad/s and ζ (Greek letter *zeta*) is the **damping factor**.

The characteristics of second-order systems with different values of ζ are illustrated in Figure 9.9. This shows the response of such systems to a step change at the input.

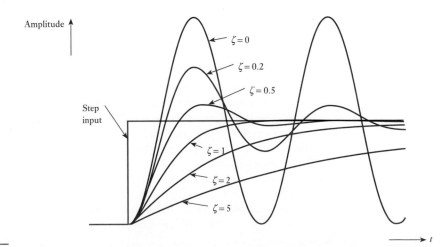

Figure 9.9
Response of second-
order systems.

Small values of the damping factor ζ cause the system to respond more rapidly, but values less than unity cause the system to **overshoot** and oscillate about the final value. When $\zeta = 1$, the system is said to be **critically damped**. This is often the ideal situation for a control system, since this condition produces the fastest response in the absence of overshoot. Values of ζ greater than unity cause the system to be **overdamped**, while values less than unity produce an **underdamped** arrangement. As the damping is reduced, the amount of overshoot produced and the **settling time** both increase. When $\zeta = 0$, the system is said to be **undamped**. This produces a continuous oscillation of the output with a natural frequency of ω_n and a peak height equal to that of the input step.

File 9C

Computer simulation exercise 9.3

Simulate the circuit of Figure 9.8, replacing the voltage source and the switch by a digital clock generator. Use values of 100 Ω, 10 mH and 100 μF for R, L and C, respectively, and set the frequency of the clock generator to 2.5 Hz. Use transient analysis to look at the output voltage over a period of 1 s.

Observe the output of the circuit and note the approximate time taken for the output to change. Increase the value of R to 200 Ω and note the effect on the output waveform. Progressively increase R up to 1 kΩ and observe the effect.

Now look at the effect of progressively reducing R below 100 Ω (down to 1 Ω or less). Estimate from your observations the value of R that corresponds to the circuit being critically damped.

9.6 Higher-order systems

Higher-order systems, that is those that are described by third-order, fourth-order or higher-order equations, often have a transient response that is similar to that of the second-order systems described in the last section. Because of the complexity of the mathematics of such systems, they will not be discussed further here.

Video 9B

Further study

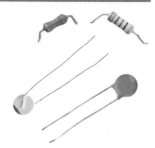

The initial and final value theorems provide a straight-forward way of modelling the behaviour of first-order systems by establishing the starting and finishing values of voltages or currents within the circuit. However, to use these theorems it is also necessary to know the time constant T of the circuit.

In the simple circuits we have considered, determining the time constant is simple, since this is given by either CR or L/R. However, in slightly more complicated circuits, the value of the time constant may not be so obvious.

To illustrate this point, determine the time constant of each of the following circuits.

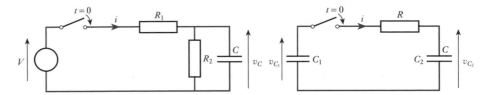

- The reaction of a circuit to instantaneous changes at its input is termed its transient response.
- The charging or discharging of a capacitor, and the energising or de-energising of an inductor, are each associated with exponential voltage and current waveforms.
- Circuits that contain resistance, and either capacitance *or* inductance, may be described by first-order differential equations and are therefore called first-order systems.
- The increasing or decreasing exponential waveforms associated with first-order systems can be found using the initial and final value theorems.
- The transient response of first-order systems can be used to determine their response to both pulse and square waveforms.
- At high frequencies, low-pass networks approximate to integrators.
- At low frequencies, high-pass networks approximate to differentiators.
- Circuits that contain both capacitance and inductance are normally described by second-order differential equations and are termed second-order systems.
- Such systems are characterised by their undamped natural frequency ω_n and their damping factor ζ. The latter determines how rapidly a system responds, while the former dictates the frequency of undamped oscillation.

Exercises

9.1 Explain the meanings of the terms 'steady-state response' and 'transient response'.

9.2 When a voltage is suddenly applied across a series combination of a resistor and an uncharged capacitor, what is the initial current in the circuit? What is the final, or steady-state, current in the circuit?

9.3 The switch in the following circuit is closed at $t = 0$. Derive an expression for the current in the circuit after this time and hence calculate the current in the circuit at $t = 4$ s.

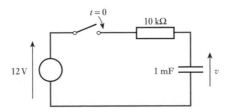

9.4 Simulate the arrangement of Exercise 9.3 and use transient analysis to investigate the current in the circuit. Use a switch element that *closes* at $t = 0$ to start the charging process, and use a second switch that *opens* at $t = 0$ to ensure that the capacitor is initially discharged (this second switch should be connected directly across the capacitor). Use your simulation to verify your answer to Exercise 9.3.

9.5 When a voltage is suddenly applied across a series combination of a resistor and an inductor, what is the initial current in the circuit? What is the final, or steady-state, current in the circuit?

9.6 The switch in the following circuit is closed at $t = 0$. Deduce an expression for the output voltage of the circuit and hence calculate the time at which the output voltage will be equal to 8 V.

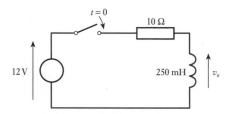

9.7 Simulate the arrangement of Exercise 9.6 and use transient analysis to investigate the output voltage of the circuit. Use a switch element that closes at $t = 0$ to start the energising process, and use your simulation to verify your answer to Exercise 9.6.

9.8 A capacitor of 25 µF is initially charged to a voltage of 50 V. At time $t = 0$, a resistance of 1 kΩ is connected directly across its terminals. Derive an expression for the voltage across the capacitor as it is discharged and hence determine the time taken for its voltage to drop to 10 V.

9.9 An inductor of 25 mH is passing a current of 1 A. At $t = 0$, the circuit supplying the current is instantly replaced by a resistor of 100 Ω connected directly across the inductor. Derive an expression for the current in the inductor as a function of time and hence determine the time taken for the current to drop to 100 mA.

9.10 What is meant by a 'first-order system', and what kind of circuits fall within this category?

9.11 Explain how the equation for an increasing or decreasing exponential waveform may be found using the initial and final values of the waveform.

9.12 The input voltage to the following CR network undergoes a step change from 20 V to 10 V at time $t = 0$. Derive an expression for the resulting output voltage.

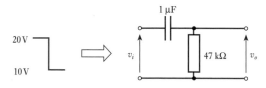

9.13 Sketch the exponential waveform $v = 5e^{-t/10}$.

9.14 For each of the following circuit arrangements, sketch the form of the output voltage when the period of the square-wave input voltage is:

(a) much greater than the time constant of the circuit;

(b) equal to the time constant of the circuit;

(c) much less than the time constant of the circuit.

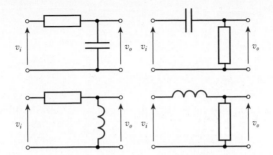

9.15 Simulate each of the circuit arrangements of Exercise 9.14, selecting component values to give a time constant of 1 ms in each case. Use a digital clock generator to apply a square-wave input voltage to the circuit and use transient analysis to observe the form of the output for input frequencies of 200 Hz, 1 kHz and 5 kHz. Compare these observations with your results for Exercise 9.14.

9.16 Under what circumstances does the behaviour of a first-order high-pass filter resemble that of a differentiator?

9.17 Under what circumstances does the behaviour of a first-order low-pass filter resemble that of an integrator?

9.18 What is meant by a 'second-order system', and what kind of circuits fall within this category?

9.19 Derive an expression for the *current* in the circuit of Figure 9.8.

9.20 Explain what is meant by the terms 'undamped natural frequency' and 'damping factor' as they apply to second-order systems.

9.21 What is meant by 'critical damping' and what value of the damping factor corresponds to this situation?

Chapter 10 Electric Motors and Generators

Objectives

When you have studied the material in this chapter, you should be able to:

- discuss the various forms of electrical machine
- explain how the interaction between a magnetic field and a rotating coil can be used to generate electricity
- explain how the interaction between a changing magnetic field and a coil can be used to generate motion
- describe the operation of various AC and DC forms of generator and motor
- discuss the use of electrical machines in a variety of industrial and domestic applications.

10.1 Introduction

An important area of electrical engineering relates to various forms of rotating **electrical machine**. These may be broadly divided into **generators**, which convert mechanical energy into electrical energy, and **motors**, which convert electrical energy into mechanical energy. In general, machines are designed to perform one of these two tasks, although in some cases a generator may also function as a motor, or vice versa, but with reduced efficiency.

Electrical machines may be divided into **DC machines** and **AC machines**, and both types operate through the interaction between a magnetic field and a set of windings. There are a great many forms of electrical machine, and this chapter does not aim to give a detailed treatment of each type. Rather, it sets out to describe the general principles involved, leaving readers to investigate further if they need more information on a particular type of motor or generator.

10.2 A simple AC generator

In Chapter 5, we looked at the nature of magnetic fields and at the effects of changes in the magnetic flux associated with conductors. You will recall that Faraday's law dictates that if a coil of N turns experiences a change in magnetic flux, then the induced voltage V is given by

$$V = N\frac{\mathrm{d}\Phi}{\mathrm{d}t} \tag{10.1}$$

where $\mathrm{d}\Phi/\mathrm{d}t$ is the rate of change of flux in webers/second.

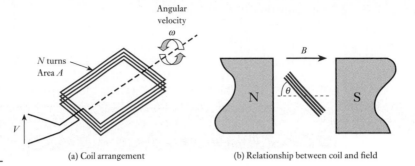

Figure 10.1
The rotation of a coil in a uniform magnetic field.

(a) Coil arrangement   (b) Relationship between coil and field

In Chapter 5, we were primarily concerned with the situation where the change in magnetic flux is caused by a change in the magnetic field associated with a stationary conductor. However, a similar effect is produced when a conductor moves within a constant field. Consider for example the situation shown in Figure 10.1. Here a coil of N turns and cross-sectional area A rotates with an angular velocity ω in a magnetic field of uniform flux density B.

If at a particular time the coil is at an angle θ to the field, as shown in Figure 10.1(b), its effective area normal to the field is $A \sin \theta$. Therefore, the flux linking the coil (Φ) is given by $BA \sin \theta$, and the rate of change of this flux ($d\Phi/dt$) is given by

$$\frac{d\Phi}{dt} = BA\frac{d(\sin \theta)}{dt} \tag{10.2}$$

Now

$$\frac{d(\sin \theta)}{dt} = \frac{d\theta}{dt}\cos \theta = \omega \cos \theta \tag{10.3}$$

since $d\theta/dt = \omega$.

Therefore, combining Equations 10.1, 10.2 and 10.3, we have

$$V = N\frac{d\Phi}{dt} = NBA\frac{d(\sin \theta)}{dt} = NBA\omega \cos \theta \tag{10.4}$$

and the induced voltage varies as the cosine of the phase angle, as shown in Figure 10.2(a). Given a constant speed of rotation, the phase angle θ varies linearly with time, and hence the output voltage varies as the cosine of time, as shown in Figure 10.2(b).

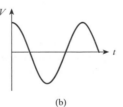

Figure 10.2
Voltage produced by the coil of Figure 10.1.

(a)   (b)

Example 10.1 │ A coil consisting of 100 turns of copper wire has an area of 20 cm². Determine the peak magnitude of the sinusoidal voltage produced across the terminals of this coil if it rotates within a magnetic field of 400 mT at a rate of 1000 rpm.

From Equation 10.4, we know that

$$V = NBA\omega \cos\theta$$

In this expression ω is the angular frequency and in this case the cyclic frequency is 1000 rpm, which is $1000/60 = 16.7$ Hz. Since $\omega = 2\pi f$, it follows that $\omega = 2 \times \pi \times 16.7 = 105$ rad/s. Therefore, substituting into the above equation gives

$$V = NBA\omega \cos\theta$$
$$= 100 \times 400 \times 10^{-3} \times 20 \times 10^{-4} \times 105 \cos\theta$$
$$= 8.4 \cos\theta$$

Therefore, the output is a sinusoidal voltage with a peak value of 8.4 V.

10.2.1   Slip rings

One problem with the arrangement of Figure 10.1 is that any wires that are connected to the coil will become tangled as the coil rotates. A solution to this problem is to use **slip rings**, which provide a sliding contact to the coil as it rotates. This idea is illustrated in Figure 10.3. Electrical contact is made to the slip rings through **brushes**, which normally take the form of graphite blocks that are held against the rings by springs.

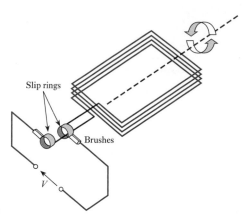

Figure 10.3
The use of slip rings.

10.3   A simple DC generator

The alternating signal produced by the arrangement of Figure 10.3 could be converted to a unidirectional signal using a rectifying circuit (as discussed in Section 2.2.6). We will look at the form of such circuits in Chapter 17. However, a more efficient (and common) approach is to rectify the output of the generator by replacing the two slip rings of Figure 10.3 with a single, split, slip ring as shown in Figure 10.4(a). A slip ring that is split in this way is referred to as a **commutator**.

The commutator is arranged so that, as the voltage produced by the coil changes polarity, the connections to the coil are reversed. Hence the voltage produced across the brushes is of a single polarity, as shown in Figure 10.4(b).

While the arrangement of Figure 10.4 produces a voltage that is unidirectional, its magnitude varies considerably as the coil rotates. This 'ripple' can be reduced by

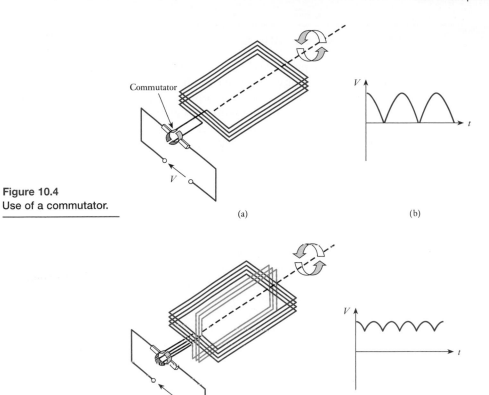

Figure 10.4
Use of a commutator.

(a)   (b)

Figure 10.5
A simple generator
with two coils.

(a)   (b)

summing the output from a number of coils that are set at different angles. This is illustrated in Figure 10.5(a), which shows an arrangement with two coils set at 90° to each other. The coils are connected in series, and the commutator now has four seg-ments to allow the connections to the coils to be changed as the coils rotate. The resultant output voltage is shown in Figure 10.5(b). This process can be extended by using additional coils to further reduce the variation in the output voltage.

The ripple voltage produced by the generator can be further reduced by winding the coils on a cylindrical iron core and by shaping the pole pieces of the magnets as shown in Figure 10.6. This has the effect of producing a high and approximately uniform magnetic field in the small air gap. The coils are wound within slots in the core, and the arrangement of coils and core is known as the **armature**. The armature of Figure 10.6 has four coils.

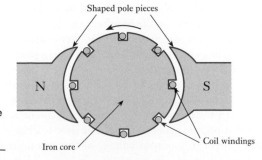

Figure 10.6
The use of an iron core
and shaped pole
pieces.

10.4 | DC generators or dynamos

Having looked at the basic principles of a simple DC generator, we are now in a position to consider a more practical arrangement. DC generators, or **dynamos** as they are sometimes called, can take a number of forms depending on the method used to produce the magnetic field. **Permanent-magnet generators** are available, but it is more common to generate the magnetic field electrically using **field coils**. The current that flows in these coils can be supplied by an external energy source (in the case of a **separately excited generator**), but it is more usual to supply this from the current produced by the generator itself (as in a **self-excited generator**). Since the field coils consume only 1 or 2 per cent of the rated output current, this loss of power is generally quite acceptable.

In addition to using multiple armature windings, it is common to use multiple poles, and Figure 10.7 shows a typical four-pole arrangement. The poles are arranged as alternating north and south poles and are held in place by a steel tube called the **stator**, which also forms the outer casing of the unit. Field coils are wound around each pole piece, these being connected in series and wired to produce the appropriate magnetic polarity. The generator in Figure 10.7 has eight slots in the armature, although many devices will have twelve or more. A typical device would produce an output ripple of about 1 or 2 per cent.

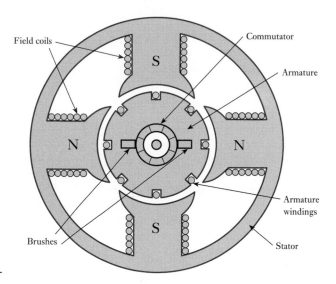

Figure 10.7
A DC generator or dynamo.

10.4.1 | Field coil excitation

In some generators the field coils are connected in series with the armature coils (a **series-wound DC generator**), while in others the field coils are connected in parallel with the armature coils (a **shunt-wound DC generator**). A third variant has two sets of field windings, one connected in series and one in parallel, and such a machine is called a **compound DC generator**.

One of the most common forms of DC generator is the shunt-wound generator. This is an example of a self-excited generator, and in this case the field coils are connected in parallel with the armature. Such an arrangement is shown in Figure 10.8. In a typical configuration, the generator might be used to charge a battery. The voltage across the armature forms the output voltage of the generator, and this is connected

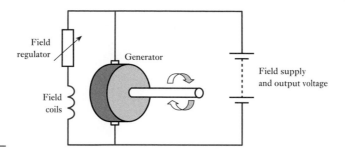

Figure 10.8
Connections for
a shunt-wound
generator.

across the battery. This voltage is also used to drive the field coils, the current in the coils being controlled by a field regulator, which might be a simple variable resistance.

10.4.2 DC generator characteristics

The various forms of DC generator have slightly different electrical characteristics, and these differences will often determine the device chosen for a particular situation. We will not look at generator characteristics in detail but simply note that the output voltage increases with the speed of rotation of the armature and that in many DC generators this is a nearly linear relationship. Generators are often run at a constant speed (although this is *not* the case in all situations), with the characteristics of the generator chosen to give the required output voltage.

The voltage produced by a generator is also affected by the current taken from the device. As the current increases the output voltage falls, as we would expect for a voltage source that has output resistance. This fall is caused partly by the resistance in the armature and partly by an effect known as **armature reaction**, where the armature current produces a magnetic flux that opposes that produced by the field coils.

Figure 10.9 shows typical characteristics for a shunt-wound DC generator. Figure 10.9(a) shows the relationship between the output voltage and the speed of rotation for zero output current; Figure 10.9(b) shows the effect of current on the output voltage; and Figure 10.9(c) shows a simple equivalent circuit of the generator.

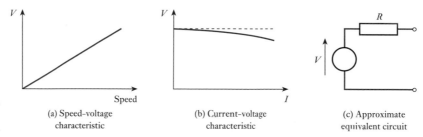

Figure 10.9
Shunt-wound generator
characteristics.

(a) Speed–voltage
characteristic

(b) Current–voltage
characteristic

(c) Approximate
equivalent circuit

10.5 AC generators or alternators

AC generators, or **alternators**, are in many ways similar to DC generators in that power is produced by the effect of a changing magnetic field on a set of coils. However, alternators do not require commutation, and this allows some simplification of their construction. Since the power required to produce the electric field is much less than that delivered by the generator, it makes sense to reverse the construction and to rotate the field coils while keeping the armature windings stationary. Note that the

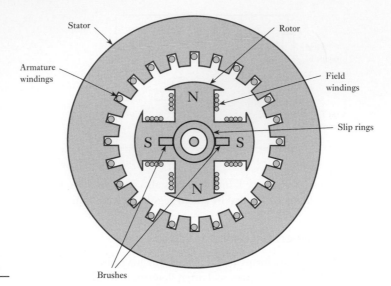

Figure 10.10
A four-pole alternator.

armature windings are the coils that produce the output e.m.f. of the generator. In the case of a DC generator, the armature windings are mounted on the rotating part of the machine (the **rotor**), so the rotor is also known as the armature. In an AC synchronous generator, the large and heavy armature coils are mounted in the stationary part of the machine (the **stator**). In this case, the field coils are mounted on the rotor and direct current is fed to these coils by a set of slip rings.

As with DC generators, multiple poles and sets of windings are used to improve efficiency. In some cases, three sets of armature windings are spaced 120° apart around the stator to form a **three-phase generator**. Figure 10.10 shows a section through a simple four-pole alternator.

In the alternator, the e.m.f. generated is in sync with the rotation of the rotor, and for this reason it is referred to as a **synchronous generator**. If the generator has a single pair of poles, then the frequency of the output will be the same as the rotation frequency. Therefore, to produce an output at 50 Hz using a two-pole generator, the speed of rotation would need to be $50 \times 60 = 3000$ rpm. Generators with additional pole pairs produce an output at a correspondingly higher frequency, since each turn of the rotor represents several cycles of the output. In general, a machine with N pole pairs produces an output at a frequency of N times the rotational frequency.

Example 10.2

A four-pole alternator is required to operate at 60 Hz. What is the required rotation speed?

A four-pole alternator has two pole pairs. Therefore the output frequency is twice the rotation speed. Therefore, to operate at 60 Hz, the required rotation speed must be $60/2 = 30$ Hz. This is equivalent to $30 \times 60 = 1800$ rpm.

10.6 ┃ DC motors

Current flowing in a conductor produces a magnetic field about it (as noted in Chapter 5). This process is illustrated in Figure 10.11(a). When the current-carrying conductor is within an externally generated magnetic field, the induced field interacts

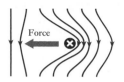

Figure 10.11
Fields and forces
associated with a
current-carrying
conductor.

(a) The magnetic field about a
current flowing into the page

(b) The effects of an
external magnetic field

with the external field and a force is exerted on the conductor. This is shown in Figure 10.11(b). If the conductor is able to move, this force will cause motion in the direction indicated in Figure 10.11(b). Therefore, if a conductor lies in a magnetic field, motion of the conductor in the field will generate an electric current, while current in the conductor will generate motion.

The reciprocal nature of the properties of conductors in magnetic fields means that many electrical machines will operate as either generators or motors. For example, the DC generators described in Section 10.4 will also function as DC motors, and the diagram of Figure 10.7 could equally well represent a four-pole DC motor. However, machines designed as motors will generally be more efficient in this role than machines designed as generators.

As in the case of generators, motors vary in the number of magnetic poles they possess and in the way that their windings are configured. Shunt-wound, series-wound and compound motors are available, and each has slightly different characteristics.

10.6.1 Shunt-wound DC motor

The shunt-wound DC motor is widely used since it has several attractive features. Its speed of rotation is largely dependent on the applied voltage, while the torque it applies is related to the current. Thus if a constant voltage is applied to such a motor the speed of rotation will tend to remain fairly constant, while the current taken by the motor will vary depending on the load applied. In practice, the speed is affected to some extent by the load applied to the motor, and therefore the speed drops slightly with increasing torque. Figure 10.12 shows typical characteristics for a shunt-wound DC motor.

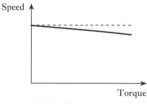

Figure 10.12
Characteristics of a
shunt-wound DC
motor.

(a) Torque–current
characteristic

(b) Speed–torque characteristic
with a constant applied voltage

10.7 AC motors

AC motors can be divided into two main forms: synchronous motors and induction motors. High-power versions of either type invariably operate from a three-phase supply, but single-phase versions are also widely used, particularly in a domestic setting. Here we will look at examples of both synchronous and induction motors, and at both three-phase and single-phase versions.

10.7.1 Synchronous motor

Just as DC generators can also be used as DC motors, the synchronous AC generator (or alternator) of Section 10.5 can also be used as a **synchronous motor**. As the name implies, such a device operates at a speed determined by the frequency of the AC input. When used with a conventional AC supply, they are therefore **constant-speed motors**.

In a three-phase synchronous motor, the three sets of stator coils produce a magnetic field that rotates around the rotor at a speed determined by the frequency of the AC supply and the number of poles in the stator. The DC field current, which is fed to the rotor through the slip rings, turns the rotor into an electromagnet (possibly with several poles), which is dragged around by the rotating magnetic field. Thus the rotor speed is determined by the speed of rotation of the magnetic field.

Single-phase motors do not have the benefit of multiple input phases to generate the rotating magnetic field. Various techniques can be used to overcome this problem, and this gives rise to several variants on the basic design. These techniques will not be discussed here.

Unfortunately, because the rotor in a synchronous motor is dragged around by the rotating magnetic field, torque is only produced when the rotor is in sync with this field. When the motor is energised from rest the rotating magnetic field is established almost immediately, but the rotor is initially stationary. Consequently, the magnetic field races past the rotor rather than causing it to turn. Basic synchronous motors therefore produce no starting torque and are not self-starting. To overcome this problem, motors incorporate some form of starting mechanism to get them moving from rest. In some cases, this involves configuring the motor to operate as an induction motor (as discussed below) until it gets up to speed. Once the motor has reached synchronism it switches to operate as a synchronous motor, which in some situations gives greater efficiency than can be gained from an induction motor.

Example 10.3

What is the speed of rotation of an eight-pole synchronous motor when used with a single-phase 50 Hz supply?

An eight-pole motor has four pole pairs, so the magnetic field rotates at four times the supply frequency, which in this case is $4 \times 50 = 200$ Hz. Since the rotor is dragged around by this rotating magnetic field, this is also the speed of rotation of the motor. Therefore, the motor turns at 200 Hz = $200 \times 60 = 12,000$ rpm.

10.7.2 Induction motors

Perhaps the most important forms of AC motor are the various types of **induction motor**. These differ from synchronous motors in that field current is not fed to the rotor through slip rings but is *induced* in the rotor by transformer action. The most common form of induction motor is the **cage rotor induction motor**, which is also known as the **squirrel-cage induction motor**. This uses a stator of a similar form to that in the synchronous motor (or the synchronous generator of Figure 10.10) but replaces the rotor and slip rings with an arrangement of the form shown in Figure 10.13. This can be seen as a series of parallel conductors that are shorted together at each end by two conducting rings.

As in the three-phase synchronous motor, in a three-phase induction motor the stator coils produce a rotating magnetic field that cycles around the rotor at a frequency determined by the supply frequency and the number of poles in the stator. A stationary conductor in the stator will see a varying magnetic field, and this changing flux will induce an e.m.f. in much the same way that an e.m.f. is induced in the

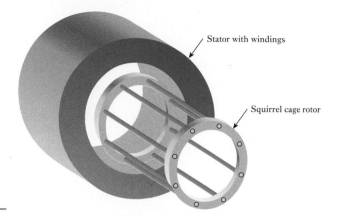

Figure 10.13
A squirrel-cage
induction motor.

secondary of a transformer. This induced e.m.f. results in a current flowing in the rotor, and this in turn produces a magnetic field. This *rotor* field interacts with the *stator* field in the same way as in a synchronous motor, and again the rotor is dragged around by the rotating field in the stator. However, in the case of an induction motor, the stator will always turn slightly *more slowly* than the stator field. This is because, if the rotor turned in sync with the stator field, there would be no change in the flux associated with the rotor, and hence no induced current. The slight speed difference is called the **slip** of the motor. This increases with the applied load and might be a few per cent at full load. Three-phase induction motors have an advantage over synchronous motors in that they are self-starting.

In domestic situations three-phase supplies are usually not available, and several techniques can be used to produce the required rotating magnetic field from a single-phase supply. This gives rise to several forms of induction motor, such as the **capacitor motor** and the **shaded-pole motor**. Such motors are inexpensive and are widely used in domestic appliances. However, further discussion of the operation of these motors is beyond the scope of this book.

10.8 Universal motors

While most motors are designed to operate on either AC or DC, some motors can operate on either. These **universal motors** resemble series-wound DC motors but are designed for both AC and DC operation. Typically operating at high speeds (usually greater than 10,000 rpm), these motors offer a high power-to-weight ratio, making them ideal for portable appliances such as hand drills and vacuum cleaners.

10.9 Stepper motors

Stepper motors, as their name implies, move in discrete steps. The motor consists of a central rotor surrounded by a number of coils (or windings). The form of a simple stepper motor is shown in Figure 10.14, and a typical motor is shown in Figure 10.15.

Diametrically opposite pairs of coils are connected together so that energising any one pair of coils will cause the rotor to align itself with that pair. By applying power to each set of windings in turn, the rotor is made to 'step' from one position to another and thus generate rotary motion. In order to reduce the number of external connections to the motor, groups of coils are connected together in sequence. In the example

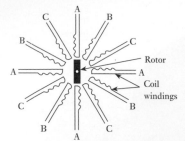

Figure 10.14
A simple stepper motor.

Figure 10.15
A typical stepper motor.

shown, every third coil is joined to give three coil sets, which have been labelled A, B and C. If initially winding A is energised, the rotor will take up a position aligned with the nearest winding in the A set. If now A is de-energised and B is activated, the motor will 'step' around to align itself with the next coil. If now B is de-energised and C is activated, the rotor will again step to the adjacent coil. If the activated coil now reverts to A, the rotor will move on in the same direction to the next coil, since this is the closest coil in the A set. In this way, the rotor can be made to rotate by activating the coils in the sequence 'ABCABCA . . .'. If the sequence in which the windings are activated is reversed (CBACBAC . . .), the direction of rotation will also reverse. Each element in the sequence produces a single step that results in an incremental movement of the rotor.

The waveforms used to activate the stepper motor are binary in nature, as shown in Figure 10.16. The motor shown in Figure 10.14 has 12 coils, and consequently 12 steps would be required to produce a complete rotation of the rotor. Typical small stepper motors have more than 12 coils and might require 48 or 200 steps to perform one complete revolution. The voltage and current requirements of the coils will vary with the size and nature of the motor.

The speed of rotation of the motor is directly controlled by the frequency of the waveforms used. Some stepper motors will operate at speeds of several tens of thousands of revolutions per minute, but all have a limited rate of acceleration determined by the inertia of the rotor. All motors have a 'maximum pull-in speed', which is the

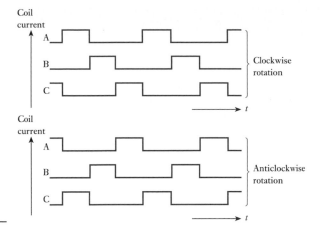

Figure 10.16
Stepper motor current waveforms.

maximum speed at which they can be started from rest without losing steps. To operate at speeds above this rate, they must be accelerated by gradually increasing the frequency of the applied waveforms. Since the movement of the rotor is directly controlled by the waveforms applied to the coils, the motor can be made to move through a prescribed angle by applying an appropriate number of transitions to the coils. This is not possible using a DC motor, since the speed of rotation is greatly affected by the applied load.

10.10 Electrical machines – a summary

It can be seen that there are a great many forms of electrical machine, and this chapter has given only a brief overview of their characteristics and variety.

While both DC and AC generators have been described, power generation is dominated by synchronous AC machines. Such devices range from small alternators used in automotive applications to the large generators used in power stations. While all synchronous generators tend to be relatively efficient, this efficiency increases with size, with large generators converting more than 98 per cent of their mechanical input power into electricity.

Both DC and AC motors are widely used, although often in different situations. Where moderate or large amounts of mechanical power are required AC motors are more common, particularly where variable-speed operation is not required. In industrial applications, three-phase induction motors are the dominant devices, while domestic appliances such as washing machines and dishwashers will normally use single-phase induction motors. DC motors are widely used in low-power applications, particularly in situations where variable-speed control is required. The simple relationship between speed and voltage in many DC motors makes them very easy to control. DC motors are also used in high-power applications that require variable speed, such as traction applications. However, the development of high-powered electronic speed controllers for AC motors has reduced their use in such applications.

While both AC and DC motors can be used in applications where their speed or position must be accurately controlled, this often requires the use of external components to monitor and then control their operation. Stepper motors, in contrast, can be easily made to produce very precise motion or a defined angle of rotation. Such motors are also often inexpensive, very rugged and can produce very high torque at low speeds. However, stepper motors are generally relatively small in size and are predominantly used in low-power applications where precision is of importance.

Video 10A

Further study

In this chapter we have looked at various forms of motor, all of which produce rotation of an output shaft.

One application where rotary motion is required is in the production of a wrist watch. Here the second hand must be made to rotate at precisely one revolution per minute, and the minute and hour hands are then turned through an appropriate combination of gears.

Consider the characteristics of the different forms of motor discussed within the chapter and decide which type of motor would be most suitable in this application.

Key points

- Electrical machines can be broadly divided into generators, which convert mechanical energy into electrical energy, and motors, which convert electrical energy into mechanical energy.
- In most cases, generators can also function as motors, and vice versa.
- Electrical machines can be divided into DC machines and AC machines.
- All electrical machines operate through the interaction between a magnetic field and a set of windings.
- The rotation of a coil in a uniform magnetic field produces a sinusoidal e.m.f. This principle is at the heart of an AC generator or alternator.
- A commutator can be used to convert the above sinusoidal e.m.f. into a unipolar form. This is the basis of a DC generator or dynamo.
- While the magnetic field in an electrical machine can be produced by a permanent magnet, it is more common to produce this electrically using field coils.
- In DC generators, the armature coils normally rotate within stationary field coils. In AC generators, the field coils normally rotate within stationary armature coils.
- DC motors are often similar in form to DC generators.
- Some forms of AC generator can also be used as motors.
- There are many forms of AC motor, the most widely used being the various types of induction motor.
- Many types of motor are not inherently self-starting, and some form of starting mechanism must be incorporated.
- Stepper motors produce motion in the form of a series of 'steps' and can be made to rotate at a precisely defined rate or through a specified angle. This makes them ideal for use in a range of control applications.

Exercises

10.1 What is meant by the term 'electrical machine'?

10.2 A coil of 50 turns with an area of 15 cm^2 rotates in a magnetic field of 250 mT at 1500 rpm. What is the peak magnitude of the sinusoidal voltage produced across its terminals?

10.3 Explain the function of slip rings.

10.4 What is the function of a commutator?

10.5 How may the ripple voltage produced by a DC generator be reduced?

10.6 How are the field coils in a DC generator normally excited?

10.7 Describe the characteristics of a shunt-wound dynamo.

10.8 What is meant by 'armature reaction'?

10.9 How does the construction of a typical alternator differ from that of a dynamo?

10.10 What is meant by the term 'armature', and what form does this take in a dynamo and in an alternator?

10.11 How does a *synchronous* generator get its name?

10.12 A six-pole alternator is required to operate at 50 Hz. What is the required rotation speed?

10.13 How does a DC motor differ from a DC generator?

10.14 Briefly describe the characteristics of a shunt-wound DC motor.

10.15 Why is a synchronous motor so called?

10.16 What is the speed of rotation of a twelve-pole synchronous motor when used with a single-phase 50 Hz supply?

10.17 How does the construction of an induction motor differ from that of a synchronous motor?

10.18 What is meant by the 'slip' of an induction motor?

10.19 Explain what is meant by a universal motor?

10.20 What sort of generator would normally be used in a power station?

10.21 What sort of motor would typically be used in a domestic washing machine?

10.22 Briefly describe the operation of a stepper motor.

10.23 How are the speed and direction of rotation of a stepper motor controlled?

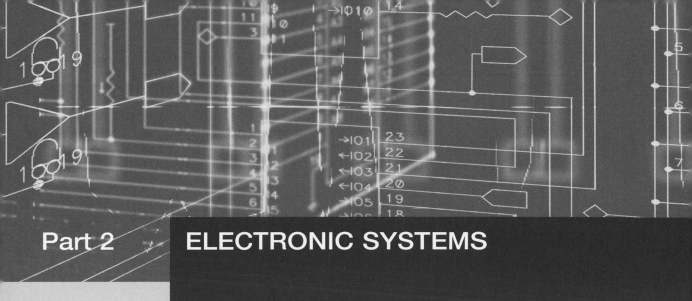

Part 2 ELECTRONIC SYSTEMS

Chapter 11

Electronic Systems

When you have studied the material in this chapter, you should be able to:

- discuss the need for electronic systems in a wide range of applications
- describe the characteristics and advantages of a 'systems approach' to engineering
- identify the inputs and outputs of an engineering system and understand the significance of the choice of system boundary
- explain the varied characteristics of physical quantities and the need to represent these quantities by electrical signals
- use block diagrams to represent complex engineering systems.

11.1 Introduction

Having looked at a range of basic electrical components and circuits in Part 1 of the text, we are now in a position to turn our attention to more complex electronic systems.

In recent years electronic systems have found their way into almost all aspects of our lives. Such systems wake us in the morning; control the operation of our cars as we drive to work; maintain a comfortable working environment in our offices and homes; allow us to communicate worldwide; provide access to information at the touch of a button; manage the provision of power to maintain our high-technology lifestyles; and provide restful entertainment after a day of 'electronically controlled' excitement.

In many cases electronic systems are used in these applications because they provide a more cost-effective solution than other available techniques. However, in many cases electronics provide the *only* solution, and the application would be impossible without its use. Our way of life increasingly depends on an ability to monitor or control our environment and to communicate efficiently. In these areas electronic systems are supreme and seem certain to remain so for the foreseeable future.

While electronic elements represent essential components of almost all complex systems, it should be noted that few, if any, engineering systems consist of entirely electronic elements. Even applications such as mobile phones or MP3 players require mechanical elements such as cases and keyboards in order to make a useable product. In practice *all* real engineering projects are interdisciplinary in nature and involve a wide range of engineering skills and techniques coming together to solve what are often quite complex problems.

Video 11A

11.2   A systems approach to engineering

Several areas of human endeavour are associated with the solution of problems that involve great complexity. These include topics as diverse as the comprehension of biological organisms, the proof of complex mathematical relationships and the rationalisation of philosophical arguments. Over the years several distinct approaches have evolved for tackling such problems, and many of these are of direct relevance to the production of complex systems within engineering.

One method is to adopt what might be termed a **systematic** approach, in which a complex problem or system is simplified by dividing it into a number of smaller elements. These elements are then themselves subdivided, the process being repeated until the various constituents have been devolved into elements that are sufficiently simple to be easily understood. This approach is widely used within engineering in what is termed **top-down design**, where a complex system is progressively divided into simpler and simpler subsystems. This results in a series of modules that are of a manageable level of complexity and size to allow direct implementation. This approach is based, to some extent, on what might be seen as a 'reductionist' view, implying that a complex system is no more than the sum of its parts.

A problem with the reductionist view is that it ignores characteristics that are features of the 'whole' rather than of individual components. These **systemic** properties are often complex in nature and may relate to several diverse aspects of the system. For example, the 'ride' and 'feel' of a car are not determined by a single module or subsystem but by the interaction of a vast number of individual components.

In recent years, modern engineering practice has evolved a more 'holistic' approach that combines the best elements of a *systematic* approach together with considerations of *systemic* issues. This results in what is called a **systems approach** to engineering.

The systems approach has its origins back in the 1960s but has gained favour within many engineering disciplines only recently. It is categorised by a number of underlying principles, which include a strong emphasis on the application of scientific methods, the use of systematic project management techniques and, perhaps most importantly, the adoption of a broad-based interdisciplinary or team approach. In addition to specialists from a range of engineering disciplines, a project might also involve experts in other fields such as artistic design, ergonomics, sociology, psychology or law. A key feature of a systems approach is that it places as much importance on identifying the relationships *between* components and events as it does on identifying the characteristics of the components and events themselves.

This text does not claim to provide a true 'systems approach' to the material it covers. In many ways such a broad-based treatment would be inappropriate for an introductory text of this kind. However, it does attempt to present information on components and techniques within the context of the systems in which they are used. For this reason, we will look in some detail at the nature and characteristics of electronic systems, before describing in detail the components in them. This allows the reader to understand *why* these components are required to have the characteristics that they have and *how* the techniques relate to the applications in which they are used.

11.3   Systems

Before looking at the nature of electronic systems, it is perhaps appropriate to make sure that we understand what we mean by the word **system**.

In an engineering context, a system can be defined as any closed volume for which all the inputs and outputs are known. This definition allows us to consider an infinite number of 'systems' depending on the volume of space that we decide to select. However, in practice, we normally select our 'closed volume' to enclose a component, or group of components, that are of interest to us. Thus we could select a volume that includes the components that control the engine of a car and call this an 'engine management system'. Alternatively, we might select a larger volume that includes the complete car and call this an 'automotive system'. A larger volume might include a complete 'transportation system', while a volume that contains the Earth might be described as an 'ecosystem'. Since we can freely select the boundaries of our systems, we can use this approach to subdivide large systems into smaller, more manageable blocks. For example, a car might be considered as a large number of smaller systems (or subsystems), each responsible for a different function.

As we change the elements within our 'system' we also change the inputs and outputs. The signals going into and out from an engine management system relate to the status, or condition, of various parts of the engine and the car. If we consider the complete car as our system, then the inputs include petrol, water and commands from the driver, while the outputs include work (in the form of movement), heat and exhaust gases. If we look at our planet as a single system, then the inputs and outputs are primarily different forms of radiation. From outside a system, only the inputs and outputs are visible. However, it may be possible to learn something of the nature of the system by observing the relationship between these inputs and outputs. One way of describing the characteristics of any system is in terms of the nature of the inputs and the outputs and the relationship between them.

In some situations only particular inputs and outputs to a system are of interest, and others may be completely ignored. For example, one input to a mobile phone might be air entering or leaving its case. An electronic engineer designing such a system might decide to ignore this form of input and to concentrate only on those inputs related to the operation of the unit. This general principle can be extended so that only particular kinds of input and output are considered. Thus a company's accounting system might consider only the flow of money into and out of the company. This concept can be extended so that the 'volume' of the system becomes nebulous, and the system is effectively defined solely by its inputs, its outputs and the relationship between them.

11.4 System inputs and outputs

Figure 11.1 represents a generalised system, together with its inputs and outputs. This diagram makes no assumptions about the form of any of its components, and this could represent a mechanical or biological system just as easily as an electrical or electronic arrangement. Thus the inputs and outputs in this case could be forces, temperatures, velocities or any other physical quantities. Alternatively, they could be electrical quantities such as voltages or currents.

When considering electronic systems, we are concerned with arrangements that generate, or manipulate, electrical energy in one form or another. However, the nature of

Figure 11.1
A generalised system.

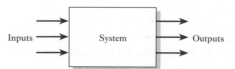

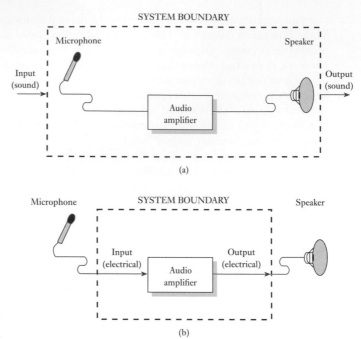

Figure 11.2
The effects of choosing system boundaries.

the inputs and outputs to such systems may depend on where we choose to draw the system's boundaries. This is illustrated in Figure 11.2, which shows an arrangement involving an audio amplifier, a microphone and a speaker. In Figure 11.2(a), we have chosen to consider the microphone and speaker as parts of our system. Here the input and output are in the form of sound waves. In Figure 11.2(b), we are considering the system to consist only of the audio amplifier itself. Now the microphone and speaker are external to the system, and the input and output are in an electrical form.

In Figure 11.2, the microphone senses variations in the external environment (in this case sound waves) and represents them electrically for processing within the electronic parts of the system. Conversely, the speaker takes the electrical outputs produced by the system and uses them to affect the external environment (again, in this case, by creating sound waves). Components that interact with the outside world in this way are referred to as *sensors* and *actuators*, and without such devices our electrical and electronic systems would be useless. We will therefore look at a range of such devices in Chapters 12 and 13. For the moment we will simply note that such elements exist and that they can be used to enable a system to interact with the world around it.

11.5   Physical quantities and electrical signals

The electrical fluctuations produced by a sensor convey information about some varying physical quantity. Such a representation is termed an electrical *signal*. In Figure 11.2, the output of the microphone is an electrical signal that represents the sounds that it detects. Similarly, the output from the amplifier is an electrical signal that represents the sounds to be produced by the speaker. Signals may take a number of forms, but before discussing these it is perhaps appropriate to look at the nature of the physical quantities that they may represent.

11.5.1 Physical quantities

The world about us may be characterised by a number of physical properties or quantities, many of which vary with time. Examples of these include temperature, humidity, pressure, altitude, position and velocity. The time-varying nature of such physical quantities allows them to be categorised into those that vary in a *continuous* manner and those that exhibit a *discontinuous* or *discrete* nature.

The vast majority of real-world physical quantities (such as temperature, pressure and humidity) vary in a continuous manner. This means that they change smoothly from one value to another, taking an infinite number of values. In contrast, discrete quantities do not change smoothly but instead switch abruptly between distinct values. Few natural quantities exhibit this characteristic (although there are some examples, such as population), but many human-made quantities are discrete.

11.5.2 Electrical signals

It is often convenient to represent a varying physical quantity by an electrical signal. This is because the processing, communication and storage of information is often much easier when it is represented electrically.

Having noted that physical quantities may be either continuous or discrete in nature, it is not surprising that the electrical signals that represent them may also be either continuous or discrete. However, there is not necessarily a direct correspondence between these forms, since it may be convenient to represent a continuous quantity by a discrete signal, or vice versa. For reasons that are largely historical, continuous signals are normally referred to as *analogue*, while discrete signals are described as *digital*.

Both analogue and digital signals can take many forms. Perhaps one of the simplest is where the voltage of a signal corresponds directly to the magnitude of the physical quantity being represented. This format is used for both the input and the output signals in Figure 11.2(b), where the voltages of the signals correspond directly to fluctuations in the input and output air pressure (sound). Most people will have seen the output of a microphone displayed on an oscilloscope and noted the relationship between the sound level and the magnitude of the displayed waveform. Figure 11.3(a) shows an example of a typical analogue signal waveform.

Although it is very common to represent the magnitude of a continuous quantity by the voltage of an electrical signal, many other forms are also used. For example, it might be more convenient to represent the value of a physical quantity by the magnitude

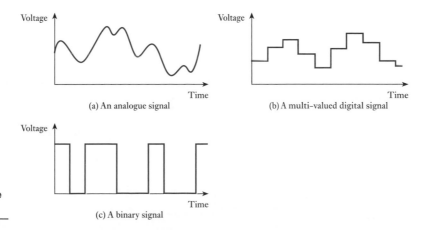

(a) An analogue signal

(b) A multi-valued digital signal

(c) A binary signal

Figure 11.3
Examples of analogue and digital signals.

of the *current* flowing in a wire (rather than by the voltage on it), or by the frequency of a sinusoidal waveform. These and other formats are used in certain situations, these being chosen to suit the application.

Digital signals may also vary in form. Figure 11.3(b) shows a signal that takes a number of discrete levels. It could be that this signal represents numerical information, such as the number of people in a building. Since the signal changes abruptly from one value to another it is digital in nature, and in many cases such signals have a limited number of allowable values. The most common forms of digital signals are those that have only two possible values, as shown in Figure 11.3(c). Such signals, which are described as *binary* signals, are widely used since they are produced by many simple sensors and can be used to control many forms of actuator. For example, a simple domestic light switch has two possible states (ON and OFF), and therefore the voltage controlled by such a switch can be seen as a binary signal representing the required state of the lights. In such an arrangement the light bulb represents the actuator of the arrangement, which in this case is used in only two possible states (again, ON and OFF). Many electrical and electronic systems are based on the use of this form of ON/OFF control and therefore all make use of binary signals of one form or another. However, binary signals are also used in more sophisticated systems (such as those based on computers), since such signals are very easy to process, store and communicate. We will return to look at these issues in later chapters.

11.6 System block diagrams

It is often convenient to represent a complex arrangement by a simplified diagram that shows the system as a set of modules or blocks. This modular approach hides unnecessary detail and often aids comprehension. An example of a typical block diagram is shown in Figure 11.4. This shows a simplified representation of an engine control unit (ECU) that might be found in a car. This diagram shows the major components of the system and indicates the flow of energy or information between the various parts. The arrows in the diagram indicate the direction of flow.

When energy or information flows from a component we often refer to that component as the *source* of that energy or information. Similarly, when energy or information flows into a component, we often say that the component represents a *load* on the arrangement. Thus in Figure 11.4 we could consider the various sensors and the power supply to be sources for the ECU and the ignition coil to be a load.

In electrical systems a flow of energy requires an electrical *circuit*. Figure 11.5 shows a simple system with a single source and a single load. In this figure the source

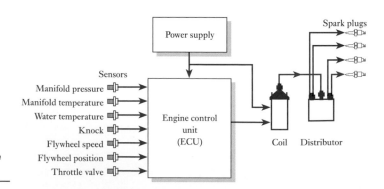

Figure 11.4
An automotive engine
control unit (ECU).

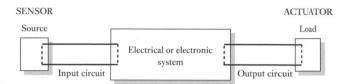

Figure 11.5
Sources and loads.

of energy is some form of sensor, and the load is some form of actuator, perhaps representing a part of the ECU of Figure 11.4. In any event the source is linked to the system by an *input circuit*, and the load is connected by an *output circuit*.

We noted earlier that we are free to choose the boundaries of our system to suit our needs. We might therefore choose to divide the system of Figure 11.5 into a number of subsystems, or modules, as shown in Figure 11.6. This process is referred to as *partitioning* and can greatly simplify the design of complex systems. It can be seen that the output of each of these subsystems represents the input of the next. Thus the output of each module represents a *source*, while the input of each module represents a *load*. In the arrangements of Figures 11.5 and 11.6, each of the various modules has a single input and a single output. In practice, modules may have multiple inputs and outputs depending on the function of that part of the system.

We noted in the early stages of this chapter that a system may be defined solely by its inputs, its outputs and the relationship between them. It follows that each of the modules in our system can be defined solely in terms of the characteristics of the sources and loads that it represents and the relationship between its input and output signals. The design of such systems therefore involves the production of modules that take signals from appropriate input devices (be they sensors, generators or other modules) and produce from them appropriate signals to drive the relevant output devices (such as actuators). Therefore, before looking at the design of these circuits, we must first know something of the nature of the signals associated with these sensors and actuators. For this reason, we will look at these devices in Chapters 12 and 13.

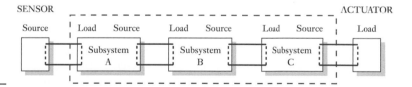

Figure 11.6
System partitioning.

Video 11B

Further study

One of the first tasks in the design of a new system is the identification of its various inputs and outputs. To investigate this process, consider the inputs and output associated with the controller of a domestic washing machine.

Identify the inputs and outputs of the unit and hence sketch a simple block diagram of the controller.

Key points

- Engineering is inherently interdisciplinary, and all engineers should have an understanding of the basic principles of electrical and electronic engineering, if only so that they can talk intelligently to others who are specialists in this area.

- Engineers often adopt a 'systems approach' to design, which combines top-down *systematic* techniques with multidisciplinary *systemic* methods.

- As far as its appearance from outside is concerned, a system can be defined solely by its inputs, its outputs and the relationship between them.

- Systems interact with the outside world through the use of *sensors* and *actuators*.

- Physical quantities may be either continuous or discrete. It is often convenient to represent physical quantities by electrical signals. These may also be either continuous or discrete. Continuous signals are normally referred to as *analogue*, while discrete signals are usually described as *digital*.

- Complex systems are often represented by block diagrams. These hide unnecessary detail and can aid comprehension.

- Energy or information flows from a *source* and flows into a *load*. Any module presents a *load* to whatever is connected to its input and represents a *source* to whatever is connected to its output.

- In order to design electrical or electronic systems, we need to understand the nature of the signals *produced* by the sensors and *used* by the actuators that form their input and output devices.

Exercises

11.1   List 10 fields of engineering that might be associated with the construction of a railway system.

11.2   Explain the distinction between a systematic approach and a systemic approach to design. Which of these methods is associated with a systems approach?

11.3   Describe briefly what is meant by a system.

11.4   Identify examples of systems that are electrical, mechanical, hydraulic, pneumatic and biological, and in each case describe the nature of the inputs and outputs.

11.5   Explain why the choice of a system's boundaries affects the form of its inputs and outputs.

11.6   Identify five naturally occurring physical quantities not mentioned in the text that are continuous in nature.

11.7   Identify five naturally occurring physical quantities not mentioned in the text that are discrete in nature.

11.8   Give an example of a situation where a continuous physical quantity is represented by a digital signal.

11.9   Give an example of a situation where a discrete physical quantity is represented by an analogue signal.

11.10 Describe what is meant by 'partitioning' with respect to the design of an electronic system.

11.11 Explain how a module in an electrical system may be described in terms of 'sources' and 'loads'.

Chapter 12 Sensors

Objectives

When you have studied the material in this chapter, you should be able to:

- discuss the role of sensors in electronic systems
- outline the requirement for a range of sensors of different types to meet the needs of varied applications
- explain the meaning of terms such as range, resolution, accuracy, precision, linearity and sensitivity, as they apply to sensors
- describe the operation and characteristics of a variety of devices for sensing various physical quantities
- give examples from the diversity of sensing devices available and outline the different characteristics of these components
- discuss the need for interfacing circuitry to make the signals produced by sensors compatible with the systems to which they are connected.

12.1 Introduction

In order to perform useful tasks electronic systems must interact with the world about them. To do this they use **sensors** to sense external physical quantities and **actuators** to affect or control them.

Sensors and actuators are often referred to as transducers. A **transducer** is a device that converts one physical quantity into another, and different transducers convert between a wide range of physical quantities. Examples include a mercury-in-glass thermometer, which converts variations in temperature into variations in the length of a mercury column, and a microphone, which converts sound into electrical signals.

In this text we are primarily interested in transducers that are used in electronic systems, so we are mainly interested in devices that produce or use electrical signals of some form. Transducers that convert physical quantities into electrical signals will normally be used to produce inputs for our system and will therefore be referred to as *sensors*. Transducers that take electrical input signals and control or affect an external physical quantity will be referred to as *actuators*. In this chapter we will look at the characteristics of sensors and in the next we will consider actuators.

Thermometers and microphones are both examples of sensors that convert one form of analogue quantity into another. Other sensors can be used with digital quantities, converting one digital quantity into another. Such systems include all forms of event counter, such as those used to count the number of people going through a turnstile.

A third class of sensors take an analogue quantity and represent it in a digital form. In some instances the output is a simple binary representation of the input, as in a thermostat, which produces one of two output values depending on whether a temperature is above or below a certain threshold. In other devices the analogue quantity at the input is represented by a multi-valued output, as in the case of a digital voltmeter, where an analogue input quantity is represented by a numerical (and therefore discrete) output. Representing an analogue quantity by a digital quantity is, by necessity, an approximation. However, if the number of allowable discrete states is sufficient, the representation may be adequate for a given application. Indeed, in many cases the error caused by this approximation is small compared with the noise or other errors within the system and can therefore be ignored.

For completeness one should say that there is a final group of sensors that take a digital input quantity and use this to produce an analogue output. However, such components are less common, and there are very few widely used examples of such devices.

Almost any physical property of a material that varies in response to some excitation can be used to produce a sensor. Commonly used devices include those whose operation is:

- resistive
- inductive
- capacitive
- piezoelectric
- photoelectric
- elastic
- thermal.

The range of sensing devices available is vast, and in this chapter we will restrict ourselves to a few examples that are widely used in electronic systems. The examples chosen have been selected to show the diversity of devices available and to illustrate some of their characteristics. These examples include sensors for a variety of physical quantities and devices that are both analogue and digital in nature. However, before we start looking at individual devices it is appropriate to consider how we quantify the performance of such components.

12.2   Describing sensor performance

When describing sensors and instrumentation systems we make use of a range of terms to quantify their characteristics and performance. It is important to have a clear understanding of this terminology, so we will look briefly at some of the more important terms.

12.2.1   Range

This defines the maximum and minimum values of the quantity that the sensor or instrument is designed to measure.

12.2.2   Resolution or discrimination

This is the smallest discernible change in the measured quantity that the sensor is able to detect. This is usually expressed as a percentage of the range of the device; for example, the resolution might be given as 0.1 per cent of the full-scale value (that is, one-thousandth of the range).

12.2.3 Error

This is the difference between a measured value and its true value. Errors may be divided into random errors and systematic errors. **Random errors** produce *scatter* within repeated readings. The effects of such errors may be quantified by comparing multiple readings and noting the amount of scatter present. The effects of random errors may also be reduced by taking the average of these repeated readings. **Systematic errors** affect all readings in a similar manner and are caused by factors such as mis–calibration. Since all readings are affected in the same way, taking multiple readings does not allow quantification or reduction of such errors.

12.2.4 Accuracy, inaccuracy and uncertainty

The term *accuracy* describes the maximum expected error associated with a measurement (or a sensor) and may be expressed as an absolute value or as a percentage of the range of the system. For example, the accuracy of a vehicle speed sensor might be given as ±1 mph or as ±0.5 per cent of the full-scale reading. Strictly speaking, this is actually a measure of its *inaccuracy*, and for this reason the term *uncertainty* is sometimes used.

12.2.5 Precision

This is a measure of the lack of random errors (scatter) produced by a sensor or instrument. Devices with high precision will produce repeated readings with very little spread. It should be noted that precision is very often confused with accuracy, which has a very different meaning. A sensor might produce a range of readings that are very consistent but that are all very inaccurate. This is illustrated in Figure 12.1, which shows the performance of three sensor systems. The figure shows the readings that might be produced by sensors of different characteristics when measuring the position of an object in two dimensions (*x* and *y*).

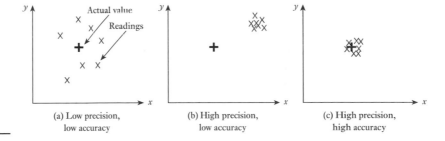

Figure 12.1
Accuracy and precision.

12.2.6 Linearity

In most situations it is convenient to have a sensor where the output is linearly proportional to the quantity being measured. If one were to plot a graph of the output of a sensor against the measured quantity, a perfectly linear device would produce a straight line going through the origin. In practice real sensors will have some non-linearity, which is defined as the maximum deviation of any reading from this straight line. Non-linearity is normally expressed as a percentage of the full-scale value.

12.2.7 Sensitivity

This is a measure of the change produced at the output for a given change in the quantity being measured. A sensor that has high sensitivity will produce a large change in its output for a given input change. The units of this measure reflect the nature of the measured quantity. For example, for a temperature sensor the sensitivity might be given as 10 mV/°C, meaning that the output would change by 10 mV for every 1 °C change in temperature.

12.3 Temperature sensors

The measurement of temperature is a fundamental part of a large number of control and monitoring systems, ranging from simple temperature-regulating systems for buildings to complex industrial process-control plants.

Temperature sensors may be divided into those that give a simple binary output to indicate that the temperature is above or below some threshold value and those that allow temperature measurements to be made.

Binary output devices are effectively temperature-operated switches, an example being the **thermostat**, which is often based on a **bimetallic strip**. This is formed by bonding together two materials with different thermal expansion properties. As the temperature of the bimetallic strip increases it bends, and this deflection is used to operate a mechanical switch.

A large number of different techniques are used for temperature measurement, but here we will consider just three forms.

12.3.1 Resistive thermometers

The electrical resistance of all conducting materials changes with temperature. The resistance of a piece of metal varies linearly with its absolute temperature. This allows temperature to be measured by determining the resistance of a sample of the metal and comparing it with its resistance at a known temperature. Typical devices use platinum wire; such devices are known as **platinum resistance thermometers** or **PRT**s.

PRTs can produce very accurate measurements at temperatures from less than −150 °C to nearly 1000 °C to an accuracy of about 0.1 °C, or 0.1 per cent. However, they have poor **sensitivity**. That is, a given change in the input temperature produces only a small change in the output signal. A typical PRT might have a resistance of 100 Ω at 0 °C, which increases to about 140 Ω at 100 °C. Figure 12.2(a) shows a typical PRT element. PRTs are also available in other forms, such as the probe shown in Figure 12.2(b).

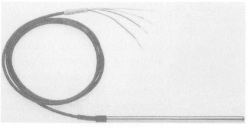

Figure 12.2
Platinum resistance thermometers (PRTs).

(a) A typical PRT element   (b) A sheathed PRT

(a) A typical disc thermistor   (b) A threaded thermistor

Figure 12.3
Thermistors.

Figure 12.4
A *pn* junction
temperature sensor.

12.3.2 Thermistors

Like PRTs, these devices also change their resistance with temperature. However, they use materials with high thermal coefficients of resistance to give much improved sensitivity. A typical device might have a resistance of 5 kΩ at 0 °C and 100 Ω at 100 °C. Thermistors are inexpensive and robust but are very non-linear and often suffer from great variability in their nominal value between devices. Figure 12.3(a) shows a typical disc thermistor, while Figure 12.3(b) shows a device incorporating a threaded section for easy attachment.

12.3.3 *pn* junctions

A *pn* junction is a semiconductor device that has the properties of a **diode**. That is, it conducts electricity in one direction (when the device is said to be *forward-biased*) but opposes the flow of electricity in the other direction (when the device is said to be *reverse-biased*). The properties and uses of semiconductor devices will be discussed in more detail later (see Chapter 17).

At a fixed current, the voltage across a typical forward-biased semiconductor diode changes by about 2 mV per °C. Devices based on this property use additional circuitry to produce an output voltage or current that is directly proportional to the junction temperature. Typical devices might produce an output voltage of 1 mV per °C, or an output current of 1 μA per °C, for temperatures above 0 °C. These devices are inexpensive, linear and easy to use but are limited to a temperature range from about −50 °C to about 150 °C by the semiconductor materials used. Such a device is shown in Figure 12.4.

12.4   Light sensors

Sensors for measuring light intensity fall into two main categories: those that generate electricity when illuminated and those whose properties (for example, their resistance) change under the influence of light. We will consider examples of both of these classes of device.

12.4.1 Photovoltaic

Light falling on a *pn* junction produces a voltage and can therefore be used to generate power from light energy. This principle is used in solar cells. On a smaller scale, **photodiodes** can be used to measure light intensity, since they produce an output voltage that depends on the amount of light falling on them. A disadvantage of this method of measurement is that the voltage produced is not related linearly to the incident light intensity. Figure 12.5(a) shows an example of a typical photodiode light sensor.

Figure 12.5
Light sensors.

(a) A photodiode   (b) A light-dependent resistor (LDR)

12.4.2 Photoconductive

Photoconductive sensors do not generate electricity, but their conduction of electricity changes with illumination. The photodiode described above as a photovoltaic device may also be used as a photoconductive device. If a photodiode is reverse biased by an external voltage source, in the absence of light it will behave like any other diode and conduct only a negligible leakage current. However, if light is allowed to fall on the device, charge carriers will be formed in the junction region and a current will flow. The magnitude of this current is proportional to the intensity of the incident light, making it more suitable for measurement than the photovoltaic arrangement described earlier.

The currents produced by photodiodes in their photoconductive mode are very small. An alternative is to use a **phototransistor**, which combines the photoconductive properties of the photodiode with the current amplification of a transistor to form a device with much greater sensitivity. The operation of transistors will be discussed in later chapters.

A third class of photoconductive device is the **light-dependent resistor** or **LDR**. As its name implies, this is a resistive device that changes its resistance when illuminated. Typical devices are made from materials such as cadmium sulphide (CdS) which have a much lower resistance when illuminated. One advantage of these devices in some applications is that they respond to different wavelengths of light in a manner similar to the human eye. Unfortunately, their response is very slow, taking perhaps 100 ms to respond to a change in illumination compared with a few microseconds, or less, for the semiconductor junction devices. A typical LDR is shown in Figure 12.5(b).

In addition to sensors that measure light intensity there are also a large number of sensors that use light to measure other quantities, such as position, motion and temperature. We will look at an example of such a sensor in Section 12.6 when we consider opto–switches.

12.5 | Force sensors

12.5.1 Strain gauge

The resistance between opposite faces of a rectangular piece of uniform electrically conducting material is proportional to the distance between the faces and inversely proportional to its cross-sectional area. The shape of such an object may be changed by applying an external force to it. The term **stress** is used to define the force per unit area applied to the object, and the term **strain** refers to the deformation produced. In a strain gauge, an applied force deforms the sensor, increasing or decreasing its length (and its cross-section) and therefore changing its resistance. Figure 12.6 shows the construction of a typical device.

Direction of sensitivity

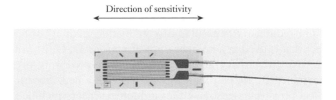

Figure 12.6
A strain gauge.

The gauge is in the form of a thin layer of resistive material arranged to be sensitive to deformation in only one direction. The long thin lines of the sensor are largely responsible for the overall resistance of the device. Stretching or compressing the gauge in the direction shown will extend or contract these lines and will have a marked effect on the total resistance. The comparatively thick sections joining these lines contribute little to the overall resistance of the unit. Consequently, deforming the gauge perpendicular to the direction shown will have little effect on the total resistance of the device.

In use, the gauge is bonded to the surface in which strain is to be measured. The fractional change in resistance is linearly related to the applied strain. If it is bonded to a structure with a known stress-to-strain characteristic, the gauge can be used to measure force. Thus it is often found at the heart of many force transducers or **load cells**. Similarly, strain gauges may be connected to diaphragms to produce **pressure sensors**.

12.5.2 Piezoelectric

Piezoelectric materials have the characteristic that they generate an electrical output when subjected to mechanical stress. Unfortunately, the output is not a simple voltage proportional to the applied force, but an amount of electric charge which is related to the applied stress. For most applications this requires some electronic circuitry to convert the signal into a more convenient voltage signal. This is not in itself difficult, but such circuits tend to suffer from 'drift', that is a gradual increase or decrease in the output. For this reason, piezoelectric transducers are more commonly used for measuring variations in forces, rather than their steady value.

12.6 | Displacement sensors

Displacement or position may be sensed using a very wide range of methods, including resistive, inductive, mechanical and optical techniques. As with many classes of sensor, both analogue and digital types are used.

12.6.1 Potentiometers

Resistive potentiometers are among the most common position transducers, and most people will have encountered them as the controls used in radios and other electronic equipment. Potentiometers may be angular or linear, consisting of a length of resistive material with an electrical terminal at each end and a third terminal connected to a sliding contact on the resistive *track*. When used as a position transducer, a potential is placed across the two end terminals and the output is taken from the terminal connected to the sliding contact. As the sliding contact moves the output voltage changes between the potentials on each end of the track. Generally, there is a linear relationship between the position of the slider and the output voltage.

12.6.2 Inductive sensors

Inductive sensors were discussed briefly in Section 5.14.

The inductance of a coil is affected by the proximity of ferromagnetic materials, an effect that is used in a number of position sensors. One of the simplest of these is the inductive **proximity sensor**, in which the proximity of a ferromagnetic plate is determined by measuring the inductance of a coil. Figure 12.7 shows examples of typical proximity sensors. Other inductive sensors include the **linear variable differential transformer** or **LVDT** (discussed in Section 5.14.2).

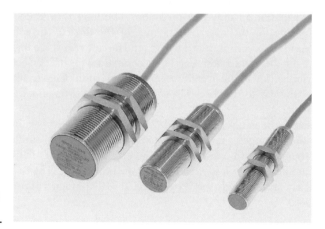

Figure 12.7
Inductive displacement or proximity sensors.

12.6.3 Switches

The simplest digital displacement sensors are mechanical switches. These are used in many forms and may be manually operated or connected to a mechanism of some kind. Manually operated switches include toggle switches, which are often used as power ON/OFF switches on electrical equipment, and momentary-action pushbutton switches as used in computer keyboards. It may not be immediately apparent that switches of this type are position sensors, but clearly they output a value dependent on the position of the input lever or surface and are therefore binary sensors.

When a switch is connected to some form of mechanism, its action as a position sensor becomes more obvious. A common form of such a device is the **microswitch**, which consists of a small switch mechanism attached to a lever or push rod, allowing it to be operated by some external force. Microswitches are often used as **limit switches**, which signal that a mechanism has reached the end of its safe travel. Such an arrangement is shown in Figure 12.8(a). Switches are also used in a number of specialised

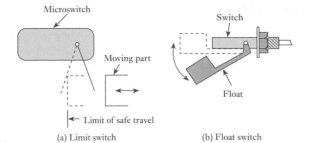

Figure 12.8
Switch position
sensors.

(a) Limit switch   (b) Float switch

position-measuring applications, such as liquid-level sensors. One form of such a sensor is shown in Figure 12.8(b). Here the switch is operated by a float, which rises with the liquid until it reaches some specific level.

12.6.4 Opto-switches

In addition to the use of mechanical switches, position can also be sensed using devices such as the **opto-switch**, which, as its name suggests, is a light-operated switch.

The opto-switch consists of a light sensor, usually a phototransistor, and a light source, usually a light-emitting diode (LEDs will be described in the next chapter), housed within a single package. Two physical arrangements are widely used, as illustrated in Figure 12.9.

Figure 12.9(a) shows a reflective device in which the light source and sensor are mounted adjacent to each other on one face of the unit. The presence of a reflective object close to this face will cause light from the source to reach the sensor, causing current to flow in the output circuit. Figure 12.9(b) shows a slotted opto-switch in which the source and sensor are arranged to face each other on either side of a slot in the device. In the absence of any object in the slot, light from the source will reach the sensor, and this will produce a current in the output circuit. If the slot is obstructed, the light path will be broken and the output current will be reduced.

Although opto-switches may be used with external circuitry to measure the current flowing and thus to determine the magnitude of the light reaching the sensor, it is more common to use them in a binary mode. In this arrangement, the current is compared with some threshold value to decide whether the opto-switch is ON or OFF. In this way, the switch detects the presence or absence of objects, the threshold value being adjusted to vary the sensitivity of the arrangement. We will consider some applications of the opto-switch later in this section.

Figure 12.9
Reflective and slotted
opto-switches.

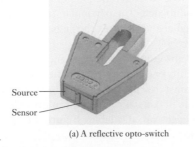

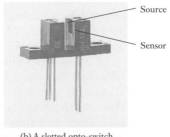

(a) A reflective opto-switch   (b) A slotted opto-switch

Video 12A

12.6.5 Absolute position encoders

Figure 12.10 illustrates the principle of a simple linear absolute position encoder. A pattern of light and dark areas is printed onto a strip and is detected by a sensor that moves along it. The pattern takes the form of a series of lines that alternate between light and dark. It is arranged so that the combination of light and dark areas on the various lines is unique at each point along the strip. The sensor, which may be a linear array of phototransistors or photodiodes, one per line, picks up the pattern and produces an appropriate electrical signal, which can be decoded to determine the sensor's position. The combination of light and dark lines at each point represents a **code** for that position. The choice of codes and their use will be discussed in more detail later (see Chapter 24).

Since each point on the strip must have a unique code, the number of distinct positions along the strip that can be detected is determined by the number of lines in the pattern. For a sensor of a given length, increasing the number of lines in the pattern increases the resolution of the device but also increases the complexity of the detecting array and the accuracy with which the lines must be printed.

Although linear absolute encoders are available, the technique is more commonly applied to angular devices. These often resemble rotary potentiometers, but they have a coded pattern in a series of concentric rings in place of the conducting track and an array of optical sensors in place of the wiper. Position encoders have excellent linearity and a long life, but they generally have poorer resolution than potentiometers and are usually more expensive.

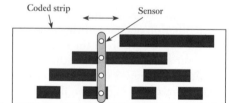

Figure 12.10
An absolute position encoder.

12.6.6 Incremental position encoders

The incremental encoder differs from the absolute encoder in that it has only a single detector, which scans a pattern consisting of a regular series of stripes perpendicular to the direction of travel. As the sensor moves over the pattern, the sensor will detect a series of light and dark regions. The distance moved can be determined by counting the number of transitions. One problem with this arrangement is that the direction of motion cannot be ascertained, as motion in either direction generates similar transitions between light and dark. This problem is overcome by the use of a second sensor, slightly offset from the first. The direction of motion may now be determined by noting which sensor is first to detect a particular transition. This arrangement is shown in Figure 12.11, which also illustrates the signals produced by the two sensors for motion in each direction.

In comparison with the absolute encoder, the incremental encoder has the disadvantage that external circuitry is required to count the transitions, and that some method of resetting this must be provided to give a reference point or datum. However, the device is simple in construction and can provide high resolution. Again, both linear and angular devices are available. Figure 12.12 shows a small angular incremental position encoder.

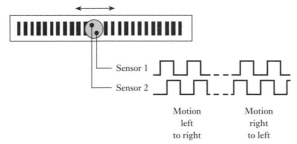

Figure 12.11
An incremental position encoder.

Figure 12.12
An angular incremental position encoder.

12.6.7 Optical gratings

The incremental encoder described above relies on counting individual lines in a pattern of stripes. To measure very small displacements these lines must be very close together, making them difficult to detect. One approach to this problem is to use optical gratings to simplify the task.

Optical gratings can take many forms, a simple version being formed by printing a pattern of opaque stripes on a transparent film. The stripes are parallel, and have lines and spaces of equal widths. If one piece of this film is placed on top of another, the pattern produced will depend on the relative positions and orientations of the stripes on the two films. If the lines of each sheet are parallel, the overall effect will depend on the relative positions of the stripes. If the lines of each coincide precisely, half the sheet will be transparent and half opaque. If, however, the lines of each sheet are side by side, the combination will be completely opaque.

Imagine now that one grating is placed on a white background and that a second grating is placed on top of the first such that the lines of each are parallel. Movement of one sheet perpendicular to the direction of the stripes will result in an overall pattern which alternates from dark to light as the stripes pass over each other. A light sensor placed above the gratings will detect these variations and by counting the number of transitions the distance moved can be determined. Since one bright 'pulse' is produced each time the lines coincide, the distance travelled is the product of the number of pulses counted and the spacing of the lines. These bands of light and dark regions are known as **moiré fringes**. The closer together the lines, the greater will be the resolution of the distance measurement.

The measurement technique thus far described suffers from the same problem as the incremental encoder described above, in that if a single sensor is used, motion in either direction generates similar signals at the sensor. This problem can be tackled in two ways. One of the gratings can be rotated slightly so that the lines of each are no longer parallel. Relative motion will now produce diagonal bands of light which move up, for motion in one direction, and down, for motion in the other direction. Alternatively, the line spacing of one of the sheets can be changed so that it is slightly different from the other. When placed together, as before, this will produce a wave of light and dark bands called **vernier fringes** in a direction perpendicular to the stripes. As the films move with respect to each other, these bands move in a direction determined by the direction of motion. In both methods a second sensor is used to allow the direction of motion to be detected, and in each case signals are produced which are similar to those produced by the incremental encoder, as shown in Figure 12.11.

Practical displacement-measuring systems use gratings which are produced photographically allowing a very high resolution. A typical application might use a linear array fixed to a static component and a small moving sensor containing a grating assembly with integrated light sensors. Line spacings down to 1 μm are readily obtainable, although line spacings of 10 μm to 20 μm are typical, with interpolation being used to obtain a measurement resolution of about 1 μm. Gratings of this type are produced in lengths of up to about 1 m, but may be joined (with some loss of accuracy) to produce greater lengths. The ability to measure distances of the order of a metre to a resolution of the order of a micron makes the use of gratings extremely attractive in some demanding applications. However, the high cost of the gratings and sensors limits their use.

12.6.8 Other counting techniques

Incremental encoders employ event counting to determine displacement. Several other techniques use this method, and Figure 12.13 shows two examples.

Figure 12.13(a) shows a technique that uses an inductive proximity sensor, as described earlier in this section. Here a ferromagnetic gear wheel is placed near the sensor; as the wheel rotates the teeth pass close to the sensor, increasing its inductance. The sensor can therefore detect the passage of each tooth and thus determine the distance travelled. A great advantage of this sensor is its tolerance to dirty environments.

Figure 12.13(b) shows a sensor that uses the slotted opto-switch discussed earlier. This method uses a disc that has a number of holes or slots spaced equally around its perimeter. The disc and opto-switch are mounted such that the edge of the disc is within the slot of the switch. As the disc rotates, the holes or slots cause the opto-switch to be periodically opened and closed, producing a train of pulses with a frequency determined by the speed of rotation. The angle of rotation can be measured by counting the number of pulses. A similar method uses an inductive proximity sensor in place of the opto-switch, and a ferromagnetic disc.

Figure 12.13
Examples of displacement sensors using counting.

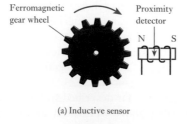

(a) Inductive sensor

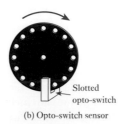

(b) Opto-switch sensor

12.6.9 Rangefinders

Measurement of large distances usually requires a non-contact method. Both passive systems (which simply observe their environment) and active systems (which send signals out into the environment) are available. Passive techniques include optical triangulation methods, in which two slightly displaced sights are aligned on a common target. The angular difference between the two sights can then be measured using one of the angular sensors described above. Trigonometry is then used to calculate the distance between the sights and the target. This method is employed in rangefinding equipment used for surveying. Active systems transmit either sound or electromagnetic energy and detect the energy reflected from a distant object. By measuring the time taken for the energy to travel to the object and back to the transmitter, the distance between them may be determined. Because the speed of light is so great, some optical systems use the phase difference between the transmitted and received signals, rather than time of travel, to determine the distance.

12.7 Motion sensors

In addition to the measurement of displacement, it is often necessary to determine information concerning the motion of an object, such as its velocity or acceleration. These quantities may be obtained by differentiation of a position signal with respect to time, although such techniques often suffer from noise, since differentiation tends to amplify high-frequency noise present in the signal. Alternatively, velocity and acceleration can be measured directly using a number of sensors.

The counting techniques described earlier for the measurement of displacement can also be used for velocity measurement. This is achieved by measuring the frequency of the waveforms produced instead of counting the number of pulses. This gives a direct indication of speed. In fact, many of the counting techniques outlined earlier are more commonly used for speed measurement than for measurement of position. In many applications the direction of motion is either known or is unimportant, and these techniques often provide a simple and inexpensive solution.

A range of other velocity sensors exist for different applications. A **tacho-generator** can be used to measure rotational speed. This is a small DC generator which produces a voltage proportional to its speed of rotation. Linear motion can be measured by converting it to a rotational movement (for example, by using a friction wheel running along a flat surface) and applying this to a tacho-generator. Alternatively, there are several methods for measuring linear motion directly, such as those employing the **Doppler effect** as used in 'radar' speed detectors. Here a beam of sound or electromagnetic radiation is directed at the moving object and the reflected radiation is detected and compared with the original transmission. The difference between the frequencies of the outgoing and reflected waveforms gives a measure of the relative speed of the object and the transducer. The velocity of fluids may be measured in many ways including pressure probe, turbine, magnetic, sonic and laser methods. These techniques are highly specialised and will not be discussed here.

Direct measurement of acceleration is made using an **accelerometer**. Most accelerometers make use of the relationship between force, mass and acceleration:

$$\text{force} = \text{mass} \times \text{acceleration}$$

A mass is enclosed within the accelerometer. When the device is subjected to acceleration the mass experiences a force, which can be detected in a number of ways. In some devices a force transducer, such as a strain gauge, is incorporated to measure the force directly. In others, springs are used to convert the force into a corresponding displacement, which is then measured with a displacement transducer. Because of the different modes of operation of the devices, the form of the output signal also varies.

12.8   Sound sensors

A number of techniques are used to detect sound. Since sound represents variations in air pressure, the objective of the microphone is to measure these variations and to represent them by some form of electrical signal (often in the form of a varying voltage). This process is illustrated in Figure 12.14.

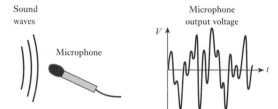

Figure 12.14
A microphone.

12.8.1   Carbon microphones

Carbon microphones are one of the oldest and simplest forms of sound detector. Sound waves are detected by a *diaphragm* which forms one side of an enclosure containing carbon particles. Sound waves striking the diaphragm cause it to move, compressing the carbon particles to a greater or lesser degree and thus affecting their resistance. Electrodes apply a voltage across the particles and the resulting current thus relates to the sound striking the device.

12.8.2   Capacitive microphones

Capacitive microphones are similar in operation to the carbon microphone described above except that movement of the diaphragm causes a variation in capacitance rather than resistance. This is achieved by arranging that motion of the diaphragm changes the separation of two plates of a capacitor, thereby varying its capacitance.

12.8.3   Moving-coil microphones

A moving-coil microphone consists of a permanent magnet and a coil connected to a diaphragm. Sound waves move the diaphragm which causes the coil to move with respect to the magnet, thus generating an electrical signal. Moving-coil devices are probably the most common form of microphone.

12.8.4   Piezoelectric microphones

The piezoelectric force sensor described earlier can also be used as a microphone. The diaphragm is made of piezoelectric material which is distorted by sound waves producing a corresponding electrical signal. This technique is often used for **ultrasonic sensors** which are used over a wide range of frequencies, sometimes up to many megahertz.

Sensor interfacing

Many electronic systems require their inputs to be in the form of electrical signals in which the voltage or current of the signal is related to the physical quantity being sensed. While some sensors produce an output voltage or current that is directly related to the physical quantity being measured, others require additional circuitry to generate such signals. The process of making the output of one device compatible with the input of another is often referred to as **interfacing**. Fortunately, the circuitry required is usually relatively simple, and this section gives a few examples of such techniques.

12.9.1 Resistive devices

In a potentiometer, the resistance between the central moving contact and the two end terminals changes as the contact is moved. This arrangement can easily be used to produce an output voltage that is directly related to the position of the central contact. If a constant voltage is placed across the outer terminals of a potentiometer, the voltage produced on the centre contact varies with its position. If the resistance of the track varies linearly, then the output voltage will be directly proportional to the position of the centre contact, and hence to the input displacement.

Many sensors represent changes in a physical quantity by changes in resistance. Examples include platinum resistance thermometers, photoconductive sensors and some forms of microphone. One way of converting a changing resistance into a changing voltage is to use the sensor in a potential divider circuit, as illustrated in Figure 12.15(a), where R_s represents the variable resistance of the sensor.

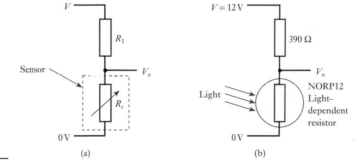

Figure 12.15
Using a resistive sensor in a potential divider.

The output voltage V_o of this arrangement is given by the expression

$$V_o = V \frac{R_s}{R_1 + R_s}$$

and clearly the output voltage V_o varies with the sensor resistance R_s. An example of the use of this arrangement is shown in Figure 12.15(b), which depicts a simple light meter based on a light-dependent resistor (LDR). Light falling on the resistor affects its resistance (as discussed in Section 12.4.2), which in turn determines the output voltage of the circuit. The LDR shown changes its resistance from about 400 Ω (at 1000 lux) to about 9 kΩ (at 10 lux), which will cause the output voltage V_o

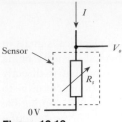

Figure 12.16
Using a resistive
sensor with a constant
current source.

to change from about 6 V to about 11.5 V in response to such changes in the light level.

While the arrangement of Figure 12.15(a) produces an output voltage that varies with the sensor resistance R_s, this is not a linear relationship. One way of producing a voltage that *is* linearly related to the resistance of a sensor is to pass a constant current through the device, as shown in Figure 12.16. From Ohm's law, the output of the circuit is given by

$$V_o = IR_s$$

and since I is constant, the output is clearly linearly related to the sensor voltage. The constant current I in the figure comes from some external circuitry – not surprisingly, such circuits are called **constant current sources**.

12.9.2 Switches

Most switches have two contacts, which are connected electrically when the switch is in one state (the closed state) and disconnected (or open circuit) when the switch is in the other state (the open state). This arrangement can be used to generate binary electrical signals simply by adding a voltage source and a resistance, as shown in Figure 12.17(a). When the switch is closed, the output is connected to the zero-volts line and therefore the output voltage V_o is zero. When the switch is open, the output is no longer connected to the zero-volts line but is connected through the resistance R to the voltage supply V. The output voltage will therefore be equal to the supply voltage minus any voltage drop across the resistance. This voltage drop will be determined by the value of the resistance R and the current flowing into the output circuit. If the value of R is chosen such that this voltage drop is small compared with V, we can use the approximation that the output voltage is zero when the switch is closed and V when it is open. The value chosen for R clearly affects the accuracy of this approximation; we will be looking at this in later chapters when we consider equivalent circuits.

One problem experienced by all mechanical switches is that of **switch bounce**. When the moving contacts in the switch come together, they have a tendency to bounce rather than to meet cleanly. Consequently, the electrical circuit is first made, then broken, then made again, sometimes several times. This is illustrated in Figure 12.17(b), which shows the output voltage of the circuit of Figure 12.17(a) as the switch is closed. The length of the oscillation will depend on the nature of the switch but might be a few milliseconds in a small switch and perhaps tens of milliseconds in a large circuit breaker. Switch bounce can cause severe problems, particularly if contact closures are being counted. Although good mechanical design can reduce this problem it cannot be eliminated, making it necessary to overcome this problem in other ways. Several

Figure 12.17
Generating a binary
signal using a switch.

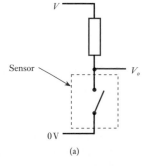

(a)

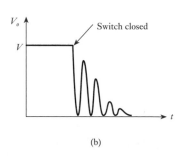

(b)

electronic solutions are possible, and it is also possible to tackle this problem using computer software techniques in systems that incorporate microcomputers.

Although the above discussion assumes the use of a mechanical switch, the circuit of Figure 12.17(a) can also be used with an opto–switch sensor. Optical switches do not produce a perfect 'closed circuit' when activated, but the effective resistance of the device does change dramatically between its ON and OFF states. Therefore, by choosing an appropriate value for the external resistance R, it can be arranged that the circuit produces a binary voltage signal that changes from approximately zero to approximately V volts, depending on the state of the switch. Optical switches do not suffer from switch bounce.

12.9.3 Capacitive and inductive devices

Sensors that change their capacitance or inductance in response to external influences normally require the use of AC circuitry. Such circuits need not be complicated, but they do involve techniques that are yet to be discussed in this text. We will therefore leave further consideration of such circuits until later.

12.9.4 Integration of sensors and signal processing

While simple sensors are used in a vast array of applications, in recent years there has been an increasing movement towards the integration of sensors with electronic circuitry that enhances their performance and functionality. In some cases this additional circuitry simply assists in the interfacing of the sensor or provides linearisation. However, in many cases the functionality provided is extensive and complex signal processing is performed.

In later chapters we will look at many of the techniques employed in the production of complex electronic circuits. In many cases, these techniques are also widely used in the production of sophisticated 'smart sensors'. For example, the methods and facilities used in the production of complex microprocessors are also used to produce devices that combine both sensing and signal-processing functions. Such devices are then used in a vast array of applications ranging from medical devices to cars, and from smart phones to aircraft.

12.10 Sensors – a summary

It is not the purpose of this chapter to provide an exhaustive list of all possible sensors. Rather, it sets out to illustrate some of the important classes of sensor that are available and to show the ways in which they provide information. It will be seen that some sensors generate output currents or voltages related to changes in the quantity being measured. In doing so they extract power from the environment and can deliver power to external circuitry (although usually the power available is small). Examples of such sensors are photovoltaic sensors and moving-coil microphones.

Other devices do not deliver power to external circuits but simply change their physical attributes, such as resistance, capacitance or inductance, in response to variations in the quantity being measured. Examples include resistive thermometers, photoconductive sensors, potentiometers, inductive position transducers and strain gauges. When using such sensors, external circuitry must be provided to convert the variation in the sensor into a useful signal. Often this circuitry is very simple, as illustrated in the last section.

Unfortunately, some sensors do not produce an output that is linearly related to the quantity being measured (for example, a thermistor). In these cases, it may be necessary to overcome the problem by using electronic circuitry or processing to compensate for any non-linearity. This process is called **linearisation**. The ease or difficulty of linearisation depends on the characteristics of the sensor and the accuracy required.

Example 12.1

Selecting an appropriate sensor for a computer mouse

In this chapter we have looked at a number of displacement and motion sensors. Armed with this information, we will select a suitable method of determining the displacement of a mouse for use as a computer pointing device. The resolution of the sensing arrangement should be such that the user can select an individual pixel (the smallest definable point within the display). A typical screen might have a resolution of 1024×768 pixels, or 2048×1536 pixels, although more sophisticated displays may have a resolution several times greater than this. Movement of the cursor from one side of the screen to the other should require a movement of the mouse of a few centimetres (the sensitivity of the mouse is often selectable using software within the computer).

Most modern mice use optical techniques to detect the motion of the mouse relative to the surface on which it is placed. These take images of the surface at regular intervals and compare them to determine how far, and in which direction, the mouse has moved. Optical mice require very sophisticated optics and advanced digital signal processing and will not be considered further here.

For the purposes of this example we will consider a non-optical mouse that senses motion using a small rubber ball that projects from its base. As the mouse is moved over a horizontal surface the ball rotates about two perpendicular axes, and this motion is used to determine the position of the cursor on a computer screen.

We have looked at several sensors that may be used to measure angular position. These include simple potentiometers and position encoders. Sensing the *absolute position* of the ball (and hence the mouse) could represent a problem, since for high-performance displays this could require a resolution of better than 1 part in 2000. Sensors with such a high resolution are often expensive and physically large. In this application, it is probably more appropriate to sense *relative* motion of the mouse. This reduces the complexity of the sensing mechanism and also means that the mouse is not tied to a fixed absolute position.

Measurement of the relative motion of the rubber ball suggests the use of some form of incremental sensor. This could use a proprietary incremental encoder, but because this is a very high-volume application, it is likely that a more cost-effective solution could be found. The diagram below shows a possible arrangement based on the use of slotted wheels and optical sensors (as described in Section 12.6).

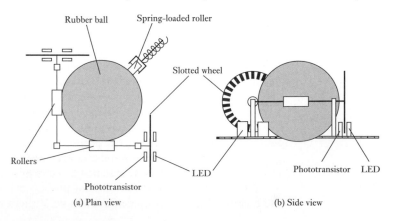

In order to resolve rotation of the ball into two perpendicular components, the ball is pressed against two perpendicular rollers by a third, spring-loaded roller. Rotation of the ball in a particular direction causes one or both of the sensing rollers to turn. Each of these rollers is connected to a slotted wheel that is placed between two slotted optical switches. The switches are positioned to allow the direction of rotation to be detected in a manner similar to that shown in Figure 12.11. The signals from the sensors are fed to the computer, which keeps track of the movement of the ball and hence determines the appropriate cursor position. This arrangement has a range limited only by the method used to count the moving slots. The sensitivity is determined by the relative sizes of the ball and the pulleys, and by the number of slots in the wheels.

Video 12B

Further study

Automotive engine management systems need to gather information about all aspects of the state and operation of the engine.

Identify a number of sensors that could be used to measure the speed of rotation of the engine, and in each case suggest the associated advantages and disadvantages. Hence identify an appropriate sensor for this application.

Key points

- A wide range of sensors are available to meet the needs of a spectrum of possible applications.
- Some sensors produce an output voltage or current that is related to the quantity being measured. They therefore supply power (albeit in small quantities).
- Other devices simply change their physical attributes, such as resistance, capacitance or inductance, in response to changes in the measured quantity.
- Interfacing circuitry may be required with some sensors to produce a signal in the desired form.
- Some sensors produce an output that is linearly related to the quantity being measured.
- Other devices are non-linear in operation.
- In some applications linearity is unimportant. For example, a proximity sensor may simply be used to detect the presence or absence of an object.
- In other applications, particularly where an accurate measurement is required, linearity is of more importance. In such applications, we will use either a sensor that has a linear characteristic or some form of linearisation to overcome non-linearities in the measuring device.

Exercises

12.1 Explain the meanings of the terms 'sensor', 'actuator' and 'transducer'.

12.2 What is meant by the resolution of a sensor?

12.3 Explain the difference between random and systematic errors.

12.4 Define the terms 'accuracy' and 'precision'.

12.5 Give an example of a digital temperature sensor.

12.6 What is the principal advantage and disadvantage of platinum resistance thermometers (PRTs) when making accurate temperature measurements?

12.7 A PRT has a resistance of $100\,\Omega$ at $0\,°C$ and a temperature coefficient of $+0.385\,\Omega$ per $°C$. What would be its resistance at $100\,°C$?

   The PRT is connected to an external circuit that measures the resistance of the sensor by passing a constant current of 10 mA through it and measuring the voltage across it. What would this voltage be at $100\,°C$?

12.8 The PRT described in the last exercise is connected as shown in the diagram below to form an arrangement where the output voltage V_o is determined by the temperature of the PRT.

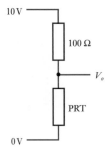

Derive an expression for V_o in terms of the temperature of the PRT.

   The resistance of the PRT is linearly related to its absolute temperature. Is V_o linearly related to temperature?

12.9 How do thermistors compare with PRTs?

12.10 *pn* junction temperature sensors are inexpensive, linear and easy to use. However, they do have certain limitations, which restrict their use. What are these limitations?

12.11 When using a photodiode as a light sensor, why might one choose to use this in a photoconductive mode rather than in a photovoltaic mode?

12.12 What is the advantage of a phototransistor light sensor in comparison with a photodiode sensor?

12.13 In what situations might one use a slow light-dependent resistor sensor in preference to a much faster photodiode or phototransistor sensor?

12.14 Explain the meanings of the terms 'stress' and 'strain'.

12.15 Suggest a suitable method for using a strain gauge to measure the vertical force applied to the end of a horizontal beam that is supported at one end.

12.16 Suggest a suitable method of employing two strain gauges to measure the vertical force applied to the end of a beam that is supported at one end. Why might this approach be used in preference to that described in the last exercise?

12.17 Describe two methods of measurement that would be suitable for a non-contact, automatic rangefinder for distances up to 10 m.

12.18 In an earlier exercise, we considered a PRT that has a resistance of 100 Ω at 0 °C and a temperature coefficient of +0.385 Ω per °C. If such a device is connected to a constant current source of 10 mA, in an arrangement as shown in Figure 12.16, what would be the output voltage of the arrangement at 0 °C?

What would be the sensitivity of this arrangement (in mV/°C) at temperatures above 0 °C?

12.19 The arrangement of Figure 12.17 produces an output of 0 V if the switch is closed and *V* if the switch is open. Devise a similar circuit that reverses these two voltages.

12.20 Suggest 10 physical quantities, not discussed in this chapter, that are measured regularly, giving in each case an application where this measurement is required.

Chapter 13 Actuators

Objectives

When you have studied the material in this chapter, you should be able to:

- discuss the need for actuators in electronic systems
- describe a range of actuators, both analogue and digital, for controlling various physical quantities
- explain the requirement for actuators with different properties for use in different situations
- describe the use of interface circuitry to match a particular actuator to the system that drives it.

13.1 Introduction

Sensors provide only half of the interaction required between an electronic system and its surroundings. In addition to being able to sense physical quantities in their environment, systems must also be able to affect the outside world in some way so that their various functions can be performed. This might require the system to move something, change its temperature or simply provide information via some form of display. All these functions are performed by **actuators**.

As with the sensors discussed in the last chapter, actuators are **transducers** since they convert one physical quantity into another. Here we are interested in actuators that take electrical signals from our system and use them to vary some external physical quantity. As one would expect, there are a large number of different forms of actuator, and it would not be appropriate to attempt to provide a comprehensive list of such devices. Rather, this chapter sets out to show the diversity of such devices and to illustrate some of their characteristics.

13.2 Heat actuators

Most heating elements may be considered as simple **resistive heaters**, which output the power that they absorb as heat. For applications requiring only a few watts of heat output, ordinary resistors of the appropriate power rating may be used. Special heating cables and elements are available for larger applications, which may dissipate many kilowatts.

Light actuators

Most lighting for general illumination is generated using conventional incandescent or fluorescent lamps. The power requirements of such devices can range from a fraction of a watt to hundreds or perhaps thousands of watts.

For signalling and communication applications, the relatively low speed of response of conventional lamps makes them unsuitable, and other techniques are required.

Video 13A

13.3.1 Light-emitting diodes

One of the most common light sources used in electronic circuits is the **light-emitting diode** or **LED**. This is a semiconductor diode constructed in such a way that it produces light when a current passes through it. A range of semiconductor materials can be used to produce infrared or visible light of various colours. Typical devices use materials such as gallium arsenide, gallium phosphide or gallium arsenide phosphide.

The characteristics of these devices are similar to those of other semiconductor diodes (which will be discussed in Chapter 17) but with different operating voltages. The light output from an LED is approximately proportional to the current passing through it; a typical small device might have an operating voltage of 2.0 V and a maximum current of 30 mA.

LEDs can be used individually or in multiple-element devices. One example of the latter is the LED **seven-segment display** shown in Figure 13.1. This consists of seven LEDs, which can be switched ON or OFF individually to display a range of patterns.

Infrared LEDs are widely used with photodiodes or phototransistors to enable short-range wireless communication. Variations in the current applied to the LED are converted into light with a fluctuating intensity, which is then converted back into a corresponding electrical signal by the receiving device. This technique is widely used in **remote control** applications for televisions and other domestic appliances. In these cases, the information transmitted is generally in a digital form. Because there is no electrical connection between the transmitter and the receiver, this technique can also be used to couple digital signals between two circuits that must be electrically isolated. This is called **opto-isolation**. Small self-contained **opto-isolators** are available that combine the light source and sensor in a single package. The input and output sections of these devices are linked only by light, enabling them to produce electrical isolation between the two circuits. This is particularly useful when the two circuits are operating at very different voltage levels. Typical devices will provide isolation of up to a few kilovolts.

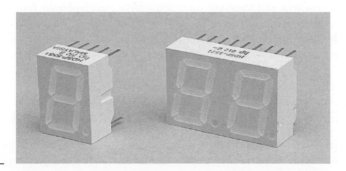

Figure 13.1
LED seven-segment displays.

13.3.2 Liquid crystal displays

Liquid crystal displays (LCDs) consist of two sheets of polarised glass with a thin layer of oily liquid sandwiched between them. An electric field is used to rotate the plane of polarisation of the liquid in certain regions, making some parts of the display opaque while others are transparent. The display segments can be arranged to create specific patterns (such as those of seven-segment displays) or in a matrix to allow any characters or images to be displayed.

A great advantage of LCDs (compared with LEDs) is that they are able to use ambient light, greatly reducing power consumption and allowing them to be used in a wide range of low-power applications. When insufficient ambient light is available they can also be backlit, although this increases their power consumption considerably. LCDs are widely used in wristwatches, mobile phones and many forms of battery-operated electronic equipment. They are also used in televisions, computer displays and other high-resolution applications. An example of a small LCD module is shown in Figure 13.2.

Figure 13.2
A liquid crystal display module.

13.3.3 Fibre-optic communication

For long-distance communication, the simple techniques used in television remote control units are not suitable as they are greatly affected by ambient light, that is, light present in the environment. This problem can be overcome by the use of a **fibre-optic cable**, which captures the light from the transmitter and passes it along the cable to the receiver without interference from external light sources. Fibres are usually made of either an optical polymer or glass. The former are inexpensive and robust, but their high attenuation makes them suitable for only short-range communications of up to about 20 metres.

Glass fibres have a much lower attenuation and can be used over several hundred kilometres, but they are more expensive than polymer fibres. For long-range communications, the power available from a conventional infrared LED is insufficient. In such applications **laser diodes** may be used. These combine the light-emitting properties of an LED with the light amplification of a laser to produce a high-power, coherent light source.

13.4   Force, displacement and motion actuators

In practice, actuators for producing force, displacement and motion are often closely related. A simple DC permanent magnet motor, for example, if opposed by an immovable object, will apply a force to that object determined by the current in the motor. Alternatively, if resisted by a spring, the motor will produce a displacement that is determined by its current and, if able to move freely, it will produce a motion related to the current. We will therefore look at several actuators that can be used to produce each of these outputs, as well as some that are designed for more specific applications.

13.4.1 Solenoids

A solenoid consists of an electric coil and a ferromagnetic slug that can move into, or out of, the coil. When a current is passed through the solenoid, the slug is attracted towards the centre of the coil with a force determined by the current in the coil. The motion of the slug may be opposed by a spring to produce a displacement output, or the slug may simply be free to move. Most solenoids are linear devices, the electric current producing a linear force/displacement/motion. However, rotational solenoids are also available that produce an angular output. Both forms may be used with a continuous analogue input, or with a simple ON/OFF (digital) input. In the latter case, the device is generally arranged so that when it is energised (that is, turned ON) it moves in one direction until it reaches an end stop. When de-energised (turned OFF) a return spring forces it to the other end of its range of travel, where it again reaches an end stop. This produces a binary position output in response to a binary input. Figure 13.3 shows examples of small linear solenoids.

Figure 13.3
Small linear solenoids.

13.4.2 Meters

Panel meters are important output devices in many electronic systems providing a visual indication of physical quantities. Although there are various forms of panel meter, one of the simplest is the **moving-iron meter**, which is an example of the rotary solenoid described above. Here a solenoid produces a rotary motion, which is opposed by a spring. This produces an output displacement that is proportional to the current flowing through the coil. A needle attached to the moving rotor moves over a fixed scale to indicate the magnitude of the displacement. Moving-iron meters can be used for measuring AC or DC quantities. They produce a displacement that is related to the magnitude of the current and is independent of its polarity.

Although moving-iron meters are used in some applications, a more common arrangement is the **moving-coil meter**. Here, as the name implies, it is the coil that moves with respect to a fixed magnet, producing a meter that can be used to determine the polarity of a signal as well as its magnitude. The deflection of a moving-coil meter is proportional to the average value of the current. AC quantities can be measured by incorporating a rectifier and applying suitable calibration. However, it should be noted that the calibration usually assumes that the quantity being measured is

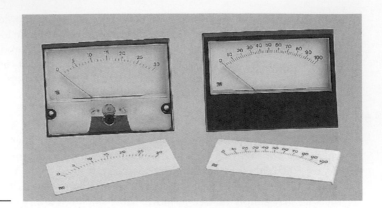

Figure 13.4
Moving-coil meters.

sinusoidal, and incorrect readings will result if other waveforms are used (as discussed in Chapter 2). Examples of typical moving-coil meters are shown in Figure 13.4.

Typical panel meters will produce a full-scale deflection for currents of 50 μA to 1 mA. Using suitable series and shunt resistances, it is possible to produce meters that will display either voltages or currents with almost any desired range (as described in Section 2.5).

13.4.3 Motors

Electric motors of various forms (as described in Chapter 10) can be used as force, displacement or motion actuators. The motors fall into three broad types: AC motors, DC motors and stepper motors.

AC motors are primarily used in high-power applications and situations where great precision is not required. Control of these motors is often by simple ON/OFF techniques, although variable power drives are also used.

DC motors are extensively used in precision position-control systems and other electronic systems, particularly in low-power applications. These motors have very straightforward characteristics, with their speed being determined by the applied voltage and their torque being related to their current. The speed range of DC motors can be very wide, with some devices being capable of speeds from tens of thousands of revolutions per minute down to a few revolutions per day. Some motors, in particular DC permanent-magnet motors, have an almost linear relationship between speed and voltage and between torque and current. This makes them particularly easy to use.

Stepper motors move in discrete steps and their speed of rotation is directly controlled by the input waveform used. Speeds of several thousands of revolutions per minute are often possible, while the discrete motion of the motor allows it to be moved through a defined angle if required. This combination of speed and controllability, together with small size and relatively low cost, makes stepper motors an attractive option in many applications.

13.5 Sound actuators

13.5.1 Speakers

Most speakers (or loudspeakers) have a fixed permanent magnet and a movable coil connected to a diaphragm. Input to the speaker generates a current in the coil, which causes it to move with respect to the magnet, thereby moving the diaphragm and generating sound. The nominal impedance of the coil in the speaker is typically in the range 4 to 15 Ω, and the power-handling capacity may vary from a few watts for a small domestic speaker to several hundreds of watts for speakers used in public address systems.

13.5.2 Ultrasonic transducers

At very high frequencies, the permanent-magnet speakers described earlier are often replaced by **piezoelectric actuators**. Such transducers are usually designed to operate over a narrow range of frequencies.

13.6 Actuator interfacing

The actuators discussed above all consume electrical power in order to vary some external physical quantity. Therefore the process of interfacing is largely concerned with the problem of enabling an electronic system to control the power in such a device.

13.6.1 Resistive devices

Where an actuator is largely resistive in nature, as in a resistive heating element, then the power dissipated in the device will be related to the voltage applied to it by the relationship

$$P = \frac{V^2}{R}$$

Here the power supplied to the actuator is simply related to the voltage applied across it. In such cases, the problems of interfacing are largely related to the task of supplying sufficient power to drive the actuator. In the case of devices requiring just a few watts (or less), this is relatively simple. However, as the power requirements increase, the task of supplying this power becomes more difficult. We will consider methods of driving high-power loads when we look at power electronic circuits (see Chapter 20).

One way of simplifying the control of high-power devices is to operate them in an ON/OFF manner. Where a device can be turned ON or OFF manually, this can be achieved using a simple mechanical switch. Alternatively, this function can be achieved under system control using an **electrically operated switch** (we will look at the operation of such circuits at a later stage).

In many cases, it is necessary to vary the power dissipated in an actuator rather than just to turn it ON and OFF. This may also be achieved using switching techniques in some cases. By repeatedly turning a device ON and OFF at high speed, it is possible to control the power dissipated in the component by altering the fraction of time for which the device is ON. Such techniques are used in conventional domestic **light dimmers**.

13.6.2 Capacitive and inductive devices

Capacitive and inductive actuators, such as motors and solenoids, create particular interfacing problems. This is particularly true when using the switching techniques described above. We will leave discussion of these problems until later (see Chapter 20) when we look at power electronic circuits.

13.7   Actuators – a summary

All the actuators we have discussed take an electrical input signal and from it generate a non-electrical output. In each case, power is taken from the input in order to apply power at the output. The power requirements are quite small in some cases, such as an LED or a panel meter, which consume only a fraction of a watt. In other cases the power required may be considerable. Heaters and motors, for example, may consume hundreds or even thousands of watts.

The **efficiency** of conversion also varies from device to device. In a heater, effectively all the power supplied by the input is converted to heat. We could say that the conversion efficiency is 100 per cent. LEDs, however, despite being one of the more efficient methods of converting electrical power into light, have an efficiency of only a few per cent, the remaining power being dissipated as heat.

Some actuators can be considered as simple resistive loads in which the current will vary in direct proportion to the applied voltage. Most heaters and panel meters would come into this category. Other devices, such as motors and solenoids, have a large amount of inductance as well as resistance, while others possess a large capacitance. Such devices behave very differently from simple resistive loads, particularly when a rapidly changing signal is applied. A third group of devices are non-linear and cannot be represented by simple combinations of passive components. LEDs and semiconductor laser diodes come into this third group. When designing electronic systems, it is essential to know the characteristics of the various actuators to be used so that appropriate interfacing circuitry can be produced.

Example 13.1 | **Controlling the power output of an electric heater.**

An electric heater operates from a 250 V, 50 Hz AC supply and produces 1 kW of heat. Some form of controller is required that will allow the heat produced by the device to be set to any value up to its maximum output.

Since an electric heater is essentially a resistive device, we can model the component by a simple resistor. The value of the resistor, which we will call R_H, can be easily found using Ohm's law, since:

$$P = \frac{V^2}{R_H}$$

$$\therefore R_H = \frac{V^2}{P} = \frac{250^2}{1000} = 62.5\ \Omega$$

One way of reducing the heat output of the device is to reduce the voltage across it, and clearly one method of achieving this is to use a potential divider as discussed in Section 1.12. Placing the heater in series with a variable resistor R_C would allow us to control the heat output by varying the value of R_C.

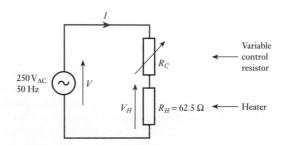

To illustrate the operation of this arrangement, consider the situation where we wish to reduce the power produced by the heater to 500 W. If we define the voltage across the heater to be V_H then using Ohm's law

$$P = \frac{V_H^2}{R_H}$$

$$\therefore V_H = \sqrt{P \times R_H} = \sqrt{500 \times 62.5} = 176.8\,V$$

From our knowledge of potential dividers it follows that:

$$V_H = V\frac{R_H}{R_H + R_C}$$

$$\therefore 176.8 = 250\frac{62.5}{62.5 + R_C}$$

$$\therefore R_C = 25.9\ \Omega$$

Thus adjusting our control resistor R_C to 25.9 Ω will reduce the power produced by the heater to 500 W and by selecting other values for the resistor we can adjust the heat output as required.

While the technique described above is simple it does have one serious drawback, and that is the power dissipated in the control resistor. A simple calculation will show that in the example given above, when 500 W is dissipated in the heater, approximately 207 W is dissipated in the control resistor. This is a tremendous waste of power and would normally be unacceptable. We therefore need a more efficient method of controlling the heater.

One way of producing more efficient control is to use a **switch-mode controller**. Here the heater is connected to the supply through a switch that can be cycled ON and OFF very quickly. The switch is driven by a repetitive waveform that varies the fraction of the time for which the heater is active. The switching rate is much faster than the response rate of the heater and so the heater effectively responds to the average value of the voltage waveform. For example, if the switch is ON all the time the heater will produce its maximum output of 1000 W, while if it is OFF all the time its power output will be zero. If the switch is ON for 50% of the time the heater will produce 50% of its maximum output (500 W), while if it is ON for 75% of the time it will produce 750 W. In this way the power output of the heater can be varied from zero to 100% of its maximum output.

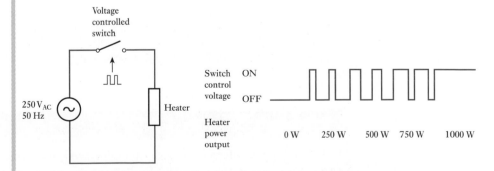

An *ideal* switch dissipates no power, since when it is switched ON it has current flowing through it but no voltage across it, and when it is switched OFF it has voltage across it but not current flowing through it. Thus in both cases the power (which is the product of the voltage and the current) is zero. Real switches do not match these idealised characteristics, but modern semiconductor switches (as discussed in later chapters) allow very rapid switching with very low power dissipation, and are a very good approximation to their idealised counterpart.

While this example has considered the control of a heater, a similar approach can be taken to the control of many high-power actuators. We will return to look at such techniques in more detail when we consider Power Electronics in more detail in Chapter 20.

Video 13B

Further study

A computer disk drive contains one or more rapidly spinning disks (or platters) that are coated with a magnetic material that allows digital information to be written and read using a magnetic head.

Clearly such drives contain several forms of actuator, but here we will concern ourselves only with those responsible for spinning the disks and for rotating the arm to position the read/write head over the appropriate part of the disk.

Consider the various actuators discussed in this chapter and decide which of these, if any, would be appropriate for use in this application.

Key points

- All useful systems need to affect their environment in order to perform their intended functions.
- Systems affect their environment using actuators.
- Most actuators take power from their inputs in order to deliver power at their outputs. The power required varies tremendously between devices.
- Some devices consume only a fraction of a watt. Others may consume hundreds or perhaps thousands of watts.
- In most cases, the energy conversion efficiency of an actuator is less than 100 per cent, and sometimes it is much less.
- Some actuators resemble resistive loads, while others have considerable capacitance or inductance. Others still are highly non-linear in their characteristics.
- The ease or difficulty of driving actuators varies with their characteristics.

Exercises

13.1 Explain the difference between a transducer and an actuator.

13.2 What form of device would normally be used as a heat actuator when the required output power is a few watts?

13.3 What form of heat actuator would be used in applications requiring a power output of several kilowatts?

13.4 Estimate the efficiency of a typical heat actuator.

13.5 What forms of light actuator would typically be used for general illumination? What would be a typical range for the output power for such devices?

13.6 Why are conventional light bulbs unsuitable for signalling and communication applications? What forms of transducer are used in such applications?

13.7 How do light-emitting diodes (LEDs) differ from conventional semiconductor diodes?

13.8 What would be a typical value for the operating voltage of an LED, and what would be a typical value for its maximum current?

13.9 From the information given for the last exercise, what would be a typical value for the maximum power dissipation of an LED?

13.10 In addition to displaying the digits 0–9, the seven-segment display of Figure 13.1 can be used to indicate some alphabetic characters (albeit in a rather crude manner). List the upper- and lower-case letters that can be shown in this way and give examples of simple status messages (such as 'Start' and 'Stop') that can be displayed using an array of these devices.

13.11 Briefly describe the operation and function of an opto-isolator.

13.12 What environmental factor causes problems for optical communication systems using conventional LEDs and photo detectors? How may this problem be reduced?

13.13 What form of optical fibre would be preferred for communication over a distance of several kilometres? What form of light source would normally be used in such an arrangement?

13.14 Explain how a single form of transducer might be used as a force actuator, a displacement actuator or a motion actuator.

13.15 Describe the operation of a simple solenoid.

13.16 Explain how a solenoid may be used as a binary position actuator.

13.17 Explain why a simple panel meter may be thought of as a rotary solenoid.

13.18 What is the most common form of analogue panel meter? What would be typical operating currents for such devices?

13.19 List three basic forms of electric motor.

13.20 What form of motor would typically be used in high-power applications?

13.21 What form of motor might be used in an application requiring precise position control?

13.22 Briefly describe the characteristics of a stepper motor that make it an attractive option in some situations.

13.23 How is the speed of a stepper motor controlled?

13.24 What would be a typical value for the impedance of the coil of a loudspeaker?

13.25 Explain how the power dissipated in an actuator may be varied using an electrically operated switch.

Chapter 14   Amplification

When you have studied the material in this chapter, you should be able to:

- explain the concept of amplification
- give examples of both active and passive amplifiers
- use simple equivalent circuits to determine the gain of an amplifier
- discuss the effects of input resistance and output resistance on the voltage gain of an amplifier and use these quantities to calculate loading effects
- define terms such as output power, power gain, voltage gain and frequency response
- describe several common forms of amplifier, including differential and operational amplifiers.

14.1   Introduction

In earlier chapters, we noted that many electronic systems are composed of one or more sensors that take information from the 'real world', one or more actuators that allow the system to output information to the real world and some form of processing that makes signals associated with the former appropriate for use with the latter. Although the form of the processing that is required varies greatly from one application to another, one element that is often required is **amplification**.

Simplistically, **amplification** means making things bigger, and the converse operation, **attenuation**, means making things smaller. These basic operations are fundamental to many systems, including both electronic and non-electronic applications.

Examples of non-electronic amplification are shown in Figure 14.1. The first shows a lever arrangement. Here, the force applied at the output is greater than that applied at the input, so we have amplified the input force. However, the distance moved by the output is less than that moved at the input. Thus, although the force has been amplified, the displacement has been reduced, or attenuated. Note that, if the positions of the input and the output of the lever were reversed, we would produce an arrangement that amplified movement but attenuated force.

The second example in Figure 14.1 shows a pulley arrangement. As in the first example, this produces a greater force at the output than is applied at the input, but the distance moved at the output is less than that at the input. Thus, again, we have a force amplifier but a movement attenuator.

In the lever arrangement shown in Figure 14.1(a), the direction of the output force is the same as that of the input. Such an amplifier is referred to as a **non-inverting amplifier**. In the pulley arrangement of Figure 14.1(b), a downward force at the input results in an upward force at the output and we have what is termed an **inverting**

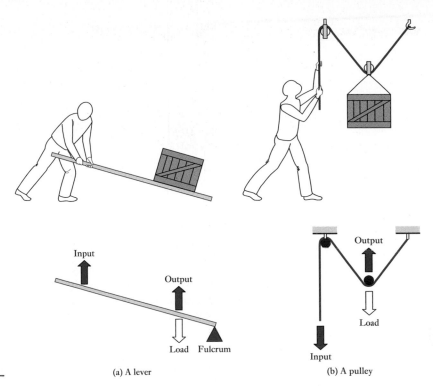

Figure 14.1
Examples of
mechanical amplifiers.

(a) A lever

(b) A pulley

amplifier. Different arrangements of levers and pulleys may be either inverting or non–inverting.

Both the examples in Figure 14.1 are **passive systems** – that is, they have no external energy source other than the inputs. For such systems, the **output power** – that is, the power delivered at the output – can never be greater than the **input power** – that is, the power absorbed by the input – and, in general, it will be less because of losses. In our examples, losses would be caused by friction at the fulcrum and pulleys.

In order to be able to provide power gain, some amplifiers are not passive but **active**. This means that they have some form of external energy source that can be harnessed to produce an output that has more power than the input. Figure 14.2 shows an

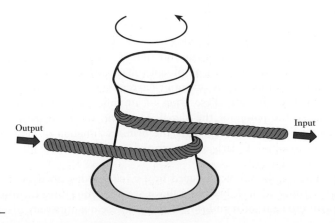

Figure 14.2
A torque amplifier.

example of such an amplifier called a **torque amplifier**. This consists of a rotating shaft with a rope or cable wound around it. The amplifier can be used as a power winch and is often found in boats and ships. One end of the rope (the output) is attached to a load and a control force is applied to the other end (the input). If no force is applied to the input, the rope will hang loosely around the rotating shaft and little force will be applied at the output. The application of a force to the input tightens the rope around the drum and increases the friction between them. This frictional force is applied to the rope and results in a force being exerted at the output. The greater the force applied to the input, the greater the frictional force experienced by the rope and the greater the force exerted at the output. We therefore have an amplifier where a small force applied at the input generates a larger force at the output. The magnitude of the amplification may be increased or decreased by changing the number of turns of the rope around the drum.

It should be noted that, as the rope is continuous, the distance moved by the load at the output is equal to the distance moved by the rope at the input. However, the force applied at the output is greater than that at the input and the arrangement therefore delivers more power at the output than it absorbs at the input. It therefore provides not only force amplification but also **power amplification**. The extra power available at the output is supplied by the rotating drum and will result in an increased drag being experienced by whatever force is causing it to rotate.

14.2 Electronic amplifiers

In electronics, there are also examples of both passive and active amplifiers. Examples of the former include a step-up transformer (as described in Section 5.12), where an alternating voltage signal applied to the input will generate a larger voltage signal at the output. Although the voltage at the output is increased, the ability of the output to provide current to an external load is reduced. The power supplied to a load will always be less than the power absorbed at the input. Thus, a transformer may provide voltage amplification but it *cannot* provide power amplification.

Although there are several examples of passive electronic amplifiers, the most important and useful electronic amplifiers are active circuits. These take power from an external energy source – usually some form of **power supply** – and use it to boost the input signal. Unless the text indicates differently, for the remainder of this book, when we use the term *amplifier*, we will be referring to an *active electronic amplifier*.

We saw earlier when looking at mechanical amplifiers that several different forms of amplification are possible. Such devices can, for example, be movement amplifiers or force amplifiers and provide power amplification or attenuation. Electronic amplifiers may also be of different types. One of the commonest is the **voltage amplifier**, the main function of which is to take an input voltage signal and produce a corresponding amplified voltage signal. Also of importance is the **current amplifier**, which takes an input current signal and produces an amplified current signal. Usually both these types of amplifier, as a result of the amplification, also increase the power of the signal. However, the term **power amplifier** is usually reserved for circuits that have the primary function of supplying large amounts of power to a load. Clearly, power amplifiers must also provide either voltage or current amplification or both.

The amplification produced by a circuit is described by its **gain**, which is often given the symbol A. From the above, we can define three quantities – namely, **voltage gain**, **current gain** and **power gain**. These quantities are given by the expressions

$$\text{voltage gain } (A_v) = \frac{V_o}{V_i} \tag{14.1}$$

$$\text{current gain } (A_i) = \frac{I_o}{I_i} \tag{14.2}$$

$$\text{power gain } (A_p) = \frac{P_o}{P_i} \tag{14.3}$$

where V_i, I_i and P_i represent the input voltage, input current and input power, respectively, and V_o, I_o and P_o represent the output voltage, output current and output power, respectively. Initially we will look at voltage amplification and leave consideration of current and power amplification until later. Note that, if the polarities of the input and output voltages of an amplifier are different, the voltage gain of a circuit will be negative. Thus, a **non-inverting amplifier** has a positive voltage gain, while an **inverting amplifier** has a negative voltage gain.

A widely used symbol for an amplifier is shown in Figure 14.3. This device has a single input and produces an amplification determined by the circuitry used. In this case, the input and output quantities are voltages and the circuit is described by its voltage gain.

Clearly the input and output voltages must be measured with respect to some reference voltage or reference point. This point is often called the **earth** of the circuit and is given the symbol shown in the diagram.

You will notice that the diagram in Figure 14.3 does not show any connection to a power source. In practice, the electronic amplifier *would* require some form of **power supply** to enable it to boost the input signal. However, we normally omit the power supply from such diagrams for reasons of clarity. The diagram represents the *functionality* of the arrangement rather than its detailed circuitry. If we were to attempt to construct this circuit, we would need to remember to include an appropriate power supply with connections to the amplifier.

In order for the amplifier to perform some useful function, something must be connected to the input to provide an **input signal** and something must be connected to the output to make use of the **output signal**. In a simple application, the input signal could come directly from one of the sensors described in Chapter 12 and the output could drive an actuator. Alternatively, the input and output could be connected to other electronic circuits. The transducer or circuitry providing an input to the amplifier is sometimes called the **source**, while the transducer or circuitry connected to the output is called the **load** of the amplifier.

An **ideal** voltage amplifier would always give an output voltage that was determined only by the input voltage and gain, irrespective of what was connected to the output (the load). Also, an ideal amplifier would not affect the signal produced by the source, implying that no current is taken from it. In fact, **real** amplifiers cannot fulfil these requirements. To understand why this is so, we need to know more about the nature of sources and loads.

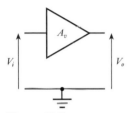

Figure 14.3
An amplifier.

14.3   Sources and loads

At the end of Chapter 11 we looked at the process of partitioning a complex system into a number of distinct modules – as illustrated in Figure 14.4. We noted that the output of each module represents a *source*, while the input of each module represents

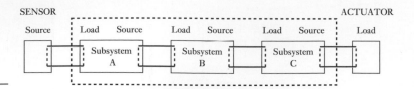

Figure 14.4
System partitioning.

a *load*. We also noted that, as far as its external appearance is concerned, a module can be described simply in terms of the nature of its inputs, its outputs and the relationship between them. This allows us to describe the operation of modules, such as amplifiers, by what is termed an **equivalent circuit**. This is a simplified representation of the module, which does not attempt to describe the internal construction of the unit, but simply to model its external characteristics.

Since an amplifier, or any other module, can be characterised by the nature of its inputs, its outputs and the relationship between them, we clearly need some way of describing these module attributes.

14.3.1   Modelling the input of an amplifier

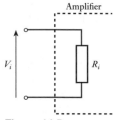

Figure 14.5
An equivalent circuit of
the input of an amplifier.

To model the characteristics of the input of an amplifier, we need to be able to describe the way in which it appears to circuits that are connected to it. In other words, we need to model how it appears when it represents the *load* of another circuit.

Fortunately, in most cases, the input circuitry of an amplifier can be modelled adequately by a single fixed resistance, which is termed its **input resistance** and given the symbol R_i. What this means is that, when an external voltage source applies an input voltage to the amplifier, the current that flows into the amplifier is related to the voltage as if the input were a simple resistance of value R_i. Therefore, an adequate model of the input of an amplifier is that shown in Figure 14.5. Such a model is termed the **equivalent circuit** of the input.

14.3.2   Modelling the output of an amplifier

To model the characteristics of the output of an amplifier, we need to be able to describe the way in which it appears to circuits that are connected to it. In other words, we need to model how it appears when it represents a *source* to another circuit.

Any real voltage source may be represented by a **Thévenin equivalent circuit** (as discussed in Section 3.8). This models the source using an ideal voltage source and a series resistor. Figure 14.6(a) shows a possible representation of this arrangement, although it is more common to use the form shown in Figure 14.6(b).

As we now have an equivalent circuit of a voltage source, we can use this to model the output of our amplifier. This is shown in Figure 14.7. Here, the voltage to be

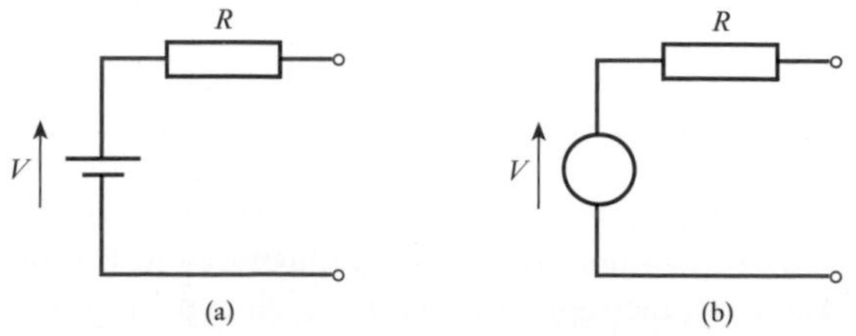

Figure 14.6
Equivalent circuits of a
voltage source.

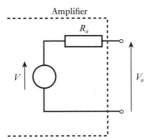

Figure 14.7
An equivalent circuit of the output of an amplifier.

produced by the amplifier is represented by the voltage V, and R_o represents the resistance associated with the output circuitry – this is termed the **output resistance** of the circuit. The actual voltage that is produced at the output of the circuit V_o is equal to V minus the voltage drop across R_o. The magnitude of this voltage drop is given by Ohm's law and is equal to the product of R_o and the current flowing through this resistance.

14.3.3 Modelling the gain of an amplifier

Having successfully modelled the input and output of our amplifier, we now need to turn our attention to the relationship between them. This is clearly the *gain* of the amplifier.

The gain of an amplifier can be modelled by means of a **controlled voltage source** – that is, a source in which the voltage is controlled by some other quantity within the circuit. Such sources are also known as **dependent voltage sources**. In this case, the controlled voltage source produces a voltage that is equal to the input voltage of the amplifier multiplied by the circuit's gain. This is illustrated in Figure 14.8, where a voltage source produces a voltage equal to $A_v V_i$.

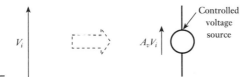

Figure 14.8
Representing gain within an equivalent circuit.

It should be noted that we are currently considering an equivalent circuit of an amplifier and *not* its physical implementation. You should not concern yourself with how a controlled voltage source operates as it is simply a convenient way to model the operation of the circuit.

Video 14A

14.4 Equivalent circuit of an amplifier

As we now have equivalent circuits for the input, output and gain of a module, we are in a position to draw an equivalent circuit for the amplifier in Figure 14.3. This is shown in Figure 14.9. The **input voltage** of the circuit V_i is applied across the **input resistance** R_i which models the relationship between the input voltage and the corresponding input current. The **voltage gain** of the circuit A_v is represented by the controlled voltage source, which produces a voltage of $A_v V_i$. The way in which the output voltage of the circuit changes with the current taken from the circuit is modelled by placing a resistor in series with the voltage source. This is the **output resistance** R_o of the

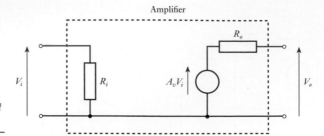

Figure 14.9
An equivalent circuit of an amplifier.

circuit. You will notice that the equivalent circuit has two input terminals and two output terminals, as in the amplifier shown in Figure 14.3. The lower terminal in each case forms a reference point, as all voltages must be measured with respect to some reference voltage. In this circuit, the input and output references are joined together. This common reference point is often joined to the **earth** or **chassis** of the system (as in Figure 14.3) and its potential is then taken as the 0 V reference.

Note that the equivalent circuit does not show any connection to a power source. The voltage source within the equivalent circuit would in practice take power from an external **power supply** of some form, but in this circuit we are only interested in its functional properties – not its physical implementation.

The advantage of an equivalent circuit is that it allows us easily to calculate the effects that external circuits (which also have input and output resistance) will have on the amplifier. This is illustrated in the following example.

Example 14.1

An amplifier has a voltage gain of 10, an input resistance of 1 kΩ and an output resistance of 10 Ω. The amplifier is connected to a sensor that produces a voltage of 2 V and has an output resistance of 100 Ω, and to a load resistance of 50 Ω. What will be the output voltage of the amplifier (that is, the voltage across the load resistance)?

First, we draw an equivalent circuit of the amplifier, sensor and load.

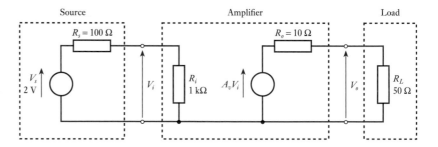

The input voltage is applied across a potential divider formed by R_s and R_i. The resulting voltage V_i is given by

$$V_i = \frac{R_i}{R_s + R_i} V_s$$

$$= \frac{1\text{ k}\Omega}{100\ \Omega + 1\text{ k}\Omega} 2\text{ V}$$

$$= 1.82\text{ V}$$

This voltage is amplified by the voltage source and the resultant voltage is applied across a second potential divider formed by R_o and R_L. Therefore, the output voltage is given by

$$V_0 = A_v V_i \frac{R_L}{R_0 + R_L}$$

$$= 10 V_i \frac{50\ \Omega}{10\ \Omega + 50\ \Omega}$$

$$= 10 \times 1.82 \frac{50\ \Omega}{10\ \Omega + 50\ \Omega}$$

$$= 15.2\ \text{V}$$

As defined in Equation 14.1, the voltage gain of an amplifier is the ratio of the output voltage to the input voltage.

Example 14.2

Calculate the voltage gain of the circuit in Example 14.1.

$$\text{voltage gain } (A_v) = \frac{V_0}{V_i} = \frac{15.2}{1.82} = 8.35$$

Examples 14.1 and 14.2 show that when an amplifier is connected to a source and a load, the resulting output voltage may be considerably less than one might have expected given the source voltage and voltage gain of the amplifier in isolation. The input voltage to the amplifier is less than the source voltage due to the effects of the source resistance. Similarly, the output voltage of the circuit is affected by the load resistance as well as the characteristics of the amplifier. This effect is known as **loading**. The *lower* the resistance that is applied across the output of a circuit, the more *heavily* it is loaded as more current is drawn from it.

File 14A

Computer simulation exercise 14.1

Simulate the circuit in Example 14.1 and experiment with different values for the various resistors, noting the effect on the voltage gain of the circuit. How should the resistors be chosen to maximise the voltage gain? In other words, under what conditions are the loading effects at a minimum?

From the above examples, it is clear that the gain produced by an amplifier is greatly affected by circuitry that is connected to it. For this reason, we should perhaps refer to the gain of the amplifier in isolation as the **unloaded amplifier gain** as it is the ratio of the output voltage to the input voltage in the absence of any loading effects. When we connect a source and a load to the amplifier, we produce **potential dividers** at the input and output that reduce the effective gain of the circuit.

Earlier in this chapter, we noted that an ideal voltage amplifier would not affect the circuit to which it was connected and would give an output that was independent of the load. This implies that it would draw no current from the source and be unaffected by loading at the output. Consideration of the analysis in Example 14.1 shows that this requires the input resistance R_i to be infinite and the output resistance R_o to be zero. Under these circumstances, the effects of the two potential dividers are removed and the voltage gain of the complete circuit becomes equal to the voltage gain of the unloaded amplifier, and is unaffected by the source and load resistance. This is illustrated in Example 14.3.

Example 14.3

An amplifier has a voltage gain of 10, an infinite input resistance and zero output resistance. The amplifier is connected, as in Example 14.1, to a sensor that produces a voltage of 2 V and has an output resistance of 100 Ω, and to a load resistance of 50 Ω. What will be the output voltage of the amplifier?

As before, we first draw an equivalent circuit of the amplifier, sensor and load.

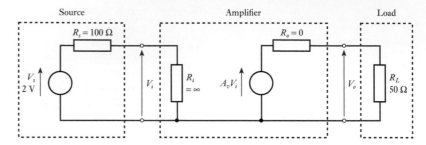

From the diagram

$$V_i = \frac{R_i}{R_s + R_i} V_s$$

When R_i is much larger than R_s, this approximates to

$$V_i = \frac{R_i}{R_s + R_i} V_s$$
$$\approx \frac{R_i}{R_i} V_s$$

In this case, R_i is infinite, so we can say

$$V_i = \frac{R_i}{R_i} V_s$$
$$= V_s$$
$$= 2 \text{ V}$$

and, therefore, the output voltage is given by

$$V_o = A_v V_i \frac{R_L}{R_o + R_L}$$
$$= 10 V_i \frac{50 \ \Omega}{0 + 50 \ \Omega}$$
$$= 10 V_i$$
$$= 10 \times 2$$
$$= 20 \text{ V}$$

The arrangement now has a voltage gain of 10 and there are no loading effects.

No real amplifier can have an infinite input resistance or zero output resistance. However, if the input resistance is *large* compared with the source resistance and the output resistance is *small* compared with the load resistance, the effects of these resistances will be small and may often be neglected. This will produce the maximum voltage gain from the circuit. For these reasons, a good voltage amplifier is characterised by a high input resistance and a low output resistance.

Often the designer does not have a totally free choice for the values of the input and output resistances of the amplifier for a given application. This may be because a particular amplifier has already been chosen or there are other constraints on the design. In this situation, an alternative is to change the values of the source and load resistances. If the source resistance could be made small compared with the input resistance of the amplifier and the load resistance made large compared with the output resistance of the amplifier, this would again mean that these resistances could be neglected and the maximum voltage gain would be achieved. Thus, when used with voltage amplifiers, it is advantageous for a signal source to have a low resistance (that is, the output resistance of the source should be low) and for a load to have a high resistance (that is, the input resistance of the load should be high). Thus, in circuits concerned with voltage amplification, the input resistances of each stage should be high and the output resistances of each stage should be low to maximise the overall voltage gain.

It should be noted that these characteristics are not beneficial in all forms of amplifier. **Current amplifiers**, for example, should have a low input resistance, so that they do not affect the current flowing into the input from a current source, and a high output resistance, so that the output current is not affected by an external load resistance.

In many applications, the voltage source supplying an input to an amplifier will be a sensor of some kind and often the load will be some form of actuator. In these circumstances, it may not be possible to change the input or output resistances of these devices to suit the application in question. Then, it will be necessary to tailor the input and output resistances of the amplifier to suit the sensors and actuators being used. It may also be necessary to allow for the effects of input and output resistance, as illustrated in Example 14.1.

14.5   Output power

In the last section, we looked at the performance of an amplifier in terms of its voltage gain and how this performance is affected by internal and external resistances. We now consider the performance of the amplifier in terms of the **power** that it can deliver to an external load.

The power dissipated in the load resistor (the **output power**, P_o) of a circuit is simply

$$P_o = \frac{V_o^2}{R_L}$$

Example 14.4 | **Calculate the output power of the circuit in Example 14.1.**

From the earlier analysis, the output voltage, V_o, of the circuit is 15.2 V and the load resistance is 50 Ω. Therefore, the output power is

$$P_o = \frac{V_o^2}{R_L} = \frac{(15.2)^2}{50} = 4.6 \text{ W}$$

In the previous section, we noted that, unless the output resistance of an amplifier is zero, the voltage gain achieved is affected by the ratio of the load resistance to the output resistance. Therefore, for a given amplifier (with a particular value of output

Table 14.1 Variation of output voltage and output power with load resistance in Example 14.1.

Load resistance, R_L (Ω)	Output voltage, V_o (V)	Output power, P_o (W)
1	1.65	2.7
2	3.03	4.6
3	4.20	5.9
10	9.10	8.3
33	14.0	5.9
50	15.2	4.6
100	16.5	2.7

resistance) and a given input voltage, the output voltage will vary with the resistance of the load connected to it. As the output power is related to the output voltage, it is clear that the output power will also vary with the load resistance. These two dependencies are illustrated in Table 14.1, which shows the output voltage and output power produced by the circuit in Example 14.1 for different values of load resistance.

You will notice that the output *voltage* increases steadily as the resistance of the load is increased. This is because the output rises as the amplifier is less heavily loaded.

However, the output *power* initially rises as the resistance of the load is increased from 1 Ω until a maximum is reached. It then drops as the load is increased further. To investigate this effect, we need to look at the expression for the output voltage of the circuit. From the analysis in Example 14.1, we know that this is

$$V_o = A_v V_i \frac{R_L}{R_o + R_L} \tag{14.4}$$

As the power dissipated in a resistance is given by V^2/R, the power dissipated in the load resistance (the output power, P_o) is given by

$$P_o = \frac{V_o^2}{R_L} = \frac{\left(A_v V_i \dfrac{R_L}{R_L + R_o}\right)^2}{R_L} = A_v^2 V_i^2 \frac{R_L}{(R_L + R_o)^2} \tag{14.5}$$

Differentiating this expression for P_o with respect to R_L gives

$$\frac{dP_o}{dR_L} = \frac{(R_L + R_o)^2 A_v^2 V_i^2 - 2 A_v^2 V_i^2 (R_L + R_o) R_L}{(R_L + R_o)^4} \tag{14.6}$$

which must equal zero for a maximum or minimum. This condition is given when the numerator is zero, which occurs when

$$(R_L + R_o) - 2R_L = 0$$

giving

$$R_L = R_o$$

Further differentiation of Equation 14.6 will confirm that this is indeed a maximum rather than a minimum value.

Substituting for the component values used in Example 14.1 shows that maximum power should be dissipated in the load when its resistance is equal to R_o, which is 10 Ω. The tabulated data in Table 14.1 confirms this result.

Thus, in circuits where the output characteristics can be adequately represented by means of a simple resistance, maximum power is transferred to the load when the load resistance is equal to the output resistance. This result holds for transfers between any two circuits and is a simplified statement of the **maximum power theorem**.

A similar analysis may be performed to investigate the transfer of power between circuits that have complex impedances rather than simple resistances. This produces the more general result that, for maximum power transfer, the impedance of the load must be equal to the complex conjugate of the impedance of the output. Thus, if the output impedance of a network has the value $R + jX$, for maximum power transfer, the load should have an impedance of $R - jX$. This implies that, if the output impedance has a capacitive component, the load must have an inductive component to obtain maximum output power. It can be seen that the simpler statement given above is a special case of this result in which the reactive component of the output impedance is zero.

The process of choosing a load to maximise **power transfer** is called **matching** and is a very important aspect of circuit design in certain areas. It should be remembered, however, that, as maximum power transfer occurs when the load and output resistances are equal, the voltage gain is far from its maximum value under these conditions. In voltage amplifiers it is more usual to attempt to maximise input resistance and minimise output resistance to maximise voltage gain. Similarly, in current amplifiers it may be advisable to have a high output impedance and a low input impedance to produce a high current transfer.

Impedance matching is of importance in circuits where the efficiency of power transfer is paramount. It should be noted, however, that, when perfectly matched, the power dissipated in the output stage of the source is equal to the power dissipated in the load. Also, when $R_L = R_o$ Equation 14.5 produces the result that

$$P_{o(\text{max})} = \frac{A_v^2 V_i^2}{4R_L}$$

In other words, the maximum output power is only one-quarter of the power that would be dissipated in the load if the output impedance were zero.

For this reason, impedance matching is seldom used in high-power amplifiers as **power efficiency** is of more importance than power transfer. In such cases, it is common to attempt to make the output resistance as small as possible to maximise the power delivered to the load for a given output voltage. We will return to this topic later (see Chapter 20) when we look in more detail at power amplifiers.

Impedance matching finds its main application in low-power **radio frequency (RF) amplifiers** where very small signals must be amplified with maximum power transfer.

File 14B

Computer simulation exercise 14.2

Use the circuit in Computer simulation exercise 14.1 to investigate the way in which the power output of the circuit is affected by the value of the load resistance R_L.

Use the simulator's sweep facility to determine the value of R_L that results in the maximum power output for a given value of R_o. Repeat this for a range of values of R_o.

Video 14B

14.6 **Power gain**

The **power gain** of an amplifier is the ratio of the power supplied by the amplifier to a load to the power absorbed by the amplifier from its source. The input power can be calculated from the input voltage and the input current or, by applying Ohm's law, from a knowledge of the input resistance and either the input voltage or current. Similarly, the output power can be determined from the output voltage and output current, or from one of these and a knowledge of the load resistance.

Example 14.5 Calculate the power gain of the circuit in Example 14.1.

To determine the power gain, first we need to calculate the input power P_i and the output power P_o.

From the example, we have that $V_i = 1.82\,\text{V}$ and $R_i = 1\,\text{k}\Omega$. Therefore

$$P_i = \frac{V_i^2}{R_i} = \frac{(1.82)^2}{1000} = 3.3\,\text{mW}$$

We also know that $V_o = 15.2\,\text{V}$ and $R_L = 50\,\Omega$. Therefore

$$P_o = \frac{V_o^2}{R_L} = \frac{(15.2)^2}{50} = 4.62\,\text{W}$$

The power gain is then given by

$$\text{power gain }(A_p) = \frac{P_o}{P_i} = \frac{4.62}{0.0033} = 1400$$

Note that when calculating the *input* power we use R_i but, when calculating the *output* power, we use R_L (*not* R_o). This is because we are calculating the power delivered to the *load*, not the power dissipated in the *output resistance*.

You will note from Example 14.5 that even a circuit with a relatively low voltage gain (in this case it is $15.2/1.82 = 8.35$) can have a relatively high power gain (in this case, over a thousand). The power gain of a modern electronic amplifier may be very high, gains of 10^6 or 10^7 being common. With these large numbers it is often convenient to use a logarithmic expression of gain rather than a simple ratio. This is normally done using **decibels** (as discussed in Section 8.3).

You will recall that the decibel (dB) is a dimensionless figure for **power gain** and is defined by

$$\text{power gain (dB)} = 10\log_{10}\frac{P_2}{P_1} \qquad (14.7)$$

where P_2 is the output power and P_1 is the input power of the amplifier or other circuit.

Example 14.6 Express a power gain of 1400 in decibels.

$$\text{power gain (dB)} = 10\log_{10}\frac{P_2}{P_1}$$

$$= 10\log_{10}(1400)$$

$$= 10 \times 3.15$$

$$= 31.5\,\text{dB}$$

14.7   Frequency response and bandwidth

All real systems have limitations in terms of the range of frequencies over which they will operate and, consequently, amplifiers are subject to such restrictions. In some cases it is also desirable to limit the range of frequencies that are amplified by a circuit. In general, an amplifier will be required to deliver a particular amount of gain over a particular range of frequencies. The gain of the circuit within this normal operating range is termed its **mid-band gain**.

The range of frequencies over which an amplifier operates is determined by its **frequency response**, which describes how the gain of the amplifier changes with frequency. The gain of all amplifiers falls at high frequencies, although the frequency at which this effect becomes apparent will vary depending on the circuits and components used. This fall in gain at high frequencies can be caused by a number of effects, including the presence of **stray capacitance** within the circuit. In some amplifiers, the gain also falls at low frequencies.

The mechanisms and circuit characteristics that determine the frequency response of electronic circuits were discussed in Chapter 8. The high- and low-frequency performance of an amplifier is described by its upper and lower cut-off frequencies, which indicate the frequencies at which the power associated with the output falls to half its mid-band value. These **half-power points** corresponds to a fall in gain of 3 dB relative to its mid-band value, which in turn represents a fall in voltage gain to $1/\sqrt{2}$ or about 0.707 of its mid-band value.

All circuits will have an **upper cut-off frequency** and some also have a **lower cut-off frequency**. This is illustrated in Figure 14.10. Figure 14.10(a) shows the frequency response of an amplifier that has both upper and lower cut-off frequencies. In this graph, the gain of the amplifier is plotted against frequency. Figure 14.10(b) shows the response of a similar amplifier, but this time plots the gain in decibels.

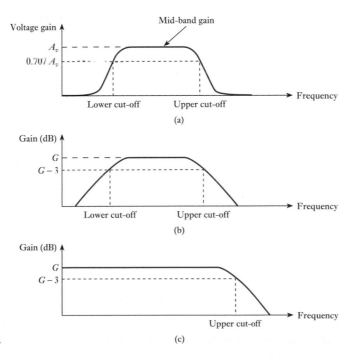

Figure 14.10
Examples of amplifier frequency response.

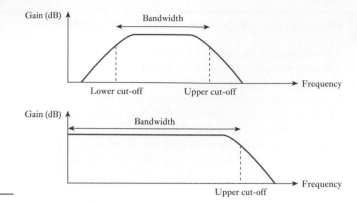

Figure 14.11
The bandwidth of an amplifier.

Figure 14.10(c) shows the response of an amplifier that has no lower cut-off. In this last circuit, the gain is constant at low frequencies.

The width of the frequency range of a circuit is termed its **bandwidth**. In circuits with both an upper and a lower cut-off frequency, this is simply the difference between these two frequencies. In circuits that have only an upper cut-off frequency, the bandwidth is the difference between this and zero hertz, so is numerically equal to the upper cut-off frequency. This is shown in Figure 14.11.

Limitations on the frequency range of an amplifier can cause distortion of the signal being amplified. For this reason, it is important that the bandwidth and the frequency range of the amplifier are appropriate for the signals being used.

Video 14C

14.8 Differential amplifiers

The amplifiers considered so far take as their input a single voltage that is measured with respect to some reference voltage, which is usually the 'ground' or 'earth' reference point of the circuit (0 V). An amplifier of this form is shown in Figure 14.3. Some amplifiers have not one but two inputs and produce an output proportional to the difference between the voltages on these inputs. Such amplifiers are called **differential amplifiers**, an example of which is shown in Figure 14.12.

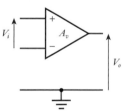

Figure 14.12
A differential amplifier.

As a differential amplifier takes as its input the difference between two input voltages, it effectively subtracts the voltage on one input from that on the other to form its input signal. Consequently, we need to differentiate between the two input terminals and, for this reason, the two inputs are labelled '+' and '−'. The former is called the **non-inverting input** because a positive voltage on this input with respect to the other input will cause the output to become positive. The latter is called the **inverting input** because a positive voltage on this input with respect to the other input will cause the output to become negative.

As a differential amplifier produces an output that is proportional to the difference between two input signals, it is clear that, if the same voltage is applied to both inputs, no output will be produced. Voltages that are common to both inputs are called **common-mode signals**, while voltage differences between the two inputs are termed **differential-mode signals**. A differential amplifier is designed to amplify differential-mode signals while ignoring (or rejecting) common-mode signals.

The ability to reject common-mode signals while amplifying differential-mode signals is often extremely useful. An example that illustrates this is the transmission of signals over great distances. Consider the situations illustrated in Figure 14.13.

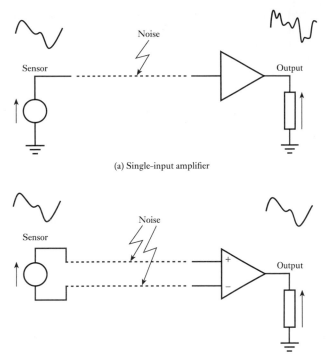

(a) Single-input amplifier

(b) Differential-input amplifier

Figure 14.13
Comparison of single
and differential input
methods.

Figure 14.13(a) shows a sensor connected to a single-input amplifier by a long cable. Any long cable is influenced by **electromagnetic interference** (EMI) and inevitably some noise will be added to the signal from the sensor. This noise will be amplified along with the wanted signal and therefore appear at the output, along with the wanted signal. Figure 14.13(b) shows a similar sensor connected by a twin conductor cable to the inputs of a differential amplifier. Again, the cable will be affected by noise, but, in this case, because of the close proximity of the two cables (which are kept as close as possible to each other), the noise picked up by each cable will be almost identical. Therefore, at the amplifier this noise appears as a common mode signal and is ignored, while the signal from the sensor is a differential-mode signal and is amplified.

In Figure 14.9 we looked at an equivalent circuit for a single-input amplifier and then went on to look at the uses of such an equivalent circuit in determining loading effects. We can also construct an equivalent circuit for a differential amplifier, and this is shown in Figure 14.14. You will note that it is very similar in form to that in Figure 14.9, but the input voltage V_i is now defined as the difference between two input voltages V_+ and V_-, and there is no connection between this second input and ground.

Figure 14.14
An equivalent circuit
for a differential
amplifier.

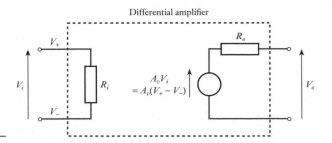

A common form of differential amplifier is the **operational amplifier** (or **op-amp**). These are constructed by fabricating all the necessary components on a single chip of silicon, forming a monolithic **integrated circuit** (IC).

In Section 14.4 when looking at voltage amplifiers, we concluded that a good voltage amplifier is characterised by a high input resistance and a low output resistance. Operational amplifiers have a very high input resistance (of perhaps several megohms or even gigohms) and a low output resistance (of a few ohms or tens of ohms). They also have a very high gain, of perhaps 10^5 or 10^6. For these reasons, operational amplifiers are extremely useful 'building blocks' for constructing not only amplifiers but also a wide range of other circuits. We will therefore look at op-amps in more detail in Chapter 16.

Unfortunately, despite their many attractive characteristics, op-amps suffer from **variability**. That is to say, their attributes (such as gain and input resistance) tend to vary from one device to another. They may also change for a particular device, with variations in temperature or over time. We therefore need techniques to overcome this variability and tailor the characteristics of the devices to match the requirements of particular applications. Techniques for achieving these goals are discussed in Chapter 15.

14.9   Simple amplifiers

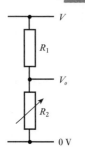

Figure 14.15
A simple potential divider.

Operational amplifiers are fairly complicated circuits that contain a number of semiconductor devices. Amplifiers may also be formed using single transistors or other active devices. Before we look at the operation of transistors, it is perhaps worth seeing how a single control device may be used to form an amplifier.

Consider the circuit in Figure 14.15. This shows a pair of resistors arranged as a potential divider. The output voltage V_o is related to the circuit parameters by the expression

$$V_o = \frac{R_2}{R_1 + R_2} V$$

If the variable resistance R_2 is adjusted to equal R_1, the output voltage V_o will clearly be half the supply voltage V. If R_2 is reduced, V_o will also reduce. If R_2 is increased, V_o will increase. If we replace R_2 with some, as yet undefined, control device that has an input voltage V_i which controls its resistance, varying V_i will vary the output voltage V_o. Figure 14.16 shows such an arrangement.

In fact, the control device does not have to be a voltage-controlled resistance. Any device in which the current is determined by a control input may be used in such an arrangement and, if the gain of the control device is suitable, it can be used as an amplifier.

Simple amplifiers of this type are used extensively within all forms of electronic circuits and we will return to them when we have considered the operation of some active components that can be used as control devices in such arrangements. Unfortunately, as with operational amplifiers, these control devices suffer from variability, and again we need a method of overcoming this problem. This is the topic of the next chapter.

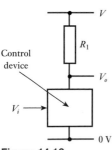

Figure 14.16
The use of a control device.

Video 14D

Further study

A particular personal audio device drives a pair of stereo headphones through a stereo output jack. The headphones supplied with the unit have an impedance of 32 Ω and each channel has a maximum output power of 30 mW.

Our task is to design a speaker unit containing two 8 Ω speakers, each capable of delivering 10 W of power. At the heart of the unit will be an audio amplifier and in later chapters we will look at circuit techniques appropriate for the design of such a unit. Here we will restrict ourselves to defining the characteristics required of such an amplifier.

From the information given above, estimate the power gain required of the amplifier (as both a simple ratio and in dBs) and also estimate the voltage gain required.

Key points

- Amplification is a fundamental part of most electronic systems.
- Amplifiers may be active or passive.
- The power delivered at the output of a passive amplifier cannot be greater than that absorbed at its input. An example of a passive electronic amplifier is a transformer.
- Active amplifiers take power from some external energy source and so can produce power amplification. Most electronic amplifiers are active.
- When designing and analysing amplifiers, equivalent circuits are invaluable. They allow the interaction of the circuit with other components to be investigated without a detailed knowledge or understanding of the internal construction of the amplifier.
- Amplifier gains are often measured in decibels (dB).
- The gain of all amplifiers falls at high frequencies. In some cases the gain also falls at low frequencies. The upper cut-off frequency and the lower cut-off frequency (if this exists) are the points at which the gain falls by 3 dB compared with its mid-band value. The difference between these two values (or between the upper cut-off frequency and zero if no lower cut-off frequency exists) defines the bandwidth of the amplifier.
- Differential amplifiers take as their input the difference between two input signals.
- Operational amplifiers (op-amps) are a common form of differential amplifier.
- In many applications, simple amplifiers, perhaps based on single transistors, may be more appropriate than more complicated circuits.

Exercises

14.1 Sketch lever arrangements that represent:

(a) a non-inverting force amplifier;

(b) a non-inverting force attenuator;

(c) an inverting force amplifier;

(d) an inverting force attenuator.

14.2 Sketch a pulley arrangement that represents a non-inverting force amplifier.

14.3 Is the torque amplifier in Figure 14.2 an inverting or a non-inverting amplifier? How could this arrangement be modified to produce an amplifier of the other type?

14.4 Conventional automotive hydraulic braking systems are an example of passive amplifiers. What physical quantity is being amplified?

Such systems may also be regarded as attenuators. What physical quantity is attenuated?

Power-assisted automotive brakes are active amplifiers. What is the source of power?

14.5 Identify examples (other than those given in the text and in earlier exercises) of both passive and active amplifiers for which the operation is mechanical, hydraulic, pneumatic, electrical and physiological. In each case, identify the physical quantity that is amplified and, for the active examples, the source of power.

14.6 If an amplifier has a voltage gain of 25, what will be the output voltage when the input voltage is 1 V?

14.7 If an amplifier has an input voltage of 2 V and an output voltage of 0.2 V, what is its voltage gain?

14.8 An amplifier has an unloaded voltage gain of 20, an input resistance of 10 kΩ and an output resistance of 75 Ω. The amplifier is connected to a voltage source of 0.5 V that has an output resistance of 200 Ω, and to a load resistor of 1 kΩ. What will be the value of the output voltage?

14.9 What is the voltage gain of the amplifier in the arrangement given in Exercise 14.8?

14.10 Calculate the input power, output power and power gain of the arrangement in Exercise 14.8.

14.11 Confirm your results for Exercises 14.8 and 14.9 using computer simulation. You may wish to start with the circuit in Computer simulation exercise 14.1.

14.12 An amplifier has an unloaded voltage gain of 500, an input resistance of 250 kΩ and an output resistance of 25 Ω. The amplifier is connected to a voltage source of 25 mV, which has an output resistance of 4 kΩ, and to a load resistor of 175 Ω. What will be the value of the output voltage?

14.13 What is the voltage gain of the amplifier in the arrangement given in Exercise 14.12?

14.14 Calculate the input power, output power and power gain of the arrangement in Exercise 14.12.

14.15 Confirm your results for Exercises 14.12 and 14.13 using computer simulation. You may wish to start with the circuit in Computer simulation exercise 14.1.

14.16 A displacement sensor produces an output of 10 mV per centimetre of movement and has an output resistance of 300 Ω. It is connected to an amplifier that has an unloaded voltage gain of 15, an input resistance of 5 kΩ and an output resistance of 150 Ω. If the output of the amplifier is connected to a voltmeter with an input resistance of 2 kΩ, what voltage will be displayed on the voltmeter for a displacement of the sensor of 1 metre?

14.17 Confirm your result for Exercise 14.16 using computer simulation. You may wish to start with the circuit in Computer simulation exercise 14.1.

14.18 An amplifier with a gain of 25 dB is connected in series to an amplifier with a gain of 15 dB and a circuit that produces an attenuation of 10 dB (that is, a gain of −10 dB). What is the gain of the overall arrangement (in dB)?

14.19 An amplifier has a mid-band gain of 25 dB. What will be its gain at its upper cut-off frequency?

14.20 An amplifier has a mid-band voltage gain of 10. What will be its voltage gain at its upper cut-off frequency?

14.21 A circuit has a lower cut-off frequency of 1 kHz and an upper cut-off frequency of 25 kHz. What is its bandwidth?

14.22 A circuit has no lower cut-off frequency but has an upper cut-off frequency of 5 MHz. What is its bandwidth?

14.23 A differential amplifier has a voltage gain of 100. If a voltage of 18.3 volts is applied to its non-inverting input and a voltage of 18.2 volts is applied to its inverting input, what will be its output voltage?

14.24 What are the minimum and maximum values of the output voltages that can be produced using the arrangement shown in Figure 14.16?

Chapter 15 Control and Feedback

Objectives

When you have studied the material in this chapter, you should be able to:

- explain the concepts of open-loop and closed-loop systems and give electronic, mechanical and biological examples of each
- discuss the role of feedback and closed-loop control in a range of automatic control systems
- identify the major components of feedback systems
- analyse the operation of simple feedback systems and describe the interaction between component values and system characteristics
- describe the uses of negative feedback in overcoming problems of variability in active devices such as transistors and integrated circuits
- explain the importance of negative feedback in improving input resistance, output resistance, bandwidth and distortion.

15.1 Introduction

Control is one of the basic functions performed by a large number of engineering systems. In simple terms, control involves ensuring that a particular operation or task is performed correctly and so it is associated with concepts such as **regulation** and **command**. While control can be associated with human activities, such as organisation and management, we are more concerned here with **control systems** that perform tasks automatically. However, the basic principles involved are very similar.

Invariably, the goal of a control system is to determine the value, or state, of one or more physical quantities. For example, a pressure regulator might aim to control the pressure within a vessel, while a climate control system might aim to determine the temperature and humidity within a building. The control system affects the various physical quantities by using appropriate actuators and, if we choose to consider our system as including these actuators, then the output of our system can be thought of as the physical quantity, or quantities, being controlled. Thus, in the example given earlier, the output of our pressure regulation system might be considered to be the actual pressure present within the vessel. The inputs to the control system will determine the value of the output that is produced, but the form of the inputs will depend on the nature of the system.

To illustrate this latter point, let us consider two possible methods for controlling the temperature within a room. The first is to use a heater that has a control that varies the heat output. The user sets the control to give a certain heat output and hopes that this will achieve the desired temperature. If the setting is too low, the room will not reach the desired value, but if the setting is too high, the temperature will rise above the desired value. If an appropriate setting is chosen, the room should stabilise at the right temperature, but will become too hot or too cold if external factors – such as the outside temperature or the level of ventilation – are changed.

An alternative approach is to use a heater equipped with a temperature controller. The user then sets the required temperature and the controller increases or decreases the heat output to achieve and then maintain this value. It does this by comparing the desired and actual temperatures and using the difference between them to determine the appropriate heat output. Such a system should maintain the temperature of the room even if external factors change.

Note that, in these two methods of temperature control, the inputs to the system are quite different. In the first, the input determines the heat to be produced by the heater, while in the second the input sets the required temperature.

15.2   Open-loop and closed-loop systems

The alternative strategies discussed above illustrate two basic forms of control, which are illustrated in Figure 15.1. In this figure the **user** is simply the person using the system, the **goal** is the desired result and the **output** of the system is the achieved result. In the example above, the goal is the *required* room temperature and the output is the *actual* room temperature. The **forward path** is the part of the system that affects the output, in response to its input. In our examples, this is the element that produces heat. In practice the forward path will also have inputs that provide it with power, but these are not normally shown in this form of diagram. As in the diagrams of amplifiers in the last chapter, we are interested in the functionality of the arrangement, not its implementation.

Figure 15.1(a) shows what is termed an **open-loop control system**, which corresponds to the first of our two heating methods. Here, the user of the system has a goal (in our example, this is achieving the required temperature) and the user uses knowledge of the characteristics of the system to select an appropriate input to the forward path. In our example, this represents the user selecting an appropriate setting for the heat control. The forward path takes this input and generates a corresponding

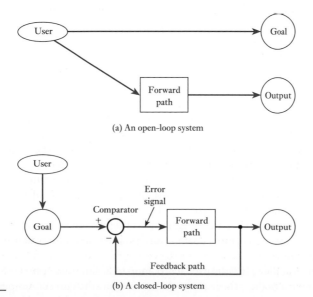

(a) An open-loop system

(b) A closed-loop system

Figure 15.1
Open-loop and
closed-loop systems.

output (which in our example is the actual room temperature). The closeness of the output to the goal will depend on how well the input has been selected. However, even if the input is chosen well, changes in the characteristics of the forward path (the heater) or in the environment (such as changes in the level of ventilation in the room) will affect the output and perhaps make it move further from the goal.

An alternative approach is shown in Figure 15.1(b). This shows a **closed-loop control system**, which corresponds to the second of our heating methods. Again, the user of the system has a goal, but in this case, the user 'inputs' this goal directly into the system. Closed-loop systems make use of a **feedback path**, via which information about the output is fed back for comparison with the goal. In this case, the difference between the output and the goal represents the current *error* in the operation of the system. Therefore, the output is subtracted from the goal to produce an **error signal**. If the output is *less* than the goal (in our example, the actual temperature is less than the required temperature), this will produce a positive error signal, which will cause the output to increase, reducing the error. If the output is *greater* than the goal (in our example, the actual temperature is higher than the required temperature), this results in a negative error signal, which will cause the output to fall, again reducing the error. In this way, the system drives the output to match the goal. One of the many attractive characteristics of this approach is that it automatically compensates for variations in the system or environment. If for any reason the output deviates from the goal, the error signal will drive the system to bring the output back to the required value.

Open-loop systems rely on knowledge of the relationship between the input and the output. This relationship may be ascertained by a process of calibration. Closed-loop systems operate by measuring the actual output of the system and using this information to drive it to the required value.

15.3   Automatic control systems

Almost all forms of automatic control are based on the use of closed-loop systems. Examples include not only a wide range of human-made systems, but also those within the natural world. Figure 15.2 shows examples of automatic control arrangements. In each case the forward path of the system includes some form of actuator that controls the output of the system. A sensor then detects the output and produces a signal that can be fed back to achieve effective control.

Figure 15.2(a) represents an arrangement similar to the example used above. Here, the forward path is the heating element and the feedback path includes a temperature sensor. The input could be in the form of an electrical signal, in which case an electrical sensor and comparator would be used. Alternatively, the input could be mechanical (in the form of the position of a dial perhaps), in which case the sensor might also produce a mechanical output, which would be compared directly with the input. Heater control can be performed using analogue techniques, but it is more usual to use a digital (usually binary) approach. In this case, the heater is turned on or off depending on whether the temperature is below or above that required.

Figure 15.2(b) shows one of many electronic automatic control systems found in cars. Here, the cruise control uses an actuator connected to the throttle to vary the power produced by the engine. This in turn affects the speed of the car, which is sensed by a speed sensor. This information is fed back to allow the cruise control to keep the speed constant, despite variations in driving conditions or the inclination of the road.

Almost all organisms use closed-loop techniques to control their various functions. In our own bodies, these are used to maintain the correct temperature, determine the

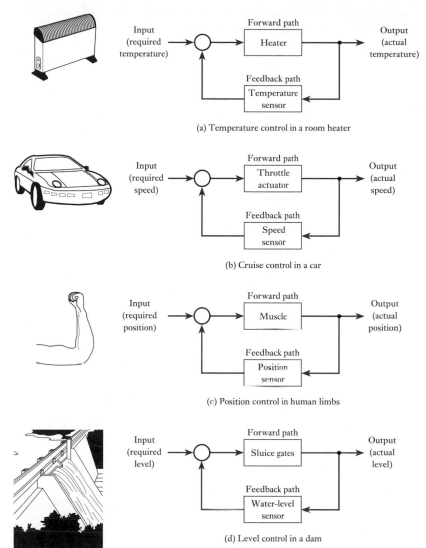

Figure 15.2
Examples of automatic control systems.

levels of chemicals or nutrients in the blood and oversee the movement of our limbs. Figure 15.2(c) illustrates the mechanism used to control the position of the arm. Here, a desire to move the arm to a particular arm position results in signals being sent to the appropriate muscles. Sensors within the arm then sense its actual position and this information is used to correct the arm's position if necessary.

The last example, shown in Figure 15.2(d), is of a level control system for a dam. Here, a measurement of the height of the water behind the dam is compared with the required level. The difference between these two values is then used to determine the rate at which water is released through the sluice gates.

In all these examples, **feedback** is used to drive the output towards the required value (the goal). In each case, the use of feedback makes the operation of the system largely independent of variations in the forward path or external environment. In the last chapter we noted that electronic amplifiers suffer from great variability in their characteristics. This suggests that such devices might benefit from the use of feedback to overcome these deficiencies.

15.4 Feedback systems

In order to understand the properties of systems that use feedback, we need to be able to analyse their behaviour. To assist us in this task, Figure 15.3 shows a block diagram of a generalised feedback system. The input and output of this system are given the symbols X_i and X_o, and these could represent physical quantities such as force, position or speed, or could represent electrical quantities such as voltage or current.

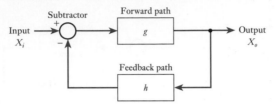

Figure 15.3
A generalised feedback system.

As in our earlier examples of closed-loop systems, the arrangement in Figure 15.3 consists of a forward path, a feedback path and a subtractor. The forward path will include within it the system or component that is to be controlled. This system is usually referred to as the **plant**. The forward path may also contain some additional elements that are added in order to drive the plant and make it easier to control. These elements are referred to as the **controller**. The combined behaviour of the controller and plant are represented in the diagram by the mathematical function g. This stands for the relationship between the input and the output of the forward path and is called its **transfer function**. The feedback path represents the sensor used to detect the output and any processing that is applied to the signals it produces. The feedback path is also represented by a transfer function, which is given the symbol h.

From a knowledge of the transfer functions g and h, it is possible to analyse the behaviour of the overall system shown in Figure 15.3 and, hence, predict its behaviour. The field of **control engineering** is largely concerned with the analysis of such systems and the design of appropriate controllers and feedback arrangements to tailor their behaviour to meet particular needs. Unfortunately, in many situations the characteristics of the physical plant are complicated, perhaps including frequency dependencies, non-linearities or time delays. Consequently, the analysis of such systems is often quite involved and so is beyond the scope of this book.

Fortunately, the feedback systems that we meet most often within electronic systems are much more straightforward. Here, we can often assume that the transfer functions of the forward and feedback paths are simple gains, which greatly simplifies analysis. This is illustrated in Figure 15.4, where the forward path is represented by a gain of A and the feedback path by a gain of B.

As the output of the forward path is X_o and its gain is A, its input must be given by X_o/A. Similarly, as the input to the feedback path is X_o and its gain is B, its output must be BX_o. From the diagram we can see that the input to the forward path (which we have just determined to be X_o/A) is actually produced by subtracting the feedback signal (which we have just found to be BX_o) from the input X_i. Therefore

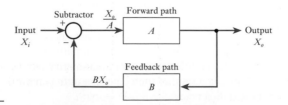

Figure 15.4
A feedback system.

$$\frac{X_o}{A} = X_i - BX_o$$

or by rearranging

$$\frac{X_o}{X_i} = \frac{A}{1 + AB}$$

The ratio of the output to the input is the gain of the feedback arrangement, which is usually given the symbol G. Therefore

$$G = \frac{A}{1 + AB} \tag{15.1}$$

This expression for the gain is also referred to as the **transfer function** of the feedback system.

It is common to refer to the forward gain A as the **open-loop gain** (as it is the gain that the circuit would have with the feedback disconnected) and to the overall gain G as the **closed-loop gain** (as this is the gain of the circuit with the feedback present). The overall characteristics of the system depend on the values of A and B or, more directly, on the product AB.

15.4.1 If *AB* is negative

If either A or B is negative (but not both), the product AB will be negative. If now the term $(1 + AB)$ is less than 1, G is greater than A. In other words, the gain of the circuit will be increased by the feedback. This is termed **positive feedback**.

A special case of positive feedback occurs when $AB = -1$. Under these circumstances

$$G = \frac{A}{1 + AB} = \frac{A}{1 - 1} = \infty \text{ (infinity!)}$$

As the gain of the circuit is infinite, it has a limited range of applications, but is useful in the production of **oscillators**.

15.4.2 If *AB* is positive

If A and B are either both positive or both negative, the term AB will be positive. Thus, the term $(1 + AB)$ must be positive and greater than 1, and G must be less than A. In other words, the gain of the circuit with feedback is less than it would be without feedback. This is **negative feedback**.

If the product AB is not only positive but also large compared with 1, the term $(1 + AB)$ is approximately equal to AB and the expression

$$G = \frac{A}{1 + AB}$$

may be simplified to

$$G \approx \frac{A}{AB} = \frac{1}{B} \tag{15.2}$$

This special case of negative feedback is of great importance as we now have a system in which the overall gain is independent of the gain of the forward path, being determined solely by the characteristics of the feedback path.

The ability to produce a system where the overall gain is independent of the gain of the forward path is of great significance. In the previous chapter we noted that devices such as transistors and operational amplifiers suffer from great variability in their characteristics. Negative feedback would seem to offer a way to tackle this problem. Therefore, for the remainder of this chapter we will concentrate on the uses and characteristics of negative feedback and will leave further discussion of positive feedback until later (see Chapter 23).

15.4.3 Notation

It should be noted that, in some textbooks, the subtractor in Figure 15.4 is replaced with an adder. This is an equally valid representation of a feedback arrangement and a similar analysis to that given above produces an expression for the overall gain of the form

$$G = \frac{A}{1 - AB}$$

This equation clearly places different requirements on A and B to achieve positive or negative feedback from the analysis given earlier.

In this text, we assume use of a subtractor in the feedback block diagram as this produces arrangements that more closely correspond to the real circuits that we will consider later. You should, however, be aware that other representations exist.

Video 15B

15.5 Negative feedback

So far, we have considered control and feedback in a generic manner. For example, in Figure 15.3 and Figure 15.4, the input and output are given the symbols X_i and X_o to indicate that they may represent any physical quantities. However, it is now appropriate to turn our attention more specifically to the use of feedback within electronic applications.

We saw in the last chapter that one characteristic of operational amplifiers is that their gain, while being large, is also variable from device to device. We also noted that their gain varies with temperature. These characteristics are common to almost all **active devices**. In contrast, passive components, such as resistors and capacitors, can be made to very high levels of precision and can be very stable as their temperature varies.

In the last section, when looking at negative feedback, we noted that a feedback circuit in which the product AB is positive and much greater than 1 has an overall gain that is independent of the forward gain A, as it is determined entirely by the feedback gain B. If we construct a negative feedback system using an active amplifier as the forward path A, and a passive network as the feedback path B, we can produce an amplifier with a stable overall gain independent of the actual value of A. This is illustrated in Example 15.1. Here an amplifier with a high, but variable, voltage gain is combined with a stable feedback network to form an amplifier with a gain of 100.

Example 15.1 | **Design an arrangement with a stable voltage gain of 100 using a high-gain active amplifier. Determine the effect on the overall gain of the circuit if the voltage gain of the active amplifier varies from 100,000 to 200,000.**

We will base our circuit on our standard block diagram.

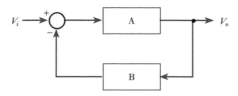

From Equation 15.2, we know that the overall gain is given by

$$G \approx \frac{1}{B}$$

and, therefore, for a gain of 100, we choose $B = 1/100$ or 0.01.

When the gain of the amplifier (A) is 100,000, the overall gain will be

$$G = \frac{A}{1 + AB} = \frac{100,000}{1 + (100,000 \times 0.01)}$$

$$= \frac{100,000}{1 + 1000} = 99.90 \approx \frac{1}{B}$$

When the gain of the amplifier (A) is 200,000, the overall gain will be

$$G = \frac{A}{1 + AB} = \frac{200,000}{1 + (200,000 \times 0.01)}$$

$$= \frac{200,000}{1 + 2000} = 99.95 \approx \frac{1}{B}$$

Notice that a change of 100 per cent in the value of the gain of the active amplifier (A) produces a change of only 0.05 per cent in the overall gain, G.

Example 15.1 shows that the large variation in gain associated with active circuits can be overcome by the use of negative feedback, provided that a stable feedback arrangement can be produced. In order to make the feedback path stable, it must be constructed using only passive components. Fortunately, this is a simple task.

We have seen that the overall gain of the feedback circuit is $1/B$. Therefore, to have an overall gain greater than 1, we require B to be less than 1. In other words, our feedback path may be a **passive attenuator**.

Construction of such a feedback arrangement using passive components is simple. If we take as an example the value used in Example 15.1, we require a passive attenuator with a voltage gain of $1/100$. This can be achieved as shown in Figure 15.5. The circuit is a simple potential divider with a ratio of 99:1. The output voltage V_o is related to the input voltage V_i by the expression

$$V_o = V_i \frac{R}{R + 99R}$$

$$\frac{V_o}{V_i} = \frac{1}{100}$$

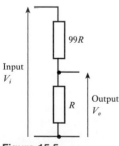

Figure 15.5
A passive attenuator with a gain of 1/100.

The resistor values of R and $99R$ are shown simply to indicate their relative magnitudes. In practice, R might be 1 kΩ and $99R$ would then be 99 kΩ. The actual values used would depend on the circuit configuration.

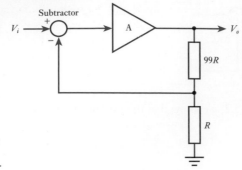

Figure 15.6
An amplifier with a gain
of 100.

Having decided that the forward path of our feedback circuit will be a high-gain active amplifier and the feedback path will be a resistive attenuator, we are now in a position to complete the circuit. Continuing with the values given in Example 15.1, we may now draw our circuit diagram, as shown in Figure 15.6. This shows an arrangement based on an active amplifier with a gain of A and a feedback network with a gain B of $1/100$. This produces an amplifier with an overall gain G of 100 (that is, $1/B$).

Our last remaining problem is that of implementing the subtractor in Figure 15.6 as this does not seem to be a standard component. Fortunately, we have already discussed a means of providing this function when we considered **operational amplifiers** in Chapter 14. Op-amps are **differential amplifiers** – that is, they amplify the difference between two input signals. We could visualise such an amplifier as a single-input amplifier with a subtractor connected to its input. Therefore, an operational amplifier may be used to replace both the amplifier and the subtractor in the circuit in Figure 15.6, as shown in Figure 15.7.

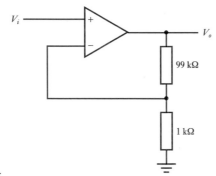

Figure 15.7
An amplifier with a gain
of 100 based on an
operational amplifier.

File 15A

Computer simulation exercise 15.1

Simulate the circuit in Figure 15.7, basing your design on one of the operational amplifiers supported by your simulation package (and remembering to include connections to appropriate power supplies). Apply a 50 mV DC input to the circuit and measure the output voltage. Then, deduce the voltage gain of the circuit and confirm that this is as expected. Experiment with different values of the input voltage and investigate how this affects the voltage gain.

Inherent in the design of our simple amplifier is the assumption that the overall gain is equal to $1/B$. From our earlier discussions we know that this assumption is only valid provided that the product AB is much greater than 1. In our circuit, the forward path is implemented using an operational amplifier and from the last chapter we know that this is likely to have a voltage gain (A) of perhaps 10^5 or 10^6. As B is $1/100$ in our example, this means that AB will have a value of about 10^3 to 10^4. As this is much greater than 1, it would seem that our assumption that the gain is equal to $1/B$ is valid. However, let us consider another example.

Example 15.2 | **Design an arrangement with a stable voltage gain of 10,000 using a high-gain active amplifier. Determine the effect on the overall gain of the circuit if the voltage gain of the active amplifier varies from 100,000 to 200,000.**

As before, we will base our circuit on our standard block diagram.

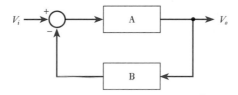

From Equation 15.2, we know that the overall gain is given by

$$G \approx \frac{1}{B}$$

and, therefore, for a gain of 10,000, we choose $B = 1/10,000$ or 0.0001.
    When the gain of the amplifier A is 100,000, the overall gain will be

$$G = \frac{A}{1 + AB} = \frac{100,000}{1 + (100,000 \times 0.0001)}$$

$$= \frac{100,000}{1 + 10}$$

$$= 9091$$

When the gain of the amplifier A is 200,000, the overall gain will be

$$G = \frac{A}{1 + AB} = \frac{200,000}{1 + (200,000 \times 0.0001)}$$

$$= \frac{200,000}{1 + 20}$$

$$= 9524$$

It can be seen that the resultant gain is *not* very close to $1/B$ (10,000) and that variations in the gain of the forward path, A, have significant effects on the overall gain.

The above example shows that, in order for the gain to be stabilised by the effects of negative feedback, the product AB must be much greater than 1. In other words

$$AB \gg 1$$

or

$$A \gg \frac{1}{B}$$

A is the open-loop gain of the active amplifier and $1/B$ is the closed-loop gain of the complete circuit. This implies that the condition required for the stabilising effects of negative feedback to be effective is that

$$\text{open-loop gain} \gg \text{closed-loop gain}$$

File 15B

> ### Computer simulation exercise 15.2
>
> Simulate the circuit in Example 15.2 using idealised gain blocks and a subtractor (these elements are available in most simulation packages). Apply a 1 V DC input to the circuit and investigate the output voltage for different values of forward gain A and feedback gain B. Then, investigate the conditions required for the overall gain to be approximately equal to $1/B$.

We have seen that negative feedback allows us to generate amplifiers with overall characteristics that are constant despite variations in the gain of the active components. In fact, negative feedback can also produce a range of other desirable effects and is widely used in many forms of electronic circuitry.

15.6   The effects of negative feedback

It is clear from the above that negative feedback has a profound effect on the gain of an amplifier – and on the consistency of that gain. However, feedback also affects other circuit characteristics. In this section we look at a range of circuit parameters.

15.6.1   Gain

We have seen that negative feedback reduces the gain of an amplifier. In the absence of feedback, the gain of an amplifier (G) is simply its **open-loop gain** A. We know from Equation 15.1 that with feedback the gain becomes

$$G = \frac{A}{1 + AB}$$

and thus the effect of the feedback is to reduce the gain by a factor of $1 + AB$.

15.6.2   Frequency response

We noted earlier (in Chapter 14) that the gain of all amplifiers falls at high frequencies and that, in many cases, it also drops at low frequencies.

From the previous discussion of gain, we know that the closed-loop gain of a feedback amplifier is largely independent of the open-loop gain of the amplifier, *provided* that the latter is considerably greater than the former. As the open-loop gain of all amplifiers falls at high frequencies (and often at low frequencies), it is clear that the closed-loop gain will also fall in these regions. However, if the open-loop gain is considerably greater than the closed-loop gain, the former will be able to fall by a considerable amount before this has an appreciable effect on the latter. Thus, the closed-loop gain

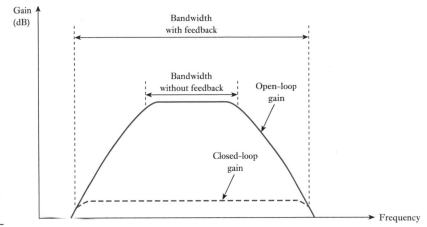

Figure 15.8
The effects of negative feedback on frequency response.

will be stable over a wider frequency range than that of the amplifier without feedback. This is illustrated in Figure 15.8.

The solid line in Figure 15.8 shows the variation of gain with frequency of an amplifier without feedback – that is, its open-loop frequency response. The addition of negative feedback (shown by the dashed line) reduces the gain of the arrangement. The resultant closed-loop gain is constant over the range of frequencies where it is considerably less than the amplifier's open-loop gain. The addition of negative feedback thus results in an increase in the bandwidth of the amplifier.

Thus, the use of negative feedback reduces gain, but increases bandwidth, allowing designers to 'trade off' one characteristic against the other. As gain is a relatively inexpensive commodity in modern electronic circuits, this is often a useful mechanism. We will see in the next chapter that, in some cases, the increase in bandwidth is directly proportional to the fall in gain. In other words, while feedback decreases the gain of an amplifier by $1 + AB$, it increases the bandwidth by $1 + AB$. In this case, the product of the gain and the bandwidth remains constant. Thus

$$\text{gain} \times \text{bandwidth} = \text{constant} \tag{15.3}$$

The value produced by multiplying these two terms is given the obvious name of the **gain–bandwidth product**.

15.6.3 Input and output resistance

One of the important characteristics of negative feedback is that it tends to keep the output contstant despite changes in its environment. One way in which this manifests itself in *voltage* amplifiers is that it tends to keep the output voltage constant, despite changes in the load applied to the amplifier. This is analogous to the way in which cruise control in a car maintains a constant speed whether going up or down a hill. From our discussions on loading in the last chapter, we know that loading effects are minimised within voltage amplifiers when the output resistance is *reduced* or when the input resistance is *increased*. It is therefore no surprise to discover that negative feedback can achieve both these desirable effects. However, the situation is slightly more complicated than this.

We noted earlier that the general representation of a feedback system shown in Figure 15.3 is applicable to a wide range of systems, and that the input and output can represent not only voltages but also currents or other physical quantities. Clearly, if our input and output quantities are temperatures, we would expect the feedback arrangement to maintain the output temperature despite changes in the environment. Similarly, if the input and output are currents (rather than voltages), we would expect the system to minimise the effects of loading on these currents. In order to do this, the circuit will *increase* the output resistance and *decrease* the input resistance – exactly the opposite action to that described above for a voltage amplifier.

In fact, negative feedback can either increase or decrease input resistance, and either increase or decrease output resistance, depending on how it is applied. The factors that determine the effects of the feedback are the way in which the output is sensed and the way in which the feedback signal is applied. If the feedback senses the output *voltage*, it will tend to make the output voltage more constant by *decreasing* the output resistance. In contrast, if the feedback senses the output *current*, it will tend to make the output current more constant by *increasing* the output resistance. Similarly, if the feedback is applied by subtracting a *voltage* related to the output from the input voltage, it will tend to make the circuit a better voltage amplifier by *increasing* the input resistance. In contrast, if the feedback is applied by subtracting a *current* from the input current, it will tend to make the circuit a better current amplifier by *decreasing* the input resistance.

To illustrate the above, let us look back at the circuit of Figure 15.7. Here the signal fed back is related to the output *voltage* (the output voltage is applied across the potential divider to produce the feedback signal) and so the feedback *reduces* the output resistance of the circuit. In this case the *voltage* from the potential divider is applied to the inverting input of the amplifier where it is effectively subtracted from the input voltage (as in Figure 15.6). Therefore the feedback *increases* the input resistance of the circuit. Thus the circuit of Figure 15.7 has a much higher input resistance, and a much lower output resistance, than the operational amplifier would have if used without feedback. We will return to look at the input and output resistance of this and other related circuits in the next chapter when we look at operational amplifiers in more detail.

So far we have noted that negative feedback can either increase or decrease input and output resistance, but we have not quantified this effect. Having noted that negative feedback decreases gain by a factor of $1 + AB$, and that it can increase bandwidth by a similar factor, it is perhaps not totally surprising to discover that the factor by which input and output resistance are increased or decreased is also $1 + AB$. The proof of this relationship is fairly straightforward but it does not seem appropriate to repeat it here.

15.6.4 Distortion

Many forms of distortion are caused by a **non-linear amplitude response** which means that the gain of the circuit varies with the amplitude of the input signal. Since negative feedback tends to make the gain more stable, it also tends to reduce distortion. Perhaps unsurprisingly, it can be shown that the factor by which distortion is reduced is again $1 + AB$.

15.6.5 Noise

The noise produced by an amplifier is also reduced by negative feedback, and again this is by a factor of $1 + AB$. However, this effect only applies to noise produced *within* the amplifier itself, not to any noise that already corrupts the input signal. The latter will be indistinguishable from the input signal and will therefore be amplified along with this signal. We will look at noise in more detail later (in Chapter 22).

15.6.6 Stability

From Equation 15.1 we know that

$$G = \frac{A}{(1 + AB)}$$

and this implies that, provided $|1 + AB|$ is greater than 1, the gain with feedback G will be less than the open-loop gain of the amplifier A.

So far in this section we have assumed that both A and B can be described by simple real gains, such that their product AB is a positive real number. Under these circumstances $|1 + AB|$ is always greater than unity. However, all amplifiers produce a phase shift in the signals passing through them, the magnitude of which varies with frequency. The result of this phase shift is that, at some frequencies, $|1 + AB|$ may be less than 1. Under these circumstances, the feedback *increases* the gain of the amplifier and is now positive, rather than negative, feedback. This can result in the amplifier becoming unstable and the output oscillating, independent of the input signal. We will return to look at the stability of circuits in Chapter 23 when we consider positive feedback in more detail.

15.7 Negative feedback – a summary

All negative feedback systems share a number of properties:

- They tend to maintain their output despite variations in the forward path or in the environment.
- They require a forward path gain that is greater than that which would be necessary to achieve the required output in the absence of feedback.
- The overall behaviour of the system is determined by the nature of the feedback path.

When applied to electronic amplifiers, negative feedback has many beneficial effects:

- It stabilises the gain against variations in the open-loop gain of the amplifying device.
- It increases the bandwidth of the amplifier.
- It can be used to increase or decrease input and output resistance as required.
- It reduces distortion caused by non-linearities in the amplifier.
- It reduces the effects of noise produced within the amplifier.

In exchange for these benefits, negative feedback reduces the gain of the amplifier. In most cases this is a small price to pay as the majority of modern amplifying devices have a high gain and are inexpensive, allowing many stages to be used if necessary. However, the use of negative feedback can have implications for the **stability** of the circuit (we will return to this issue in Chapter 23).

Negative feedback plays a vital role in electronic circuits for a wide range of applications. We will look at a few examples of such circuits in the next chapter when we look in more detail at operational amplifiers.

Video 15C

Further study

The world in which we live is full of various forms of control system including both open-loop and closed-loop systems. For example, air conditioning units often have controls that allow the user to set the required temperature. The cooling system is then turned on and off automatically to achieve this temperature. This is an example of a closed-loop control system where the input temperature represents the 'goal' of the system and the 'output' is the actual temperature achieved. While such systems are common, simpler air conditioning systems are also used (particularly within cars) where the user determines the amount of cooling to be provided and adjusts this to achieve a comfortable temperature. This is clearly an example of an open-loop approach.

Identify other common control systems – both natural and man-made – and consider whether these are open-loop or closed-loop in nature. In each case determine the major components of the system and relate these to the appropriate elements of Figure 15.1. Suitable examples might include other automotive systems, as well as domestic and industrial systems. You might also like to consider systems such as the temperature control mechanism of the human body, a stock control system within a warehouse and an aircraft autopilot.

Key points

- Feedback systems form an essential part of almost all automatic control systems, be they electronic, mechanical or biological.

- Feedback systems may be divided into two types. In negative feedback systems the feedback tends to *reduce* the input to the forward path. In positive feedback systems the feedback tends to *increase* the input to the forward path.

- If the gain of the forward path is A, the gain of the feedback path is B and the feedback signal is subtracted from the input, then the overall gain G of the system is given by

$$G = \frac{A}{(1 + AB)}$$

- If the loop gain AB is positive we have negative feedback. If the loop gain is also large compared with 1, the expression for the gain simplifies to $1/B$. In these circumstances, the overall gain is independent of the gain of the forward path.

- If the loop gain AB is negative and less than 1 we have positive feedback. For the special case where AB is equal to -1, the gain is infinite. This condition is used in the production of oscillators.

- Negative feedback tends to increase the bandwidth of an amplifier at the expense of a loss of gain. In many cases the factor by which the bandwidth is increased and the gain is reduced is $(1 + AB)$. Thus the gain–bandwidth product remains constant.

- Negative feedback also tends to improve the input resistance, output resistance, distortion and noise of an amplifier. In each case, the improvement is generally by a factor of $(1 + AB)$.

- While negative feedback brings many benefits, it can also bring with it problems of instability.

Exercises

15.1   What is meant, in engineering, by the term 'control'?

15.2   Give three examples of common control systems.

15.3   In each of the control systems identified in your answer to the last exercise, what constitutes the input and the output to the system?

15.4   In a car's cruise control, what is the input and what is the output?

15.5   In a stock control system in a warehouse, what is the input and what is the output?

15.6   Explain the meanings of the terms 'user', 'goal', 'output', 'forward path', 'feedback path' and 'error signal', as they relate to an open- or closed-loop system.

15.7   From where does the forward path obtain its power?

15.8   Sketch a block diagram of a generalised feedback system and derive an expression for the output in terms of the input and the gains of the forward path A and feedback path B.

15.9   In the expression you derived in Exercise 15.8, what range of values for AB corresponds to positive feedback?

15.10 In the expression you derived in Exercise 15.8, what range of values for AB corresponds to negative feedback?

15.11 In the expression you derived in Exercise 15.8, what range of values for AB would be used to produce an oscillator?

15.12 Explain why the characteristics of active devices encourage the use of negative feedback.

15.13 Design an arrangement with a stable voltage gain of 10 using a high-gain active amplifier. Determine the effect on the overall gain of the circuit if the voltage gain of the active amplifier varies from 100,000 to 200,000.

15.14 What is the voltage gain of the following arrangement?

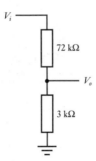

15.15 Design a passive attenuator with a gain of 1/10.

15.16 Determine the voltage gain of the following amplifier.

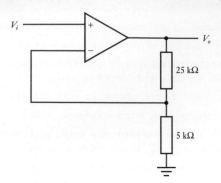

15.17 Confirm your results for Exercise 15.16 using computer simulation. You may wish to start with the circuit in Computer simulation exercise 15.1.

15.18 Design an amplifier with a gain of 10 based on an operational amplifier.

15.19 Confirm your results for Exercise 15.18 using computer simulation. You may wish to start with the circuit in Computer simulation exercise 15.1.

15.20 Design an arrangement with a stable voltage gain of 20,000 using a high-gain active amplifier. Determine the effect on the overall gain of the circuit if the voltage gain of the active amplifier varies from 100,000 to 200,000.

15.21 Confirm your results for Exercise 15.20 using computer simulation. You may wish to start with the circuit in Computer simulation exercise 15.2.

15.22 By what factor is the gain of an amplifier changed by the use of negative feedback? Is this an increase or a decrease in gain?

15.23 How is the bandwidth of an amplifier changed by the use of negative feedback?

15.24 Under what circumstances might it be advantageous to increase the input resistance of an amplifier? How might this be achieved?

15.25 Under what circumstances might it be advantageous to decrease the input resistance of an amplifier? How might this be achieved?

15.26 Under what circumstances might it be advantageous to decrease the output resistance of an amplifier? How might this be achieved?

15.27 Under what circumstances might it be advantageous to increase the output resistance of an amplifier? How might this be achieved?

15.28 How does negative feedback affect the distortion produced by an amplifier?

15.29 How does negative feedback affect noise present in an input signal to an amplifier?

Chapter 16 Operational Amplifiers

Objectives

When you have studied the material in this chapter, you should be able to:

- outline the uses of operational amplifiers in a range of engineering applications
- describe the physical form of a typical op-amp and its external connections
- explain the concept of an ideal operational amplifier and describe the characteristics of such a device
- draw and analyse a range of standard circuits for performing functions such as the amplification, buffering, addition and subtraction of signals
- design op-amp circuits to perform simple tasks, including the specification of appropriate passive components
- describe the ways in which real operational amplifiers differ from ideal devices
- explain the importance of negative feedback in tailoring the characteristics of operational amplifiers to suit a particular application.

16.1 Introduction

Operational amplifiers (or **op-amps**) are among the most widely used building blocks for the construction of electronic circuits. One of the reasons for this is that they are nearly ideal voltage amplifiers, which greatly simplifies design. As a result, op-amps tend to be used not only by specialist electronic engineers but also by other engineers who want a simple solution to an instrumentation or control problem. Thus, a mechanical engineer who wishes to display the speed of rotation of an engine or a civil engineer who needs to monitor the stress on a bridge would be very likely to use an operational amplifier to construct the instrumentation required.

Operational amplifiers are a form of **integrated circuit** (IC) – that is, they are constructed by integrating a large number of electronic devices into a single semiconductor component. In later chapters we will look at the operation of such components, but for the moment we are concerned simply with their characteristics and how they are used.

A typical op-amp takes the form of a small plastic package with an appropriate number of pins for carrying signals and power in and out of the device. Figure 16.1 shows two common forms. Figure 16.1(a) shows an eight-pin 'dual in line' or DIL package. Such a component would often accommodate a single op-amp and such devices might be used when manually producing a prototype circuit or perhaps when manufacturing systems in very small quantities. Larger versions of this package, with a greater number of pins, can house two or four op-amps within a single component. Figure 16.1(b) shows a surface-mounted technology (SMT) component. These components have the advantage of a much smaller physical size, permitting

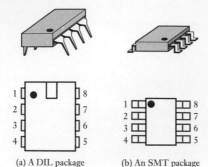

Figure 16.1
Typical operational
amplifier packages.

(a) A DIL package   (b) An SMT package

a much greater circuit density. Because of their size these components are invariably assembled using computer-controlled 'pick and place' machines, and this is by far the most common form of device used in the mass-production of electronic circuits.

In both packages, the pins are numbered anticlockwise when viewed from the top. Pin number 1 is usually marked by a dot or notch or both. The way in which the pins are connected internally is termed the 'pin-out' of the device and Figure 16.2 illustrates typical pin-outs for a range of components. In this figure, the connections labelled V_{pos} and V_{neg} represent the positive and negative power supply voltages respectively. You will remember that these connections are normally omitted from circuit diagrams to aid clarity, but must be connected within a physical circuit to provide a power source for the circuit. Typical values for these quantities might be +15 V and −15 V, but these values will be discussed later in this chapter when we look at real devices in Section 16.5.

One of the many attractive features of operational amplifiers is that they can be configured easily to produce a wide range of electronic circuits. These include not only amplifiers of various forms but also circuits with more specialised functions, such as adding, subtracting or modifying signals. We will look at a few basic circuits in this chapter, but will meet a range of other op-amp circuits later in the text.

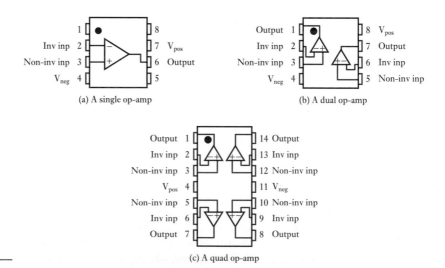

Figure 16.2
Typical operational
amplifier pin-outs.

| 16.2 | **An ideal operational amplifier** |

We noted in the last section that operational amplifiers are nearly ideal voltage amplifiers. Design is often much simpler when using idealised components and so it is common initially to assume that our components are perfect, then to investigate the effects of any non-ideal characteristics. In order to do this in the case of operational amplifiers, we need first to have an idea of how an ideal component would behave.

In Chapter 14 we looked briefly at ideal voltage amplifiers and deduced that these would have an infinite input resistance and a zero output resistance. Under these circumstances, the amplifier would draw no current from the source and its output voltage would be unaffected by the value of the load. Therefore, there would be no **loading effects** when using such an amplifier.

While it is relatively easy to deduce the input and output resistance of an ideal amplifier, it is less obvious what the gain of such a device would be. Clearly, the gain required of a circuit will differ with the application and it is perhaps not clear that one particular gain is 'ideal' for all situations. However, we saw in the last chapter that negative feedback can be used to tailor the gain of an amplifier to any particular value (which is determined by the feedback gain), provided that the open-loop gain is sufficiently high. Therefore, when using negative feedback, it is advantageous to have as high an open-loop gain as possible. Thus, an ideal operational amplifier would have an infinite open-loop gain.

Therefore, an ideal op-amp would have an infinite input resistance, zero output resistance and infinite voltage gain. We are now in a position to draw an equivalent circuit for such a device, starting with the diagram given in Figure 14.14 and modifying the parameters accordingly. This is shown in Figure 16.3. Note that the infinite input resistance means that no current flows into the device and the input terminals appear to be unconnected. Similarly, the zero output resistance means that there is no output resistor. The output voltage is equal to the voltage produced by the controlled voltage source, which is A_v times the differential input voltage V_i. In this case, the voltage gain A_v is infinite.

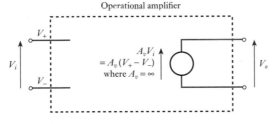

Figure 16.3
Equivalent circuit of an ideal operational amplifier.

Video 16A

| 16.3 | **Some basic operational amplifier circuits** |

Before we look at some simple amplifier circuits, we should perhaps make sure that we are clear on some of the terminology. Electronic amplifiers can be either non-inverting or inverting (see Chapter 14). If the input to a *non-inverting* amplifier is a positive voltage, then the output will also be positive. If a similar input signal is applied to an *inverting* amplifier, the output will be negative. When the input is not a fixed voltage, but an alternating waveform, then the output voltage of either amplifier will also alternate. The effects of these two forms of amplification on an alternating waveform are illustrated in Figure 16.4. In this figure the non-inverting amplifier has a gain of +2, while the inverting amplifier has a gain of −2.

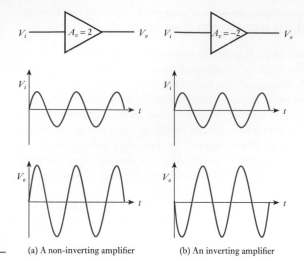

Figure 16.4
Non-inverting and
inverting amplifiers.

(a) A non-inverting amplifier   (b) An inverting amplifier

In the last chapter, we derived the circuit of a non-inverting amplifier from 'first principles'. That is, we started with a generalised block diagram of a feedback system and devised elements to implement the forward and feedback paths. While that circuit was based on an operational amplifier, this is not the process we normally adopt when using op-amps. Commonly, we start with a standard or 'cookbook' circuit and adapt it to suit our needs. Often, this adaptation requires no more than selecting appropriate component values. There are a wide range of these cookbook circuits available to perform a wide range of tasks. We begin by looking at a few well-known examples.

16.3.1 A non-inverting amplifier

The first of our standard circuits is that derived in the last chapter for a **non-inverting amplifier**. This is shown in Figure 16.5(a). Also shown, in Figure 16.5(b), is the same circuit redrawn in a different orientation. This latter form is electrically identical to the earlier circuit and has the same characteristics. It is important that readers can recognise and use this circuit in either form.

Rather than analyse the circuit from first principles (as in Chapter 15) we will look at the operation of the circuit assuming that it contains an ideal op-amp. You will see that this makes the analysis very straightforward.

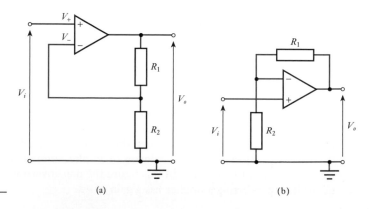

Figure 16.5
A non-inverting
amplifier.

(a)   (b)

First, as the gain of the op-amp is infinite, if the output voltage is finite, the input voltage to the op-amp ($V_+ - V_-$) must be zero. Therefore

$$V_- = V_+ = V_i$$

As the op-amp has an infinite input resistance, its input current must be zero. Therefore, V_- is determined simply by the output voltage and the potential divider formed by R_1 and R_2. Thus

$$V_- = V_o \frac{R_2}{R_1 + R_2}$$

Therefore, as $V_- = V_i$,

$$V_i = V_o \frac{R_2}{R_1 + R_2}$$

and the overall gain of the circuit is given by

$$G = \frac{V_o}{V_i} = \frac{R_1 + R_2}{R_2} \qquad\qquad (16.1)$$

which is consistent with the analysis in Chapter 15 of the circuit of Figure 15.7.

Example 16.1 | **Design a non-inverting amplifier with a gain of 25 based on an operational amplifier.**

We start with our standard circuit.

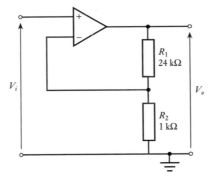

From Equation 16.1 we have

$$G = \frac{V_o}{V_i} = \frac{R_1 + R_2}{R_2}$$

Therefore, if $G = 25$

$$\frac{R_1 + R_2}{R_2} = 25$$

$$R_1 + R_2 = 25R_2$$

$$R_1 = 24R_2$$

As it is the ratio of the resistor values that determines the gain, we are free to choose the actual values. Here we will choose $R_2 = 1 \text{ k}\Omega$, which means that R_1 must be 24 kΩ. When using *ideal* op-amps, the actual values of the resistors are unimportant – it is only the ratio of the values that is significant. However, when we use *real* components, there are factors that affect our choice of component values. These are discussed in Section 16.6.

File 16A

> ### Computer simulation exercise 16.1
>
> Simulate the circuit in Example 16.1 using one of the operational amplifiers supported by your simulation package. Apply a 100 mV DC input to the circuit and measure the output voltage. Then, deduce the voltage gain of the circuit and confirm that this is as expected. Experiment with different values for the two resistors and see how this affects the voltage gain. Experiment with different values for the input voltage (including both positive and negative values) and confirm that the circuit behaves as you expect.

16.3.2 An inverting amplifier

The second of our standard circuits is that of an **inverting amplifier**. This is shown in Figure 16.6. As in the previous circuit, because the gain of the op-amp is infinite, if the output voltage is finite, the input voltage to the op-amp ($V_+ - V_-$) must be zero. Therefore

$$V_- = V_+ = 0$$

As the op-amp has an infinite input resistance, its input current must be zero. Therefore, the currents I_1 and I_2 must be equal and opposite. By applying Ohm's law to the two resistors, we see that

$$I_1 = \frac{V_o - V_-}{R_1} = \frac{V_o - 0}{R_1} = \frac{V_o}{R_1}$$

and

$$I_2 = \frac{V_i - V_-}{R_2} = \frac{V_i - 0}{R_2} = \frac{V_i}{R_2}$$

Therefore, because

$$I_1 = -I_2$$

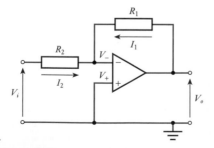

Figure 16.6
An inverting amplifier.

then

$$\frac{V_o}{R_1} = -\frac{V_i}{R_2}$$

and the gain G is given by the very simple result

$$G = \frac{V_o}{V_i} = -\frac{R_1}{R_2} \qquad (16.2)$$

Note the minus sign in the expression for the gain, showing that this is indeed an inverting amplifier.

In this circuit, the negative feedback maintains the voltage on the inverting input (V_-) at zero volts. This may be understood by noting that if V_- becomes more positive than the voltage on the non-inverting input (in this case zero volts), this will cause the output of the op-amp to become negative, which will drive V_- negative through R_1. If, on the other hand, V_- becomes negative with respect to zero volts, the output will become positive, which will tend to make V_- more positive. Thus, the circuit will act to keep V_- at zero, even though this terminal is not physically connected to earth. Such a point within a circuit is referred to as a **virtual earth** and this kind of amplifier is called a **virtual earth amplifier**.

Example 16.2

Design an inverting amplifier with a gain of −25 based on an operational amplifier.

We start with our standard circuit.

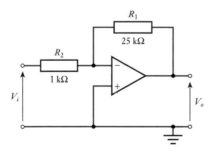

From Equation 16.2, we have

$$G = \frac{V_o}{V_i} = -\frac{R_1}{R_2}$$

Therefore, if $G = -25$

$$-\frac{R_1}{R_2} = -25$$

$$R_1 = 25R_2$$

As it is the ratio of the resistor values that determines the gain, we are free to choose the actual values. Here, we will choose $R_2 = 1$ kΩ, which means that R_1 must be 25 kΩ. As discussed above, we will leave consideration of the choice of component values until Section 16.6.

Computer simulation exercise 16.2

Simulate the circuit in Example 16.2 using one of the operational amplifiers supported by your simulation package. Apply a 100 mV DC input to the circuit and measure the output voltage. Then, deduce the voltage gain of the circuit and confirm that this is as expected. Experiment with different values for the two resistors and see how this affects the voltage gain. Experiment with different values for the input voltage (including both positive and negative values) and confirm that the circuit behaves as you expect.

You will note that the assumption that we are using an ideal operational amplifier greatly simplifies the analysis of these circuits.

Video 16B

| 16.4 | **Some other useful circuits** |

Having seen how we can use operational amplifiers to produce simple non-inverting and inverting amplifiers, we will now look at a few other standard circuits.

16.4.1 A unity gain buffer amplifier

This is a special case of the non-inverting amplifier discussed in Section 16.3.1 with R_1 equal to zero and R_2 equal to infinity. The resulting circuit is shown in Figure 16.7.

File 16C

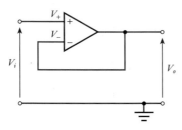

Figure 16.7
A unity gain buffer amplifier.

From Equation 16.1, we know that the gain of a non-inverting amplifier circuit is given by

$$G = \frac{R_1 + R_2}{R_2}$$

This may be rearranged to give

$$G = \frac{R_1}{R_2} + 1$$

If we substitute appropriate values for R_1 and R_2 we get

$$G = \frac{0}{\infty} + 1 = 1 \qquad (16.3)$$

and we therefore have an amplifier with a gain of 1 (unity).

At first sight this may not seem a very useful circuit as the voltage at the output is the same as that at the input. However, one must remember that voltage is not the only important attribute of a signal. The importance of this circuit is that it has a very high input resistance and a very low output resistance, making it very useful as a **buffer**. We will look at input and output resistance later in this chapter.

16.4.2 A current-to-voltage converter

Some sensors operate such that the physical quantity being measured is represented by the magnitude of the *current* produced at its output, rather than by the magnitude of a voltage (see Chapter 12). This illustrates one of many situations where we may wish to convert a varying current into a corresponding varying voltage. A circuit to perform this transformation is shown in Figure 16.8.

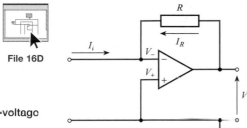

File 16D

Figure 16.8
A current-to-voltage converter.

The analysis of this circuit is similar to that of the inverting amplifier of Section 16.3.2. Again, the inverting input to the op-amp is a **virtual earth point** and the voltage at this point (V_-) is zero. As the currents into the virtual earth point must sum to zero and the input current to the op-amp is zero, it follows that

$$I_i + I_R = 0$$

and

$$I_i = -I_R$$

Now, as V_- is zero, I_R is given by

$$I_R = \frac{V_o}{R}$$

and therefore

$$I_i = -I_R = -\frac{V_o}{R}$$

or rearranging

$$V_o = -I_i R \tag{16.4}$$

Thus, the output voltage is directly proportional to the input current. The minus sign indicates that an input current that flows in the direction of the arrow in Figure 16.8 will produce a negative output voltage.

16.4.3 A differential amplifier (subtractor)

A common requirement within signal processing is the need to subtract one signal from another. A simple circuit for performing this task is shown in Figure 16.9. As no current flows into the inputs of the op-amp, the voltages on the two inputs are determined simply by the potential dividers formed by the external resistors. Thus

$$V_+ = V_1 \frac{R_1}{R_1 + R_2}$$

$$V_- = V_2 + (V_o - V_2)\frac{R_2}{R_1 + R_2}$$

As in earlier circuits, the negative feedback forces V_- to equal V_+ and therefore

$$V_+ = V_-$$

and

$$V_1 \frac{R_1}{R_1 + R_2} = V_2 + (V_o - V_2)\frac{R_2}{R_1 + R_2}$$

Multiplying through by $(R_1 + R_2)$ gives

$$V_1 R_1 = V_2 R_1 + V_2 R_2 + V_o R_2 - V_2 R_2$$

which may be arranged to give

$$V_o = \frac{V_1 R_1 - V_2 R_1}{R_2}$$

and hence the output voltage V_o is given by

$$V_o = (V_1 - V_2)\frac{R_1}{R_2} \tag{16.5}$$

Thus, the output voltage is simply the differential input voltage $(V_1 - V_2)$ times the ratio of R_1 to R_2. Note that if $R_1 = R_2$, the output is simply $V_1 - V_2$.

File 16E

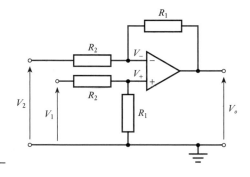

Figure 16.9
A differential amplifier or subtractor.

16.4.4 An inverting summing amplifier (adder)

As well as subtracting one signal from another, we often need to add them together. Figure 16.10 shows a simple circuit for adding together two input signals V_1 and V_2. This circuit may be easily expanded to sum any number of signals, simply by adding further input resistors.

File 16F

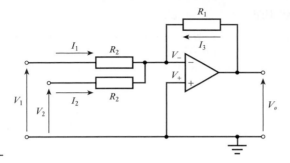

Figure 16.10
An inverting summing
amplifier or adder.

The circuit is similar in form to the inverting amplifier of Section 16.3.2, with the addition of an extra input resistor. As for the earlier circuit, the inverting input to the op-amp forms a virtual earth and therefore V_- is zero. This makes the various currents in the circuit easy to calculate

$$I_1 = \frac{V_1}{R_2}$$

$$I_2 = \frac{V_2}{R_2}$$

$$I_3 = \frac{V_o}{R_1}$$

As no current flows into the op-amp, the external currents flowing into the virtual earth must sum to zero. Therefore

$$I_1 + I_2 + I_3 = 0$$

or rearranging

$$I_3 = -(I_1 + I_2)$$

Substituting for the various currents then gives

$$\frac{V_o}{R_1} = -\left(\frac{V_1}{R_2} + \frac{V_2}{R_2}\right)$$

and the output voltage V_o is given by

$$V_o = -(V_1 + V_2)\frac{R_1}{R_2} \qquad (16.6)$$

The output voltage is determined by the sum of the input voltages $(V_1 + V_2)$ and the ratio of the resistors R_1 and R_2. The minus sign in the expression for the gain indicates that this is an inverting adder. Note that if $R_1 = R_2$ the output is simply $-(V_1 + V_2)$.

This circuit can be easily modified to add more than two input signals. Any number of input resistors may be used and, provided that they are all of value R_2, the output will become

$$V_o = -(V_1 + V_2 + V_3 + \cdots)\frac{R_1}{R_2} \qquad (16.7)$$

16.4.5 An integrator

Replacing R_1 in the inverting amplifier of Figure 16.6 with a capacitor produces a circuit that acts as an integrator. This is shown in Figure 16.11. Again, V_- is a virtual earth point and the currents into this point must sum to zero. Thus

$$I_C + I_R = 0$$

$$I_C = -I_R = -\frac{V_i}{R}$$

As V_- is zero, the output voltage V_o is simply the voltage across the capacitor. The voltage across any capacitor is proportional to its charge and inversely proportional to its capacitance. In turn, the charge is equal to the integral of the current into the capacitor. Thus

$$V_o = \frac{q}{C} = \frac{1}{C} \int_0^t I_C \, dt + \text{constant}$$

where the constant represents the initial voltage on the capacitor at $t = 0$. If we assume that initially there is no charge on the capacitor, its voltage will be zero and substituting for I_C gives

$$V_o = -\frac{1}{C} \int_0^t \frac{V_i}{R} \, dt$$

or

$$V_o = -\frac{1}{RC} \int_0^t V_i \, dt \tag{16.8}$$

Therefore the output voltage is proportional to the integral of the input voltage, the constant of proportionality being determined by a **time constant** equal to the product of R and C.

One problem associated with the use of integrators is that any DC component in the input is integrated to produce a continuously increasing output, which eventually results in the output saturating at one of the supply rails. A common cause of such a DC component is an *input offset voltage*, which is discussed later in this chapter. To overcome this problem, the circuit is usually modified by adding a resistor in parallel with the capacitor to reduce its DC gain.

File 16G

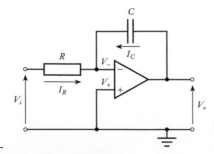

Figure 16.11
An integrator.

16.4.6 A differentiator

Exchanging the position of the resistor and the capacitor in the integrator produces a differentiating circuit. This is shown in Figure 16.12. As before, V_- is a virtual earth point and the currents into this point must sum to zero. Therefore

$$I_C + I_R = 0$$

and thus

$$I_C = -I_R = -\frac{V_o}{R}$$

As V_- is zero, the voltage across the capacitor is simply the input voltage V_i, and therefore

$$V_i = \text{voltage across capacitor}$$

$$= \frac{1}{C} \int_0^t I_C \, dt + \text{constant}$$

and, differentiating both sides with respect to t,

$$\frac{dV_i}{dt} = \frac{I_C}{C}$$

Substituting for I_C gives

$$\frac{dV_i}{dt} = -\frac{V_o}{RC}$$

and rearranging gives

$$V_o = -RC \frac{dV_i}{dt} \qquad\qquad (16.9)$$

Therefore, the output voltage is proportional to the derivative of the input voltage with respect to time. In fact, the circuit given above is rarely used in this form as it greatly amplifies high-frequency noise and unwanted spikes in the signal and is inherently unstable. The addition of a resistor in series with the capacitor reduces the undesirable amplification of noise at the expense of a slightly less precise differentiation.

File 16H

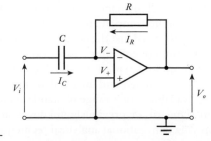

Figure 16.12
A differentiator.

File 16C
File 16D
File 16E
File 16F
File 16G
File 16H

Computer simulation exercise 16.3

Simulate the various circuits described in Section 16.4 using one of the operational amplifiers supported by your simulation package. Apply appropriate input signals and confirm that the circuits operate as expected. Experiment with different values for the various resistors and capacitors in your circuits and note any effects on the circuit's operation.

16.4.7 Active filters

In Section 8.13.3 we looked at the use of **active filters** and several forms of these can be constructed using operational amplifiers.

Figure 16.13 shows four filters, each constructed around a non-inverting amplifier. The circuits shown are **two-pole filters**, but several such stages may be cascaded to form higher-order filters. Circuits of this type are referred to as **Sallen–Key filters**. By choosing appropriate component values, these circuits can be designed to produce the characteristics of various forms of filter, such as Bessel, Butterworth or Chebyshev. In general, the cascaded stages will not be identical, but are designed such that the combination has the required characteristics. The component values shown produce Butterworth filters with $f_0 = 1/(2\pi RC)$ in each case. Other combinations of components produce other types of filter and the cut-off frequency may be slightly above or slightly below this value. The resistors R_1 and R_2 define the overall gain of each circuit, as in the non-inverting amplifier circuit described in Section 16.3.1. The gain, in turn, determines the Q of the circuit.

File 16J
File 16K
File 16L
File 16M

Computer simulation exercise 16.4

Simulate the filters in Figure 16.13 using $R = 16$ kΩ, $C = 10$ nF, $R_1 = 5.9$ kΩ and $R_2 = 10$ kΩ. Plot the frequency response of each arrangement and note the general shape of the response and its cut-off frequency (in the case of the high-pass and low-pass filters) or centre frequency (in the case of the band-pass and band-stop filters).

16.4.8 Further circuits

We have seen that operational amplifiers can be used to produce many useful circuits using only a small number of additional components. In later chapters we will meet many other op-amp circuits for performing other functions. You will find that these are often equally simple in design and are equally straightforward to analyse.

While the analysis of these circuits is generally relatively simple, in many situations we do not need to analyse them at all. In many cases, we can simply take a standard cookbook circuit and select appropriate component values to customise the circuit to our needs. In such instances, we often need just the circuit diagram and an equation relating the circuit's function to the component values (as in Equations 16.1 to 16.9 above). Appendix C gives some examples of typical cookbook circuits for use in a range of situations. Some of these are discussed and analysed within the text and others are not. However, despite the availability of a range of standard circuits, it is very useful to be able to analyse non-standard, or previously unknown, arrangements. This can generally be done by applying conventional analytical techniques.

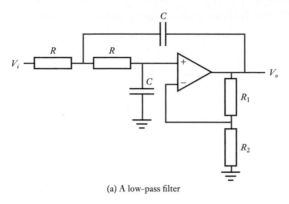

(a) A low-pass filter

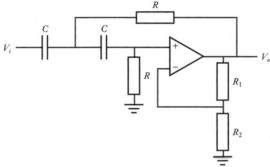

(b) A high-pass filter

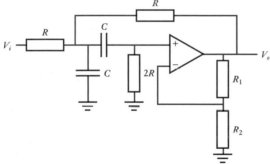

(c) A band-pass filter

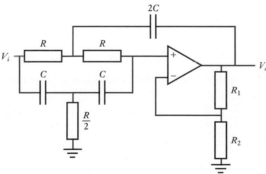

Figure 16.13
Operational amplifier
filter circuits.

(d) A band-stop filter

Example 16.3 Calculate the voltage gain of the following circuit.

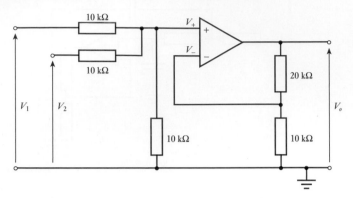

The analysis is similar to that used in earlier examples. First we note that, as in earlier circuits, the negative feedback forces V_- to equal V_+, and therefore

$$V_- = V_+$$

Since no current flows into the inputs of the op-amp, V_- and V_+ are determined by the potential dividers formed by the resistors.

V_- is easy to calculate and is given by

$$V_- = V_o \frac{10\ \text{k}\Omega}{10\ \text{k}\Omega + 20\ \text{k}\Omega} = \frac{V_o}{3}$$

V_+ is slightly more complicated to compute, since it is determined by the two input voltages. However, applying the principle of superposition, we know that the voltage on V_+ will be equal to the sum of the voltages that would be generated if each input voltage were applied separately.

If V_1 is applied while V_2 is set to zero, then the resistor connected to V_2 effectively goes to ground and is in parallel with the existing 10 kΩ resistor that goes from V_+ to ground. Therefore,

$$V_+ = V_1 \frac{10\ \text{k}\Omega//10\ \text{k}\Omega}{10\ \text{k}\Omega//10\ \text{k}\Omega + 10\ \text{k}\Omega} = V_1 \frac{5\ \text{k}\Omega}{5\ \text{k}\Omega + 10\ \text{k}\Omega} = \frac{V_1}{3}$$

If now V_2 is applied while V_1 is set to zero, we have a directly equivalent situation and clearly, because of the symmetry of the circuit,

$$V_+ = V_2 \frac{10\ \text{k}\Omega//10\ \text{k}\Omega}{10\ \text{k}\Omega//10\ \text{k}\Omega + 10\ \text{k}\Omega} = V_2 \frac{5\ \text{k}\Omega}{5\ \text{k}\Omega + 10\ \text{k}\Omega} = \frac{V_2}{3}$$

Therefore if both inputs are applied simultaneously we have

$$V_+ = \frac{V_1}{3} + \frac{V_2}{3}$$

Now since

$$V_- = V_+$$

we have

$$\frac{V_o}{3} = \frac{V_1}{3} + \frac{V_2}{3}$$

and

$$V_o = V_1 + V_2$$

Thus the circuit is a non-inverting adder. This circuit can be extended to have any number of inputs (see Appendix C).

File 16N

Computer simulation exercise 16.5

Use circuit simulation to investigate the circuit of Example 16.3 using one of the operational amplifiers supported by your simulation package. Apply appropriate input signals and confirm that the circuit operates as expected.

We have seen in the various arrangements discussed above that the functionality of the circuit usually depends on the *relative* values of the various components rather than their *absolute* values. For example, in the case of the inverting amplifier circuit, the gain is given by the ratio of R_1 to R_2. This would seem to suggest that we are free to choose any values for the various resistors provided that the ratio of their magnitudes is appropriate. This assumption would be correct if we were able to use ideal operational amplifiers within our circuits. Unfortunately, when we use real components these impose restrictions on how we must choose component values. In order to understand these restrictions, we need to know something about the nature of real devices.

16.5 Real operational amplifiers

In Section 16.2 we looked at the characteristics that we would require of an ideal operational amplifier. In that section we deduced that an ideal device would be characterised by an infinite voltage gain, infinite input resistance and zero output resistance.

No real op-amp can satisfy these requirements and it is important to recognise the limitations of physical components and how these influence the design and performance of physical circuits. In this section, we look at various characteristics of operational amplifiers and see how these compare with those of an ideal component.

One of the problems in comparing real and ideal components is that there are a great many operational amplifiers available and the characteristics of these devices vary considerably. One of the best-known general-purpose op-amps is the **741**. This device was one of the first widely-used operational amplifiers and not surprisingly it is now far from 'state of the art'. However, since it is so widely-known, it is common to judge the characteristics of more modern components by comparing them with the 741. When studying and experimenting with op-amps it is also convenient to use these devices since they are very inexpensive and because they are modelled within almost all circuit simulation packages. In many situations the performance and characteristics of a 741 are perfectly adequate to achieve the desired result. However, in real industrial situations it is more common to use components that are tailored to a specific class of applications. For example, some components are optimised for use in situations requiring low power consumption, while others are designed to produce low levels of noise.

The characteristics of op-amps are clearly determined by the circuitry used to implement them and a wide range of techniques are used to achieve a range of features. The 741 uses **bipolar transistors**, while other devices use **field-effect transistors**

(FETs). A third group of op-amps use a combination of these two forms of transistor and these are referred to as **BiFET** or **Bimos** devices. In later chapters we will look at the characteristics of both bipolar and field-effect transistors and investigate some of the circuit techniques used within various forms of operational amplifier. For the moment, we will put aside the implementation of these devices and concentrate on their external characteristics. Here, we look at the characteristics of general-purpose devices, such as the 741, but also consider the range of performance achieved by other devices.

16.5.1 Voltage gain

Most operational amplifiers have a gain of between 100 and 140 dB (a voltage gain of between 10^5 and 10^7). The 741 has a gain of about 106 dB (a voltage gain of about 2×10^5), while some components have gains of 160 dB (a voltage gain of about 10^8) or more. While these gains are clearly not infinite, in many situations they are 'high enough' and gain limitations will not affect circuit operation. Unfortunately, the gain, though high, is normally subject to great variability. The gain will often vary tremendously from one device to another and with temperature.

16.5.2 Input resistance

The typical input resistance of a 741 is 2 MΩ, but again this quantity varies considerably from device to device and may be as low as 300 kΩ. This value is low for modern op-amps and it is not uncommon for devices that use bipolar transistors (like the 741) to have input resistances of 80 MΩ or more. In many applications, this value will be very large compared with the source resistance and may be considered to be high enough for loading effects to be ignored. In applications where higher input resistances are required it is common to use devices that use field-effect transistors (FETs) in their input stages. These have a typical input resistance of about 10^{12} Ω. When using these devices, loading effects can almost always be ignored. Field-effect and bipolar transistors are discussed later (see Chapters 18 and 19).

16.5.3 Output resistance

The 741 has a typical output resistance of 75 Ω, this being a typical figure for bipolar transistor op-amps. Some low-power components have a much higher output resistance, perhaps up to several thousand ohms. Often of more importance than the output resistance of a device is the maximum current that it will supply. The 741 will supply 20 mA, with values in the range 10 to 20 mA being typical for general-purpose op-amps. Special high-power devices may supply output currents of an amp or more.

16.5.4 Output voltage range

With voltage gains of several hundred thousand times it would seem that, if 1 volt were applied to the input of an operational amplifier, one would have to keep well clear of the output! However, in practice, the output voltage is limited by the supply voltage. Most op-amps based on bipolar transistors (like the 741) produce a maximum output voltage swing that is slightly less than the difference between the two supply voltages. An amplifier connected to a positive supply of +15 V and a negative supply of −15 V, for example (a typical arrangement), might produce an output voltage range of about ±13 V. Op-amps based on field-effect transistors can often produce output voltage swings that go very close to both supply voltages. These are often referred to as 'rail-to-rail' devices.

16.5.5 Supply voltage range

A typical arrangement for an operational amplifier might use supply voltages of +15 V and −15 V, although a wide range of supply voltages are usually possible. The 741, for example, may be used with supply voltages in the range ±5 V to ±18 V, this being fairly typical. Some devices allow higher voltages to be used – perhaps up to ±30 V – while others are designed for low-voltage operation – perhaps down to ±1.5 V.

Many amplifiers allow operation from a single voltage supply, which may be more convenient in some applications. Typical voltage ranges for a single supply might be 4 to 30 V, though devices are available that will operate down to 1 V or less.

16.5.6 Common-mode rejection ratio

An ideal operational amplifier would not respond to common-mode signals. In practice, all amplifiers are slightly affected by common-mode voltages, though in good amplifiers the effects are very small. A measure of the ability of a device to ignore common-mode signals is its **common-mode rejection ratio** or **CMRR**. This is the ratio of the response produced by a differential-mode signal to the response produced by a common-mode signal of the same size. The ratio is normally expressed in decibels.

Typical values for CMRR for general-purpose operational amplifiers are between 80 and 120 dB. High-performance devices may have a ratio of up to 160 dB or more. The 741 has a typical CMRR of 90 dB.

16.5.7 Input currents

For an operational amplifier to work correctly, a small input current is required into each input terminal. This current is termed the **input bias current** and must be provided by external circuitry. The polarity of this current will depend on the input circuitry used in the amplifier and, in most situations, it is so small that it can be safely ignored.

Typical values for this current in bipolar op-amps range from a few microamps down to a few nanoamps or less. For the 741 this value is typically 80 nA. Operational amplifiers based on FETs have much smaller input bias currents, with values of a few picoamps being common and values down to less than a femtoamp (10^{-15} A) being possible.

16.5.8 Input offset voltage

One would expect that, if the input voltage of the amplifier were zero, the output would also be zero. In general, though, this is not the case. The transistors and other components in the circuit are not precisely matched and a slight error is usually present, which acts like a voltage source added to the input. This is the **input offset voltage** V_{ios}, which is defined as the small voltage required at the input to make the output zero.

The input offset voltage of most op-amps is generally in the range of a few hundred microvolts up to a few millivolts. For the 741, a typical value is 2 mV. This may not seem very significant, but remember that this is a voltage added to the *input*, so it is multiplied by the gain of the amplifier. Fortunately, the offset voltage is approximately constant, so its effects can be reduced by subtracting an appropriate voltage from the input. The 741, in common with many operational amplifiers, provides connections to allow an external potentiometer to 'trim' the offset to zero. Some op-amps are **laser trimmed** during manufacture to produce a very low offset voltage without the need for manual adjustment. Unfortunately, the input offset voltage varies with temperature by a few microvolts per degree centigrade, making it generally impossible to remove the effects of the offset voltage completely by trimming alone.

16.5.9 Frequency response

Operational amplifiers have no lower cut-off frequency and the gain mentioned earlier is therefore the gain of the amplifier at DC.

All amplifiers have an **upper cut-off frequency** (as discussed in Section 14.7) and one would perhaps imagine that, to be generally useful, operational amplifiers would require very high upper cut-off frequencies. In fact, this is not the case and in many devices the gain begins to roll off above only a few hertz. Figure 16.14 shows a typical frequency response for the 741 op-amp.

The magnitude of the gain of the amplifier is constant from DC up to only a few hertz. Above this frequency, it falls steadily until it reaches unity at about 1 MHz. Above this frequency, the gain falls more rapidly. The upper cut-off frequency is introduced intentionally by the designer to ensure the stability of the system. We will return to the question of **stability** in Chapter 23.

The frequency range of an operational amplifier is usually described by the frequency at which the gain drops to unity (this is called the **transition frequency**, f_T) or by its **unity gain bandwidth**. The latter is the bandwidth over which the gain is greater than unity and it is clear that for an operational amplifier these two measures are equal. From Figure 16.14 it can be seen that the 741 has an f_T of about 1 MHz. Typical values for f_T for other general-purpose operational amplifiers vary from a few hundred kilohertz up to a few tens of megahertz. However, a high-speed device may have an f_T of several gigahertz.

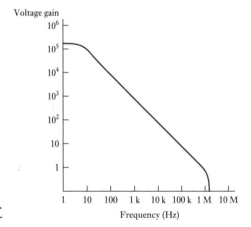

Figure 16.14
Typical gain against frequency characteristic for a 741.

16.5.10 Slew rate

While the bandwidth determines the ability of an operational amplifier to respond to rapidly changing small signals, when large signals are used, it is often the **slew rate** that is the limiting factor. This is the maximum rate at which the output voltage can change and typically is a few volts per microsecond. The effects of the slew rate are most obvious when an amplifier is required to output a large amplitude square or pulsed wave. Rather than a rapid transition from one level to another, the signal 'ramps' between the two values at a rate determined by the slew rate. The limitations of slew rate may also affect sinusoidal or other analogue signals of a large amplitude and high frequency.

16.5.11 Noise

All operational amplifiers add noise to the signals that pass through them. Noise is generated by a number of mechanisms and these have different frequency characteristics. Some produce essentially **white noise**, meaning that it has equal power density at all frequencies (that is, the noise power within a given bandwidth is equal at all frequencies). Others produce more power in some parts of the frequency spectrum than others. For this reason, it is difficult to describe accurately the noise performance of a given device without being specific about the frequency range over which it is being used. Clearly, as noise is present at all frequencies, the amount of noise detected will depend on the bandwidth over which measurements are made. Manufacturers normally give a figure indicating the noise voltage divided by the square root of the bandwidth of measurement.

Low-noise op-amps are likely to have noise voltages of about $3 \text{ nV}/\sqrt{\text{Hz}}$. General-purpose devices may have noise voltages several orders of magnitude greater. We will consider noise in more detail in Chapter 22.

16.6 Selecting component values for op-amp circuits

Earlier in this chapter we derived expressions for the gain of a range of op-amp circuits. The analysis assumed the use of an ideal amplifier and resulted in simple expressions, usually involving ratios of the values of circuit components. This implies that the absolute values of the components are unimportant. This would suggest that an inverting amplifier with a gain of 10 could be formed using resistors of $1 \, \Omega$ and $10 \, \Omega$, $1 \, \text{k}\Omega$ and $10 \, \text{k}\Omega$, or $1 \, \text{G}\Omega$ and $10 \, \text{G}\Omega$. While this would be true if we were using an ideal op-amp, it is certainly *not* true when we use real components.

In our analysis we assumed that our operational amplifier had an infinite gain, an infinite input resistance and zero output resistance. However, from the discussion in the last section, we know that this is not true for real op-amps. Therefore, in order for our analysis to represent a reasonable model of the operation of a real circuit, we need to select external components such that the assumptions made during the calculations are reasonable. We will therefore look at each of our assumptions in turn to see what restrictions they impose on the circuit's design.

Our first assumption was that the gain of the op-amp was infinite. We used this assumption when we assumed that the input voltage to the op-amp was zero. We know that one of the requirements of effective negative feedback is that the closed-loop gain must be much less than the open-loop gain (as discussed in Chapter 15). In other words, the gain of the complete circuit with feedback must be much less than the gain of the operational amplifier without feedback.

Our second assumption was that the input resistance of the operational amplifier was infinite. We used this assumption when we assumed that the input current to the op-amp was zero. This will be a reasonable approximation provided that the currents flowing in the external components are large compared with the current into the op-amp. This will be true provided that the resistors forming the external circuitry are much smaller than the input resistance of the op-amp.

Our final assumption was that the output resistance of our operational amplifier was zero. This was used when we assumed that there would be no loading effects. This will be a reasonable assumption if the external resistors are much larger than the output resistance of the op-amp.

Therefore, these three assumptions will be reasonable provided that:

- we limit the gain of our circuits to a value much less than the open-loop gain of the op-amp;
- the external resistors are small compared with the input resistance of the op-amp;
- the external resistors are large compared with the output resistance of the op-amp.

From the last section we know that the gain of our op-amp is likely to be greater than 10^5. Therefore, the assumption that the gain of the op-amp is infinite will be a reasonable approximation, provided that the gain of our complete circuit is *much* less than this. Therefore, we should limit the gain of any individual circuit to 10^3 or less.

A typical value for the input resistance of a bipolar operational amplifier is in the $1 \, \text{M}\Omega$ to $100 \, \text{M}\Omega$ range, and a typical value for the output resistance might be $10 \, \Omega$ to $100 \, \Omega$. Therefore, for circuits using such devices, resistors in the $1 \, \text{k}\Omega$ to $100 \, \text{k}\Omega$ range would be appropriate.

Operational amplifiers based on FETs have a much higher input resistance – of the order of $10^{12} \, \Omega$ or more. Circuits using these devices may therefore use higher-value resistors – of the order of $1 \, \text{M}\Omega$ or more, if desired. However, resistors in the $1 \, \text{k}\Omega$ to $100 \, \text{k}\Omega$ range will generally produce satisfactory results with all forms of op-amps.

File 16P

> **Computer simulation exercise 16.6**
>
> Simulate the non-inverting amplifier of Example 16.1 using a 741 operational amplifier and measure its gain. Modify your circuit by replacing R_1 with a resistor of $24 \, \Omega$ and R_2 with a resistor of $1 \, \Omega$ and again measure its gain. Repeat this exercise, replacing the two resistors with values of $24 \, \text{M}\Omega$ and $1 \, \text{M}\Omega$ and hence confirm the design rules given above. Repeat this process for the inverting amplifier of Example 16.2.

16.7 The effects of feedback on op-amp circuits

We know that the use of negative feedback has a dramatic effect on almost all of the characteristics of an amplifier (as discussed in Chapter 15). All the circuits discussed in this chapter make use of negative feedback, so it is appropriate to look briefly at its effects on various aspects of a circuit's operation.

16.7.1 Gain

Negative feedback reduces the gain of an amplifier from A to $A/(1 + AB)$. It therefore reduces the gain by a factor of $(1 + AB)$. In return for this loss of gain, feedback gives consistency as, providing the open-loop gain is much greater than the closed-loop gain, the latter is approximately equal to $1/B$.

We have seen that an additional benefit of the use of negative feedback is that it simplifies the design process. Standard cookbook circuits can be used and these can be analysed without needing to consider the detailed operation of the operational amplifier itself.

Video 16C

16.7.2 Frequency response

In Chapter 15, we looked at the effects of negative feedback on the frequency response and bandwidth of an amplifier. At that time, we noted that negative feedback tends to increase the bandwidth of an amplifier by keeping its closed-loop gain constant, despite a fall in its open-loop gain.

The effects of negative feedback on a typical operational amplifier, a 741, are shown in Figure 16.15. The figure shows the frequency response of the circuit without feedback (its open-loop response) and the response of amplifiers with different amounts of feedback. Without feedback, the amplifier has a gain of about 2×10^5 and a bandwidth of about 5 Hz. However, if feedback is used to reduce the gain to 1000, then the bandwidth increases to about 1 kHz. Decreasing the gain to 100 increases the bandwidth to about 10 kHz, while decreasing the bandwidth to 10 increases the bandwidth to about 100 kHz. It can be seen that this behaviour illustrates the relationship discussed in Chapter 15, where we noted that in many cases

$$\text{gain} \times \text{bandwidth} = \text{constant} \tag{16.10}$$

In this case, the **gain–bandwidth product** is about 10^6 Hz. Note that the gain of the op-amp falls to unity at about 10^6 Hz and so, in this case, the gain–bandwidth product is equal to the **unity gain bandwidth**.

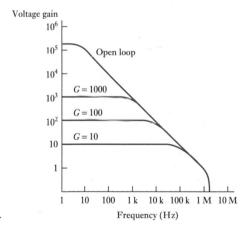

Figure 16.15
Gain against frequency characteristics for a 741 with feedback.

Example 16.4

An audio amplifier is to be produced using a 741 op-amp. What is the maximum gain that can be achieved using this arrangement if the amplifier must have a bandwidth of 20 kHz?

For a 741

$$\text{gain} \times \text{bandwidth} = 10^6$$

Therefore, if the bandwidth required is 2×10^4, then the maximum gain is given by

$$\text{gain} = \frac{10^6}{\text{bandwidth}} = \frac{10^6}{2 \times 10^4} = 50$$

High-speed op-amps may have unity gain bandwidths of a gigahertz or more, allowing the production of wide-bandwidth amplifiers that also have high gain. However, not all op-amps have a frequency response of the form shown in Figure 16.15 and then the relationship between gain and bandwidth is not so straightforward.

File 16Q

> ### Computer simulation exercise 16.7
>
> Simulate a non-inverting amplifier with a gain of 10 based on a 741 operational amplifier, plot its frequency response and measure its bandwidth. Repeat this process with your circuit modified to produce gains of 1 and 100. In each case, calculate the product of the gain and the bandwidth and investigate the relationship of Equation 16.10. Also, compare the gain–bandwidth product with the unity gain bandwidth.

Video 16D

16.7.3 Input and output resistance

Negative feedback can be used to either increase or decrease both the input and the output resistance of a circuit (as discussed in Section 15.6.3). We have also seen that the amount by which the resistance is changed is given by the expression $(1 + AB)$. As this is also the factor by which the gain is reduced, we can determine the value of this expression simply by dividing the open-loop gain by the closed-loop gain. For example, if an op-amp with an open-loop gain of 2×10^5 is used to produce an amplifier with a gain of 100, then $(1 + AB)$ must be equal to $2 \times 10^5/100 = 2 \times 10^3$. Note that, when using negative feedback, the factor $(1 + AB)$ will always be positive. If we use an op-amp to produce an inverting amplifier with a gain of -100, we are effectively using the op-amp in a configuration where its gain is -2×10^5 and so $(1 + AB)$ is equal to $-2 \times 10^5/-100 = 2 \times 10^3$ as before.

In order to determine whether the feedback increases or decreases the output resistance, we need to see whether it is the output voltage or the output current that is being used to determine the feedback quantity. In all the circuits discussed in this chapter, it is the output voltage that is being used to determine the feedback so, in each case, the feedback *reduces* the output resistance.

In order to determine whether the input resistance is increased or decreased, we need to determine whether it is a voltage or a current that is being subtracted at the input. In the case of the non-inverting amplifier of Section 16.3.1, it is a *voltage* that is subtracted from the input voltage to form the input to the op-amp. Thus, in this circuit, the feedback *increases* the input resistance by a factor of $(1 + AB)$. In the case of the inverting amplifier of Section 16.3.2, it is a *current* that is subtracted from the input current to form the input to the op-amp and therefore the feedback *decreases* the input resistance. In this particular circuit, the resistor R_2 goes from the input to the virtual earth point. Therefore, the input resistance is simply equal to R_2.

When considering other circuits, one needs to look at the quantity being fed back and the quantity being subtracted from the input to determine the effects of the feedback on the input and output resistance.

Example 16.5 | **Determine the input and output resistance of the following circuit, assuming that the operational amplifier is a 741.**

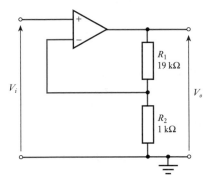

The open-loop gain of a 741 is typically 2×10^5 and the closed-loop gain of this circuit is 20. Therefore $(1 + AB) = (2 \times 10^5)/20 = 10^4$.

The output resistance of a 741 is typically about 75 Ω and in this circuit the output *voltage* is fed back. Thus the feedback *reduces* the output resistance by a factor of $(1 + AB)$ which becomes $75/10^4 = 7.5$ mΩ.

The input resistance of a 741 is typically about 2 MΩ and in this circuit a feedback *voltage* is subtracted from the input voltage. Thus, the feedback *increases* the input resistance by a factor of $(1 + AB)$, which becomes $2 \times 10^6 \times 10^4 = 2 \times 10^{10} = 20$ GΩ.

Example 16.6 | **Determine the input and output resistance of the following circuit, assuming that the operational amplifier is a 741.**

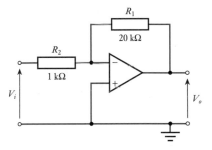

The open-loop gain of a 741 is typically 2×10^5 and the magnitude of the closed-loop gain of this circuit is 20. Therefore $(1 + AB) = (2 \times 10^5)/20 = 10^4$.

The output resistance of a 741 is typically about 75 Ω and in this circuit the output *voltage* is fed back. Thus, the feedback *reduces* the output resistance by a factor of $(1 + AB)$, which becomes $75/10^4 = 7.5$ mΩ.

The input resistance of a 741 is typically about 2 MΩ and in this circuit a feedback *current* is subtracted from the input current. Thus, the feedback *decreases* the input resistance. In this case, the input is connected to a virtual earth point by the resistance R_2 and so the input resistance is equal to R_2, which is 1 kΩ.

Examples 16.5 and 16.6 illustrate the very dramatic effects that feedback can have on the characteristic of a circuit. While the input and output resistance of an operational amplifier make it a *good* voltage amplifier, the use of feedback can turn it into an *excellent* one. This is most striking in the case of the buffer amplifier of Section 16.4.1, where feedback produces such a high input resistance – and such a low output resistance – that loading effects can almost always be ignored. This is shown in Example 16.7.

Example 16.7 | Determine the input and output resistance of the following circuit, assuming that the operational amplifier is a 741.

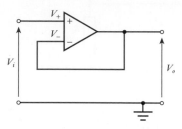

The open-loop gain of a 741 is typically 2×10^5 and the closed-loop gain of this circuit is 1. Therefore $(1 + AB) = (2 \times 10^5)/1 = 2 \times 10^5$.

The output resistance of a 741 is typically about 75 Ω and in this circuit the output *voltage* is fed back. Thus the feedback *reduces* the output resistance by a factor of $(1 + AB)$, which becomes $75/(2 \times 10^5) \approx 400\ \mu\Omega$.

The input resistance of a 741 is typically about 2 MΩ and in this circuit a feedback *voltage* is subtracted from the input voltage. Thus, the feedback *increases* the input resistance by a factor of $(1 + AB)$, which becomes $(2 \times 10^6) \times (2 \times 10^5) = 4 \times 10^{11} = 400\ \text{G}\Omega$.

While it is clear that negative feedback can dramatically improve the input and output resistance of a circuit, it should be remembered that this improvement is brought about at the expense of a loss in gain. As the open-loop gain of the operational amplifier changes with frequency (as shown in Figure 16.14), so will the input and output resistance. The various calculations and examples above use the low-frequency open-loop gain of the op-amp, therefore the values obtained represent the resistances at very low frequencies. As the frequency increases, the gain of the op-amp falls and the improvement brought about by feedback will be reduced.

16.7.4 Stability

While negative feedback can be used to tailor the characteristics of an operational amplifier for a given application, its use does have implications for the stability of the circuit. We will return to look at considerations of stability in Chapter 23.

Video 16E

Further study

While we are normally faced with the task of designing circuits to perform a particular function, it is also useful to be able to analyse given circuits to determine their characteristics.

Consider the following circuits and in each case determine the function of the arrangement and the relationship between the inputs and the output.

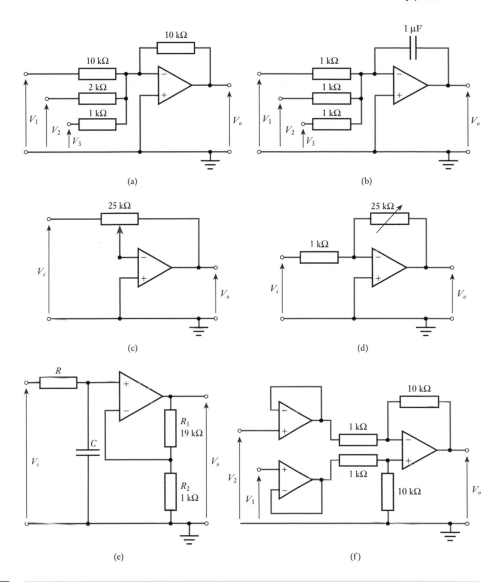

(a)

(b)

(c)

(d)

(e)

(f)

- Operational amplifiers are among the most widely used building blocks for the construction of electronic circuits.

- Op-amps are small integrated circuits that typically take the form of a plastic package containing one or more amplifiers.

- Although they are often omitted from circuit diagrams, op-amps require connections to power supplies (typically +15 V and −15 V) in order to function.

- An ideal operational amplifier would have an infinite voltage gain, infinite input resistance and zero output resistance.

- Designers often base their designs on a number of standard cookbook circuits. Analysis of these circuits is greatly simplified if we assume the use of an ideal op-amp.

- Standard circuits are available for various forms of amplifiers, buffers, adders, subtractors and many other functions.

- Real operational amplifiers have several non-ideal characteristics. However, if we choose component values appropriately, these factors should not affect the operation of our cookbook circuits.
- When designing op-amp circuits, we normally use resistors in the range 1 kΩ to 100 kΩ.
- Feedback allows us to increase dramatically the bandwidth of a circuit by trading off gain against bandwidth.
- Feedback allows us to tailor the characteristics of an op-amp to suit a particular application. We can use feedback to overcome problems associated with the variability of the gain of the op-amp and can also either increase or decrease the input and output resistance depending on our requirements.

Exercises

16.1 What is meant by the term 'integrated circuit'?

16.2 Explain the acronyms DIL and SMT as applied to IC packages.

16.3 What are typical values for the positive and negative supply voltages of an operational amplifier?

16.4 Outline the characteristics of an 'ideal' op-amp.

16.5 Sketch an equivalent circuit of an ideal operational amplifier.

16.6 Determine the gain of the following circuit.

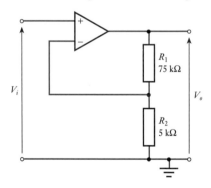

16.7 Sketch the circuit diagram of a non-inverting amplifier with a gain of 30.

16.8 Use circuit simulation to investigate your solution to the last exercise. Use one of the operational amplifiers supported by your simulation package and apply a DC input voltage of 100 mV. Then, confirm that the circuit works as expected.

16.9 Determine the gain of the following circuit.

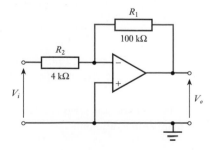

16.10 Sketch the circuit diagram of an inverting amplifier with a gain of −30.

16.11 Use circuit simulation to investigate your solution to the last exercise. Use one of the operational amplifiers supported by your simulation package and apply a DC input voltage of 100 mV. Then, confirm that the circuit works as expected.

16.12 Sketch a circuit that takes two input signals, V_A and V_B, and produces an output equal to $10(V_B - V_A)$.

16.13 Sketch a circuit that takes four input signals, V_1 to V_4, and produces an output equal to $-5(V_1 + V_2 + V_3 + V_4)$.

16.14 Derive an expression for the output V_o of the following circuit in terms of the input voltages V_1 and V_2 and hence determine the output voltage if $V_1 = 1\,\text{V}$ and $V_2 = 0.5\,\text{V}$.

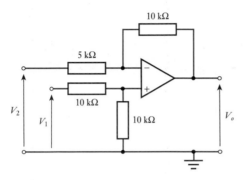

16.15 Derive an expression for the output V_o of the following circuit in terms of the input voltages V_1 and V_2 and hence determine the output voltage if $V_1 = 1\,\text{V}$ and $V_2 = 0.5\,\text{V}$.

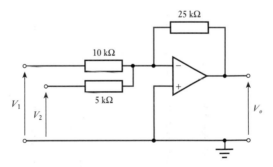

16.16 Derive an expression for the output voltage V_o of the following circuit in terms of the input voltages V_1, V_2 and V_3 and the component values.

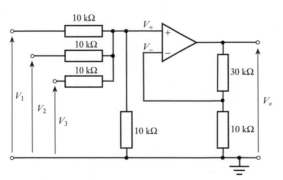

16.17 Simulate the circuit of Exercise 16.16 using one of the operational amplifiers supported by your simulation package. Apply appropriate input signals and hence confirm your answer to this exercise.

16.18 What are typical ranges for the open-circuit voltage gain, input resistance and output resistance of general-purpose operational amplifiers?

16.19 What are typical ranges for the supply voltages of general-purpose operational amplifiers?

16.20 What is meant by the term 'common-mode rejection ratio'? What would be a typical CMRR for a general-purpose op-amp?

16.21 Explain the term 'input bias current'.

16.22 Define the term 'input offset voltage' and give a typical figure for this quantity. How may the effects of the input offset voltage be reduced?

16.23 Sketch a typical frequency response for a 741 op-amp. What is its upper cut-off frequency? What is its lower cut-off frequency?

16.24 Give a typical value for the gain–bandwidth product of a 741. How does this relate to the unity gain bandwidth?

16.25 If an amplifier with a gain of 25 is constructed using a 741, what would be a typical value for the bandwidth of this circuit?

16.26 What is meant by the 'slew rate' of an op-amp? What would be a typical value for this parameter?

16.27 What range of resistor values would normally be used for circuits based on a bipolar operational amplifier?

16.28 Estimate the gain, input resistance and output resistance of the following circuits at low frequencies, assuming that each is constructed using an operational amplifier that has an open-loop gain of 10^6, input resistance of 10^6 Ω and output resistance of 100 Ω.

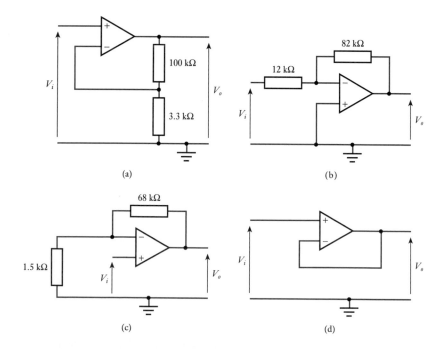

Semiconductors and Diodes

When you have studied the material in this chapter, you should be able to:

- explain the basic function of diodes within electrical circuits and describe the characteristics of an ideal diode
- describe the electrical characteristics of conductors, insulators and semiconductors
- discuss the doping of semiconductor materials and the construction of semiconductor diodes
- describe the characteristics of a typical diode and sketch its current–voltage characteristics
- outline the use of several forms of special-purpose semiconductor devices, including Zener, tunnel and varactor diodes
- design a range of circuits that exploit the characteristics of semiconductor diodes.

17.1 Introduction

So far, we have considered 'black box' amplifiers and operational amplifiers, but have not yet looked in detail at the operation of the devices at the heart of these systems. In many applications we may ignore the internal operation of these components and look simply at their external characteristics. However, it is sometimes necessary to look at the construction of the active components of our system to gain more insight into their characteristics and operation.

Most modern electronic systems are based on **semiconductor devices**, such as diodes and transistors. In this chapter we will look at the nature of semiconductor materials and consider their use in diodes. We will then move on to look at transistors in later chapters.

17.2 Electrical properties of solids

Solid materials may be divided, with respect to their electrical properties, into three categories:

- conductors
- insulators
- semiconductors.

The different characteristics of these groups are produced by the atomic structure of the materials and, in particular, by the distribution of electrons in the outer orbits of their atoms. These outermost electrons are termed **valence electrons** and they play a major part in determining many of the properties of the material.

17.2.1 Conductors

Conductors such as copper or aluminium have a cloud of free electrons at all temperatures above absolute zero. This is formed by the weakly bound 'valence' electrons in the outermost orbits of their atoms. If an electric field is applied across such a material, electrons will flow, causing an electric current.

17.2.2 Insulators

In insulating materials, such as polythene, the valence electrons are tightly bound to the nuclei of the atoms and very few are able to break free to conduct electricity. The application of an electric field does not cause a current to flow as there are no mobile charge carriers.

17.2.3 Semiconductors

At very low temperatures, semiconductors have the properties of an insulator. However, at higher temperatures, some electrons are free to move and the materials take on the properties of a conductor – albeit a poor one. Nevertheless, semiconductors have some useful characteristics that make them distinct from both insulators and conductors.

17.3 Semiconductors

Semiconductor materials have very interesting electrical properties that make them extremely useful in the production of electronic devices. The most commonly used semiconductor material for such applications is **silicon**, but **germanium** is also used, as are several more exotic materials, such as **gallium arsenide**. Many metal oxides have semiconducting properties (for example, the oxides of manganese, nickel and cobalt).

17.3.1 Pure semiconductors

At temperatures near absolute zero, the valence electrons in a semiconductor are tightly bound to their nuclei and the material has the characteristics of an insulator. The reasons for this effect may be understood by considering the structure of a typical semiconductor. Figure 17.1 shows a two-dimensional representation of a crystal of silicon. Silicon is a tetravalent material – that is, it has four valence electrons. The outermost electron shell of each atom can accommodate up to eight electrons and the atom is most stable when the shell is fully populated. In a crystal of pure silicon, each atom shares its valence electrons with its four neighbouring atoms so that each atom has a part-share in eight valence electrons rather than sole ownership of four. This is a very stable arrangement that is also found in materials such as diamond. This method of atomic bonding is called **covalent bonding**.

At low temperatures, the tight bonding of the valence electrons in semiconductor materials leaves no electrons free to conduct electricity, resulting in the insulating properties described above. However, as the temperature rises, thermal vibration of the crystal lattice results in some of the bonds being broken, generating a few **free**

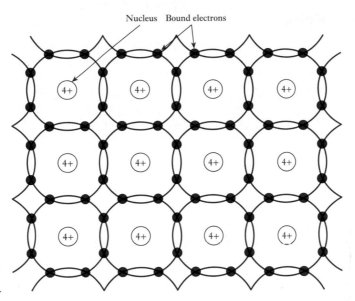

Figure 17.1
The atomic structure of
silicon.

electrons that are able to move throughout the crystal. This also leaves behind **holes** that accept electrons from adjacent atoms and therefore also move about. Electrons are negative charge carriers and will move against an applied electric field, generating an electric current. Holes, being the absence of an electron, act like positive charge carriers and will move in the direction of an applied electric field, so will also contribute to current flow. This process is illustrated in Figure 17.2.

At normal room temperatures, the number of charge carriers present in pure silicon is small and consequently it is a poor conductor. This form of conduction is called **intrinsic conduction**.

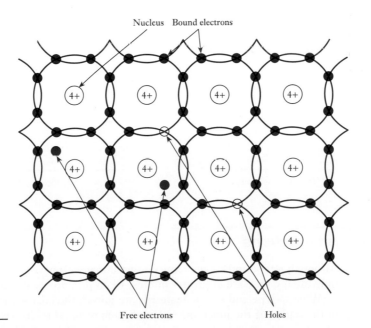

Figure 17.2
The effects of thermal
vibration on the
structure of silicon.

17.3.2 Doping

The addition of small amounts of impurities to a semiconductor can drastically affect its properties. This process is known as **doping**. Of particular interest are impurities of materials that can fit within the crystal lattice of the semiconductor, but have a different number of valence electrons. An example of such an impurity is the presence of **phosphorus** in silicon. Phosphorus is a pentavalent material – that is, it has five valence electrons in its outer electron shell. When a phosphorus atom is present within the lattice of a piece of silicon, four of its valence electrons are tightly bound by the covalent bonding described earlier. However, the fifth electron is only weakly bound and is therefore free to move within the lattice and contribute to an electric current. Materials such as phosphorus are known as **donor impurities** because they produce an excess of free electrons. Semiconductors containing such impurities are called **n-type semiconductors** as they have free *negative* charge carriers.

Boron has three valence electrons and is thus a trivalent material. When a boron atom is present within a silicon crystal, the absence of an electron in the outer shell leaves a space (a hole) that can accept an electron from an adjacent atom to complete its covalent bonds. This hole moves from atom to atom and acts as a mobile positive charge carrier in exactly the same manner as the holes generated in the intrinsic material by thermal vibration. Materials such as boron are known as **acceptor impurities** because they accept electrons to produce holes. Semiconductors containing such impurities are called **p-type semiconductors** as they have free *positive* charge carriers.

It is important to remember that a piece of doped semiconductor in isolation will be electrically neutral. Therefore, the presence of *mobile* charge carriers of a particular polarity must be matched by an equal number of *fixed* (or *bound*) charge carriers of the opposite polarity. Thus, in an *n*-type semiconductor, the free electrons produced by the doping will be matched by an equal number of positive charges bound within the atoms in the lattice. Similarly, in a *p*-type semiconductor, free holes are matched by an equal number of bound negative charges. This process is illustrated in Figure 17.3.

Both *n*-type and *p*-type semiconductors have much greater conductivities than that of the intrinsic material, the magnitude depending on the doping level. This is called **extrinsic conductivity**. The dominant charge carriers in a doped semiconductor (that is, electrons in an *n*-type material and holes in a *p*-type material) are called the **majority charge carriers**. The other charge carriers are called the **minority charge carriers**.

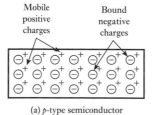
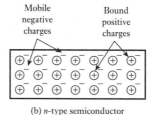

Figure 17.3
Charges within doped semiconductors.

(a) *p*-type semiconductor   (b) *n*-type semiconductor

17.4 *pn* junctions

Although *p*-type and *n*-type semiconductor materials have some useful characteristics individually, they are of greater interest when they are used together.

When *p*-type and *n*-type materials are joined, the charge carriers in each interact in the region of the junction. Although each material is electrically neutral, each has

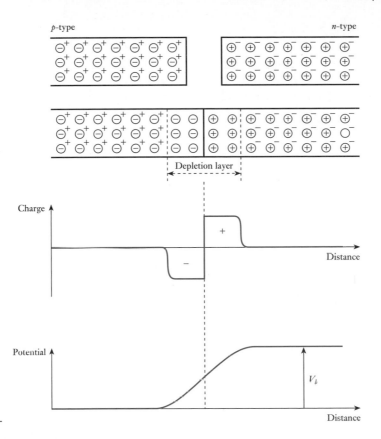

Figure 17.4
A *pn* junction.

a much higher concentration of majority charge carriers than minority charge carriers. Thus, on the *n*-type side of the junction there are far more free electrons than on the *p*-type side. Consequently, electrons diffuse across the junction from the *n*-type side to the *p*-type side where they are absorbed by recombination with free holes that are plentiful in the *p*-type region. Similarly, holes diffuse from the *p*-type side to the *n*-type side and combine with free electrons.

This process of diffusion and recombination of charge carriers produces a region close to the junction that has very few mobile charge carriers. This region is referred to as a **depletion layer** or, sometimes, a **space-charge layer**. The diffusion of negative charge carriers in one direction and positive charge carriers in the other generates a net charge imbalance across the junction. The existence of positive and negative charges on either side of the junction produces an electric field across it. This produces a **potential barrier** that charge carriers must overcome to cross the junction. This process is illustrated in Figure 17.4.

Only a small number of *majority* charge carriers have sufficient energy to surmount this barrier and these generate a small **diffusion current** across the junction. However, the field produced by the space-charge region does not oppose the movement of *minority* charge carriers across the junction; rather, it assists it. Any such charge carriers that stray into the depletion layer, or are formed there by thermal vibration, are accelerated across the junction, forming a small **drift current**. In an isolated junction, a state of dynamic equilibrium exists in which the diffusion current exactly matches the drift current. This situation in shown in Figure 17.5(a). The application of an external potential across the device will affect the height of the potential barrier and change the state of dynamic equilibrium.

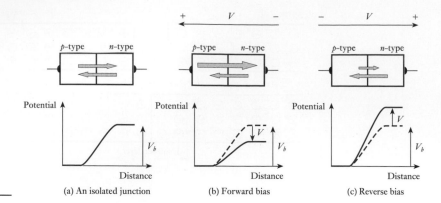

Figure 17.5
Currents in a *pn*
junction.

17.4.1 Forward bias

If the *p*-type side of the device is made positive with respect to the *n*-type side, the applied potential neutralises some of the space charge and the width of the depletion layer decreases. The height of the barrier is reduced and a larger proportion of the majority carriers in the region of the junction now have sufficient energy to surmount it. The diffusion current produced is therefore much larger than the drift current and a net current flows across the junction. This situation is shown in Figure 17.5(b).

17.4.2 Reverse bias

If the *p*-type side of the device is made negative with respect to the *n*-type side, the space charge increases and the width of the depletion layer is increased. This produces a larger potential barrier and reduces the number of majority carriers that have sufficient energy to surmount it, reducing the diffusion current across the junction. This situation is shown in Figure 17.5(c).

Even a small negative bias – of perhaps 0.1 V – is sufficient to reduce the diffusion current to a negligible value. This leaves a net imbalance in the currents flowing across the junction, which are now dominated by the drift current. As the magnitude of this current is determined by the rate of thermal generation of minority carriers in the region of the junction, it is not related to the applied voltage. At normal room temperatures, this reverse current is very small – typically a few nanoamps for silicon devices and a few microamps for germanium devices. It is, however, exponentially related to temperature and doubles for a temperature rise of about 10 °C. Reverse current is proportional to the junction area and so is much greater in large-power semiconductors than in small, low-power devices.

17.4.3 Forward and reverse currents

The current flowing through a *pn* junction can be approximately related to the applied voltage by the expression

$$I = I_s(e^{eV/\eta kT} - 1)$$

where I is the current through the junction, I_s is a constant called the **reverse saturation current**, e is the electronic charge, V is the applied voltage, k is Boltzmann's constant, T is the absolute temperature and η (Greek letter *eta*) is a constant in the range 1 to 2, determined by the junction material. Here, a positive applied voltage represents a forward-bias voltage and a positive current a forward current.

The constant η is approximately 1 for germanium and about 1.3 for silicon. However, for our purposes, it is reasonable to use the approximation that

$$I \approx I_s(e^{eV/kT} - 1) \tag{17.1}$$

and we will make that assumption for the remainder of this text.

At room temperatures, if V is less than about -0.1 V, the exponential term within the brackets in Equation 17.1 is small compared with 1 and I is given by

$$I \approx I_s(0 - 1) = -I_s \tag{17.2}$$

Similarly, if V is greater than about $+0.1$ V, the exponential term is much greater than 1 and I is given by

$$I \approx I_s(e^{eV/kT})$$

At normal room temperatures, e/kT has a value of about $40\,\text{V}^{-1}$ and so we can make the approximation

$$I \approx I_s(e^{eV/kT}) \approx I_s e^{40V} \tag{17.3}$$

We therefore have a characteristic for which the reverse-bias current is approximately constant at $-I_s$ (which explains why this quantity is called the *reverse saturation current*) and the forward-bias current rises exponentially with the applied voltage.

In fact, the expressions in Equations 17.1 to 17.3 are only approximations of the junction current in a real device as effects such as **junction resistance** and **minority carrier injection** tend to reduce the current flowing. However, this analysis gives values that indicate the form of the relationship and are adequate for our purposes. Figure 17.6 shows the current–voltage characteristic of a *pn* junction.

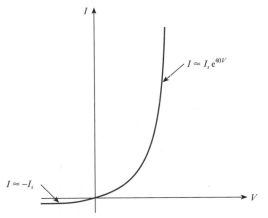

Figure 17.6
Current–voltage characteristics of a *pn* junction.

17.5 Diodes

Simplistically, a **diode** is an electrical component that conducts electricity in one direction but not the other. One could characterise an **ideal diode** as a component that conducts no current when a voltage is applied across it in one direction, but appears as a short circuit when a voltage is applied in the opposite direction. One could picture such a device as an electrical equivalent of a hydraulic non-return valve, which allows water to flow in one direction but not the other.

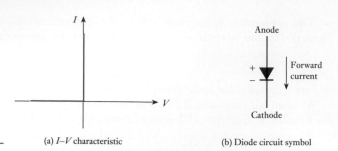

Figure 17.7
An ideal diode.

(a) I–V characteristic   (b) Diode circuit symbol

The characteristics of an ideal diode are shown in Figure 17.7(a), while Figure 17.7(b) shows the circuit symbol for a diode. A diode has two electrodes, called the **anode** and the **cathode**. The latter diagram also shows the polarity of the voltage that must be applied across the diode in order for it to conduct. It can be seen that the symbol for a diode resembles an arrow that points in the direction of current flow.

Diodes have a wide range of applications, including the **rectification** of alternating voltages. This process is illustrated in Figure 17.8. Here the diode conducts for the positive half of the input waveform, but opposes the flow of current during the negative half-cycle. When diodes are used in such circuits, they are often referred to as **rectifiers** and the arrangement of Figure 17.8 would be described as a **half-wave rectifier**. We will return to look at this circuit in Section 17.8, when we consider some applications of semiconductor diodes.

In practice, no real component has the properties of an ideal diode, but semiconductor diodes, in the form of *pn* junctions, can produce a good approximation to these characteristics.

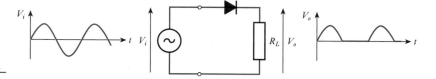

Figure 17.8
A diode as a rectifier.

17.6 ## Semiconductor diodes

A *pn* junction is not an ideal diode, but it does have a characteristic that approximates to such a device. When viewed on a large scale, the relationship between the current and the applied voltage is as shown in Figure 17.9. When forward biased, a *pn* junction exhibits an exponential current–voltage characteristic. A small forward voltage is required to make the device conduct, but then the current increases rapidly as this

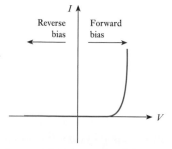

Figure 17.9
Forward and reverse currents in a semiconductor diode.

voltage is increased. When reverse biased, the junction passes only a small reverse current which is almost always negligible. The *pn* junction, therefore, represents a reasonable approximation to an ideal diode and is widely used in diode applications.

17.6.1 Diode characteristics

While the graph shown in Figure 17.9 provides an overview of diode behaviour, we often need a more detailed view of the component's characteristics. As we have seen, when reverse biased, a semiconductor diode passes only a very small current – the **reverse saturation current**. For silicon devices, this saturation current is typically 1 nA and is negligible in almost all applications. The reverse current is approximately constant as the reverse voltage is increased to a critical voltage called the **reverse breakdown voltage**, V_{br}. If the negative voltage is increased beyond this point, the junction breaks down and begins to conduct. This limits the useful voltage range of the diode. The reverse characteristics of a typical silicon diode are shown in Figure 17.10(a). The value of the reverse breakdown voltage will depend on the type of diode and may have a value from a few volts to a few hundred volts.

When a semiconductor diode is forward biased, a negligible current will flow for a small applied voltage, but this increases exponentially as the voltage is increased. *When viewed on a large scale*, it appears that the current is zero until the voltage reaches a so-called **turn-on voltage** and as the voltage is increased beyond this point the junction begins to conduct and the current increases rapidly. This turn-on voltage is about 0.5 V for a silicon junction. A further increase in the applied voltage causes the junction current to increase rapidly. This results in the current–voltage characteristic being almost vertical, showing that the voltage across the diode is approximately constant, irrespective of the junction current. The characteristic of a typical silicon diode is shown in Figure 17.10(a). However, in many applications, it is reasonable to approximate the characteristic by a straight-line response, as shown in Figure 17.10(b). This simplified form represents the forward characteristic of the diode by a forward voltage drop (of about 0.7 V for silicon devices) combined with a forward resistance. The latter results in the slope of the characteristic above the turn-on voltage. In many cases, the forward resistance of the diode may be ignored and it can be considered simply as a near ideal diode with a small forward voltage drop. This voltage drop is termed the **conduction voltage** of the diode. As the current through the diode increases, the voltage across the junction also increases. At 1 A, the conduction voltage might be about 1 V for a silicon diode, rising to perhaps 2 V at 100 A. In practice, most diodes would be destroyed long before the current reached such large values.

So far, we have considered only diodes constructed from **silicon** and have seen that such devices have a turn-on voltage of about 0.5 V and a conduction voltage of

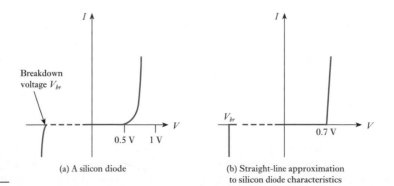

Figure 17.10
Semiconductor diode characteristics.

about 0.7 V. While silicon is the usual material for the manufacture of semiconductor diodes, many other materials are also used. Examples include **germanium** (which has a turn-on voltage of about 0.2 V and a conduction voltage of about 0.25 V) and **gallium arsenide** (which has a turn-on voltage of about 1.3 V and a conduction voltage of about 1.4 V).

Diodes are used for a number of purposes within electronic circuits. In many cases, relatively low voltages and currents are present and devices for such applications are usually called **signal diodes**. A typical device might have a maximum forward current of 100 mA and reverse breakdown voltage of 75 V. Other common applications for diodes include their use within power supplies to convert alternating currents into direct currents. Such diodes will usually have a greater current-handling capacity (usually measured in amperes or tens of amperes) and are generally called **rectifiers** rather than diodes. Reverse breakdown voltages for such devices will vary with the application but are typically hundreds of volts.

Diodes and rectifiers can be made using a variety of semiconductor materials and may use other techniques in place of simple *pn* junctions. This allows devices to be constructed with a wide range of characteristics in terms of current-handling capability, breakdown voltage and speed of operation.

File 17A

Computer simulation exercise 17.1

Use simulation to investigate the relationship between the current and the applied voltage in a small signal diode (such as a 1N4002). Measure the current while the applied voltage is swept from 0 to 0.8 volts and plot the resulting curve.

Look at the behaviour of the device over different voltage ranges, including both forward- and reverse-bias conditions. Estimate from these experiments the reverse breakdown voltage of the diode.

17.6.2 Diode equivalent circuits

It is often convenient to represent a diode by a simple **equivalent circuit** that embodies its basic characteristics. As with many devices, several equivalent circuits are used, these differing in their level of sophistication. Figure 17.11 shows three forms of equivalent circuit for a diode, together with an indication of their characteristics. These arrangements do not model the reverse breakdown of the device.

Figure 17.11(a) shows the simplest form of equivalent circuit, where the diode is represented by an ideal device. Such a model may be appropriate when using relatively large voltages and small currents, when the voltage across the conducting diode may be small compared with the other voltages within the circuit. Figure 17.11(b) shows a more elaborate model that includes not only an ideal diode but also a voltage source, representing the conduction voltage of the diode, V_{ON}. The voltage source *opposes* forward voltages across the diode, so conduction will only occur when the input voltage is greater than this conduction voltage. Note that the internal voltage source cannot be used to produce current in an external circuit as it is in series with an ideal diode that opposes such current flow. Including the conduction voltage within the equivalent circuit provides a much more realistic representation of the operation of the device, allowing it to be used in a wide range of situations. Figure 17.11(c) is the most sophisticated of the three models shown. This includes not only an ideal diode and a voltage source but also a resistor, representing the *on* resistance of the device r_{ON}. The inclusion of the diode's resistance is of particular importance in high-current applications where

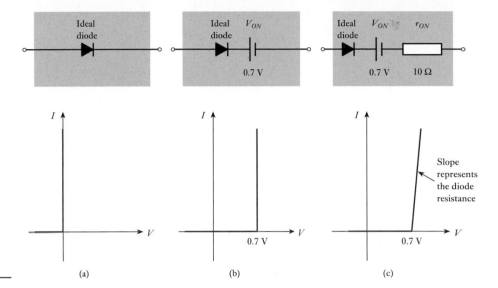

Figure 17.11
Diode equivalent circuits.

the voltage drop across the diode may be considerable. The value of r_{ON} will vary from one component to another, but a typical value might be 10 Ω. Appropriate values for V_D and r_{ON} may be obtained by plotting the diode's characteristic and drawing a straight-line approximation to the exponential curve. Where this line crosses the horizontal axis gives V_D and the reciprocal of the slope of this line gives r_{ON}. Clearly, the values obtained for these two quantities will depend on how the line is chosen. It is normal to draw the line tangential to the curve in the region of the expected quiescent current in the device as this optimises the model in this region of operation.

17.6.3 Diode circuit analysis

The non-linear characteristics of diodes complicate the analysis of circuits in which they are used. Consider, for example, the circuit in Figure 17.12. If the diode in this circuit were replaced with a resistor, then calculating the current in the circuit would be a case of simply dividing the voltage E by the sum of the two resistances. However, with a diode in the circuit, the calculation is more complicated. Applying Kirchhoff's voltage law around the circuit shows that

$$E = V_D + V_R$$

and therefore

$$E = V_D + IR \tag{17.4}$$

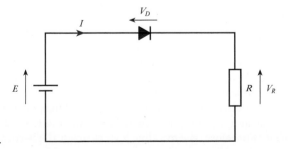

Figure 17.12
A simple diode circuit.

From Equation 17.3, we know that the relationship between the current in the diode and the voltage across it is given by

$$I \approx I_s \, e^{40V_D} \tag{17.5}$$

where I_s is the reverse saturation current of the diode.

Determining the current I in the circuit requires that we solve the two simultaneous equations given in Equations 17.4 and 17.5. This can be achieved using programs such as Mathcad, but also by using a graphical approach.

Load lines

To understand the graphical method, we need to look at the form of the two equations. Equation 17.5 describes the characteristics of the diode, which can be represented as shown in Figure 17.13(a). The graph shows the voltage across the diode for a given current. Equation 17.4 may be rearranged to give

$$V_D = E - IR$$

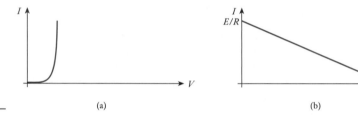

Figure 17.13
Graphical representations of Equations 17.4 and 17.5.

(a) (b)

This is also the relationship between the voltage across the diode and the current, which can be plotted as shown in Figure 17.13(b). If these two figures are plotted on identical axes they may be superimposed as shown in Figure 17.14, and the point of intersection of the two lines represents the solution of the two simultaneous equations. This is termed the **operating point** of the circuit. The straight line shown in Figure 17.14 represents the *load* applied to the circuit and, for this reason, is referred to as a **load line**. From the graph it is possible to read off the current in the circuit and the voltage across the diode. As the voltage across the resistor is equal to $E - V_D$, the distance along the voltage axis from the operating point to E represents the voltage across the resistor.

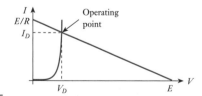

Figure 17.14
Use of a load line.

The construction of a load line is very straightforward. First, the diode characteristic is drawn on appropriate axes and then the load line is added. The load line is defined by two points – one on each axis. If the current in the diode were zero, then the current in the resistor would also be zero and $E - IR = E$. Thus, the load line goes through the horizontal axis at $V = E$. If the voltage across the diode were zero, then $E - IR = 0$ and $I = E/R$. Therefore, the load line goes through the vertical axis at $I = E/R$.

Example 17.1   **Determine the current flowing in the following circuit and the voltage across the diode, given that the characteristics of the diode are as shown.**

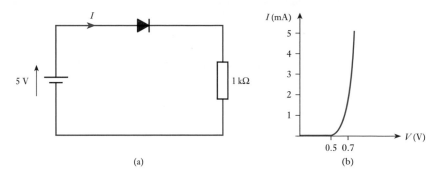

(a)   (b)

To determine the current, we first draw the diode characteristic with an extended voltage axis, then superimpose the load line. The load line is defined by the point on the *voltage* axis where V is equal to the applied voltage (5 V) and the point on the *current* axis where $I = E/R = 5\,\text{V}/1\,\text{k}\Omega = 5\,\text{mA}$.

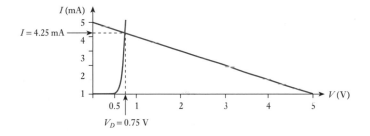

The point of intersection of the two lines shows the current in the circuit (4.25 mA) and the voltage across the diode (0.75 V).

Load lines are used not only in diode circuits but also in the analysis of other non–linear components, such as transistors. We will therefore return to this topic in later chapters.

Analysis using simplified equivalent circuits

We saw earlier how a diode could be represented by a range of simplified equivalent circuits. The use of these equivalent circuits greatly simplifies analysis, as it removes the need to solve complicated simultaneous equations.

One way to use equivalent circuits is to replace the diode characteristics used in Figure 17.14 with the simpler characteristics of Figure 17.11. This is illustrated in Figure 17.15, which shows the use of load lines for two of the simplified models

Figure 17.15
Using load lines with simplified diode models.

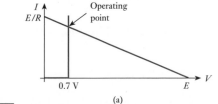
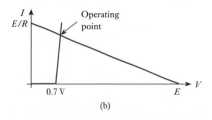

(a)   (b)

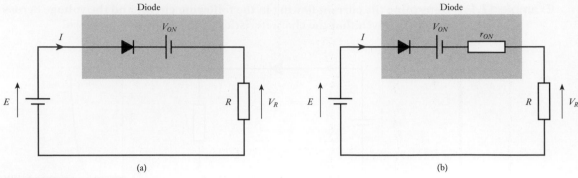

Figure 17.16
Use of simplified equivalent circuits.

shown in Figure 17.11. The use of straight-line approximations to the diode characteristics greatly simplifies the task of constructing these diagrams – at the expense of some loss of accuracy.

While equivalent circuits can be used in the construction of load lines, in practice this is rarely done because the use of simplified diode models removes the need for a graphical approach. This may be illustrated using the circuits shown in Figure 17.16. The arrangement of Figure 17.16(a) shows the simple circuit of Figure 17.12, but the diode has been replaced with the equivalent circuit of Figure 17.11(b). As the diode in this arrangement is 'ideal', it has no voltage across it when forward biased (as here), so the current in the circuit can be obtained directly by applying Kirchhoff's voltage law around the circuit, which shows that

$$E - V_{ON} = V_R = IR$$

or rearranging

$$I = \frac{E - V_{ON}}{R}$$

The circuit in Figure 17.16(b) uses the slightly more sophisticated model of Figure 17.11(c) and a similar analysis of the circuit shows that the current is given by

$$I = \frac{E - V_{ON}}{R + r_{ON}}$$

Thus, the use of simplified equivalent circuits removes the non-linear element from the circuit and greatly simplifies its analysis, at the expense of some loss of accuracy.

Example 17.2   Repeat the investigation of Example 17.1 using a simplified model of the diode.

We have looked at three simplified models for the diode.

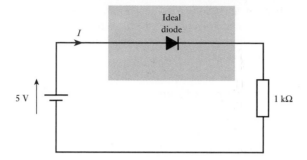

If we assume that the diode is 'ideal', there is no voltage drop across the diode and the current in the circuit is given by $I = E/R = 5\,\text{V}/1000\,\Omega = 5$ mA.

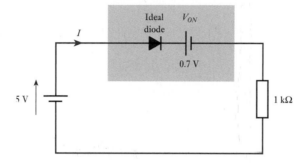

If we include the effect of the conduction voltage of the diode, then the current is given by $I = (E - V_D)/R = (5 - 0.7)/1000 = 4.3$ mA.

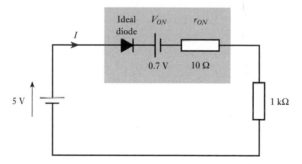

If we include the effect of the conduction voltage and internal resistance of the diode, then the current is given by $I = (E - V_D)/(R + r_{ON}) = (5 - 0.7)/(1000 + 10) = 4.26$ mA.

17.6.4 Effects of temperature

From Equation 17.1, we have

$$I \approx I_s(e^{eV/kT} - 1)$$

Clearly, for a given value of diode current I the voltage across the junction V is inversely proportional to the absolute junction temperature T. For silicon devices, the junction voltage *decreases* by about 2 mV/°C rise in temperature.

The diode current is also affected by the reverse saturation current I_s. We noted earlier that this current is related to the production of minority carriers as a result of thermal vibration. As the temperature rises, the number of minority carriers produced increases and the reverse saturation current goes up; I_s approximately doubles for an increase in temperature of about 10 °C. This corresponds to an increase of about 7 per cent/°C.

17.6.5 Reverse breakdown

The reverse breakdown of a diode may be brought about by one of two phenomena. In devices with heavily doped p- and n-type regions, the transition from one to the other is very abrupt and the depletion region is often only a few nanometres thick. Under these circumstances, junction voltages of only a few volts produce fields across the junction of several hundreds of megavolts per metre. Such high field strengths result in electrons being pulled from their covalent bonds, producing additional charge carriers and a large reverse current. The current produced by this **Zener breakdown** must be limited by external circuitry to prevent damage to the diode. The voltage at which Zener breakdown occurs is determined by the energy gap of the semiconductor used and is thus largely independent of temperature. However, it *decreases* very slightly with increasing temperature. Zener breakdown normally occurs at voltages below 5 V.

Diodes in which one, or both, of the semiconductor regions are lightly doped have a less abrupt transition and consequently a wider depletion layer. In such devices, the field generated by an applied voltage is usually insufficient to produce Zener breakdown, but does accelerate current carriers within the depletion layer. As these carriers are accelerated, they gain energy that they may lose by colliding with atoms within the lattice. If the carriers gain sufficient energy, they may ionise these atoms by freeing their electrons. This will generate additional carriers that are themselves accelerated by the applied field. At some point, the applied field is large enough to produce an 'avalanche' effect in which the current increases dramatically. This gives rise to **avalanche breakdown**. For high-voltage operation, it is possible to construct devices with breakdown voltages of several thousands of volts. Alternatively, it is possible to arrange that breakdown occurs at only a few volts. The voltage at which avalanche breakdown occurs *increases* with junction temperature.

Generally, if a diode suffers reverse breakdown at a voltage of less than 5 V, it is likely to be caused by Zener breakdown. If it occurs at a voltage of greater than 5 V, it is likely to be the result of avalanche breakdown.

17.7 Special-purpose diodes

We have already encountered several forms of semiconductor diode within this text. These include the *signal diode* and *rectifier*, discussed in the last section, and the *pn junction temperature sensor*, *photodiode* and *light-emitting diode* (*LED*) (described in Chapter 12). There are several other forms of diode that are widely used, each having its own unique characteristics and applications. We will look briefly at some of the more popular forms.

17.7.1 Zener diodes

When the reverse breakdown voltage of a diode is exceeded, the current that flows is generally limited only by external circuitry. If steps are not taken to limit this current, the power dissipated in the diode may destroy it. However, if the current is limited by the circuitry connected to the diode, the breakdown of the junction need not cause any damage to the device. This effect is utilised in special–purpose devices called **Zener diodes**, although it should be noted that the name is largely historical and is used to describe devices the operation of which may depend on either Zener or avalanche breakdown. From Figure 17.10 it is clear that when the junction is in the breakdown region, the junction voltage is approximately constant irrespective of the reverse current flowing. This allows the device to be used as a **voltage reference**. In such devices, the breakdown voltage is often given the symbol V_Z. Zener diodes are available with a variety of breakdown voltages to allow a wide range of reference voltages to be produced.

A typical circuit using a Zener diode is shown in Figure 17.17, which also shows the symbol used for a Zener diode. Here a poorly regulated voltage V is applied to a series combination of a resistor and a Zener diode. The diode is connected so that it is reverse biased by the positive applied voltage. If V is greater than V_Z the diode junction will break down and conduct, drawing current from the resistance R. The diode prevents the output voltage going above its breakdown voltage V_Z and thus generates an approximately constant output voltage irrespective of the value of the input voltage, provided it remains greater than V_Z. If V is less than V_Z the diode will conduct negligible current and the output will be approximately equal to V. In this situation, the Zener diode has no effect on the circuit.

If circuitry is connected to the output of the arrangement shown in Figure 17.17, this will also draw current through the resistance R. The value of R must be chosen so that the voltage drop across it caused by this current is not great enough to reduce the voltage across the Zener diode to below its breakdown voltage. This requirement must be balanced against the fact that the power dissipated in the diode and in the resistor increases as R is reduced.

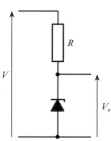

Figure 17.17
A simple voltage reference using a Zener diode.

Example 17.3

Design a voltage reference of 3.6 V capable of driving a load of 200 Ω. The reference voltage is to be produced from a supply voltage that can vary between 4.5 and 5.5 V.

A suitable circuit would be as shown here.

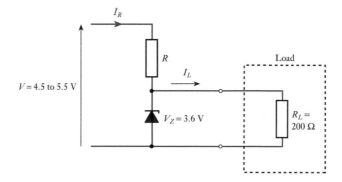

Clearly a Zener diode with a breakdown voltage V_Z of 3.6 V is used, but calculations must be performed to determine the value required for the resistor R and the power ratings required for the resistor and diode.

For power dissipation considerations, we would like R to be as high as possible. The maximum value of R is determined by the requirement that the voltage drop across R must not take the voltage at the output (and across the diode) below the required output voltage. This condition is most critical when the input voltage is at its lowest, so the value of R is determined by calculating the value for which the output would be equal to the required value (3.6 V) when the input voltage is at its lowest permissible value (4.5 V). We can ignore the effects of the Zener diode at this point as all the current flowing through R will be flowing into the load.

The current flowing in the load I_L can be calculated from

$$I_L = \frac{V_Z}{R_L}$$
$$= \frac{3.6 \text{ V}}{200 \ \Omega} = 18 \text{ mA}$$

The voltage drop across the resistance R caused by I_L must be less than the difference between the minimum supply voltage (4.5 V) and the Zener voltage (3.6 V), therefore

$$I_L R < 4.5 - 3.6 \text{ V}$$

and

$$R < \frac{4.5 - 3.6 \text{ V}}{I_L}$$
$$< \frac{0.9 \text{ V}}{18 \text{ mA}}$$
$$< 50 \ \Omega$$

So a standard value of 47 Ω would probably be chosen.

Maximum power dissipation in the various components is generated when the input voltage is at its maximum value (5.5 V). As the output voltage is fixed, the voltage across the resistance is easy to calculate – it is simply $V - 3.6$ V. Therefore, the maximum power dissipation in the resistance is simply

$$P_{R(\text{max})} = \frac{V^2}{R} = \frac{(5.5 - 3.6)^2}{47} \text{ W}$$
$$= 77 \text{ mW}$$

The power dissipated in the Zener diode P_Z is also simple to calculate. The voltage across it is fixed (3.6 V) and the current through it is simply the current through the resistance I_R minus the current taken by the load I_L. This is also at a maximum when the supply voltage has its maximum value.

Thus

$$P_{Z(\text{max})} = V_Z I_{Z(\text{max})} = V_Z (I_{R(\text{max})} - I_L)$$
$$= 3.6 \left(\frac{5.5 - 3.6}{47} - 0.018 \right) = 81 \text{ mW}$$

Thus, the Zener diode must have a power rating of greater than 81 mW.

It should be noted that although the voltage across a Zener diode is *approximately* constant in its breakdown region, irrespective of the current passing through it, it is not *completely* constant. You will observe from Figure 17.10 that the characteristic is not vertical above the breakdown voltage, but has a finite slope. This slope represents an effective output resistance that is typically a few ohms up to a few hundred ohms, causing the output voltage to vary slightly with current. For a high-precision voltage reference it is necessary to pass a fairly constant current through the Zener diode to produce a more constant output voltage than would otherwise be the case. We also noted earlier that the breakdown voltages of these devices vary slightly with temperature. Typical devices have **temperature coefficients** for their breakdown voltages of between 0.001 per cent/°C and 0.1 per cent/°C.

File 17B

Computer simulation exercise 17.2

The D1N750 is a 4.7 V Zener diode. Simulate the circuit of Figure 17.17 using this diode and a suitable resistor.

Apply a swept DC input voltage to the circuit and plot the output voltage against the input voltage for a range of values of R. Investigate the effect of connecting a load resistor to the circuit.

17.7.2 Schottky diodes

Unlike conventional pn junction diodes that are formed at the junction of two layers of doped semiconductor material, **Schottky diodes** are formed by a junction between a layer of metal (such as aluminium) and a semiconductor. The rectifying contact formed relies only on majority charge carriers and so is much faster in operation than pn junction devices, which are limited in speed by the relatively slow recombination of minority charge carriers.

Schottky diodes also have a low forward voltage drop of about 0.25 V. This characteristic is used to great effect in the design of high-speed logic gates (as described in Chapter 26).

17.7.3 Tunnel diodes

The **tunnel diode** uses high doping levels to produce a device with a very narrow depletion region. This region is so thin that a quantum mechanical effect known as *tunnelling* can take place. This results in charge carriers being able to cross the depletion layer, even though they do not have sufficient energy to surmount it. The combination of the tunnelling effect and conventional diode action produces a characteristic as shown in Figure 17.18.

This rather strange characteristic finds application in a number of areas. Of particular interest is the fact that, for part of its operating range, the voltage across the device falls for an increasing current. This corresponds to a region where the incremental resistance of the device is *negative*. This property is utilised in high-frequency oscillator circuits in which the negative resistance of the tunnel diode is used to cancel losses within passive components.

Figure 17.18
Characteristic of a
tunnel diode.

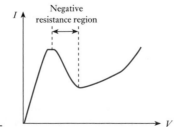

17.7.4 Varactor diodes

A reverse-biased diode has two conducting regions of *p*- and *n*-type semiconductor separated by a depletion region. This structure resembles a capacitor, with the depletion region forming the insulating dielectric. Small silicon signal diodes have a capacitance of a few picofarads, which changes with the reverse-bias voltage as this varies the thickness of the depletion region.

This effect is used by **varactor diodes** which act as voltage-dependent capacitors. A typical device might have a capacitance of 160 pF at 1 V, falling to about 9 pF at 10 V. Such devices are used at the heart of many automatic tuning arrangements, where the varactor is used within an *LC* or *RC* tuned circuit. The capacitance of the device – and therefore the frequency characteristics of the circuit – may then be varied by the applied reverse-bias voltage.

17.8   Diode circuits

In this section we look at just a few of the many circuits that make use of diodes of one form or another.

Video 17A

17.8.1 A half-wave rectifier

One of the commonest uses of diodes is as a rectifier within a power supply to generate a direct voltage from an alternating supply. A simple arrangement to achieve this is the half-wave rectifier (discussed briefly in Section 17.5), shown in Figure 17.19. While the input voltage is greater than the turn-on voltage of the diode, the diode conducts and the input voltage (minus the small voltage drop across the diode) appears across the load. During the part of the cycle in which the diode is reverse biased, no current flows in the load.

To produce a steadier output voltage, a **reservoir capacitor** is normally added to the circuit, as shown. This is charged while the diode is conducting and maintains the output voltage when the diode is turned off by supplying current to the load. This current gradually discharges the capacitor, causing the output voltage to decay. One effect of adding a reservoir capacitor is that the diode conducts for only short periods of time. During these periods, the diode currents are thus very high. The magnitude of the *ripple* in the output voltage is affected by the current taken by the load, the size of the capacitor and the frequency of the incoming signal. Clearly, as the supply frequency is increased the time for which the capacitor must maintain the output is reduced.

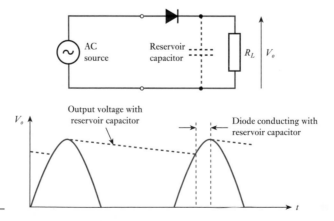

Figure 17.19
A half-wave rectifier.

Example 17.4

A half-wave rectifier connected to a 50 Hz supply generates a peak voltage of 10 V across a 10 mF reservoir capacitor. Estimate the peak ripple voltage produced if this arrangement is connected to a load that takes a constant current of 200 mA.

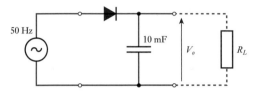

The voltage V across a capacitor is related to its charge q and its capacitance C by the expression

$$V = \frac{q}{C}$$

Differentiating with respect to time gives

$$\frac{\mathrm{d}V}{\mathrm{d}t} = \frac{1}{C}\frac{\mathrm{d}q}{\mathrm{d}t} = \frac{i}{C}$$

where i is the current into, or out of, the capacitor. In this example, the current is constant at 200 mA and the capacitance is 10 mF, so

$$\frac{\mathrm{d}V}{\mathrm{d}t} = \frac{i}{C} = \frac{0.2}{0.01} = 20 \text{ V/s}$$

Therefore, the output voltage will fall at a rate of 20 volts per second.

During each cycle, the capacitor is discharged for a time almost equal to the period of the input, which is 20 ms. Therefore, during this time, the voltage on the capacitor (the output voltage) will fall by 20 ms $\times$ 20 V/s = 0.4 V.

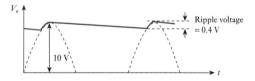

File 17C

Computer simulation exercise 17.3

Simulate the circuit in Figure 17.19 and investigate the behaviour of the circuit.

While typical half-wave rectifier arrangements might have input voltages of several hundred volts, the operation of the circuit is more apparent if smaller voltages are used so that the turn-on voltage of the diode is more easily observed.

Simulate the circuit with and without a reservoir capacitor and use transient analysis to study the circuit's behaviour. Plot the output voltage and current through the diode and see how these relate to the input voltage. Vary the input frequency and see how this affects the output.

Note the peak voltage at the output and the ripple voltage for a given set of circuit parameters (for use in the next simulation exercise).

17.8.2 A full-wave rectifier

One simple method for effectively increasing the frequency of the waveform applied to the capacitor in the previous circuit is to use a full-wave rectifier arrangement, as shown in Figure 17.20. When terminal A of the supply is positive with respect to terminal B, diodes D2 and D3 are forward biased and diodes D1 and D4 are reverse biased. Current therefore passes from terminal A, through D2, through the load R_L, and returns to terminal B through D3. This makes the output voltage V_o positive. When terminal B is positive with respect to terminal A, diodes D1 and D4 are forward biased and D2 and D3 are reverse biased. Current now flows from terminal B through D4, through the load, R_L, and returns to terminal A through D1. As the direction of the current in the output resistor is the same, the polarity of the output voltage is unchanged. Thus, both positive and negative half-cycles of the supply produce positive output peaks and the time during which the capacitor must maintain the output voltage is reduced.

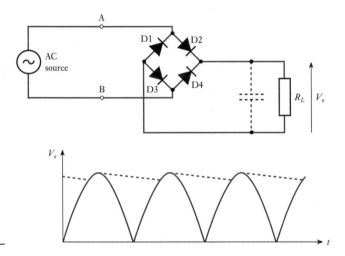

Figure 17.20
A full-wave rectifier.

Example 17.5

Determine the effect on the ripple voltage of replacing the half-wave rectifier in Example 17.4 with a full-wave arrangement, assuming that the reservoir capacitor and load remain the same.

As the capacitor and load are unchanged, the rate of change of the voltage on the capacitor is also unchanged at $20\,\text{V/s}$. However, in this case the time between successive peaks in the output voltage is equal to half the period of the input, which is 10 ms. Hence, the ripple voltage is now $10\,\text{ms} \times 20\,\text{V/s} = 0.2\,\text{V}$. Thus, the ripple voltage is halved.

File 17D

Computer simulation exercise 17.4

Repeat the investigations of Computer simulation exercise 17.3 for the full-wave rectifier circuit in Figure 17.20. Compare the peak voltage at the output and the ripple voltage with those obtained using the earlier circuit for similar circuit parameters.

17.8.3 A voltage doubler

The voltage doubler circuit in Figure 17.21 produces an output voltage considerably greater than the peak voltage of the input. To understand its operation, consider initially the half-cycle when the input terminal A is negative with respect to terminal B. Diode D1 is forward biased by the input voltage and so conducts, charging the capacitor C_1 to close to the peak voltage of the input waveform. During the next half-cycle, input terminal A is positive with respect to terminal B and diode D1 is reverse biased and so does not conduct. As terminal A becomes more positive with respect to terminal B the voltage across D1 increases, being equal to the input voltage plus the voltage across C_1. When the input reaches its peak value the voltage across D1 is nearly twice the input voltage. This forward biases diode D2, charging C_2 to close to twice the peak input voltage. This forms the output voltage of the arrangement.

If greater output voltages are required, several stages of voltage doubling can be cascaded to produce progressively higher voltages. These **voltage multiplier** circuits are ideal for applications that require high voltages at relatively low currents. Common applications include the extra-high-tension (EHT) supplies of cathode ray tubes (CRTs) and photomultiplier tubes.

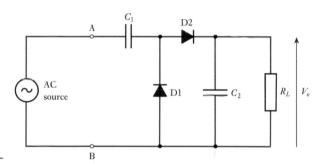

Figure 17.21
A voltage doubler.

17.8.4 A signal rectifier

Video 17B

A common use of signal diodes is in the rectification (also called **demodulation** or **detection**) of modulated signals, such as those used for radio frequency broadcasting. Such signals often use **full amplitude modulation** (**full AM**) which produces a waveform of the type shown in Figure 17.22. The signal consists of a high-frequency carrier component, the amplitude of which is modulated by a lower-frequency signal. It is this low-frequency signal that conveys the useful information and that must be recovered by **demodulating** the signal. We will look at both modulation and

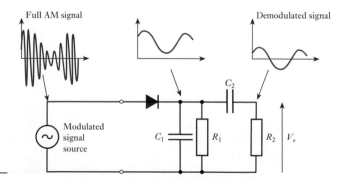

Figure 17.22
A signal rectifier.

demodulation in Chapter 29, but for the moment will just look at a simple method of demodulation.

The demodulator works in a similar manner to the half-wave rectifier described earlier. The modulated signal is passed through a diode that applies only the positive half of each cycle to a parallel RC network formed by R_1 and C_1. This behaves as a low-pass filter and the values of R_1 and C_1 are chosen such that they produce a high-frequency cut-off that is above the signal frequency but below that of the radio frequency carrier. Thus, the carrier is removed, leaving only the required signal plus a DC component. This direct component is removed by a second capacitor C_2 that applies the demodulated signal to R_2. This second RC network is effectively a high-pass filter that removes the DC component but has a cut-off frequency sufficiently low to pass the signal frequency.

The output voltage developed across R_2 represents the envelope of the original signal. For this reason, the circuit is often called an **envelope detector**. This arrangement forms the basis of most AM radio receivers, from simple **crystal sets**, which consist largely of the detector and a simple frequency-selective network, to complicated **superheterodyne receivers**, which use sophisticated circuitry to select and amplify the required signal. We will look in more detail at such arrangements in Chapter 29.

17.8.5 Signal clamping

Diodes may be used in a number of ways to change the form of a signal. Such arrangements come under the general heading of **wave-shaping circuits** and Figure 17.23 shows a few examples.

Figure 17.23(a) shows a simple arrangement for limiting the negative excursion of a signal. When the input signal is positive the diode is reverse biased and has no effect. However, when the input is negative and larger than the turn-on voltage of the diode,

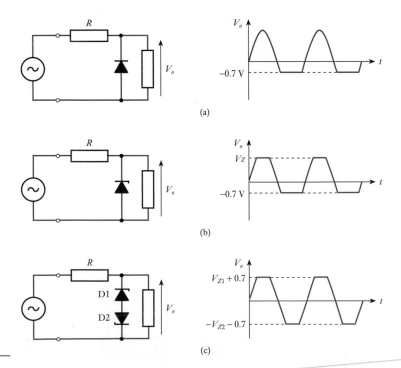

Figure 17.23
Signal-clamping circuits.

the diode conducts, clamping the output signal. This prevents the output from going more negative than the turn-on voltage of the diode (about 0.7 V for a silicon device). If a second diode is added in parallel with the first but connected in the opposite sense, the output will be clamped to ±0.7 V.

If the diode of Figure 17.23(a) is replaced with a Zener diode, as shown in Figure 17.23(b), the waveform is clamped for both positive *and* negative excursions of the input. If the input goes more positive than the breakdown voltage of the Zener diode V_Z, breakdown will occur preventing the output from rising further. If the input goes negative by more than the forward turn-on voltage of the Zener, it will conduct, again clamping the output. The output will therefore be restricted to the range $+V_Z > V_o > -0.7$ V.

Two Zener diodes may be used, as shown in Figure 17.23(c), to clamp the output voltage to any chosen positive and negative voltages. Note that the voltages at which the output signal is clamped are the sums of the breakdown voltage of one of the Zener diodes V_Z and the turn-on voltage of the other.

File 17E
File 17F
File 17G

Computer simulation exercise 17.5

Use simulation to investigate the behaviour of the various circuits in Figure 17.23. Apply a sinusoidal input voltage of 10 volts peak and use various combinations of simple diodes and Zener diodes. Suitable components might include 1N4002 signal diodes and D1N750 4.7 V Zener diodes.

Use transient analysis to look at the relationship between the input and output waveforms.

17.8.6 Catch diodes

Many actuators are inductive in nature (as discussed in Chapter 13). Examples include relays and solenoids. One problem with such actuators is that a large back e.m.f. is produced if they are turned off rapidly. This effect is used to advantage in some automotive ignition systems in which a circuit breaker (the 'points') is used to interrupt the current in a high-voltage coil. The large back e.m.f. produced is used to generate the spark required to ignite the fuel in the engine. In other electronic systems, these reverse voltages can do serious damage to delicate equipment if they are not removed. Fortunately, in many cases, the solution is very simple. This involves placing a **catch diode** across the inductive component, as shown in Figure 17.24, to reduce the magnitude of this reverse voltage.

The diode is connected so that it is normally reverse biased by the applied voltage and so normally is non-conducting. However, when the supply voltage is removed, any back e.m.f. produced by the inductor will forward bias the diode, which then conducts and dissipates the stored energy. The diode must be able to handle a current equal to the forward current flowing before the supply is removed.

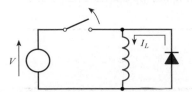

Figure 17.24
Use of a catch diode.

Video 17C

Further study

A common application for diodes is in the production of power supplies.

Design a unit that operates from a mains supply of 240 V AC at 50 Hz. The unit is required to drive an appliance that requires a fairly constant input of 12 V and takes a current that varies from 100 to 200 mA.

Key points

- Semiconductor materials are used at the heart of a multitude of electronic devices.
- The electrical properties of materials are brought about by their atomic structure.
- At very low temperatures semiconductors have the properties of an insulator. At higher temperatures thermal vibration of the atomic lattice leads to the generation of mobile charge carriers.
- Pure semiconductors are poor conductors, even at high temperatures. However, the introduction of small amounts of impurities dramatically changes their properties.
- Doping of semiconductors with appropriate materials can lead to the production of *n*-type or *p*-type materials.
- A junction between *n*-type and *p*-type semiconductors (a *pn* junction) has the properties of a diode.
- Semiconductor diodes approximate ideal diodes, but have a conduction voltage. Silicon diodes have a conduction voltage of about 0.7 V.
- In addition to conventional *pn* junction diodes, there are a wide variety of more specialised diodes, such as Zener, Schottky, tunnel and varactor diodes.
- Diodes are used in a range of applications, in both analogue and digital systems, including rectification, demodulation and signal clamping.

Exercises

17.1   Describe briefly the electrical properties of conductors, insulators and semiconductors.

17.2   Name three materials commonly used for semiconductor devices. Which material is most widely used for this purpose?

17.3   Outline the effect of an applied electric field on free electrons and holes.

17.4   Explain, with the aid of suitable diagrams where appropriate, what is meant by the terms 'tetravalent material', 'covalent bonding' and 'doping'.

17.5   What are meant by the terms 'intrinsic conduction' and 'extrinsic conduction'? What form of charge carriers is primarily responsible for conduction in doped semiconductors?

17.6   Explain what is meant by a depletion layer and why this results in a potential barrier.

17.7   Explain the diode action of a *pn* junction in terms of the effects of an external voltage on the drift and diffusion currents.

17.8   Sketch the current–voltage characteristic of a silicon diode for both forward- and reverse-bias conditions.

17.9   Sketch the current–voltage characteristic of an ideal diode.

17.10   What is the difference between a diode and a rectifier?

17.11   What is meant by the reverse saturation current of a diode?

17.12   Explain what is meant by the turn-on voltage and the conduction voltage of a diode. What are typical values of these quantities for a silicon diode?

17.13   What are typical values for the turn-on voltage and conduction voltage of diodes formed from germanium and gallium arsenide?

17.14   Explain the use of equivalent circuits of diodes and give examples of diode equivalent circuits of different levels of sophistication.

17.15   Use computer simulation to plot the characteristic of a typical silicon diode, such as a 1N4002, and use this to determine appropriate values for V_{ON} and r_{ON}. Assume that the diode is to be used in a circuit where the quiescent current through the diode will be approximately 20 mA, then construct an equivalent circuit of the diode when used in such a circuit.

17.16   Repeat the previous exercise using a different diode, such as a 1N914, and compare your results.

17.17   Explain, with the aid of a suitable sketch, how you would use load line analysis to determine the current I in the following circuit.

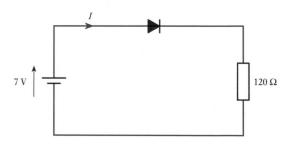

17.18 Estimate the current in the circuit of the previous exercise, assuming that the diode can be adequately represented by an equivalent circuit consisting of an ideal diode and a fixed voltage source.

17.19 Explain the terms 'Zener breakdown' and 'avalanche breakdown'.

17.20 Sketch a simple circuit that uses a Zener diode to produce a constant output voltage of 5.6 V from an input voltage that may vary from 10 to 12 V. Select appropriate component values, such that the circuit will deliver a current of at least 100 mA to an external load, and estimate the maximum power dissipation in the diode.

17.21 A half-wave rectifier is connected to a 50 Hz supply and generates a peak output voltage of 100 V across a 220 μF reservoir capacitor. Estimate the peak ripple voltage produced if this arrangement is connected to a load that takes a constant current of 100 mA.

17.22 What would be the effect on the ripple voltage calculated in the last exercise of replacing the half-wave rectifier with a full-wave rectifier of similar peak output voltage?

17.23 Sketch the output waveforms of the following circuits. In each case, the input signal is a sine wave of ±5 V peak.

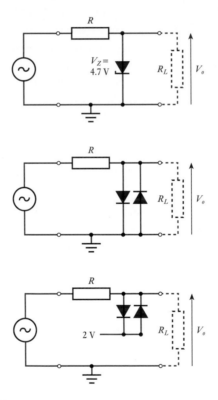

17.24 Use circuit simulation to verify your answers to the last exercise. How does the value of R_L affect the operation of the circuits?

17.25 Design a circuit that will pass a signal unaffected, except that it limits its excursion to the range $+10.4\,\text{V} > V > -0.4\,\text{V}$.

17.26 Use circuit simulation to verify your solution to the last exercise.

Chapter 18   Field-effect Transistors

When you have studied the material in this chapter, you should be able to:

- describe the construction and operation of the major forms of field-effect transistors (FETs)
- derive and use equivalent circuits of such devices to describe their behaviour and characteristics
- outline the use of FETs in the formation of amplifiers and design and analyse simple circuits
- explain the distinction between 'small-signal' and 'large-signal' aspects of a circuit's design
- describe the limitations imposed on the frequency response of FET circuits by device capacitances
- discuss the use of FETs in a range of amplifier circuits
- suggest a number of other uses of FETs in both analogue and digital systems.

18.1   Introduction

Field-effect transistors, or FETs, are probably the simplest form of transistor to understand and are widely used in both analogue and digital applications. They are characterised by a very high input resistance and small physical dimensions and can be used to create circuits with low power consumption, making them ideal for use in **very large-scale integration** (VLSI) circuits. There are two main forms of field-effect transistor – namely, the **insulated-gate FET** and the **junction-gate FET**. We will look at both forms in this chapter.

Continuing with our 'top-down' approach, we will begin by looking at the general behaviour of FETs, then turn our attention to their physical construction and operation. We will then consider the characteristics of these devices and look briefly at some simple circuits. The chapter ends by looking at the use of FETs in a range of analogue and digital applications.

18.2   An overview of field-effect transistors

While there are many forms of FETs, the general operation and characteristics of these devices are essentially the same. In each case, a voltage applied to a control input produces an electric field, which affects the current that flows between two of the terminals of the device.

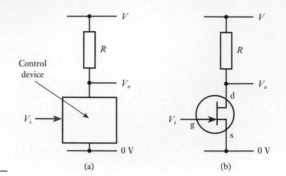

Figure 18.1
An FET as a control device.

(a)   (b)

In Chapter 14, when considering simple amplifiers, we looked at an arrangement using an unspecified 'control device' to produce amplification. This arrangement is repeated in Figure 18.1(a). Here, the control device varies the current flowing through a resistor in response to some input voltage. The output voltage V_o of this arrangement is equal to the supply voltage V minus the voltage across the resistor R. Therefore $V_o = V - IR$ where I is the current flowing through the resistor. The resistor current I is equal to the current flowing into the control device (ignoring any current flowing through the output terminal) so the output voltage is directly affected by the control device. If the current flowing in the control device is, in turn, determined by its input voltage V_i, then the output voltage of the circuit is controlled by this input voltage. Given an appropriate 'gain' in the control device, this arrangement can be used to create voltage amplification.

In simple electronic circuits, the control device shown in Figure 18.1(a) will often be some form of **transistor** – generally either a *field-effect transistor*, as discussed in this chapter, or a *bipolar transistor*, as discussed in the next chapter. Figure 18.1(b) shows a simple amplifier based on an FET. This diagram shows the circuit symbol for a junction–gate FET, although other forms of device could also be used. The input voltage to the FET controls the current that flows through the device and hence determines the output voltage as described above.

FETs have three terminals: the **drain**, **source** and **gate** – labelled d, s and g, respectively, in Figure 18.1(b). It can be seen that the gate represents the **control input** of the device and a voltage applied to this terminal will affect the current flowing from the drain to the source.

18.2.1 Notation

When describing FET circuits, we are often interested in the voltages between their various terminals and the current flowing into these terminals. We normally adopt a notation whereby voltages are given symbols of the form V_{XY} where X and Y correspond to the symbols for two of the device's terminals. This symbol then represents the voltage on X with respect to Y. For example, V_{GS} would be used to represent the voltage on the *gate* with respect to the *source*.

Device currents are given labels to represent the associated terminal. For example, the current into the drain would be labelled I_D. We normally use *upper-case* letters for steady voltages and currents and *lower-case* letters for varying quantities. For example, V_{GS} and I_D represent steady quantities, while v_{gs} and i_d represent varying quantities.

A special notation is used to represent the power supply voltages and currents in FET circuits. The voltage and current associated with the supply line that is connected (directly or indirectly) to the drain of the FET are normally given the labels V_{DD} and

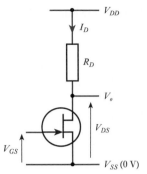

Figure 18.2
Labelling of voltages
and currents in FET
circuits.

I_{DD}. Similarly, the corresponding labels for the supply connected to the source are V_{SS} and I_{SS}. In many cases, V_{SS} is taken as the zero-volts reference (or ground) of the circuit. Passive components connected to the various terminals are often given corresponding labels, so that a resistor connected to the gate of an FET might be labelled R_G. This notation is illustrated in Figure 18.2.

18.3 Insulated-gate field-effect transistors

In FETs, conduction between the drain and source electrodes takes place through a **channel** of semiconductor material. Insulated-gate FETs are so called because their metal gate electrode is *insulated* from the conducting channel by a layer of insulating oxide. Such devices are often referred to as **insulated-gate field-effect transistors** (IGFETs), but are more commonly described as **metal oxide semiconductor field-effect transistors** (MOSFETs). Digital circuits constructed using such techniques are usually described as using **MOS technology**. Here we refer to insulated-gate devices as MOSFETs.

The channel in a MOSFET can be made of either *n*-type or *p*-type semiconductor material, leading to two polarities of transistor, termed *n*-channel and *p*-channel devices. Figure 18.3 illustrates the construction of each form of device. The behaviour of these two forms is similar, except that the polarities of the various currents and voltages are reversed. To avoid duplication, we will concentrate on *n*-channel devices in this section.

An *n*-channel device is formed by taking a piece of *p*-type semiconductor material (the **substrate**) and forming *n*-type regions within it to represent the drain and the source. Electrical connections are made to these regions, forming the drain and source

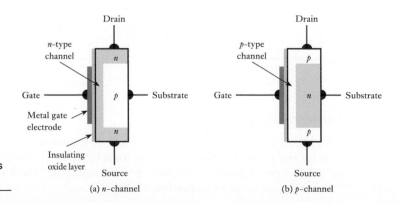

Figure 18.3
Insulated-gate
field-effect transistors
– MOSFETs.

(a) *n*-channel

(b) *p*-channel

electrodes. A thin *n*-type channel is then formed to join these two regions. This channel is covered by an insulating oxide layer and then by a metal gate electrode. Electrical connections are made to the gate and substrate, although the latter is often internally connected to the source to form a device with three external connections.

18.3.1 MOSFET operation

The channel between the drain and the source represents a conduction path between these two electrodes and therefore permits a flow of current. With zero volts applied to the gate, a voltage applied between the drain and the source will produce an electric field that will cause the mobile charge carriers in the channel to flow, thus creating a current. The magnitude of this current will be determined by the applied voltage and the number of charge carriers available in the channel.

The application of a voltage to the gate of the device will affect the number of charge carriers in the channel and hence the flow of electricity between the drain and source. The metal gate electrode and the semiconductor channel represent two conductors separated by an insulating layer. The construction therefore resembles a capacitor and the application of a voltage to the gate will induce the build-up of charge on each side. If the gate of an *n*-channel MOSFET is made *positive* with respect to the channel, this will attract electrons to the channel region, increasing the number of mobile charge carriers and increasing the apparent thickness of the channel. Under these circumstances the channel is said to be **enhanced**. If the gate is made *negative*, this will repel electrons from the channel, reducing its thickness. Here, the channel is said to be **depleted**. Thus, the voltage on the gate directly controls the effective thickness of the channel and the resistance between the drain and the source. This process is illustrated in Figure 18.4.

Note that the junction between the *p*-type substrate and the various *n*-type regions represents a *pn* junction, which will have the normal properties of a semiconductor diode. However, in normal operation the voltages applied ensure that this junction is always reverse biased so that no current flows. For this reason, the substrate can be largely ignored when considering the operation of the device.

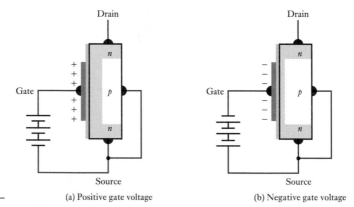

Figure 18.4
The effect of gate voltage on a MOSFET.

(a) Positive gate voltage

(b) Negative gate voltage

18.3.2 Forms of MOSFET

The MOSFETs shown in Figure 18.3 and described above can be used with both positive and negative gate voltages, which can be used to enhance or deplete the channel. Such a device is called a **depletion–enhancement MOSFET** or **DE MOSFET** – or sometimes simply a **depletion MOSFET**.

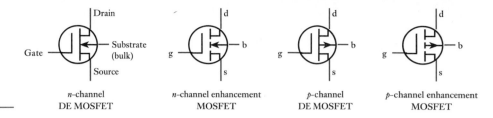

**Figure 18.5
MOSFET circuit
symbols.**

Other forms of MOSFETs are constructed in a similar manner, except that no channel is formed during manufacture. In the absence of a gate voltage there is no conduction path between the drain and the source and no current will flow. However, the application of a positive voltage to the gate (of an *n*-channel device) will attract electrons and repel holes from the region around the gate to form a conducting *n*-type channel. This region is called an inversion layer as it represents an *n*-type layer in a *p*-type material. A device of this type can be used in an enhancement mode, as with the depletion–enhancement MOSFET, but cannot be used in a depletion mode. For this reason, such devices are called **enhancement MOSFETs**.

To avoid confusion, different circuit symbols are used for the various types of MOSFETs, which are shown in Figure 18.5. The vertical line in the centre of the symbols represents the channel and is shown solid in a DE MOSFET (as it is present even in the absence of a gate voltage) and dashed in an enhancement MOSFET (as it is present only when an appropriate gate voltage is applied). The arrow on the substrate indicates the polarity of the device. This represents the *pn* junction between the substrate and the channel and points in the same direction as the symbol of an equivalent diode. That is, it points *towards* an *n*-type channel and *away from* a *p*-type channel. Note that, within the circuit symbols, the substrate is given the label 'b' (which stands for *bulk*) to avoid confusion with the source. As noted earlier, the substrate is often joined to the source internally to give a device with only three terminals. If the significance of the various elements of the symbols is considered, then it is relatively easy to remember the symbols for the various types of devices.

18.4 Junction-gate field-effect transistors

As in a MOSFET, conduction in a junction-gate transistor takes place through a channel of semiconductor material. However, in this case, the conduction within the channel is controlled not by an insulated gate but by a gate formed by a reverse-biased *pn* junction. Junction-gate FETs are sometimes referred to as **JUGFETs**, but here we will use another widely used abbreviation – **JFET**.

The form of a JFET is illustrated in Figure 18.6, which shows both *n*–channel and *p*-channel versions. As with MOSFETs, the operation of *n*-channel and *p*-channel devices is similar, except for the polarity of the voltages and currents involved. Here, we concentrate on *n*-channel devices.

Figure 18.6(a) shows an *n*-channel JFET. Here, a substrate of *n*-type material is used, with electrical connections at each end to form the drain and the source. A region of *p*-type material is now added between the drain and the source to form the gate. The fusion of the *n*-type and *p*-type materials forms a *pn* junction, which has the electrical properties of a semiconductor diode. If this junction is forward biased (by making the gate *positive* with respect to the other terminals of the device) current will flow across this junction. However, in normal operation, the gate is kept *negative* with

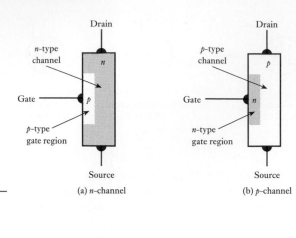

Figure 18.6
Junction field-effect transistors – JFETs.

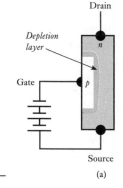

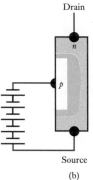

Figure 18.7
An *n*-channel JFET with a negative gate bias.

respect to the rest of the device, reverse biasing the junction and preventing any current from flowing across it. This situation is shown in Figure 18.7.

In the last chapter, we noted that reverse biasing a *pn* junction creates a **depletion layer** about the junction in which there are very few mobile charge carriers. Such a region is effectively an insulator, so the formation of such a region about the gate in a JFET reduces the cross-sectional area of the channel and increases its effective resistance. This is shown in Figure 18.7(a). As the magnitude of the reverse-bias voltage is increased, the thickness of the depletion layer increases and the channel is further reduced, as shown in Figure 18.7(b). Thus, the resistance of the channel is controlled by the voltage applied to the gate.

The circuit symbols used for *n*-channel and *p*-channel JFETs are shown in Figure 18.8. As in the MOSFET, the arrow shows the polarity of the device and as before it points *towards* an *n*-type channel and *away from* a *p*-type channel.

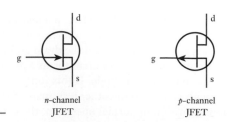

Figure 18.8
JFET circuit symbols.

FET characteristics

While MOSFETs and JFETs operate in somewhat different ways, their characteristics are in many ways quite similar. When considering amplifiers in Chapter 14, we characterised them in terms of the nature of their input and their output, and the relationship between them. We can consider the characteristics of FETs in a similar manner.

18.5.1 Input characteristics

Both MOSFETs and JFETs have a very high input resistance. In a MOSFET, the gate electrode is insulated from the rest of the device by an oxide layer, which prevents any current flowing into this terminal. In a JFET, the gate takes the form of a *pn* junction that is kept in a reverse-biased state. The current across such a reverse-biased junction is negligible in almost all cases (see Chapter 17). Thus, in both MOSFETs and JFETs, the gate is effectively insulated from the remainder of the device.

18.5.2 Output characteristics

The output characteristics of a device describe how the output voltage affects the output current. When considering amplifiers in Chapter 14, we saw that, in many cases, this relationship can be adequately represented by a simple, fixed output resistance. However, in transistors the situation is slightly more complicated and we need to look in more detail at how the device operates when connected to an external power supply.

In most circuits that use an *n*-channel FET, a voltage is applied across the device such that the drain is positive with respect to the source, so that V_{DS} is positive. The applied voltage produces a **drain current** I_D through the channel between the drain and the source. As the current flows through the channel, its resistance produces a potential drop such that the potential gradually falls along the length of the channel. As a result, the voltage between the gate and the channel varies along the length of the channel.

The effect of this variation in gate-to-channel voltage on a MOSFET is shown in Figure 18.9(a). Here, the voltage on the gate is positive with respect to the source and a correspondingly larger positive voltage is applied to the drain. In this arrangement, the potential in the channel at the end adjacent to the drain terminal makes it more positive than the gate terminal and the channel in this area is *depleted*. At the other end of the channel, the potential is approximately equal to that of the source, which is negative with respect to the gate, and the channel is *enhanced*. Thus, the apparent thickness of the channel changes along its length, being narrower near the drain and wider nearer the source. In this example the gate voltage is positive with respect to the

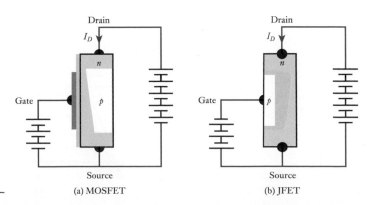

Figure 18.9
Typical FET circuit configurations.

source, but in a DE MOSFET the gate voltage could alternatively be zero or negative with respect to the source. This would reduce the average thickness of the channel, but the tapering effect of the variation in channel potential would still be present.

Figure 18.9(b) shows the corresponding situation in a JFET. In this case, the gate-to-source voltage is negative and the drain-to-source voltage is positive. Here, the gate-to-channel voltage is negative along the length of the channel (ensuring that the gate junction is reverse biased throughout its length). However, at the drain end of the channel, this reverse-bias voltage is much greater than at the source end, so the depletion layer is much thicker near the drain than it is near the source. This results in a tapered channel, as shown in the figure, where the channel is much narrower at the drain end than at the source end.

It can be seen that, in both the MOSFET and the JFET, the thickness of the channel is controlled by the voltage applied to the gate, but it is also influenced by the drain-to-source voltage. For small values of V_{DS}, as the gate-to-source voltage V_{GS} is made more positive (for n-channel FETs), the channel thickness increases and the effective resistance of the channel is decreased. The behaviour of the channel resembles that of a resistor, with the drain current I_D being proportional to the drain voltage V_{DS}. The value of this effective resistance is controlled by the gate voltage V_{GS}, the resistance reducing as V_{GS} is made more positive. This is referred to as the **ohmic region** of the device's operation.

As V_{DS} is increased the channel becomes more tapered and eventually the channel thickness is reduced to approximately zero at the end near the drain. The channel is now said to be **pinched off** and the drain-to-source voltage at which this occurs is called the **pinch-off voltage**. This does not stop the flow of current through the channel, but it does prevent any further increase in current. Thus, as the drain voltage is increased above the pinch-off voltage, the current remains essentially constant. This is referred to as the **saturation region** of the characteristic.

If the gate voltage is held constant and the drain voltage is gradually increased, the current initially rises linearly with the applied voltage (in the ohmic region) and then, above the pinch-off voltage, becomes essentially constant (in the saturation region). This behaviour is shown in Figure 18.10(a). Varying the voltage on the gate changes the effective resistance of the channel in the ohmic region and also the value of the steady current produced in the saturation region. This is illustrated in Figure 18.10(b) by drawing the output characteristic of Figure 18.10(a) for a range of different gate voltages. These **output characteristics** are also called the **drain characteristics** of the device. This figure shows a generic set of curves that could represent any form of FET. The range of values of V_{GS} will depend on the type of device concerned, as discussed in earlier sections.

From the output characteristics, it is clear that FETs have two distinct regions of operation – namely, the ohmic region and the saturation region. In the first of these,

Figure 18.10
FET output characteristics.

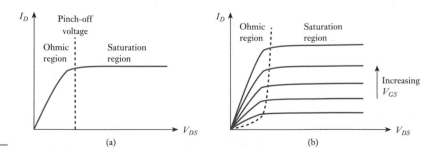

the device resembles a **voltage-controlled resistance** and there are several applications that make use of this characteristic. In the second region (saturation) the output current I_D is largely independent of the applied voltage and is controlled by the input voltage V_{GS}. It is this second region that is normally used in the creation of amplifiers. When used in this region it is the slight slope of the output characteristic that indicates how the output current changes with the output voltage. This slope therefore represents the **output resistance** of the device in that operating region.

MOSFET output characteristics

While the general form of the output characteristics is similar for all forms of FET (as illustrated in Figure 18.10(b)), there are differences in the voltage ranges involved. We noted in Section 18.3.2 that DE MOSFETs are used with both positive and negative gate voltages, while enhancement MOSFETs use only positive gate voltages (for n-channel devices). In each case, the gate voltage V_{GS} at which the device starts to conduct is called the **threshold voltage** V_T and for DE MOSFETs this will have a negative value of a few volts, while for enhancement MOSFETs it will have a positive value of a few volts. The actual value of the threshold voltage will depend on the form of the transistor and will therefore vary from one device to another.

Figure 18.11(a) shows the output characteristic of a typical n-channel DE MOSFET. It can be seen that when the drain voltage is greater than the pinch-off voltage the drain current I_D is approximately constant, the magnitude of this current being determined by the gate-to-source voltage V_{GS}. When V_{GS} is equal to zero, the value of the drain current is referred to as the **drain-to-source saturation current** I_{DSS}. As V_{GS} becomes more negative the drain current is reduced, until it is zero for all values of V_{DS}. The value of the gate voltage at which the channel stops conducting is termed the **threshold voltage** V_T or sometimes the **gate cut-off voltage** $V_{GS(OFF)}$. The device shown in Figure 18.11(a) has a threshold voltage of $-5\,V$. It can be seen that the voltage at which the channel becomes pinched off varies with V_{GS}. The boundary between the ohmic and saturation regions of the characteristic is defined by the locus of pinch-off, which is given by

$$V_{DS}\,(\text{at pinch-off}) = V_{GS} - V_T \tag{18.1}$$

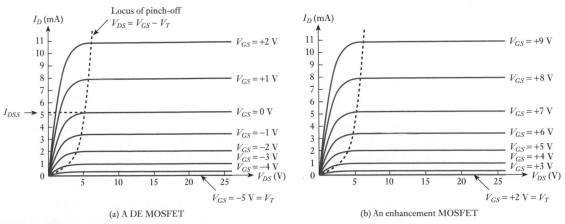

(a) A DE MOSFET

(b) An enhancement MOSFET

Figure 18.11
Typical MOSFET output characteristics.

Figure 18.11(b) shows the output characteristic of a typical *n*-channel enhancement MOSFET. It can be seen that this is of a similar form to that of a DE MOSFET, except that the useful range of the gate-to-source voltage is different. Here the gate must be made positive by a few volts to turn on the device and making the gate more positive progressively increases the drain current. The gate-to-source voltage at which the device begins to turn on is again termed the threshold voltage, which in an *n*-channel enhancement MOSFET is positive.

The device shown in Figure 18.11(b) has a threshold voltage of +2 V. As an enhancement MOSFET is turned off when $V_{GS} = 0$, the saturation current I_{DSS} is *not* defined for such a device.

JFET output characteristics

The output characteristics of a JFET are similar to those of MOSFETs except that the useful range of V_{GS} is different. Figure 18.12 shows the characteristics of a typical *n*-channel JFET and, as in the case of the DE MOSFET, the drain current produced for $V_{GS} = 0$ is termed the **drain-to-source saturation current I_{DSS}**. However, the gate voltage at which the channel stops conducting is now termed the **pinch-off voltage V_P** (or sometimes the **gate cut-off voltage $V_{GS(OFF)}$**). The device shown in Figure 18.12 has a pinch-off voltage of −7 V.

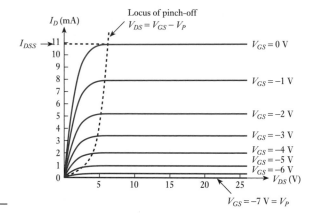

Figure 18.12
Typical JFET output characteristics.

18.5.3   **Transfer characteristic**

Having considered the input and the output characteristics of FETs, we can now turn our attention to the relationship between the input and the output. This is often termed the **transfer characteristic** of the device.

In Chapter 14, when we looked at voltage amplifiers, we represented the relationship between the input and the output by the *gain* of the circuit, this being equal to V_o/V_i. However, in the case of an FET, the input quantity is the gate *voltage*, while the output quantity is the drain *current*. It is also clear from Figures 18.10–18.12 that there is no linear relationship between these two quantities. If, however, we arrange that the device remains within the saturation region, we can plot the relationship between the input voltage V_{GS} and the output current I_D, and this is shown in Figure 18.13 for various forms of FET. It can be seen that the basic form of the transfer function is similar for each device, although the characteristics are offset with respect to each other because of their different gate voltage ranges. For each device, the relationship between I_D and V_{GS} is approximately parabolic.

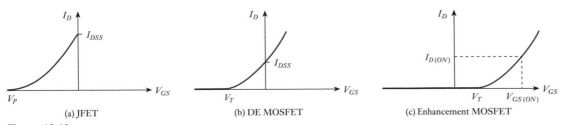

Figure 18.13
FET transfer characteristics.

MOSFET transfer characteristics

The characteristics of a DE MOSFET are normally described by defining the **threshold voltage** V_T and the **drain-to-source saturation current** I_{DSS}, as shown in Figure 18.13(b). In an enhancement device, the characteristic is normally described by giving V_T and some particular value of the drain current, $I_{D(ON)}$, corresponding to a specified value of the gate voltage, $V_{GS(ON)}$, as shown in Figure 18.13(c). In both cases, the relationship between I_D and V_{GS} in the saturation region is approximately parabolic (a square law) and is described by the simple expression

$$I_D = K(V_{GS} - V_T)^2 \qquad (18.2)$$

where K is a constant that depends on the physical parameters and geometry of the device.

JFET transfer characteristics

The characteristics of a JFET are normally described by defining the **pinch-off voltage** V_P and the **drain-to-source saturation current** I_{DSS}, as shown in Figure 18.13(a). Again, in the saturation region the relationship between the drain current and the gate voltage obeys a square law. In this case, the relationship is

$$I_D = I_{DSS}\left(1 - \frac{V_{GS}}{V_P}\right)^2 \qquad (18.3)$$

which may be rearranged into the form

$$I_D = K'(V_{GS} - V_P)^2$$

which is clearly of the same form as Equation 18.2. As in the earlier equation, K' is a constant related to the physical parameters and geometry of the device.

18.5.4 FET operating ranges

While the characteristics of Figure 18.13 are clearly not linear, over a small range of values of V_{GS} they may be said to approximate to a linear relationship. Thus, if we restrict V_{GS} to small fluctuations about a particular mean value (termed the **operating point**), the relationship between variations in V_{GS} and the resultant variations in I_D is *approximately* linear. Figure 18.14 illustrates the operating points and normal operating ranges for the three forms of FET.

When the device is constrained to operation about its operating point, then the transfer characteristics of an FET are described by the *change* in the output that is

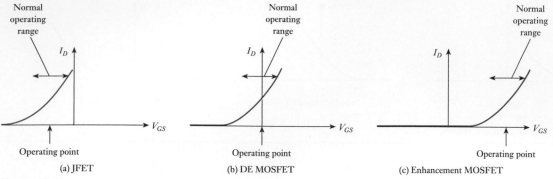

Figure 18.14
Normal operating ranges for FETs.

produced by a corresponding *change* in the input. This corresponds to the slope of the curve at the operating point in the graphs of Figure 18.14. This quantity has the units of current/voltage, which is the reciprocal of resistance, or **conductance**. As this quantity describes the transfer properties of the FET, it is given the name **transconductance** and, like conductance, it has the units of siemens. The symbol given to transconductance is g_m.

It should be noted that g_m represents the slope of the transfer characteristic at the operating point, *not* the ratio of the drain current to the gate voltage at this point. Thus, if a small change in the gate voltage ΔV_{GS} produces a small change in the drain current ΔI_D, then

$$g_m = \frac{\Delta I_D}{\Delta V_{GS}} \tag{18.4}$$

$$g_m \neq \frac{I_D}{V_{GS}}$$

From Figure 18.14, it is clear that the slope of the transfer function varies along the curve, so g_m is not constant for a given device. Clearly, in the limit g_m is given by

$$g_m = \frac{dI_D}{dV_{GS}}$$

From Equation 18.3 we know that for a JFET

$$I_D = I_{DSS}\left(1 - \frac{V_{GS}}{V_P}\right)^2$$

and we can therefore find g_m by differentiation. This gives

$$g_m = -\frac{2I_{DSS}}{V_P}\left(1 - \frac{V_{GS}}{V_P}\right)$$

$$= -2\frac{\sqrt{I_{DSS}}}{V_P} \times \sqrt{I_D} \tag{18.5}$$

Thus, in a JFET, g_m is proportional to the square root of the drain current. A similar analysis can be performed to obtain a similar result for the MOSFET.

18.5.5 Equivalent circuit of an FET

In Chapter 14, we noted the usefulness of equivalent circuits when describing the characteristics of amplifiers. We are now in a position to construct an equivalent circuit for an FET and this is shown in Figure 18.15. The circuit shows no input resistor, as the input resistance is so high that it can normally be considered to be infinite. As the output of an FET is normally considered to be the drain *current*, the equivalent circuit models the output using a Norton equivalent circuit rather than the Thévenin arrangement used in Chapter 14. The equivalent circuit describes the behaviour of the device when its input fluctuates by a small amount about its normal operating point, rather than describing its behaviour in response to constant (DC) voltages. For this reason, it is referred to as a **small-signal equivalent circuit**.

Figure 18.15
A small-signal equivalent circuit of an FET.

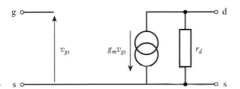

The equivalent circuit represents the transfer characteristics of the FET using a dependent current source, which produces a current of $g_m v_{gs}$, where v_{gs} is the fluctuating, or small-signal, input voltage. This flows *downwards* as the drain current i_d is assumed to flow *into* rather than *out of* the device. The resistor r_d models the way in which the output current is affected by the output voltage. If the lines in the output characteristic of Figure 18.10(b) were completely horizontal, then the output current would be independent of the output voltage. In practice, these have a slight slope, giving rise to r_d, which is termed the small-signal **drain resistance**. This resistance is given by the slope of the output characteristic and, for this reason, is also known as the output **slope resistance**.

The small-signal equivalent circuit is a very useful model for representing the behaviour of a device in response to small changes in its input signal. It must, however, be used in conjunction with data on the DC characteristics of the device – that is, the behaviour of the device in response to steady DC voltages. As we have seen, the DC characteristics of MOSFETs and JFETs are not the same since they require different bias voltages to place them in their normal operating regions. However, their small-signal characteristics and small-signal equivalent circuits are similar. The design of circuits using FETs must take both of these considerations into account.

18.5.6 FETs at high frequencies

In Figure 18.15 we looked at a small-signal equivalent circuit for an FET. This circuit is sufficient for most purposes but does not adequately describe the behaviour of these devices at high frequencies.

The MOSFET consists of two conducting regions – the gate and the channel – separated by an insulator. This construction forms a capacitor, with the insulating layer forming the dielectric. In the JFET, the insulator is replaced by a depletion layer, which has the same effect. In both cases, capacitance is present between the gate and the channel and, as the channel is joined to the drain at one end and the source at the

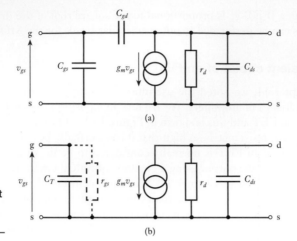

Figure 18.16
Small-signal equivalent circuits for an FET at high frequencies.

other, capacitance is present between the gate and each of the other terminals. The reverse-biased junction between the channel and the substrate also acts as a capacitor, the depletion layer separating the two conducting regions. This produces capacitance between the drain and the substrate and, thus, between the drain and source (as the latter is normally joined to the substrate). There is, therefore, capacitance between each pair of terminals of the device.

At low frequencies, the effects of these capacitances are small and can normally be neglected (as in Figure 18.15). However, at high frequencies their effects become more significant and it is necessary to include them in the small-signal equivalent circuit, as shown in Figure 18.16(a). Each of the capacitors shown in this figure has a magnitude of the order of 1 pF.

The presence of C_{gd} makes analysis of this circuit much more complicated. Fortunately, it is possible to represent the effects of this capacitance by increasing the magnitude of the capacitance between the gate and the source. In fact, the capacitance required between the gate and the source to give the same effect as C_{gd} is $(A + 1)C_{gd}$ where A is the voltage gain between the drain and the gate. This apparent increase in capacitance is brought about by a phenomenon known as the **Miller effect**. Thus, although C_{gd} and C_{gs} are of the same order of magnitude, it is C_{gd} that tends to dominate the high-frequency performance of the device.

It is therefore possible to represent the FET by the equivalent circuit shown in Figure 18.16(b) where the effects of both C_{gs} and C_{gd} are combined into a single capacitance, C_T, which represents the total input capacitance. From the discussion on low-pass RC networks in Section 8.6, it is clear that the presence of this capacitance will produce a fall in gain at high frequencies and give rise to a high-frequency cut-off, at a frequency determined by the value of the capacitance and impedance of the input to ground. This impedance will almost always be dominated by the source resistance. However, in some cases, it may be appropriate to include in the equivalent circuit a resistance r_{gs} representing the **small-signal gate resistance** of the device.

The effects of capacitance greatly reduce the performance of FETs at high frequencies. The presence of capacitance across the input reduces the input impedance from several hundreds of megohms at low frequencies to perhaps a few tens of kilohms at frequencies of the order of 100 MHz. There is also a reduction in g_m at high frequencies.

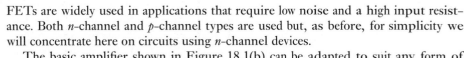

18.6 FET amplifiers

FETs are widely used in applications that require low noise and a high input resistance. Both *n*-channel and *p*-channel types are used but, as before, for simplicity we will concentrate here on circuits using *n*-channel devices.

The basic amplifier shown in Figure 18.1(b) can be adapted to suit any form of FET by adding additional circuitry to 'bias' the gate to the appropriate operating point for the FET used. This is illustrated in Figure 18.17, which shows examples of amplifiers based on different forms of FET. We noted in Figure 18.14 that when using a DE MOSFET it is normal to bias the circuit so that the operating point is at zero volts. In other words, the circuit is designed so that, in the absence of any input signal, the voltage on the gate (with respect to the source) is zero. This can be achieved simply by connecting a resistor from the gate to ground, as shown in Figure 18.17(a). A **coupling capacitor** (also called a **blocking capacitor**) is then used to couple the input signal to the amplifier while blocking any DC component from upsetting the biasing of the FET.

When using an *n*-channel enhancement MOSFET, it is normal to bias the gate to an appropriate *positive* voltage. This can be done by replacing the single gate resistor in Figure 18.17(a) with a pair of resistors forming a potential divider between V_{DD} and V_{SS}, as shown in Figure 18.17(b). The values of these resistors are chosen to produce a bias voltage appropriate to the FET used.

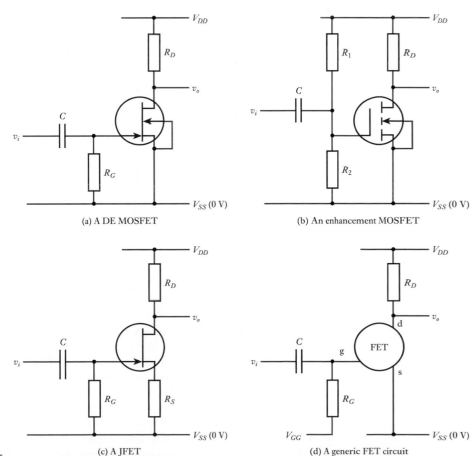

(a) A DE MOSFET

(b) An enhancement MOSFET

(c) A JFET

(d) A generic FET circuit

Figure 18.17
Simple FET amplifiers.

The normal operating point for an *n*-channel JFET requires a *negative* bias voltage. This can be produced by connecting the gate resistor to a negative voltage supply (if this is available) or by connecting the gate to zero volts and adding a resistor between the source and V_{SS} as shown in Figure 18.17(c). Current flowing through this source resistor will make the source positive with respect to V_{SS} and hence with respect to the gate. The gate will therefore be negative with respect to the source, as required. This technique is called **automatic bias**. By choosing an appropriate value for the source resistor, the gate-to-source voltage can be set to the desired value. One disadvantage of this approach is that, although the source resistor correctly sets the bias voltage of the circuit, it also reduces the small-signal gain of the amplifier. We will return to look at this problem, and a method of overcoming it, when we look at feedback amplifiers in more detail later in this section.

In addition to the circuits shown in Figures 18.17(a)–(c), it is possible to construct a single circuit that will work with any form of FET. Such an arrangement is shown in Figure 18.17(d). This can be configured to operate with any form of *n*-channel FET simply by choosing an appropriate value for the gate supply voltage V_{GG}. In each case, this is chosen to match the operating point appropriate for the device, as shown in Figure 18.14. In practice, it is often inconvenient to use a separate gate supply voltage, so the circuits of Figures 18.17(a)–(c) are often more attractive. However, the generic circuit of Figure 18.17(d) does provide a way of looking at FET amplifier circuits in general terms without being concerned about which form of device we are using.

The circuit in Figure 18.17(d) (and the other circuits in Figure 18.17) cannot amplify DC signals because of the presence of the capacitor C. However, AC signals applied to the input will be coupled through the capacitor and will change the gate voltage. The circuit is therefore an **AC-coupled amplifier** or simply an **AC amplifier**. In the absence of any input voltage, the circuit is said to be in its *quiescent* state. The drain current flowing though the FET under these conditions is termed the **quiescent drain current**. This current will flow through the drain resistor and the resulting voltage drop will determine the **quiescent output voltage**. When an input signal causes the gate to become more *positive*, this will increase the current through the FET (and hence through R_D), increasing the voltage drop across the resistor and making the output more *negative*. Similarly, when the input makes the gate more *negative*, this will reduce the current in the FET and the resistor and make the output more *positive*. The arrangement is therefore an **inverting amplifier**.

In the circuit of Figure 18.17(d), the input signal is applied between the gate and the source of the FET and the output is measured between the drain and the source. The source is therefore common to both the input circuit and the output circuit. For this reason, amplifiers of this general form are called **common-source amplifiers**. The circuits of Figures 18.17(a)–(c) are also common-source amplifiers.

18.6.1 Equivalent circuit of an FET amplifier

It is often useful to represent an amplifier using an equivalent circuit (see Chapter 14). In Figure 18.15 we looked at a small-signal equivalent circuit for an FET and it is quite simple to extend this to encompass the complete amplifier circuit. This is illustrated in Figure 18.18, which shows an equivalent circuit for a common-source amplifier.

It is important to remember that as the supply lines are constant voltages they do not fluctuate with respect to each other and thus *there is no small-signal voltage between the supply lines*. Therefore, as far as small-signal (AC) signals are concerned, the rails V_{DD}, V_{SS} and V_{GG} may be considered to be joined together. If this concept seems

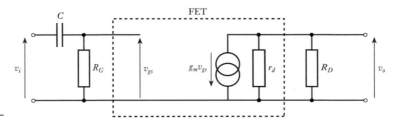

Figure 18.18
A small-signal
equivalent circuit of an
FET amplifier.

strange, remember that it is normal to place a large **reservoir capacitor** across the output of a power supply (see Section 17.8.1) to reduce ripple. At all but the lowest frequencies this capacitor resembles a short circuit between the supply rails.

As V_{DD} and V_{SS} are effectively joined as far as AC signals are concerned, the small-signal drain resistance r_d and external drain resistor R_D appear in parallel. Similarly, as V_{GG} and V_{SS} are effectively joined for small signals, the gate resistor R_G appears from the gate to ground, the common reference point. If a single resistor is used to generate the gate bias, this resistor simply appears as R_G, as shown. If the gate is biased by two resistors forming a potential divider, both resistors will appear in parallel. It should be noted that any biasing arrangement consisting of a combination of resistors and voltage sources can be modelled by its **Thévenin equivalent circuit**, as described in Section 3.8. This allows the arrangement to be replaced with a single resistor and a voltage source, as shown in Figure 18.17(d), which can then be represented by a single resistor in the small-signal equivalent circuit as constant voltages are not shown. Normally, C would be chosen to have a negligible effect at the frequencies of interest. If this is the case, its effects may be ignored.

It should be remembered that the small-signal model of Figure 18.18 is appropriate for all types of FET. The differences between the various types of device affect their biasing arrangements rather than their small-signal behaviour.

18.6.2 Small-signal voltage gain

Having derived a small signal equivalent circuit for the amplifier, we are now in a position to determine its small-signal voltage gain.

From Figure 18.18 it is clear that if we ignore the effects of the input capacitor C, the voltage on the gate of the FET is the same as that at the input, v_i. The output voltage is determined by the current generator and the effective resistance of the parallel combination of the small-signal drain resistance r_d and resistance R_D. We often use a shorthand notation for 'the parallel combination of', which is simply two parallel lines. Thus, 'the parallel combination of R_1 and R_2' would be written as $R_1//R_2$. Therefore, the output voltage is given by

$$v_o = -g_m v_{gs}(r_d//R_D)$$
$$= -g_m v_i(r_d//R_D)$$

and thus

$$\frac{v_o}{v_i} = -g_m(r_d//R_D)$$

The minus sign in the expression for the output voltage reflects the fact that the output voltage falls as the output current increases, so the output voltage is inverted with respect to the input. We therefore have an inverting amplifier.

The voltage gain is thus given simply by the product of the transconductance of the FET g_m and the effective resistance of the parallel combination of r_d and R_D. This can be expanded to give

$$\text{voltage gain} = \frac{v_o}{v_i} = -g_m \frac{r_d R_D}{r_d + R_D} \qquad (18.6)$$

It is also straightforward to calculate the **small-signal input resistance** and the **small-signal output resistance** of the amplifier from the equivalent circuit. The input resistance is simply equal to the gate resistance R_G. Because of the very high input resistance of the FET, the gate resistance can normally be chosen to be as high as necessary to suit a particular application. The output resistance is given by the parallel combination of r_d and R_D. The input and output resistances calculated from the small-signal equivalent circuit are small-signal resistances and are given the symbols r_i and r_o. They are the relationship between small-signal voltages and small-signal currents – they do not relate to the DC voltages and currents within the circuit.

Example 18.1 | Determine the small-signal voltage gain, input resistance and output resistance of the following circuit, given that $g_m = 2$ mS and $r_d = 100$ kΩ.

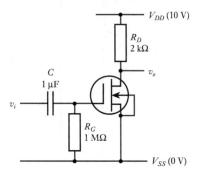

The first step in this problem is to determine the small-signal equivalent circuit of the amplifier.

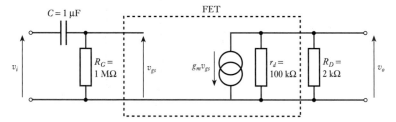

Clearly, from the equivalent circuit

$$\frac{v_o}{v_i} = -g_m(r_d // R_D)$$

$$= -g_m \frac{r_d R_D}{r_d + R_D}$$

$$= -2 \times 10^{-3} \frac{100 \times 10^3 \times 2 \times 10^3}{100 \times 10^3 + 2 \times 10^3}$$

$$= -3.9$$

The minus sign indicates that the amplifier is an inverting amplifier. The small-signal input resistance of the amplifier is simply R_G, thus

$$r_i = R_G$$

$$= 1\ \text{M}\Omega$$

The small-signal output resistance is given by

$$r_o = r_d // R_D$$

$$= \frac{r_d R_D}{r_d + R_D}$$

$$= \frac{100 \times 10^3 \times 2 \times 10^3}{100 \times 10^3 + 2 \times 10^3}$$

$$\approx 2.0\ \text{k}\Omega$$

This example considers a circuit containing an n-channel DE MOSFET; a similar calculation could be performed for a circuit incorporating another type of FET.

A typical value for the small-signal drain resistance r_d would be in the range 50 to 100 kΩ, which will generally be much larger than R_D. Under these circumstances the effects of r_d may usually be neglected and the characteristics of the amplifier may be approximated by the expressions

$$r_i \approx R_G$$
$$r_o \approx R_D$$
$$\frac{v_o}{v_i} \approx -g_m R_D$$

Example 18.2

Determine the low-frequency cut-off produced by the coupling capacitor in the circuit in Example 18.1.

The low-frequency cut-off is determined by the value of C and the input resistance. As the input resistance is approximately equal to R_G it follows (from Equation 8.11) that

$$f_c = \frac{1}{2\pi C R_G} = \frac{1}{2 \times \pi \times 10^{-6} \times 10^6} = 0.16\ \text{Hz}$$

Clearly, by changing the values of R_D we change the small-signal voltage gain of the amplifier, but we must remember that this will also affect the direct current – that is, the steady current I_D that flows even in the absence of any input, which in turn affects the value of g_m. We must therefore turn our attention to the DC aspects of the amplifier.

18.6.3 Biasing considerations

The **biasing** arrangement of an amplifier determines the operation of the circuit in the absence of any input signal. This is said to be the **quiescent** state of the circuit. Of central importance in the amplifiers being considered is the **quiescent drain current** $I_{D(quiescent)}$, which in turn determines the **quiescent output voltage** $V_{o(quiescent)}$. As the output voltage is equal to the supply voltage minus the voltage across R_D, it follows that

$$V_{o(quiescent)} = V_{DD} - I_{D(quiescent)}R_D$$

In many cases, V_{DD} will be fixed and the design will require a particular value of the quiescent output voltage. This means that an appropriate combination of $I_{D(quiescent)}$ and R_D will need to be used to produce the required output voltage.

Taking as an example the circuit of Figure 18.17(a), it is clear that the quiescent drain current is affected by the value of the drain resistor R_D and by the voltage–current characteristics of the FET. If the relationship between the drain current and the drain voltage in the FET were linear, as in a resistor, it would be simple to calculate the value of the resistor required to give an appropriate current. The ratio of the resistance of the FET to that of R_D could then be set to give an appropriate quiescent output voltage. However, we know from Figure 18.10 that the relationship between drain current and drain voltage is *not* linear. Indeed, in the section of the characteristic in which we wish to operate (the saturation region), the drain current is largely independent of the drain voltage. This makes determination of the quiescent conditions somewhat more complicated.

Use of a load line

One solution to this problem is to use a **load line** (as discussed in Chapter 17). It is clear that, although the currents flowing through R_D and the FET are not easily determined, the voltages across these two devices must sum to the voltage between the supply rails ($V_{DD} - V_{SS}$). The voltage across the FET is determined by its characteristics and the bias voltage V_{GS}. From Figures 18.11 and 18.12 it can be seen that the basic form of FET characteristics is the same for all forms of devices. Figure 18.19(a) shows generalised FET output characteristics that could apply to any *n*-channel FET.

When a current flows through the FET it also flows through R_D producing a voltage drop across it. The voltage on the drain of the FET is simply the supply voltage V_{DD} minus the voltage drop across the resistor, which is $I_D R_D$. Figure 18.19(b) shows the voltage on the drain of the FET for different values of drain current. When the drain current is zero there is no voltage drop across the resistor and the drain voltage is simply the supply voltage V_{DD}. As I_D increases, V_{DS} decreases, the slope of the line being the inverse of the drain resistance R_D.

Figures 18.19(a) and 18.19(b) each represent the relationship between the drain current and the drain voltage in our simple amplifier. Clearly, the actual operating condition must satisfy both these relationships. To determine this condition, we simply plot both characteristics on a single graph, as shown in Figure 18.20.

The straight line in this graph is a **load line** as it indicates the effect of the load resistance on the drain voltage. The intersection of this line with one of the output characteristic lines represents a point at which both relationships are satisfied. Consider, for example, point A on this line. The graph indicates that, if V_{GS} is set to

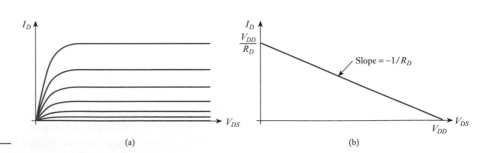

Figure 18.19
Current–voltage
relationships for an
FET and for R_D.

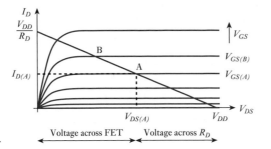

Figure 18.20
A load line for the
simple FET amplifier.

$V_{GS(A)}$, the drain current will be $I_{D(A)}$ and the drain voltage (which is also the output voltage of the amplifier) will be $V_{DS(A)}$.

It may help to visualise the significance of the load line to note that since the voltage across the FET plus the voltage across R_D must equal the supply voltage V_{DD}, the distance from zero to $V_{DS(A)}$ represents the voltage across the FET, while the distance from this point to V_{DD} represents the voltage across R_D. If, now, the gate voltage is increased to $V_{GS(B)}$, the drain current will increase and the drain voltage decrease, as indicated by point B on the characteristic. Thus, the load line shows how the drain current and drain voltage vary for different values of the gate voltage.

It should be remembered that the various lines of the output characteristic are only a representative few of the infinite number of lines that could be plotted. For simplicity, we plot a small number of lines and estimate the behaviour of the device between them.

The graph in Figure 18.20 shows the characteristics of an amplifier with a given value of R_D. If this value were changed, the slope of the load line would change, thereby affecting the characteristics of the amplifier. In practice, the designer is normally faced with the problem of selecting a value for R_D for optimum performance. In doing this, the designer defines an **operating point** that corresponds to the position on the characteristic under quiescent conditions. This point is also known as the **working point**. The designer therefore starts with the output characteristics of the FET to be used, but not knowing the value of the load resistor. In order to determine the value of this resistor, the designer must select the ideal operating point for the system.

If we assume that the designer chooses the point corresponding to position A in Figure 18.20, a line would then be drawn through this point to the V_{DD} position on the horizontal axis and this would form the load line. The value of R_D required can then be found by measuring the slope of this line.

When the operating point is known, the required gate voltage V_{GS} will be known and then the necessary biasing circuitry can be designed, as discussed earlier. The operating point determines the quiescent state of the circuit, and so defines the quiescent drain current and output voltage. When a small–signal input is applied to the circuit, the variations in the gate voltage cause the circuit to move along the load line on either side of the operating point. If the input signal is sufficiently large, this could cause the circuit to enter the ohmic region or *limit* as the output reaches the supply rail. Either of these conditions will **distort** the output signal. If a large output swing is required, it is important that the operating point is chosen appropriately.

18.6.4 Choice of operating point

In Figure 18.10, we divided the output characteristic of FETs into two regions – the ohmic and saturation regions – and noted that, for normal operation, the ohmic region is avoided. In fact, other areas of the characteristic are also normally avoided when using such a device as a linear amplifier. Figure 18.21 shows these regions.

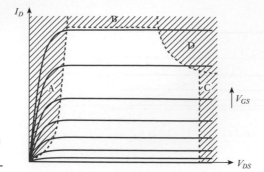

Figure 18.21
The forbidden operating regions of an FET.

Region A is the ohmic region discussed earlier. It is not used because the drain current is heavily dependent on the drain voltage in this area. When constructing a linear amplifier, we wish the drain current to be controlled by the input signal, *not* by the voltage across the device.

Region B may be caused by one of two factors, depending on the type of FET being used. For all devices, there is a maximum allowable drain current before the device is damaged and the designer must ensure that the device is not taken into this region. For JFETs there is also a limit imposed by the fact that the gate voltage must not forward bias the gate junction. One or other of these constraints limits the maximum drain current, or gate voltage, that may be used.

Constraints are also imposed by the **breakdown voltage** of the device, as shown by region C. If this voltage is exceeded, permanent damage to the device may result.

Finally, a fourth prohibited region is imposed by **power dissipation** considerations. The power dissipated by the FET is given by the product of the drain current and the drain voltage (as the gate current is negligible) and results in the generation of heat. This heat causes the temperature of the device to rise and operation is limited by the allowable temperature of the junction. The region of operation that satisfies the power dissipation conditions is bounded by a hyperbola (the locus of the point where current times voltage equals a constant), as shown by region D. These four regions define an allowable area in which the device must operate.

In selecting the operating point for an amplifier, the designer must ensure that the transistor is kept within safe limits and within its normal operating region. This will normally require that the supply voltage is less than the breakdown voltage of the device and the maximum current and power limits are not infringed. For maximum voltage swing, the operating point is normally placed approximately halfway between the supply voltage and the lower end of the saturation region, as shown in Figure 18.22. This allows for maximum transition of the input before the signal distorts.

Once the operating point has been selected, the load line is drawn from the supply voltage point on the horizontal axis, through the operating point. The value of V_{GS} corresponding to the operating point determines the biasing circuit that must be used on the gate, while the slope of the load line determines the required value for the load resistor. The position of the operating point shows the **quiescent output voltage** and the **quiescent drain current** for the circuit. A small-signal input causes the circuit to move along the load line on either side of the operating point – as illustrated in Figure 18.22 – and results in changes to the drain current and the output voltage. By comparing the magnitudes of a small-signal input voltage and the resultant small-signal output voltage, it is possible to deduce the small-signal voltage gain of the arrangement from this diagram.

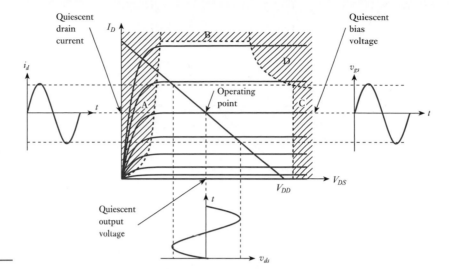

Figure 18.22
Choice of operating point in an FET amplifier.

18.6.5 Device variability

One problem associated with the use of FETs – in common with the use of most active devices – is the large spread in the characteristics of devices that are supposedly of the same type. This variability is illustrated for a JFET in Figure 18.23, which shows the spread of characteristics that might be obtained from devices from a single batch of components of the same type.

When choosing component values for a circuit using device A in Figure 18.23, the designer might choose to set the gate bias voltage to V_{GGA} to obtain the maximum output range. Similarly, when using device B, V_{GGB} might be chosen. However, when choosing component values, the designer will not know the characteristic of the device, only that it should lie within the range indicated by these two devices. Clearly, if the gate bias voltage is set to V_{GGA} and the component used has a characteristic similar to device B, the usable range of the amplifier will be very limited. The designer is therefore forced to assume the extreme case and choose that the gate voltage be V_{GGB} in all cases, as shown in Figure 18.24. One effect of this is that the **quiescent drain current** $I_{D(quies)}$ is ill defined. In Figure 18.24, it may be anywhere between I_{DA} and I_{DB}.

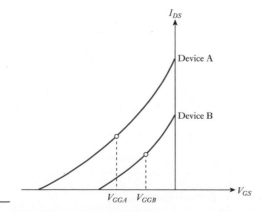

Figure 18.23
Typical transfer characteristics of JFETs.

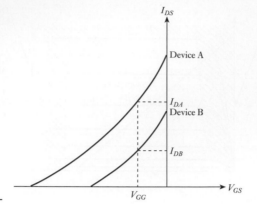

Figure 18.24
Selection of V_{GG} for a
JFET amplifier.

The variability of the quiescent drain current may be reduced by the use of **negative feedback**. In fact, the **automatic bias** network described earlier, and shown in Figure 18.17(c), uses negative feedback to stabilise the drain current. The operation of this arrangement is illustrated in Figure 18.25.

As the current flowing through the FET increases, the voltage drop across the source resistor also increases. As this voltage drop is determining the gate bias voltage,

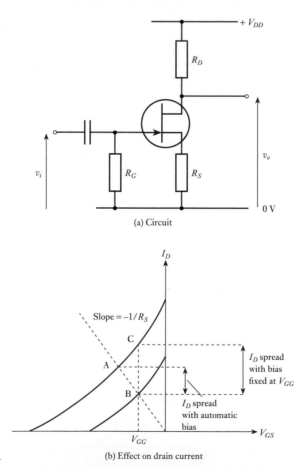

Figure 18.25
The effect of automatic
bias on drain current.

it has the effect of making the gate-to-source voltage more negative as the current increases, which tends to reduce the current. This therefore constitutes *negative* feedback. Using suitable component values, the circuit in Figure 18.25 would operate at point B using a device with characteristics corresponding to the lower line, and at point A with a device corresponding to the upper line. Devices with intermediate characteristics would operate at some point along the line from A to B. If a fixed bias voltage V_{GG} were used, as described earlier, the operating point of such a circuit would move from point B to point C across the spectrum of devices. Thus, the variation of I_D is much less with automatic biasing than with a fixed bias voltage. The circuit in Figure 18.25 uses a JFET, but this technique may also be used with other forms of FETs.

It might seem that an alternative method of solving the problems of device variability would be to **measure** the characteristics of the actual device to be used and to tailor the circuit accordingly. This approach has many disadvantages. First, measuring devices is a slow process if performed manually, so it is expensive. It also implies that component values need to be individually calculated, which may be time consuming. A second disadvantage is that, should the component fail in the field, it would need to be replaced and there is no guarantee that the replacement would exactly match the original.

An alternative arrangement is to **select** components that match a particular characteristic. This again implies measuring each device, although this is often performed automatically by the manufacturer, who divides products into distinct ranges of performance. Almost all active components are subject to manufacturing spread, so the more precisely they are characterised into distinct groups, the more expensive they become. In general, good designs are tolerant of a broad spread of active component characteristics, rather than needing more expensive selected components.

Example 18.3   A 2N5486 *n*-channel JFET has $V_P = -6$ V and $I_{DSS} = 8$ mA. Use the device's transfer characteristic to design a biasing arrangement such that, when used with a supply voltage of 15 V and a drain resistor of 2.5 kΩ, the amplifier has a quiescent output voltage of 10 V.

A suitable circuit is given below.

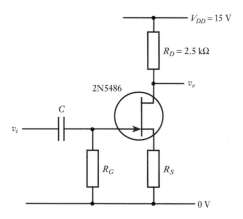

From Equation 18.3 we know that

$$I_D = I_{DSS}\left(1 - \frac{V_{GS}}{V_P}\right)^2$$

which, using the figures given for V_P and I_{DSS}, may be plotted to give

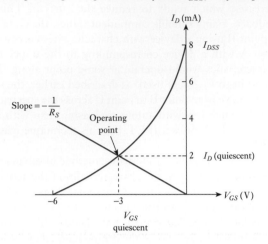

The quiescent output voltage $V_{o(quiescent)}$ is given by

$$V_{o(quiescent)} = V_{DD} - V_R$$

where V_R is the voltage drop across the drain resistor R_D. Therefore, the required value of V_R is given by

$$V_R = V_{DD} - V_{o(quiescent)} = 15 - 10 = 5\,\text{V}$$

and the required quiescent drain current $I_{D(quiescent)}$ is

$$I_{D(quiescent)} = \frac{V_R}{R_D} = \frac{5\,\text{V}}{2.5\,\text{k}\Omega} = 2\,\text{mA}$$

From the transfer characteristic, this value of drain current corresponds to a gate-to-source voltage of −3 V. As the gate is at ground potential, this gate-to-source voltage must be obtained by a voltage drop across R_S of +3 V. Thus, the value of R_S is given by

$$R_S = \frac{V_{GS}}{I_D} = \frac{3\,\text{V}}{2\,\text{mA}} = 1.5\,\text{k}\Omega$$

The value of R_G is not critical as it is simply required to bias the gate to zero volts. It would normally be chosen to give a high input resistance, but must not be so high that the voltage drop caused by the effects of the gate current (a few nanoamps) becomes significant. A value of 470 kΩ would be suitable.

Example 18.4

Perform the design of Example 18.3 numerically rather than using graphs.

As before

$$I_{D(quiescent)} = \frac{V_R}{R_D} = \frac{5\,\text{V}}{2.5\,\text{k}\Omega} = 2\,\text{mA}$$

From Equation 18.3

$$I_D = I_{DSS}\left(1 - \frac{V_{GS}}{V_P}\right)^2$$

or, by rearranging,

$$V_{GS} = V_P\left(1 - \sqrt{\frac{I_D}{I_{DSS}}}\right)$$

$$= -6\left(1 - \sqrt{\frac{2}{8}}\right)$$

$$= -3 \text{ V}$$

as before. Thus, R_S is 1.5 kΩ, as above.

Video 18C

18.6.6 A negative feedback amplifier

Negative feedback can be used to stabilise not only the biasing conditions of an amplifier but also its voltage gain. Consider, for example, the circuit in Figure 18.26, which is based on an enhancement MOSFET. The source resistor R_S in this circuit provides negative feedback, as described in the last section, thus stabilising the quiescent operating conditions of the circuit. However, the presence of the source resistor also stabilises the small-signal gain of the arrangement.

From the definition of g_m we know that

$$g_m = \frac{i_d}{v_{gs}}$$

therefore

$$i_d = g_m v_{gs} = g_m(v_g - v_s)$$

The source voltage v_s is given by

$$v_s = R_S i_d$$

and combining this with the previous equation and rearranging gives

$$v_s = \frac{R_S g_m}{1 + R_S g_m} v_g = \frac{1}{\dfrac{1}{R_S g_m} + 1} v_g$$

If $1/R_S g_m \ll 1$, then $v_s \approx v_g$. In other words, the source voltage tends to *follow* the gate voltage (the input).

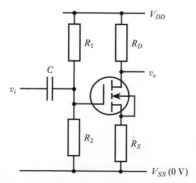

Figure 18.26
A negative feedback
amplifier.

The source voltage appears across the source resistor R_S and is related to the source current i_s by the expression

$$i_s = \frac{v_s}{R_S}$$

As the gate current is negligible, the drain current is equal to the source current, so

$$i_d = i_s = \frac{v_s}{R_S}$$

The voltage drop across the drain resistor R_D is equal to $i_d R_D$, so the small-signal output voltage is given by

$$v_o = 0 - i_d R_D = -\frac{v_s}{R_S} R_D$$

The zero in this expression represents the small-signal voltage on the V_{DD} supply which, being a constant voltage, has no small-signal component. As $v_s \approx v_g = v_i$ (from above), this may be rearranged to give

$$\frac{v_o}{v_i} \approx -\frac{R_D}{R_S} \tag{18.7}$$

and the small-signal voltage gain is given by the ratio of the drain resistor to the source resistor. The minus sign in this expression shows that we have an inverting amplifier.

An important implication of this expression is that the gain of the feedback amplifier is determined by the resistor values (which are relatively constant) rather than the characteristics of the FET (which are quite variable). Comparing this circuit with the earlier circuits, we can also deduce values for the input and output resistance and usually it is reasonable to assume that for the circuit in Figure 18.26

$$r_i = R_1 // R_2$$
$$r_o \approx R_D$$
$$\frac{v_o}{v_i} \approx -\frac{R_D}{R_S}$$

Example 18.5 | Estimate the input resistance, output resistance, small-signal voltage gain and lower cut-off frequency of the following circuit, given that $g_m = 72$ mS.

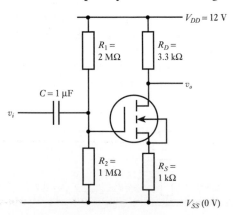

Much of the analysis is similar to that of earlier examples. From the above we have

$$r_i \approx R_1 // R_2 = 1 \text{ M}\Omega // 2 \text{ M}\Omega = 667 \text{ k}\Omega$$

$$r_o \approx R_D = 3.3 \text{ k}\Omega$$

$$\frac{v_o}{v_i} \approx -\frac{R_D}{R_S} = -\frac{3.3 \text{ k}\Omega}{1 \text{ k}\Omega} = -3.3$$

The lower cut-off frequency is given by

$$f_o \approx \frac{1}{2\pi r_i C} = \frac{1}{2 \times \pi \times 667 \times 10^3 \times 1 \times 10^{-6}} = 0.24 \text{ Hz}$$

In fact, the relationship given in Equation 18.7 provides only an approximate value for the gain of the circuit in Figure 18.26. A more detailed analysis yields the relationship that

$$\frac{v_o}{v_i} \approx -\frac{g_m R_D}{1 + g_m R_S + \dfrac{R_D + R_S}{r_d}} \tag{18.8}$$

Provided R_D and R_S are each small compared with the small-signal drain resistance of the FET r_d, this simplifies to give

$$\frac{v_o}{v_i} \approx -\frac{g_m R_D}{1 + g_m R_S} \tag{18.9}$$

When $g_m R_S \gg 1$, Equation 18.9 gives the same result as the simpler expression in Equation 18.7.

Example 18.6

For the circuit in Example 18.5, compare the values obtained for the gain of the circuit when using the expressions of Equations 18.7, 18.8 and 18.9, assuming that $g_m = 72$ mS and $r_d = 50$ kΩ.

From Equation 18.7 we have

$$\frac{v_o}{v_i} \approx -\frac{R_D}{R_S} = -\frac{3.3 \text{ k}\Omega}{1 \text{ k}\Omega} = -3.3$$

From Equation 18.8 we have

$$\frac{v_o}{v_i} \approx -\frac{g_m R_D}{1 + g_m R_S + \dfrac{R_D + R_S}{r_d}} = -\frac{72 \times 10^{-3} \times 3.3 \text{ k}\Omega}{1 + 72 \times 10^{-3} \times 1 \text{ k}\Omega + \dfrac{3.3 \text{ k}\Omega + 1 \text{ k}\Omega}{50 \text{ k}\Omega}} = -3.251$$

From Equation 18.9 we have

$$\frac{v_o}{v_i} \approx -\frac{g_m R_D}{1 + g_m R_S} = -\frac{72 \times 10^{-3} \times 3.3 \text{ k}\Omega}{1 + 72 \times 10^{-3} \times 1 \text{ k}\Omega} = -3.255$$

It can be seen that, in this case, the simple expression in Equation 18.7 produces a value that agrees quite closely with the values obtained using the more complete expressions.

File 18A

Computer simulation exercise 18.1

Simulate the feedback amplifier of Example 18.5 using an enhancement MOSFET, such as the IRF150. If your simulation package does not support this particular MOSFET, then you may need to experiment with the component values.

Apply a 1 kHz sinusoidal input voltage at 1 V peak and use transient analysis to investigate the relationship between the input and output waveforms. What happens to the output of your amplifier if the magnitude of the input is progressively increased?

Perform an AC analysis on your circuit to determine its frequency response. Compare the low-frequency behaviour of your circuit with that predicted in Example 18.5.

18.6.7 Using a decoupling capacitor

While the use of negative feedback in the circuit in Figure 18.26 produces a gain that is stable in spite of variations in the characteristics of the active device, this is achieved at the expense of a considerable loss in gain. An alternative approach is to use negative feedback to stabilise the *biasing* conditions of the circuit, but leave the small-signal performance unchanged. Such an arrangement is shown in Figure 18.27, which differs from the earlier circuit by the addition of a source capacitor C_S. This capacitor has no effect on the direct voltages and currents within the circuit, so the DC biasing of the circuit is unchanged. However, the presence of the source capacitor produces a low impedance for alternating signals and so removes small-signal voltages from the source. A capacitor used in this way is termed a **decoupling capacitor** or, sometimes, a **bypass capacitor** and its value should be such that *at the frequencies of interest* its impedance is small compared with the impedances within the circuit between the source and ground.

In the circuit shown in Figure 18.27, the source is effectively connected to ground as far as alternating signals are concerned and, thus, the small-signal behaviour of the circuit is similar to that of the circuit in Figure 18.17(b). So, for the circuit in Figure 18.27

$$r_i \approx R_1 // R_2$$

$$r_o \approx R_D$$

$$\frac{v_o}{v_i} \approx -g_m R_D$$

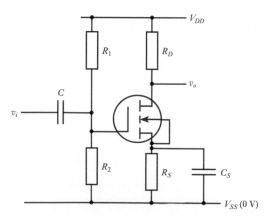

Figure 18.27
Use of a decoupling capacitor.

Example 18.7 | Estimate the input resistance, output resistance, small-signal voltage gain and lower cut-off frequency of the following circuit, given that $g_m = 72$ mS. You may assume that C_S has been chosen so that its impedance is 'small' at the frequencies of interest.

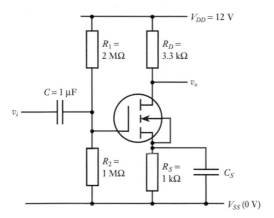

Because of the similarity between the circuits, the input resistance and output resistance of the circuit are the same as in the circuit in Example 18.5. Similarly, as the input resistance and the size of the coupling capacitor are the same as in the earlier example, the lower cut-off frequency is also the same as in Example 18.5.

The gain of the circuit is given by

$$\frac{v_o}{v_i} \approx -g_m R_D = -72 \times 10^{-3} \times 3.3 \times 10^3 = -238$$

File 18B

Computer simulation exercise 18.2

Simulate the feedback amplifier of Example 18.7 using an enhancement MOSFET such as the IRF150. Choose a large value for C_S, such as 1 F. If your simulation package does not support this particular MOSFET, then you may need to experiment with the component values.

Apply a 1 kHz sinusoidal input voltage at 10 mV peak and use transient analysis to investigate the relationship between the input and the output waveforms. What happens to the output of your amplifier if the magnitude of the input is progressively increased?

Perform an AC analysis on your circuit to determine its frequency response. Compare the low-frequency behaviour of your circuit with that predicted in Example 18.7.

Choosing the decoupling capacitor

So far we have assumed that the decoupling capacitor is chosen so that its impedance is 'small' at the frequencies of interest, but how do we determine the required size of this component?

In order to perform its required function, the impedance of the capacitor must be small compared with the effective resistance connected across it. This resistance is not simply R_S because of the effective resistance seen 'looking into' the source of the FET. We will see in the next section that the resistance looking into the device is approximately $1/g_m$ which is generally much smaller than R_S and therefore tends to determine the effective resistance seen in parallel with C_S.

At high frequencies, C_S will tend to look like a 'short circuit' to small signals, so the gain of the circuit will be approximately $-g_m R_D$ as discussed above. However, at low frequencies, the impedance of the capacitor will increase and the gain will fall. The point at which the impedance of the capacitor matches the effective resistance across it determines the **low-frequency cut-off** of the circuit. Therefore

$$f_o \approx \frac{1}{2\pi R C_S} \qquad (18.10)$$

where R is the effective resistance in parallel with the decoupling capacitor, which is generally approximately equal to $1/g_m$. Typical values of $1/g_m$ will be in the range of a few ohms to a few hundred ohms, so very large coupling capacitors may be needed for amplifiers that require good low-frequency performance.

At very low frequencies, the impedance of the capacitor becomes large compared with the other resistances in the circuit and has little effect. Therefore, at low frequencies, the gain is approximately $-R_D/R_S$ which is the gain the circuit would have without a decoupling capacitor (as discussed in Section 18.6.6 above when we looked at a negative feedback amplifier).

In the circuit in Figure 18.27 each of the capacitors produces a low-frequency cut-off and their effects on the gain of the circuit are cumulative. Where a particular lower cut-off frequency is required, it is normal to pick one of the capacitors to produce this effect and to choose the other to produce a cut-off at a much lower frequency. This simplifies the analysis of the circuit by making one of the capacitors dominant, allowing the effects of the other to be largely ignored.

File 18C

Computer simulation exercise 18.3

Use your circuit from Computer simulation exercise 18.2 to investigate the effect of the decoupling capacitor. Perform an AC analysis on your circuit and note the effect of changes to the decoupling capacitor on the frequency response.

Use the gain of your circuit to estimate the g_m of the FET in your circuit, then use Equation 18.10 to estimate the cut-off frequency produced by a particular value of decoupling capacitor. Measure the cut-off frequency produced within your simulation and compare this with your calculated value.

18.6.8 Source followers

In previous sections we have considered a series of common-source amplifier circuits. Some other widely used FET amplifier configurations are shown in Figure 18.28. In these circuits, the *drain* terminal is common to the input and the output circuits (remember that V_{DD} is effectively joined to ground for small signals). Consequently, they are called **common-drain amplifiers**.

The analysis of these circuits is similar to that of the feedback circuit in Section 18.6.6, but with fewer stages. From the definition of g_m we know that

$$g_m = \frac{i_d}{v_{gs}}$$

therefore

$$i_d = g_m v_{gs} = g_m(v_g - v_s)$$

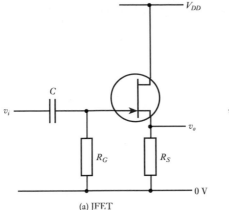

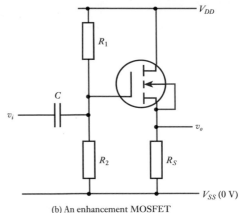

Figure 18.28
Source follower
amplifiers.

(a) JFET

(b) An enhancement MOSFET

As the source voltage v_s is given by

$$v_s = R_S i_d$$

then

$$v_s = R_S \times g_m(v_g - v_s)$$
$$= R_S g_m v_g - R_S g_m v_s$$

and rearranging gives

$$v_s = \frac{R_S g_m}{1 + R_S g_m} v_g = \frac{1}{\dfrac{1}{R_S g_m} + 1} v_g$$

If $1/R_S g_m \ll 1$ then $v_s \approx v_g$. In this circuit, the source voltage v_s represents the *output* of the circuit and so the output tends to follow the gate voltage (which is the *input*). For this reason, these circuits are often called **source followers** and, as the output follows the input, circuits of this type are **non-inverting amplifiers**.

As the small-signal output of the source follower is very nearly the same as the small-signal input, the gain of the amplifier v_s/v_g is approximately unity. In most cases, these circuits are used because of their very high input resistance and comparatively low output resistance. The input resistance is defined by the gate resistance R_G and the output resistance by the characteristics of the FET.

Source follower output resistance

To determine the output resistance of the circuit, we wish to know how the output voltage v_s changes with the output current i_s in the absence of any change at the input. Thus the output resistance r_o is v_s/i_s with $v_g = 0$.

From our earlier analysis we have

$$i_d = g_m v_{gs} = g_m(v_g - v_s)$$

and substituting $v_g = 0$ gives

$$i_d = g_m v_{gs} = g_m(0 - v_s) = -g_m v_s$$

As gate currents are negligible, the magnitude of the source current is equal to that of the drain current. However, currents are conventionally considered to flow *into* the device, so $i_s = -i_d$. Therefore

$$i_s = -i_d = g_m v_s$$

and

$$r_o = \frac{v_s}{i_s} = \frac{1}{g_m}$$

As g_m varies with current, so will the output resistance, but it is typically a few tens or hundreds of ohms for currents of a few milliamps.

Source followers do not have as low an output resistance as similar circuits constructed using bipolar transistors (these are discussed in the next chapter) but their very high input resistance makes them extremely useful as **unity gain buffer amplifiers**.

For the circuits in Figure 18.28

$$r_i \approx R_G \text{ or } R_1 // R_2$$
$$r_o \approx 1/g_m$$
$$\frac{v_o}{v_i} \approx 1$$

Example 18.8

Estimate the input resistance, output resistance, small-signal voltage gain and lower cut-off frequency of the following circuit, given that $g_m = 72$ mS.

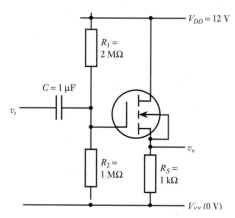

Much of the analysis is similar to that of earlier examples. From the above we have

$$r_i \approx R_1 // R_2 = 1 \text{ M}\Omega // 2 \text{ M}\Omega = 667 \text{ k}\Omega$$
$$r_o \approx 1/g_m = 1/72 \times 10^{-3} \approx 14 \ \Omega$$
$$\frac{v_o}{v_i} \approx 1$$

The lower cut-off frequency is given by

$$f_o \approx \frac{1}{2\pi r_i C} = \frac{1}{2 \times \pi \times 667 \times 10^3 \times 1 \times 10^{-6}} = 0.24 \text{ Hz}$$

18.6.9 Differential amplifiers

In Section 14.8 we discussed the use of differential amplifiers, which produce an output proportional to the *difference* between two input signals and ignore signals that are common to both inputs, the latter property being known as **common-mode rejection**.

A common form of differential amplifier is the **long-tailed pair**, which is often used in the input stage of operational amplifiers. Such a circuit is shown in Figure 18.29. In this arrangement, two FET amplifiers share a common-source resistor R_S and have similar gate resistors and drain resistors. The transistors are chosen to have matched characteristics so that the circuit is as symmetrical as possible. The circuit has two inputs, v_1 and v_2, and two output signals, v_3 and v_4. A small-signal equivalent circuit for this arrangement is shown in Figure 18.30.

The input and output voltages are measured to the common reference point (ground). The gate resistors are normally chosen to be sufficiently high in value as to have little effect on the operation of the circuit, other than to set the correct DC-bias conditions for the FETs. They are therefore omitted from the small-signal equivalent circuit. We will assume that the devices are perfectly matched so that the transconductance g_m and the drain resistance r_d of both are equal.

As the input voltages v_1 and v_2 are measured with respect to ground, the voltages applied across the gate-to-source junction of each device are simply

$$v_{gs1} = v_1 - v_s$$

and

$$v_{gs2} = v_2 - v_s$$

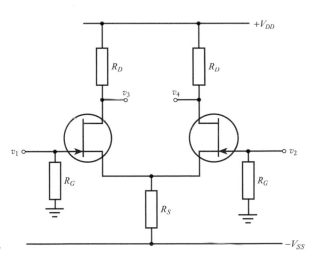

Figure 18.29
An FET differential amplifier.

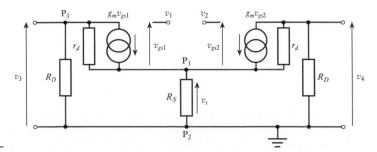

Figure 18.30
Equivalent circuit for the FET differential amplifier.

From **Kirchhoff's law**, we know that the currents flowing into any node of a circuit must sum to zero. We can apply this principle to a number of points in the circuit to produce a series of simultaneous equations.

If we consider point P_1, we see that

$$g_m v_{gs1} + \frac{(v_3 - v_s)}{r_d} + g_m v_{gs2} + \frac{(v_4 - v_s)}{r_d} - \frac{v_s}{R_S} = 0$$

which, substituting for v_{gs1} and v_{gs2}, gives

$$g_m(v_1 - v_s) + \frac{(v_3 - v_s)}{r_d} + g_m(v_2 - v_s) + \frac{(v_4 - v_s)}{r_d} - \frac{v_s}{R_S} = 0 \qquad (18.11)$$

Applying the same principle to point P_2 yields

$$\frac{v_3}{R_D} + \frac{v_4}{R_D} + \frac{v_s}{R_S} = 0 \qquad (18.12)$$

and to point P_3 gives

$$\frac{v_3}{R_D} + \frac{(v_3 - v_s)}{r_d} + g_m(v_1 - v_s) = 0 \qquad (18.13)$$

From these equations, it is possible to deduce an expression for the outputs of the circuit v_3 and v_4 in terms of the inputs, but the analysis is fairly involved. We can simplify the process by making an assumption about the term v_s/R_s in Equation 18.12. This term represents the *small-signal* current in the source resistor R_S – that is, the fluctuations in this current as a result of the varying inputs. For reasons that will become apparent later, it is advantageous to assume that this term is very small, so its effects can be neglected. Making this approximation implies that the current through R_S is constant and it is acting as a **constant current generator**.

If the term v_s/R_S is negligible, Equation 18.12 becomes

$$\frac{v_3}{R_D} + \frac{v_4}{R_D} = 0 \qquad (18.14)$$

which simplifies to give

$$v_3 = -v_4$$

Combining this result with Equations 18.11 and 18.13 we can obtain an expression for the output signals, which is

$$v_3 = -v_4 = (v_1 - v_2)\frac{-g_m}{2\left(\dfrac{1}{r_d} + \dfrac{1}{R_D}\right)} \qquad (18.15)$$

Thus, the output signals are equal and opposite and their magnitude is determined by the difference between the two input signals. We therefore have a *differential amplifier*.

The *differential* output voltage of this arrangement v_o is given by $v_3 - v_4$ and, as v_3 and v_4 are equal and opposite, the differential voltage gain of the circuit is simply

$$\text{differential voltage gain} = \frac{v_o}{v_i} = \frac{v_3 - v_4}{v_1 - v_2} = \frac{-g_m}{\left(\dfrac{1}{r_d} + \dfrac{1}{R_D}\right)}$$

We noted in Section 18.6.2 that r_d is usually much greater than R_D. If we make this assumption, we can simplify the above expression to

$$\text{differential voltage gain} \approx -g_m R_D \qquad\qquad (18.16)$$

which is identical to the expression derived earlier for the voltage gain of a simple common-source amplifier.

It should be noted that the input signals are not required to be symmetrical to produce symmetrical outputs. Note also that the magnitude of the output signals is not affected by the actual value of the input signals, only by the *difference* between them. Thus, common-mode signals are ignored.

File 18D

Computer simulation exercise 18.4

Simulate the differential amplifier of Figure 18.29. A suitable circuit might use 2N3819 JFETs with $V_{DD} = 12$ V, $V_{SS} = -12$ V, $R_D = 2.2$ kΩ and $R_S = 3.3$ kΩ. If your simulation package does not support this particular JFET, you may need to experiment with component values.

Initially, apply a 1 kHz sinusoidal input voltage of about 50 mV peak to one input to the amplifier and a constant (DC) voltage to the other. Use transient analysis to look at the form of the outputs, then experiment with different magnitudes and combinations of input signals.

Common-mode rejection ratio of the long-tailed pair amplifier

The **common-mode rejection ratio** (CMRR) is the ratio of the gain for *difference* input signals to that for *common-mode* input signals. That is

$$\text{CMRR} = \frac{\text{differential-mode gain}}{\text{common-mode gain}}$$

From Equation 18.15 it would seem that the CMRR for the long-tailed pair amplifier is infinite as the output is unaffected by common-mode signals. It should be remembered, however, that this expression was derived by making the simplifying assumption that the current through the source resistor was constant. If the variation of this current is taken into account, it can be shown that the common-mode rejection ratio varies directly with the value of R_S, such that

$$\text{CMRR} \approx g_m R_S$$

The higher the value of R_S, the more closely the circuit represents a constant current source and the higher the CMRR. However, in practice it is not possible to use a very high value for R_S without upsetting the DC conditions of the circuit. For circuits requiring a high CMRR, R_S is normally replaced with a **constant current source** that behaves like a resistor of very high value without affecting the DC operation of the circuit. Using such techniques, it is possible to produce amplifiers with CMRRs of more than 100 dB. Such arrangements are discussed in Section 18.7.1.

In our analysis, we have assumed a perfectly matched pair of transistors. In practice, the devices will never be identical, which will lead to a reduction in common-mode rejection. When the devices are fabricated within a single integrated circuit, as in an operational amplifier, the manufacturing tolerances will tend to affect both devices equally and a good match will generally be obtained. Because of their close proximity, the devices will also generally operate at the same temperature. With circuits constructed using discrete devices the problems are more severe as the devices will generally be less well matched and their operating temperatures will not always be equal. These problems may be reduced by using a matched pair of transistors within a single package. This ensures a close correspondence of device characteristics and a similar operating temperature.

18.7 Other FET applications

18.7.1 An FET as a constant current source

Provided that the drain-to-source voltage is greater than the pinch-off voltage, the drain current of an FET is controlled by the gate-to-source voltage. Therefore, a very simple constant current source can be formed by applying a constant voltage to the gate. For JFETs and DE MOSFETs, the simplest forms of such an arrangement are shown in Figures 18.31(a) and 18.31(b). In these circuits the gate is simply connected to the source giving a drain current of I_{DSS}. The current produced by these arrangements is determined by the characteristics of the device and is generally in the range 1 mA to 5 mA. Commercially available 'constant current sources' are often simply FETs with their source and gate joined internally to produce two-terminal devices that are then selected for different current ranges.

It is also possible to produce a variable constant current source by using the automatic bias technique discussed in Section 18.6. Such an arrangement is shown in Figure 18.31(c). Current through the device produces a voltage drop across the resistor and generates a bias voltage between the gate and source. The value of this resistor is adjusted to produce the required current.

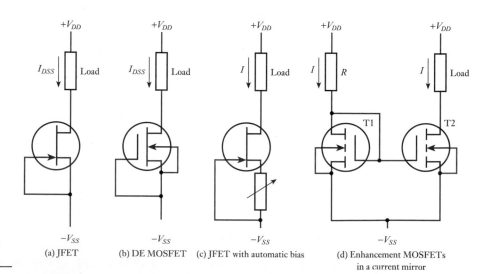

Figure 18.31
FET constant current sources.

(a) JFET   (b) DE MOSFET   (c) JFET with automatic bias   (d) Enhancement MOSFETs in a current mirror

Another widely used form of constant current source is the **current mirror** shown in Figure 18.31(d). Here two similar devices are connected with their source and gate terminals joined. In this arrangement, the gate-to-source voltage of each device is identical and, if the devices have similar characteristics, this will produce identical drain currents. In the diagram, the current I in T1 is set by the resistor R resulting in a similar current I flowing through T2 and the load. Additional transistors can be connected in parallel with T2 to produce identical currents in several parts of a circuit. Current mirrors are widely used in integrated circuits where it is possible to achieve very good matching of the transistors and the close proximity of the components ensures that the devices are at approximately the same temperature.

FET constant current sources are often used to produce the source current for long-tailed pair amplifiers, as described in Section 18.6.9. An example of such an arrangement is shown in Figure 18.32.

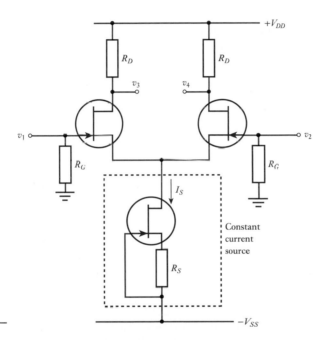

Figure 18.32
A long-tailed pair amplifier with an FET current source.

File 18E

Computer simulation exercise 18.5

Use circuit simulation to investigate the properties of the circuit in Figure 18.31(a). Use a 15 V supply and a 2N3819 JFET. Look at the current through the load as it is varied between 100 Ω and 1 kΩ. What happens if the load is increased to 2 kΩ? Can you explain this?

18.7.2 An FET as a voltage-controlled resistance

From Figure 18.10 it is clear that, for *small values of drain-to-source voltage*, FETs have a characteristic that can be described as *ohmic*, in that the drain current increases linearly with the drain voltage. The value of the effective resistance (which corresponds to the slope of the characteristic) is controlled by the gate voltage. This allows the device to be used as a **voltage-controlled resistance** (VCR). The range of resistance that can be produced varies from a few tens of ohms (or less for power devices) up to several gigohms.

A common application of this arrangement is within **automatic gain control** circuits. Here, the voltage-controlled resistance is used in a potential divider arrangement with a fixed resistor to form a **voltage-controlled attenuator**, as shown in Figure 18.33. The attenuator is used within the feedback path of an amplifier to vary its gain. The voltage fed to the FET to control its resistance is derived from the output signal of the amplifier and is arranged so that, as the magnitude of the output voltage increases, the amount of negative feedback is increased, thereby reducing the amplifier's gain. This allows the output amplitude to be maintained at some fixed value independent of the magnitude of the input signal. This technique is used, for example, to keep the volume of a radio receiver constant, even if the strength of the radio signal changes.

Another use for voltage-controlled attenuators is in the design of oscillators. The automatic gain control arrangement described above can be used to stabilise the gain of an oscillator circuit without producing distortion of the output. We will return to this topic in Chapter 23.

The voltage-controlled attenuator described above may be used with DC or AC input signals as the FET is essentially symmetrical in its operation (although the characteristics of the device for input signals of different polarities are usually very different). However, to avoid excessive distortion the magnitude of the input signals must be restricted to a few tens of millivolts.

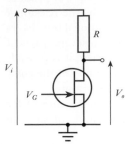

Figure 18.33
A voltage-controlled attenuator.

18.7.3 An FET as an analogue switch

By applying a suitable voltage to the gate of an FET, the effective drain-to-source resistance can be varied from a few tens of ohms or less (effectively, a short circuit in many applications) to a value so high that the channel can almost always be considered to be an open circuit. The resistances of the device in these two states are called the **on resistance** and the **off resistance** of the FET. The ability to turn the device 'on' and 'off' in this way allows it to be used as a switch, as illustrated in Figure 18.34.

Figure 18.34(a) shows a JFET being used as a *series* switch. MOSFETs can be used in a similar manner. When the device is turned on, the resistance between the input and output is small. Provided the resistances of the source and load are large compared with the on resistance of the FET, the device will resemble a short circuit. When the device is turned off, the resistance between the source and load will be equal to the off resistance of the FET. Provided this is large compared with the resistances within the circuit, this will represent an open circuit. Because of the many orders of magnitude difference between the on and off resistances of the FET, it is usually easy to satisfy these conditions, allowing the FET to be used as a very efficient switch. Figure 18.34(b) shows an FET used in a *shunt* arrangement. Here the series resistor R is chosen to be large compared with R_{ON} and small compared with R_{OFF}. The potential divider produces an output voltage close to V_i when the device is turned off, but close to zero when it is turned on.

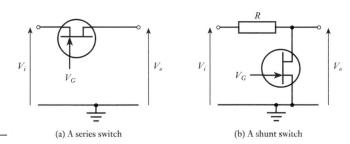

Figure 18.34
The FET as an analogue switch.

(a) A series switch   (b) A shunt switch

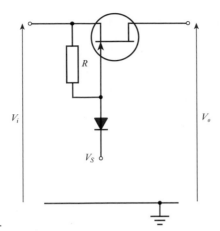

Figure 18.35
A series switch with
automatic gate bias.

When using FETs as analogue switches, care must be taken to ensure that the operating conditions of the device are correct. It is obviously essential to ensure that the breakdown voltage of the gate is not exceeded, but it is also necessary to guarantee that the gate is taken to an appropriate voltage to turn the device either completely on or completely off.

For n-channel MOSFETs the gate can be taken to a large positive voltage to turn the device on and must be made negative with respect to the input voltage by an appropriate amount to turn it off.

In the case of JFETs, the situation is slightly more complicated – particularly when used in the series arrangement – as the gate junction must not be forward biased. A simple circuit that overcomes this problem is shown in Figure 18.35. When the switching voltage V_S is more positive than the input voltage V_i the diode is reverse biased and the gate voltage is set equal to V_i by the resistor R, thus turning the FET on. If V_S is taken negative, the diode conducts, taking the gate negative with respect to the source and turning the FET off.

18.7.4 An FET as a logical switch

In addition to their use within analogue circuits, FETs (particularly MOSFETs) are widely used in digital applications. Such circuits (as we will see in Chapter 26) usually adopt a two-state, or **binary**, arrangement in which all signals are constrained to be within one of two voltage ranges, one range representing one state (for example, the ON state) and the other representing a second state (for example, the OFF state). These ranges are often referred to as 'logical 1' and 'logical 0'. Within circuits using MOSFETs it is common for voltages close to zero to represent a logical 0 and for voltages close to the positive supply voltage to represent a logical 1.

One of the simplest logic circuits is the **logical inverter**, which is required to generate a voltage corresponding to a logical 1 if the input corresponds to a logical 0, and vice versa. A simple circuit to realise this function is shown in Figure 18.36(a). The circuit uses an n-channel enhancement MOSFET and a resistor. It is similar in appearance, and operation, to the amplifier described in Section 14.9 and discussed earlier in this chapter.

When used as a logical inverter, the input voltage will be either close to zero (logical 0) or close to the supply voltage V_{DD} (logical 1). When the input voltage is close to zero volts, the enhancement MOSFET is turned off as this device needs a positive voltage on the gate to produce a channel between the drain and the source (see Section 18.3.2).

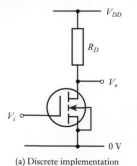
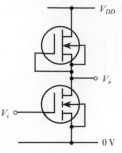

(a) Discrete implementation   (b) Integrated circuit implementation

Figure 18.36
Logical inverters using
MOSFETs.

The drain current is thus negligible and there is no appreciable voltage drop across the resistor R_D. The output voltage is therefore approximately equal to the supply voltage V_{DD} (logical 1). When the input voltage is close to the supply voltage the MOSFET is turned on and current flows through R_D, dropping the output voltage to close to ground (logical 0). Thus, when the input is high the output is low and, conversely, when the input is low the output is high. We therefore have the function of an inverter.

The arrangement shown in Figure 18.36(a) is quite acceptable when discrete components are to be used, but is unattractive when an integrated circuit is to be produced. One of the reasons for MOSFETs being so widely used in digital integrated circuits is that each transistor requires a very small area of silicon, allowing a very large number of devices to be fabricated on a single chip. Resistors, on the other hand, occupy a proportionately larger area, making them components to be avoided wherever possible. Therefore, when producing logic inverters using MOSFETs, it is common to use the circuit shown in Figure 18.36(b). Here, a second MOSFET is used as an **active load**, greatly reducing the area of silicon required. We will return to the use of active loads in Chapter 20.

The arrangement shown in Figure 18.36(b) uses n-channel MOSFETs. Circuits of this kind are often called **NMOS** circuits, the letters being an abbreviation of **n-channel metal oxide semiconductor**. Similarly, circuits based on p-channel devices are referred to as **PMOS** circuits. We will return to look at other forms of NMOS and PMOS logic circuitry in Chapter 26.

18.7.5 CMOS circuits

In the NMOS and PMOS circuits described above, the value of the resistance R_D (or the effective resistance of the MOSFET used in its place) affects the output resistance of the circuit when the output is high and the power dissipation of the gate when the output is low.

When the input is low, the switching MOSFET is turned off and the output is pulled high by the resistor R_D. As the output resistance of the device is determined by the value of R_D, to achieve a low output resistance, R_D should be as small as possible.

When the input is high, the switching MOSFET is turned on and the output is pulled low. As the switching MOSFET has a low ON resistance, the output resistance is low, enabling the circuit to sink a high current from an external load. In this state, almost the entire supply voltage is applied across R_D, producing a large current and consequently a high power consumption. To minimise this power consumption, the resistor should be as large as possible.

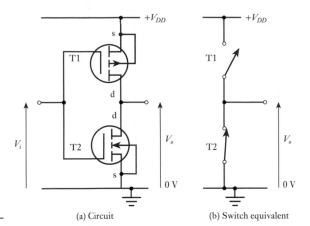

(a) Circuit   (b) Switch equivalent

Figure 18.37
A CMOS logic inverter.

Clearly, the requirements of a low output resistance and low power consumption place conflicting requirements on the value of R_D. This problem can be overcome by using the arrangement shown in Figure 18.37. Here, both NMOS and PMOS devices are combined within a single circuit which is now described as complementary MOS or CMOS logic. When the input voltage is close to zero, the n-channel device T2 is turned off, but the p-channel device T1 is turned on. When the input voltage is close to the supply voltage, the position is reversed, with T1 off and T2 on. Thus, with the input in either state, one of the transistors is on and the other off.

The circuit in Figure 18.37(a) may be represented by the arrangement in Figure 18.37(b). With switch T1 closed and T2 open, the output is pulled high and the output resistance is low, being determined by the on resistance of T1. With T2 closed and T1 open, the output is pulled low and the output resistance is also low, now being determined by the on resistance of T2. In both cases, as one of the switches is turned off, the only supply current flowing is that drawn by the load. If the load is another circuit of the same form, this will be negligible because of the high input resistance of MOSFETs. Thus, in either state, the output resistance is very low and the power consumption extremely small. In fact, when static (that is, when in one state or the other), the power consumption is generally negligible. In practice, the power consumed by a CMOS circuit is determined by the small amount of current that flows as the devices switch from one state to the other. This is because for a short period of time both transistors are conducting, producing a short burst of current from the supply to ground.

CMOS circuits are the most widely used form of digital circuitry and so we will discuss CMOS logic in more detail later (in Chapter 26). CMOS techniques are also used in analogue applications (and we will look at such circuits in Chapter 21).

18.8 FET circuit examples

18.8.1 FET input buffer for an operational amplifier

A pair of FETs may be used in a **long-tailed pair amplifier** to improve the performance of an operational amplifier, as shown in Figure 18.38.

If used with a bipolar op-amp, the FET buffer can be used to provide a very high input resistance. If FETs T1 and T2 are matched, they may also improve the CMRR of the amplifier. Usually these two transistors are in the form of a matched pair within

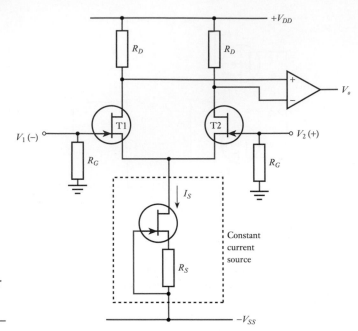

Figure 18.38
An FET input buffer for
an operational
amplifier.

a single package. This arrangement would typically produce a CMRR in excess of 120 dB – somewhat more than most general-purpose operational amplifiers.

In common with most operational amplifiers, the arrangement above would usually be used with some form of feedback. The circuit may be treated as a single amplifying block, with V_1 and V_2 being the inverting and non-inverting inputs and V_o the output.

18.8.2 An integrator with reset

In Chapter 16 we considered the use of an operational amplifier as an integrator. Very often, it is necessary to have some method for zeroing the output of the circuit by removing the charge on the capacitor. One of the simplest ways to achieve this is to place some form of electrically activated switch across the capacitor. FETs are an obvious choice for this application. Figure 18.39 shows a simple arrangement based on the use of an enhancement MOSFET.

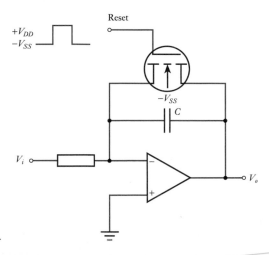

Figure 18.39
An integrator with
reset.

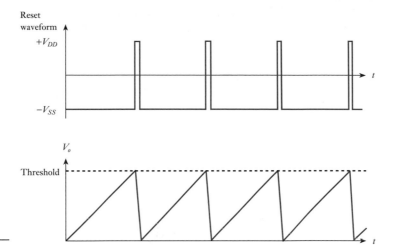

Figure 18.40
Waveforms from a
sawtooth generator.

The FET switch is controlled by a gate signal that is switched between the two supply voltages. The substrate of the device is connected to the most negative voltage in the system $-V_{SS}$ to ensure that the voltages on the source and drain are never more negative than the substrate.

When the gate input signal is equal to $-V_{SS}$ the gate is at the same potential as the substrate and, therefore, being an enhancement device, the channel is turned off. Under these conditions, the presence of the FET has no effect and the circuit acts as a simple integrator.

When the gate input voltage is equal to $+V_{DD}$ the channel is turned on and the capacitor is effectively shorted out. Any charge on the capacitor will quickly be removed and the output voltage of the circuit will be clamped near to ground until the FET is again turned off.

If the FET is turned off, a fixed DC voltage at the input V_i will produce a steady increase in the output voltage. If a voltage–level detector is connected to the output of the integrator and made to generate a reset pulse whenever the output reaches a particular threshold value, this will produce a **sawtooth waveform**, as shown in Figure 18.40.

18.8.3 Sample and hold gate

In essence, a sample and hold gate can be simply a capacitor and a switch, as shown in Figure 18.41. When the switch is closed, the capacitor quickly charges or discharges so that its voltage – and hence the output voltage – equals the input voltage. If the switch is now opened, the capacitor simply holds its existing charge and its voltage remains constant. The circuit is used to take a *sample* of a varying voltage by closing the switch, then *hold* that value by opening the switch.

In practice, the simple circuit in Figure 18.41 has a couple of weaknesses. First, when the switch is closed the capacitor represents a very low impedance to the source, so loads it heavily, possibly distorting the input value. If the source has a fairly high output

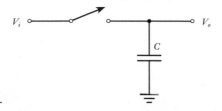

Figure 18.41
A simple sample and
hold gate.

resistance it may take some time for the capacitor to charge up, reducing the speed at which samples can be taken. Second, in practice the capacitor will be connected to a load that will tend to discharge the capacitor because of its finite input resistance.

To overcome these problems buffer amplifiers are normally used – as shown in Figure 18.42. This arrangement uses two operational amplifiers as unity gain buffers (as described in Section 16.4.1). The first amplifier provides a high input resistance but a low output resistance to provide rapid charging of the capacitor. The second amplifier would normally be an FET input device which takes very little current from the capacitor, allowing the sampled voltage to be held for a considerable time. The near-ideal properties of FET switches make them an obvious choice as the switching device in such circuits, creating the circuit shown in Figure 18.43.

The performance of the sample and hold circuit shown in Figure 18.43 is limited by the leakage current of the FET switch and any input bias current of the operational amplifiers. More advanced circuits reduce the effects of these currents.

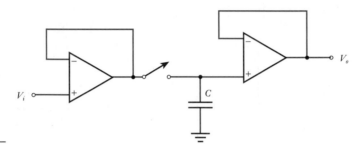

Figure 18.42
A sample and hold gate with input and output buffers.

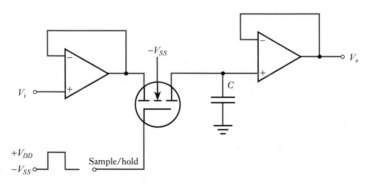

Figure 18.43
A sample and hold gate using an FET as the switching device.

Video 18D

Further study

A particular industrial application requires a control unit that can be configured using three push-buttons. The buttons are required to vary the voltage gain of an amplifier within the unit and set this to 1, 10 or 100 depending on which button is selected.

Design a switchable gain amplifier for use in this application. The unit should take three control signals (coming from the three switches) and use these to configure the behaviour of the circuit.

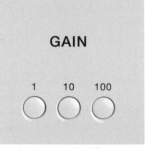

Key points

- FETs are widely used in both analogue and digital applications.

- The term FET describes a range of components that may be divided into *insulated-gate* devices, also called MOSFETs, and *junction* devices, which are also called JFETs.

- Insulated-gate FETs are further divided into *enhancement* and *depletion* types.

- The various forms of FET can be produced in both *n*-channel and *p*-channel versions.

- Although the characteristics of the various classes of FET are slightly different, their principles of operation are similar.

- All are characterised by very high input resistance. In most cases, their input currents are so low that they may be considered to be negligible. For this reason, they are often used within the input stage of amplifiers to provide a high input impedance.

- In addition to their use in the construction of amplifiers, FETs can be used as voltage-controlled resistances, constant current sources and both analogue (linear) and digital (logical) switches.

- In switching applications, their very high off resistance and very low on resistance make them nearly ideal switches.

- Another important characteristic of FETs is their small physical size when implemented in integrated circuit form. This, combined with their excellent switching properties, has led to their extensive use in digital very large-scale integration (VLSI) circuitry. The majority of microprocessors, memories and associated components are implemented using FETs.

Although the myriad variations of FET devices may be a little daunting, designing circuits using FETs is not as complicated as it might at first appear. In most cases the circuit differences required to cater for the various device types relate simply to the biasing arrangements. The small-signal characteristics of all kinds of FETs are of the same basic form. Thus, when the general principles have been mastered, these may be applied to all types of FETs.

Exercises

18.1 Why are field-effect transistors so called?

18.2 What characteristics of FETs make them ideal for use in integrated circuits?

18.3 Name the three terminals common to all FETs. Which of these constitutes the control input?

18.4 What is meant by the symbols V_{DS}, I_D, V_{GS}, v_{gs}, R_S, V_{DD} and V_{SS}?

18.5 What name is given to the conductive path between the drain and the source in an FET?

18.6 Explain the difference between an IGFET and a MOSFET.

18.7 What is the difference between an *n*-channel FET and a *p*-channel FET? How do their characteristics compare?

18.8 What is meant by the 'substrate' of a MOSFET? What character is used to represent this terminal in its circuit symbol?

18.9 Explain what is meant by 'enhancement' of a channel. In an n-channel MOSFET, what polarity of gate-to-source voltage will cause the channel to be enhanced?

18.10 Explain what is meant by 'depletion' of a channel. In an n-channel MOSFET, what polarity of gate-to-source voltage will cause the channel to be depleted?

18.11 Explain the difference between a DE MOSFET and an enhancement MOSFET.

18.12 What is used in a JFET to replace the insulated gate found in a MOSFET?

18.13 In a circuit based on an n-channel JFET, what would be the normal polarity of the gate-to-source voltage? Why?

18.14 Explain the significance of the depletion layer in a JFET.

18.15 How is the polarity of an FET (that is, n-channel or p-channel) indicated in its circuit symbol?

18.16 Describe the input characteristics of a MOSFET and a JFET.

18.17 Sketch typical output characteristics for an FET, indicating the ohmic region, saturation region and pinch-off voltage.

18.18 In which region of its characteristic does an FET resemble a voltage-controlled resistance?

18.19 Explain the concept of an operating point.

18.20 Define the term 'transconductance' and give its units.

18.21 Sketch a small-signal equivalent circuit for an FET.

18.22 Explain the function of the capacitor C in the circuits shown in Figure 18.17. What effect will this capacitor have on the frequency response of the circuit?

18.23 Repeat the calculations of Example 18.1 using a device with a g_m of 2.5 mS and a load resistor R_L of 3.3 kΩ.

18.24 An n-channel JFET has a pinch-off voltage of $-4\,$V and a drain-to-source saturation current of 6 mA. Calculate the transconductance of this device at drain currents of 1, 2 and 4 mA.

18.25 What is meant by the quiescent output voltage of a circuit?

18.26 An amplifier is required to operate from a supply of 25 V with a load resistance of 4.7 kΩ and a quiescent output voltage of 15 V. Use a graphical method to design a biasing arrangement to satisfy these conditions, assuming the design uses a JFET as described in Exercise 18.24.

18.27 Determine the input resistance, output resistance, small-signal voltage gain and low-frequency cut-off of the following circuit, given that $g_m = 3$ mS.

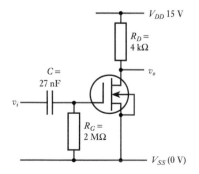

18.28 Determine the input resistance, output resistance, small-signal voltage gain and low-frequency cut-off of the following circuit.

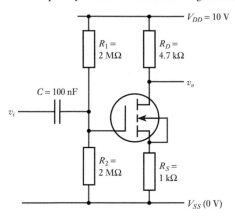

18.29 Repeat Computer simulation exercise 18.1 and note the effect of increasing the input voltage until the output becomes grossly distorted. What are the maximum and minimum values of the output voltage obtainable from the circuit? What causes these limitations?

18.30 Determine the input resistance, output resistance and small-signal voltage gain of the following circuit, assuming that $g_m = 24$ mS. You may assume that C_S has been chosen so that its impedance is 'small' at the frequencies of interest.

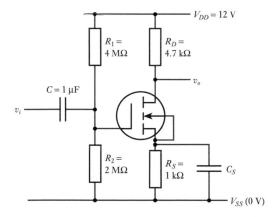

18.31 What is the voltage gain of a source follower amplifier?

18.32 Use simulation to estimate the common-mode rejection ratio (CMRR) of the circuit in Figure 18.29. Take as your starting point the circuit in Computer simulation exercise 18.4.

18.33 Modify your circuit for the previous exercise by replacing the source resistor with a constant current source formed using a JFET and a resistor, as shown in Figure 18.31(c). Use a 2N3819 JFET and an appropriate value resistor. *Hint:* before modifying your circuit, place a current probe in series with the source resistor to determine the source current. Replace the source resistor with the FET and a resistor of a few hundred ohms and measure the resulting source current. Adjust the value of the resistor until the source current is approximately the same as before – the biasing of the circuit will now be similar to the original circuit.

When the circuit is operating correctly, measure the CMRR of the new circuit and compare this with the value obtained in the previous exercise.

18.34 Explain the use of an FET in an automatic gain control circuit.

18.35 Discuss the use of an FET as an analogue switch.

18.36 How would the circuit in Figure 18.35 be modified if a p-channel device were used?

18.37 Explain why the circuits in Figure 18.36 use enhancement MOSFETs rather than depletion devices for the switching transistor.

18.38 Explain why the circuit in Figure 18.36(b) might be more attractive than that in Figure 18.36(a) in some situations.

18.39 Explain briefly why CMOS circuitry is superior to NMOS and PMOS circuitry for the production of logic gates.

Chapter 19

Bipolar Junction Transistors

Objectives

When you have studied the material in this chapter, you should be able to:

- explain the importance of bipolar transistors in modern electronic circuits
- describe the construction, operation and characteristics of bipolar transistors
- analyse simple amplifier circuits based on transistors and determine their operating conditions and voltage gain
- use a range of small-signal equivalent circuits for bipolar transistors, including the hybrid-parameter and hybrid-π models
- discuss the importance of negative feedback in overcoming variability within transistor circuits
- design a range of transistor circuits using simple design rules.

19.1 Introduction

Bipolar transistors are one of the main 'building blocks' in electronic systems and are used in both analogue and digital applications. The devices incorporate two *pn* junctions and are also known as **bipolar junction transistors** (BJTs). It is common to refer to bipolar transistors simply as 'transistors', the term FET being used to identify the field–effect transistor. Bipolar transistors get their name from the fact that current is carried by both polarities of charge carriers (that is, by electrons and by holes), unlike FETs, which are *unipolar*. Bipolar transistors generally have a higher gain than FETs and can often supply more current. However, they have a lower input resistance than FETs, are more complicated in operation and often consume more power.

We will start by looking at the general behaviour of bipolar transistors before turning our attention to their physical construction, operation and characteristics. We will then go on to see how such devices can be used in amplifier circuits and a range of other applications.

19.2 An overview of bipolar transistors

In Chapter 14, we considered the use of a 'control device' in the construction of an amplifier, and in Chapter 18 we saw how an FET could be used in such an arrangement. Bipolar transistors may be used in a similar manner, as shown in Figure 19.1. The devices have three terminals – called the **collector**, **base** and **emitter** – that are given the symbols c, b and e. The base is the control input and signals applied to this terminal affect the flow of current between the collector and emitter. As in the corresponding

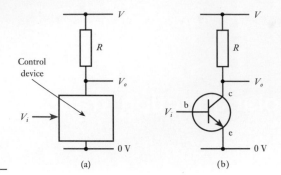

Figure 19.1
A bipolar transistor as a control device.

FET circuit, variations in the input to the transistor alter the current flowing through the resistor R and hence determine the output voltage, V_o. However, the behaviour of bipolar transistors differs from that of FETs, producing circuits with somewhat different characteristics.

While FETs are 'voltage-controlled' devices, bipolar transistors are often considered to be **current-controlled** components. When a control *current* is supplied to the base of a transistor, this causes a larger current to flow from the collector to the emitter (provided that external circuitry is able to supply this current). When used in this way the transistor acts as an almost linear **current amplifier** where the output current (the current flowing into the collector) is directly proportional to the input current (the current flowing into the base). This relationship is illustrated in Figure 19.2. The **current gain** produced by a transistor might be a hundred or more and is relatively constant for a given device (although it will vary with temperature).

While it is common to view the transistor as a current-controlled device, an alternative view is to consider the input as the *voltage* applied to its base. This voltage produces an input current, which is then amplified to produce an output current. The behaviour of the device is then described by the relationship between the output current and the input voltage (the transconductance), as in the FET. Unfortunately, the relationship between the input voltage and the input current is not linear, so, unlike the current gain, the transconductance is not constant but varies with the magnitude of the output current. We look at this relationship in Section 19.3.

Regardless of which 'model' is adopted, an input signal applied to the base controls the current flowing into the collector of the transistor. Therefore, when used in a circuit of the form shown in Figure 19.1(b), variations in the input will cause variations in the current through the resistor, which will control the output voltage.

Figure 19.2
The relationship between the collector current and base current in a bipolar transistor.

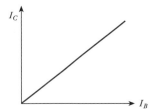

19.2.1 Construction

Bipolar transistors are formed from three layers of semiconductor material and two device polarities are possible. The first is formed by placing a thin layer of *p*-type

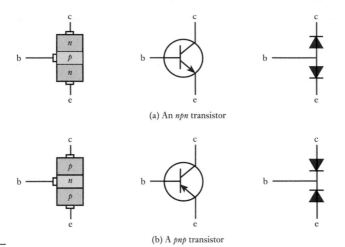

(a) An *npn* transistor

(b) A *pnp* transistor

Figure 19.3
npn and *pnp*
transistors.

semiconductor between two layers of *n*-type material to form an ***npn* transistor**. The second is formed by placing a thin layer of *n*-type material within two layers of *p*-type material to give a ***pnp* transistor**. Both types of device are widely used and circuits often combine components of these two polarities. The operation of the two forms is similar, differing mainly in the polarities of the voltages and currents (and in the polarities of the charge carriers involved). Figure 19.3 shows the form of each kind of transistor together with its circuit symbol. It can be seen that in each case the sandwich construction produces two *pn* junctions (diodes). However, the operation of the transistors is very different from that of two connected diodes.

As the operation of *npn* and *pnp* transistors is similar, in this chapter we concentrate on the former to avoid duplication. In general, the operation of *pnp* devices is similar to that of *npn* devices if the polarities of the various voltages and currents are reversed.

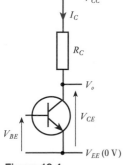

Figure 19.4
Labelling of voltages
and currents in bipolar
transistor circuits.

19.2.2 Notation

The notation used to represent the various voltages and currents in bipolar transistor circuits is similar to that used in FET circuits. For example, V_{CE} would be used to represent the voltage on the *collector* with respect to the *emitter* and the current into the base would be labelled I_B. Again, we use upper-case letters for steady voltages and currents, but lower-case letters for varying quantities. The voltage and current associated with the supply line that is connected (directly or indirectly) to the collector of the transistor are normally given the labels V_{CC} and I_{CC}, while the corresponding labels for the supply connected to the emitter are V_{EE} and I_{EE}. In circuits using *npn* transistors, V_{CC} is normally positive and V_{EE} is taken as the zero-volts reference (or ground). This notation is illustrated in Figure 19.4.

19.3 Bipolar transistor operation

The *npn* or *pnp* structure produces two *pn* junctions connected 'back to back', as shown in Figure 19.3. If a voltage is connected across the device between the collector and the emitter, with the base open circuit, one or other of these junctions is reverse biased, so negligible current will flow. If a transistor was nothing more than two 'back-to-back diodes', it would have little practical use. However, the construction of the device – in particular, the fact that the base region is very thin – allows the base to act as a control input. Signals applied to this electrode can be used to produce, and control, currents between the other two terminals. To see why this is so, consider the circuit configuration shown in Figure 19.5(a).

The normal circuit configuration for an *npn* transistor is to make the collector more positive than the emitter. Typical voltages between the collector and the emitter (V_{CE}) might be a few volts. With the base open circuit, the only current flowing from the collector to the emitter will be a small **leakage current** I_{CEO}, the subscript specifying that it is the current from the *C*ollector to the *E*mitter with the base *O*pen circuit. This leakage current is small and can normally be neglected. If the base is made positive with respect to the emitter this will forward bias the base–emitter junction, which will behave in a manner similar to a diode (as described in Chapter 17). For small values of the base-to-emitter voltage (V_{BE}) very little current will flow, but as V_{BE} is increased beyond about 0.5 V (for a silicon device), the base current begins to rise rapidly.

The fabrication of the device defines that the emitter region is heavily doped, while the base is lightly doped. The heavy doping in the emitter region results in a large number of majority charge carriers, which are electrons in an *npn* transistor. The light doping in the base region generates a smaller number of holes, which are the majority carriers in the *p*-type base region. Thus, in an *npn* transistor, the base current is dominated by electrons flowing from the emitter to the base. In addition to being lightly doped, the base region is very thin. Electrons that pass into the base from the emitter as a result of the base–emitter voltage become minority charge carriers in the *p*-type base region. As the base is very thin, electrons entering the base find themselves close to the space-charge region formed by the reverse bias of the base–collector junction. While the reverse-bias voltage acts as a barrier to majority charge carriers near the junction, it actively propels minority charge carriers across it. Thus, any electrons entering the junction area are swept across into the collector and give rise to a collector current. Careful design of the device ensures that the majority of the electrons entering the base are swept across the junction into the collector. Thus, the flow of electrons from the emitter to the collector is many times greater than the flow from the emitter to the base. This allows the transistor to function as a current-amplifying device, with a small base current generating a larger collector current. As conventional current flow is in the opposite direction to the flow of the negatively charged electrons, a flow of electrons from the emitter to the collector represents a flow of conventional current in the opposite direction, as shown in Figure 19.5(b). This phenomenon of current amplification is referred to as **transistor action**.

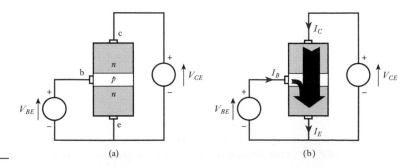

Figure 19.5
Transistor operation.

(a)   (b)

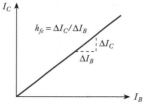
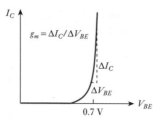

Figure 19.6
Characteristics of a
typical silicon bipolar
transistor.

(a) Relationship between output
current and input current

(b) Relationship between output
current and input voltage

The relationship between the collector current I_C and base current I_B for a typical silicon bipolar transistor is shown in Figure 19.6(a).

Because of a slight non-linearity in the relationship between I_B and I_C, there are two ways that the current gain of the device can be specified. The first is the **DC gain**, h_{FE} or β, which is found simply by dividing the collector current by the base current. This is usually given at a particular value of I_C. Because of the slight non-linearity of the characteristic, h_{FE} has slightly different values at different values of I_C. The DC current gain is used in large-signal calculations and, therefore

$$ I_C = h_{FE} I_B \tag{19.1} $$

When considering small signals, we need to know the relationship between a small change in I_B (ΔI_B) and the corresponding change in I_C (ΔI_C). The ratio $\Delta I_C/\Delta I_B$ is called the small-signal current gain and is given the symbol h_{fe}. It is also called the **AC gain** of the device. The value of h_{fe} may be obtained from the slope of the characteristic given in Figure 19.6(a) and

$$ i_c = h_{fe} i_b \tag{19.2} $$

For most practical purposes, h_{FE} and h_{fe} can be considered to be equal. A typical value for a general-purpose silicon transistor would be in the range 100 to 300, but the current gain of bipolar transistors varies considerably with temperature and operating conditions. There is also a considerable spread of characteristics between devices of the same nominal type and even within the same batch.

The characteristics of a bipolar transistor may also be described by the relationship between the output current I_C and input voltage V_{BE} as shown in Figure 19.6(b). As the base–emitter junction resembles a simple pn junction, the input current I_B is exponentially related to the input voltage V_{BE} (as illustrated in Figure 17.9). As the output current is approximately linearly related to the input current (by the current gain, h_{FE}), the relationship between I_C and V_{BE} has the same shape (although with correspondingly larger values of current). The slope of this curve at any point is given by the ratio $\Delta I_C/\Delta V_{BE}$, which represents the **transconductance** of the device g_m (you may like to compare this with the similar discussion of the FET given in Section 18.5). In the limit

$$ g_m = \frac{\mathrm{d}I_C}{\mathrm{d}V_{BE}} \tag{19.3} $$

Unlike h_{FE}, which is approximately constant for a given device, g_m varies with the collector current (and, therefore, the emitter current) at which the circuit is operated.

Video 19A

A simple amplifier

In Figure 19.1(b) we looked at a basic amplifier based on a bipolar transistor. Varying the voltage applied to the input will vary the current flowing into the base of the transistor. This base current will be amplified by the transistor to produce a correspondingly larger collector current. This collector current, flowing through the collector resistor, will determine the output voltage.

While the arrangement in Figure 19.1(b) has some interesting properties, it does not represent a useful amplifier. The relationship between the base current and the collector current is relatively linear (as shown in Figure 19.2), but the device can only be used with positive base currents. If we wish to operate with a bipolar signal (that is, one that goes negative as well as positive), we must offset the input from zero in order to amplify the complete signal. This is done by **biasing** the input of the amplifier.

A simple amplifier with a biasing arrangement is shown in Figure 19.7. The base resistor R_B applies a positive voltage to the base of the transistor, forward biasing the base–emitter junction and producing a base current. This in turn produces a collector current, which flows through the collector load resistor R_C, producing a voltage drop and making the output voltage V_o less than the collector supply voltage V_{CC}. When no input is applied to the circuit it is said to be in its quiescent state. The value of the base resistor R_B will determine the **quiescent base current**, which in turn will determine the **quiescent collector current** and the **quiescent output voltage**. The values of R_B and R_C must be chosen carefully to ensure the correct operating point for the circuit.

If a *positive* voltage is applied to the input of the amplifier, this will tend to increase the voltage on the base of the transistor, thus increasing the base current. This in turn will raise the collector current, increasing the voltage drop across the collector resistor R_C and *decreasing* the output voltage V_o. If a *negative* voltage is applied to the input, this will decrease the current through the transistor and thus *increase* the output voltage. We therefore have an **inverting amplifier**. As with the MOSFET amplifiers described in the previous chapter, a **coupling capacitor** is used to prevent input voltages from affecting the mean voltage applied to the base. Therefore, the circuit cannot be used to amplify DC signals and so is an **AC-coupled amplifier**.

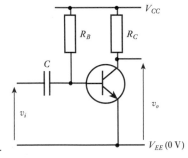

Figure 19.7
A simple amplifier.

File 19A

Computer simulation exercise 19.1

Simulate the circuit in Figure 19.7 using an appropriate bipolar transistor. A suitable arrangement might use a 2N2222 transistor with $V_{CC} = 10$ V, $R_B = 910$ kΩ, $R_C = 2.7$ kΩ and $C = 1$ μF. If your simulation package does not provide this particular transistor then you may need to experiment with other parts, adjusting the resistor values to suit.

Apply a small alternating voltage to the input and observe the resultant variations at the output.

Before we can look in detail at the operation of amplifier circuits, we need to know a little more about the characteristics of bipolar transistors.

19.5 Bipolar transistor characteristics

19.5.1 Transistor configurations

Bipolar transistors are very versatile components and can be used in a number of circuit configurations. These arrangements differ in the way that signals are applied to the device and how the output is produced. In each configuration, control signals are applied to the transistor by an 'input circuit' and a controlled quantity is sensed by an 'output circuit'.

In Figure 19.7, we considered a circuit where the input takes the form of a voltage applied to the base (with respect to the emitter) and the output is represented by the voltage on the collector (with respect to the emitter). This arrangement is represented in Figure 19.8. It can be seen that the emitter terminal is common to both the input and the output circuits. For this reason, this arrangement is known as a **common-emitter** circuit. It should perhaps be noted at this point that the 'E' or 'e' in h_{FE} and h_{fe} each stands for 'emitter' as these are the current gains of the transistor when used in a common-emitter configuration. Common-collector and common-base circuits are also used and these have different current gains and characteristics. This allows us to select from a range of characteristics to suit our needs. In this section, we will concentrate on the characteristics associated with common-emitter circuits.

In the previous chapter, when looking at FETs, we noted that the behaviour of such a transistor may be understood by considering its input characteristics, its output characteristics and the relationship between the input and the output (the transfer characteristics). We now consider these three aspects of the behaviour of a bipolar transistor.

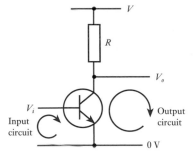

Figure 19.8
A common-emitter
arrangement.

19.5.2 Input characteristics

From Figure 19.5(a), it is clear that the input of the transistor takes the form of a forward-biased pn junction and the input characteristics are therefore similar to the characteristics of a semiconductor diode. The input characteristics of a typical silicon device are shown in Figure 19.9.

In Chapter 17, we deduced that the current in a semiconductor diode is given by

$$I \approx I_S \, e^{40V}$$

where the '40' represents an approximate value for e/kT and V is the voltage across the device.

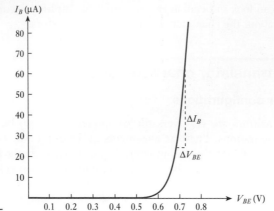

Figure 19.9

Input characteristics of a bipolar transistor in the common-emitter configuration.

In this case, the current I is the base current of the transistor I_B, and the junction voltage V is the base-to-emitter voltage V_{BE}. We therefore have

$$I_B \approx I_{BS}\, e^{40V_{BE}} \tag{19.4}$$

where I_{BS} is a constant determined by the base characteristics. This equation represents the **input characteristic** of the device.

The slope at any point of the line in Figure 19.9 represents the relationship between a small change in the base-to-emitter voltage ΔV_{BE} and the corresponding change in the base current ΔI_B. The slope therefore indicates the **small-signal input resistance** of the arrangement, which is given the symbol h_{ie}. Clearly, the magnitude of the input resistance varies with position along the characteristic. The value at any point may be found by differentiating Equation 19.4 with respect to I_B, which gives the simple result that

$$h_{ie} = \frac{\mathrm{d}V_{BE}}{\mathrm{d}I_B} \approx \frac{1}{40 I_B}\,\Omega \tag{19.5}$$

As a typical value for I_B might be a few tens of microamps, a typical value for h_{ie} might be a few kilohms.

The input characteristics of a bipolar transistor in the common-emitter configuration may therefore be described by very simple expressions for the base current and the small-signal input resistance. However, it should be noted that the 'e' subscript in h_{ie} stands for 'emitter' and that h_{ie} is the small-signal input resistance in the *common-emitter configuration*. Other circuit configurations will have a different input resistance.

19.5.3 Output characteristics

Figure 19.10 shows the relationship between the collector current I_C and collector voltage V_{CE}, for a typical device for various values of the base current. These two quantities represent the output current and output voltage of the transistor in the common-emitter configuration. The relationship between them is often referred to as the common-emitter **output characteristic**. You might like to compare this characteristic with that obtained for FETs in Figure 18.10.

For a given base current, the collector current initially rises rapidly with the collector voltage as this increases from zero. However, it soon reaches a steady value and any further increase in the collector voltage has little effect on the collector current. The

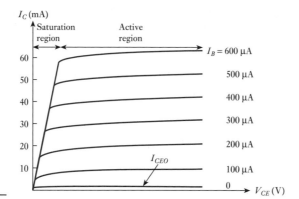

Figure 19.10
A typical common-
emitter output
characteristic.

value of collector current at which the characteristic stabilises is determined by the base current. The ratio between this steady value of collector current and the value of the base current represents the DC gain of the device, h_{FE}.

The section of the characteristic over which the collector current is approximately linearly related to the base current is referred to as the **active region**. Most linear amplifier circuits operate in this region. The section of the characteristic close to the origin where this linear relationship does not hold is called the **saturation region**. The saturation region is generally avoided in linear circuits, but widely used in non-linear arrangements, including digital circuitry. It is important to note that the term 'saturation' has a different meaning when applied to bipolar transistors from that when discussing FETs. Saturation occurs in bipolar transistors when V_{CE} is very low, because the efficiency of the transistor action is reduced and many charge carriers pass from the emitter to the base without being swept into the collector region.

In an ideal bipolar transistor, the various lines in the output characteristic would be horizontal, indicating that the output current was completely independent of the collector voltage. In practice, this is not the case and all real devices have a slight gradient, as shown in Figure 19.10. The slope of these lines indicates the change in output current with output voltage and is therefore a measure of the **output resistance** of the arrangement. A typical value for this resistance might be of the order of 100 kΩ. It is often convenient to consider the reciprocal of the output resistance, namely the **output admittance** h_{oe}, where again the 'e' subscript indicates that this value is for a device in the common-emitter configuration.

If the nearly horizontal portions of the output characteristic are extended 'back-wards' (to the left in Figure 19.10) they converge at a point on the negative portion of the horizontal axis. This point is referred to as the **Early voltage**, after J. M. Early of Bell Laboratories. The Early voltage is given the symbol V_A and has a typical value of between 50 and 200 volts.

Notice in Figure 19.10 that the collector current is not zero when the base current is zero. This is because of the presence of the **leakage current** I_{CEO}. The effect of I_{CEO} is magnified in the figure to allow it to be visible. In silicon devices its effects are usually negligible.

19.5.4 Transfer characteristics

Figure 19.6 represents the transfer characteristics of a bipolar transistor in a common–emitter configuration. From this figure, it is clear that the characteristics can be described in two ways: the first, in terms of the current gain of the device; the second, in terms of its transconductance.

For a given device, the current gain tends to be relatively constant regardless of how the device is used (although the gain *will* vary with temperature and from one device to another). In contrast, the transconductance varies with the operating conditions of the circuit, as shown in Figure 19.6.

From Figure 19.5(b) it is clear that the emitter current I_E must be given by the sum of the collector current I_C and the base current I_B. Thus

$$I_E = I_C + I_B$$

and as

$$I_C = h_{FE}I_B$$

it follows that

$$I_E = h_{FE}I_B + I_B = (h_{FE} + 1)I_B$$

As h_{FE} is usually much greater than unity, we may make the approximation that

$$I_E \approx h_{FE}I_B = I_C \qquad (19.6)$$

Combining the results of Equations 19.3, 19.4 and 19.6 it can be shown that

$$g_m \approx 40I_C \approx 40I_E \text{ siemens} \qquad (19.7)$$

It is important to note that in bipolar transistors g_m is proportional to I_E, whereas in FETs, g_m is proportional to the square root of the drain current (see Section 18.5.4). As g_m is directly controlled by the quiescent collector (emitter) current, the voltage gain of an amplifier formed using the device will also be related to this current. The choice of quiescent circuit conditions thus plays a major part in determining the performance of an amplifier.

The value of g_m represents the ratio of changes in the collector (emitter) current to changes in the base voltage. The inverse of this quantity has the units of resistance and is the ratio of the small-signal base-to-emitter voltage to the corresponding change in the emitter current. This is termed the **emitter resistance**, r_e. It follows from Equation 19.7 that

$$r_e = \frac{1}{g_m} \approx \frac{1}{40I_C} \approx \frac{1}{40I_E} \Omega \qquad (19.8)$$

From Equation 19.5

$$h_{ie} \approx \frac{1}{40I_B}$$

it follows that

$$h_{ie} \approx h_{fe}r_e \qquad (19.9)$$

Note that here we use h_{fe} rather than h_{FE} as both h_{ie} and r_e are small-signal quantities. We noted earlier that h_{ie} might have a typical value of a few kilohms, so r_e will have a typical value of a few ohms or tens of ohms.

19.5.5 Equivalent circuits for a bipolar transistor

We are now in a position to create equivalent circuits for a bipolar transistor and Figure 19.11 shows two such circuits. You might wish to compare these models with

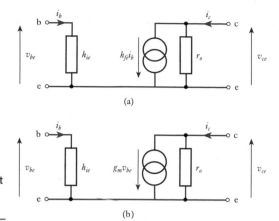

Figure 19.11
Small-signal equivalent circuits for a bipolar transistor.

that shown in Figure 18.15 for an FET. As in the earlier chapter, the models in Figure 19.11 are small-signal equivalent circuits.

It can be seen that the two models in Figure 19.11 are very similar, differing only in the magnitude of the current produced by their current generator. In Figure 19.11(a) the behaviour of the device is modelled by its current gain h_{fe}, which for general-purpose small-signal transistors would normally have a value of between 100 and 300 (but could be much lower in power transistors). In Figure 19.11(b) the transistor is modelled by its transconductance, g_m. The value of g_m depends on the circuit configuration and an estimate of its value can be obtained from Equation 19.7. The input and output resistance of each model are the same (as one would expect, given that these are alternative representations of the same device). The magnitude of the input resistance h_{ie} depends on the circuit arrangement and an estimate of its value can be obtained from Equation 19.5. The magnitude of the output resistance corresponds to the slope of the output characteristic and would typically be between 10 kΩ and 1 MΩ. This corresponds to values of the output admittance h_{oe} of between 1 and 100 μS.

It is worth noting at this point that h_{fe} varies considerably from one device to another, but g_m is defined by the physics of the materials used and can be calculated from the collector (or emitter) current. It must be remembered, however, that we are using an approximation for g_m that ignores, for example, the effects of η in the diode equations (see Section 17.4.3).

The equivalent circuits given in Figure 19.11 are perfectly acceptable for most applications, but are not the only ways to model a device. Various other representations are used, each using different refinements to provide a better model of the operation of the device.

The hybrid-parameter model

You will have noticed that two of the symbols used to describe the characteristics of the bipolar transistor in the simple equivalent circuit of Figure 19.11(a), namely h_{fe} and h_{ie}, have a similar form and that a third quantity, the output resistance, can be represented as $1/h_{oe}$. There is, in fact, a fourth commonly used member of the set h_{re}, which is the **reverse voltage transfer ratio**. This describes how the output voltage affects the current in the input circuit. To understand the nature of this *feedback* effect, we need to look back at the input characteristic of the device.

In Figure 19.9, we looked at the variation of base current with base-to-emitter voltage. This figure gives a single characteristic, indicating a unique correspondence

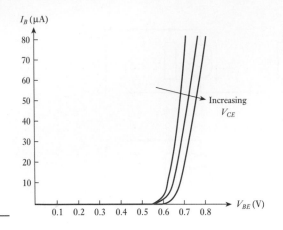

Figure 19.12
Variation of V_{BE} with collector voltage.

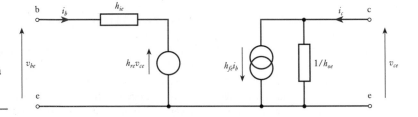

Figure 19.13
The hybrid-parameter equivalent circuit for a common-emitter transistor.

between one and the other. In practice, the position of this characteristic depends on the collector voltage, as shown in Figure 19.12. As V_{CE} increases, the value of V_{BE} required to produce a given base current also increases. The reverse voltage transfer ratio h_{re} is the rate of change of V_{BE} with V_{CE} for a given I_B. This effect is fairly small and h_{re} has a typical value of about 10^{-3} to 10^{-5}. The interaction between V_{BE} and V_{CE} is brought about because a change in the collector voltage alters the width of the base region. This **base width modulation** results in variations of V_{BE} with V_{CE} at constant I_C and is called the **Early effect**. An equivalent circuit that incorporates this additional feature is given in Figure 19.13. An additional voltage generator has been added that is controlled by the output voltage. You will also notice that the output resistance is now shown as $1/h_{oe}$ (that is, the reciprocal of the output admittance), which is the normal notation used in this form of equivalent circuit.

The parameters used in the equivalent circuit shown in Figure 19.13 have different forms: h_{ie} is an impedance, h_{oe} is an admittance, h_{fe} is a current ratio and h_{re} is a voltage ratio. For this reason, the model is called the **hybrid-parameter model**, often abbreviated to the **_h_-parameter model**, and this accounts for the 'h' in the symbols used for the parameters. Each parameter is identified by a two-letter suffix. The second letter of each is an 'e', indicating that these quantities relate to the common-emitter configuration. The first suffix indicates the nature of the parameter: h_{ie} describes the characteristics of the _input_, h_{re} refers to the _reverse_ effects, h_{fe} gives the _forward_ current gain and h_{oe} relates to the _output_ of the device.

These _h_-parameters are frequently given in manufacturers' data sheets and the hybrid model is widely used.

The hybrid-π model

The hybrid-parameter model described above provides a good description of the operation of transistors at low frequencies, but does not take into account the effects

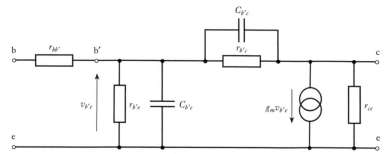

Figure 19.14
The hybrid-π equivalent circuit for a common-emitter transistor.

of device capacitances, which greatly alter a device's performance at high frequencies. A more sophisticated model can be used to take these effects and some other additional physical properties of the device into account.

Figure 19.14 shows the **hybrid-π model** for a bipolar transistor. The circuit still has three nodes – e, b and c – representing terminals of the device, but now another node b' is added that represents the base *junction*. The ohmic resistance between the base terminal and the base junction is represented by $r_{bb'}$, which is typically in the range 5 to 50 Ω for general-purpose devices and a few ohms for high-frequency types. The parameter $r_{b'e}$ is the base–emitter junction resistance. A similar analysis to that given in Section 19.5.4 shows that this is given approximately by the expression

$$r_{b'e} \approx h_{fe} r_e \approx \frac{h_{fe}}{40 I_E} \tag{19.10}$$

This is the expression previously derived for h_{ie}, which is the total resistance seen looking into the base. From this analysis it is apparent that a more accurate expression for h_{ie} is

$$h_{ie} = r_{bb'} + r_{b'e} \approx r_{bb'} + h_{fe} r_e \approx r_{bb'} + \frac{h_{fe}}{40 I_E} \tag{19.11}$$

If we take typical values for h_{fe} and I_E of 200 and 5 mA, respectively, this gives a value for $r_{b'e}$ of approximately 1 kΩ. As this is large compared with typical values for $r_{bb'}$ (5 to 50 Ω), the approximation that $h_{ie} \approx r_{b'e}$ is normally acceptable.

The effects of variations in the output voltage on the input are represented in the hybrid-π model by a resistor $r_{b'c}$ rather than by a voltage generator as in the hybrid-parameter model. Capacitors $C_{b'e}$ and $C_{b'c}$ are included to represent the capacitance across the base–emitter junction and the capacitance from the base to the collector, respectively. The current produced by the current source in the circuit is proportional to the current that passes through the base–emitter junction resistance, $r_{b'e}$. However, because of the presence of $r_{b'c}$ and the two capacitors, this is not equal to the current flowing into the base terminal. The current generator is therefore made equal to g_m times the voltage across the base–emitter junction $v_{b'e}$.

The hybrid-π model agrees more closely with actual device performance than the simpler models described earlier. However, it is much more difficult to analyse. The inclusion of device capacitances allows the high-frequency performance of the device to be predicted, although the model breaks down at frequencies considerably below those at which the gain falls to unity.

19.5.6 Bipolar transistors at high frequencies

A detailed analysis of the high-frequency performance of bipolar transistors, as predicted by the hybrid-π model, is beyond the scope of this text. However, looking at the model, it is clear that the presence of $C_{b'e}$ will produce a fall in gain at high frequencies, as described in Section 8.6. $C_{b'c}$ also affects the frequency response of the device as it produces feedback from the output to the input. As the device inverts the input signal, this is **negative feedback**. As the impedance of the capacitor decreases as the frequency increases, the amount of negative feedback increases with frequency, reducing the gain at high frequencies. The combined effects of the two capacitances produce a frequency response of the general form shown in Figure 19.15, where $h_{fe(0)}$ represents the low-frequency value of h_{fe}.

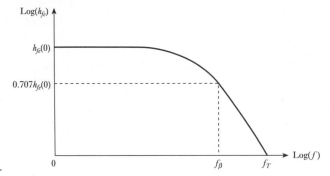

Figure 19.15
Variation of current gain with frequency for a bipolar transistor.

The frequency at which the gain falls to 0.707 (that is, $1/\sqrt{2}$) of its low-frequency value represents the **bandwidth** of the device and is given the symbol f_β. In Section 8.6 we deduced that the upper cut-off frequency of an RC network is given by

$$f_c = \frac{1}{2\pi CR}\,\text{Hz}$$

It can be shown that f_β is given by the expression

$$f_\beta \approx \frac{1}{2\pi(C_{b'e} + C_{b'c})r_{b'e}}$$

In general, the effects of $C_{b'e}$ dominate those of $C_{b'c}$ and so it is usually possible to make the approximation

$$f_\beta \approx \frac{1}{2\pi C_{b'e}r_{b'e}} \tag{19.12}$$

The frequency at which the gain drops to unity is termed the **transition frequency** and is given the symbol f_T. It can be shown that this is simply related to f_β by the expression

$$f_T = h_{fe(0)}f_\beta \tag{19.13}$$

From Equation 19.10 we know that $r_{b'e}$ varies with the emitter current I_E, and from Equations 19.12 and 19.13 it follows that f_β and f_T also vary with current. By substituting from Equation 19.10 we can obtain expressions indicating this relationship. These are

$$f_\beta \approx \frac{40I_E}{h_{fe(0)}2\pi(C_{b'e} + C_{b'c})} \approx \frac{40I_E}{h_{fe(0)}2\pi C_{b'e}}$$

and

$$f_T \approx \frac{40I_E}{2\pi(C_{b'e} + C_{b'c})} \approx \frac{40I_E}{2\pi C_{b'e}}$$

Thus, the bandwidth and transition frequency of the transistor are both directly proportional to the emitter current. This result has implications for high-bandwidth, low-power-consumption amplifiers.

19.5.7 Leakage currents

In Section 19.3, we discussed the presence of a leakage current I_{CEO} that flows from the collector to the emitter in the absence of any base current. This current was represented in Figure 19.10.

A leakage current is also present between the collector and the base across the reverse-biased collector–base junction. This is given the symbol I_{CBO} as it is the current that flows from the *C*ollector to the *B*ase with the third terminal (the emitter) *O*pen circuit.

An understanding of the relationship between these two currents and their significance within transistor circuits can be gained by remembering that the transistor can be seen as two back-to-back *pn* junctions. One of these, the collector–base junction, is reverse biased in normal operation. We know (from the discussions in Chapter 17) that reverse-biased junctions have a reverse saturation current that is approximately constant for all but the smallest reverse voltages. This accounts for I_{CBO} which flows across the reverse-biased collector–base junction.

When the leakage current I_{CBO} enters the base region, it produces a similar effect to current entering from the base terminal. The leakage current is amplified by the current gain of the device, producing a much larger current from the collector to the emitter. This is the collector-to-emitter leakage current I_{CEO}. Therefore

$$I_{CEO} \approx h_{FE}I_{CBO}$$

and the collector current is given by

$$I_C = h_{FE}I_B + I_{CEO}$$

In some semiconductors – for example, those manufactured using germanium – the effects of leakage currents are very significant, particularly at high temperatures (in the next section we will discuss the effects of temperature on leakage currents). However, in silicon devices the leakage currents are extremely small and can, with good design, normally be neglected.

19.5.8 Temperature effects

In Section 17.6.4 we looked at how temperature affects the behaviour of semiconductor diodes. The junctions within bipolar transistors are also affected by temperature, which changes their current gain, base–emitter voltage and leakage currents.

Details of how h_{FE} increases with temperature for a particular device can be found in its data sheet. Typical silicon devices might show an increase in gain of about 15 per cent when taken from 0 to 25 °C, but a doubling in gain when the temperature increases from 25 to 75 °C.

The value of V_{BE} shows a similar variation with temperature to that seen in semiconductor diodes. In silicon devices this corresponds to a *fall* in V_{BE} of about 2 mV per °C rise in temperature.

The transistor leakage currents I_{CBO} and I_{CEO} increase with temperature. I_{CBO} shows a similar rate of increase to that of semiconductor diodes, doubling for an increase of about 10 °C (an increase of about 7 per cent/°C). As I_{CEO} is related to I_{CBO} by the expression

$$I_{CEO} \approx h_{FE} I_{CBO}$$

this also increases with temperature, but more rapidly as h_{FE} also increases with temperature. With silicon devices, leakage currents are generally of the order of a few nanoamps and their effects are almost always negligible. Consequently, the effects of temperature on these currents are not usually of great concern.

Video 19B

19.6 Bipolar amplifier circuits

Having looked at the characteristics of bipolar transistors, we are now in a position to look in more detail at amplifier circuits based on these devices. We will begin by returning to look at the simple amplifier of Section 19.4, and then progress to more sophisticated arrangements.

The circuit of Figure 19.7 is an AC-coupled amplifier and in such circuits it is convenient to consider its DC (or quiescent) behaviour separately from its AC (or small-signal) behaviour.

19.6.1 DC analysis of a simple amplifier

The DC analysis of our simple amplifier is very straightforward.

Example 19.1

Determine the quiescent collector current and the quiescent output voltage of the following circuit, given that the h_{FE} of the transistor is 100.

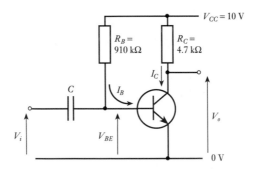

The base-to-emitter junction of the transistor resembles a forward-biased *pn* junction, therefore we will assume that the base-to-emitter voltage V_{BE} is 0.7 V.

From a knowledge of V_{BE} we also know the voltage across R_B as this is simply $V_{CC} - V_{BE}$, which in turn enables us to calculate the base current I_B. In this case

$$I_B = \frac{V_{CC} - V_{BE}}{R_B} = \frac{10 - 0.7 \text{ V}}{910 \text{ k}\Omega} = 10.2 \text{ μA}$$

The collector current I_C is now given by

$$I_C = h_{FE} I_B = 100 \times 10.2 \text{ μA} = 1.02 \text{ mA}$$

The quiescent output voltage is simply the supply voltage minus the voltage drop across R_C and is therefore

$$V_o = V_{CC} - I_C R_C = 10 - 1.02 \times 10^{-3} \times 4.7 \times 10^3 \approx 5.2\,\text{V}$$

Thus, the circuit has a quiescent collector current of about 1 mA and a quiescent output voltage of approximately 5.2 V.

Note: In this circuit, the quiescent collector current and the quiescent output voltage are both determined by the value of h_{FE}, which varies considerably between devices (and with temperature). For this reason, this simple circuit arrangement is rarely used. However, it is useful to look at the disadvantages of this arrangement before progressing to more suitable techniques.

19.6.2 Small-signal analysis of a simple amplifier

In order to determine the small-signal behaviour of our simple amplifier, we first need to construct a small-signal equivalent circuit of the arrangement. In Section 19.5.5 we derived equivalent circuits for a bipolar transistor, and it is a simple matter to extend this model to represent the complete amplifier shown in Figure 19.7. When doing this we must remember that the supply line V_{CC} is effectively joined to the earth line (0 V) as far as small signals are concerned, so the equivalent circuit is as shown in Figure 19.16. Clearly, a more sophisticated model could be produced using a full hybrid-parameter or hybrid-π model, but a simple circuit is sufficient for our present needs.

Both R_B and h_{ie} could be replaced by a single resistor representing the resistance of the parallel combination. This could also be done with $1/h_{oe}$ and R_C if desired. It should be remembered that as h_{oe} is an admittance it must be inverted to give an equivalent resistance. The input capacitor C is present to prevent any incoming signal from affecting the bias conditions of the transistor. Normally its value is chosen so that it has a negligible effect at the frequencies of interest. However, the presence of C will introduce a **low-frequency cut-off** into the frequency characteristic of the circuit, as discussed in Section 8.6.

From the small-signal equivalent circuit for the amplifier we can deduce its small-signal voltage gain. If we assume that C can be ignored, it is clear that

$$v_{be} = v_i \tag{19.14}$$

and that

$$v_o = -g_m v_{be} \left(\frac{1}{h_{oe}} // R_C \right)$$

where $(1/h_{oe})//R_C$ simply means the effective resistance of the parallel combination of $1/h_{oe}$ and R_C. Combining these two expressions gives

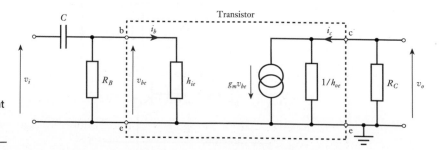

Figure 19.16
Small-signal equivalent circuit of a simple transistor amplifier.

$$v_o = -g_m v_i \left(\frac{1}{h_{oe}} // R_C \right)$$

and so the voltage gain is given by

$$\text{voltage gain} = \frac{v_o}{v_i} = -g_m \left(\frac{1}{h_{oe}} // R_C \right) = -g_m \frac{R_C}{h_{oe} R_C + 1} \tag{19.15}$$

The negative polarity of the gain indicates that this is an **inverting amplifier** and a positive input voltage will produce a negative output voltage. You might like to compare this expression with that obtained in Equation 18.6 for the voltage gain of an amplifier using an FET.

In many cases, the collector resistance R_C will be much smaller than the output resistance of the transistor, so $(1/h_{oe})//R_C$ is approximately equal to R_C. *In this case,* Equation 19.15 may be approximated by

$$\text{voltage gain} = \frac{v_o}{v_i} \approx -g_m R_C \tag{19.16}$$

It is interesting to note that, as the emitter resistance r_e is equal to $1/g_m$, the voltage gain is also given by

$$\text{voltage gain} = \frac{v_o}{v_i} \approx -\frac{R_C}{r_e} \tag{19.17}$$

Note also that, as $g_m \approx 40 I_E$, the voltage gain may be expressed as

$$\text{voltage gain} \approx -g_m R_C \approx -40 I_E R_C \approx -40 I_C R_C \approx -40 V_{RC}$$

where V_{RC} is the voltage across R_C. Thus, the voltage gain of the amplifier is related to the large-signal voltage across R_C. The voltage gain is therefore determined by the quiescent conditions of the circuit as set by the biasing arrangement. We noted in Example 19.1 that the quiescent conditions of this arrangement are affected by the current gain of the device, which varies considerably from one device to another. It is therefore clear that the voltage gain of the circuit will also vary greatly with the current gain of the device used. These characteristics make this a very poor circuit and, in practice, transistors are invariably used with negative feedback to stabilise these parameters.

From the equivalent circuit, it is also straightforward to determine the **small-signal input resistance** and the **small-signal output resistance** of the amplifier. This is illustrated in the following example.

Example 19.2

Determine the small-signal voltage gain, input resistance and output resistance of the following circuit, given that $h_{fe} = 100$ and $h_{oe} = 10\ \mu S$.

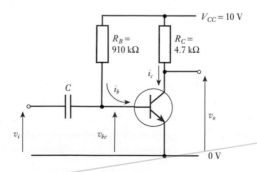

The first step in this problem is to determine the small–signal equivalent circuit of the amplifier.

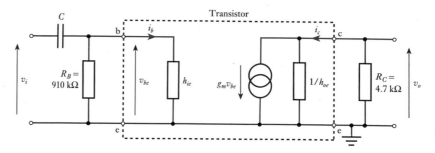

Voltage gain

In order to determine the behaviour of the circuit, we need to establish the values of g_m and h_{ie}. To do this, we must know the DC operating conditions as both are affected by the quiescent current. Fortunately, we have already investigated the DC conditions of the circuit in Example 19.1, from which we know that I_C is 1.02 mA. Therefore, as $I_E \approx I_C$, it follows that $I_E \approx 1.02$ mA and thus

$$g_m \sim 40 I_E \approx 40.8 \text{ mS}$$

and

$$h_{ie} \approx \frac{h_{fe}}{40 I_E} \approx \frac{100}{40 \times 1.02 \times 10^{-3}} \approx 2.45 \text{ k}\Omega$$

From Equation 19.15 we have

$$\text{voltage gain} = \frac{v_o}{v_i} = -g_m \frac{R_C}{h_{oe} R_C + 1}$$

and substituting for the component values gives

$$\text{voltage gain} = -40.8 \times 10^{-3} \frac{4700}{10 \times 10^{-6} \times 4700 + 1} \approx -183$$

If we consider that $1/h_{oe}$ is large compared with R_C and assume that the voltage gain is equal to $-g_m R_C$, this gives a value of -192. Given the inaccuracies in our calculations, this is probably a reasonable approximation. Therefore

$$\text{voltage gain} = \frac{v_o}{v_i} \approx -g_m R_C$$

Input resistance

From the equivalent circuit it is clear that the **small-signal input resistance** is simply $R_B // h_{ie}$. As $R_B \gg h_{ie}$, it is reasonable to say

$$r_i = R_B // h_{ie} \approx h_{ie} \approx 2.4 \text{ k}\Omega$$

Output resistance

The **small-signal output resistance** is the resistance seen 'looking into' the output terminal of the circuit. As the idealised current generator has an infinite internal resistance, the output resistance is simply the parallel combination of R_C and $1/h_{oe}$. Thus

$$r_o = R_C // \frac{1}{h_{oe}} = 4700 // 100,000 \approx 4.5 \text{ k}\Omega$$

and again it is reasonable to use the approximation that $r_o \approx R_C$.

It can be seen that, for the simple common-emitter amplifier of Examples 19.1 and 19.2,

$$r_i \approx h_{ie}$$

$$r_o \approx R_C$$

$$\frac{v_o}{v_i} \approx -g_m R_C \approx -\frac{R_C}{r_e}$$

The effect of a load resistor

In Chapter 14 we looked at the effects of loading on amplifiers and noted that this reduces their output voltage. If we add a load resistor R_L to the amplifier in Figure 19.7 (or Example 19.2), this will also reduce the gain of the circuit. In order that the load does not upset the DC operation of the circuit, this must be connected to the output of the amplifier using a **coupling capacitor** (as at the input) and again this will introduce a low-frequency cut-off. If we choose a capacitor of an appropriate size then, at the frequencies of interest, the effects of the capacitor can be ignored and, in the small-signal equivalent circuit of Figure 19.16, the load resistor R_L will appear in parallel with R_C. A similar analysis to that given above will then show that the voltage gain is given by

$$\text{voltage gain} = \frac{v_o}{v_i} \approx -g_m (R_C // R_L) = -\frac{R_C // R_L}{r_e} \tag{19.18}$$

If R_L is much greater than R_C then the parallel combination $R_C // R_L$ will be approximately equal to R_C and the gain will be unaffected. This corresponds to the situation where the load resistance is much greater than the output resistance of the amplifier. However, where the resistance of the load is comparable with the output resistance of the circuit, then loading effects must be considered.

19.6.3 Large-signal considerations

The large-signal considerations in the design of a transistor amplifier are concerned with the **quiescent operation** of the circuit. In other words, the voltages and currents in the circuit for zero input. These voltages and currents are determined by the biasing arrangements of the circuit and the characteristics of the device itself.

In Section 19.6.1 we looked at the quiescent conditions of a simple amplifier and found them easy to calculate. The quiescent currents and voltages in the amplifier are determined by the various resistors in the circuit and the current gain of the transistor. We also saw in Section 19.6.2 that calculating the various small-signal characteristics of a circuit is also straightforward. Unfortunately, although it is simple to calculate the steady-state voltages and currents within a given circuit, *designing* a circuit to give a desired set of characteristics is slightly more complicated. This is because the collector resistor R_C determines not only the quiescent conditions of the circuit but also the small-signal gain. To overcome this problem, we can adopt the technique used earlier when dealing with FET amplifiers – that is, use of a **load line** (see Section 18.6.3).

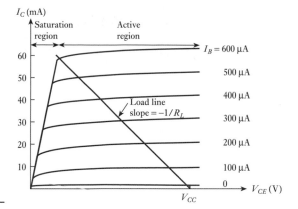

Figure 19.17
A load line.

Figure 19.17 shows a typical **output characteristic** for a bipolar transistor in a common-emitter configuration with a load line superimposed. The function and nature of the load line are identical to those of the load line used for the FET amplifier in Section 18.6. The load line passes through V_{CC} (the collector supply voltage) and has a slope equal to the reciprocal of the load resistance. The various lines of the output characteristic represent the relationship between the collector current and the collector-to-emitter voltage for different values of base current. The load line represents the relationship between the collector current and the voltage at the junction of the load resistor R_C and the transistor. The intersection of the load line and one of the lines of the output characteristic represents a possible solution that satisfies both these relationships simultaneously. The point of intersection of the load line and the characteristic line represents the base current to be used and is termed the **operating point**. This corresponds to the quiescent state of the circuit.

The load line can be used in a number of ways. If the desired operating point is known, a load line can simply be drawn from this point to the V_{CC} position on the voltage axis. The slope of this line then gives the value of the required load resistor. Alternatively, if the value of R_C is fixed, the load line can be drawn with a slope equal to $-1/R_C$ through the V_{CC} point on the voltage axis and the operating point chosen appropriately. Choice of the operating point then determines the required quiescent base current. Thus the choice of operating point is of great importance as it directly affects not only the quiescent conditions of the circuit but also the output voltage range.

19.6.4 Choice of operating point

In Section 18.6.4 we considered the choice of operating point for an FET amplifier and noted that there are several forbidden zones within the characteristic of the FET that must be avoided. These forbidden regions also exist for bipolar transistors, as shown in Figure 19.18.

Region A represents the **saturation region** which is normally avoided in linear applications as it is highly non-linear. However, digital circuits often make use of this part of the characteristic (we will discuss the reasons for this in Chapter 26). All devices have a maximum collector current $I_{C(max)}$ that must not be exceeded. This results in forbidden zone B. Devices also have a maximum collector voltage $V_{CE(max)}$ resulting in zone C. If this voltage is exceeded, the device will exhibit **avalanche breakdown** at the collector–base junction, which can lead to damage (see Section 17.6.5). The last region that must be avoided – area D – is caused by a need to restrict

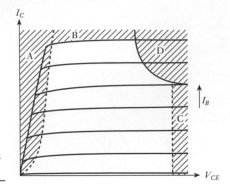

Figure 19.18
The forbidden
operating regions for a
bipolar transistor.

the power dissipated within the device to safe limits. As the power dissipation is the product of the collector voltage and collector current (the base current can normally be neglected), this area is bounded by a hyperbola $P = P_{max}$.

To ensure safe operation of a circuit, the operating point and the various signals within the circuit must be chosen to ensure that the device remains within the allowed regions of the characteristic. Clearly, the output voltage cannot exceed the supply voltage V_{CC} and, to maintain linearity, the device should not be operated in the saturation region. Therefore, for a maximum output voltage range the operating point should be positioned halfway between V_{CC} and the saturation region. For maximum collector current swing the operating point should be chosen at half $I_{C(max)}$. To satisfy both these requirements, the operating point would be positioned close to the centre of the allowable region, as shown in Figure 19.19. In many applications neither maximum voltage range nor maximum current swing are required. In such cases, the operating point may be chosen to simplify circuit design, providing that adequate voltage and current ranges are produced.

The current and voltage corresponding to the operating point are the **quiescent collector current** and the **quiescent output voltage** of the circuit, respectively. Small-signal input voltages to the amplifier will cause corresponding small-signal changes in the base current, causing the circuit to traverse the load line on either side of the operating point.

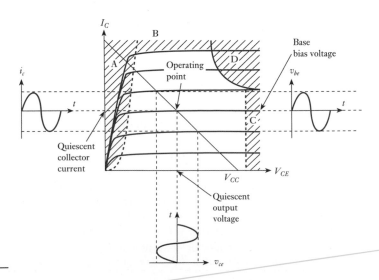

Figure 19.19
Choice of operating
point in a bipolar
transistor amplifier.

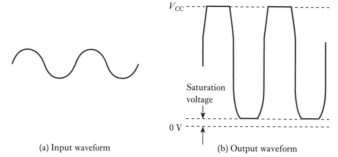

Figure 19.20
Clipping of a sinusoidal
signal.

(a) Input waveform   (b) Output waveform

In normal operation, the input signal should be limited to ensure that the transistor remains within its linear region. If the input is increased so that the device leaves this region, the output waveform will be distorted. This is illustrated in Figure 19.20, which shows the effects of applying an excessive sinusoidal input signal to an amplifier. The output voltage is **clipped** as it attempts to exceed the supply voltage and drives the transistor into its saturation region. This clipping is a form of **amplitude distortion**. When driven into saturation, the collector voltage drops to its **saturation voltage** and then remains relatively constant as the base current is increased further. When in this saturated state, the collector current is determined predominantly by the external circuit. A typical value for the saturation voltage of a small general-purpose silicon transistor would be 0.2 V. Note that this value is less than the voltage between the base and the emitter V_{BE}, so the collector is negative with respect to the base and the base–collector junction is forward biased by a small amount.

19.6.5 Device variability

Bipolar transistors, in common with almost all active devices, suffer from considerable device variability during their production. Devices from the same batch of components may have parameters that differ by several hundred per cent. This leads to considerable problems in the design of circuits using these components and greatly limits the usefulness of the simple amplifier circuits discussed so far.

To illustrate this point, let us consider the simple amplifier in Examples 19.1 and 19.2 when used with a typical small-signal bipolar transistor. In the examples we assumed that $h_{FE} = 100$. Typical general-purpose transistors of a given type might have a spread of current gain in the range 80 to 350. If we repeat the calculations performed in the examples for different values of current gain, we get a range of results, as illustrated in Table 19.1.

From the table of results, it is clear that the performance of the amplifier is greatly affected by the current gain of the transistor used. You will notice that the characteristics

Table 19.1 The effects of device variability on amplifier characteristics.

Current gain	Amplifier characteristics		
	Quiescent collector current (mA)	Quiescent output voltage (V)	Voltage gain
80	0.82	6.2	−147
100	1.02	5.2	−183
150	1.53	2.8	−275
200	2.04	0.4	−366

are only tabulated for values of current gain up to 200, even though the device may have a current gain up to 350. The reason for this will become clear shortly. If we look first at the quiescent collector current, we see that this rises linearly with the current gain. This is because the base current I_B is fixed and the collector current is simply the current gain times the base current. As the collector current rises, the quiescent output voltage drops. You will remember that this is given by the expression

$$V_o = V_{CC} - I_C R_C$$

As I_C increases, the voltage drop across the load resistor increases and the output voltage falls. From the previous section we know that the available output swing of the amplifier is limited by the distance of the operating point from the supply voltage and the saturation region. With a device with a current gain of 100, the operating point is positioned midway between these two limits. The quiescent output voltage is 5.2 V, allowing it to swing by nearly 5 V in either direction. When a device with a current gain of 80 is used, the quiescent output voltage rises to 6.2 V. This allows the output to go positive by only a little under 4 V before it is limited by the supply rail. If the transistor has a current gain of 150, the quiescent output voltage is only 2.8 V, permitting a maximum negative-going excursion of only about 2.5 V before the device enters saturation. The extreme case is reached with a device with a current gain of 200. Here, the quiescent output voltage is only 0.4 V, which is at the edge of the non-linear saturation region, allowing no useful output swing in the negative direction. For values of current gain above 200, the transistor is driven hard into saturation and the circuit cannot be used as a linear amplifier.

One apparent solution to these problems would be to measure the current gain of the transistor to be used and design the circuit appropriately. This is not an attractive solution in a mass-production environment in which it would be totally impractical to design each circuit uniquely to match individual components. It would also cause severe problems when a device failed and needed to be replaced. An alternative would be to select devices within a particular close range of parameters. This is possible but expensive, particularly if a very narrow spread is required.

The only practical solution to this problem is to design circuits that are not greatly affected by changes in the current gain of the transistors within them. Fortunately, we have already looked at techniques to achieve this objective – namely, the use of feedback.

Video 19C

19.6.6 The use of feedback

We have seen that amplifiers based on bipolar transistors suffer from problems associated with the variability of their gain. In Chapter 15 we saw how negative feedback could be used to overcome the problems associated with the variability of gain in amplifier circuits, and in Chapter 16 we looked at the use of such techniques with operational amplifiers. We also noted that this leads to circuits that are easier to design and understand. In Chapter 18 we discussed the use of feedback in overcoming variability in the characteristics of FETs and saw how this simplifies the design of biasing networks. It will perhaps come as no surprise to find that feedback can also be used in circuits using bipolar transistors and you will be pleased to hear that here too it produces simpler circuits.

An amplifier using negative feedback

Consider the circuit of Figure 19.21(a).

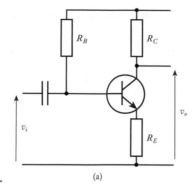 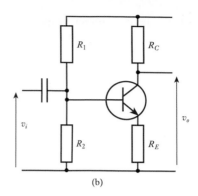

Figure 19.21
Amplifiers with
negative feedback.

You will notice that an emitter resistor has been added. The voltage across this resistor is clearly proportional to the emitter current, which is (almost exactly) equal to the collector current. As the input voltage is applied to the base, the voltage across the emitter resistor is effectively subtracted from the input to produce the voltage that is applied to the transistor. Therefore, we have **negative feedback** with a voltage proportional to the output current being subtracted from the input. From the discussions in Section 15.6.3 it is clear that this will improve the constancy of the output current by *increasing the output resistance* and will improve the performance of the circuit as a voltage amplifier by *increasing the input resistance*. The voltage gain will also be *stabilised*, making it affected less by variations in the current gain of the device.

As the input is no longer applied directly between the base and the emitter, it is no longer appropriate to describe this circuit as a common-emitter amplifier. A more accurate description would be a **series feedback amplifier**, although the term does not describe this circuit uniquely.

A further development of this circuit is shown in Figure 19.21(b). Here, *two* resistors are used to provide base bias rather than the single resistor used in the earlier circuit. This arrangement produces good stabilisation of the DC operating conditions of the circuit and also gives an arrangement that is very easy to analyse. To illustrate this point, let us consider both the large-signal (DC) and small-signal (AC) characteristics of the circuit in Figure 19.21(b). It is normally the designer's task to choose component values to obtain a given set of characteristics rather than to analyse an existing circuit. However, we will first look at the behaviour of a given circuit, then turn our attention to the problem of designing for a given specification.

DC analysis of a negative feedback amplifier

In performing the DC analysis of the circuit, it is useful to make two simplifying assumptions. These are that:

- the DC gain h_{FE} is large and therefore, in normal operation, the base current I_B can be neglected;
- the DC base–emitter voltage V_{BE} is constant. Here we will assume that it has a value of 0.7 V.

We will discuss the validity of these assumptions later.

In analysing the DC characteristics of a circuit, the two quantities of greatest interest are the quiescent output voltage and quiescent collector current. However, it is easier in this case to start by determining values for the quiescent base and emitter voltages as these lead directly to the quantities required.

Example 19.3 Determine the quiescent voltages and currents in the following feedback circuit.

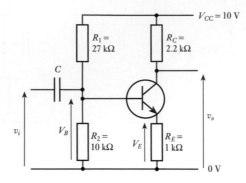

Quiescent base voltage

If we assume that the base current is negligible, as no constant current can flow through the input capacitor, the **quiescent base voltage** is determined simply by the supply voltage V_{CC} and potential divider formed by R_1 and R_2. Hence

$$V_B \approx V_{CC} \frac{R_2}{R_1 + R_2}$$

Therefore in our example

$$V_B \approx 10 \frac{10\ \text{k}\Omega}{27\ \text{k}\Omega + 10\ \text{k}\Omega} \approx 2.7\ \text{V}$$

Quiescent emitter voltage

As the base-to-emitter voltage V_{BE} is assumed to be constant, it is simple to determine the emitter voltage from the base voltage. Thus, the **quiescent emitter voltage** is simply

$$V_E = V_B - V_{BE}$$

and in our circuit

$$V_E = 2.7 - 0.7 = 2.0\,\text{V}$$

Quiescent emitter current

Knowing the voltage across the emitter resistor and its value gives us the emitter current

$$I_E = \frac{V_E}{R_E}$$

and therefore

$$I_E = \frac{2.0\ \text{V}}{1\ \text{k}\Omega} = 2\ \text{mA}$$

Quiescent collector current

If the base current is negligible, it follows that the collector current is equal to the emitter current

$$I_C \approx I_E$$

Therefore in our circuit

$$I_C \approx I_E = 2 \text{ mA}$$

Quiescent collector (output) voltage

In this circuit the output voltage is simply the collector voltage. This is determined by the supply voltage V_{CC} and voltage across the collector resistor R_C. The voltage across R_C is simply the product of its resistance and the collector current, so

$$V_{o(quiescent)} = V_C = V_{CC} - I_C R_C$$

In this case

$$V_{o(quiescent)} = 10 \text{ V} - 2 \text{ mA} \times 2.2 \ \Omega = 5.6 \text{ V}$$

It is clear from Example 19.3 that the DC conditions of such circuits can be determined quickly and easily. It is interesting to note that at no time in the analysis did we need to know the value of the current gain of the transistor. The fact that we have determined the quiescent conditions of the circuit without a knowledge of the current gain indicates that the gain does not directly affect the DC operation of the circuit. However, we did have to make some assumptions in order to perform the analysis in this way and it is perhaps useful at this point to look back at these assumptions in the light of the results obtained to see if they appear justifiable.

The first assumption made was that the DC gain h_{FE} is large and that the base current may therefore be neglected. This assumption is required at two stages within the analysis. The first is when determining the base voltage V_B. By neglecting the effects of I_B, the determination of V_B becomes trivial and independent of the actual value of I_B. This assumption would seem justifiable, provided that the current flowing through the base resistors R_1 and R_2 is large compared with the base current. If we look at the actual circuit values in this example, we see that the current I_{bias} flowing through the potential divider formed by R_1 and R_2 is given by

$$I_{bias} = \frac{V_{CC}}{(R_1 + R_2)} = \frac{10 \text{ V}}{(27 \text{ k}\Omega + 10 \text{ k}\Omega)} \approx 270 \ \mu\text{A}$$

The base current is given by I_E / h_{FE} and therefore varies depending on the current gain of the transistor used. For a typical general-purpose transistor, h_{FE} might be in the range 80 to 350 and, as I_E was calculated to be 2 mA, this gives a range for I_B of approximately 6 to 25 μA. Thus, the base current will be small compared with the current through the potential divider for all values of the current gain within the given range and the base current can safely be ignored.

The other stage of the analysis that required the assumption that the base current could be neglected was when it was assumed that the collector current was approximately equal to the emitter current. Again, for typical values for h_{FE} of between 80 and 350, this assumption is reasonable.

The second assumption made in the analysis was that the base–emitter voltage V_{BE} was constant. This assumption was used when determining the emitter voltage from the base voltage. From Figure 19.6, we know that, in practice, V_{BE} is not constant but varies with the bias current. However, the figure also shows that the voltage is approximately constant for small fluctuations in the base current. Provided the variation of V_{BE} is small compared with the magnitude of the base voltage it will have little effect on the value of the emitter voltage. In our example, V_B is about 2.7 V, which is large compared with likely fluctuations of V_{BE} and the assumption is therefore reasonable.

We have looked at our two simplifying assumptions with respect to the circuit used in the example and find that they are justified. In other circuits using different component values they may not be and a more complicated analysis would be required. Generally, when designing circuits of this form we aim to ensure that these assumptions are valid, because if they are not, the effects of the negative feedback will not be fully utilised. In this circuit, this means that the values of R_1 and R_2 should be chosen to ensure that the current in the potential divider chain is large compared with the base current, and V_B should be chosen to be large compared with likely fluctuations of V_{BE}. In meeting these requirements, there may well be other factors to be considered, such as the AC characteristics described below. Inevitably design is a compromise between a number of conflicting requirements.

File 19B

> ### Computer simulation exercise 19.2
>
> Simulate the circuit in Example 19.3 using an appropriate bipolar transistor (for example, a 2N2222) and a coupling capacitor of 1 μF. Measure the quiescent voltages and currents in the circuit and compare these with the values calculated in the example.

AC analysis of a series feedback amplifier

As before, the analysis of the circuit can be simplified if we make some assumptions. These are that:

- the small–signal current gain h_{fe} is large, so the small–signal base current i_b may be neglected;
- the large–signal base-to-emitter voltage V_{BE} is approximately constant, so the small-signal base-to-emitter voltage v_{be} is very small.

It can be seen that these assumptions are directly equivalent to those made in the DC analysis earlier.

Clearly, a small–signal equivalent circuit can be constructed and used to determine the AC characteristics of the circuit. However, even without the equivalent circuit it is fairly easy to determine the small–signal gain of the arrangement, just using our simplifying assumptions. Example 19.4 shows such an analysis, while Example 19.5 shows the use of an equivalent circuit to derive additional information about the AC behaviour of the circuit.

Example 19.4 Determine the small-signal behaviour of the following feedback circuit.

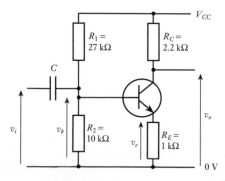

This circuit is identical to that in Example 19.3, but we are now considering the small-signal (or AC) behaviour.

Small-signal voltage gain

From the circuit diagram it is clear that the input signal is applied to the base of the transistor through the coupling capacitor C. Normally C would be chosen to have negligible impedance at the frequencies of interest and so it can be ignored. We will discuss later how to determine the effects of C if these are not negligible, but for the moment we will assume that the base voltage v_b is equal to the input voltage v_i.

From the second of our assumptions, v_{be} is considered to be very small, so the *small-signal* voltage on the emitter is effectively equal to that on the base. That is,

$$v_e \approx v_b \approx v_i \tag{19.19}$$

Now, from Ohm's law we know that

$$i_e = \frac{v_e}{R_E}$$

and as

$$i_c \approx i_e$$

it follows that

$$v_o = -i_c R_C \approx -i_e R_C = -\frac{v_e}{R_E} R_C$$

where the minus sign reflects the fact that the output voltage goes down when the current increases. If you expected to see V_{CC} in this expression, remember that the supply rail has no small-signal voltages on it.

Substituting from Equation 19.19 gives

$$v_o = -v_e \frac{R_C}{R_E} \approx -v_i \frac{R_C}{R_E}$$

and so the voltage gain of the circuit is given by

$$\text{voltage gain} = \frac{v_o}{v_i} \approx -\frac{R_C}{R_E} \tag{19.20}$$

For the component values used this gives

$$\text{voltage gain} \approx -\frac{2.2 \text{ k}\Omega}{1.2 \text{ k}\Omega} \approx -2.2$$

We therefore have a very simple expression for the voltage gain of the circuit that relies only on the values of the passive components. It is interesting to compare the results of Equation 19.20 with the expression derived for the gain of a simple common-emitter amplifier given in Equation 19.17, from which we have

$$\text{voltage gain (common emitter)} \approx -\frac{R_C}{r_e}$$

Thus, *with* feedback the gain is approximately $-R_C/R_E$, and *without* feedback it is approximately $-R_C/r_e$. An important difference between these two arrangements is that with feedback the gain is determined by the ratio of two stable and well-defined passive components. Without feedback, the gain is controlled by r_e which varies with the transistor's operating conditions.

File 19C

> ### Computer simulation exercise 19.3
>
> Simulate the circuit of Example 19.4 using an appropriate bipolar transistor (for example, a 2N2222) and a coupling capacitor of 1 µF. Apply an input voltage of 50 mV peak at 1 kHz. Use transient analysis to measure the voltage gain of the circuit. Compare the measured value with that calculated in the example.

Example 19.5

Determine the small-signal behaviour of the circuit in Example 19.4 using a small-signal equivalent circuit.

For a more detailed view of the AC characteristics of the circuit, we turn again to a small-signal equivalent circuit.

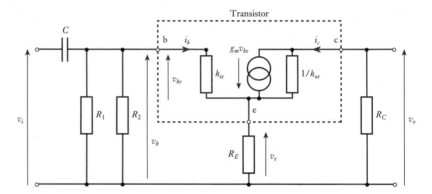

The presence of the emitter resistor makes the equivalent circuit more complicated than that given in Figure 19.16. The emitter is connected to ground through the emitter resistor R_E, but the base resistors R_1 and R_2 and the collector resistor R_C are shown connected to ground as these resistors are connected either to ground or to V_{CC} (which is at ground potential for AC signals). The small-signal voltage gain and the input and output resistance can be calculated directly from the equivalent circuit as follows.

Small-signal voltage gain

The analysis of Example 19.4 can be performed with reference to the equivalent circuit, producing an identical result. The equivalent circuit perhaps illustrates more clearly why the voltage gain of the amplifier is negative. If one ignores the effects of i_b and considers a current flowing from the current source, through R_E and then back through R_C, it is clear that the magnitude of the current flowing in each resistor is equal. However, as the current is flowing in opposite directions in the two resistors, the polarity of the voltage across each will be reversed. As the voltage across R_E is approximately equal to the input voltage, the output voltage will be inverted with respect to the input. Also, as the same current flows through both R_C and R_E, it is logical that the voltage gain will be the ratio of the resistor values. Therefore

$$\text{voltage gain} = \frac{v_o}{v_i} \approx -\frac{R_C}{R_E}$$

as before.

From this expression for the voltage gain of the amplifier it would seem that using a very small value for R_E would produce a very high gain. Taking this to its logical conclusion, we might expect that taking a vanishingly small value for R_E would result in an extremely high gain circuit. A moment's thought should make it clear that the gain of the resulting circuit cannot exceed that of the common–emitter amplifier of Figure 19.7, which is simply $-g_m R_C$, so a more complete expression for the gain is required.

If we ignore the effects of the output resistance $1/h_{oe}$ (this is normally large compared with R_E and R_C) and the base current (which is normally small compared with I_E) then the output voltage is given by

$$v_o = -g_m v_{be} R_C$$

as the current in R_C is equal to that produced by the current generator. The negative sign here reflects the fact that a positive current from the current generator produces a negative output voltage.

The input voltage to the amplifier is equal to the sum of the voltages across the base–emitter junction v_{be} and the emitter resistor v_e. Therefore

$$v_i = v_{be} + v_e$$

If we again ignore the effects of h_{oe} and i_b, then v_e is given by

$$v_e = g_m v_{be} R_E$$

and combining this with the previous expression gives

$$v_i = v_{be} + g_m v_{be} R_E$$

This may be rearranged to give

$$v_{be} = \frac{v_i}{1 + g_m R_E}$$

and combining this with the expression above for v_o gives

$$v_o = -\frac{g_m v_i R_C}{1 + g_m R_E}$$

Thus, the voltage gain of the amplifier is given by

$$\text{voltage gain} = \frac{v_o}{v_i} = -\frac{g_m R_C}{1 + g_m R_E}$$
$$= -\frac{R_C}{R_E + 1/g_m}$$

It can now be seen that, if R_E is zero, the gain becomes equal to $-g_m R_C$, as for the common-emitter amplifier. However, when R_E is much greater than $1/g_m$ the gain tends to $-R_C/R_E$, as before.

Note that $1/g_m$ is equal to r_e, so the gain may be written as

$$\text{voltage gain} = -\frac{R_C}{R_E + r_e}$$

and the gain is approximately equal to $-R_C/R_E$ when $R_E \gg r_e$ and equal to $-R_C/r_e$ in the special case where $R_E = 0$ (when we have the common-emitter circuit).

Small-signal input resistance

From the equivalent circuit, it is apparent that the input resistance is formed by the parallel combination of R_1, R_2 and the resistance seen looking into the base of the transistor. This last term is not simply the sum of h_{ie} and R_E because of the effect of the current generator. When a current i_b enters the base of the transistor, the current that flows through R_E is the sum of the base current and the collector current. As the collector current is equal to $h_{fe}i_b$, the emitter current is given by

$$i_e = i_b + h_{fe}i_b = (h_{fe} + 1)i_b$$

When a current flows through a resistor, a voltage drop is produced and the ratio of the voltage to the current determines the resistance of the component. When i_b flows into the base of the transistor, a much greater current flows in the emitter resistor, producing a proportionately larger voltage drop. Therefore, the emitter resistor *appears* much larger when viewed from the base. In fact, the emitter resistor appears to be increased by a factor of $(h_{fe} + 1)$. The input resistance h_{ie} is not amplified in this way as the current passing through this is simply i_b. Therefore, the **effective input resistance** seen looking into the base of the transistor is

$$r_b = h_{ie} + (h_{fe} + 1)R_E \tag{19.21}$$

and the input resistance of the amplifier is

$$r_i = R_1 // R_2 // r_b \tag{19.22}$$

It is interesting to look at the relative magnitudes of the three components of the input resistance, as defined in Equation 19.22. R_1 and R_2 are base-bias resistors and we noted when considering the DC performance of the circuit that these should be chosen such that the current flowing through them is large compared with the base current. This limits their maximum values, which typically might be a few kilohms or a few tens of kilohms. The magnitude of r_b is given in Equation 19.21. We know from Section 19.5.2 that a typical value for h_{ie} might be a few kilohms. As R_E will also typically be a few kilohms and h_{fe} will be perhaps 80 to 350, it would be reasonable to use the approximation

$$r_b = h_{ie} + (h_{fe} + 1)R_E \approx h_{fe}R_E$$

From this it can be seen that r_b will generally be several hundred kilohms and will usually be large compared with R_1 and R_2. Therefore, as the three resistors are in parallel, the effects of r_b will often be negligible and it will be the parallel combination of R_1 and R_2 that will determine the input resistance of the amplifier.

If this is so

$$r_i \approx R_1 // R_2$$

and in this circuit

$$r_i \approx 27 \text{ k}\Omega // 10 \text{ k}\Omega \approx 7.3 \text{ k}\Omega$$

From this it is clear that, to achieve a high input resistance, it is desirable to make R_1 and R_2 as high as possible. This is in conflict with the biasing considerations discussed earlier, as it is necessary to ensure that the current flowing through R_1 and R_2 is large compared with the base current to ensure that the effects of the latter may be ignored. These opposing requirements are resolved by a compromise, where R_2 is typically chosen to be about 10 times the value of R_E.

An advantage of this circuit – in comparison with the simple common-emitter amplifier – is that the input resistance is determined by the passive components within the circuit rather than by the transistor. This makes the circuit much more predictable and less affected by the characteristics of the active device used.

Small-signal output resistance

In Example 19.2 we considered the output resistance of a simple common-emitter amplifier and deduced that this was equal to the parallel combination of R_C and $1/h_{oe}$. The addition of the emitter resistor R_E in the series feedback amplifier places an extra resistance in series with $1/h_{oe}$, as shown in the equivalent circuit. The addition of this resistance makes calculation of the output resistance somewhat more complicated. However, as $1/h_{oe}$ is normally much greater than R_C, it follows that $1/h_{oe}$ in series with an additional resistance will also be greater than R_C and the resistance of the parallel combination is dominated by R_C. Therefore, in this example

$$r_o \approx R_C = 2.2 \text{ k}\Omega$$

As with the input resistance, the output resistance is determined by the passive components within the circuit rather than the transistor. This improves the predictability of the circuit by reducing its dependence on device characteristics.

Therefore, for the series feedback amplifier in Examples 19.3, 19.4 and 19.5, our analysis suggests that

$$r_i \approx R_1 // R_2$$

$$r_o \approx R_C$$

$$\frac{v_o}{v_i} \approx -\frac{R_C}{R_E}$$

In the light of the results of these examples, we can now re-examine the assumptions made and see if they appear appropriate for the AC analysis of the circuit.

The assumption that h_{fe} is large is used in determining the small-signal voltage gain of the arrangement. Here, we assumed that the current flowing in R_E is the same as that flowing in R_C. For a general-purpose device with a gain of between 80 and 350 this assumption would seem valid.

The second assumption, that v_{be} is very small, was also used in determining the gain, where it was assumed that the small-signal voltage across the emitter resistor v_e was equal to the input voltage v_i. To investigate this assertion it is useful to consider the derivation of the input resistance of the circuit given in Example 19.5. Here it was observed that the input of the circuit appears as if it were the input resistance of the transistor h_{ie} in series with a resistance that is approximately h_{fe} times the emitter resistance R_E. This is illustrated in Figure 19.22.

The two resistors form a potential divider and the voltage v_e is given by the expression

$$v_e = v_b \frac{h_{fe}R_E}{h_{ie} + h_{fe}R_E}$$

and the assumption that v_{be} may be neglected is valid provided that $h_{fe}R_E \gg h_{ie}$.

In our example R_E is 1.0 kΩ, so $h_{fe} \times R_E$ will be in the range 80 to 350 kΩ. This will always be much greater than h_{ie} which will be of the order of 1 kΩ. Therefore, the effects of h_{ie} can be neglected and

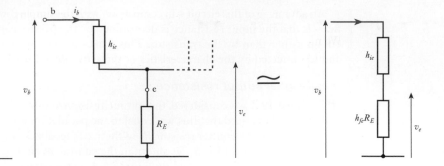

Figure 19.22
An equivalent circuit for the input of the feedback amplifier.

$$v_e = v_b \frac{h_{fe}R_E}{h_{ie} + h_{fe}R_E} \approx v_b \frac{h_{fe}R_E}{h_{fe}R_E} \approx v_b$$

Having looked at the large-signal and small-signal characteristics of the series feedback amplifier, we have found that the circuit can be analysed very easily provided that a few simplifying assumptions are made. The assumptions required are effectively the same for both the DC and AC cases and may be summarised as follows:

- the current gains h_{FE} and h_{fe} are high and therefore the base currents I_B and i_b may be neglected;
- the steady-state base–emitter voltage V_{BE} is approximately constant (at say $0.7\,\text{V}$) and therefore the small-signal base–emitter voltage v_{be} is very small.

It should be remembered that these assumptions will not always be justified. It is up to the designer to verify at the end of the design process that the approximations made are appropriate.

File 19D

Computer simulation exercise 19.4

Simulate the circuit of Example 19.3 using an appropriate bipolar transistor (for example, a 2N2222) and a coupling capacitor of 1 μF. Apply an input voltage of 50 mV peak at 1 kHz. Measure the small-signal input and output resistance of the circuit and compare these with the predicted values.

Hint: measure the input resistance by inserting a resistor in series with the input and comparing the *small-signal* voltage across this resistor with the voltage across the input to the amplifier. Measure the output resistance by connecting resistors of different values from the output to ground and measuring the change in the *small-signal* output voltage.

The effects of a load resistor

As with the simple common-emitter amplifier discussed in Section 19.6.2, applying a load to the series feedback amplifier reduces its gain. In the common-emitter amplifier, we noted that the gain was reduced from $-R_C/r_e$ to $-(R_C//R_L)/r_e$ and in the feedback amplifier it is reduced from $-R_C/R_E$ to $-(R_C//R_L)/R_E$. As before, if R_L is large compared with R_C, then the loading effects will be small.

The effects of coupling capacitors

We noted earlier that if **coupling capacitors** are used at the input or at the output their values are normally chosen so that their effects are negligible at the frequencies of interest. Coupling capacitors are used to prevent external circuitry from upsetting the biasing of the transistor and we noted in Section 8.6 that their use produces a **low-frequency cut-off**. The frequency of this cut-off was shown to be given by

$$f_c = \frac{1}{2\pi CR}$$

where C is the value of the coupling capacitor and R is the resistance in series with this capacitance. For the coupling capacitor used at the input of an amplifier, this resistance will be the sum of the input resistance of the amplifier and the source resistance. For the coupling capacitor used at the output, it will be the sum of the output resistance and the load resistance. It is worthy of note that in this case the use of negative feedback increases the input resistance of the amplifier, reducing the size of the capacitor required to provide a given lower cut-off frequency.

File 19E

Computer simulation exercise 19.5

Having estimated the input resistance of the circuit used in Computer simulation exercise 19.4, calculate the low-frequency cut-off produced by the coupling capacitor, assuming that it is fed from a low-resistance source. Use an AC sweep to measure the frequency response of the amplifier and compare the cut-off frequency with the calculated value.

Now consider the use of a coupling capacitor of 100 nF to couple the output of the amplifier to a load of 10 kΩ. Calculate the effect of this arrangement on the mid-band gain of the amplifier and the frequency of the low-frequency cut-off produced. Now simulate this arrangement and compare the values obtained with those you have predicted.

Design of an amplifier to meet a given specification

So far, we have looked at the analysis of existing circuits and used this to increase our understanding of the operation of these circuits. We now turn our attention to the process of designing an amplifier for a given task. The circuit used in Examples 19.3 to 19.5 gives us a good blueprint for a design, but we need to choose component values to suit our purpose. The process involved in this choice will vary from one application to another, as the specification will differ in terms of the parameters that are specified. In some cases, the supply voltage and quiescent output voltage may be given, while in others the quiescent current may be of importance. An example of the design process is given in Example 19.6.

Example 19.6 Design a single-stage amplifier with a small-signal voltage gain of −4 and a maximum output swing of 10 V peak to peak (when used with a high-impedance load) that operates from a 15 V supply line. The amplifier should be AC coupled, but have a gain that is approximately constant down to 100 Hz.

We will use an amplifier with an emitter resistor and potential divider biasing as in earlier examples.

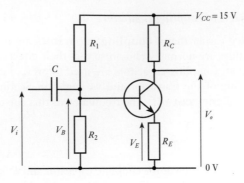

Designing circuits is not a precise art. There is no unique ideal solution and often we use rules of thumb to simplify component choice.

Quiescent output voltage and collector current

One of our first tasks is to decide on appropriate values for the quiescent output voltage and the quiescent collector current. The first of these is constrained by the relatively large required output swing. In order to produce 10 V peak-to-peak output, it must be able to go above and below its quiescent value by at least 5 V. In order to leave a reasonable voltage across the emitter resistor (to increase stability), let us choose to make the quiescent output voltage about 5.5 V below V_{CC}, therefore

$$V_{C(quiescent)} \approx V_{CC} - 5.5 = 9.5 \text{ V}$$

The choice of the quiescent collector current is fairly arbitrary, as the load impedance is high. Let us choose a value of 1 mA.

This immediately allows us to calculate an appropriate value for R_C as

$$V_{C(quiescent)} = V_{CC} - I_{C(quiescent)}R_C$$

and thus

$$R_C = \frac{V_{CC} - V_{C(quiescent)}}{I_{C(quiescent)}} = \frac{15.0 \text{ V} - 9.5 \text{ V}}{1 \text{ mA}} = 5.5 \text{ k}\Omega$$

As 5.5 kΩ is not a standard resistor value and the value is not critical, we choose the closest standard value, which is 5.6 kΩ. Therefore

$$R_C = 5.6 \text{ k}\Omega$$

Small-signal voltage gain

From our discussions in Example 19.5 we know that

$$\text{voltage gain} = -\frac{R_C}{R_E + r_e}$$

Therefore, in order that the gain is determined by the passive components in the circuit we require $R_E \gg r_e$ and, in order to obtain a gain of −4, we require $R_C = 4R_E$.

With a quiescent collector current of 1 mA, the emitter current is also 1 mA and the emitter resistance is given by

$$r_e \approx \frac{1}{40I_E} \approx \frac{1}{40 \times 10^{-3}} \approx 25 \ \Omega$$

If R_C is equal to 5.6 kΩ then R_E should be 5.6/4 = 1.4 kΩ to give a gain of −4. This value satisfies the condition that $R_E \gg r_e$.

Again, 1.4 kΩ is not a standard resistor value. If it is vital that the gain is very close to −4, this value could easily be achieved by combining two high-tolerance resistors of appropriate values. Alternatively, we choose the closest standard value and accept that the actual gain will be slightly above or below the specified value. In this case, let us choose R_E = 1.3 kΩ. This gives a value for the gain of

$$\text{voltage gain} = -\frac{R_C}{R_E + r_e} \approx -\frac{5.6 \text{ k}\Omega}{1.3 \text{ k}\Omega + 25 \text{ k}\Omega} \approx -4.2$$

Base-bias resistors

If the quiescent emitter current is to be about 1 mA and R_E is 1.3 kΩ, the quiescent emitter voltage must be given by

$$V_{E(quiescent)} = I_{E(quiescent)} \times R_E = 10^{-3} \times 1.3 \times 10^3 = 1.3 \text{ V}$$

To achieve this emitter voltage the base must be biased to 1.3 + 0.7 = 2.0 V. The ratio of R_1 to R_2 must therefore be determined by the relationship

$$\frac{R_2}{R_1 + R_2} V_{CC} = 2.0 \text{ V}$$

Choice of the absolute values of the base resistors is a compromise between high values, which give a high input resistance to the circuit, and low values, which make the current through the bias resistors large compared with the base current. As mentioned earlier, a common rule of thumb solution is to choose R_2 as approximately 10 times R_E. Therefore, R_2 becomes 13 kΩ. Rearranging the above expression, we have

$$R_1 = \frac{R_2(V_{CC} - 2.0)}{2.0} = -\frac{13 \text{ k}\Omega(15.0 - 2.0)}{2.0} = 84.5 \text{ k}\Omega$$

The nearest standard value for R_1 is 82 kΩ. The use of this value raises the base voltage slightly above 2.0 V, which in turn increases the emitter current and reduces the quiescent collector voltage. Calculation of these values is left as an exercise for the reader.

Input resistance and the choice of C

From our earlier discussions, we know that the input resistance is given approximately by the parallel combination of R_1 and R_2. Therefore in this case

$$\text{input resistance} \approx R_1 /\!/ R_2 = 13 \text{ k}\Omega /\!/ 82 \text{ k}\Omega \approx 11.2 \text{ k}\Omega$$

From Section 8.6 we know that, if the effects of the source resistance are neglected, the presence of the coupling capacitor will produce a low-frequency cut-off at a frequency given by

$$f_c = \frac{1}{2\pi CR}$$

where R is the input resistance of the amplifier.

In this example, we require the gain to be approximately constant down to 100 Hz. We therefore choose a lower cut-off frequency of 100/10 = 10 Hz, which gives a value for C of

$$C = \frac{1}{2\pi f_c R} = \frac{1}{2 \times \pi \times 10 \times 11.2 \times 10^3} = 1.4\ \mu F$$

Therefore, a non-polarised capacitor of more than 1.4 μF would be used – for example, a 2.2 μF polyester type.

Thus our final design is as follows.

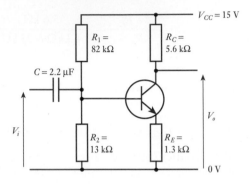

File 19F

Computer simulation exercise 19.6

Simulate the circuit designed in Example 19.6 and compare its performance with the original specification.

19.6.7 Use of a decoupling capacitor

A comparison of the results of Example 19.2, which looked at a simple common-emitter amplifier, and Example 19.4, which was concerned with a series feedback amplifier, indicates that the use of feedback has several advantages. These include the stabilisation of circuit parameters, such as voltage gain and input and output resistance, which become less affected by changes in the transistor's characteristics. However, this is achieved at the expense of a considerable reduction in voltage gain. This fall in gain is a direct result of the negative feedback incorporated. In some applications this reduction in gain is unacceptable and a **decoupling capacitor** is used to reduce the amount of AC negative feedback while maintaining DC feedback. This increases the small-signal gain of the circuit but does not affect the DC feedback, which provides stability to the bias conditions of the circuit.

The use of a decoupling capacitor is illustrated in Figure 19.23. The capacitor C_E is placed across the emitter resistor R_E, providing a low-impedance path for alternating signals from the emitter to ground, but having no effect on the steady bias voltages. As the emitter is now effectively connected to ground for small-signal inputs, we can accurately describe this amplifier as a **common-emitter amplifier** as the emitter is common to both the input and output circuits.

The decoupling capacitor does not change the DC performance of the circuit as the capacitor is effectively an infinite impedance at DC. Therefore, the quiescent conditions of the circuit are unaffected.

For AC signals, the capacitor represents a low impedance. Its value would normally be chosen so that, at the frequencies to be used by the circuit, the capacitor looks like a short circuit. A small-signal equivalent circuit for the amplifier of Figure 19.23 is shown in Figure 19.24.

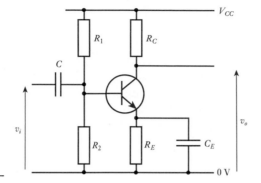

Figure 19.23
Use of a decoupling
capacitor.

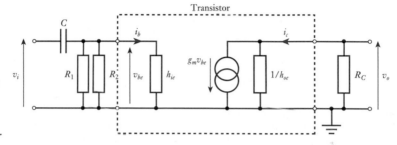

Figure 19.24
Small-signal equivalent
circuit of an amplifier
using a decoupling
capacitor.

The presence of the decoupling capacitor effectively removes R_E from the small-signal equivalent circuit, producing an arrangement that is almost identical to that of the simple common-emitter amplifier shown in Figure 19.16. The only difference between these two circuits is the number of resistors that connect the base to ground. It follows that the AC analysis of the two circuits is similar with regard to their voltage gain and input and output resistance. The differences in the calculations are simply that R_B in the analysis of the common-emitter amplifier is replaced by the resistance of the parallel combination of R_1 and R_2 when using the capacitively decoupled amplifier.

Although the decoupling capacitor will normally be chosen to provide a low impedance at the frequencies of interest, at low frequencies its effects will diminish and the small-signal gain will fall. The point at which this occurs will be determined by the frequency at which the impedance of the decoupling capacitor becomes appreciable compared with the resistance within the circuit across the capacitor – that is, between the emitter and ground. This resistance is given by the parallel combination of the emitter resistor R_E and the resistance seen looking into the emitter, r_e. The point at which the impedance of the capacitor matches the effective resistance across it determines the **low-frequency cut-off**. Therefore

$$f_{co} = \frac{1}{2\pi C_E (R_E // r_e)}$$

From Equation 19.8 we know that

$$r_e \approx \frac{1}{40 I_E}$$

It follows that this will have a typical value of a few ohms and will generally be much smaller than R_E. We can therefore approximate f_c using the expression

$$f_c = \frac{1}{2\pi C_E r_e}$$

(19.23)

This expression can be used to calculate the size of decoupling capacitor that will be required to produce a sufficiently low cut-off frequency for a given application. Note that the low value of r_e means that the capacitor must be considerably larger than would be required if the cut-off frequency were determined by the time constant $C_E R_E$. This can prove inconvenient in applications where good low-frequency response is required as large capacitors may be both expensive and bulky.

Below the cut-off frequency, the gain drops at 6 dB/octave, as described in Section 8.6, until the impedance of C_E becomes comparable with R_E. Below this point, the gain levels out as R_E tends to dominate the response. The frequency at which this occurs is given by

$$f_1 = \frac{1}{2\pi C_E R_E}$$

(19.24)

The resultant frequency response of the amplifier is shown in Figure 19.25, which also shows the equivalent responses for a simple common-emitter amplifier and series feedback amplifier, using the same transistor and component values (with appropriate biasing arrangements). For simplicity, the figure shows the responses of amplifiers that are *not* fitted with coupling capacitors. The addition of coupling capacitors to these circuits would introduce additional attenuation at low frequencies, determined by the size of the coupling capacitors used.

From Figure 19.25(a), it is apparent that the usable bandwidth of the simple common-emitter amplifier extends down to DC and is limited at its upper end by the frequency response of the transistor. The **mid-band gain** of the amplifier is approximately $-R_C/r_e$ and, as r_e is a function of I_E, will vary considerably with the h_{FE} of the transistor used.

Figure 19.25(b) shows the response of the series feedback amplifier with the response of the simple common-emitter amplifier shown as a dashed line for comparison. The response of the series feedback amplifier also extends down to DC, but the gain is considerably less than the common-emitter circuit at $-R_C/R_E$. The gain, however, is determined by stable and well-defined passive components rather than the characteristics of the transistor. The upper cut-off frequency of the amplifier is increased from f_2 to f_3 by the negative feedback and the bandwidth of the circuit is thus increased.

The response of the common-emitter amplifier using an emitter resistor and a decoupling capacitor is shown in Figure 19.25(c). This has a gain of $-R_C/R_E$ at low frequencies, which then rises to $-R_C/r_e$ as the effects of the decoupling capacitor come into play. This produces a low-frequency cut-off frequency at f_{co}. The high-frequency response would appear, at first sight, to be unaffected by the addition of the emitter resistor and decoupling capacitor. However, from Equation 19.8 we see that

$$r_e \approx \frac{1}{40 I_E}$$

Therefore, as the presence of the feedback resistor stabilises the emitter current, it also stabilises the gain of the amplifier.

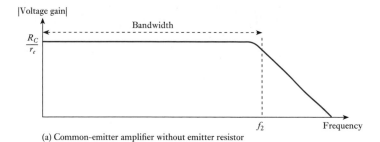

(a) Common–emitter amplifier without emitter resistor

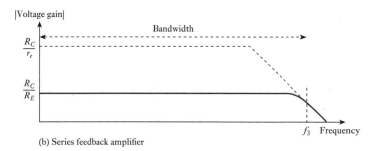

(b) Series feedback amplifier

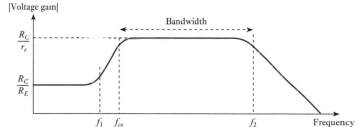

Figure 19.25
A comparison of the
frequency responses of
various amplifiers.

(c) Common–emitter amplifier with emitter resistor and decoupling capacitor

Example 19.7   **Perform a DC analysis on the following circuit.**

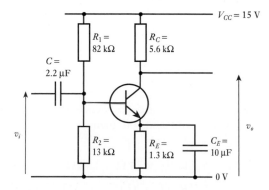

The circuit shown is similar to that designed in Example 19.6, but with the addition of a decoupling capacitor. As the capacitors may be considered to be open circuit for DC signals, we can analyse the circuit as if they were not present. Therefore, assuming that the gain of the transistor is reasonably high, so that the base current may be neglected, we have

$$V_B = V_{CC} \frac{R_2}{R_1 + R_2} = 15.0 \times \frac{13\ k\Omega}{13\ k\Omega + 82\ k\Omega} = 2.05\ V$$

$$V_E = V_B - V_{BE} = 2.05 - 0.7 = 1.35\ V$$

$$I_E = \frac{V_E}{R_E} = \frac{1.35\ V}{1.3\ k\Omega} = 1.04\ mA$$

$$I_C = I_E = 1.04\ mA$$

$$V_{o(quiescent)} = V_{CC} - I_C R_C = 15\ V - 1.04\ mA \times 5.6\ k\Omega = 9.2\ V$$

Example 19.8

Determine the small-signal voltage gain, input resistance and output resistance of the circuit in Example 19.7.

Small-signal voltage gain

The mid-band gain of the amplifier is approximately $-R_C/r_e$ where r_e is given by

$$r_e \approx \frac{1}{40 I_E} \approx \frac{1}{40 \times 1.04 \times 10^{-3}} \approx 24\ \Omega$$

Therefore

$$\text{small-signal voltage gain} \approx -\frac{R_C}{r_e} = -\frac{5.6\ k\Omega}{24\ \Omega} \approx -233$$

Small-signal input resistance

The input resistance of the circuit is given by the parallel combination of R_1, R_2 and the resistance seen looking into the base of the transistor. In the circuit in Example 19.6, the resistance seen looking into the base of the transistor is approximately equal to $h_{fe}R_E$, so it is normally so high that it may be ignored. However, the presence of the decoupling capacitor in the circuit in Example 19.7 removes the effect of R_E for small signals. The resistance seen looking into the base of the transistor is now h_{ie}, which is likely to be of similar magnitude to the parallel combination of R_1 and R_2, so cannot be ignored. From Equation 19.9, we know that

$$h_{ie} \approx h_{fe}r_e$$

In this case r_e is about 24 Ω, so h_{ie} will be a few kilohms (depending on the value of h_{fe}) and the input resistance of the complete amplifier r_i will be $R_1 // R_2 // h_{ie}$. For a given type of transistor, we can determine the range of h_{fe} from the data sheet and so deduce the likely range of h_{ie} and hence the input resistance. For example, if we know that h_{fe} is between 100 and 400, then we can compute a range for the input resistance. If $h_{fe} = 100$

$$h_{ie} \approx h_{fe}r_e = 100 \times 24\ \Omega = 2.4\ k\Omega$$

$$r_i = R_1 // R_2 // h_{ie} = 82\ k\Omega // 13\ k\Omega // 2.4\ k\Omega = 2.0\ k\Omega$$

If $h_{fe} = 400$

$$h_{ie} \approx h_{fe}r_e = 400 \times 24\ \Omega = 9.6\ k\Omega$$

$$r_i = R_1 // R_2 // h_{ie} = 82\ k\Omega // 13\ k\Omega // 9.6\ k\Omega = 5.2\ k\Omega$$

Thus, the small-signal input resistance for such a circuit, using such a transistor, would be in the range 2.0–5.2 kΩ.

Small-signal output resistance

The output resistance of the circuit is similar to that of a simple common-emitter amplifier, which we have previously shown to be approximately equal to R_C. Therefore

$$\text{small-signal output resistance} \approx R_C = 5.6 \text{ k}\Omega$$

Therefore, for the amplifier in Examples 19.7 and 19.8, our analysis suggests that

$$r_i \approx R_1 // R_2 // h_{ie}$$

$$r_o \approx R_C$$

$$\frac{v_o}{v_i} \approx -\frac{R_C}{r_e}$$

From Examples 19.7 and 19.8 it is clear that the use of a decoupling capacitor dramatically increases the gain of the circuit. It can also be seen that the DC conditions, voltage gain and output resistance of the resulting circuit are largely independent of the current gain of the transistor. However, the use of a decoupling capacitor reduces the input resistance of the circuit and makes it dependent on the characteristics of the transistor. We will see in the following example that the low-frequency characteristics of the circuit are also affected by the current gain of the transistor.

Example 19.9 | **Determine the low-frequency effects of the coupling and decoupling capacitors in the circuit of Example 19.7.**

The coupling capacitor C

As in Example 19.6, the coupling capacitor introduces a low-frequency cut-off given by

$$f_c = \frac{1}{2\pi CR}$$

where R is the input resistance of the amplifier. In this case, the input resistance is dependent on the current gain of the transistor. If, as in Example 19.8, we know that the input resistance is in the range 2.0–5.2 kΩ, then we can determine the range of the low-frequency cut-off.

If $R = 2.0$ kΩ

$$f_c = \frac{1}{2\pi CR} = \frac{1}{2 \times \pi \times 2.2 \times 10^{-6} \times 2.0 \times 10^3} = 36 \text{ Hz}$$

If $R = 5.2$ kΩ

$$f_c = \frac{1}{2\pi CR} = \frac{1}{2 \times \pi \times 2.2 \times 10^{-6} \times 5.2 \times 10^3} = 14 \text{ Hz}$$

We would therefore expect the low-frequency cut-off produced by the coupling capacitor to be in the range 14–36 Hz.

The decoupling capacitor C_E

From Equation 19.23, we have

$$f_c \approx \frac{1}{2\pi C_E r_e} = \frac{1}{2 \times \pi \times 10 \times 10^{-6} \times 24} = 663 \text{ Hz}$$

Note that although C_E is larger than C it produces a much higher low-frequency cut-off because of the low value of r_e. Circuits that are required to work at low frequencies require very large, and expensive, decoupling capacitors.

File 19G

> ### Computer simulation exercise 19.7
>
> Simulate the circuit of Example 19.7 using a suitable transistor, such as a 2N2222. Measure the DC and AC characteristics of this circuit and compare these with those predicted in Examples 19.7–19.9.

Split emitter resistors

A variant of the circuit in Figure 19.23 uses two resistors in series in place of R_E, with a decoupling capacitor connected across only the resistor connected to ground, as shown in Figure 19.26. This use of **split emitter resistors** allows the total emitter resistance to be tailored to suit the biasing requirements of the circuit, while permitting only part of this resistance to be decoupled to produce the required small-signal performance.

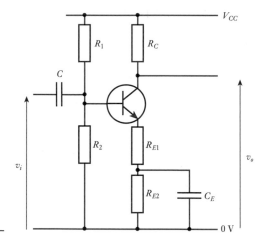

Figure 19.26
Use of split emitter resistors.

Decoupled and non-decoupled amplifiers

The use of an emitter resistor and a decoupling capacitor produces an amplifier that has a high gain for AC signals, but uses negative feedback to stabilise the DC operating conditions.

A disadvantage of this approach is that the arrangement has a frequency response that may be inconvenient in some situations (as shown in Figure 19.25(c)). Circuits that must operate at low frequencies will require very large decoupling capacitors to produce a sufficiently low cut-off frequency. In many applications, it is preferable to use several stages of amplification, relying on feedback, rather than a single stage that requires a large decoupling capacitor. In discrete circuits, transistors are much less expensive, and smaller, than large capacitors. In integrated circuits, transistors require much less chip area than small capacitors, while large capacitors are completely impractical.

It is very important to distinguish between the functions of coupling and decoupling capacitors:

- **Coupling capacitors** are used to couple one stage of a circuit to the next by passing the AC component of a signal while preventing the DC component from one stage affecting the biasing conditions of the next.
- **Decoupling capacitors** remove the AC component of a signal by shorting it to ground while leaving the DC component unchanged. They therefore decouple a node of the circuit from the AC signal.

19.6.8 Amplifier configurations

So far, we have concentrated on circuits in which the input is applied to the base of the transistor and the output is taken from the collector. The common-emitter amplifier is the simplest of these. It is also possible to construct circuits using other transistor configurations in which the input and output are applied to other nodes. Examples of these configurations are shown in Figure 19.27.

Of these arrangements, the common-emitter configuration is by far the most widely used, which is why the text so far has concentrated on such circuits. However, the other configurations are of interest as they have different characteristics that are utilised for particular applications.

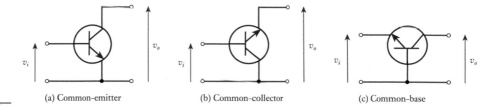

Figure 19.27
Transistor circuit configurations.

(a) Common-emitter   (b) Common-collector   (c) Common-base

Video 19D

Common-collector amplifiers

Figure 19.28 shows the circuit of a simple common-collector amplifier. The circuit is similar to the series feedback circuit used earlier, except that the output is taken from the emitter rather than the collector, eliminating the need for a collector resistor. The collector is connected directly to the positive supply, which is at earth potential for AC signals as there are no AC voltages between the supply and ground. Input signals are

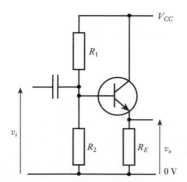

Figure 19.28
A common-collector amplifier.

applied between the base and ground, while output signals are measured between the emitter and ground. As the collector is at ground potential, for small signals the collector is common to both the input *and* output circuits, hence the arrangement is a **common-collector** amplifier.

As with the series feedback amplifier considered earlier, this circuit uses **negative feedback**. However, in this case, the voltage that is subtracted from the input is related to the output *voltage* rather than the output *current* as the emitter voltage *is* the output voltage of the emitter follower. From the discussion in Section 15.6.3 it is clear that this will *increase the input resistance*, as before, but *decrease the output resistance*. Analysis of the circuit is very much simpler than that of the series feedback amplifier and, having done the groundwork, it is easy to write down its characteristics.

The DC analysis of the circuit is similar to that of the series feedback amplifier, except that fewer steps are required. Once the quiescent base voltage has been determined, the quiescent emitter voltage is found by subtracting the (constant) base-to-emitter voltage V_{BE}, which gives the quiescent output voltage directly. The quiescent emitter current is found by dividing the emitter voltage by the value of the emitter resistor R_E.

The AC analysis is also straightforward. One of the assumptions made when looking at the series feedback amplifier was that the small-signal base-to-emitter voltage v_{be} was negligible. This implies that the emitter tracks the base with some constant offset voltage (typically about 0.7 V). Thus the voltage gain of the common-collector amplifier is approximately unity and the emitter simply follows the input signal. For this reason, this form of amplifier is often called an **emitter follower** amplifier. The small-signal input resistance is calculated in the same manner as for the series feedback amplifier (see Example 19.5) and thus

$$r_i = R_1 // R_2 // r_b$$

where r_b is the resistance seen looking into the base. As before, r_b is given by

$$r_b = h_{ie} + (h_{fe} + 1)R_E$$

so r_b is again likely to be large compared with R_1 and R_2 and its effects may be ignored. Therefore, as for the series feedback amplifier

$$r_i \approx R_1 // R_2 \tag{19.25}$$

The output resistance of the amplifier is given by the parallel combination of the emitter resistor R_E and the resistance seen looking back into the emitter r_e. From Equation 19.8, we know that

$$r_e \approx \frac{1}{40I_E}$$

which means that r_e will usually be of the order of a few ohms or tens of ohms and will dominate the output resistance of the circuit. Therefore

$$r_o \approx r_e \approx \frac{1}{40I_E} \tag{19.26}$$

Example 19.10 | Determine the quiescent output voltage, the small-signal voltage gain, the input resistance and the output resistance of the following circuit.

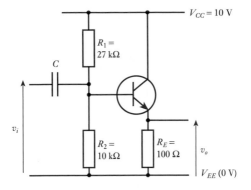

Quiescent output voltage

If we assume that the base current is negligible, as no constant current can flow through the input capacitor, the **quiescent base voltage** is determined simply by the supply voltage V_{CC} and by the potential divider formed by R_1 and R_2. Thus

$$V_B \approx V_{CC} \frac{R_2}{R_1 + R_2}$$

Therefore in this case

$$V_B \approx 10 \frac{10 \text{ k}\Omega}{27 \text{ k}\Omega + 10 \text{ k}\Omega} \approx 2.7 \text{ V}$$

The **quiescent emitter voltage** is then given by

$$V_E = V_B - V_{BE} = 2.7 - 0.7 = 2.0 \text{ V}$$

However, in this circuit V_E represents our output voltage, so the **quiescent output voltage** is given by

$$V_{o(quiescent)} = V_E = 2.0 \text{ V}$$

Small-signal voltage gain

As discussed earlier, the input signal is applied to the base of the transistor through the coupling capacitor C, and therefore

$$v_e \approx v_b \approx v_i$$

However, in this circuit v_e is the output voltage and as the output is equal to the input

$$\text{voltage gain} \approx 1$$

Input resistance

From Equation 19.25

$$r_i \approx R_1 // R_2$$
$$= 27 \text{ k}\Omega // 10 \text{ k}\Omega \approx 7.3 \text{ k}\Omega$$

Output resistance

From Equation 19.26

$$r_o \approx r_e \approx \frac{1}{40 I_E}$$

In this case, $I_E = V_E / R_E = 2.0\,\text{V} / 100\,\Omega = 20\,\text{mA}$. Therefore

$$r_o \approx \frac{1}{40 I_E} = \frac{1}{40 \times 20 \times 10^{-3}} = 1.25\,\Omega$$

Therefore, for the emitter follower amplifier of Example 19.10

$$r_i \approx R_1 // R_2$$

$$r_o \approx r_e$$

$$\frac{v_o}{v_i} \approx 1$$

The fact that the emitter follower has a voltage gain of approximately unity might at first sight appear to make it of little use. However, it is its relatively high input impedance and low output impedance that make it of interest as a **unity gain buffer amplifier**. You might like to compare the characteristics of the emitter follower with those of the source follower described in Section 18.6.8. In doing so, remember that in Section 19.5.4 we noted that for the bipolar transistor $r_e = 1/g_m$.

Common-base amplifiers

The common-base configuration is the least widely used transistor arrangement, but does have characteristics that make it of interest. Figure 19.29 shows a simple common-base arrangement.

The circuit is similar to the series feedback amplifier used earlier, except that the base is connected to ground through a capacitor, placing it at earth potential for AC signals. The input is applied between the emitter and ground and the output is measured between the collector and ground. As the base is at ground potential as far as small signals are concerned, the input and output are effectively measured with respect to the base and the base is thus common to the input and output circuits. This arrangement is therefore a common-base amplifier.

As the presence or absence of AC signals makes no difference to the quiescent state of the circuit, the DC analysis is identical to that of the series feedback amplifier performed in Example 19.4.

We found, when considering the emitter follower circuit, that the output resistance was very low ($r_e // R_E$). It is perhaps no surprise that, if we use a similar arrangement

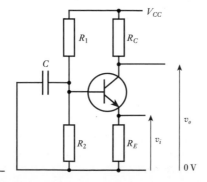

Figure 19.29
A common-base
amplifier.

with the emitter as the input terminal, this will produce a very low *input* resistance. In fact, the input resistance is also equal to $r_e // R_E$ and, as before, r_e dominates, giving

$$r_i \approx r_e$$

The output arrangement is similar to that used in the common–emitter amplifier and produces a similar (though not identical) output resistance, which is dominated by the collector resistor. Therefore

$$r_o \approx R_C$$

As the base is at earth potential for AC signals, it follows that making the input (the emitter) more positive *reduces* the voltage across the base–emitter junction and reduces the emitter current. This in turn reduces the collector current, which increases the output voltage. Thus, this amplifier is a **non-inverting amplifier**. Analysis shows that the gain is given by

$$\text{voltage gain} \approx g_m R_C \approx \frac{R_C}{r_e}$$

This is of the same magnitude as the gain of the simple common–emitter amplifier (not the series feedback amplifier, which it more closely resembles), but with opposite polarity. As the emitter and collector currents are almost identical, the current gain of the common-base amplifier is approximately unity.

Therefore, for the common-base amplifier of Figure 19.29

$$r_i \approx r_e$$

$$r_o \approx R_C$$

$$\frac{v_o}{v_i} \approx g_m R_C \approx \frac{R_C}{r_e}$$

Common-base amplifiers are characterised by a low input resistance and high output resistance. These characteristics are not generally associated with good voltage amplifiers, but make them useful as **transimpedance amplifiers** – that is, amplifiers that take an input *current* and produce a related output *voltage*. The common-base configuration is also often used in **cascode amplifiers** (*not* cascade amplifiers), where it is combined with a common-emitter amplifier. The common-base stage then provides voltage gain and the common-emitter stage gives current gain.

Amplifier configurations – a summary

So far, we have considered three basic circuit configurations for bipolar transistors, namely common–emitter, common–collector and common–base. These arrangements are shown in Figures 19.23, 19.28 and 19.29, respectively.

It is interesting to compare these three configurations in terms of a few key characteristics. Table 19.2 shows approximate maximum values for these characteristics for the circuits given in the figures.

19.6.9 Cascaded amplifiers

If more gain is required than can be obtained from a single amplifier, many stages can be **cascaded** by connecting them in series. Indeed, we have seen that it is often better to use many stages, each with a relatively low gain, than to use one high-gain stage.

Table 19.2 A comparison of amplifier configurations.

	Common-emitter	Common-collector	Common-base
Input terminal	Base	Base	Emitter
Output terminal	Collector	Emitter	Collector
Voltage gain, A_v	$-g_m R_C$ (high)	≈ 1 (unity)	$g_m R_C$ (high)
Current gain, A_i	$-h_{fe}$ (high)	h_{fe} (high)	≈ -1 (unity)
Power gain, A_p	$A_v A_i$ (very high)	$\approx A_i$ (high)	$\approx A_v$ (high)
Input impedance	$R_1 // R_2$ (moderate)	$R_1 // R_2$ (moderate)	$\approx r_e$ (very low)
Output impedance	$\approx R_C$ (high)	$\approx r_e$ (very low)	$\approx R_C$ (high)
Phase shift (mid-band)	$180°$	$0°$	$0°$

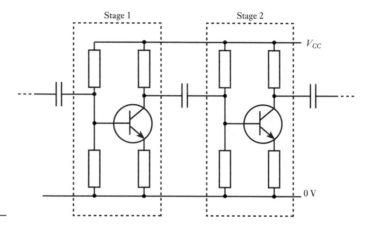

Figure 19.30
Capacitive coupling
between amplifier
stages.

When connecting the output of one stage to the input of the next, it is important to ensure that the bias conditions of the circuits are not affected. One way to ensure this is to use **coupling capacitors** between each stage. These pass AC signals, but prevent DC voltages from one stage upsetting the bias conditions of the next. Figure 19.30 shows an example of such an arrangement.

The bias conditions of each stage can be analysed separately as they are effectively isolated as far as DC signals are concerned. The small-signal performance of the combination may be determined by calculating the input and output resistance and the gain of each stage, then combining these (as described in Section 14.4). If the output resistance of one stage is low compared with the input resistance of the next, then the effects of **loading** can normally be ignored.

The use of capacitive coupling, although simple in concept, does have some disadvantages. First, each coupling capacitor produces a **low-frequency cut-off** which limits the low-frequency response of the amplifier as described in Section 8.5. Second, the presence of capacitors makes the circuit more expensive and less suitable for the production of integrated circuits as capacitors require a large amount of 'chip' area.

If care is taken with the design of the circuit, it is possible to dispense with the use of coupling capacitors between the stages of cascaded amplifiers by ensuring that the quiescent output voltage of one stage represents the correct biasing voltage for the next. This not only removes the need for coupling capacitors but also reduces the complexity of the biasing circuitry required. An example of an amplifier of this form is shown in Figure 19.31.

A major advantage of this technique is that it removes the frequency limitations introduced by capacitive coupling, allowing the production of amplifiers that can be

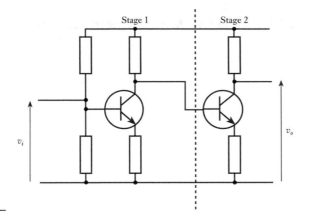

Figure 19.31
A two-stage DC-coupled amplifier.

used at frequencies down to DC. Amplifiers that have a low-frequency cut-off as a result of coupling capacitors are often called **AC-coupled amplifiers**. Circuits that have no capacitive coupling, and are therefore able to amplify signals down to DC, are referred to as **DC-coupled** or **directly coupled** amplifiers. In applications where the signals to be amplified include DC values, directly coupled amplifiers are essential.

Analysis of directly coupled amplifiers is very straightforward, as is illustrated in Example 19.11.

Example 19.11 | Calculate the quiescent output voltage and the small-signal voltage gain of the following circuit.

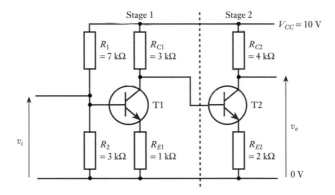

Quiescent output voltage

If we adopt the notation that V_{B1} is the voltage on the base of T1, V_{C2} is the voltage on the collector of T2 and so on, it follows that

$$V_{B1} = V_{CC}\frac{R_1}{R_1 + R_2} = 10\frac{3\ \text{k}\Omega}{7\ \text{k}\Omega + 3\ \text{k}\Omega} = 3.0\ \text{V}$$

$$V_{E1} = V_{B1} - V_{BE} = 3.0 - 0.7 = 2.3\ \text{V}$$

$$I_{C1} \approx I_{E1} = \frac{V_{E1}}{R_{E1}} = \frac{2.3\ \text{V}}{1\ \text{k}\Omega} = 2.3\ \text{mA}$$

and therefore

$$V_{C1} = V_{CC} - I_{C1}R_{C1} = 10\,\text{V} - 2.3\,\text{mA} \times 3\,\text{k}\Omega = 3.1\,\text{V}$$

V_{C1} forms the bias voltage V_{B2} for the second stage, thus

$$V_{E2} = V_{B2} - V_{BE}$$
$$= 3.1 - 0.7$$
$$= 2.4\,\text{V}$$

and

$$I_{C2} \approx I_{E2} = \frac{V_{E2}}{R_{E2}}$$
$$= \frac{2.4\,\text{V}}{2\,\text{k}\Omega}$$
$$= 1.2\,\text{mA}$$

Therefore

$$\text{quiescent output voltage} = V_{C2} = V_{CC} - I_{C2}R_{C2}$$
$$= 10\,\text{V} - 1.2\,\text{mA} \times 4\,\text{k}\Omega$$
$$= 5.2\,\text{V}$$

Voltage gain

Calculation of the voltage gain of the amplifier is also straightforward. In the absence of additional base-bias resistors, the input resistance of the second stage is at least h_{fe} times R_{E2} (see Example 19.5). This is likely to be greater than 100 kΩ and is certainly large compared with the output resistance of the previous stage, which must be less than R_{C1} (3 kΩ). Therefore, loading effects can be ignored and the gain of the combination is simply the product of the gains of the two stages when considered separately. These gains are given simply by the ratios of the collector and emitter resistors (see Example 19.4). Therefore

$$\text{overall gain} = \text{gain of stage 1} \times \text{gain of stage 2}$$
$$= -\frac{R_{C1}}{R_{E1}} \times -\frac{R_{C2}}{R_{E2}}$$
$$= -\frac{3\,\text{k}\Omega}{1\,\text{k}\Omega} \times -\frac{4\,\text{k}\Omega}{2\,\text{k}\Omega}$$
$$= -3 \times -2 = 6$$

The process of *designing* multi-stage amplifiers is similar to that outlined in Example 19.6, except that it must be performed over each section in turn, beginning with the output stage. The specified output swing and output current of the final stage will generally determine its design and the required biasing conditions of this stage then determine the form of the preceding section. This process is then repeated, working back towards the input.

File 19H

Computer simulation exercise 19.8

Simulate the circuit of Example 19.11 and apply an input signal of 0.1 volts peak at 1 kHz.

Determine the DC and AC characteristics of this arrangement and compare the measured values with those predicted in the text.

19.6.10 Darlington transistors

An interesting method of combining two or more transistors is the **Darlington connection** illustrated in Figure 19.32(a).

The current gain of the first transistor is multiplied by that of the second to produce a combination that acts like a single transistor with an h_{fe} equal to the product of the gains of the two transistors. The two devices are often available within single packages, which are called **Darlington transistors**.

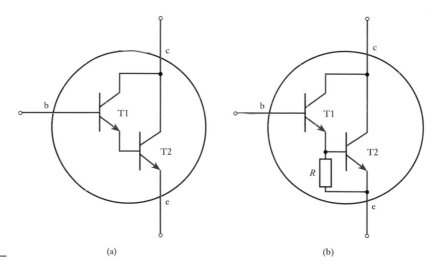

Figure 19.32
The Darlington connection.

(a)   (b)

Looking at the form of the device it is clear that in operation the offset voltage between the input (base) and the output (emitter) of the circuit will be twice that of a single transistor at about 1.4 V. The saturation voltage for the device is also greater than that of a single transistor, at slightly more than the conduction voltage of a single junction (about 0.7 V). This is because the emitter of the input transistor must be at least this voltage above the emitter voltage of the output transistor in order for the latter to be turned on. The collector of the input transistor cannot then be below the voltage on its own emitter.

One disadvantage of the arrangement shown in Figure 19.32(a) is that, if T1 is turned off rapidly, it stops conducting and there is no path by which the stored charge in the base of T2 can be quickly removed. T2 thus responds relatively slowly. This problem can be resolved by adding a resistor across the base–emitter junction of T2, as shown in Figure 19.32(b). If now T1 is turned off, R provides a conduction path to remove the charge stored in T2. R also prevents the small leakage current from T1 being amplified by T2 to produce a significant output current. R is chosen such that the voltage drop across it caused by the leakage current is smaller than the turn-on voltage of T2. Typical values for R might be a few kilohms in a small-signal Darlington down to a few hundred ohms in a power device. R is normally included within the Darlington package.

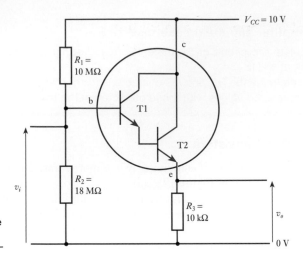

Figure 19.33
A high-input-resistance
buffer amplifier.

The very high gain of Darlingtons – which may be 10^4 to 10^5 or more – has a number of uses, including the production of circuits with very high input resistance. For example, the arrangement in Figure 19.33 has a gain of unity and an input resistance approximately equal to the parallel combination of its base resistors (about 6.4 MΩ). This is because the resistance seen looking into the base of the Darlington pair (approximately R_3 times the product of the gains of the two transistors) is so large that its effects may be neglected. The circuit is useful as a **unity gain buffer amplifier**.

Another common form of the Darlington configuration uses transistors of opposite polarities, as shown in Figure 19.34. This is known as the **complementary Darlington connection**, or sometimes as the **Sziklai connection**.

This arrangement also behaves as a single high-gain transistor and has the advantage that its base-to-emitter voltage is equal to that of a single transistor (although its saturation voltage is still greater than the conduction voltage of a single junction).

A common application of the Darlington connection is in the production of high-gain power transistors. Conventional high-power devices have relatively low gains, of perhaps 10 to 60. A typical Darlington power device might have a minimum gain of 1000 at 10 A.

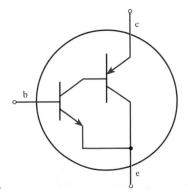

Figure 19.34
The complementary
Darlington connection.

Bipolar transistor applications

19.7.1 A bipolar transistor as a constant current source

In Section 18.7.1, we looked at the use of FETs as constant current sources. It is also possible to use bipolar transistors for this purpose, as illustrated in Figure 19.35.

Figure 19.35(a) shows a circuit using an *npn* transistor. Here a pair of base resistors form a potential divider that applies a constant voltage to the base of the transistor. The constancy of the base-to-emitter voltage results in a fixed emitter voltage, which therefore produces a constant emitter current. This in turn draws a collector current equal to the emitter current through the load.

Figure 19.35(b) works in the same manner, but uses a *pnp* transistor and produces a current to ground rather than a current from the supply.

Strictly speaking, the first of these two circuits is a **current sink** and the second a **current source**. However, it is common to use the general term **current source** to refer to both types of circuit. The circuits may be refined by using a **Zener diode** in place of R_2 to improve the constancy of the emitter voltage. The circuit can be made into a variable current source by using a variable resistance in place of R_E.

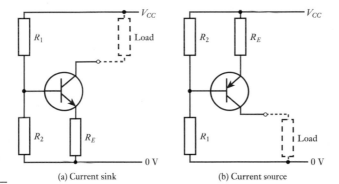

Figure 19.35
Current sources using bipolar transistors.

(a) Current sink   (b) Current source

19.7.2 A bipolar transistor as a current mirror

A current mirror may be considered as a form of constant current source where the current produced is equal to some input current. The principle is illustrated in Figure 19.36(a). Two transistors are joined at the base and emitter, and so have identical base-to-emitter voltages. If the transistors are identical, this will produce equal collector currents in each device. The current in T1 represents the input current, which in this example is

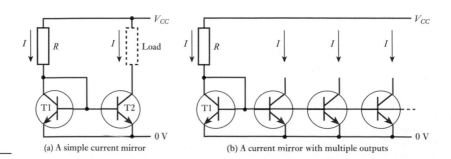

Figure 19.36
Current mirrors.

(a) A simple current mirror   (b) A current mirror with multiple outputs supplied by the resistor R. The circuit acts as a current sink, the current through T2 being drawn through the load.

A major use of this technique is in creating a number of equal currents within a circuit. This can be achieved by connecting a number of output transistors to a single input transistor, as shown in Figure 19.36(b). Current mirrors are widely used in integrated circuits where it is possible to achieve very good matching of the transistors. The close proximity of the transistors within the circuit also ensures that they are at approximately the same temperature. This is important because of the variation in the current–voltage characteristics of the transistors with temperature.

19.7.3 Bipolar transistors as differential amplifiers

In Section 18.6.9 we looked at the use of FETs in differential amplifiers. Bipolar transistors are also widely used in such circuits and, again, a common form is the **long-tailed pair**. Figure 19.37 shows an example of such an arrangement in which, for simplicity, the base-bias circuitry has been omitted. You might like to compare this with the FET amplifier shown in Figure 18.29.

The circuit resembles two series feedback amplifiers that share a common emitter resistor. The emitter resistor acts as a current source and the two transistors share the current that it supplies. With perfectly matched transistors and resistors, the circuit is symmetrical. If the input voltages, v_1 and v_2, are identical, the outputs will be equal, with the current from R_E being shared equally between the two devices. The voltage V_E will be simply V_{BE} less than the voltage on the inputs. If now v_1 is made slightly positive with respect to v_2, V_E will rise with v_1, while the base–emitter voltage on T2 will be reduced, tending to turn it off. Thus the current supplied by R_E will be split unevenly, with more of the current going through T1. This will reduce v_3 and increase v_4. If v_2 is made slightly positive with respect to v_1, the same process occurs in reverse.

Thus, if we consider the input signal to be $v_1 - v_2$ and the output signal to be $v_3 - v_4$, we have an inverting differential amplifier.

It can be shown that the voltage gain of this arrangement is given simply by

$$\text{differential voltage gain} = -g_m R_C = -\frac{R_C}{r_e}$$

You might like to compare this result with that obtained earlier for the common-emitter amplifier and for the FET differential amplifier in Section 18.6.9.

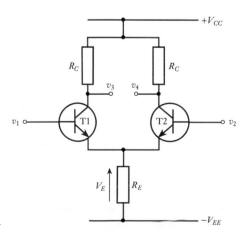

Figure 19.37
A long-tailed pair
amplifier.

One of the main advantages of this form of amplifier is its great linearity for small differential input voltages (approximately ±25 mV), which is considerably better than can be achieved using a single transistor. For this reason, this type of circuit forms the basis of most bipolar operational amplifiers and is used even where a non-differential input is required (simply by earthing the unused input). The temperature stability of the arrangement is also good as the temperature drifts of the two transistors tend to cancel each other out.

A useful measure of the performance of a differential amplifier is its **common-mode rejection ratio** (CMRR). That is, the ratio of the differential-mode gain to the common-mode gain. It can be shown that for this amplifier

$$ \text{CMRR} = g_m R_E = \frac{R_E}{r_e} $$

and again you might like to compare this with the performance of the equivalent FET circuit.

To obtain a good CMRR it is necessary to use a high value for R_E. One way to increase the effective value of R_E, without upsetting the DC conditions of the circuit, is to replace it with a **constant current source**, as described earlier in this section. An ideal current source has an infinite internal resistance, giving the maximum possible CMRR.

Figure 19.38 shows a circuit using a current mirror as a constant current source for a long-tailed pair amplifier. Circuits of this type often form the basis of the input stage of bipolar operational amplifiers. Note that, as with the FET long-tailed pair amplifier, the two output signals are symmetrical even if the inputs are not. Thus the two outputs may be used individually as single-ended (that is, non-differential) outputs if required.

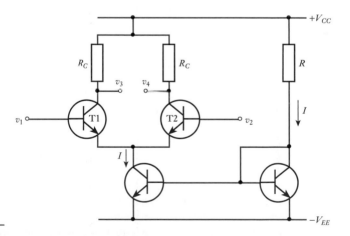

Figure 19.38
A long-tailed pair amplifier using a current mirror.

19.8 Circuit examples

19.8.1 A phase splitter

Consider the circuit in Figure 19.39. This produces two output signals that are inverted with respect to each other. By taking outputs from both the collector and the emitter, we have combined an inverting negative feedback amplifier and a non-inverting emitter follower into a single stage.

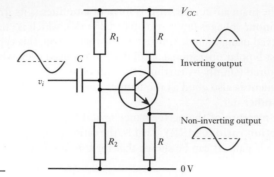

Figure 19.39
A phase splitter.

From Equation 19.20, we know that the voltage gain of the negative feedback amplifier is equal to $-R_C/R_E$. In this circuit the two resistors are equal giving a gain of -1. The emitter follower has a gain of $+1$, so the two signals are identical in size but of opposite polarity.

It should be remembered that the output resistances of the two outputs are very different. Therefore, the signals should be fed into high-input-resistance circuits if the correct relative signal magnitudes are to be maintained. It is also worth noting that the DC quiescent output voltages are different on the two outputs.

File 19J

Computer simulation exercise 19.9

Simulate the circuit in Figure 19.39 using a suitable *npn* transistor (for example, a 2N2222). A suitable choice of components would be $R = 1\ k\Omega$, $R_1 = 6.7\ k\Omega$, $R_2 = 3.3\ k\Omega$ and $C = 1\ \mu F$. Use a supply voltage $V_{CC} = 10\ V$ and apply a 1 kHz sinusoidal input of 500 mV peak.

Display the two output signals and compare their relative magnitude and phase.

19.8.2 A bipolar transistor as a voltage regulator

The emitter follower configuration discussed earlier produces a voltage that is determined by its input voltage (with an offset of about 0.7 V). It also has a very low output resistance, meaning that its output voltage is not greatly affected by the load connected to it. These two characteristics allow this circuit to form the basis of a **voltage regulator**, as shown in Figure 19.40(a). The resistor and Zener diode form a constant voltage reference V_Z (as discussed in Section 17.7.1), which is applied to the base of the transistor. The output voltage is equal to this voltage minus the approximately constant base-to-emitter voltage of the transistor.

The circuit in Figure 19.40(a) is shown redrawn in Figure 19.40(b) which illustrates a more common way of representing the circuit. The arrangement as shown provides a considerable amount of regulation but, for large fluctuations in output current, suffers from the variation of V_{BE} with current. More effective regulator circuits will be discussed in Chapter 20.

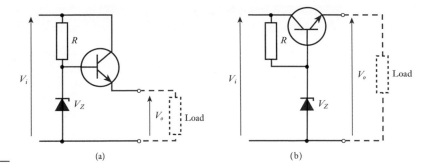

Figure 19.40
A simple voltage
regulator.

(a)   (b)

19.8.3 A bipolar transistor as a switch

We saw (in Sections 18.7.3 and 18.7.4) that FETs can be used as both analogue and logical switches. Bipolar transistors are not usually used in analogue switching applications, but can be used as logical switching elements.

Figure 19.41 shows a simple form of **logical inverter** based on a bipolar transistor. The circuit resembles the common-emitter amplifier considered earlier and indeed its operation is similar. The main difference between this circuit and a linear amplifier is that the inputs are restricted to two distinct ranges. Input voltages close to zero (representing logical '0') are insufficient to forward-bias the base of the transistor, so it is cut off. Negligible collector current flows and the output voltage is therefore close to the supply voltage. Input voltages close to the supply voltage (representing a logical '1') forward bias the base junction, turning on the transistor. The resistor R_B is chosen such that the base current is sufficient to saturate the transistor, producing an output voltage equal to the transistor's collector saturation voltage (normally about 0.2 V). Thus an input of logical '0' produces an output of logical '1', and vice versa. The circuit is therefore a logical inverter.

If the transistor were an **ideal switch**, it would have an infinite resistance when turned off, zero resistance when turned on and would operate in zero time. In fact, the bipolar transistor is a good, but not an ideal, switch. When turned off, the only currents that are passed are small leakage currents, which are normally negligible. When turned on, the device has a low ON resistance, but a small saturation voltage, as described above. The speed of operation of bipolar transistors is very fast, with devices able to switch from one state to another in a few nanoseconds (or considerably faster for high-speed devices). We will return to look in more detail at the characteristics of transistors as switches in Chapter 26.

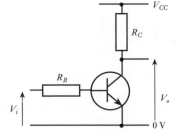

Figure 19.41
A logical inverter based
on a bipolar transistor.

Video 19E

Further study

There are many situations that call for the use of a phase splitter (as discussed in Section 19.8.1). However, the simple design shown in that section suffers from the fact that the output resistances of its two outputs are very different. This means that if the circuit is used to drive two similar circuits (as is often the case) the magnitude of the signal fed to each channel will be different unless the input resistance of each channel is relatively high.

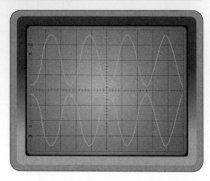

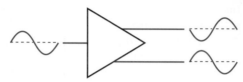

Design a phase splitting circuit that does not suffer from this problem.

Key points

- Bipolar transistors are one of the most important forms of electronic component and a clear understanding of their operation and use is essential for anyone working in this area.
- They are used in a wide variety of both analogue and digital circuits.
- Bipolar transistors can be considered as either current-controlled or voltage-controlled devices.
- If we choose to view them as current-controlled devices, we depict them as current amplifiers and describe their performance by their current gain.
- If we choose to view them as voltage-controlled devices, we depict them as transconductance amplifiers and describe their behaviour using their trans-conductance, g_m.
- If the base–emitter junction is forward biased and the base–collector junction is reverse biased, the collector current is related to the base current by the current gain.
- Two forms of current gain are used. The DC gain, h_{FE}, describes the relationship between large-signal base and collector currents, while the AC gain, h_{fe}, describes the relationship between the corresponding small-signal currents.
- For most purposes these two current gains may be considered to be equal.
- For high-power transistors, the current gain may be as low as 10, while for high-gain devices it may be as high as 1000. Compound transistors, such as Darlington devices, may have gains of 100,000 or more.
- Wherever possible, we attempt to design circuits in which the actual value of the gain is unimportant.

■ The g_m of a transistor is determined by the physics of the device and its operating conditions and does not vary significantly between components.

■ The current gain of a transistor varies with temperature and from one component to another.

■ One of the main uses of bipolar transistors is in the production of amplifiers. When designing such circuits, it is necessary to consider two main areas:

– the quiescent or DC aspects

– the small-signal or AC aspects.

■ An invaluable tool in the design and analysis of transistor circuits is a small-signal equivalent circuit. Models of varying complexity are available, though in most cases the simplest representations are quite adequate.

■ Simple circuits, such as common-emitter amplifiers, can be analysed fairly easily, but their characteristics tend to vary greatly with the parameters of the transistors.

■ Feedback can be used to stabilise the characteristics of the circuits so that they are less affected by the devices used.

■ This stabilisation is achieved at the expense of a fall in gain, but, as transistors are inexpensive, adding more gain is not usually a problem.

■ A major advantage of the use of feedback is that, by making the circuit less dependent on the characteristics of the transistor, the analysis of the circuit is made much simpler.

■ Usually it is possible to simplify the analysis of transistor circuits that use feedback by making a few simple assumptions. These are that:

– the gain of the transistor is so high that base currents can be neglected;

– the base–emitter voltage is approximately constant and therefore the small-signal voltage between the base and the emitter is negligible.

■ Although the majority of transistor circuits apply the input signal to the base and take the output signal from the collector, other configurations are also used.

■ The common-collector (emitter follower) mode produces a unity gain amplifier with a high input resistance and a low output resistance, which is good in a buffer amplifier.

■ The common-base configuration produces an amplifier with a low input resistance and a high output resistance, which makes it a good transimpedance amplifier.

■ Several transistor amplifiers may be cascaded using coupling capacitors to carry the AC signals from one stage to the next, while blocking any DC component that would upset the biasing. Unfortunately, the use of capacitors restricts the low-frequency performance of the arrangement and produces a complicated and bulky circuit.

■ An alternative method is to use direct coupling between stages. This requires fewer components and permits operation down to DC.

■ Bipolar transistors are used in a wide range of applications in addition to their use as simple amplifiers. More complicated circuits, such as operational amplifiers and other integrated circuits, are often constructed using bipolar transistors.

Exercises

19.1 Why are bipolar transistors so called?

19.2 Is a bipolar transistor a voltage-controlled or a current-controlled device?

19.3 Sketch the relationship between the base current and the collector current in a bipolar transistor (when the device is in its normal operating environment).

19.4 Sketch the construction of the two polarities of bipolar transistor.

19.5 What is meant by the symbols V_{CC}, V_{CE}, V_{BE}, v_{be} and i_c?

19.6 Explain what is meant by 'transistor action'.

19.7 Explain the terms h_{FE} and h_{fe} and describe their relative magnitudes.

19.8 Sketch the relationship between the base voltage and the collector current in a bipolar transistor (when the device is in its normal operating environment).

19.9 What is meant by the transconductance of a transistor? How does this quantity relate to the characteristic described in the previous exercise?

19.10 Determine the quiescent collector current and the quiescent output voltage of the following circuit, given that the h_{FE} of the transistor is 100.

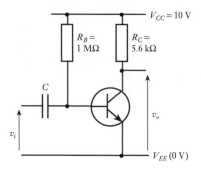

19.11 Repeat Exercise 19.10, but replace the transistor with one that has a current gain of 200. Is this circuit useful as an amplifier?

19.12 Derive a simple small-signal equivalent circuit for the following circuit, then deduce the small-signal voltage gain, input resistance and output resistance, given that $h_{FE} \approx h_{fe} = 175$ and $h_{oe} = 15$ µS. How is the small-signal voltage gain related to the quiescent voltage across R_C?

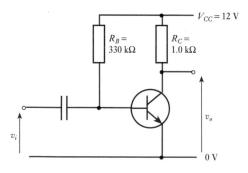

19.13 Repeat the calculations of Exercise 19.12, this time assuming that the transistor has $h_{oe} = 330$ µS.

19.14 Calculate the quiescent collector current, quiescent output voltage and small–signal voltage gain of the following circuit.

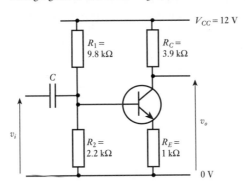

19.15 For the circuit of Exercise 19.14, estimate the small-signal input and output resistance.

19.16 For the circuit of Exercise 19.14, estimate the effect on the frequency response of the circuit of using a coupling capacitor C of 1 µF.

19.17 Use simulation to confirm your answers to Exercises 19.14 and 19.16.

19.18 If the circuit of Exercise 19.14 were modified by the addition of an emitter decoupling capacitor of 10 µF, estimate the quiescent output voltage, small-signal voltage gain and low-frequency cut-off of the resulting circuit. What would be a suitable value for the decoupling capacitor if the amplifier were required for use with signals down to 100 Hz?

19.19 Use simulation to confirm your answers to Exercise 19.18.

19.20 Using a circuit of the form shown in Exercise 19.14, design an amplifier with a small-signal voltage gain of −3, quiescent output voltage of 7 V, supply voltage of 12 V and collector load resistance of 2.2 kΩ.

19.21 Use simulation to confirm the operation of your solution to Exercise 19.20.

19.22 For the amplifier designed in Exercise 19.20, calculate the small-signal input resistance and then determine an appropriate value for the input capacitor to allow satisfactory operation at frequencies down to 50 Hz.

19.23 Calculate the quiescent collector current, quiescent output voltage and small–signal voltage gain of the following circuit.

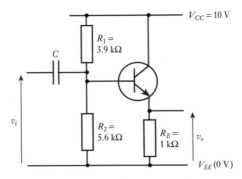

19.24 Estimate the input resistance of the circuit in Exercise 19.23, then determine an appropriate value for a coupling capacitor to allow satisfactory operation down to 50 Hz.

19.25 Use simulation to confirm your answers to Exercises 19.23 and 19.24.

19.26 Determine the quiescent output voltage, voltage gain and input and output resistance of the following circuit. You may find it helpful to redraw the circuit in a more familiar form. You may assume that the capacitor C has a negligible impedance at the frequencies of interest.

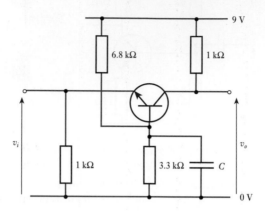

19.27 Design a two-stage, direct-coupled amplifier with a voltage gain of 10. The circuit should operate from a 15 V supply and have a maximum output swing of at least 4 volts peak to peak.

19.28 Use simulation to investigate the performance of your solution to the previous exercise.

19.29 Design a long-tailed pair amplifier based on the circuit of Figure 19.38 and use simulation to determine the small-signal voltage gain of the circuit and its CMRR.

Chapter 20 Power Electronics

Objectives

When you have studied the material in this chapter, you should be able to:

- describe a range of power amplifier circuits based on bipolar transistors
- discuss methods for reducing distortion in power amplifiers and coping with problems of temperature instability
- explain the various classes of amplifier circuit and outline the distinctions between these classes
- explain the operation of special-purpose switching devices, such as thyristors and triacs, and describe their use in AC power control
- sketch circuits of simple unregulated and regulated power supplies
- discuss the need for voltage regulation in power supplies and describe the operation of both conventional and switching voltage regulators.

20.1 Introduction

We have seen in earlier chapters how transistors may be used to produce various types of amplifiers. Such circuits usually deliver more power to their load than they absorb from their input, so they provide some degree of power amplification. However, the term **power amplifier** is normally reserved for circuits whose main function is to deliver large amounts of power to a load. Power amplifiers are used in audio systems to drive the speakers and are also used in a wide range of other applications. In some cases, as in audio amplifiers, we require a linear (or nearly linear) relationship between the input of the amplifier and the power delivered to the load. In other cases – for example, when we are controlling the power dissipated in a heater – linearity is less important. In such situations, our main concern is often the efficiency of the control process. Linear amplifiers often dissipate a lot of power (in the form of heat) and this is usually undesirable. Where linearity is unimportant, we often use techniques that improve the efficiency of the control process at the expense of some distortion of the amplified signal. In some cases, we are concerned only with the amount of power that is delivered and the nature of the waveform is unimportant. In such cases, we often use switching techniques, which can control large amounts of power and offer great efficiency.

In this chapter, we will look at power electronic circuits using both linear and switching techniques. Linear circuits can be produced using either FETs or bipolar transistors, but here we will concentrate on the latter. Although switching circuits can be constructed using transistors, special-purpose components can often perform such tasks more efficiently. Here, we will look at several devices that are specifically designed for switching applications, including **thyristors** and **triacs**.

20.2 Bipolar transistor power amplifiers

When designing a power amplifier, we normally require a low output resistance so that the circuit can deliver a high output current. We looked at various transistor configurations (in Chapter 19) and noted that the common–collector amplifier has a very low output resistance. You may recall that this arrangement is also known as an emitter follower because the voltage on the emitter follows the input voltage. While the emitter follower does not produce any voltage gain, its low output resistance makes it attractive in high-current applications and it is often used in power amplifiers.

20.2.1 Current sources and current sinks

In many cases, the load applied to a power amplifier is not simply resistive but has an impedance that includes inductive or capacitive components. For example, a speaker connected to an audio amplifier has inductance as well as resistance, while a long cable will add capacitance to a load. When driving reactive loads, an amplifier needs to *supply* current to the load at some times (when it is acting as a **current source**), but *absorb* current from the load at other times (when it is acting as a **current sink**). These two situations are illustrated in Figure 20.1, which shows an emitter follower connected to a capacitive load.

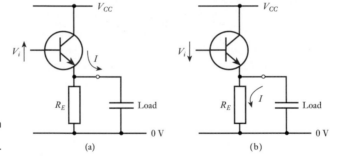

Figure 20.1
An emitter follower with a capacitive load.

Consider initially the situation shown in Figure 20.1(a), where the input is becoming more positive. The transistor drives the output more positive by passing its emitter current into the load, the low output impedance of the transistor allowing the capacitor to be charged quickly. If the input becomes more negative, as shown in Figure 20.1(b), charge must be removed from the capacitor. This cannot be done by the transistor, which can only source current in this configuration. Therefore, the charge must be removed by the emitter resistor R_E, which will be considerably greater than the output resistance of the amplifier. Therefore, this arrangement can charge the capacitor quickly but is slow to discharge it.

Replacing the transistor with a *pnp* device, as shown in Figure 20.2, produces an arrangement that can discharge the load quickly but is slow to charge it. Clearly, the

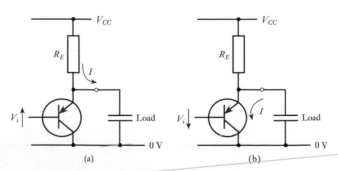

Figure 20.2
An emitter follower using a *pnp* transistor.

rate at which the capacitor can be charged or discharged through the resistor R_E can be increased by decreasing the value of the resistor. However, decreasing R_E increases the current flowing through the transistor, increasing its power dissipation. In high-power applications, this can cause serious problems.

20.2.2 Push–pull amplifiers

One approach to this problem – often used in the high-power output stage of an amplifier – is to use two transistors in a **push–pull** arrangement, as shown in Figure 20.3.

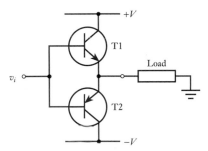

Figure 20.3
A simple push–pull amplifier.

Here, one transistor is able to source and the other to sink current, so the load can be driven from a low-resistance output in either direction. This arrangement is commonly used with a **split power supply** – that is, a supply that provides both positive and negative voltages, with the load being connected to earth. For *positive* input voltages, transistor T1 will be conducting but T2 will be turned off, as its base junction will be reverse biased. Similarly, for *negative* input voltages, T2 will be conducting and T1 will be turned off. Thus, at any time, only one of the transistors is turned on, reducing the overall power consumption.

A possible method for driving the push–pull stage is shown in Figure 20.4(a). Here, a conventional common-emitter amplifier is used to drive the bases of the two output transistors. This circuit is shown redrawn in Figure 20.4(b), which is a more conventional method of drawing the arrangement, but is electrically identical.

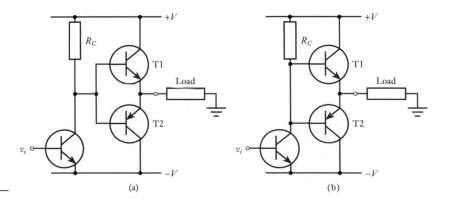

Figure 20.4
Driving a push–pull output stage.

Distortion in push–pull amplifiers

A problem with the simple push–pull arrangement described above is that, for small values of the base voltage on either side of zero, both of the output transistors are turned off. This gives rise to an effect known as **crossover distortion**, as shown in

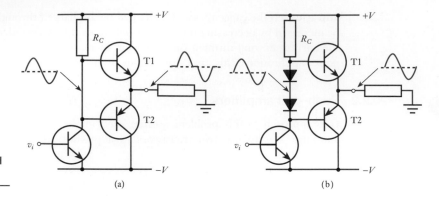

Figure 20.5
Tackling crossover
distortion in push–pull
amplifiers.

Figure 20.5(a). One solution to this problem is shown in Figure 20.5(b). Here, two
diodes are used to apply different voltages to the two bases. When conducting, the
voltage across each diode is approximately equal to the base-to-emitter voltage of each
transistor. Consequently, one transistor should turn on precisely when the other turns
off, greatly reducing the distortion produced.

A slight problem with the arrangement in Figure 20.5(b) is that the current passing through the output transistors is considerably greater than that passing through
the diodes. Consequently, the conduction voltage of the diodes is less than the V_{BE} of
the transistors and a small dead band still remains. Thus, crossover distortion is
reduced but not completely eliminated. More effective methods for biasing the output
transistors are available and we will consider several approaches later in this chapter
when we look at power amplifiers in more detail.

File 20A
File 20B

Computer simulation exercise 20.1

Simulate the arrangement of Figure 20.5(a). Suitable transistors would be a
2N2222 for T1 and a 2N2907A for T2. Use +15 V and –15 V supplies, a value
of 10 kΩ for R_C and a 10 Ω load. Any *npn* transistor may be used for the drive
transistor, although a suitable biasing arrangement must be added to set the
quiescent voltage on the bases of the output transistors to about zero. You
may need to experiment with component values to achieve this. Alternatively,
the drive transistor can be replaced with a sinusoidal current generator. If a
current generator is used, this should be configured to give an offset current of
1.5 mA so that the quiescent voltage on the bases of the transistors is close to
zero (this current flowing through the 10 kΩ resistor will give a voltage drop of
15 V, making the quiescent base voltage equal to about zero).

Apply a sinusoidal input to the circuit to produce a sinusoidal voltage on the
bases of the transistors of 5 V peak at 1 kHz and observe the form of the output. You should see that the output suffers from crossover distortion. Display
the frequency spectrum of the output using the fast Fourier transform (FFT)
feature of the simulator and observe the presence of harmonics of the input
signal. Note which harmonics are present in this signal.

Modify the circuit by adding two diodes, as in Figure 20.5(b). You may use any
conventional small-signal diodes – for example, 1N914 devices. The drive transistor biasing arrangement or the offset of the current generator must now be
adjusted to return the quiescent voltage on the junction of the two diodes to zero.

Again, apply a sinusoidal input and observe the output voltage. Display the
spectrum of this waveform and notice the effect of the diodes on the crossover
distortion.

20.2.3 Amplifier efficiency

An important aspect in the choice of a technique to use in a power output stage is its efficiency. We may define the efficiency of an amplifier as

$$\text{efficiency} = \frac{\text{power dissipated in the load}}{\text{power absorbed from the supply}} \tag{20.1}$$

Efficiency is important as it determines the power dissipated by the amplifier itself. The **power dissipation** of an amplifier is important for a number of reasons. One of the least important, except in battery-powered applications, is the actual cost of the electricity used, as this is generally negligible. Power dissipated by an amplifier takes the form of **waste heat** and the production of excess heat requires the use of larger, and more expensive, power transistors to dissipate this heat. It might also require the use of other methods of heat dissipation, such as **heat sinks** or **cooling fans**. It may also be necessary to increase the size of the power supply to deliver the extra power required by the amplifier. All these factors increase the cost and size of the system.

Of great importance in determining the efficiency of an amplifier is its **class**, a term that describes its mode of operation. The main classes of operation are described in the following section.

20.3 Classes of amplifier

All amplifiers can be allocated to one of a number of classes, depending on the way in which the active device is operated.

Video 20B

20.3.1 Class A

In class A amplifiers, the active device (for example, a bipolar transistor or an FET) conducts during the complete period of any input signal. An example of such a circuit would be a conventionally biased single-transistor amplifier of the type shown in Figure 20.6.

It can be shown that, for conventional class A amplifiers, maximum efficiency is achieved for a sinusoidal input of maximum amplitude, when it reaches only 25 per cent. With more representative inputs, the efficiency is very poor.

The efficiency of class A amplifiers can be improved by coupling the load using a transformer. The primary replaces the load resistor, while the load is connected to the secondary to form a **transformer-coupled amplifier**. This arrangement enables efficiencies approaching 50 per cent to be achieved, but it is unattractive because of the disadvantages associated with the use of inductive components, including their cost and bulk.

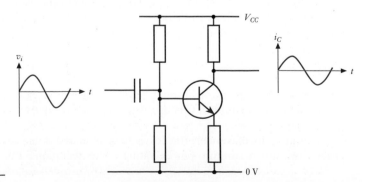

Figure 20.6
A class A amplifier.

20.3.2 Class B

In a class B amplifier, the output active devices conduct for only half the period of an input signal. These are normally push–pull arrangements, in which each transistor is active for half of the input cycle. Such circuits were discussed in Section 20.2.2, and Figure 20.7 illustrates the currents flowing in the two output transistors of such a circuit in response to a sinusoidal input signal.

Class B operation has the advantage that no current flows through the output transistors in the quiescent state, so the overall efficiency of the system is much higher than in class A. If one assumes the use of ideal transistors, it can be shown that the maximum efficiency is about 78 per cent.

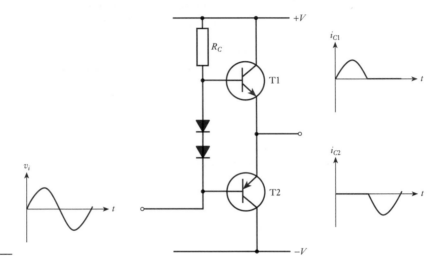

Figure 20.7
A class B amplifier.

20.3.3 Class AB

Class AB describes an amplifier that lies partway between classes A and B. The active device conducts for more than 50 per cent of the input cycle, but less than 100 per cent.

A class AB amplifier can be formed from a standard push–pull stage by ensuring that both devices conduct simultaneously for part of the input waveform. One way to achieve this is shown in Figure 20.8, where a third diode has been added to the class B amplifier of Figure 20.7 to increase the voltage difference between the bases of the two transistors. This ensures that for input voltages close to zero both devices are turned on, greatly reducing the amount of crossover distortion produced. This arrangement produces lower distortion than is normally associated with class B operation, without the efficiency penalty imposed by the use of class A. The efficiency of a class AB amplifier will lie between those of class A and class B designs and will depend on the bias conditions of the circuit.

20.3.4 Class C

Following on from the definitions of classes A and B, it is perhaps not surprising that the definition of class C is that the active device conducts for *less* than half of the input cycle.

Class C is used to enable the device to be operated at its peak current limit without exceeding its maximum power rating. The technique can produce efficiencies

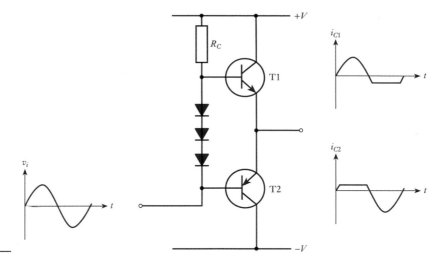

Figure 20.8
A class AB amplifier.

approaching 100 per cent, but results in gross distortion of the waveform. For these reasons, class C is used only in fairly specialised applications. One such use is in the output stage of radio transmitters, where inductive filtering is used to remove the distortion. A possible circuit for a class C amplifier is given in Figure 20.9. Often the collector resistor in this circuit is replaced by an RC tuned circuit.

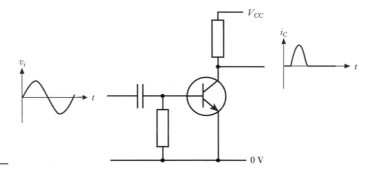

Figure 20.9
A class C amplifier.

20.3.5 Class D

In class D amplifiers, the active devices are used as switches and are either completely on or completely off. A perfect switch has the characteristics of having infinite resistance when open and zero resistance when closed. If the devices used for the amplifier were perfect switches, this would result in no power being dissipated in the amplifier itself, because, when a switch was on, it would have current flowing through it but no voltage across it and, when it was off, it would have voltage across it but no current flowing through it. As power is the product of voltage and current, the dissipation in both states would be zero. Although no *real* device is an ideal switch, transistors do make very good switching devices and amplifiers based on power transistors are both efficient and cost effective.

Amplifiers of this type are often called **switching amplifiers** or **switch–mode amplifiers**. Class D amplifiers may use single devices or push–pull pairs. In the latter case, only one of the two devices is on at any time.

Switching amplifiers may be used to provide continuous control of power by switching the output voltage on and off repeatedly at high speed. The power delivered to the load is controlled by varying the fraction of time for which the output is turned on. This process is referred to as **pulse-width modulation** (PWM) and we consider this process in more detail later in this chapter when we look at switch-mode power supplies.

20.3.6 Amplifier classes – a summary

Class A, B, AB and C arrangements are linear amplifiers, while class D circuits are switching amplifiers. The distinctions between the linear amplifiers may be seen as the differences between the quiescent currents flowing through the output devices. This, in turn, is determined by the biasing arrangements of the circuit.

File 20C
File 20D
File 20E
File 20F

Computer simulation exercise 20.2

Simulate the circuits in Figures 20.6, 20.7, 20.8 and 20.9. In each case, apply a sinusoidal input signal and use transient analysis to investigate the nature of the current flowing in the output transistors.

Video 20C

20.4 Power amplifiers

20.4.1 Class A

In class A circuits the quiescent current in the output device is greater than the maximum output current. This allows the current through the device to decrease by an amount equal to the peak output current without 'bottoming out'. This leads to low distortion but consumes a great deal of power. As the quiescent current is large, considerable power is dissipated even in the absence of an input.

Class A techniques are occasionally used in power amplifiers, but only where the very low distortion produced by these techniques can justify the high cost. Examples of such applications include high-performance audio amplifiers.

20.4.2 Class B

In class B amplifiers the quiescent current is zero, giving a much lower power consumption and greater efficiency. We have already met this form of amplifier in our discussions of **push–pull amplifiers** and Figure 20.10 shows one of the simplest push–pull arrangements.

The amplifier in Figure 20.10 is, strictly speaking, a class C amplifier as there is a dead band of $\pm V_{BE}$ for which neither transistor is conducting. Each device therefore conducts for less than half of the input cycle. However, this arrangement is often considered to be a poorly designed class B amplifier as, for large input signals, each device conducts for nearly half the cycle. The limitations of this design are clear. Small fluctuations of the input will produce no output change, while large input swings will be distorted as they traverse the dead band. This leads to the generation of large amounts of **crossover distortion**.

To convert the amplifier in Figure 20.10 into a true class B amplifier we need to arrange that one transistor turns on precisely when the other turns off. To do this, the

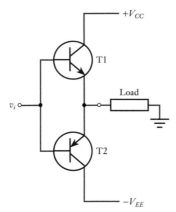

Figure 20.10
A simple push–pull
arrangement.

bases of the transistors must be biased so that they are separated by twice their normal
base-to-emitter voltage. In Section 20.2.2 we looked at a simple method for achieving
this result and two variants of this technique are shown in Figure 20.11.

We noted in Section 20.2.2 that a problem with the arrangements of Figure 20.11
is that the current passing through the output transistors is considerably greater than
that passing through the diodes. Consequently, the voltage across each of the diodes
is less than the V_{BE} of the transistors and a small dead band still remains. Crossover
distortion is reduced but not completely eliminated.

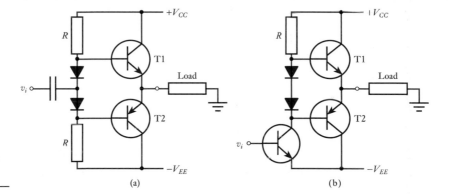

Figure 20.11
The use of diodes for
base biasing.

Several approaches are used in an attempt to produce the correct offset between
the bases of the output transistors. One of the simplest is to add a small preset variable
resistance between the bases, in series with the two diodes, as shown in Figure 20.12.

Current flowing through the biasing network produces a voltage across the preset
resistance, increasing the voltage between the bases. If this voltage is increased sufficiently, both transistors will conduct at the same time, producing a quiescent current
through the output transistors. For true class B operation, the resistance should be set
to a value that just prevents quiescent current from flowing through the transistors.

A problem associated with this arrangement is that of **temperature instability**,
caused by changes in the temperature of the transistors. As the output devices warm
up, because of the power that is being dissipated in them, their V_{BE} drops in comparison with the voltages across the relatively cool diodes. If the voltage between the
bases of the two transistors becomes greater than the sum of the turn-on voltages of
the transistors, a quiescent current will flow, further increasing the power dissipation
and thence the temperature of the output transistors. In extreme cases, this process

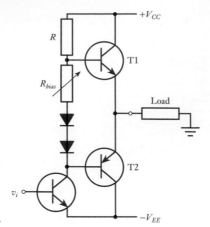

Figure 20.12
The use of a preset
resistance to set the
bias voltage.

can lead to **thermal runaway** and the ultimate destruction of the circuit. If the diodes of this circuit are placed close to the output transistors, they will provide **temperature compensation** as variations in the V_{BE} of the output transistors, caused by temperature changes, will be matched by similar variations in the voltages across the diodes.

Unfortunately, the improvement in efficiency of class B amplifiers as compared with that of class A comes at a price – there is always some non-linearity associated with the transition from one transistor being turned on to the other being turned on, which inevitably leads to a small amount of crossover distortion.

20.4.3 Class AB

The distortion produced at the crossover point can be reduced by allowing a small quiescent current to flow through the output devices. This smooths the transition from one transistor to the other and thus reduces crossover distortion. This technique has the effect of increasing the quiescent power consumption of the system and reducing the overall efficiency. However, the reduction in distortion makes this approach extremely attractive for applications such as **audio amplifiers**. As current flows in both transistors when no input is applied, each output transistor is conducting for more than 50 per cent of the input cycle. This arrangement is therefore class AB rather than B. The circuit in Figure 20.12 can be used as a class AB amplifier simply by choosing an appropriate value for the preset resistor. This circuit, and those that follow, may therefore be class B or AB depending on the value set for the quiescent current. If appropriately adjusted, the amplifiers could be used in class A too, by ensuring that the quiescent current is so large that both transistors keep conducting throughout the input cycle. However, use of class A in this form of power output stage is unusual. In practice, the circuits are normally adjusted to suit the application, the quiescent current being set to zero to minimise power consumption (class B operation) or set for a quiescent current that minimises crossover distortion (class AB operation).

20.4.4 Output stage techniques

The stability of the quiescent conditions of the circuit in Figure 20.12 can be improved by the addition of small emitter resistors, as shown in Figure 20.13. The voltage drop across these resistors is subtracted from the base–emitter voltage of the

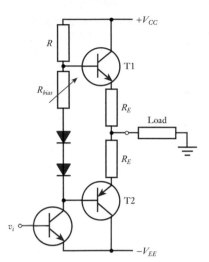

Figure 20.13
The use of emitter resistors.

circuit. They therefore provide series negative feedback by reducing the voltage across the base–emitter junction as the current increases, thus stabilising the quiescent current.

Typical values for these resistors might be a few ohms for low-power applications and much less for higher-power circuits. The power dissipated in these resistors is thus fairly small. The small voltage drop across the resistors is compensated for by the setting of the bias resistor, R_{bias}.

We noted in the last chapter that the current gain of high-power transistors is relatively low. For this reason, it is often useful in power output stages to use a **Darlington** arrangement, as shown in Figure 20.14. This figure shows two widely used configurations, the first using Darlington pairs of the same type of transistor, the second using **complementary Darlington** transistors.

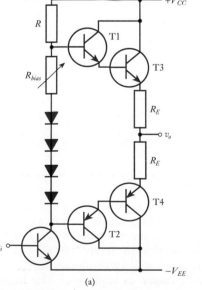
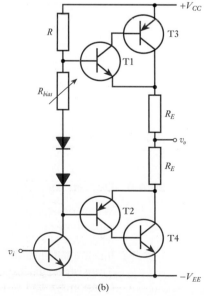

Figure 20.14
The use of Darlington transistors in power output stages.

The first circuit in Figure 20.14 uses four diodes. These are required as a voltage of $4V_{BE}$ is needed between the bases of T1 and T2 to turn on the four output transistors. In the second circuit, only two diodes are required as a voltage of only $2V_{BE}$ between the bases of T1 and T2 will cause all four output transistors to conduct.

An alternative to the use of a chain of diodes in the bias network is shown in Figure 20.15(a). Here, a single transistor is used with a pair of resistors to provide its base bias. As, for a high-gain transistor, the base current is generally negligible, the base–emitter voltage V_{BE} is determined by the collector–emitter voltage V_{CE}, and the ratio of the resistors R_1 and R_2. Therefore

$$V_{BE} = V_{CE}\frac{R_1}{R_1 + R_2}$$

and rearranging

$$V_{CE} = V_{BE}\frac{R_1 + R_2}{R_1}$$

The voltage across the network is thus determined by V_{BE} and the ratio of the resistors, R_1 and R_2. By adjusting the relative values of the two resistors, the voltage across the network can be set to equal any multiple of V_{BE}, as required. For this reason, the circuit is called a V_{BE} **multiplier**. As with biasing diodes, it is usual to mount the transistor close to the output transistors to achieve good **temperature compensation**. Figure 20.15(b) shows an output stage using a V_{BE} multiplier to set the quiescent current.

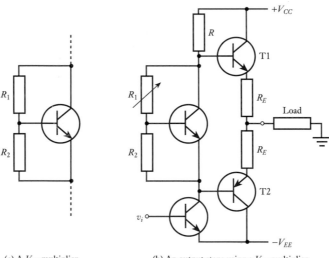

Figure 20.15
The V_{BE} multiplier arrangement.

(a) A V_{BE} multiplier   (b) An output stage using a V_{BE} multiplier

20.4.5 Design for integration

When designing amplifiers for construction using discrete components (resistors, capacitors, transistors and so on), it is natural to use low-cost passive components wherever possible to minimise the total cost of the circuit. When designing circuits for implementation in **integrated form**, the rules of economics are different. Transistors, diodes and other active components require less chip area than passive components and so are 'cheaper' to implement. This leads us to rethink the way in

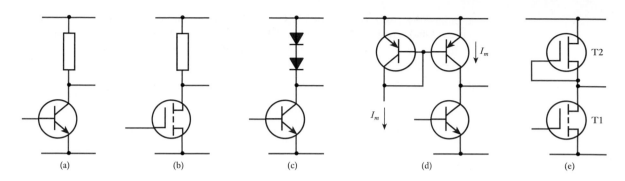

(a) (b) (c) (d) (e)

Figure 20.16
A comparison of
passive and active load
arrangements.

which we design circuits in an attempt to minimise the use of passive components and
replace them with active parts wherever possible.

Active loads

One common use of passive components within conventional circuits is the use of
load resistors, as shown in Figures 20.16(a) and (b).

In circuits based on bipolar transistors, it is possible to replace the load resistor
with an active load consisting of a series of diodes, as shown in Figure 20.16(c). This
produces an effective load equal to the sum of the slope resistances of the diodes.
Unfortunately, the resistance contributed by each diode is small and the number of
diodes that can be used is limited by the fact that each contributes a voltage drop of
about 0.7 V.

An alternative arrangement for bipolar transistor circuits is to use a **current
mirror**, as shown in Figure 20.16(d). As a *constant* current I_m is supplied by the
current mirror, any small-signal variations in the collector current appear directly
at the output. This arrangement is economical as a number of load transistors can be
driven from one current mirror (as shown in Figure 19.36(b)).

An FET can also be used as an active load, as shown in Figure 20.16(e). We have
already come across this circuit in Section 18.7.4, where we considered the use of an
FET as a logical switch or inverter. The arrangement in Figure 20.16(e) can be

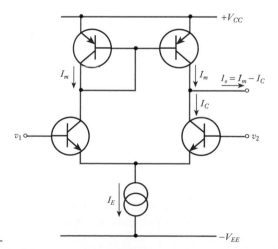

Figure 20.17
A long-tailed pair
amplifier with current
mirror load.

modified by replacing the DE MOSFET used for the load transistor T2 with an enhancement device that has its gate connected to its drain rather than its source. This gives a more linear transfer characteristic that is more suitable for analogue applications. However, when used in digital circuits (such as the logical inverter in Figure 18.36(b)) a DE MOSFET is preferred.

In **long-tailed pair amplifiers** two load resistors are required. In bipolar transistor circuits this can conveniently be achieved using both arms of a current mirror, as shown in Figure 20.17. This arrangement is not only very economical in terms of space, using no resistors, but also gives a very high voltage gain, of the order of several thousand.

20.5 Four-layer devices

Although transistors make excellent logical switches, they have limitations when it comes to switching high currents at high voltages. For example, to make a bipolar transistor with a high current gain requires a thin base region, which produces a low breakdown voltage. An alternative approach is to use one of a number of devices designed specifically for use in such applications. These components are not transistors, but their construction and mode of operation have a great deal in common with those of bipolar transistors.

20.5.1 The thyristor

The thyristor is a **four-layer device** consisting of a *pnpn* structure as shown in Figure 20.18(a). The two end regions have electrical contacts called the anode (*p* region) and the cathode (*n* region). The inner *p* region also has an electrical connection, called the gate. The circuit symbol for the thyristor is shown in Figure 20.18(b).

In the absence of any connection to the gate, the thyristor can be considered to be three diodes in series formed by the *pn*, *np* and *pn* junctions. As two of these diodes are in one direction and one is in the other, any applied voltage must reverse bias at least one of the diodes and no current will flow in either direction. However, when an appropriate signal is applied to the gate, the device becomes considerably more useful.

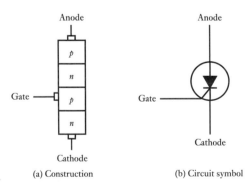

Figure 20.18
A thyristor.

(a) Construction (b) Circuit symbol

Thyristor operation

The operation of the thyristor is most readily understood by likening it to two interconnected transistors, as shown in Figure 20.19(a), which can be represented by the circuit of Figure 20.19(b).

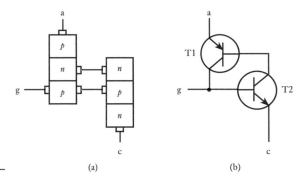

Figure 20.19
Thyristor operation.

(a)   (b)

Let us consider initially the situation in which the anode is positive with respect to the cathode but no current is flowing in either device. As T1 is turned off, no current flows into the base of T2, which is therefore also turned off. As T2 is turned off, no current flows from the base of T1, so this transistor remains off. This situation is stable and the circuit will stay in this state until external events change the condition of the circuit.

Consider now the effect of a positive pulse applied to the gate. When the gate becomes positive, T2 will turn on, causing current to flow from its collector to its emitter. This current will produce a base current in T1, turning it on. This in turn will cause current to flow through T1, producing a base current in T2. This base current will tend to increase the current in T2, which in turn will increase the current in T1 and the cycle will continue until both devices are saturated. The current flowing between the anode and the cathode will increase until it is limited by external circuitry. This process is said to be *regenerative* in that the current flow is self-increasing and self-maintaining. Once the thyristor has been 'fired', the gate signal can be removed without affecting the current flow.

The thyristor will only function as a control device in one direction so, if the anode is made negative in relation to the cathode, the device simply acts like a reverse-biased diode. This is why the thyristor is also called a **silicon-controlled rectifier** (SCR).

The thyristor may be thought of as a very efficient electrically controlled switch, with the rather unusual characteristic that when it is turned on, by a short pulse applied to the gate, it will stay on as long as current continues to flow through the device. If the current stops or falls below a certain **holding current**, the transistor action stops and the device automatically turns off. In the OFF state, only leakage currents flow and breakdown voltages of several hundreds or thousands of volts are common. In the ON state, currents of tens or hundreds of amperes can be passed with an ON voltage of only a volt or so. The current needed to turn on the device varies from about 200 μA for a small device to about 200 mA for a device capable of passing 100 A or so. The switching times for small devices are generally of the order of 1 μs, but are somewhat longer for larger devices.

It should be remembered that, even though the thyristor is a very efficient switch, power is dissipated in the device as a result of the current flowing through it and the voltage across it. This power produces heat which must be removed to prevent the device from exceeding its maximum working temperature. For this reason, all but the smallest thyristors are normally mounted on heat sinks, which are specifically designed to dispel heat.

The thyristor in AC power control

Although the thyristor can be used in DC applications, it is most often found in the control of AC systems. Consider the arrangement shown in Figure 20.20(a). Here, a

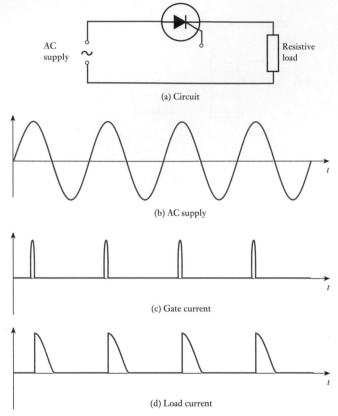

(a) Circuit

(b) AC supply

(c) Gate current

Figure 20.20
Use of a thyristor in AC
power control.

(d) Load current

thyristor is connected in series with a resistive load to an AC supply. External circuitry senses the supply waveform, shown in Figure 20.20(b), and generates a series of gate trigger pulses, as shown in Figure 20.20(c). Each pulse is positioned at the same point within the phase of the supply, so the thyristor is turned on at the same point in each cycle. Once turned on, the device continues to conduct until the supply voltage and the current through the thyristor drop to zero, producing the output waveform shown in Figure 20.20(d). In the illustration, the thyristor is fired approximately halfway through the positive half-cycle of the supply, so the thyristor is on for approximately one-quarter of the cycle. Therefore, the power dissipated in the load is approximately one-quarter of what it would be if the load were connected directly to the supply. By varying the phase angle at which the thyristor is fired, the power delivered to the load can be controlled from 0 to 50 per cent of full power. Such control is called half-wave control.

The gate current pulse is generated by applying a voltage of a few volts between the gate electrode and the cathode. As the cathode is within the supply circuit, it is common to use **opto-isolation** (as described in Section 13.3.1) to insulate the electronics used to produce these pulses from the AC supply. To achieve full-wave control using thyristors requires the use of two devices connected in inverse parallel, as shown in Figure 20.21. This allows power to be controlled from 0 to 100 per cent of full power, but unfortunately requires duplication of the gate pulse-generating circuitry and isolation network.

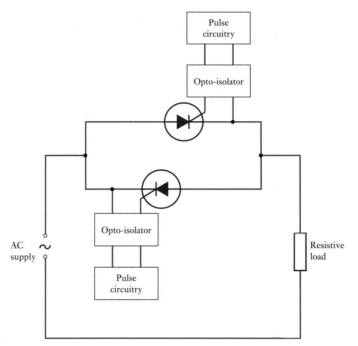

Figure 20.21
Full-wave power
control using thyristors.

20.5.2 The triac

A more elegant solution to full-wave control of AC power is to use a **triac**. This is effectively a bidirectional thyristor that can operate during both halves of the supply cycle. It resembles two thyristors connected in inverse parallel, but has the advantage that gate pulses can be supplied by a single isolated network. Gate pulses of either polarity will trigger the triac into conduction throughout the supply cycle. As the device is effectively symmetrical, its two electrodes are simply given the names MT1 and MT2, where MT simply stands for 'main terminal'. Voltages applied to the gate of the device are applied with respect to MT1. The circuit symbol for a triac is shown in Figure 20.22(a).

Gate trigger pulses in triac circuits are often generated using another four-layer device – the **bidirectional trigger diode**, or **diac**. The diac resembles a triac but without any gate connection. It has the property that, for small applied voltages, it passes no current, but if the applied voltage is increased above a certain point – termed the **breakover voltage** – the device exhibits **breakdown** and begins to

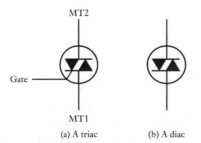

Figure 20.22
Symbols for a triac and
a diac.

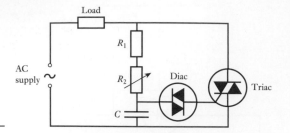

Figure 20.23
A simple lamp dimmer
using a triac.

conduct. Typical values for the breakover voltage for a diac are 30 to 35 V. The device operates in either direction and is used to produce a burst of current into the gate when a control voltage, derived from the supply voltage, reaches an appropriate value. The circuit symbol for a diac is shown in Figure 20.22(b).

Triacs are widely used in applications such as lamp dimmers and motor speed controllers. A circuit for a simple domestic lamp dimmer is given in Figure 20.23. Operation of the circuit is very straightforward: as the supply voltage increases at the beginning of the cycle, the capacitor is charged through the resistors and its voltage increases. When it reaches the breakover voltage of the diac (about 30 V), the capacitor discharges through the diac, producing a pulse of current, which fires the triac. The phase angle at which the triac is triggered is varied by changing the value of R_2, which controls the charging rate of the capacitor. R_1 is present to limit the minimum resistance of the combination to prevent excessive dissipation in the variable resistor. Once the triac has been fired, it is maintained in its ON state by the load current flowing through it, while the voltage across the resistor–capacitor combination is limited by the ON voltage of the triac, which is of the order of 1 V. This situation is maintained until the end of the present half-cycle of the supply. At this point, the supply voltage falls to zero, reducing the current through the triac below its holding current, turning it OFF. The supply voltage then enters its next half-cycle, the capacitor voltage again begins to rise (this time in the opposite sense) and the cycle repeats. If the component values are chosen appropriately, the output can be varied from zero to nearly full power by adjusting the setting of the variable resistor.

The simple lamp dimmer circuit of Figure 20.23 controls the power in the lamp by varying the phase angle of the supply at which the triac is fired. Not surprisingly, this method of operation is called **phase control** or sometimes **duty-cycle control**. Using this technique, large transients are produced as the triac switches on partway through the supply cycle.

These transients can cause problems of **interference**, either by propagating noise spikes through the supply lines or by producing electromagnetic interference (EMI). An alternative method of control is to turn the triac on for complete half-cycles of the supply, varying the ratio of on to off cycles to control the power. Switching occurs when the voltage is zero, so interference problems are removed. This technique is called **burst firing** and is useful for controlling processes with a relatively slow speed of response. It is not suitable for use with lamp control as it can give rise to flickering.

20.6   Power supplies and voltage regulators

In Section 17.8 we saw how semiconductor diodes can be used to rectify alternating voltages and in Section 19.8.2 we looked at the use of bipolar transistors in voltage regulation. These techniques can be brought together and developed to form a range of power supplies.

20.6.1 Unregulated DC power supplies

A basic low–voltage unregulated supply takes the form of a step–down transformer, a full–wave rectifier and a **reservoir capacitor** – also called a **smoothing capacitor**. A typical circuit is shown in Figure 20.24.

The output voltage of an unregulated supply is determined by the input voltage and the step–down ratio of the transformer. As the load current is increased, the ripple voltage increases (as discussed in Section 17.8) and the mean output voltage falls. Unregulated supplies are used in applications where a constant output voltage is not required and ripple is acceptable (for example, when charging a battery).

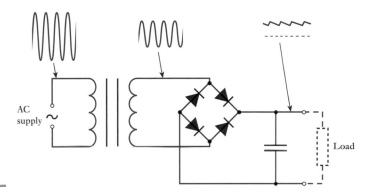

Figure 20.24
A typical unregulated power supply.

20.6.2 Regulated DC power supplies

When a more constant output voltage is required, an unregulated supply can be combined with a voltage regulator to form a regulated supply, as shown in Figure 20.25.

We considered simple voltage regulators in Section 19.8.2, where we looked at the circuit shown in Figure 20.26(a). You will recall that this is basically an emitter follower circuit where the output voltage is given by the base voltage of the transistor (which is set by the Zener diode) minus the fairly constant base–emitter voltage.

In practice, it is common to use a slightly more sophisticated arrangement, such as that shown in Figure 20.26(b). This is similar to the earlier circuit, except that the output voltage is now sampled (using the potential divider of R_3 and R_4) and a fraction of this voltage is used to provide negative feedback. The voltage on the emitter of T2 is held constant by the Zener diode. If the voltage on the base of T2 rises to a point where the base–emitter voltage is greater than the turn-on voltage of the transistor,

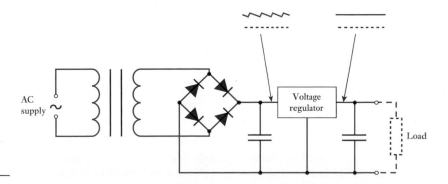

Figure 20.25
A regulated power supply.

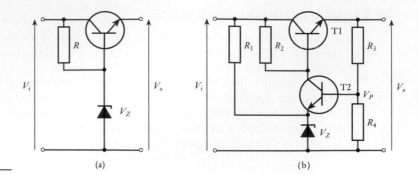

Figure 20.26
Voltage regulators.

(a) (b)

this will produce a collector current that flows through R_2. This will reduce the voltage on the base of T1 and so reduce the output voltage. The circuit will therefore stabilise at a point where the voltage at the midpoint of the potential divider V_P is approximately equal to V_Z plus the base–emitter voltage of T2 (about 0.7 V). As V_P is determined by the output voltage and the ratio of R_3 and R_4, it follows that

$$V_P = V_o \frac{R_4}{R_3 + R_4} = V_Z + 0.7 \text{ V}$$

and therefore

$$V_o = (V_Z + 0.7 \text{ V}) \frac{R_3 + R_4}{R_4} \tag{20.2}$$

Example 20.1

Determine the output voltage of the following regulator (assuming that the input voltage is sufficiently high to allow normal operation).

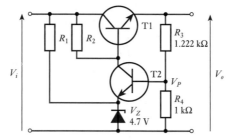

The voltage V_P on the base of T2 will be $V_Z + 0.7$. Therefore, the output voltage will be this value multiplied by $(R_3 + R_4)/R_4$. Therefore

$$V_o = (V_Z + 0.7 \text{ V}) \frac{R_3 + R_4}{R_4}$$

$$= (4.7 + 0.7) \frac{1.222 \text{ k}\Omega + 1 \text{ k}\Omega}{1 \text{ k}\Omega}$$

$$= 12.0 \text{ V}$$

Voltage regulators normally take the form of a dedicated integrated circuit. These invariably use more complicated circuits than that shown in Figure 20.26(b), often replacing T2 with an operational amplifier to give greater gain and better regulation. They also generally include additional circuitry to provide **current limiting** to prevent the circuit from being damaged by excessive current flow. Fixed-voltage regulator ICs are available for a wide range of standard voltages (for example, +5, +15 and −15 V), while other components allow the output voltage to be set using external components (usually a couple of resistors).

Power dissipation

A disadvantage of the power supplies described so far is that they are relatively inefficient. This can be appreciated by considering the regulator in Figure 20.26(b). In order to provide a constant output voltage, the magnitude of the unregulated input voltage (V_i) must be somewhat larger than the regulated output voltage (V_o). Therefore, the voltage across the output transistor is $V_o - V_i$, and the power dissipated in the transistor is equal to this voltage times the output current. In many cases, the power dissipated in the regulator is comparable with that delivered to the load and, in high-power supplies, this results in the production of large amounts of heat. A further disadvantage of these supplies is that the transformers are heavy and bulky as the low frequency of the AC supply requires a large inductance.

Example 20.2

Compare the power dissipated in the load with that dissipated in the output transistor of the regulator when the circuit of Figure 20.26(b) is connected to a load $R_L = 5\ \Omega$, given that $V_i = 15$ V and $V_o = 10$ V.

The output voltage $V_o = 10$ V and the load resistance, R_L, is 5 Ω, so the output current is

$$I_o = \frac{V_o}{R_L} = \frac{10}{5} = 2\ \text{A}$$

Therefore, the power delivered to the load is

$$P_o = V_o I_o = 10 \times 2 = 20\ \text{W}$$

The current through the output transistor (T1) is equal to output current I_o and the voltage across the transistor is given by the difference between the input voltage and the output voltage. Therefore, the power dissipated in the output transistor P_T is given by

$$P_o = (V_i - V_o)I_o = (15 - 10) \times 2 = 10\ \text{W}$$

Thus the power dissipated in the output transistor is half that of the power delivered to the load.

20.6.3 Switch-mode power supplies

One way to tackle the problems associated with high power consumption in power supplies is to use a **switching regulator**. The basic configuration of such a regulator is shown in Figure 20.27(a). The unregulated voltage is connected to a switch that is opened and closed at a rate of about 20 kHz (or more). While the frequency remains constant, the **duty cycle** (that is, the ratio of the on time to the off time) is varied. If the switch is closed for a relatively short period during each cycle, the average value of the output will be low, as shown in Figure 20.27(b). However, if the switch is closed for a larger proportion of each cycle, the average value will be higher, as in Figure 20.27(c).

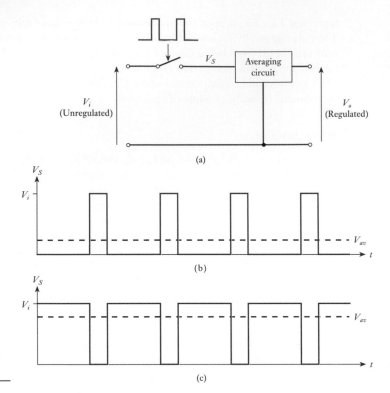

Figure 20.27
A switching regulator
arrangement.

By varying the duty cycle of the switching waveform, the average value of the output voltage can be varied from zero up to the input voltage.

A great advantage of switching regulators is that their power dissipation is very low. When an ideal switch is off, the *current* through it is zero, but when it is on, the *voltage* across it is zero. Therefore, in either state, the power dissipated in the switch is zero. Transistors are not ideal switches, but both bipolar transistors and MOSFETs have very good switching characteristics. When the transistor is turned off, it passes negligible current, while, when it is turned on, the voltage across it is small. Thus, in either state, the switch (and the regulator as a whole) consume very little power.

The averaging circuit normally uses an inductor–capacitor arrangement, as shown in Figure 20.28. When the switch is first closed, current starts to flow through the inductor and into the capacitor. The diode is reverse biased by the applied voltage and so passes no current. Because of the nature of inductance, the current builds slowly in the circuit as energy is stored in the inductor. When the switch is now opened, the energy stored in the inductor produces an e.m.f., which acts to

Figure 20.28
An *LC* averaging
circuit.

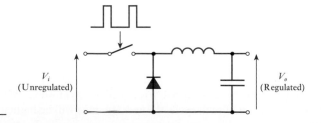

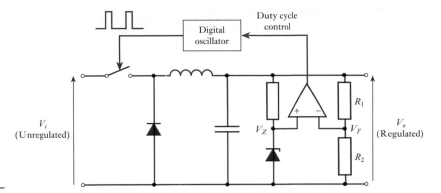

Figure 20.29
The use of feedback in
a switching regulator.

continue this current. This has the effect of forward biasing the diode and further charging the capacitor as the current decays. Current is taken out of the capacitor by the load and the circuit soon reaches equilibrium, where the voltage on the capacitor is equal to the average value of the switching waveform plus a small ripple voltage.

The duty cycle of the switch in the switching regulator is controlled using feed-back from the output and Figure 20.29 shows a possible arrangement to achieve this. The potential divider formed by R_1 and R_2 produces a voltage V_F that is related to the output voltage by the expression $V_F = V_o \times R_2/(R_1 + R_2)$. This is compared with a reference voltage V_Z, produced using a Zener diode. The output from the comparator is used to vary the duty cycle of a digital oscillator, which then controls the switch. If V_F falls below V_Z the duty cycle of the switch will be increased to raise the output voltage, while if V_F rises above V_Z the output will be reduced. In this way the feedback maintains the output such that V_F is equal to V_Z and thus

$$V_F = V_o \frac{R_2}{R_1 + R_2} = V_Z$$

and rearranging

$$V_o = V_Z \frac{R_1 + R_2}{R_2} \tag{20.3}$$

Switching regulators can be used to replace conventional regulators in arrangements of the form shown in Figure 20.25. This reduces the power consumed in the regulator and may cut its size and weight (by avoiding the need for a large heat sink). In some cases it may also be possible to achieve additional weight savings by removing the transformer. Here the AC supply is simply rectified and applied to the switching regulator. A power supply that makes use of a switching regulator is often referred to as a **switch-mode power supply** or **switch-mode power unit** (SMPU).

While it is possible to construct switching regulators from discrete components, it is more usual to use IC elements. These contain all of the active components within a single device. Combining such an IC with a handful of external components allows a complete switch-mode power supply to be constructed easily. Details of appropriate circuits are normally given in the data sheet for the particular component. Alternatively, complete switch-mode power supplies can be purchased as ready-made modules.

Video 20D

Further study

When considering four-layer devices in Section 20.5 we predominantly looked at their uses within AC circuits. However, devices such as thyristors are also extensively used in DC applications.

High-power DC motors pass large currents and controlling these using electromechanical switches may necessitate the use of switches that are both large and cumbersome. Design a more elegant controller using a thyristor. The arrangement should have two push-buttons. Pushing one should turn the motor ON and pushing the other should turn it OFF.

Key points

- Power amplifiers are designed to deliver large amounts of power to their load.
- Power amplifiers can be constructed using either FETs or bipolar transistors. Bipolar circuits often make use of emitter follower circuits as these have low output resistance.
- Many power amplifiers use a push–pull arrangement with a split power supply. Where distortion is of importance, care must be taken in the design of the biasing arrangement of such circuits to reduce crossover distortion.
- The efficiency of an amplifier is greatly affected by its class of operation:
  - in class A amplifiers, the active device conducts all the time;
  - in class B amplifiers, the active device conducts for half of the period of the input;
  - in class AB amplifiers, the active device conducts for more than half of the period of the input;
  - in class C amplifiers, the active device conducts for less than half of the period of the input;
  - class D amplifiers are switching circuits and the active device is always either fully on or fully off.
- While transistors make excellent logical switches, they are less good when dealing with high voltages and high currents. In such cases, we often use special-purpose devices, such as thyristors or triacs.
- A transformer, rectifier and capacitor can be combined to form a simple unregulated power supply. Unfortunately, such circuits suffer from variability in the output voltage and output voltage ripple.
- A more constant output voltage can be produced by adding a regulator. Conventional regulators are cheap and easy to use, but very inefficient. Switching regulators provide much higher efficiency, but require a more complicated circuit.
- Switch-mode power supplies can provide very high efficiency combined with low volume and weight.

Exercises

20.1 What is meant by the term 'power amplifier'?

20.2 Why is efficiency of importance in power amplifiers?

20.3 Which bipolar transistor configuration is most often used in power amplifier circuits? Why is this?

20.4 Explain the operation of the simple push–pull amplifier in Figure 20.3. Why does this circuit produce crossover distortion?

20.5 How may the crossover distortion in a push–pull amplifier be reduced?

20.6 Explain what is meant by the efficiency of an amplifier.

20.7 What happens to the power that is absorbed from the supply but is not supplied to the load?

20.8 Outline the distinction between the various classes of amplifier. Which forms are linear amplifiers?

20.9 What is the maximum efficiency of a class A amplifier? Under what circumstances is this maximum efficiency achieved?

20.10 Into which class does the simple push–pull amplifier of Figure 20.3 fall?

20.11 Why are class C amplifiers rarely used?

20.12 Explain the use of pulse-width modulation in power control.

20.13 What class of amplifier (that is, class A, class B and so on) does the following circuit represent?

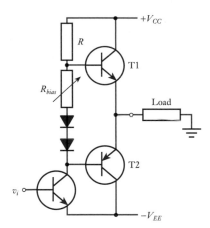

20.14 Explain what is meant by temperature instability and describe how this problem could be tackled.

20.15 Why is class AB generally used in preference to class B in audio amplifiers? What are the disadvantages of class AB in comparison with class B?

20.16 Explain the operation of a V_{BE} multiplier arrangement.

20.17 Explain the advantages of 'active loads' within integrated circuits.

20.18 Why is a bipolar transistor not ideal for switching high currents at high voltages?

20.19 Explain briefly the operation of a thyristor.

20.20 Explain the need for opto-isolation in AC control using thyristors.

20.21 Why is a thyristor not ideal for AC power control?

20.22 How does a triac differ from a thyristor?

20.23 What is meant by 'phase control' of a triac circuit? Describe some potential problems with this approach.

20.24 What is meant by 'burst firing' of a triac circuit? Describe some potential problems with this approach.

20.25 Sketch the circuit of a simple unregulated power supply, explaining the function of each component. What limits the usefulness of such a circuit?

20.26 Sketch a simple voltage regulator that uses feedback to stabilise the output voltage. Explain the operation of your circuit.

20.27 Modify the circuit in Example 20.1 to produce an output voltage of 15 V.

20.28 A voltage regulator of the form shown in Figure 20.26(a) is connected to a load of 10 Ω. Calculate the power dissipated in the output transistor if the input voltage is 25 V and the output voltage is 15 V.

20.29 Explain what is meant by a switching regulator. What are the advantages of this form of regulator?

20.30 Explain the operation of the switching regulator in Figure 20.29.

Chapter 21   Internal Circuitry of Operational Amplifiers

Objectives

When you have studied the material in this chapter, you should be able to:

■ describe a range of circuit techniques that are commonly used in the construction of operational amplifiers

■ sketch the basic form of a rudimentary op-amp based on bipolar transistors, and explain the operation of its various elements

■ identify and understand the operation of a range of circuit elements within the circuits of commercial bipolar op-amps

■ outline the form of a rudimentary op-amp based on CMOS circuitry, and explain the operation of its various elements

■ recognise key elements within the design of commercial CMOS op-amps

■ extract the key parameters of op-amp performance from the data sheets of commercial devices

■ explain the distinguishing characteristics, and unique circuit elements, of BiFET and BiMOS operational amplifiers.

21.1   Introduction

In most situations we view operational amplifiers as 'black boxes' and have little interest in the circuitry used to implement them. However, the circuits used within these integrated circuits employ many of the circuit elements that we have studied in earlier chapters, and therefore represent an excellent way of extending our understanding of these techniques. They also give us an opportunity to consider more complex circuits. While it is unlikely that many readers of this text will be faced with the task of designing a commercial operational amplifier, an appreciation of the circuitry used within such devices does give us a better understanding of their external characteristics (as discussed in Section 16.5).

While we have looked at a range of operational amplifier applications, we have assumed that the op-amps used are 'conventional', general-purpose devices. Various specialised forms of amplifier are available including devices with current-mode (Norton) inputs or with differential outputs. Here we will concentrate on the most widely used forms of op-amp, which are differential voltage amplifiers with a single output.

Op-amps may be constructed with either bipolar transistors or field-effect transistors, or a combination of the two. We will look at all these forms, but will start by looking at circuits based on bipolar transistors since these are the most widely used.

Video 21A

21.2   Bipolar operational amplifiers

21.2.1   A simple differential amplifier

In Figure 19.37 we looked at a differential amplifier based on bipolar transistors in the form of a long-tailed pair. This provides differential inputs and high gain within a single circuit and, when used with a constant current source, also gives a high common–mode rejection ratio. In Section 19.7.2 we looked at the use of a current mirror as a constant current source, and in Figure 19.38 we saw how a current mirror could be incorporated into a long-tailed pair amplifier. In addition to a differential amplifier, an op-amp also needs an output stage. The need to both source and sink current requires the use of a push–pull amplifier, and in Figure 20.5 we looked at simple push–pull arrangements. Combining the differential amplifier of Figure 19.38 with a push–pull output stage of the form shown in Figure 20.5(b) produces a rudimentary op-amp, as shown in Figure 21.1.

In the circuit in Figure 21.1, transistors T1 and T2 form the long-tailed pair differential amplifier with collector resistors R_1 and R_2. In fact, R_1 is unnecessary as the output is taken as a single–ended signal from T2. The constant current generator for the input amplifier is provided by the current mirror of T3 and T4. The output from T2 is amplified by the series negative feedback amplifier formed by T5 (which, in this case, is a *pnp* transistor rather than the *npn* transistor shown in Figure 20.5(b)), which drives the output transistors T6 and T7. Diodes D1 and D2 are present to provide differential bias to the bases of the output transistors in order to reduce crossover distortion.

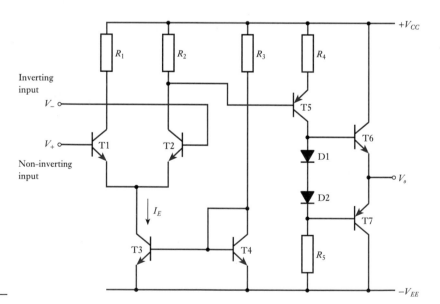

Figure 21.1
A rudimentary bipolar operational amplifier.

21.2.2 An improved amplifier

The design outlined in the last section can be improved in a number of ways using techniques already discussed:

- the use of emitter resistors to improve the stability of the quiescent conditions of the output transistors;
- an improvement in the biasing arrangements of the circuit to allow accurate setting of the quiescent current in the output devices;
- the use of active, rather than passive, loads to make the circuit more suitable for integration.

An improved circuit is shown in Figure 21.2.

The emitter resistors R_4 and R_5 improve the stability of the quiescent conditions of the output stage by applying series negative feedback (as discussed in Section 20.4.4). For such a low-power application, the resistors would be about 25 to 50 Ω.

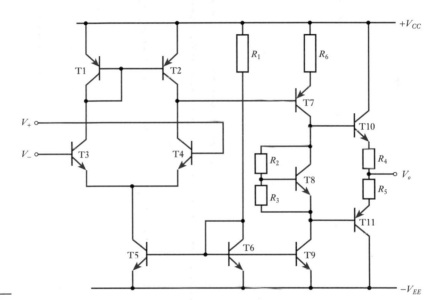

Figure 21.2
An improved bipolar operational amplifier.

T8 forms a V_{BE} multiplier arrangement with R_2 and R_3 to provide a stable bias voltage between the bases of T10 and T11 (as discussed in Section 20.4.4). The values of the resistors would be of the order of a few kilohms, the ratio being chosen to select the desired quiescent current through the output transistors.

Transistors T1 and T2 form a current mirror, which is used as an active load for the long-tailed pair amplifier formed by T3 and T4 (as discussed in Section 20.4.5). The output of this load is fed to T7, which also has an active load in the form of T9. Again, this is part of a current mirror, which this time shares a transistor (T6) with the current mirror used to provide the constant current into the emitters of the input differential amplifier.

21.2.3 Real bipolar operational amplifiers

While the circuit in Figure 21.2 provides many useful features, real op-amps usually include additional refinements that are outside the scope of this text. These include such features as short–circuit protection, to safeguard the device against excessive output current, and offset null adjustment, to remove the effects of the input offset voltage. Figure 21.3 shows the circuit diagram of a basic, general-purpose op-amp – the 741. While this circuit is considerably more complicated than that shown in Figure 21.2, many similarities can be seen in the circuit elements used.

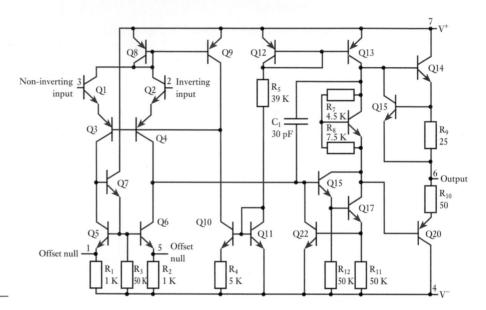

Figure 21.3
The circuit diagram of a 741 operational amplifier.

Figure 21.4 shows an extract from the data sheet of a 741 operational amplifier. The data sheet in fact covers a family of components with similar characteristics. The first part of the extract (shown in Figure 21.4(a)) gives an overview of the characteristics of the devices and shows the pin-outs of the various packages. It can be seen that some of the components contain a single op-amp, while others contain two. This section of the data sheet also indicates the operating temperature range of the devices. The second part of the extract (shown in Figure 21.4(b)) shows a table giving values for a wide range of parameters such as the input offset voltage, the input bias current, and the input and output resistances. (The significance of many of these parameters was discussed in Section 16.5.) Note that, in most cases, the table shows *typical* values for the various parameters along with an indication of the maximum or minimum expected values. This allows designers to assess the likely variation in device characteristics across a range of components. The information given in Figure 21.4 represents only a small part of the data provided within the data sheet, which runs to some 21 pages.

μA741, μA741Y
GENERAL-PURPOSE OPERATIONAL AMPLIFIERS

SLOS094B – NOVEMBER 1970 – REVISED SEPTEMBER 2000

- **Short-Circuit Protection**
- **Offset-Voltage Null Capability**
- **Large Common-Mode and Differential Voltage Ranges**
- **No Frequency Compensation Required**
- **Low Power Consumption**
- **No Latch-Up**
- **Designed to Be Interchangeable With Fairchild μA741**

description

The μA741 is a general-purpose operational amplifier featuring offset-voltage null capability.

The high common-mode input voltage range and the absence of latch-up make the amplifier ideal for voltage-follower applications. The device is short-circuit protected and the internal frequency compensation ensures stability without external components. A low value potentiometer may be connected between the offset null inputs to null out the offset voltage as shown in Figure 2.

The μA741C is characterized for operation from 0 °C to 70°C. The μA741I is characterized for operation from –40°C to 85°C. The μA741M is characterized for operation over the full military temperature range of –55°C to 125°C.

symbol

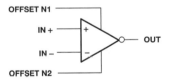

μA741M . . . J PACKAGE
(TOP VIEW)

NC	1	14	NC
NC	2	13	NC
OFFSET N1	3	12	NC
IN–	4	11	V_{CC+}
IN+	5	10	OUT
$V_{CC–}$	6	9	OFFSET N2
NC	7	8	NC

μA741M . . . JG PACKAGE
μA741C, μA741I . . . D, P, OR PW PACKAGE
(TOP VIEW)

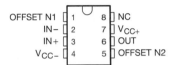

μA741M . . . U PACKAGE
(TOP VIEW)

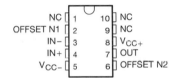

μA741M . . . FK PACKAGE
(TOP VIEW)

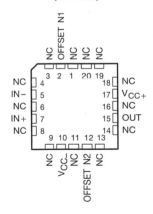

NC – No internal connection

TEXAS INSTRUMENTS
POST OFFICE BOX 655303 ● DALLAS, TEXAS 75265

1

Figure 21.4(a)
An extract from a data sheet for a 741 operational amplifier.

µA741, µA741Y
GENERAL-PURPOSE OPERATIONAL AMPLIFIERS

SLOS094B ± NOVEMBER 1970 ± REVISED SEPTEMBER 2000

electrical characteristics at specified free-air temperature, V$_{CC+}$ = ± 15 V (unless otherwise noted)

PARAMETER		TEST CONDITIONS	T$_A$†	µA741C			µA741I, µA741M			UNIT
				MIN	TYP	MAX	MIN	TYP	MAX	
V$_{IO}$	Input offset voltage	V$_O$ = 0	25°C		1	6		1	5	mV
			Full range			7.5			6	
ΔV$_{IO(adj)}$	Offset voltage adjust range	V$_O$ = 0	25°C		+15			±15		mV
I$_{IO}$	Input offset current	V$_O$ = 0	25°C		20	200		20	200	nA
			Full range			300			500	
I$_{IB}$	Input bias current	V$_O$ = 0	25°C		80	500		80	500	nA
			Full range			800			1500	
V$_{ICR}$	Common-mode input voltage range		25°C	±12	±13		±12	±13		V
			Full range	±12			±12			
V$_{OM}$	Maximum peak output voltage swing	R$_L$ = 10 kΩ	25°C	±12	±14		±12	±14		V
		R$_L$ ≥ 10 kΩ	Full range	±12			±12			
		R$_L$ = 2 kΩ	25°C	±10	±13		±10	±13		
		R$_L$ ≥ 2 kΩ	Full range	±10			±10			
A$_{VD}$	Large-signal differential voltage amplification	R$_L$ ≥ 2 kΩ	25°C	20	200		50	200		V/mV
		V$_O$ = ±10 V	Full range	15			25			
r$_i$	Input resistance		25°C	0.3	2		0.3	2		MΩ
r$_o$	Output resistance	V$_O$ = 0, See Note 5	25°C		75			75		Ω
C$_i$	Input capacitance		25°C		1.4			1.4		pF
CMRR	Common-mode rejection ratio	V$_{IC}$ = V$_{ICR}$min	25°C	70	90		70	90		dB
			Full range	70			70			
k$_{SVS}$	Supply voltage sensitivity (ΔV$_{IO}$/ΔV$_{CC}$)	V$_{CC}$ = +9 V to ± 15 V	25°C		30	150		30	150	µV/V
			Full range			150			150	
I$_{OS}$	Short-circuit output current		25°C		±25	±40		±25	±40	mA
I$_{CC}$	Supply current	V$_O$ = 0, No load	25°C		1.7	2.8		1.7	2.8	mA
			Full range			3.3			3.3	
P$_D$	Total power dissipation	V$_O$ = 0, No load	25°C		50	85		50	85	mW
			Full range			100			100	

† All characteristics are measured under open-loop conditions with zero common-mode input voltage unless otherwise specified. Full range for the µA741C is 0°C to 70°C, the µA741I is ±40°C to 85°C, and the µA741M is −55°C to 125°C.
NOTE 5:   This typical value applies only at frequencies above a few hundred hertz because of the effects of drift and thermal feedback.

operating characteristics, V$_{CC±}$ = ±15 V, T$_A$ = 25°C

PARAMETER		TEST CONDITIONS		µA741C			µA741I, µA741M			UNIT
				MIN	TYP	MAX	MIN	TYP	MAX	
t$_r$	Rise time	V$_I$ = 20 mV,	R$_L$ = 2 kΩ,		0.3			0.3		µs
	Overshoot factor	C$_L$ = 100 pF,	See Figure 1		5%			5%		
SR	Slew rate at unity gain	V$_I$ = 10 V,	R$_L$ = 2 kΩ,		0.5			0.5		V/µs
		C$_L$ = 100 pF,	See Figure 1							

5

Figure 21.4(b)
An extract from a data sheet for a 741 operational amplifier.

Video 21B

21.3 CMOS operational amplifiers

21.3.1 A simple differential amplifer

While many general-purpose op-amps are based on bipolar transistors, FETs are often used to produce devices with more specialised characteristics. FETs can be used to produce op-amps with very low power consumption that are ideal for battery-powered operation. They can also produce devices in which the input and output voltages are able to 'swing' all the way between the supply voltages – so-called **rail-to-rail operation**.

In Section 18.6.9 we looked at the construction of differential amplifiers using FETs, and in Section 18.7.1 we saw how FETs could be used to form constant current sources. Figure 18.32 shows a long-tailed pair amplifier with a single FET as a current source but, within operational amplifiers, it is common to use current mirrors as constant current sources, as in the bipolar circuits discussed earlier. Figure 21.5 shows a rudimentary op-amp based on enhancement MOSFETs. The circuit uses both *n*-channel and *p*-channel devices and is therefore a **CMOS** circuit (as discussed in Section 18.7.5).

In Figure 21.5 transistors T1 and T2 form a long-tailed pair, with transistors T3 and T4 acting as an active (current mirror) load. A constant current is fed to the long-tailed pair by the current mirror formed by T5 and T6, the value of this current being determined by the current flowing through the resistor *R*. The output from the differential amplifier is taken as a single sided output from the drain of T2. This goes to a common-source amplifier formed by T7. The current mirror of T5 and T6 is extended by T8 to form an active load for T7. You might like to compare the circuit in Figure 21.5 with the bipolar circuit in Figure 21.2, which uses many of the same circuit techniques. Notice that the circuit in Figure 21.5 uses a single supply line, which is common in CMOS op-amps.

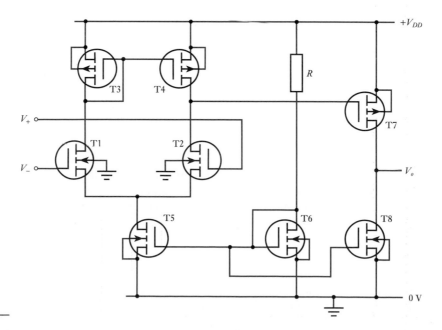

Figure 21.5
A rudimentary CMOS operational amplifier.

21.3.2 Real CMOS operational amplifiers

The circuit in Figure 21.5 illustrates some of the elements of a typical CMOS op-amp, but real devices are invariably more sophisticated and therefore more complicated.

Figure 21.6 shows an example of a low-power-consumption CMOS device – the TLC271. When comparing Figures 21.5 and 21.6 it should be noted that the circuitry in the latter is largely an 'upside-down' version of the former. The circuit in Figure 21.6 uses p-channel devices for the long-tailed pair and the current source that feeds it, and n-channel devices for the active load current mirror. The circuit in Figure 21.5 uses MOSFETs of the opposite polarities, in keeping with the discussions in earlier chapters. However, the basic operation of the two circuits is the same.

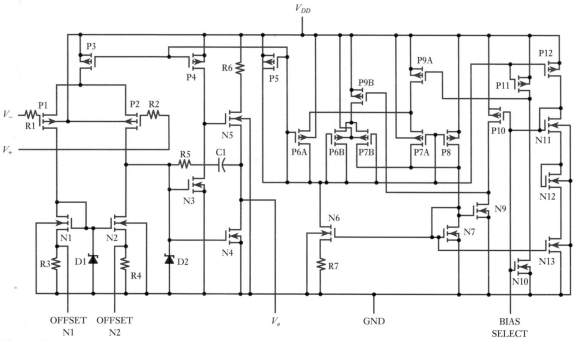

Figure 21.6
The circuit diagram of a TLC271 operational amplifier.

Figure 21.7 shows an extract from the data sheet of a TLC271 op-amp. The first part of the extract (shown in Figure 21.7(a)) gives an overview of the characteristics of the device and shows its pin-out. Data is also given on the operating temperature range of the devices, and on the noise performance. The second part of the extract (shown in Figure 21.7(b)) gives values for parameters such as the input offset voltage, the input bias current and the common-mode rejection ratio. The complete data sheet for this part runs to some 84 pages.

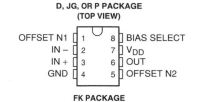

TLC271, TLC271A, TLC271B
LinCMOS™ PROGRAMMABLE LOW-POWER
OPERATIONAL AMPLIFIERS
SLOS090D – NOVEMBER 1987 – REVISED MARCH 2001

- Input Offset Voltage Drift . . . Typically 0.1 µV/Month, Including the First 30 Days
- Wide Range of Supply Voltages Over Specified Temperature Range:
  0°C to 70°C . . . 3 V to 16 V
  −40°C to 85°C . . . 4 V to 16 V
  −55°C to 125°C . . . 5 V to 16 V
- Single-Supply Operation
- Common-Mode Input Voltage Range Extends Below the Negative Rail (C-Suffix and I-Suffix Types)
- Low Noise . . . 25 nV/√Hz Typically at f = 1 kHz (High-Bias Mode)
- Output Voltage Range Includes Negative Rail
- High Input Impedance . . . 10^{12} Ω Typ
- ESD-Protection Circuitry
- Small-Outline Package Option Also Available in Tape and Reel
- Designed-In Latch-Up Immunity

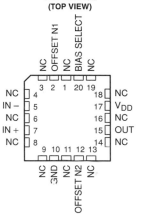

D, JG, OR P PACKAGE
(TOP VIEW)

OFFSET N1	1	8	BIAS SELECT
IN −	2	7	V_{DD}
IN +	3	6	OUT
GND	4	5	OFFSET N2

FK PACKAGE
(TOP VIEW)

NC – No internal connection

description

The TLC271 operational amplifier combines a wide range of input offset voltage grades with low offset voltage drift and high input impedance. In addition, the TLC271 offers a bias-select mode that allows the user to select the best combination of power dissipation and ac performance for a particular application. These devices use Texas Instruments silicon-gate LinCMOS™ technology, which provides offset voltage stability far exceeding the stability available with conventional metal-gate processes.

AVAILABLE OPTIONS

		PACKAGE			
T_A	V_{IO}max AT 25°C	SMALL OUTLINE (D)	CHIP CARRIER (FK)	CERAMIC DIP (JG)	PLASTIC DIP (P)
0°C to 70°C	2 mV	TLC271BCD			TLC271BCP
	5 mV	TLC271ACD	−	−	TLC271ACP
	10 mV	TLC271CD			TLC271CP
−40°C to 85°C	2 mV	TLC271BID			TLC271BIP
	5 mV	TLC271AID	−	−	TLC271AIP
	10 mV	TLC271ID			TLC271IP
−55°C to 125°C	10 mV	TLC271MD	TLC271MFK	TLC271MJG	TLC271MP

The D package is available taped and reeled. Add R suffix to the device type (e.g., TLC271BCDR).

Please be aware that an important notice concerning availability, standard warranty, and use in critical applications of Texas Instruments semiconductor products and disclaimers thereto appears at the end of this data sheet.

LinCMOS is a trademark of Texas Instruments.

TEXAS
INSTRUMENTS
POST OFFICE BOX 655303 • DALLAS, TEXAS 75265

1

Figure 21.7(a)
An extract from a data sheet for a TLC271 operational amplifier.

TLC271, TLC271 A, TLC271B
LinCMOS™ PRQGRAMMABLE LOW-POWER
OPERATIONAL AMPLIFIERS
SLOS094B – NOVEMBER 1997 – REVISED MARCH 2001

HIGH-BIAS MODE

electrical characteristics at specified free-air temperature (unless otherwise noted)

PARAMETER			TEST CONDITIONS	T_A†	TLC271C, TLC271C, TLC271BC						UNIT
					$V_{DD} = 5$ V			$V_{DD} = 10$ V			
					MIN	TYP	MAX	MIN	TYP	MAX	
V_{IO}	Input offset current	TCL271C	$V_O - 1.4V$, $V_{IC} - 0V$, $R_S - 50\Omega$ $R_L - 10\Omega$	25°C		1.1	1.0		1.1	10	mV
				Full range	25°C		1.2	±15		12	
		TCL271AC		25°C		0.9	5		0.9	5	
				Full range			6.5			6.5	
		TCL271BC		25°C		0.34	2		0.39	2	
				Full range			3			3	
V_{VID}	Average temperature boeiliclent or input offset woltage			25°C la 70°C		1.8			2		V
I_{IO}	Imput offset current (see Note 4)		$V_O - V_{DD}/2$, $V_{IC} - V_{DD}/2$	25°C		0.1	60		0.1	60	PA
				70°C		7	300		7	300	
I_{IB}	Imput btas ourrent (see Note 4)		$V_O - V_{DD}/2$, $V_{IC} - V_{DD}/2$	25°C		0.6	60		0.7	60	PA
				70°C		40	600		50	500	
V_{ICR}	Common-mode input vottage range (see Note 5)			25°C	−0.2 10 4	−0.3 10 4.2		−0.2 10 9	−0.3 10 9.2		W
				Fut range	−0.2 10 3.5			−0.2 10 8.9			W
V_{OH}	High-level output voltage		$V_{ID} - 100$ mV, $R_L - 10$ kΩ	Full range	3.2	3.8		8	8.5		dB
				0°C	3	3.6		7.8	8.5	150	W
				Full range	3	3.6		7.8	8.4	150	W
V_{OL}	Law-tever otput voltage		$V_{ID} - -100$ mV, $I_{OL} - 0$	25°C		±25	50		0	50	mA
				25°C		0	50		0	50	
				Full range		0	50		0	50	
A_{VO}	Large-aignal differential wovtage amgilitic agaion		$R_L - 10$ kΩ, See Note 5	25°C	5	23		10	36		V/mV
				Full range	4	27		7.5	42		
				70°C	4	20		7.5	32		
CVRR	Common-made revection ratio		$V_{IC} - V_{ICR}min$	25°C	65	80		65	65		dB
				0°C	60	84		60	88		
				70°C	60	85		60	88		
I_{SVR}	Supply-citage refection ratio ($\Delta V_{DD}/\Delta V_{IO}$)		$V_{DD} - 5$ V to 10 V $V_O - 1.4$ W	25°C	65	95		65	85		dB
				0°C	60	94		60	94		
				70°C	60	96		60	96		
$I_{I(SEL)}$	Input current (BIAS SELECT)		$VI_{(SEL)} - a$	25°C		−1.4			−1.9		μA
I_{DD}	Supply current		$V_O - V_{DD}/2$, $V_{IC} - V_{DD}/2$, No load	25°C		675	1600		950	2000	μA
				0°C		775	1600		1125	2200	
				70°C		575	1300		750	1700	

†Full range is 0°C to 70°C.
NOTES   4.   The typical values of Input blas current and Input offset below 5 pA were detemined mathematicially.
        5.   This range also appies to each Input Individually.
        6.   At $V_{DD} = 5V$, $V_O = 0.25$ V to 2 V; at $V_{DD} - 10$ V $V_O - 1$ V to 6 V

TEXAS INSTRUMENTS
POST OFFICE BOX 655303 • DALLAS, TEXAS 75265

5

Figure 21.7(b)
An extract from a data sheet for a TLC271 operational amplifier.

21.4 | **BiFET operational amplifiers**

In addition to circuits based entirely on bipolar transistors – or entirely on FETs – there are also several circuit techniques that use a combination of the two. One such example is the BiFET op–amp, which uses largely bipolar circuitry, but increases the input resistance of the arrangement by using a JFET input stage. Figure 21.8 illustrates this approach.

The basic operation of the BiFET circuit is similar to that of the corresponding bipolar arrangement, but the use of JFETs increases the input resistance substantially.

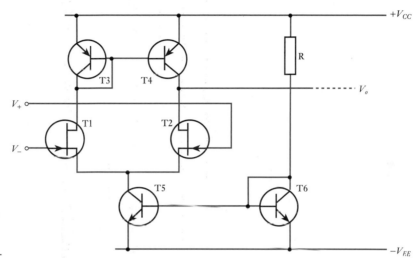

Figure 21.8
A BiFET input stage for an operational amplifier.

21.5 | **BiMOS operational amplifiers**

Another circuit technique that combines bipolar transistors and FETs is the BiMOS arrangement, which combines bipolar transistors with MOSFETs. This can produce circuits with input resistances that are even higher than can be achieved using BiFET techniques. The principles of a BiMOS input stage are illustrated in Figure 21.9.

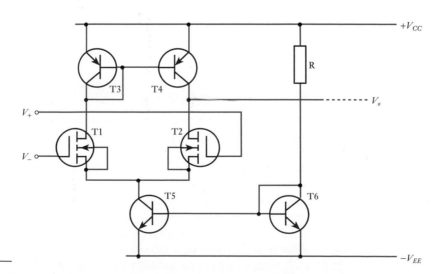

Figure 21.9
A BiMOS input stage for an operational amplifier.

Video 21C

Further study

Figure 21.3 shows the circuit of a general purpose operational amplifier – the 741.

Clearly the circuitry used in this device is considerably more complex than that shown in the earlier figures in this chapter, and a full explanation of the operation of this circuit is beyond the scope of this text. However, close inspection of the circuit shows that it does use several of the techniques that have been discussed in this and earlier chapters.

Study the circuit and attempt to identify circuit elements and techniques that have been described within the text. Hence attempt to understand the overall operation of the device.

Key points

- Operational amplifiers can be constructed using bipolar transistors, field effect transistors, or a combination of the two.
- While the circuitry used in real operational amplifiers is often fairly complex, many of the techniques used are relatively easy to understand.
- Most op-amps, both bipolar and CMOS, use a long-tailed pair amplifier at their input.
- Most op-amps, both bipolar and CMOS, make use of current mirrors as constant current sources.
- Real devices often use quite elaborate circuitry to stabilise the quiescent conditions of the circuit.
- Device data sheets give a considerable amount of information on the performance of the device, and how this performance varies with temperature and between devices.
- While conventional, general-purpose op-amps can be used in a wide range of applications, more specialised devices are available for applications that have special requirements. These include situations that require a very low power consumption, rail-to-rail operation or a very high input resistance.

Exercises

21.1 While the most widely used forms of operational amplifiers are differential voltage amplifiers with a single output, suggest some other forms of op-amps that are available.

21.2 In bipolar op-amps, what form of circuitry is normally used to implement the differential input amplifier?

21.3 What form of circuit is normally used to provide a constant current to the input amplifier of a bipolar op-amp?

21.4 What form of amplifier is normally used for the output stage of a bipolar op-amp? What are the advantages of this circuit approach?

21.5 Explain the function of transistor T4 in Figure 21.1.

21.6 Explain the function of diodes D1 and D2 in Figure 21.1.

21.7 In the circuit of Figure 21.1 what would be the effect of connecting the base of T5 to the collector of T1 instead of the collector of T2?

21.8 In Figure 21.1, the output of the differential amplifier formed by T1 and T2 is taken from the collector of T2. What voltage gain is produced by the remaining stages of the amplifier? In other words, what is the relationship between V_o and the voltage on the collector of T2?

21.9 Explain the function of T8 in Figure 21.2. What determines the collector–emitter voltage across this device?

21.10 In Figure 21.2, T8 is an *npn* transistor; could this function be performed by a *pnp* device? If so, what changes would need to be made to the remainder of the circuit?

21.11 In Figure 21.2, how does the use of T8 (rather than the use of two diodes as in Figure 21.1) affect the voltage gain of the circuit?

21.12 Explain the functions of R_4 and R_5 in the circuit of Figure 21.2.

21.13 The μA741 op-amp (as described in Figure 21.4) is available in several variants. Over what temperature range would the commercial μA741C normally be used? How does this compare with the μA741M military part?

21.14 From the data given in Figure 21.4, how is the input offset voltage of the 741 affected by temperature? Does the commercial form of this device differ from the military form in this regard?

21.15 What is meant by rail-to-rail operation?

21.16 In CMOS op-amps, what form of circuitry is normally used to implement the differential input amplifier?

21.17 What form of circuit is normally used to provide a constant current to the input amplifier of a CMOS op-amp?

21.18 In Figure 21.5, how is the drain current of T5 related to that of T8?

21.19 Explain the function of the resistor R in Figure 21.5.

21.20 For the TLC271 op-amp described in Figure 21.7, what would be a typical figure for the input impedance?

21.21 In Figure 21.7(b), data is given for many parameters. In some cases a figure is given for a typical value and for a *maximum* value. In other cases a figure is given for a typical value and for a *minimum* value. Why is this?

21.22 Why might a BiFET or BiMOS op-amp be chosen in preference to a bipolar device in some situations?

21.23 Explain the function of T3 and T4 in the circuit of Figure 21.8.

Chapter 22

Noise and Electromagnetic Compatibility

When you have studied the material in this chapter, you should be able to:

- discuss the problems associated with noise within electronic systems
- identify the major causes of noise and interference within electronic systems and outline the characteristics of the noise that they produce
- describe ways of representing noise sources within electronic circuits in order to model their behaviour
- use appropriate measures to quantify the effects of noise on electronic signals such as the signal-to-noise ratio and the noise figure
- discuss a range of techniques appropriate for designing systems for low-noise applications
- outline the basic principles of electromagnetic compatibility (EMC) and adopt appropriate techniques to reduce EMC-related problems.

22.1 Introduction

All electrical signals are affected by **noise**. This is a random fluctuation of a signal that is produced either by variations in components within the system or by external effects of the environment. Noise has a number of causes, but since *all* real physical devices produce noise, it is always present within electronic systems.

In most cases the amount of noise produced within a system is not related to the magnitude of the signals present. For this reason, the relative size of the noise will be greatest when the signal is small. This effect is readily experienced when listening to a radio or personal music system. Noise, experienced as a background 'hiss', is more apparent during quiet, rather than loud, passages. We can quantify the 'quality' of a signal with respect to noise by giving the **signal-to-noise ratio**, which describes the relative magnitudes of the signal and the noise. We will look at this quantity in more detail later in this chapter.

Noise is present in all forms of electrical signal and affects both analogue and digital systems. This is illustrated in Figure 22.1, which shows the effects of relatively small amounts of noise on an analogue and a digital (binary) signal. It is clear that in each case the noise corrupts the original signal, but it is also clear that in the case of the binary signal it is still apparent which parts of the signal represent the higher voltage and which the lower. This suggests that it might be possible to extract the original signal and thereby remove the effects of the noise. When noise affects analogue signals it may be possible to use filtering to remove it within particular frequency ranges, but it is not normally possible to remove noise that is of a similar frequency to the signal itself. This is an important distinction between analogue and digital signals that is

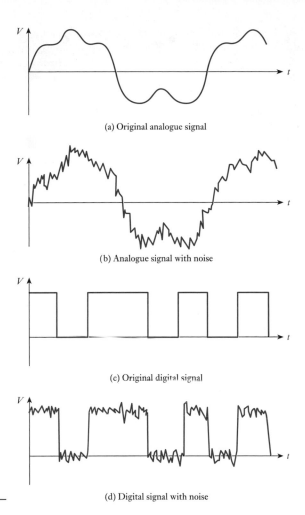

(a) Original analogue signal

(b) Analogue signal with noise

(c) Original digital signal

Figure 22.1
The effects of noise
on analogue and
digital signals.

(d) Digital signal with noise

developed further in later chapters. The presence of noise often limits the ultimate performance of electronic systems and one of the major tasks of electronic design is that of reducing the magnitude of these effects.

One of the sources of noise that we will consider within this chapter is *interference* from external noise sources. This in turn will lead on to the study of **electromagnetic compatibility** (EMC), which is concerned with the ability of a system to work correctly in the presence of interference, and also its ability to work without itself causing excessive amounts of interference.

22.2 Noise sources

Video 22A

22.2.1 Thermal noise

All electronic components that possess resistance (in practice this means all real components) generate what is called **thermal noise** (which is also called **Johnson noise**) as a result of the random, thermally induced motion of their atoms.

Thermal noise has components at all frequencies with equal noise power in all parts of the spectrum. For this reason it is often described as **white noise** by analogy with white light. Although theoretically the noise has an infinite **bandwidth** – that is, an infinite frequency range – within a given application, only the noise that is within the bandwidth of the operation of the system will have any effect. It can be shown that the **noise power** P_n that results from this form of noise is constant for any resistance and is related to its absolute temperature T and the bandwidth B of the measuring system by the expression

$$P_n = 4kTB \tag{22.1}$$

where k is Boltzmann's constant ($k \approx 1.3805 \times 10^{-23}$ J/K).

For a given resistance of value R, the power dissipated in it is related to the root-mean-square (r.m.s.) value of the voltage V across it by the expression

$$P = \frac{V^2}{R}$$

which may be rearranged to give

$$V = (PR)^{1/2}$$

Combining this result with that of Equation 22.1 gives an expression for the r.m.s. voltage produced by a resistance of value R as a result of thermal noise. This is

$$V_n(\text{r.m.s.}) = (4kTBR)^{1/2} \tag{22.2}$$

Thus, although the noise *power* generated by all resistors is equal, the noise *voltage* increases with the value of the resistance.

The noise obeys a **Gaussian amplitude distribution**, but, as it is random in nature, it is not possible to predict its instantaneous value. However, the expression above may be used to determine the r.m.s. noise voltage for a given resistance.

It is interesting to look at the nature of this relationship. In most systems the operating temperature is close to ambient. As this is expressed as an absolute temperature, taking an approximate value of 20 °C (68 °F) will normally suffice. We may then expand the expression as

$$V_n(\text{r.m.s.}) = (4kT)^{1/2}(BR)^{1/2} = 1.27 \times 10^{-10} \times (BR)^{1/2}$$

If we consider, for example, a system with a bandwidth of 20 kHz (which is typical of a high-quality audio amplifier), we find that a resistance of 1 kΩ has an open-circuit noise voltage of about 500 nV and that a resistance of 1 MΩ has an open-circuit noise voltage of about 18 μV. Note that these are open-circuit noise voltages and that a 'noisy' resistance of R can be modelled by a voltage source V_n in series with an ideal noiseless resistance of R. This is illustrated in Figure 22.2.

Figure 22.2
Representation of a noisy resistor.

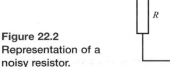

(a) A 'noisy' resistor

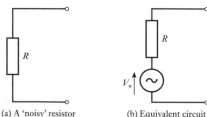

(b) Equivalent circuit

As any signal source has a resistive component to its output impedance, it follows that all sources give rise to thermal noise. The noise voltage calculations above, if applied to this source resistance, thus give an indication of the ultimate limit to the **signal-to-noise ratio** of a system. In practice, thermal noise will also be produced by the various resistors within the electronic system itself and noise will be produced by other noise sources, as described below. Good design techniques will reduce the noise produced by the *system*, but the noise produced by the *source* is normally beyond the control of the designer.

22.2.2   Shot noise

The current flowing within an electronic circuit is made up of large numbers of individual **charge carriers**. With large currents, the averaging effect gives the impression of a continuous and constant stream. However, for smaller flow rates, the granular nature of the current becomes more apparent. The statistical variation of the flow gives rise to a noise current, the magnitude of which is given by

$$I_n(\text{r.m.s.}) = (2eBI)^{1/2} \tag{22.3}$$

where e is the electronic charge (1.6×10^{-19} coulombs), B is the bandwidth over which the noise is measured and I is the mean value of the current. You may care to compare this expression with that of Equation 22.2. You will notice that thermal noise produces a *noise voltage*, whereas shot noise produces a *noise current*.

As with thermal noise, the magnitude of shot noise increases with the bandwidth of the measuring system. Again, it is both white and Gaussian in nature. The noise current also increases with the mean value of the current (as you would expect), but, as it increases with the *square root* of the mean current, the relative size of the noise decreases as the current increases.

To illustrate this effect, let us consider the shot noise associated with currents flowing in a circuit with a bandwidth of 20 kHz. For a current of 1 A, the r.m.s. noise current would be about 80 nA, or 0.000008 per cent of the mean current. With a current of 1 µA, the noise current falls to 80 pA, but this now represents 0.008 per cent of the mean current. Decreasing the current to 1 pA produces a noise current of about 8 per cent of the mean current.

Shot noise is generated by the random flow of charge carriers across potential barriers, such as *pn* junctions. Within high-gain transistors, base currents are small, making the effects of this form of noise more significant.

22.2.3   1/f noise

1/f noise is caused by not one but a variety of noise sources. It is so called because the power spectrum of the noise is inversely proportional to frequency. This means that the power in any octave (or decade) of frequency is the same. This represents a halving of the power for a doubling of the frequency, and thus corresponds to a fall in power of 3 dB/octave. Clearly, most of the power of this form of noise is concentrated at low frequencies. 1/f noise is often known as **pink noise** to distinguish it from white noise, which has a uniform spectrum.

One of the most important forms of 1/f noise is **flicker noise**, which is caused by random variations in the diffusion of charge carriers within devices. Other forms of 1/f noise include the current-dependent fluctuations of resistance exhibited by all real resistors. This noise is in addition to any thermal noise and proportional to the mean current flowing through the device.

22.2.4 Interference

Another source of noise within an electronic system is interference from external signal sources. This can take many forms and enter the system at any stage.

Common noise sources include radio transmitters, AC power cables, lightning, switching transients in nearby equipment, mechanical vibration (particularly in mechanical sensors), ambient light (particularly in optical sensors) and unintentional coupling within systems (perhaps caused by stray capacitance or inductance). Interference will be discussed in more detail in Section 22.9, when we look at electromagnetic compatibility.

22.3 Representing noise sources within equivalent circuits

In earlier chapters we have looked at equivalent circuits for a range of sensors, amplifiers and devices. Having now established that all real devices produce noise, it is necessary to consider how we may adapt our equivalent circuits to incorporate the effects of noise.

The output of any sensor or circuit may be modelled using either a Thévenin or a Norton equivalent circuit, and we can represent the presence of noise simply by adding a voltage generator V_n or a current generator I_n as in Figure 22.3.

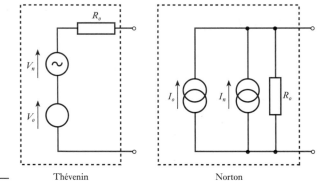

Figure 22.3
Equivalent circuits of output networks with noise.

Thévenin   Norton

A typical amplifier circuit will contain many resistors, each contributing thermal noise to the signal. Clearly, noise introduced near the input of the amplifier is going to have more effect than that introduced close to the output as the former will be amplified by the gain of the amplifier. For this reason, in applications where a low noise level is required, great emphasis is placed on producing a low-noise input stage.

Rather than consider a large number of separate noise sources within an amplifier, it is common to combine their effects and represent them using a single equivalent noise source. As the noise produced at the output of an amplifier is affected by its gain, it is common to represent noise by an equivalent noise source at the input, as shown in Figure 22.4.

The input voltage V_i is applied across the series combination of the input resistance R_i and the noise source V_n. The voltage across the input resistance (which in turn determines the output voltage) is thus $V_i - V_n$. As the noise is random, the polarity of the noise is unimportant.

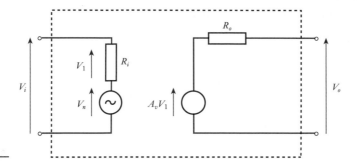

Figure 22.4
Representation of
noise in an amplifier.

22.4 Noise in bipolar transistors

Bipolar transistors suffer from noise that comes from a variety of sources. The semiconductor materials used to produce the devices clearly have resistance that leads to the production of **thermal noise** throughout the device. **Shot noise** is also produced by the random traversal of charge carriers across the junctions. Fluctuations in the diffusion process throughout the device lead to **flicker noise**.

At low frequencies (up to a few kilohertz), flicker noise is the dominant noise source. However, it becomes less significant at higher frequencies, while the effects of thermal and shot noise become more significant. Both noise voltages and noise currents are produced. These can be represented in a small-signal equivalent circuit, as shown in Figure 22.5.

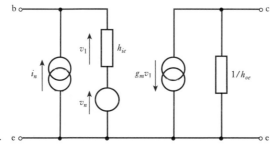

Figure 22.5
Representation of
noise voltages and
currents in an
equivalent circuit for
a bipolar transistor.

22.5 Noise in FETs

Noise in FETs is normally dominated by flicker noise (particularly at low frequencies) and thermal noise resulting from the resistance of the channel. Shot noise is generally insignificant in MOSFETs as there is no junction and in JFETs the only currents across the gate junction are those caused by leakage.

22.6 Signal-to-noise ratio

It is often useful to have a quantitative method for describing the quality of a signal in terms of its corruption by noise. This may be done by measuring its **signal-to-noise ratio** (S/N ratio) – that is, the ratio of the magnitude of the signal to that of the noise. This is most commonly expressed as the ratio of the signal power P_s

to the noise power P_n. As this is a power ratio, it may be expressed in decibels, as discussed earlier.

The signal-to-noise ratio may therefore be defined as

$$\text{S/N ratio} = 10 \log_{10}\left(\frac{P_s}{P_n}\right) \text{dB} \tag{22.4}$$

As both the signal and the noise are present at the same point within a circuit, they are applied to the same impedance. Thus, the power ratio may be expressed as a ratio of the r.m.s. signal voltage V_s to the r.m.s. noise voltage V_n:

$$\text{S/N ratio} = 10 \log_{10}\left(\frac{V_s}{V_n}\right)^2 \text{dB}$$

$$= 20 \log_{10}\left(\frac{V_s}{V_n}\right) \text{dB} \tag{22.5}$$

The S/N ratio within a given system will vary with the magnitude of the signal. If the signal becomes very small, the relative size of the noise will increase and the S/N ratio will become smaller. For this reason, it is advantageous to represent quantities with the largest possible signals to produce the highest S/N ratio. It is often useful to define the maximum achievable S/N ratio for a given system. This is given by the ratio of the largest possible signal to the noise present. When expressed as a ratio of r.m.s. voltages, this is

$$\text{max S/N ratio} = 20 \log_{10}\left(\frac{V_{s(\text{max})}}{V_n}\right) \text{dB}$$

Example 22.1 | At a particular point in a circuit, a signal of 2.5 volts r.m.s. is corrupted by 10 mV r.m.s. of noise. What is the S/N ratio at this point?

From the above

$$\text{S/N ratio} = 20 \log_{10}\left(\frac{V_s}{V_n}\right) = 20 \log_{10}\left(\frac{2.5}{0.01}\right) = 48 \text{ dB}$$

In many cases, the noise voltage will be made up of a number of different components. As noise voltages are random in nature, they cannot simply be added together to obtain their combined effect. Instead, we must add the squares of the r.m.s. voltages (which are related to the noise power), then take the square root of the result to obtain the r.m.s. voltage of the combination. Thus, for two noise sources

$$V_n = \sqrt{(V_{n1}^2 + V_{n2}^2)} \tag{22.6}$$

22.7 Noise figure

S/N ratio calculations can be used to indicate the quality of a signal, but not to describe how well an amplifier, or other circuit, performs regarding noise. Simply measuring the S/N ratio at the output of a circuit is not an indication of its performance as this will be affected by the nature of the input signal. We have seen that any real signal source has noise associated with it, which must be taken into account when determining the noise performance of any circuit to which it is connected.

One method that can be used to describe how well an amplifier performs is to give the ratio of the noise produced at its output to that which would be present at the output of an ideal 'noiseless' amplifier of the same gain when both are connected to the same input. This comparison will give different results depending on the nature of the input signal used. A common method of comparison is to measure this ratio when the input is simply the thermal noise from a resistor of a specified value. The ratio measured under these conditions is termed the **noise figure** (NF) of the system, where

$$\text{NF} = 10\log_{10}\frac{\text{noise output power from amplifier}}{\text{noise output power from noiseless amplifier}} \tag{22.7}$$

$$\text{NF} = 20\log_{10}\frac{\text{r.m.s. noise output voltage from amplifier}}{\text{r.m.s. noise output voltage from noiseless amplifier}} \tag{22.8}$$

We noted earlier that it is often convenient to represent all the noise sources within a network by a single noise source at its input. A 'noiseless' amplifier has no noise sources and so the output noise is the same as the noise from the source. If we represent all the noise sources in our amplifier under test by a noise source of r.m.s. magnitude $V_{n(total)}$ then

$$\text{NF} = 20\log_{10}\frac{V_{n(total)}}{V_{ni}}\,\text{dB} \tag{22.9}$$

where V_{ni} is the r.m.s. noise voltage of the source.

The NF gives a way of comparing the performance of amplifiers or other circuits regarding noise. However, it should be remembered that the NF depends on the value of the source resistance and, as many noise sources are frequency dependent, varies with frequency. A perfect noiseless amplifier has an NF of 0 dB. Good low-noise amplifiers will have NFs of 2 to 3 dB.

22.8 Designing for low-noise applications

In most cases, it is the noise generated by the first stage of an amplifier that determines the overall noise performance of the system. This is because noise generated near the input of the circuit is amplified, along with the signal, by all later stages, whereas noise generated near the output receives relatively little amplification. Therefore, to achieve a good low-noise design, particular attention must be paid to the all-important first stage.

22.8.1 Source resistance

From our considerations of thermal noise, it would seem advantageous to use a source with as low an internal resistance as possible. In fact, when the noise of the first stage is considered, this is not the best condition. It can be shown that the optimum value for the source resistance R_s is given by the expression

$$R_s = \sqrt{\frac{V_n^2}{I_n^2}} \tag{22.10}$$

where V_n and I_n are the r.m.s. noise voltage and current, respectively. These are combined in this way, rather than as a simple ratio, because they are uncorrelated quantities. The value of this optimum source resistance will vary from circuit to circuit. A typical value might be a few kilohms or a few tens of kilohms. In many cases, the designer is not able to choose the resistance of the source as this is determined by another system or a particular sensor. In such cases, this must be taken into account in the design.

22.8.2 Bipolar transistor amplifiers

In bipolar amplifiers, both noise voltages and noise currents increase with collector current. Low-noise designs therefore use low quiescent currents of a few microamps. Suitable transistors combine low flicker noise and high current gain at low collector currents. Bipolar transistor amplifiers can produce good low-noise circuits with a wide range of source resistances, from a few hundred ohms to a few hundred kilohms.

22.8.3 FET amplifiers

The dominant noise voltage source in JFETs is thermal noise caused by the resistance of the channel, which decreases with increasing drain current. Low-noise designs therefore use a fairly high drain current. MOSFETs generally have a poorer noise performance than JFETs at frequencies up to several hundred kilohertz. FETs provide good noise performance for source resistances from a few tens of kilohms to several hundred megohms.

22.8.4 A comparison of bipolar and FET amplifiers

Bipolar transistors and FETs can both be used to produce excellent low-noise amplifiers with noise figures of 1 dB or better (given suitable care in design). Generally, bipolar transistors are preferable for low source resistances and will produce good results down to a few hundred ohms. FETs are superior with high source resistances and can be used with sources of 100 MΩ or more.

22.8.5 Interference in low-noise applications

In many applications, interference plays a dominant role in determining the overall noise performance. The susceptibility of an electronic system to interference is affected not only by the circuit used, but also by its construction, location and use. Such considerations come within the very important topic of electromagnetic compatibility, which is discussed next.

Video 22B

22.9 Electromagnetic compatibility

Electromagnetic compatibility (EMC) is concerned with the ability of a system to operate in the presence of interference from other electrical equipment and not to interfere with the operation of other equipment or other parts of itself.

Examples of problems associated with EMC are illustrated in Figure 22.6. The diagram in Figure 22.6(a) shows an external electromagnetic noise source interfering with the operation of a system. Figure 22.6(b) shows an arrangement where the operation of one part of a system adversely affects the functioning of another part. Figure 22.6(c) illustrates the situation where a system produces radiation that interferes with other equipment.

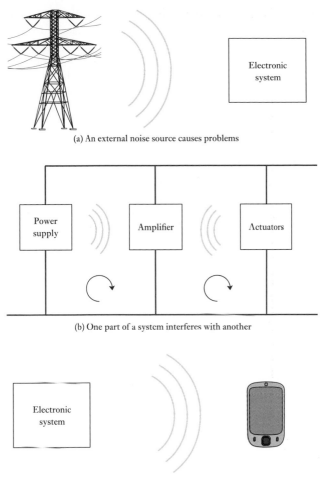

(a) An external noise source causes problems

(b) One part of a system interferes with another

Figure 22.6
Examples of EMC
problems.

(c) A system interferes with the operation of other equipment

22.9.1 Sources of electromagnetic interference

All **electromagnetic (E-M) waves** consist of an electric field and a magnetic field that are at right angles to each other. The properties of such waves vary with their frequency, but all travel at the speed of light. Examples include visible light, X-rays and radio waves.

E-M waves are used as the basis of many forms of communication and information processing and many systems rely on their production or detection. However, problems may arise when unwanted electromagnetic signals interfere with the operation of a system. Sources of electromagnetic interference may be natural or human-made.

Natural sources of interference

There are several natural phenomena that generate electromagnetic disturbances. The most significant of these are lightning, solar emissions and cosmic radiation.

The frequency of **lightning** varies dramatically between different parts of the world, being commonplace in certain areas at some times of the year. A direct lightning strike on a piece of equipment can produce voltages of the order of hundreds of thousands of volts and can deliver currents of hundreds of thousands of amps. It is therefore extremely difficult to design systems that can withstand such an incident. In most places, direct strikes are extremely rare, except in the case of large, exposed conductors, such as overhead power cables.

Unfortunately, the adverse effects of lightning are not restricted to those of a direct hit. Thunderstorms result in very intense electric fields in the region of several kilovolts per metre at ground level. At the instant of a lightning strike the intensity of the field falls rapidly as a result of the electrical discharge. This can result in large transients being induced in any nearby conductor. Lightning also produces high-frequency emissions up to about 100 MHz. These can propagate over a considerable area and are a major source of atmospheric noise.

Variations in radiation from the Sun cause fluctuations in the ionosphere, which in turn affect the way in which it reflects or transmits radio waves. For this reason, variations in **solar emissions** have a great effect on radio communications, particularly in the 2–30 MHz range, and on satellite communication at higher frequencies. **Cosmic radiation** produces significant background noise in the 100–1000 MHz range.

Human-made sources of interference

The commonest sources of interference are various electrical and electronic systems that radiate energy as a direct result of their operation. Interference can also be caused by electrostatic discharges or an electromagnetic pulse.

Electrical and electronic systems may generate interference through a number of mechanisms. For example, noise may be caused by the radiation of high-frequency signals used within the circuit (such as the oscillations produced within a superheterodyne radio receiver), the pulsed currents within digital circuits such as computers, or switching transients such as those of an electric light switch. Examples of systems that are common sources of interference include:

■ automotive ignition systems	■ switching power supplies
■ electric motors	■ power distribution systems
■ industrial plant	■ circuit breakers/contactors
■ mobile telephones	■ computers.

Automotive systems in general are common sources of interference, with noise being generated by the ignition system, alternator, electric switches and electrostatic discharges caused by friction in the brakes.

On a smaller scale, all conductors within electronic circuits are potential sources of electromagnetic interference. The various conductors within a circuit – be they component leads, wires or printed circuit tracks – all represent small **antennas**. The mechanisms by which these antennas radiate and receive energy can be understood by considering two simplified models – namely, the *short-wire* and the *small-loop* models.

The **short-wire** model is used to describe the radiation from individual wires within a circuit. When a current flows through a wire, it forms a **monopole antenna**, which radiates energy in all directions. This form of antenna has a very high impedance, which results in the production of an electromagnetic field in which the electrical component is much greater than the magnetic component. For this reason, such an antenna is often referred to as an **electric dipole**.

Combinations of conductors and components also form loops within circuits. When a fluctuating current flows in such a loop, it produces a **small-loop** antenna that will again result in the generation of an E-M wave. This form of antenna has a relatively low impedance, which results in an electromagnetic field in which the magnetic component is larger than the electrical component. Such an antenna is called a **magnetic dipole**.

From the above, it is clear that all useful electronic circuits have the potential to produce electromagnetic interference as a consequence of the currents within them. The extent to which a given circuit radiates energy is determined by a large number of factors, including the magnitude and frequency of the signals involved and construction of the circuit. In general, this form of radiation tends to be more of a problem at high frequencies (perhaps above 30 MHz) or in circuits that have transients with components at such frequencies. Unfortunately, just as a radio antenna can receive as well as transmit signals, so the electric and magnetic dipoles within circuits provide a mechanism whereby circuits are *affected* by E–M waves as well as *producing* them.

Electromagnetic interference may also be produced by equipment that would seem to have little to do with electricity. Most of us have experienced a mild electric shock as a result of touching an earthed conductor (perhaps a metal hand rail) after walking on a non-conducting carpet. This is an example of an **electrostatic discharge** (ESD). Such discharges can also be experienced when taking off clothing made of a poorly conducting synthetic material. A build-up of static charge can be produced via friction between solids or fluids and can produce voltages of several thousand volts. In addition to giving an uncomfortable shock, the rapid discharging of the stored energy can damage electronic components and produce radiated transient signals.

Another human-made source of electromagnetic interference, though thankfully a fairly rare one, is the **electromagnetic pulse** (EMP) produced as a result of a nuclear explosion. Though of primary interest to those designing military systems that must continue to operate in the area of a nuclear explosion, it should be noted that the effects of a nuclear explosion in the upper atmosphere can affect electronic systems over an extremely large area – far greater than that directly affected by the blast.

Many electromagnetic noise sources have an extremely wide bandwidth and may produce perturbations of very large magnitudes. For example, a conventional 220 V AC domestic supply will normally have wide-band noise with a bandwidth of perhaps hundreds of megahertz. It is also likely to have occasional transients of over 1000 volts. In contrast, some noise sources have a very narrow and well-defined bandwidth. An example of such a narrow-band noise source is a mobile phone that may produce high levels of interference as a result of its transmitted signal.

22.9.2 Electromagnetic susceptibility

All electronic circuits are, to some extent, affected by electromagnetic interference. The **electromagnetic susceptibility** of a circuit is related to the extent to which it is sensitive to such disturbances.

Interference enters a system either by **conduction** or **radiation**. The importance of these two mechanisms varies with frequency. Generally, at frequencies up to 30 MHz, conduction is the dominant mechanism, while at frequencies above 30 MHz, radiation tends to be the more significant.

Conducted interference often enters a system via input or output cables or power supply leads. Radiated interference can enter a system via its casing and act directly on the internal circuitry. We will see in Section 22.10 that it is common to use an earthed metal case to screen sensitive electronic equipment to reduce the effects of radiated interference. However, radiated energy may induce noise in external cables and so enter the unit by conduction.

The electromagnetic susceptibility of a system is determined by the ease with which noise is able to enter it and the amount of noise that it can tolerate before its operation is seriously affected. These factors may also be expressed in terms of the **electromagnetic immunity** of the system, which is the ability of a system to function correctly in the presence of electromagnetic interference.

22.9.3 Electromagnetic emission

EMC is concerned not only with the ability of a system to operate correctly in the presence of electromagnetic interference, but also with its ability to operate without itself generating noise that might interfere with other equipment.

As with energy entering a system, it is clear that electromagnetic energy may also *leave* a system by either conduction or radiation. Again, the importance of these two mechanisms varies with frequency and, as one might expect, the 'routes' by which interference leaves a system are similar to those by which it can enter. Conducted energy tends to exit through input and output leads and power supply lines, while radiated energy may radiate directly through the case. The use of an earthed metal case can reduce the amount of radiated energy leaving a system, although energy may radiate from noise that is conducted out of the unit through cables.

Because of the similarity between the mechanisms by which interference enters and leaves a system, it follows that many of the methods used to tackle these problems are also similar. We will look at some of these techniques in Section 22.10.

22.9.4 Electromagnetic coupling between stages

Another important aspect of EMC is related to the ways in which one section of a system may interfere with the operation of another. Because of the close proximity of the various parts of the system, high-frequency energy can easily *radiate* from one section to another. Also, because of the interconnections between these sections, energy can be *conducted* between them through signal, power supply or ground leads. Examples of internal EMC coupling are shown in Figure 22.7.

Earlier we noted that all electronic circuits are potential sources of electromagnetic interference and that all circuits can be affected by such radiation. These effects produce unintentional coupling between the various sections of a system that can greatly affect its operation. High-frequency circuits are particularly sensitive to this form of coupling, and consequently great care is required in their design.

Common sources of internal noise include the power supply unit (PSU). Noise on the incoming AC supply is often passed by the regulating circuitry and appears on the DC supply lines. High-voltage transients are a particular problem as these will often produce spikes on the DC supply or cause the PSU to radiate noise to other parts of the system. In addition to noise that enters the system from the AC supply, the power supply unit may also produce its own interference. With simple linear supplies, the fluctuating field from the transformer can induce noise currents at the supply

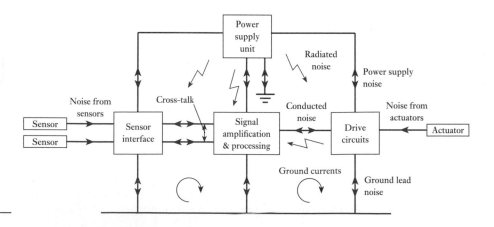

Figure 22.7
Examples of internal
EMC coupling.

frequency, while with **switch-mode power supplies** the high–frequency switching currents cause transients that can propagate throughout the system.

In addition to transmitting noise from the PSU, the power supply lines also propagate noise from one stage to another because of their impedance. When the current taken by a component or module changes, the voltage on the power line fluctuates and this variation in supply voltage is coupled to other parts of the system. Similar inter stage coupling can occur because of the impedance of the system's earth connection. Variations in the current taken by a module may cause fluctuations of the module's earth potential. If another module shares this earth connection, then its earth voltage will also be affected. In this way, the operation of one module can directly affect that of another.

Any lines that are routed near to each other may be susceptible to **cross-talk**, where signals on one line affect those on another. This phenomenon derives its name from the early days of the telephone industry where users would tend to overhear conversations on lines that were routed close to their own. Because of this problem, the industry developed several very effective methods of reducing coupling between adjacent lines, many of which are widely used today. An example of such a technique is the use of **twisted pairs** of cables, as described in the next section.

Digital systems have particular problems with unintended coupling between stages. This topic will be discussed in more detail in Chapter 26 when we consider noise in digital systems.

22.10 Designing for EMC

The EMC performance of a system is affected by almost all aspects of its design. Key factors include the frequency range of the signals involved (high frequencies cause more problems than low frequencies) and the magnitudes of the voltages and currents used. Unfortunately, these factors are often dictated by the functional requirements of the system and it may not be possible for the designer to select these parameters freely.

The circuits used within a system, and the components used within these circuits, also play a large part in determining EMC performance. In general, the designer has much more control over the selection of these aspects of the design and it is important that EMC issues are considered along with functional considerations within the circuit design process.

Other issues of very great importance to EMC are the physical layout and construction of the system. We have seen that conductors within a circuit act as small antennas that both radiate and receive electromagnetic interference. Keeping such conductors short and avoiding the formation of large loops can greatly improve EMC performance. Careful design of the layout of a system and use of appropriate grounding and shielding techniques can reduce the emission and susceptibility of a system by several orders of magnitude.

A detailed treatment of EMC requires a study of many disciplines and is not within the scope of this book. However, it is perhaps useful to consider some aspects of design that are of particular importance.

22.10.1 Analogue *vs* digital systems

Because of the wide variety of both analogue and digital systems, it is extremely difficult to make definitive statements about the characteristics of these two forms of circuitry.

Analogue systems are often characterised by restricted bandwidths and small signal amplitudes. The former is generally an advantage in EMC terms as this reduces the range of frequencies that are likely to interfere with the system. However, the latter characteristic is a disadvantage as it limits the S/N ratio that can be achieved. In analogue systems it is normal to attempt to maximise signal magnitudes to try to reduce the effects of noise.

We will see in later chapters that digital circuits are affected by noise in a different way from analogue circuits. We will also see that they are generally associated with high-frequency signals that require a very wide bandwidth. The use of a wide bandwidth tends to increase the problems associated with EMC as this implies sensitivity to noise over a greater range of frequencies. Fortunately, digital circuits tend to be less affected by noise than their analogue counterparts and, in general, digital systems are more immune to outside disturbances than analogue ones. Conversely, digital systems tend to produce more electrical interference than analogue systems.

Within the remainder of this section we will look at issues that are of relevance to all forms of electronic circuitry and leave consideration of problems that are unique to digital systems until we have looked at digital circuits in the following chapters. We will return to design considerations for EMC in Chapter 26 when we look in more detail at noise in digital systems.

22.10.2 Circuit design

We have already seen that conductors within circuits act as small unintentional antennas. In most circuits, the majority of currents flow within loops formed by an outward and a return path. Such loops can be formed by any conductive route and may include wires, printed circuit board tracks and circuit components. Figure 22.8 shows examples of current loops. Each current loop forms a magnetic dipole that radiates energy with a magnitude that is proportional to the current within the loop, the loop area and the square of the frequency. Emissions can be decreased by reducing any of these factors. Emissions due to current loops are oriented such that the electric field is at a maximum in the plane of the loop and a minimum along its axis.

While most currents flow within a loop, within a given region currents may take on a unidirectional nature, to form a *short-wire* antenna. This may occur, for example, in an isolated cable or ground lead. In such cases the current flow generates a monopole or electric dipole. At a given distance from such a source, the magnitude of the radiation is proportional to the magnitude of the current, the length of the conductor and the frequency. The radiation is not oriented about the source.

It can be seen that the radiated emission of a circuit is affected by the frequency of the signals used and the magnitudes of the currents. Wherever possible, these factors

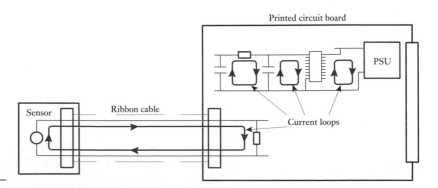

Figure 22.8
Examples of current loops within circuits.

should be reduced in order to improve EMC performance. Current loop antennas are particularly sensitive to frequency and anything that can be done to reduce the high-frequency content of a signal will be very beneficial.

22.10.3 Circuit layout

The radiation from both current-loop and short-wire antennas is greatly affected by the layout of the circuit. When laying out **printed circuit boards** (PCBs), every effort should be made to minimise track lengths and the areas of any current loops. These precautions are of particular importance when routing high-frequency signals, such as digital clock lines.

Particular attention needs to be given to the layout of the power supply lines of a PCB. Figure 22.9(a) shows a common method of laying out the power supply lines on double-sided PCBs that have many digital logic devices. Here, the positive supply voltage (often +5 V) is fed along one edge of one side of the board, while the 0 V return line is fed down the opposite edge of the other side of the board. Tracks are then fed from these two 'rails' to each of the logic devices. The resultant 'comb-like' arrangement has the advantage of being very easy to design, but is very poor from an EMC viewpoint. The power lines form a huge current loop that has an area almost equal to that of the board and in which large currents flow.

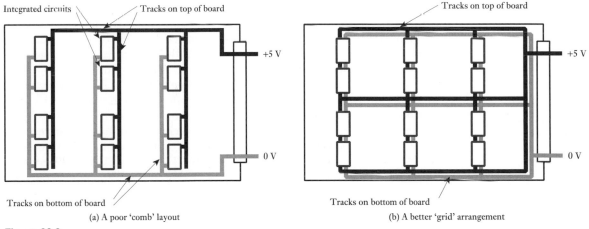

(a) A poor 'comb' layout

(b) A better 'grid' arrangement

Figure 22.9
Examples of power supply routing methods.

A better arrangement is shown in Figure 22.9(b). Here, the supply and return lines are again fed to opposite sides of the board, but in this case they are placed on top of each other to minimise the area of any loops formed and to produce capacitance between the two lines to reduce the effects of transients. In this arrangement, several lines are used in a grid to reduce power line and ground impedances.

22.10.4 Multi-layer PCBs

The grid arrangement described above can be extended by the use of multi-layer PCBs where separate planes are used for the power supply and ground. This produces a very low impedance for both the supply and ground. Tracks on alternate layers are normally arranged to be perpendicular to each other, to reduce mutual coupling (and facilitate routing).

The parallel-plate arrangement of multi-layer PCBs produces a low-impedance **transmission line** effect that reduces coupling between circuits. However, when using such boards, discontinuities, such as those at right-angled bends, should be avoided as they create high fields at the corner. It is better to use curves rather than sharp corners, although in practice 90° corners are normally broken into pairs of 45° bends.

22.10.5 Device packaging

From an EMC standpoint, surface-mounted devices have several advantages over dual in line (DIL) parts. Many of these advantages come directly from the reduced size of the parts, which leads to a greater board density. This, in turn, results in shorter lead runs and a reduction in parasitic capacitance. Most DIL packages have the power connections on opposing corner pins, thereby maximising the distance between them. This not only increases the parasitic inductance of the device, but tends to produce long tracks and large current loops.

22.10.6 Circuit partitioning and grounding

Most electrical circuits operate with one or more power supply 'rails' and a single 'ground' connection. Typically, the ground is connected to a metal case or enclosure or to the earth return of the AC power lines. Where a system has a number of components or modules, each section is normally connected to this common ground to facilitate the passing of signals. The way in which the various parts of a system are connected to the ground can have a large effect on its EMC performance.

The importance of grounding methods stems from the fact that any connection or conduction path has a certain impedance. The significance of this impedance can be seen by looking at the simple grounding scheme shown in Figure 22.10. This figure shows three modules – A, B and C – using a **series grounding arrangement**. Module A is connected to the system ground by a lead that has an impedance of Z_A.

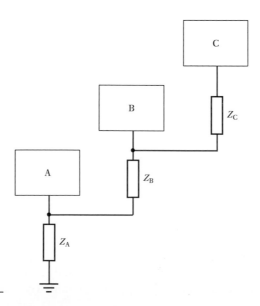

Figure 22.10
A simple grounding scheme.

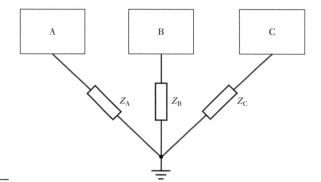

Figure 22.11
A single-point
grounding scheme.

Module B has its earth connection joined not to the system ground, but to the ground of module A. The impedance of this connection is Z_B. Similarly, module C is joined to the ground connection of module B by a lead with an impedance of Z_C.

This grounding scheme is often used because it simplifies wiring and is therefore cheaper. However, its disadvantage is that the common impedances produce coupling between the various modules. Variations in the currents taken by one of the modules will cause the ground potentials of the other modules to fluctuate.

An alternative grounding scheme is shown in Figure 22.11. Here, each module is connected directly to the system ground. This **single-point grounding** arrangement is often referred to as a **star connection** scheme.

The advantage of this method is that ground connection impedances are not shared and fluctuations in the ground potential of one module (as a result of variations in its ground current) will not be coupled to other sections of the system. This arrangement is the preferred method at relatively low frequencies and it is common for separate grounds to be used for sections where mutual interference is likely. An example of such an arrangement is shown in Figure 22.12. Here, a system is partitioned into three distinct regions to minimise interaction. Analogue and digital circuitry are separated, and a third section is used to isolate particularly 'noisy' parts of the system. This last section might include drive electronics for high-power actuators or eletromagnetic components, such as relays. The ground connections of these three sections are joined off the board at the common earth point of the system.

When arranging the grounding of the various parts of a system, it is important to ensure that sections do not have multiple ground paths. The presence of two or more

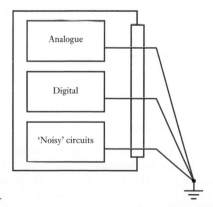

Figure 22.12
System partitioning to
reduce EMC problems.

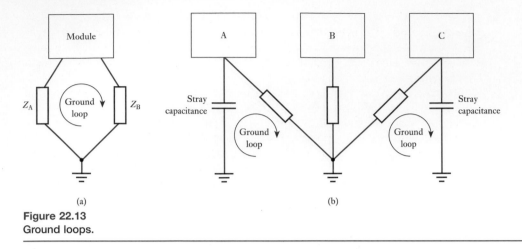

Figure 22.13
Ground loops.

ground connections to a given module will result in the formation of one or more **ground loops** (or **earth loops**), as shown in Figure 22.13(a). Such loops act like other forms of current loop and couple electromagnetic fields into the ground currents.

Unfortunately, ground loops may also be formed by stray capacitance at high frequencies, as shown in Figure 22.13(b). The single-point grounding arrangement tends to result in long ground leads for those parts of the circuit that are furthest from the common earthing point. At high frequencies, stray capacitance may provide earth routes that are comparable in impedance with the intended grounding path, forming a ground loop. For this reason, it is more common to use a low-impedance **ground plane** at high frequencies, as shown in Figure 22.14. In this **multi-point grounding** arrangement, individual components and modules are each connected directly to the ground plane using leads that are as short as possible. The technique relies on having a very low impedance within the ground plane to prevent impedance coupling problems. This is often achieved by using a plane within a multi-layer PCB.

Typically, single-point grounding techniques are used for systems with operating frequencies of up to 1 MHz, while systems operating at above 10 MHz will use a multi-point method. Systems operating between these values often adopt a hybrid approach.

Figure 22.14
A multi-point grounding scheme.

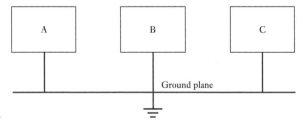

22.10.7 Enclosures and cable shielding

One of the major weapons used to combat radiative coupling is shielding. This will often include the use of a grounded conductive case around the system and shielded connecting cables. Shielding may also be used within the system to reduce coupling between its various sections.

A shielded case acts as a form of Faraday cage and can greatly reduce the field produced inside the enclosure as a result of an external electromagnetic field. It may also dramatically reduce radiative emissions. The effectiveness of the shielding provided by a case depends on a number of factors, including the material used, its thickness, construction and the frequency and nature of the interference concerned. Enclosures are usually made of metal, although metal-plated plastic cases are also used. They are normally connected directly to the primary grounding point of the system in order to achieve a low ground impedance. Earthed metal cases not only provide EMC screening but also perform a safety function.

Shielding efficiency is greatly affected by any apertures in the conductive surface, such as those caused by displays, cooling fans or cable ports. Cables entering an enclosure provide a potential route for the entry of noise, either by conduction through the cable itself or by radiation through the aperture.

External signal cables have the potential to pick up radiative noise and conduct it within the protective enclosure. Once inside, it can then be distributed within the system, either by conduction or radiation. For this reason, external cables are often screened to reduce their susceptibility to noise. There are many forms of screened cable and numerous ways of interconnecting the signal and screen conductors. A few examples are given in Figure 22.15, although these should *not* be taken as recommended configurations.

Figure 22.15(a) shows a signal being passed from one location to another using a single, unscreened conductor (wire). The return path for this signal is through the ground.

Figure 22.15(b) shows an improved arrangement where a screened coaxial cable is used, the screen being joined to ground at either end. This cable has a central

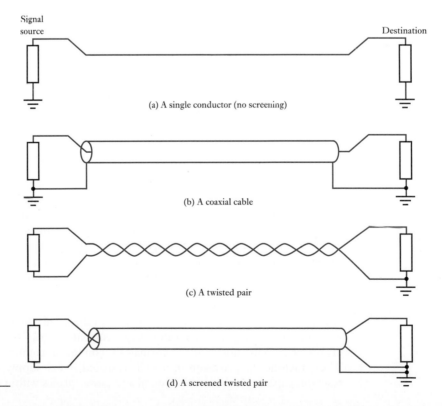

Figure 22.15
Cable screening
techniques.

conductor that is surrounded by a layer of insulating dielectric and then by a ring of copper braid that forms a conducting screen along the length of the cable. The screen is then itself surrounded by an insulating layer. The effectiveness of this arrangement will depend on many factors, but one might expect an improvement of several orders of magnitude in the amount of noise detected at the destination.

A further improvement might be achieved by means of a differential arrangement, as shown in Figure 22.15(c). In Section 14.8 we noted that noise pick-up could be reduced by connecting a signal source to a pair of wires and using a differential amplifier to detect the differential signal while rejecting the common-mode noise (see Figure 14.13). To make this technique as effective as possible, we require the two wires to pick up exactly the same amount of noise so that the differential noise is at a minimum. It has been found that this can best be achieved by twisting the two wires around each other to form a **twisted pair cable**. One might expect that the noise picked up by this arrangement would be several orders of magnitude less than that of the coaxial cable discussed earlier.

A further improvement could be achieved by using a **screened twisted pair cable**, as shown in Figure 22.15(d). It is likely that the noise delivered to the destination of this final arrangement would be more than a million times less than that expected when using a single unscreened cable.

With any cabling arrangement, great attention must be paid to the connections between the cable and the rest of the system. A short length of unscreened cable where a conductor is joined to a connector can dramatically compromise screening efficiency, as can a **pigtail** (a short length of single wire used to connect onto the outer shield of a cable). For maximum effectiveness, screened cables should be terminated within screened connectors that maintain the shielding integrity.

22.10.8 Supply line filtering and decoupling

We noted earlier that the domestic AC supply is a major source of electrical noise. For this reason, it is normal to use a **mains filter** to remove high-frequency noise from the supply in noise-sensitive applications. The filter must be fitted as close as possible to the point where the supply leads enter the system's enclosure to prevent the unfiltered leads from radiating noise within the case. To reduce this problem, some filters are incorporated into shielded mains sockets that can be fitted directly into an aperture in an enclosure. In this way, the supply is filtered as it enters the enclosure.

Mains filters tend to reduce the amount of high-frequency noise entering the system, but are less effective at dealing with high-voltage transients. Special **transient suppressors** can be used to counter these potentially dangerous events. Suppressors may be of several types, including *pn* junction devices that resemble Zener diodes. Typically, such a device would be installed across the output of a DC supply and be chosen to have a breakdown voltage somewhat greater than the nominal supply voltage. If a transient causes the voltage to exceed the breakdown voltage of the device, the suppressor will conduct, preventing the supply voltage from rising. A typical device might respond in about a nanosecond and could pass tens or perhaps hundreds of amps.

Power supply line transients and noise can also be produced by the action of circuit components. This is a particular problem in digital circuits where the switching of logic gates can cause large current surges. To minimise the effects of power supply noise, it is normal to employ a combination of techniques. These include the provision of a well-regulated power supply with low-inductance reservoir capacitors, the use of ground and power planes within a multi-layer PCB to minimise the supply line impedance and the fitting of power supply **decoupling capacitors**.

Within digital circuits, it is normal to fit a decoupling capacitor adjacent to every integrated circuit to supply the surge of current it requires during switching. This capacitor reduces the voltage transient produced on the supply line and minimises the area in which the transient current flows. A typical circuit would use ceramic capacitors (or other low-inductance ones) of between 10 and 100 nF, fitted as close as possible to each integrated circuit. Each board would then be fitted with a larger bulk decoupling capacitor of perhaps 100 μF fitted at the point where the power line enters the board. This would typically be an electrolytic or tantalum capacitor.

22.10.9 Isolation

In Section 20.5 we saw how opto-isolators could be used to safeguard electronic circuits in the presence of high voltages. These devices can also be used to pass information between circuits while preventing the passage of electrical noise. We will return to this use of opto-isolation when we look at the effects of noise in digital systems (in Chapter 26).

22.10.10 Achieving good EMC performance

Good EMC performance cannot be achieved simply through good design. In fact, EMC considerations affect all aspects of the design, development, construction, testing, installation, use and maintenance of a system. While a good design is essential to achieving good EMC characteristics, other issues related to **quality** are also very important. One could produce a design that was highly optimised to minimise both the sensitivity and emission of a system, only to have one's efforts nullified by the inadvertent fitting of plastic rather than metal washers when fitting a screen or by failing to refit an earth strap during maintenance. Good EMC performance requires commitment throughout all the phases of a system's development and use.

22.10.11 EMC and the law

While there are clear commercial advantages to producing systems that have good EMC characteristics, there are also legal requirements to be considered. Many countries have legislation that places restrictions on the amount of interference that electrical equipment may produce. Within Europe, legislation imposes strict rules on the performance of electrical products and equipment with regard to EMC. The directive covers the whole range of electrical and electronic equipment capable of producing interference or being affected by it. The 'essential requirements' of the directive are that:

The apparatus shall be so constructed that:

(a) The electromagnetic disturbance it generates does not exceed a level allowing radio and telecommunications equipment and other apparatus to operate as intended.
(b) The apparatus has an adequate level of intrinsic immunity to electromagnetic disturbance enabling it to operate as intended.

EMC issues clearly have both commercial and legal implications and all engineers need to be well acquainted with this very important area.

Video 22C

Further study

In this chapter we have looked at various aspects of electromagnetic compatibility. While EMC is of great importance in a wide range of applications, it is of particular significance in situations where failure of a system could have safety implications. Electronic systems within vehicles invariably come within this category.

Identify some of the safety-related systems within a modern, high-performance car and consider the EMC related factors that could affect their operation. What design measures could be taken to maximise the dependability of these systems in relation to EMC issues?

Key points

- Noise in electronic systems may be categorised into a number of classes:
  - thermal (Johnson) noise produced by any component that possesses resistance;
  - shot noise produced by the random nature with which charge carriers cross junctions;
  - 1/f noise, which is produced by a number of sources, including flicker noise as a result of fluctuations in the diffusion process;
  - interference from external signals or events.
- Bipolar transistors and FETs both suffer from all these types of noise to a lesser or greater extent. High-performance, low-noise amplifiers can be constructed using both forms of transistor.
- Electromagnetic compatibility (EMC) is concerned with the ability of a system to operate in the presence of interference from other electrical equipment and not interfere with the operation of other equipment or other parts of itself.
- There are many natural and human-made sources of electromagnetic interference.
- Interference may enter or leave a system by either conduction or radiation.
- Electromagnetic coupling between the various stages of a system can also be through conduction or radiation. The power supply is a particular problem area.
- Circuit layout plays a large part in determining EMC performance. All wires should be kept as short as possible and loop sizes should be minimised.
- Grounding is of great importance, with different techniques being appropriate at high and low frequencies.
- The shielding of cables is vital and can reduce the amount of interference entering the system by many orders of magnitude.
- Good EMC performance cannot be achieved simply by means of good design alone. It requires attention to detail throughout all phases of the development, use and maintenance of a system.

Exercises

22.1 What is electronic noise?

22.2 Why is electronic noise often more apparent at times when the input signal is small?

22.3 Give three everyday examples of the effects of electronic noise.

22.4 Give one reason why noise might be a greater problem in an analogue system than in a digital one.

22.5 Why is thermal noise produced by all real components?

22.6 How does Johnson noise differ from thermal noise?

22.7 Estimate the noise voltage produced by thermal noise in a 47 kΩ resistor at normal room temperature (≈ 300 K) when measured over a bandwidth of 20 kHz (the typical audio spectrum).

22.8 Calculate the thermal noise voltage produced by a resistor of 10 kΩ when connected to a system with a bandwidth of 5 MHz at normal room temperature.

22.9 Why is shot noise a particular problem in high-gain bipolar transistors?

22.10 Calculate the percentage fluctuation caused by shot noise current, of a 1 nA signal, when measured by a system with a bandwidth of 5 MHz at normal ambient temperature.

22.11 Explain what is meant by the terms 'white noise' and 'pink noise'.

22.12 What causes flicker noise in semiconductor devices?

22.13 What are the major causes of noise within bipolar transistors? Which of these forms of noise dominate at low frequencies?

22.14 What are the dominant causes of noise in field-effect transistors? Why is shot noise normally insignificant in such devices?

22.15 Give an expression for the signal-to-noise ratio of a signal in terms of the signal and noise voltages.

22.16 In a particular circuit, a signal of 100 mV r.m.s. is corrupted by 200 μV r.m.s. of noise. What is the S/N ratio of the signal?

22.17 A transducer may be represented by an ideal, noiseless voltage source of 1 mV r.m.s., in series with a resistor of 1 kΩ. If the resistor is at a temperature of 300 K, and the bandwidth of the measuring system is 100 kHz, what is the signal-to-noise ratio of the output signal?

22.18 Which type of transistor (bipolar or FET) is better for the production of low-noise systems?

22.19 List three natural and three human-made sources of electromagnetic interference.

22.20 Explain briefly what is meant by an 'electric dipole'.

22.21 Explain briefly what is meant by a 'magnetic dipole'.

22.22 Give an example of a noise source which is narrow band and one that is wide band.

22.23 In what frequency range is 'conduction' the dominant mechanism by which interference enters a system?

22.24 In what frequency range is 'radiation' the dominant mechanism by which interference enters a system?

22.25 How may an electronic system be protected from radiated interference?

22.26 How may an electronic system be protected from conductive interference?

22.27 What are the primary mechanisms of electromagnetic coupling between the various parts of an electronic system?

22.28 Describe the importance of the layout of a circuit in relation to its EMC performance.

22.29 Why is the choice of a system's grounding method affected by its operating frequency?

22.30 Discuss the importance of quality in relation to EMC.

Chapter 23

Positive Feedback, Oscillators and Stability

Objectives

When you have studied the material in this chapter, you should be able to:

- describe the use of positive feedback in the production of both sine wave and digital oscillators
- explain the conditions required for a circuit to oscillate
- sketch simple circuits of sine wave and digital oscillators
- discuss the problems of amplitude stability in such circuits
- describe the use of crystals in producing highly stable oscillators
- explain the effects of positive feedback on the stability of a circuit.

23.1 Introduction

In Chapter 15 we looked at feedback in general terms and spent some time looking at the use and characteristics of negative feedback. In this chapter, we will continue our study in this area by looking at some of the features of positive feedback.

Positive feedback is used in a range of both analogue and digital circuits to produce a variety of effects. In this chapter, we will concentrate on the most common use of positive feedback, which is in the production of **oscillators**.

While positive feedback is often used intentionally to achieve particular circuit characteristics, it can also occur *unintentionally* in circuits. This is particularly common in circuits that make use of negative feedback. In this situation, the presence of feedback can adversely affect the operation of the circuit and have dramatic effects on its stability.

Video 23A

23.2 Oscillators

When considering generalised feedback arrangements (in Chapter 15) we derived an expression for the gain of a feedback system of the form shown in Figure 23.1. You will recall that this is referred to as the **closed-loop** gain G of the system and is given by the expression

$$G = \frac{A}{(1 + AB)}$$

where A represents the **forward gain** and B the **feedback gain** of the arrangement. If the **loop gain** AB is negative and its magnitude is less than or equal to 1, then the overall gain is greater than the forward gain and we have positive feedback. You will

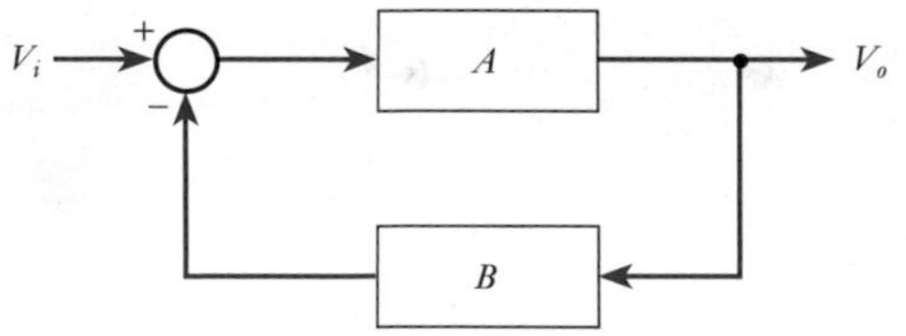

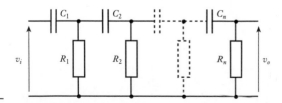

Figure 23.1
A generalised feedback
arrangement.

note that if $AB = -1$ the closed-loop gain is theoretically infinite. Under these circumstances, the system will generally produce an output even in the absence of any input. This situation is used in the production of oscillators and $AB = -1$ represents the condition needed for oscillation to occur.

When looking at feedback (in Chapter 15), we considered A and B as representing simple (real) voltage ratios. Using this approach, a non-inverting amplifier has a positive gain and an inverting amplifier has a negative gain. Thus, the condition $AB = -1$ can be satisfied if the magnitude of B is equal to $1/A$ and if either A or B (but not both) is 'inverting'. The inversion of a sine wave represents a phase shift of $180°$ and an alternative way to describe the condition for oscillation is that the product AB must have a magnitude of 1 and a phase angle of $180°$ (or π radians).

These requirements are expressed by the **Barkhausen criterion**, which, using our notation, says that the condition needed for oscillation to occur is that:

■ the magnitude of the loop gain AB must be equal to 1;
■ the phase shift of the loop gain AB must be $180°$ or $180°$ plus an integer multiple of $360°$.

The second condition is slightly more complicated than our original requirement as it acknowledges that shifting a sine wave by a complete cycle leaves it unchanged. Thus, if a phase shift of $180°$ will cause oscillation, then a phase shift of $180°$ plus any multiple of $360°$ will have the same effect.

In order to make a useful oscillator, a frequency-selective element is added to ensure that the condition for oscillation is met at only a single frequency. The circuit then oscillates continuously at that frequency.

23.2.1 The *RC* or phase-shift oscillator

A simple way to produce a phase shift of $180°$ at a single frequency is to use an RC ladder network, as shown in Figure 23.2. Here, several RC stages are cascaded, each producing an additional high-frequency cut-off. From the discussion in Section 8.5, we know that a single RC stage of this type produces a maximum phase shift of $90°$, but this maximum value is achieved only at infinite frequency. We therefore require at least three stages to produce a phase shift of $180°$ at any non-infinite frequency.

If we adopt a ladder with three identical RC stages, then standard circuit analysis reveals that the ratio of the output voltage to the input voltage is given by the expression

Figure 23.2
An *RC* ladder network.

$$\frac{v_o}{v_i} = \frac{1}{1 - \dfrac{5}{(\omega CR)^2} - j\left(\dfrac{6}{\omega CR} - \dfrac{1}{(\omega CR)^3}\right)} \tag{23.1}$$

The magnitude and phase angle of this ratio is clearly dependent on the angular frequency, ω. We are interested in the condition where the phase shift is 180°. This implies that the gain is negative and real and the imaginary part of the ratio is zero. This condition is met when

$$\frac{6}{\omega CR} = \frac{1}{(\omega CR)^3}$$

or

$$6 = \frac{1}{(\omega CR)^2}$$

This can be rearranged to give

$$\omega = \frac{1}{CR\sqrt{6}}$$

and therefore

$$f = \frac{1}{2\pi CR\sqrt{6}}$$

Substituting $\omega CR = 1/\sqrt{6}$ in Equation 23.1 gives

$$\frac{v_o}{v_i} = \frac{1}{1 - 5 \times 6} = -\frac{1}{29}$$

Therefore, at the frequency where the phase shift is equal to 180°, the gain of the ladder network is 1/29. If we use the RC ladder network as our feedback path, it is clear that $B = -1/29$. In order for the loop gain, AB, to be equal to -1, we therefore require the forward gain of the arrangement A to be +29. Oscillators based on this principle are called **RC oscillators** or sometimes **phase-shift oscillators** and Figure 23.3 shows their basic form.

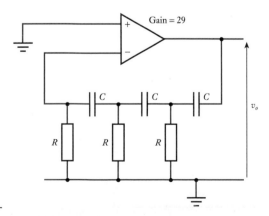

Figure 23.3
An RC or phase-shift oscillator.

It can be seen from Figure 23.3 that the phase-shift oscillator consists of an invert-ing amplifier (the input is applied to the inverting input) and a feedback network with a phase shift of 180°. The same result may be achieved using a non-inverting amplifier and a feedback network with a phase shift of 0°. This approach is used in the Wien-bridge oscillator.

23.2.2 Wien-bridge oscillator

The Wien-bridge oscillator uses a series/parallel combination of resistors and capac-itors for the feedback network, as shown in Figure 23.4. If we consider that R_1 and C_1 together constitute an impedance $\mathbf{Z}_1$, and that R_2 and C_2 represent an impedance $\mathbf{Z}_2$, it is clear that the output of the network is related to the input by the expression

$$\frac{v_o}{v_i} = \frac{\mathbf{Z}_2}{\mathbf{Z}_1 + \mathbf{Z}_2}$$

As

$$\mathbf{Z}_1 = R_1 + \frac{1}{j\omega C_1}$$

and

$$\mathbf{Z}_2 = \frac{1}{\dfrac{1}{R_2} + j\omega C_2}$$

if we make $R_1 = R_2$ and $C_1 = C_2$, it is relatively straightforward to show that

$$\frac{v_o}{v_i} = \frac{1}{3 - j\left(\dfrac{1 - \omega^2 R^2 C^2}{\omega CR}\right)} \tag{23.2}$$

In order for the phase shift of this network to be zero, the imaginary part must also be zero. This is true when

$$\omega^2 R^2 C^2 = 1$$

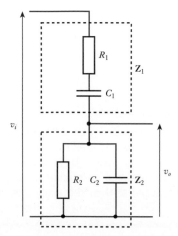

Figure 23.4
The Wien-bridge network.

that is, when

$$\omega = \frac{1}{RC}$$

Substituting for ω in Equation 23.2 gives

$$\frac{v_o}{v_i} = \frac{1}{3}$$

Thus, at the selected frequency the network has a phase shift of zero and a gain of 1/3. Further investigation of Equation 23.2 shows that the gain is a maximum at this point and that this is therefore the resonant frequency of the circuit. To form an oscillator, this network must be combined with a non-inverting amplifier with a gain of 3, making the magnitude of the loop gain unity. Figure 23.5 shows a possible arrangement using the non-inverting amplifier circuit discussed in Chapter 16.

Figure 23.5(a) shows the basic non-inverting amplifier circuit with the resistors chosen to give a gain of 3. Figure 23.5(b) shows the same circuit redrawn in a more convenient form. Figure 23.5(c) shows the oscillator formed by adding the feedback network. From the above, it is clear that the frequency of oscillation of the circuit is given by

$$f = \frac{1}{2\pi CR} \qquad (23.3)$$

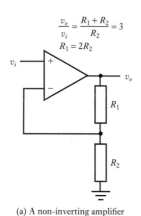

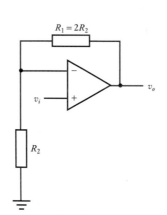

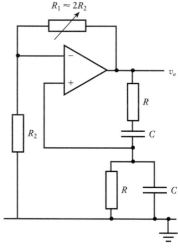

(a) A non-inverting amplifier

(b) Redrawn circuit

(c) The complete oscillator

Figure 23.5
A Wien-bridge oscillator.

File 23A

Computer simulation exercise 23.1

Simulate the Wien-bridge network of Figure 23.4 and measure its gain and phase response using the component values $R = 1$ kΩ and $C = 1$ μF. Measure the response over a frequency range from 10 Hz to 10 kHz and determine the frequency at which the output reaches its maximum amplitude. Measure the voltage gain and the phase angle at this frequency and compare these with the values given above.

23.2.3 Amplitude stabilisation

In the phase-shift and Wien-bridge circuits discussed above, the loop gain of the circuit is determined by component values in the oscillator. If the gain set is too low, the oscillations will die; if it is too high, the oscillations will grow until limited by circuit constraints.

In Figure 23.5(c), R_2 has been shown as a variable resistor to allow it to be adjusted to the correct value. In practice, the gain must be set such that the magnitude of the loop gain is slightly greater than unity to ensure that any oscillation grows rather than decays and to allow for any downward fluctuation in the gain of the amplifier.

Several methods exist for limiting the magnitude of the oscillation. In the circuit shown in Figure 23.5(c), the amplitude is restricted simply by the limitations on the output swing of the amplifier. Fortunately for this application, operational amplifiers have non-linear gain characteristics and the gain tends to drop as the amplitude approaches the supply rails. Thus, if the gain is set to slightly greater than that required to maintain the oscillation for small signals, as the signal amplitude increases it will enter a region where the gain falls and the magnitude will stabilise at that value. While this is a simple method, it does produce some distortion because the amplifier is being used in its non-linear region.

A possible solution is to replace the variable resistor R_1 in the circuit of Figure 23.5(c) with a suitable thermistor (as discussed in Section 12.3.2). The resistor values are chosen such that, when the thermistor is at normal room temperature, the gain is slightly greater than that required for oscillation, so the amplitude of the output increases. This increases the power dissipated in the thermistor, causing it to heat up. The increase in temperature causes the resistance of the thermistor to fall, reducing the gain of the circuit. The amplitude of the oscillation therefore stabilises at a point where the magnitude of the loop gain is exactly unity. This limits the amplitude of the output signal without causing distortion.

Although the use of a thermistor is a possible solution to the problem, there are several more elegant solutions. However, the detailed design of oscillators is beyond the scope of this book and we will not discuss this further.

Video 23B

23.2.4 Digital oscillators

The oscillators considered so far are intended to produce a sinusoidal output (although, as we have seen, this is often slightly distorted due to amplitude stabilisation problems). Positive feedback is also widely used in a range of digital applications. These include a range of digital oscillator circuits.

A simple digital oscillator is illustrated in Figure 23.6, which shows a **relaxation oscillator**. To understand the operation of this circuit, imagine that initially (when power is applied to the circuit) the inputs have a slight bias, such that the non-inverting input of the operational amplifier is more positive than the inverting input. This offset will be amplified by the op-amp and its output will become large and positive. The actual voltage produced will depend on the nature of the op-amp (as discussed in Section 16.5.4), but for our current purposes it is sufficient to assume that the output will be close to the positive supply voltage, V_{pos}. The potential divider formed by the two resistors sets the voltage on the non-inverting input and as these are of equal value (R_1) the voltage on the non-inverting input will be about $V_{pos}/2$. If we assume that initially the capacitor is uncharged, the voltage on the inverting input will be zero and the voltage difference between the two op-amp inputs will maintain the output at its maximum positive value.

If the output is positive, a positive voltage is applied across the RC combination, which will charge the capacitor and cause the voltage across it to increase exponentially

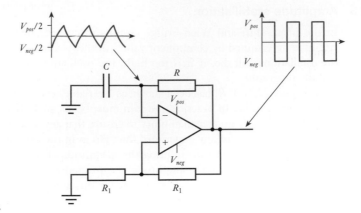

Figure 23.6
A relaxation oscillator.

towards V_{pos}. The voltage across the capacitor sets the voltage on the inverting input, so this also increases. However, when this voltage becomes greater than the voltage on the non-inverting input (which is $V_{pos}/2$), the polarity of the input voltage to the op-amp will be reversed and its output will become close to the negative supply voltage, V_{neg}. This in turn will change the voltage on the non-inverting input to $V_{neg}/2$, which will tend to force the output to become even more negative. The voltage across the RC network is now reversed and the capacitor will start to charge exponentially towards V_{neg}. This continues until the voltage on the inverting input reaches $V_{neg}/2$, when the output will reverse once more and the cycle will start again. This produces a continuous oscillation of the output between the two supply voltages. The frequency of oscillation is determined by the rate at which the capacitor charges and this is set by the time constant of the arrangement, which is equal to CR.

In practice, the output produced by the relaxation oscillator in Figure 23.6 is not a perfect square wave as the slew rate of the op-amp limits the speed at which the output can change. The slew rate of an op-amp was discussed in Section 16.5.10.

File 23B

Computer simulation exercise 23.2

Simulate the relaxation oscillator of Figure 23.6 using an appropriate operational amplifier. Suitable values would be $R = 1$ kΩ, $C = 1$ μF and $R_1 = 10$ kΩ. Observe both the output of the circuit and the voltage across the capacitor and confirm that these are as expected.

23.2.5 Crystal oscillators

The **frequency stability** of an oscillator is largely determined by the ability of the feedback network to select a particular operating frequency. In a **resonant circuit**, this ability is described by its quality factor or Q, which determines the ratio of its resonant frequency to its bandwidth (this topic was discussed in Section 8.12 when we looked at resonance). A circuit with a very high Q will be very frequency selective and therefore tend to have a stable frequency. Networks based on resistors and capacitors have relatively low values of Q. Those based on combinations of inductors and capacitors are better in this respect, with Q values of up to several hundred. These are suitable for most purposes, but are not adequate for some demanding applications, such as the measurement of time. In such cases, it is normal to use a frequency-selective network based on a crystal.

Some materials have a **piezoelectric** property in that deformation of the substance causes them to produce an electrical signal. The converse is also true: an applied electric field will cause the material to deform. A result of these properties is that if an alternating voltage is applied to a crystal of one of these materials it will vibrate. The mechanical resonance of the crystal, caused by its size and shape, produces an electrical resonance with a very high Q. Resonant frequencies from a few kilohertz to many megahertz are possible with a Q as high as 100,000.

These piezoelectric resonators are commonly referred to simply as **crystals** and are most commonly made from **quartz** or some form of **ceramic** material. The circuit symbol for a crystal is shown in Figure 23.7(a). Functionally, the device resembles a series RLC resonant circuit (with a small amount of parallel capacitance, C_P), and Figure 23.7(b) shows a simple equivalent circuit. The devices have a pair of resonant frequencies: at one (the parallel resonant frequency) the impedance approaches infinity, while at the other (the series resonant frequency) it drops almost to zero. Over the remainder of the frequency range, the device looks like a capacitor. The parallel resonance occurs at a slightly higher frequency than the series resonance, but the frequency difference is normally so small that it may be ignored. The presence of these two forms of resonance allows the device to be used in a number of different circuit configurations.

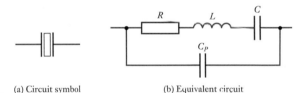

Figure 23.7
A crystal.

(a) Circuit symbol   (b) Equivalent circuit

Crystal oscillators are widely used in a range of analogue and digital applications. They form the basis of the time measurement in digital watches and clocks and are used to generate the timing reference (clock) in most computers. Figure 23.8 shows the circuit of a simple **digital oscillator** based on a crystal. This is a form of **Pierce oscillator**, where a logical inverter is used to provide a high-gain inverting amplifier and the crystal provides positive feedback at its resonant frequency. The second inverter in this circuit 'squares up' the output from the oscillator and also acts as a buffer, increasing the circuit's ability to drive a load.

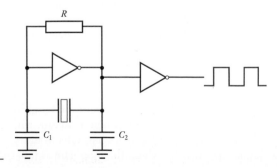

Figure 23.8
A crystal oscillator.

<table>
<tr><td>**23.3**</td><td></td></tr>
</table>

23.3 Stability

In the previous section, we used a general expression for the gain of a feedback network, namely

$$G = \frac{A}{(1 + AB)}$$

and considered the conditions necessary to produce oscillation. However, so far we have only considered the situation where a circuit designer intentionally sets out to produce this effect. Unfortunately, oscillation sometimes occurs unintentionally as a result of unwanted positive feedback in a circuit. To see how this can occur, we need to look in more detail at the nature of the *gains* in the circuit.

In our earlier discussions, we have assumed that the forward and feedback gains of our circuit (A and B) can be represented by simple, real numbers. When we design a circuit to make use of negative feedback (as in most amplifiers), the selected component values will set A and B, such that $|1 + AB|$ is positive and greater than unity. This gives the various advantages associated with the use of negative feedback (as discussed in Section 15.6). However, from our study of frequency response in Chapter 8, we know that the gain of an amplifier has not only a *magnitude* but also a *phase angle*. We also know that the gain of all amplifiers falls at high frequencies and that associated with this fall in gain is an increasing phase shift. In almost all cases, the phase shift will become greater than 180° as the frequency is increased. As a phase shift of 180° corresponds to an inversion of a sine wave, the effective gain of the circuit changes polarity. Thus, a circuit designed to take advantage of negative feedback may, at high frequencies, see the effects of positive feedback instead. Not only does this remove the beneficial effects of negative feedback (such as its effects on gain, frequency response and input/output impedance) but it may also result in the circuit becoming unstable and starting to oscillate. Thus, when designing a feedback circuit, one must consider not only its performance within its operational frequency range but also its stability.

The stability of an amplifier is determined by the term $(1 + AB)$. If this term is positive, we have negative feedback and the stability of the circuit is assured. If, as a result of a phase shift, the term $(1 + AB)$ becomes less than 1 (because AB becomes negative), the feedback becomes positive and all the advantages of negative feedback are lost. In the extreme case, if $(1 + AB)$ becomes equal to 0, the closed-loop gain of the arrangement is infinite and the system becomes unstable and will oscillate. This corresponds to the circuit satisfying the conditions for oscillation set out in the Barkhausen criterion discussed earlier.

The condition that $(1 + AB) = 0$ represents the case where $AB = -1$, or, in other words, where the loop gain has a magnitude of 1 and a phase of 180°. Indeed, the amplifier will remain stable even if the phase shift is equal to 180° provided that the magnitude of the loop gain is less than unity. The task of the designer is thus to ensure that the loop gain of the amplifier falls below unity *before* the phase shift reaches 180°.

23.3.1 Gain and phase margins

In practice, it is advisable to allow some margin for variability in the phase and gain values. This leads to the concept of the **phase margin**, which is the angle by which the phase is less than 180° when the loop gain falls to unity, and the **gain margin**, which is the amount (in dB) by which the loop gain is less than 0 dB (that is, unity gain) when the phase reaches 180°. These quantities may be illustrated using a Bode diagram, as shown in Figure 23.9.

The above discussion makes it clear why designers of op–amps (such as the 741 described in Chapter 16) choose to add a single dominant time constant to the amplifier to roll off the gain (as shown in Figure 16.14). This ensures that the gain falls to less than 0 dB well before the phase shift reaches 180°. This produces large gain and phase margins and ensures good stability.

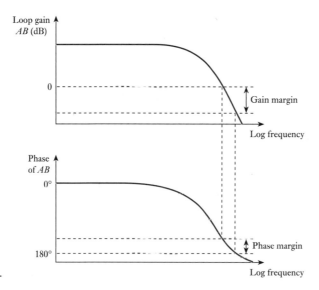

Figure 23.9
Gain and phase margins.

23.3.2 Nyquist diagrams

An alternative method of investigating the stability of a circuit is the use of a **Nyquist diagram**, which illustrates the relationship between gain and phase in a single plot. The diagram is essentially a plot of the real and imaginary parts of the loop gain AB for all frequencies. An example of a Nyquist diagram for an amplifier with a single low-frequency cut-off and a single high-frequency cut-off is shown in Figure 23.10.

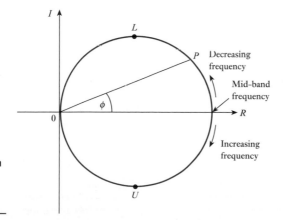

Figure 23.10
Nyquist diagram for an amplifier with single upper and lower cut-offs.

The diagram is formed by the locus of P, where the distance of P from the origin (the magnitude of OP) represents the magnitude of AB and the angle ϕ represents its phase. At mid-band frequencies, the phase of the output is zero and the gain is simply real. As the frequency is reduced, the lower cut-off frequency causes the magnitude of the gain to fall and produces a positive phase angle. When the frequency is equal to the lower cut-off frequency, the gain will have dropped to 0.707 of its mid-band value and its phase will be $+45°$. Thus, the lower cut-off frequency corresponds to point L. As the frequency goes to very low values, the magnitude of the gain tends to zero and the phase tends to $+90°$. The locus of P thus approaches the origin along the positive imaginary axis. At high frequencies, the upper cut-off frequency reduces the magnitude of the gain and produces a phase lag. At the upper cut-off frequency the magnitude of the gain has again dropped to 0.707 of its mid-band value and the phase angle is $-45°$. This corresponds to point U. As the frequency increases, the magnitude of the gain falls towards zero and the phase angle tends to $-90°$. The locus therefore approaches the origin along the negative imaginary axis. For this idealised case (an amplifier having one upper and one lower cut-off frequency), the Nyquist diagram is a circle.

Figure 23.11 shows some examples of Nyquist diagrams for a range of amplifiers. Figure 23.11(a) represents an amplifier with no low-frequency cut-off (a DC-coupled amplifier). The gain therefore stays constant at low frequencies rather than falling to zero. This example is for an amplifier having a single high-frequency cut-off. The maximum phase shift is therefore $-90°$ and the locus approaches the origin along the negative imaginary axis.

Figure 23.11(b) represents a DC amplifier with two high-frequency cut-offs. This has a maximum phase shift of $180°$ and thus approaches the origin along the negative real axis. Figure 23.11(c) shows a system with three high-frequency cut-offs and the locus therefore approaches the origin along the positive imaginary axis. Figure 23.11(d)

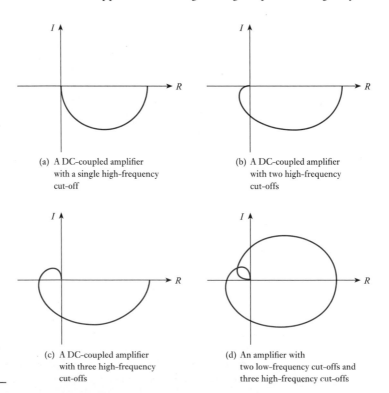

(a) A DC-coupled amplifier with a single high-frequency cut-off

(b) A DC-coupled amplifier with two high-frequency cut-offs

(c) A DC-coupled amplifier with three high-frequency cut-offs

(d) An amplifier with two low-frequency cut-offs and three high-frequency cut-offs

Figure 23.11
Examples of Nyquist diagrams.

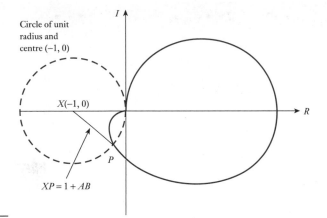

Figure 23.12
Investigations of
stability using a
Nyquist diagram.

shows the response of an amplifier with two low-frequency cut-offs and three high-frequency cut-offs. This has a maximum phase shift of +180° at low frequencies and -270° at high frequencies.

The stability of an amplifier is determined by the magnitude of the term $1 + AB$, which may be represented on the Nyquist diagram by a line drawn from the point $(-1, 0)$ to the locus, P. This is shown in Figure 23.12.

If a circle of unit radius is drawn with its centre at $(-1, 0)$, whenever P lies within this circle it represents a point at which $1 + AB$ is less than unity. Under these circumstances, the feedback is positive rather than negative. This implies that the gain of the amplifier is greater than A and all the advantages of negative feedback are lost. If the locus passes through the point $(-1, 0)$ this represents the condition that $1 + AB$ is equal to 0. This means that the gain of the amplifier is infinite and it is thus unstable and will oscillate.

The Nyquist diagram can therefore be used to investigate the stability of a system, the general principles of which are summed up in the **Nyquist stability criterion**. This may be paraphrased as:

- if the locus of P does not enter the unit circle centred on $(-1, 0)$ the circuit is stable and has negative feedback;
- if the locus enters the unit circle, the feedback is positive within that region;
- if the locus encircles the point $(-1, 0)$, the amplifier will oscillate.

It can be seen that amplifiers with no more than a single upper and lower cut-off will always be stable as the locus of P is always to the right of the origin. Amplifiers with two upper or two lower cut-offs can enter a region where the feedback is positive but cannot encircle the point $(-1, 0)$ and so are always stable. Systems with more than two upper or lower cut-offs may be unstable if the locus of P encircles the point $(-1, 0)$.

23.3.3 Unintentional feedback

Stability can also be affected by unintended feedback in a circuit. For example, the presence of stray capacitance or stray inductance in a circuit may introduce additional feedback paths that do not form part of the original design. If these represent positive feedback, then they can cause instability in a similar manner to that described above. These problems are more severe in high-frequency applications, where small amounts of capacitance can have a dramatic effect. Such problems must be tackled by careful design to minimise these spurious effects.

Video 23C

Further study

A particular electronic application requires a sine wave signal with a frequency of 1 kHz and an amplitude of several volts. It is important that this signal represents a reasonably pure sine wave with minimal distortion. The application also requires an indicator light in the form of a low-voltage incandescent light bulb.

When discussing sine wave oscillators we noted that one problem with simple circuits is that they tend to produce distorted waveforms due to problems associated with amplitude stabilisation. We also noted that one approach to solving this problem is to use a thermistor within the circuit to stabilise the amplitude without producing distortion.

Incandescent light bulbs have a relatively low resistance when they are cold that increases rapidly when the bulb is turned on and heats up. Since our system has need of such a bulb it would be convenient to use this bulb to stabilise the output of our oscillator without the need for a thermistor. Sketch the circuit of such an arrangement and consider how the various component values would be determined.

Key points

- Positive feedback is used in a range of both analogue and digital circuits.
- One of the primary uses of positive feedback is in the production of oscillators.
- The requirements for oscillation are that the loop gain AB must have a magnitude of 1 and a phase of 180° (or 180° plus some integer multiple of 360°).
- This condition can be satisfied by using an arrangement that produces a phase shift of 180° (at a particular frequency) in association with an inverting amplifier – as in a phase-shift oscillator.
- Alternatively, it can be satisfied using an arrangement that produces a phase shift of 0° in association with a non-inverting amplifier – as in the Wien-bridge oscillator.
- Sine wave oscillators present the problem of maintaining the gain of the circuit at 1 without causing distortion of the output.
- Positive feedback is also used in digital applications, such as the production of digital oscillators.
- Where good frequency stability is required, circuits normally make use of crystals.
- While positive feedback is a useful tool in circuit design, it can also pose problems. At high frequencies, negative feedback arrangements can exhibit positive feedback, which may lead to instability. Unwanted feedback can also cause problems.

Exercises

23.1 State the Barkhausen criterion for oscillation.

23.2 As an RC network can produce up to 90° of phase shift, why can a phase-shift oscillator not use just two stages in its ladder network?

23.3 Calculate the frequency of oscillation of a phase-shift oscillator that uses a three-stage ladder network, each with $R = 1$ kΩ and $C = 1$ μF.

23.4 Simulate the RC ladder network of Figure 23.2 using three stages, each with $R = 1$ kΩ and $C = 65$ nF, and measure the gain and frequency response of this arrangement. Measure the frequency at which the phase shift is equal to 180° and the gain at this frequency and confirm that these are as expected.

23.5 A Wien-bridge oscillator of the form shown in Figure 23.5 is constructed using $R = 100$ kΩ and $C = 10$ nF. Calculate the frequency of oscillation.

23.6 Why does amplitude stabilisation present a problem in simple sine wave oscillators?

23.7 Explain how a thermistor might be used to stabilise the output amplitude of an oscillator.

23.8 A relaxation oscillator of the form shown in Figure 23.6 is constructed with $R = 1$ kΩ, $C = 10$ μF and $R_1 = 10$ kΩ. Estimate the frequency of oscillation of this circuit by considering the charging rate of the capacitor.

23.9 Use simulation to confirm your answer to the previous exercise. You may find it useful to start with the circuit of Computer simulation exercise 23.2.

23.10 Why are crystal oscillators used in digital watches rather than circuits based on RC or RL techniques?

23.11 Calculate the percentage accuracy required in the frequency of oscillation of a clock that must keep time to within 1 second per month.

23.12 Why does the phase response of an amplifier have implications for its stability at high frequencies?

23.13 Explain what is meant by the gain margin and the phase margin of a circuit.

23.14 How can stray capacitance affect the stability of a circuit?

23.15 The Nyquist diagrams below represent four circuits. In each case, determine the number of low-frequency and high-frequency cut-offs present and whether or not the circuit is stable.

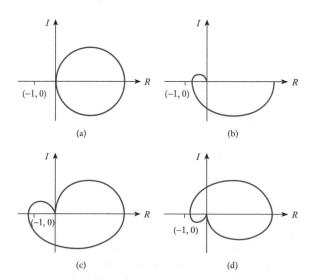

Chapter 24 Digital Systems

Objectives

When you have studied the material in this chapter, you should be able to:

- define terms such as binary variable, logic state, logic gates and logic operators such as AND, OR, NOT and Exclusive OR

- use truth tables and Boolean algebra to represent the operation of simple logic functions

- design arrangements of standard logic gates to perform particular functions that are specified in words or symbolically

- perform addition and subtraction using binary arithmetic and convert numbers between a range of number bases

- describe codes used for the representation of numeric and non-numeric quantities such as alphabetic characters.

24.1 Introduction

In Chapter 11 we noted that some information sources produce discrete – in other words, digital – signals, and in Chapters 12 and 13 we looked at a range of digital sensors and actuators. It is now time to look at the techniques used to process digital signals and the design of digital systems.

Although digital signals can take many forms, in this chapter we are primarily concerned with **binary** signals as these are the most common form of digital information. Binary signals may be used individually, perhaps to represent the state of a single switch, or in combination to represent more complicated quantities. We will start by looking at the processing of individual binary quantities and then move on to more complicated arrangements.

Video 24A

24.2 Binary quantities and variables

A **binary quantity** is one that can take only two states. Examples include a switch that can be only on or off, a hydraulic valve that can be only open or closed and an electric heater that can be only on or off. It is common to represent such quantities using **binary variables**, which are simply symbolic names for the quantities.

Figure 24.1 illustrates a simple binary arrangement involving a battery, a switch and a lamp. If the state of the switch is represented by the binary variable S and the state of the lamp by the binary variable L, we can represent the relationship between these two variables symbolically using a table

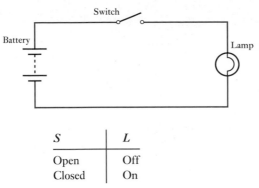

Figure 24.1
A simple binary
arrangement.

S	L
Open	Off
Closed	On

We can also use a symbolic name for the state of each variable, so that rather than using terms such as open and closed or on and off, we can use *symbols* for the states such as '0' and '1'. If we use the symbol '0' to represent the switch being open and the lamp being off, our table becomes

S	L
0	0
1	1

The mapping between on and off and '0' and '1' is arbitrary, but the user must know what the relationship is. It is common to use '1' to represent the ON state, a switch being closed or a statement being true. It is common to use '0' for the OFF state, a switch being open or a statement being false.

The table lists on the left all the possible states of the switch and indicates, on the right, the corresponding states of the lamp. Such a table is called a **truth table** and it defines the relationship between the two variables. The order in which the possible states are listed is normally *ascending binary order*. If you are not aware of the meaning of this phrase, it will become clear when we look at number systems and binary arithmetic later in this chapter.

Figure 24.2(a) shows an arrangement incorporating two switches in series. Here, it is necessary for *both* switches to be *closed* in order for the lamp to light. The relationships between the positions of the switches and the state of the lamp are given in the truth table shown in Figure 24.2(b). Notice that the table now has four rows to represent all the possible combinations of the two switches. Alternatively, we could express this relationship in words as 'the lamp will be illuminated if, and only if, switch *S1* is closed AND switch *S2* is closed'. We can abbreviate this statement as

$$L = S1 \text{ AND } S2$$

This AND relationship is very common in electronic systems and is found in a variety of everyday applications. For example, automotive brake lamps are often only illuminated if the foot brake is depressed, closing a switch, AND the ignition switch is on.

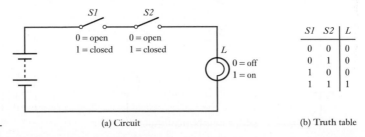

Figure 24.2
Two switches in series.

(a) Circuit

S1	S2	L
0	0	0
0	1	0
1	0	0
1	1	1

(b) Truth table

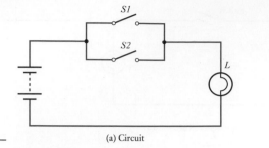

S1	S2	L
0	0	0
0	1	1
1	0	1
1	1	1

Figure 24.3
Two switches in parallel.

(a) Circuit

(b) Truth table

Figure 24.3(a) shows an arrangement that has two switches in parallel. In this configuration, the lamp will light if either of the switches is closed. This function is described in the truth table of Figure 24.3(b), where the meanings of '0' and '1' are as for the previous example. We can express this relationship in words as 'the lamp will be illuminated if, and only if, switch *S1* is closed OR switch *S2* is closed (or if both are closed)' or, in the abbreviated form,

$$L = S1 \text{ OR } S2$$

An example of the OR function, again an automotive application, might be the courtesy light, which is illuminated if the driver's door is open (closing a switch) OR if the passenger's door is open. This function is sometimes called the 'Inclusive OR' function, as it includes the case where both inputs are true (that is, in this case, when both switches are closed).

Our examples of the AND and OR functions can be extended to the use of three or more switches, as illustrated in Figures 24.4 and 24.5. These figures show three switches, but the process can be expanded, allowing any number of switches to be connected in series or in parallel. Note that there are eight possible combinations of the positions of three switches, leading to eight rows in the truth tables.

Consider now the circuit in Figure 24.6. Here, two switches are connected in parallel and this combination is in series with a third switch. This produces an arrangement

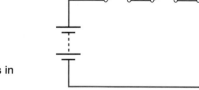

S1	S2	S3	L
0	0	0	0
0	0	1	0
0	1	0	0
0	1	1	0
1	0	0	0
1	0	1	0
1	1	0	0
1	1	1	1

Figure 24.4
Three switches in series.

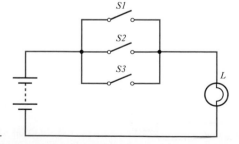

S1	S2	S3	L
0	0	0	0
0	0	1	1
0	1	0	1
0	1	1	1
1	0	0	1
1	0	1	1
1	1	0	1
1	1	1	1

Figure 24.5
Three switches in parallel.

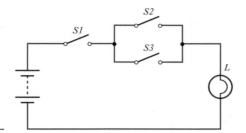

Figure 24.6
A series/parallel
configuration.

S1	S2	S3	L
0	0	0	0
0	0	1	0
0	1	0	0
0	1	1	0
1	0	0	0
1	0	1	1
1	1	0	1
1	1	1	1

that can be described by the truth table shown or by the statement 'the lamp will be illuminated if, and only if, *S1* is closed AND either *S2* OR *S3* is closed'. This can again be given in an abbreviated form as

$$L = S1 \text{ AND } (S2 \text{ OR } S3)$$

Notice the use of brackets to make the meaning of the expression clear and avoid ambiguity.

In the examples so far considered, we have started with a combination of switches and represented them in a truth table and verbal description. In practice, we will generally need to perform the process in reverse, being given a function and being required to devise an arrangement to produce this effect. In such cases, we might be given either a truth table or a verbal description of the required system. Consider the following truth table:

S1	S2	S3	L
0	0	0	0
0	0	1	0
0	1	0	0
0	1	1	1
1	0	0	0
1	0	1	1
1	1	0	1
1	1	1	0

This represents an arrangement with three switches – *S1*, *S2* and *S3* – and a lamp *L*. We would consider the switches to be the three *inputs* to the network and the lamp to be the *output*.

It is not immediately obvious what arrangement of switches would correspond to this truth table and perhaps it is not clear if *any* combination of the three switches can produce the desired results. In such cases, it is often useful to consider the desired arrangement as a 'black box' with the various switches as inputs and the lamp as an output. Such an arrangement is shown in Figure 24.7.

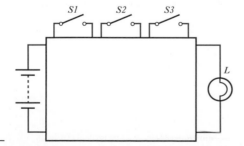

Figure 24.7
Representation of an
unknown network.

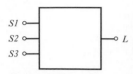

Figure 24.8
Symbolic
representation of an
unknown network.

The diagram in Figure 24.7 makes no assumptions concerning the method of interconnection of the switches and lamp. It may be that, in order to produce the desired function, we will need some form of electronic circuitry in our 'black box'. As the three switches and lamp represent simple binary devices, we could produce a more general arrangement by showing these as simple binary variables without defining their type. You will remember that, when representing 'black box' amplifiers, such as operational amplifiers, we often omit the connections to the power supply. If we adopt a similar scheme here, we arrive at a diagram of the form shown in Figure 24.8.

We now have a symbolic representation of a network with three inputs and one output that makes no assumptions as to the form of the inputs or output. This can clearly be extended to represent systems with any number of inputs and outputs. The inputs could represent switches, as in the earlier examples, but they could equally well be signals from binary sensors, such as thermostats, level sensors or proximity switches. Similarly, the output devices could be lamps but could equally well be heaters or solenoids. To implement our digital system, we need to take the various input signals and use them to produce appropriate output signals. The normal building blocks used to achieve this are **logic gates**.

Video 24B

24.3 Logic gates

A logic gate is an element that takes one or more binary input signals and produces an appropriate binary output, depending on the state(s) of the input(s). There are three elementary gate types, two of which – the AND and the OR functions – we have already met. These elementary gates can be combined to form more complicated gates, which in turn may be connected to produce any required function. Each type of gate has its own **logic symbol**, which allows complicated functions to be represented by a **logic diagram**. The function of each gate can also be represented by a mathematical notation known as **Boolean notation**. This allows complicated functions to be manipulated, and perhaps simplified, by means of **Boolean algebra**.

24.3.1 Elementary logic gates

The AND gate

The output of an AND gate is true (1) if, and only if, all of the inputs are true. The gate can have any number of inputs. The logic symbol and truth table for a two-input AND gate are given in Figure 24.9. The labelling of the inputs and outputs is arbitrary. The Boolean notation for the AND function is a dot. For example, the gate in Figure 24.9 could be described by the expression $C = A \cdot B$. In practice, it is common to omit the dot, so the expression is often written as $C = AB$.

Figure 24.9
A two-input AND gate.

(a) Circuit symbol

A	B	C
0	0	0
0	1	0
1	0	0
1	1	1

(b) Truth table

$C = A \cdot B$

(c) Boolean expression

The OR gate

The output of an OR gate is true (1) if, and only if, at least one of its inputs is true. It is also called the 'Inclusive OR' gate, for the reasons discussed earlier. The gate can have any number of inputs. The logic symbol and truth table for a two-input OR gate are given in Figure 24.10. The Boolean notation for the OR function is '+'. For example, the gate in Figure 24.10 could be described by the expression $C = A + B$.

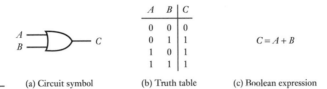

Figure 24.10
A two-input OR gate.

(a) Circuit symbol   (b) Truth table   (c) Boolean expression

The NOT gate

The output of a NOT gate is true (1) if, and only if, its single input is false. This gate has the function of a **logical inverter** as the output is the **complement** of the input. The gate is sometimes referred to as an **invert gate** or simply as an **inverter**. The circuit symbol and truth table for a NOT gate are shown in Figure 24.11. In Boolean notation, inversion is represented by putting a line (a bar) above the expression for the signal. The operation of the gate in Figure 24.11 can be written as $B = \overline{A}$ which is read as 'B equals NOT A' or as 'B equals A bar'.

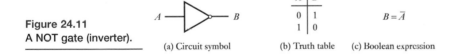

Figure 24.11
A NOT gate (inverter).

(a) Circuit symbol   (b) Truth table   (c) Boolean expression

The circle in the symbol for an inverter represents the process of inversion. The triangular symbol without the circle would represent a function in which the output state was identical to the input. This function is called a **buffer**. The presence of a buffer does not affect the state of a logic signal. However, when we come to consider the implementation of gates using electronic circuits, we will see that a buffer can be used to change the electrical properties of a logic signal. It is interesting to note that the symbol for a buffer is similar to that used for a single-input analogue amplifier. As the buffer does not produce any logical function, it is usually not considered to be an elementary gate. However, for completeness, its logic symbol and truth table are given in Figure 24.12. Clearly, in this case, $B = A$.

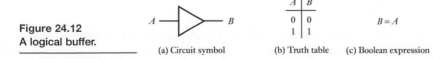

Figure 24.12
A logical buffer.

(a) Circuit symbol   (b) Truth table   (c) Boolean expression

24.3.2 Compound gates

The elementary gates described above can be combined to form any desired logic function. However, it is often more convenient to work with slightly larger building blocks. Several compound gates are used that are simple arrangements of these elementary gates.

The NAND gate

The NAND gate is functionally equivalent to an AND gate followed by an inverter – the name is an abbreviation of Not AND.

Following the example set with the symbol for an inverter, the logic symbol for a NAND gate is simply that for an AND gate with a circle at the output. The truth table for the NAND gate is similar to that for an AND gate with the output state inverted. A NAND gate can have any number of inputs. The logic symbol and truth table for a two-input NAND gate are shown in Figure 24.13. This function would be written as

$$C = \overline{A \cdot B}$$

or simply

$$C = \overline{AB}$$

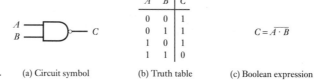

A	B	C
0	0	1
0	1	1
1	0	1
1	1	0

$$C = \overline{A \cdot B}$$

Figure 24.13
A two-input NAND gate.

(a) Circuit symbol   (b) Truth table   (c) Boolean expression

The NOR gate

The NOR gate is functionally equivalent to an OR gate followed by an inverter – the name being an abbreviation of Not OR.

The logic symbol is that of an OR gate with a circle at the output to indicate an inversion. A NOR gate can have any number of inputs. Figure 24.14 shows the logic symbol and truth table of a two-input NOR gate. This function would be written as $C = \overline{A + B}$.

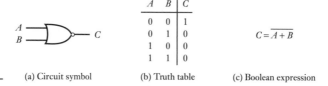

A	B	C
0	0	1
0	1	0
1	0	0
1	1	0

$$C = \overline{A + B}$$

Figure 24.14
A two-input NOR gate.

(a) Circuit symbol   (b) Truth table   (c) Boolean expression

The Exclusive OR gate

The output of an Exclusive OR gate is true (1) if, and only if, one or other of its two inputs is true, but not if both are true.

The gate gets its name from the fact that it resembles the Inclusive OR gate, except that it *excludes* the case where both inputs are true. An Exclusive OR gate always has exactly two inputs. The logic symbol and truth table for an Exclusive OR gate are given in Figure 24.15. The Exclusive OR function has its own Boolean symbol, which is '⊕'. The arrangement in Figure 24.15 would therefore be written as $C = A \oplus B$.

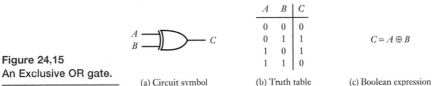

Figure 24.15
An Exclusive OR gate.

(a) Circuit symbol   (b) Truth table   (c) Boolean expression

The Exclusive NOR gate

The last member of our group of compound gates is the Exclusive NOR gate, which, as its name suggests, is the inverse of the Exclusive OR gate. This may be considered to be an Exclusive OR gate followed by an inverter. This gate gives a true output when both inputs are 0 or when both are 1. It therefore gives a true output when the inputs are equal. For this reason, this gate is also known as an **equivalence** or **equality gate**. The logic symbol and a truth table for the Exclusive NOR gate are shown in Figure 24.16. This function would be written as $C = \overline{A \oplus B}$.

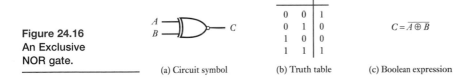

Figure 24.16
An Exclusive
NOR gate.

(a) Circuit symbol   (b) Truth table   (c) Boolean expression

24.3.3 Using logic gates

The various logic gates are summarised in Table 24.1, which shows their circuit symbols, Boolean expressions and truth tables. The table shows two circuit symbols for each gate. The first is the 'distinctive shape' symbol given earlier, the second is an alternative symbol defined in international standard IEC 617. Both forms are widely used, but, in this text, we adopt the distinctive shape symbols as these are probably more widely used in engineering courses at this level.

Using appropriate combinations of these gates, it is possible to implement any required relationship between a set of binary inputs and outputs. Applications might range from the use of a handful of gates to produce a simple control mechanism to the use of perhaps millions of gates to produce a complete microcomputer.

Table 24.1 Logic gates.

Function	Symbol	Alternative symbol	Boolean expression	Truth table
Buffer			$B = A$	$A \mid B$ 0 $\mid$ 0 1 $\mid$ 1
NOT			$B = \bar{A}$	$A \mid B$ 0 $\mid$ 1 1 $\mid$ 0
AND			$C = A \cdot B$	$A \quad B \mid C$ 0 $\quad$ 0 $\mid$ 0 0 $\quad$ 1 $\mid$ 0 1 $\quad$ 0 $\mid$ 0 1 $\quad$ 1 $\mid$ 1
OR			$C = A + B$	$A \quad B \mid C$ 0 $\quad$ 0 $\mid$ 0 0 $\quad$ 1 $\mid$ 1 1 $\quad$ 0 $\mid$ 1 1 $\quad$ 1 $\mid$ 1
NAND			$C = \overline{A \cdot B}$	$A \quad B \mid C$ 0 $\quad$ 0 $\mid$ 1 0 $\quad$ 1 $\mid$ 1 1 $\quad$ 0 $\mid$ 1 1 $\quad$ 1 $\mid$ 0
NOR			$C = \overline{A + B}$	$A \quad B \mid C$ 0 $\quad$ 0 $\mid$ 1 0 $\quad$ 1 $\mid$ 0 1 $\quad$ 0 $\mid$ 0 1 $\quad$ 1 $\mid$ 0
Exclusive OR			$C = A \oplus B$	$A \quad B \mid C$ 0 $\quad$ 0 $\mid$ 0 0 $\quad$ 1 $\mid$ 1 1 $\quad$ 0 $\mid$ 1 1 $\quad$ 1 $\mid$ 0
Exclusive NOR			$C = \overline{A \oplus B}$	$A \quad B \mid C$ 0 $\quad$ 0 $\mid$ 1 0 $\quad$ 1 $\mid$ 0 1 $\quad$ 0 $\mid$ 0 1 $\quad$ 1 $\mid$ 1

24.4 Boolean algebra

Boolean algebra defines constants, variables and functions to describe binary systems. It also defines a number of theorems that can be used to manipulate, and perhaps simplify, logic expressions.

24.4.1 Boolean constants

Boolean constants consist of '0' and '1'. The former represents the false state and the latter the true state.

24.4.2 Boolean variables

Boolean variables are quantities that can take different values at different times. They may represent input, output or intermediate signals and are given names, usually consisting of alphabetic characters, such as 'A', 'B', 'X' or 'Y'. Boolean variables may only take the values '0' or '1'.

24.4.3 Boolean functions

Each of the elementary logic functions (such as AND, OR and NOT) are represented by unique symbols (such as '+', '·' and '−'). These various symbols were introduced in the previous section.

24.4.4 Boolean theorems

Boolean algebra has a set of rules that define how it can be used. These consist of a set of **identities** and a set of **laws**, which are summarised in Table 24.2. Many of these rules are self-evident (given a little thought about the meaning of the relevant expression), while others are less obvious. These various rules may be used to simplify algebraic expressions or simply change their form to aid implementation. We will look at algebraic simplification later in this chapter.

Table 24.2 Summary of Boolean algebra identities and laws.

Boolean identities

AND function	OR function	NOT function
$0 \cdot 0 = 0$	$0 + 0 = 0$	$\bar{0} = 1$
$0 \cdot 1 = 0$	$0 + 1 = 1$	$\bar{1} = 0$
$1 \cdot 0 = 0$	$1 + 0 = 1$	$\bar{\bar{A}} = A$
$1 \cdot 1 = 1$	$1 + 1 = 1$	
$A \cdot 0 = 0$	$A + 0 = A$	
$0 \cdot A = 0$	$0 + A = A$	
$A \cdot 1 = A$	$A + 1 = 1$	
$1 \cdot A = A$	$1 + A = 1$	
$A \cdot A = A$	$A + A = A$	
$A \cdot \bar{A} = 0$	$A + \bar{A} = 1$	

Boolean laws

Commutative law	Absorption law
$AB = BA$	$A + AB = A$
$A + B = B + A$	$A(A + B) = A$
Distributive law	**De Morgan's law**
$A(B + C) = AB + AC$	$\overline{A + B} = \bar{A} \cdot \bar{B}$
$A + BC = (A + B)(A + C)$	$\overline{A \cdot B} = \bar{A} + \bar{B}$
Associative law	**Note also**
$A(BC) = (AB)C$	$A + \bar{A} B = A + B$
$A + (B + C) = (A + B) + C$	$A(\bar{A} + B) = AB$

24.5 Combinational logic

Digital systems can be divided into two broad categories. In the first, the outputs are determined solely by the current states of the inputs to the circuit – such arrangements are described as **combinational logic**. In the second form of system, the outputs are determined not only by the current inputs but also by the sequence of inputs that has led to the current state. Such systems are known as **sequential logic**. In this section we will look at the design of combinational logic circuits (leaving sequential logic until Chapter 25).

We have seen that logic functions can be described in a number of ways. For example, we can describe the required operation of a system using a Boolean expression, words or a truth table. Therefore, in order to be able to design and use logic gates effectively, we need to be able to take descriptions in any of these forms and from them generate a circuit diagram of an arrangement that will perform that function. It is also useful to be able to perform the reverse operation – that is, taking a circuit diagram and generating from it a description of its functionality.

24.5.1 Implementing a logic function from a Boolean expression

One of the many advantages of using Boolean algebra is that it produces an unambiguous description of a system that can be easily converted into a circuit diagram. As a Boolean expression combines its various terms using AND, OR or NOT operations, these can be implemented directly as a logic circuit using the corresponding logic gates. This process is illustrated in Example 24.1.

Example 24.1 | Implement the function $X = A + B\overline{C}$.

This expression has one output (X) and three inputs (A, B and C). X is formed by ORing together two components, A and $B\overline{C}$. The first of these is one of the inputs, while the second is formed by ANDing together B and the inverse of C (remember that $B\overline{C}$ is a shorthand notation for $B \cdot \overline{C}$). Therefore, the circuit diagram is

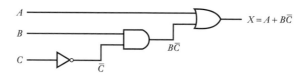

This process can also be used for more complicated expressions, as illustrated in Example 24.2.

Example 24.2 | Implement the function $Y = \overline{\overline{AB} + C\overline{D}}$.

This expression has one output (Y) and four inputs (A, B, C and D). In this example, two terms are ORed together and the result is inverted. These two operations may be combined by the use of a NOR gate. Y is therefore formed by NORing together two components, $\overline{A}B$ and $C\overline{D}$. These components in turn are formed by ANDing together signals derived from the inputs.

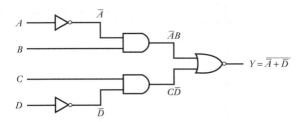

It can be seen that a simple way to implement a Boolean expression is to identify the major elements in the expression and note how these are combined. This defines the logic gate that will generate the output signal. We then work backwards to identify the nature of the inputs to this gate, to see how these are formed from other components. This process is then repeated until the required signals can be generated directly from the input signals. Where systems have more than one output, each output will be represented by a separate Boolean expression that can be implemented separately.

File 24A
File 24B

Computer simulation exercise 24.1

Simulate the logic circuits of Examples 24.1 and 24.2 and confirm that they produce the required outputs for all possible combinations of the inputs.

24.5.2 Generating a Boolean expression from a logic diagram

It is sometimes necessary to reverse the process described above and to generate a Boolean expression to describe an existing logic circuit. Fortunately, this is very straightforward. Perhaps the easiest way is to annotate the logic diagram by starting at the inputs and moving towards the outputs, writing the Boolean expression on the output of each gate. This, in turn, gives you the input to the next gate and so on until you reach the output.

Example 24.3 Derive a Boolean expression for the following circuit.

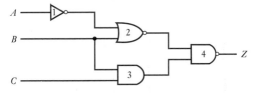

The derivation is performed by working across the circuit, starting with the inputs and working towards the output, writing the output of each gate on the circuit diagram. If we start with gate 1, we can see that its output is simply $\overline{A}$, so we write this against its output on the circuit. We then know both inputs to gate 2 and can write its output on the diagram. If we repeat this for gate 3 and then gate 4, we end up with the diagram below, which gives us a Boolean expression for the output, Z.

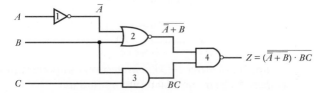

24.5.3 Implementing a logic function from a description in words

In many cases, our starting point in the design of a logic system is not a Boolean expression but a description of the required function in words. Often the simplest approach is to produce a Boolean expression from the original description and to implement it as before. Provided that the description of the required function is clear and unambiguous, this is normally not too difficult.

Example 24.4

Implement the function of an Exclusive OR gate.

From the discussion in the last section, we can describe the required operation of an Exclusive OR gate as

The output should be true if either of its inputs are true, but not if both inputs are true.

If we consider a gate with inputs A and B, we can rephrase this as

The output is true if A OR B is true, AND if A AND B is NOT true.

We can express this in Boolean notation as

$$X = (A + B) \cdot (\overline{AB})$$

and implement it as below.

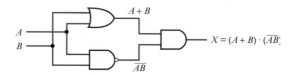

It is worth noting at this point that, while the implementation of Example 24.4 is correct, it is not the only way that this function can be produced. This can be understood by noting that there are other ways to express the operation of an Exclusive OR gate that, therefore, generate alternative ways in which to implement it. This is illustrated in Example 24.5, which uses an alternative definition of the gate.

Example 24.5

Implement an Exclusive OR gate using an alternative configuration.

From the truth table of the Exclusive OR gate, we can see that

The output is true if A is true AND B is NOT true, OR if A is NOT true AND B is true.

This leads to the Boolean expression

$$X = A\bar{B} + \bar{A}B$$

which can be implemented as

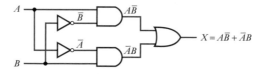

Examples 24.4 and 24.5 illustrate a very important property of Boolean algebra, which is that Boolean expressions are not unique.

Computer simulation exercise 24.2
Simulate the Exclusive OR gates of Example 24.4 and 24.5 and confirm that they each produce the required output for all possible combinations of the inputs, so proving the equivalence of these two implementations.

24.5.4 Implementing a logic function from a truth table

If our required system is defined by a truth table, we again produce a Boolean expression from the table and implement it as before. The task of producing an expression from a truth table can be easily understood if we bear in mind what the table actually represents and here an example might be useful. Consider the following truth table, which is for an Exclusive NOR gate:

A	B	C
0	0	1
0	1	0
1	0	0
1	1	1

The table lists, on the right-hand side, the output for each possible combination of the values of the inputs. Where the output is a '1', this corresponds to a set of inputs for which the output is true. Therefore, if we list these combinations, we have a list of all the conditions for which the output is true. The function can then be described by saying that the output will be true if, and only if, the inputs correspond to one or other of the combinations in this list. In the example above, the only combinations for which the output is true are when both A and B are '0' and when they are both '1'. If A and B are both '0', then $\overline{A}$ and $\overline{B}$ must be equal to '1'. Therefore, the first condition corresponds to $\overline{A}$ AND $\overline{B}$ being equal to '1' and the second to A AND B being equal to '1'. The function can therefore be described as

$$C = \overline{A}\overline{B} + AB$$

A more formal description of the process might be:

- a **minterm** is generated for each row in which a '1' appears in the output column of the truth table;
- the minterm contains each input variable in turn, the input being non-inverted if it is a '1' in the truth table and inverted if it is a '0';
- the overall expression for the logic function is then the sum of the minterms.

This process can be applied to a truth table of any size. For example, this table

A	B	C	D	minterms
0	0	0	0	
0	0	1	1	$\overline{A}\overline{B}C$
0	1	0	0	
0	1	1	0	
1	0	0	1	$A\overline{B}\overline{C}$
1	0	1	0	
1	1	0	0	
1	1	1	1	ABC

corresponds to the expression

$$D = \overline{A}\overline{B}C + A\overline{B}\overline{C} + ABC$$

It is important to remember when writing down the minterms that $\overline{A}\overline{B} \neq \overline{AB}$ and therefore, for example, $\overline{A}\overline{B}C \neq \overline{AB}C$.

Looking back at the truth table, it can be seen that there is a very simple relationship between the combinations of inputs for which the output is '1' and the resulting Boolean expression. This makes it very easy to write down the Boolean expression directly from the truth table. Once we have this expression, we can implement the function as before.

Example 24.6 | Implement the function of the following truth table.

A	B	C	X
0	0	0	0
0	0	1	1
0	1	0	0
0	1	1	0
1	0	0	0
1	0	1	1
1	1	0	1
1	1	1	0

There are three combinations of inputs for which the output is true, so the expression will have three minterms ORed together. By inspection, the expression is

$$X = \overline{A}\overline{B}C + A\overline{B}C + AB\overline{C}$$

This can be implemented as

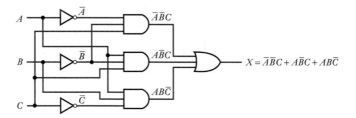

It can be seen that, as circuit diagrams grow in complexity, it becomes increasingly difficult to follow the interconnections. This problem can be reduced by using labels to indicate connections rather than drawing interconnecting lines. Using this approach, the diagram above becomes that shown below, which is much easier to understand.

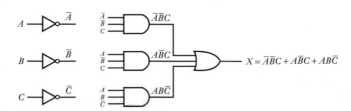

24.6 Boolean algebraic manipulation

The various functions and laws of Boolean algebra allow us to manipulate expressions to simplify them or make them easier to implement. For example, the associative law allows us to combine similar operations within a single gate. This process is illustrated in Figure 24.17, which shows that several AND or OR operations can be performed within a single device.

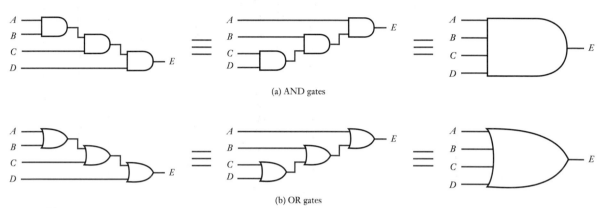

(a) AND gates

(b) OR gates

Figure 24.17
The physical significance of the associative law.

Algebraic manipulation also allows us to select the form of the gates that we use to implement a function. We will see in Chapter 26 that, when using a particular device technology to implement a circuit, it is often easier, and less expensive therefore, to produce one form of gate rather than another. It is therefore important to be able to manipulate logic expressions so that they may be implemented in a suitable form. In several logic families, NAND gates are the simplest form of gate, so it is useful to be able to construct circuits using only these gates.

Modifying a circuit to use only NAND gates

We saw in Section 24.5.4 that the function of *any* truth table (and therefore of any combinational arrangement) can be expressed as the sum of a series of minterms. In other words, we can implement any system by an arrangement of the general form shown in Figure 24.18.

However, from De Morgan's theorem, we know that

$$A + B + C = \overline{\overline{A}\,\overline{B}\,\overline{C}}$$

and therefore the OR gate in the arrangement of Figure 24.18 can be replaced with a series of inverters and a NAND gate, as shown in Figure 24.19.

It follows that the functions implemented by the AND and OR gates in Figure 24.18 can be replaced by an arrangement using only NAND gates, as shown in Figure 24.20.

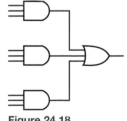

Figure 24.18
Implementation of a system using a combination of AND and OR gates.

Figure 24.19
Functional equivalence of an OR gate.

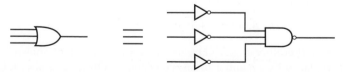

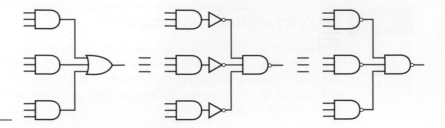

Figure 24.20
Implementation
of a generalised
combinational
arrangement using
only NAND gates.

Example 24.7 Implement the function of the following circuit using only NAND gates.

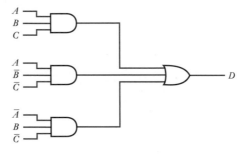

Converting this to use only NAND gates, as shown in Figure 24.20, gives

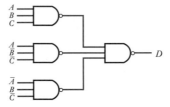

This transformation can also be achieved using Boolean algebraic manipulation directly. First, we derive the Boolean expression for the circuit (as discussed in Section 24.5.4) which gives

$$D = ABC + A\overline{B}\overline{C} + \overline{A}B\overline{C}$$

and then we use De Morgan's theorem to transform this to a suitable form. This gives

$$D = ABC + A\overline{B}\overline{C} + \overline{A}B\overline{C} = \overline{\overline{ABC} \cdot \overline{A\overline{B}\overline{C}} \cdot \overline{\overline{A}B\overline{C}}}$$

which is in a form suitable for direct implementation using NAND gates.

Modifying a circuit to use only NOR gates

When using some logic families, NOR gates are the simplest. A similar manipulation to that given above can be used to implement functions using only these gates.
    Again, using De Morgan's theorem we know that

$$A \cdot B \cdot C = \overline{\overline{A} + \overline{B} + \overline{C}}$$

and therefore the AND gates in the arrangement shown in Figure 24.18 can be replaced by a number of inverters and an OR gate, as shown in Figure 24.21.

Figure 24.21
Functional
equivalence of
an AND gate.

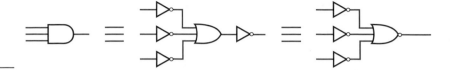

It follows that the functions implemented by the AND and OR gates of Figure 24.18 can be replaced with an arrangement using only NOR gates, as shown in Figure 24.22. In this case the final circuit requires more gates than the original implementation but, provided that the inverses of the various input signals are available, only one extra gate is required.

Example 24.8 | **Implement the function of the following circuit using only NOR gates.**

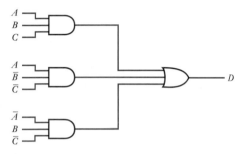

This circuit is the same as that in Example 24.7. Converting this to use only NOR gates, as shown in Figure 24.22, gives the following arrangement. Note that the inputs to this circuit are inverted in relation to the original circuit.

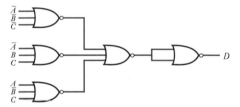

As in the previous example, this transformation can also be achieved using Boolean algebraic manipulation directly. First, we derive the Boolean expression for the circuit which, as before, is

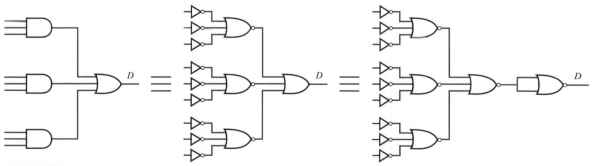

Figure 24.22
Implementation of a generalised combinational arrangement using only NOR gates.

$$D = ABC + A\overline{B}\overline{C} + \overline{A}B\overline{C}$$

and then we use De Morgan's theorem to transform this to a suitable form. This gives

$$D = ABC + A\overline{B}\overline{C} + \overline{A}B\overline{C} = \overline{\overline{(\overline{A} + \overline{B} + \overline{C})} + \overline{(\overline{A} + B + B)} + \overline{(A + \overline{B} + C)}}$$

which is in a form suitable for direct implementation using NOR gates.

Video 24D

24.7 Algebraic simplification

One of the main aims of a logic designer is to produce a simple system. Logic minimisation aims to take an algebraic expression and reduce it to a form that is easier to implement. In practice, circuits that contain the smallest number of gates are not necessarily the cheapest to construct as other factors, such as the type of gates used, may affect the overall cost. However, reducing the complexity of a circuit is a good first step in reducing cost.

The usual form of Boolean expression is the **sum-of-products** form. This is the format obtained in Example 24.6, consisting of a number of ANDed terms (**minterms** or **products**) that are ORed together. Each product consists of a combination of some or all of the input variables, each in a complemented or uncomplemented form. Examples of sum-of-products expressions are

$$A\overline{B} + \overline{A}B$$

$$XYZ + \overline{X}Y\overline{Z} + X\overline{Y}Z$$

$$AB\overline{C}D + A\overline{B}C + \overline{A}BCD + ABC\overline{D}$$

The sum-of-products form must *not* include inversions of a series of terms, as in

$$\overline{AB\overline{C}D} + \overline{A}BCD$$

Various techniques are available to simplify Boolean algebraic expressions in order to reduce the complexity of the circuitry required to implement them. These include analytical, graphical and computer-based methods. We will begin by looking at algebraic simplification, which is performed by taking the sum-of-products expression for a function and using the identities and rules of Boolean algebra to combine terms, progressively reducing its complexity. This is best illustrated by the use of examples.

Example 24.9 | **Implement the Boolean expression**

$$X = ABC + \overline{A}BC + AC + A\overline{C}$$

This can be implemented directly as

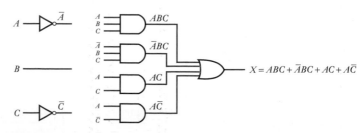

Alternatively, it can be rearranged using the commutative and distributive laws to give

$$X = BC(A + \overline{A}) + A(C + \overline{C})$$

$$= BC + A$$

which can be implemented as

$$X = BC + A$$

File 24D

Computer simulation exercise 24.3

Simulate the two implementations in Example 24.9 and confirm that they each produce the required output for all possible combinations of the inputs, so proving the equivalence of these two implementations.

Example 24.10

Implement the Boolean expression

$$E = B\overline{C}\overline{D} + \overline{A}BD + ABD + BC\overline{D} + \overline{B}CD + \overline{A}\overline{B}CD + A\overline{B}CD$$

This can be implemented directly, as in the last example, but would use a large number of gates. Alternatively, it may be simplified as follows.
  Combining terms 2 and 3, 1 and 4, and 6 and 7

$$E = BD(A + \overline{A}) + B\overline{D}(C + \overline{C}) + \overline{B}CD + \overline{B}CD(A + \overline{A})$$

$$= BD + B\overline{D} + \overline{B}CD + \overline{B}CD$$

and combining the new terms 1 and 2, and 3 and 4

$$E = B(D + \overline{D}) + \overline{B}D(C + \overline{C})$$

$$= B + \overline{B}D$$

which further combines to give

$$E = B + D$$

This expression can be implemented by means of a single gate.

$$E$$

Example 24.11

Simplify the Boolean expression

$$E = A\overline{B}\overline{C} + \overline{A}CD + ABD + BCD$$

It is not immediately obvious how to combine these terms to reduce its complexity. In fact, it is necessary to *expand* the terms first, as follows:

$$E = A\overline{B}\overline{C} + \overline{A}CD + ABD + BCD$$

$$= A\overline{B}\overline{C}(D + \overline{D}) + \overline{A}CD(B + \overline{B}) + ABD(C + \overline{C}) + BCD(A + \overline{A})$$

$$= A\overline{B}\overline{C}D + A\overline{B}\overline{C}\overline{D} + \overline{A}BCD + \overline{A}\overline{B}CD + ABCD + AB\overline{C}D$$

$$\quad + ABCD + \overline{A}BCD$$

Combining terms 5 and 7 produces

$$E = A\bar{B}\bar{C}D + A\bar{B}CD + \bar{A}B\bar{C}D + \bar{A}BCD + ABCD + AB\bar{C}D + \bar{A}BCD$$

and duplicating terms 1, 3 and 6 gives

$$E = A\bar{B}\bar{C}D + A\bar{B}\bar{C}D + A\bar{B}CD + \bar{A}B\bar{C}D + \bar{A}B\bar{C}D$$
$$+ \bar{A}BCD + ABCD + AB\bar{C}D + AB\bar{C}D + \bar{A}BCD$$

Combining terms 1 and 3, 5 and 6, 4 and 8, 2 and 9, and 7 and 10, we get

$$E = A\bar{B}\bar{C}(D + \bar{D}) + \bar{A}\bar{C}D(B + \bar{B}) + B\bar{C}D(A + \bar{A})$$
$$+ \bar{A}CD(B + \bar{B}) + BCD(A + \bar{A})$$
$$= A\bar{B}\bar{C} + \bar{A}\bar{C}D + B\bar{C}D + \bar{A}CD + BCD$$

and combining terms 2 and 4, and 3 and 5, gives

$$E = A\bar{B}\bar{C} + \bar{C}D(A + \bar{A}) + BD(C + \bar{C})$$
$$= A\bar{B}\bar{C} + \bar{C}D + BD$$

This represents a considerable simplification of the original expression.

It can be seen from Examples 24.9 to 24.11 that this form of algebraic simplification can greatly reduce the complexity of Boolean expressions, thereby greatly reducing the cost of their implementation. However, it is clear from the third of these examples that the process of simplification is not always straightforward. Manipulation of this kind is often inspired guesswork and suffers from the problem that one is never sure whether or not an optimum solution has been obtained. For example, can the expressions in Example 24.11 be further simplified?

It is important to understand the process of algebraic simplification as it is a useful technique for simple functions. However, for more complicated expressions, we normally resort to more powerful methods.

24.8 Karnaugh maps

The **Karnaugh map** is a graphical method for representing the information within a truth table.

In a truth table, each possible combination of the inputs is represented by a unique line in the table, while the value of the output corresponding to that pattern of inputs is shown in the output column. In a Karnaugh map, each possible combination of the inputs is represented by a box within a grid and the corresponding value of the output is written within that box. An example of this arrangement is shown in Figure 24.23 for a function with two inputs, A and B.

The usefulness of the Karnaugh map stems from the arrangement of the boxes within the grid. For a system with two inputs, the grid has four boxes, corresponding to the four combinations of the inputs. The boxes are positioned so that, in going from one box to another, either horizontally or vertically, only one of the variables associated with that box changes. For example, in going from the top left-hand box to the top right-hand box, A changes from '0' to '1', but B remains unchanged. In moving from the top left-hand box to the bottom left-hand box, B changes from '0' to '1' but A remains unchanged. This principle is maintained as more boxes are added to represent systems with a greater number of inputs. Figure 24.24 shows Karnaugh maps for systems with

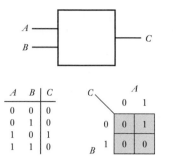

Figure 24.23
The truth table and
Karnaugh map for a
system with two inputs.

(a) Truth table

(b) Karnaugh map

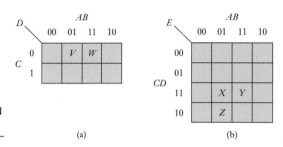

Figure 24.24
Karnaugh maps for
systems with three and
four inputs.

(a)

(b)

three and four inputs. This technique can be extended to systems with a greater
number of inputs, if required.

For a system with three input variables, a four-by-two array is used, as in Figure 24.24(a).
Two of the variables are associated with the four *columns* and the remaining variable
with the two *rows*. The distribution of the variables is shown above and to the side of the
map. The values of the appropriate variables are shown adjacent to each row and column.
Thus, box V corresponds to AB being '01' and to C being '0'. Box V is therefore $\overline{A}B\overline{C}$.
Similarly, box W is $AB\overline{C}$.

Systems with four inputs require a four-by-four array, as shown in Figure 24.24(b).
Here, box X corresponds to $\overline{A}BCD$, box Y to $ABCD$ and box Z to $\overline{A}BC\overline{D}$.

You will notice that the grid is *not* labelled in simple binary order. The sequence,
in fact, corresponds to the **Gray code**, which will be discussed later in this chapter. For
the moment, it is sufficient to notice that the sequence is such that adjacent squares,
both horizontally and vertically, differ in terms of the state of only one variable. For
example, X and Y differ only in terms of the state of A, while X and Z differ only in
terms of the state of D. The importance of this property will become apparent shortly.
Note that this relationship does not apply to boxes that are linked diagonally, as with
Y and Z, which differ in terms of the states of both A and D.

To examine the use of Karnaugh maps, consider the maps shown in Figure 24.25.
The figure shows two Karnaugh maps for functions E and F. We can extract an

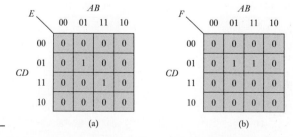

Figure 24.25
Simple Karnaugh
maps.

(a)

(b)

algebraic expression for the functions directly from the maps in a manner similar to that used in Example 24.6 to extract this information from a truth table. Clearly E is given by the expression

$$E = \overline{A}B\overline{C}D + ABCD$$

and F is given by

$$F = \overline{A}B\overline{C}D + AB\overline{C}D$$

It is interesting at this point to attempt to simplify these expressions algebraically using the techniques described earlier. If this is done, it will be found that E cannot be simplified, but that F can be reduced by noting that

$$F = \overline{A}B\overline{C}D + AB\overline{C}D$$

$$= B\overline{C}D(A + \overline{A})$$

$$= B\overline{C}D$$

The fact that F can be simplified while E cannot is evident if we look again at the maps in Figure 24.25. In the map for F, the two '1's are adjacent. Thus, from the characteristics of the Karnaugh map, we know that these correspond to boxes representing combinations of inputs that only vary in the state of one input. Therefore *any* two adjacent elements containing '1's can be represented by an expression of the form

$$WXYZ + WXY\overline{Z}$$

where W, X, Y and Z each represent one of the input variables or its inverse. This will always simplify, giving

$$WXYZ + WXY\overline{Z} = WXY(Z + \overline{Z}) = WXY$$

removing the effect of one of the variables. The variable removed from the expression is the term that is present in both its normal and inverted forms. This can be understood by noting that this means the output will be true whether this variable is '1' or '0'. It is thus independent of this variable and the variable no longer appears within the expression.

Given a Karnaugh map containing two adjacent '1's, it is normal to draw a loop around the pair to indicate their union. The combination can then be represented by a term that indicates the input states that are constant for that group. In the example shown in Figure 24.25(b), B is '1', C is '0' and D is '1' for both elements that are '1'. However, A takes a value of '0' for one element and '1' for the other, so does not appear in the expression for the combination, which is simply $B\overline{C}D$. Some examples of combinations of two elements are shown in Figure 24.26. Note that, for the purposes

Figure 24.26
Combination of adjacent pairs of '1's in a Karnaugh map.

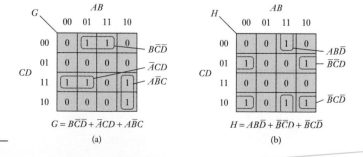

$$G = B\overline{C}\overline{D} + \overline{A}CD + A\overline{B}C$$

(a)

$$H = AB\overline{D} + \overline{B}\overline{C}D + \overline{B}C\overline{D}$$

(b)

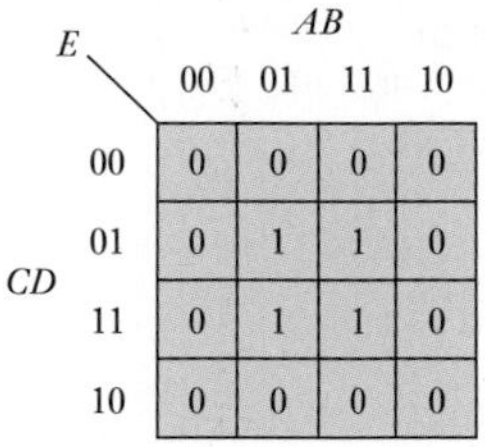

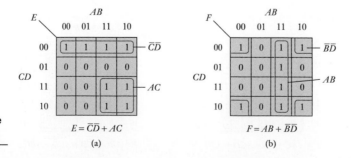

Figure 24.27
A Karnaugh map with
an array of four '1's.

of combining '1's, the top and bottom rows are considered to be adjacent, as are the right and left columns. It is as if the map were cylindrical in each plane, such that the top and bottom and the two sides are touching.

Consider the map in Figure 24.27. Here, four '1's are arranged in a square. The map may be described by the expression

$$E = \overline{A}B\overline{C}D + AB\overline{C}D + \overline{A}BCD + ABCD$$

This can be simplified by combining terms 1 and 2, and 3 and 4

$$E = B\overline{C}D(A + \overline{A}) + BCD(A + \overline{A})$$

$$= B\overline{C}D + BCD$$

This simplification is evident from the map in that we could draw loops round these two pairs of elements. However, the two resulting terms may themselves be combined

$$E = BD(C + \overline{C})$$

$$= BD$$

The fact that the four terms may be combined into one can be understood by noting that

$$E = \overline{A}B\overline{C}D + AB\overline{C}D + \overline{A}BCD + ABCD$$

$$= BD(\overline{A}\,\overline{C} + A\overline{C} + \overline{A}C + AC)$$

The term within the brackets represents all the possible combinations of the variables A and C. Thus, E is true if BD is true, independent of the values of A and C.

Other groups of four elements of a Karnaugh map can be combined provided that they represent all the possible combinations of any two variables. Some examples of allowed groupings are shown in Figure 24.28 and some illegal groupings in Figure 24.29.

Figure 24.28
Examples of allowable
groupings of four '1's.

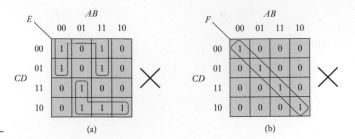

Figure 24.29
Examples of illegal groupings of four '1's.

Having observed that we may form groups of two or four '1's, it is perhaps not surprising to note that we may also form groups of eight elements, provided that we follow certain rules. In fact, we can form groups of 2^n elements, provided that they represent all the possible states of n variables. This will be true if the groups are rectangles with each side having 2^m elements. Therefore, allowable groups will have sides of 1, 2, 4, 8, ... elements, so permitted groupings are 1×1 (a single element), 1×2, 2×1, 2×2, 1×4, 2×4, 4×4 and so on. The size of any rectangle will clearly be limited by the size of the map. Figure 24.30 shows some examples of such groupings. In each case, the algebraic expression representing the group is obtained by noting which variable states are constant for all elements in the group.

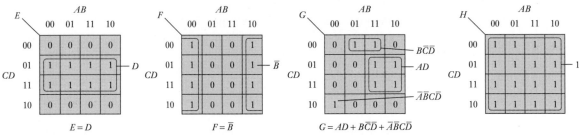

Figure 24.30
Examples of permitted groupings of 2^n elements.

The form of G in Figure 24.30 illustrates the fact that not all the '1's within a map may combine with other terms. It should also be noted that it is possible to combine some '1's within more than one grouping. This is shown in Figure 24.31 which shows two maps for the same function E, where the element $\overline{A}\overline{B}\overline{C}D$ is used within two groups. Also within this figure we note that the element $ABCD$ can be combined in two ways. This produces two different algebraic expressions for the function E. These two expressions are equally valid representations of E and are equally simple. This illustrates that *the simplified expressions produced by this technique are not unique.*

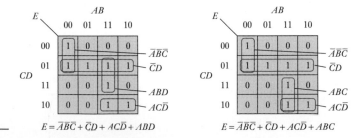

Figure 24.31
Alternative groupings of map elements.

As it seems that there may be a number of ways to group the '1's in a Karnaugh map, we require some rules to ensure that we perform the groupings in such a way that the expression produced is as simple as possible. These rules may be expressed as follows:

1   The largest possible groups of cells should be constructed first, each group containing 2^n elements.
2   Progressively smaller groups should be added until every cell containing a '1' has been included at least once.
3   Any redundant groups should then be removed, even if these are large groups, to avoid duplication.

Use of these rules is illustrated in the following examples.

Example 24.12 | Simplify the following Boolean expression using a Karnaugh map.

$$D = \overline{A}B\overline{C} + AB\overline{C} + A\overline{B}\overline{C} + \overline{A}\overline{B}C + \overline{A}BC + ABC$$

This may be represented by a Karnaugh map as

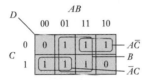

This gives an expression for D of the form

$$D = B + A\overline{C} + \overline{A}C$$

Note that the term $A\overline{B}\overline{C}$ is grouped with $AB\overline{C}$ even though the latter is contained within another group. This also applies to $\overline{A}\overline{B}C$ and $\overline{A}BC$. We always make the largest possible groups.

Example 24.13 | Simplify the following Boolean expression using a Karnaugh map.

$$E = \overline{A}B\overline{C}\overline{D} + AB\overline{C}\overline{D} + A\overline{B}\overline{C}\overline{D} + \overline{A}B\overline{C}D + \overline{A}B\overline{C}D + AB\overline{C}D$$

$$+ A\overline{B}\overline{C}D + \overline{A}BCD + ABCD + \overline{A}BC\overline{D} + ABC\overline{D}$$

This can be represented by a Karnaugh map as

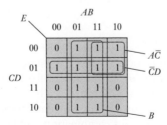

Here one element, $A\overline{B}CD$, is contained within three groups to obtain the simplest form. This is

$$E = B + A\overline{C} + \overline{C}D$$

Example 24.14 | Simplify the following Boolean expression using a Karnaugh map.

$$E = \overline{A}B\overline{C}\overline{D} + \overline{A}B\overline{C}D + AB\overline{C}D + A\overline{B}CD + \overline{A}\overline{B}CD + \overline{A}BCD$$

$$+ ABCD + ABC\overline{D}$$

This produces the following map:

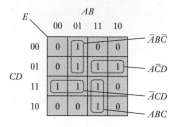

Here, the largest possible grouping – formed by joining the four innermost cells – is not used. This is because the four groups shown are all essential to include the four outer '1's. This makes the inner group of four unnecessary as all its components are already included in other groupings. This is an application of the third of our rules, which says that we must remove any redundant cells.

Karnaugh maps may also be used to simplify expressions with more than four input variables. The same basic rules apply, but the topology of the map becomes more complicated. This is illustrated in Figure 24.32, which shows a five-variable map.

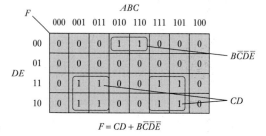

Figure 24.32
A Karnaugh map for a system with five input variables.

$$F = CD + B\overline{C}\overline{D}\overline{E}$$

As the number of input variables becomes larger, this method of simplification becomes unwieldy and automated computer techniques are usually adopted. These are discussed briefly later in this chapter.

24.8.1 Don't care conditions

Don't care conditions occur when the state of an input or output variable is unimportant. In the case of an input variable, this means that the output is the same whether the input is a '0' or a '1'. When an output variable has a don't care state, this means that the output state for that combination of inputs is unimportant. Often this is because that input combination will never occur. A don't care condition is represented within a truth table or a Karnaugh map by an 'X', as illustrated in Figure 24.33.

Figures 24.33(a) and 24.33(b) show the use of don't care conditions for input variables. When used in this way, they represent a shorthand method of representing

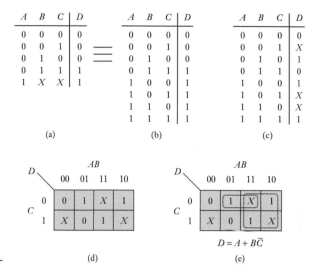

Figure 24.33
The representation of 'don't care' conditions.

a number of input combinations within a single line. This simplifies the truth table and aids comprehension. The truth table of Figure 24.33(a) has precisely the same meaning as that of Figure 24.33(b), but is easier to assimilate.

When used for output variables, don't care conditions are of greater significance. Here, they mean that the output state is unimportant, leaving the designer free to choose which state the system should adopt. When using a truth table to represent an expression, as shown in Figure 24.33(c), it is not obvious how the don't care states should be assigned. The designer could simply choose to make them *all* '0' or *all* '1'. Alternatively, *some* could be set to '0' and *some* to '1'. However, it is not clear which choice will produce the simplest implementation.

One of the great strengths of Karnaugh maps is their ability to deal sensibly with don't care conditions. Figure 24.33(d) shows the Karnaugh map representation of the expression given in the truth table of Figure 24.33(c). From this it is apparent which don't care terms should be considered as '0' and which as '1' to produce the simplest expression. In this case, $\overline{A}\overline{B}C$ should be chosen as a '0' and the remainder as '1's to produce the simplest grouping, as shown in Figure 24.33(e).

24.9 Automated methods of minimisation

Although Karnaugh maps can be used fairly easily with up to six variables, with larger numbers it becomes impractical and other methods are required. For such problems, it is normal to use a tabular method, such as that developed by McCluskey from an original technique proposed by Quine. This approach is universally known as **Quine–McCluskey minimisation**.

The process of minimisation is performed on a table of the input products (minterms) that is systematically reduced by examining each pair of terms to see if the Boolean simplification $AB + A\overline{B} = A$ can be applied. This process is applied exhaustively until a minimised expression is produced. The technique can also cater for don't care conditions in a similar manner to that used with Karnaugh maps. The Quine–McCluskey method can handle systems with any number of inputs and can be performed by hand or, as is more usual, by computer. The algorithm is very simple to program and computers are now frequently used for all but the simplest minimisation tasks.

24.10   Propagation delay and hazards

So far we have considered logic gates purely from a functional viewpoint and have ignored any issues relating to their implementation. In practice, physical logic gates take a finite time to respond to input signals and there is a delay between the time when the input signals change and when the output responds. This delay is termed the **propagation delay time**. When using modern electronic components this time is very short (often less than a nanosecond) and in many cases will be unimportant. However, in some situations this delay can affect the operation of the circuit and must be taken into account in the design. The problems associated with this phenomenon are illustrated in Figure 24.34(a).

The circuit of Figure 24.34(a) represents the function

$$C = A \cdot B = A \cdot \overline{A} = 0$$

and thus the output C should always be 0.

Figure 24.34(b) shows the response of the arrangement to a transition from 0 to 1 at the input. Because of the delay caused by the inverter, the AND gate sees a '1' on both inputs for a brief period and may, if it is sufficiently fast, produce an output pulse of logic 1, despite the fact that the logic function predicts that it will always be 0. This is an example of a **hazard** – a transient effect that generates unwanted transitions of the output.

The effects of hazards are further illustrated by the circuit in Figure 24.34(c). Here, the output represents the function $A\overline{B} + BC$, which may be represented by the Karnaugh map in Figure 24.34(d). Let us consider the situation in which, initially, all three inputs are at 1. Under these circumstances, both inputs to gate Y are at 1 and, consequently, the output of Y is 1 and D is 1. It might be expected that taking B to 0 would have no effect on the circuit as, with B at 0, both inputs to gate X are at 1 and again D is 1. However, the delay caused by the inverter

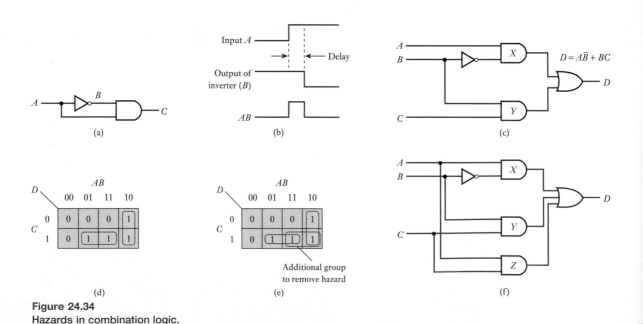

Figure 24.34
Hazards in combination logic.

results in the output of Y going to 0 before the output of X goes to 1. Consequently, for a brief period, both inputs to the OR gate are 0 and the output D pulses low.

The solution to this problem can be understood by looking at the Karnaugh map in Figure 24.34(d). The conditions under which the output is high are represented by the two ringed groups. Each group is implemented by one of the AND gates in Figure 24.34(c). If either produces an output of 1, the output of the complete circuit will be high. As the pattern of inputs changes, we move about the Karnaugh map, producing a high or low output as appropriate. The hazard condition represents the situation in which the input combination jumps from one ring to another. While in either state the output is high but, during the brief period of transition, the output is incorrect. The problem can be overcome by bridging the gap between the rings, as shown in Figure 24.34(e). The addition of an extra redundant term in no way changes the logic function, but it does ensure that unwanted transitions do not occur as the inputs change between those representing the two groups. The resultant logic circuit is shown in Figure 24.34(f).

The importance of the presence of hazards within combinational logic differs between applications. In some situations they are unimportant as the system being driven is not sufficiently fast to respond to the transients produced. However, in many cases the elimination of hazards is essential, particularly when driving circuits that respond to transitions rather than levels (as in many of the sequential circuits discussed in the next chapter). The addition of extra gates clearly removes hazards at the expense of increased complexity. Some examples of hazard removal are given in Example 24.15.

| **Example 24.15** | **Removal of hazard conditions.** |

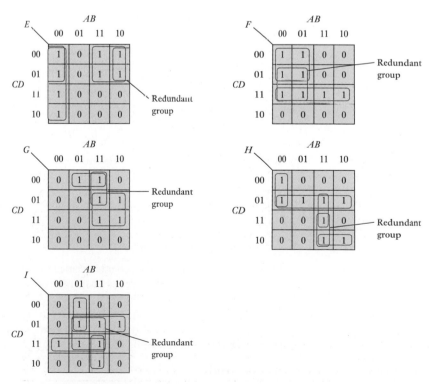

24.11 | Number systems and binary arithmetic

So far we have dealt with simple binary signals, such as those produced by switches and those required to turn lamps on or off. Sometimes groups of binary signals are combined to form binary words. These words can be used to represent various forms of information, the most common being *numeric* and *alphabetic* data. When numerical information is represented, this permits arithmetic operations to be performed on the data.

24.11.1 Number systems

The decimal number system

In everyday arithmetic, we use numbers with a base of 10, this choice being almost certainly related to the fact that we have 10 fingers and thumbs. This system requires 10 symbols to represent the values that each digit may take, for which we use the symbols 0, 1, 2, . . . , 9. Our numbering system is 'order dependent' in that the significance of a digit in a number depends on its position. For example, the number

$$1234$$

means 1 *thousand*, plus 2 *hundreds*, plus 3 *tens*, plus 4 *units*. Each column of the number represents a power of 10, starting with units on the right-hand side ($10^0 = 1$) and moving to increasing powers of 10 as we move to the left. Thus our number is

$$1234 = (1 \times 10^3) + (2 \times 10^2) + (3 \times 10^1) + (4 \times 10^0)$$

Digits at the left-hand side of the number are of much greater significance than those on the right-hand side. For this reason, the left-hand digit is termed the **most significant digit (MSD)**, while that on the right is termed the **least significant digit (LSD)**.

The numbering system can be extended to represent magnitudes that are not integer quantities by extending the sequence below the units column, a decimal point being placed to the right of the units column to indicate its position. Thus

$$1234.56 = (1 \times 10^3) + (2 \times 10^2) + (3 \times 10^1) + (4 \times 10^0)$$
$$+ (5 \times 10^{-1}) + (6 \times 10^{-2})$$

Numbers of any size can be represented by using a sufficiently large number of digits. Leading zeros have no effect on the magnitude of the number, provided that they are to the left of the decimal point. Similarly, trailing zeros have no effect if they are to the right of the decimal point.

The binary number system

Binary numbers have similar characteristics to decimal numbers, except that they have a base of 2. As each digit may now take only two values, only two symbols are required. These are usually 0 and 1. One advantage of this system is that digits can be represented by any binary quantity, such as a switch position or a lamp being ON or OFF.

As binary quantities use symbols that are also used in decimal numbers, it is common to identify the notation being used by adding a subscript indicating the base. Thus

$$1101_2$$

is a binary number, while

$$1101_{10}$$

is a decimal number. In many cases, the base is known or obvious, in which case the subscript is usually omitted.

Like their decimal counterparts, the digits in a binary word are also position dependent. As before, the digits represent ascending powers of the base, such that

$$1101_2 = (1 \times 2^3) + (1 \times 2^2) + (0 \times 2^1) + (1 \times 2^0)$$

Therefore, rather than having units, tens, hundreds and thousands columns, as in decimal numbers, we have 1s, 2s, 4s, 8s, 16s . . . columns.

Fractional parts may also be represented, as

$$1101.01_2 = (1 \times 2^3) + (1 \times 2^2) + (0 \times 2^1) + (1 \times 2^0) + (0 \times 2^{-1}) + (1 \times 2^{-2})$$

The position of the units column is now indicated by a **binary point** (rather than a decimal point) and columns to the right of this point represent magnitudes of $1/2$, $1/4, \ldots, 1/2^n$. The term *binary digit* is often abbreviated to **bit**. Thus, a binary number consisting of eight digits would be referred to as an 8-bit number.

Other number systems

Although there are clear reasons for using both decimal and binary numbers, any integer may be used as the base for a number system. For reasons that are unimportant at this stage, common numbering systems include those using bases of 8 (octal) and 16 (hexadecimal – or simply hex). Octal numbers require eight symbols and use $0, 1, \ldots, 7$. Hexadecimal numbers require 16 symbols and use $0, 1, \ldots, 9$, A, B, C, D, E and F. From the above discussion of decimal and binary numbers, it is clear that

$$123_8 = (1 \times 8^2) + (2 \times 8^1) + (3 \times 8^0)$$

and that

$$123_{16} = (1 \times 16^2) + (2 \times 16^1) + (3 \times 16^0)$$

Table 24.3 gives the numbers 0 to 20_{10} in decimal, binary, octal and hexadecimal.

24.11.2 Number conversion

Most scientific calculators can perform conversions between number bases automatically, but it is perhaps useful to see how this process can be performed 'manually'. This is instructive, if only because it gives an insight into the relationships between the various number systems.

Conversion from binary to decimal

Converting binary numbers into decimals is straightforward. It is achieved simply by adding up the decimal values of each '1' in the number.

For small numbers, this conversion can be performed quite simply using mental arithmetic. Larger numbers take a little longer. Numbers with fractional parts can be converted in the same manner by adding the decimal equivalent of each term.

Table 24.3 Number representations.

Decimal	Binary	Octal	Hexadecimal
0	0	0	0
1	1	1	1
2	10	2	2
3	11	3	3
4	100	4	4
5	101	5	5
6	110	6	6
7	111	7	7
8	1000	10	8
9	1001	11	9
10	1010	12	A
11	1011	13	B
12	1100	14	C
13	1101	15	D
14	1110	16	E
15	1111	17	F
16	10000	20	10
17	10001	21	11
18	10010	22	12
19	10011	23	13
20	10100	24	14

Example 24.16 Convert 11010_2 to decimal.

$$11010_2 = (1 \times 2^4) + (1 \times 2^3) + (0 \times 2^2) + (1 \times 2^1) + (0 \times 2^0)$$

$$= 16 + 8 + 0 + 2 + 0$$

$$= 26_{10}$$

Conversion from decimal to binary

Conversion from decimal to binary is effectively the reverse of the above process, although the similarity is not at first apparent. It is achieved by repeatedly dividing the number by 2 and noting any remainder. This procedure is repeated until the number vanishes.

Example 24.17 Convert 26_{10} to binary.

	Number	Remainder
Starting point	26	
÷ 2	13	0
÷ 2	6	1
÷ 2	3	0
÷ 2	1	1
÷ 2	0	1

read number from this end
= 11010

Thus

$$26_{10} = 11010_2$$

Numbers with fractional parts are converted in parts, the integer part being converted as above and the fractional part being converted by repeated multiplication by 2, noting, and then discarding, the overflow beyond the binary point after each multiplication. As with fractional parts in decimal numbers, the number of places used to the right of the binary point depends on the accuracy required.

Example 24.18

Convert 34.6875_{10} to binary.

First, the whole number part (34) is converted as before

	Number	Remainder
Starting point	34	
$\div 2 =$	17	0
$\div 2 =$	8	1
$\div 2 =$	4	0
$\div 2 =$	2	0
$\div 2 =$	1	0
$\div 2 =$	0	1

read number from this end
$= 100010$

then the fraction part (0.6875) is converted

	Overflow	Number	
read number from top		.6875	$\times 2 =$
	1	.375	$\times 2 =$
$= 0.1011$	0	.75	$\times 2 =$
	1	.5	$\times 2 =$
	1	.0	

Thus

$$34.6875_{10} = 100010.1011_2$$

Conversion from hexadecimal to decimal

Conversion from hexadecimal to decimal is similar to the conversion from binary to decimal, except that powers of 16 are used in place of powers of 2.

Example 24.19

Convert $A013_{16}$ to decimal.

$$A013_{16} = (A \times 16^3) + (0 \times 16^2) + (1 \times 16^1) + (3 \times 16^0)$$

$$= (10 \times 4096) + (0 \times 256) + (1 \times 16) + (3 \times 1)$$

$$= 40,960 + 0 + 16 + 3$$

$$= 40,979_{10}$$

Conversion from decimal to hexadecimal

Conversion from decimal to hexadecimal is likewise similar to the conversion from decimal to binary, except that divisions and multiplications are by 16 rather than by 2.

Example 24.20 **Convert 7046_{10} to hexadecimal.**

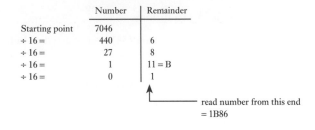

	Number	Remainder
Starting point	7046	
$\div 16 =$	440	6
$\div 16 =$	27	8
$\div 16 =$	1	$11 = B$
$\div 16 =$	0	1

read number from this end
$= 1B86$

Thus

$$7046_{10} = 1B86_{16}$$

Conversions between other bases

Conversions from other number bases to decimal are similar to those described for binary and hexadecimal numbers, using the appropriate power or multiplication factor. Conversion from one non-decimal number base to another can be achieved using decimal as an intermediate stage. This involves converting a number in one number system into decimal form and then converting it from decimal into the target number system.

It is also possible to convert directly from one number system to another, although this is sometimes tricky as most of us are strongly bound to thinking in decimal numbers. Examples of direct conversions that are easy to achieve include translations from binary to hexadecimal and vice versa. These conversions are straightforward because each hexadecimal digit corresponds to exactly four binary digits (4 bits). This allows each hexadecimal digit to be converted directly. All that is required is a knowledge of the binary equivalent for each of the 16 hexadecimal digits (as given in Table 24.3) and the translation is easy.

Example 24.21 **Convert $F851_{16}$ to binary.**

$$F851_{16} = (1111)(1000)(0101)(0001)$$

$$= 1111100001010001_2$$

Example 24.22 **Convert 1111100001010001_2 to hexadecimal.**

$$111011011000100_2 = (0111)(0110)(1100)(0100)$$

$$= 76C4_{16}$$

Note that, when arranging binary numbers into groups of four for conversion to hexadecimal, the grouping begins with the right most digit (the LSD) and extra leading zeros are added at the left-hand side as necessary.

From the above example, it is clear that large binary numbers are unwieldy and difficult to remember. Because it is so easy to convert from binary to hexadecimal, it is very common to use the latter in preference to the former for large numbers. Clearly 76C4 is much easier to write and remember than 111011011000100.

Video 24F

24.11.3 Binary arithmetic

One of the many advantages of using binary rather than decimal representations of numbers is that arithmetic is much simpler. To see why this is true one only has to consider that, in order to perform decimal long multiplication, one needs to know all the products of all possible pairs of the 10 decimal digits. To perform binary long multiplication, however, one only needs to know that $0 \times 0 = 0$, $0 \times 1 = 1 \times 0 = 0$ and that $1 \times 1 = 1$. This simplicity is a characteristic of all forms of binary arithmetic, but for the moment we will consider only addition and subtraction.

Binary addition

The addition of two single-digit binary quantities is a very simple task, the rules of which may be summarised as

$$0 + 0 = 0$$

$$0 + 1 = 1$$

$$1 + 0 = 1$$

$$1 + 1 = 10$$

It can be seen that the addition of two single-digit numbers can give rise to a two-digit number. For reasons that will become apparent shortly, an arrangement to perform this function is termed a **half adder**.

The half adder

The characteristics of a half adder are shown in Figure 24.35. The block diagram of Figure 24.35(a) shows an arrangement with two inputs, A and B, and two outputs, C (the **carry**) and S (the **sum**).

The truth table of Figure 24.35(b) shows the states of the two outputs for all possible combinations of the inputs. This is the first example we have met of a system with more than one output. As you can see, the truth table is similar to those we have used before, except that it has two output columns rather than one. The system could be described by two independent truth tables, one for each output, but it is easier to combine them.

Figure 24.35
A binary half adder.

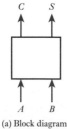

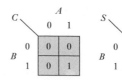

(a) Block diagram   (b) Truth table   (c) Karnaugh maps

From the truth table we can obtain Boolean expressions for the two outputs in terms of the input signals. These are

$$C = A \cdot B$$

and

$$S = \overline{A}B + A\overline{B}$$

You may recognise the expression for S as the Exclusive OR function, so we could say

$$S = A \oplus B$$

We can represent the data contained in the truth table of Figure 24.35(b) by means of two Karnaugh maps, as shown in Figure 24.35(c), in an attempt to simplify the expressions obtained. However, as can be readily seen, the '1's cannot be combined and so no simplification is possible.

The half adder can be implemented using simple gates, as shown in Figure 24.36(a), or using an Exclusive OR gate, as shown in Figure 24.36(b). While the circuit in Figure 24.36(a) provides the correct logic functions for a half adder, it has some problems when one considers its speed of operation. All electronic circuits take a finite time to respond to changes in their inputs. When considering logic gates, the time taken for the output to change as a result of a change in an input signal is termed the **propagation delay time** of the circuit (as discussed in Section 24.10). This delay time varies for different logic gates, but it is perhaps obvious that the time taken for a system to respond will depend, to a large extent, on the number of gates through which the signal passes on its route through the system. This leads to the concept of **logical depth**, which is the *maximum* number of simple gates through which a signal will pass between the input and output.

If we look at the circuit in Figure 24.36(a), we see that the carry output C is separated from the inputs by only a single gate (a logical depth of 1), whereas the sum output S is controlled by signals that pass through three gates (a logical depth of 3). This difference in logical depth means that the carry output will respond before the sum output, which may cause problems in some situations. It might seem at first sight that the circuit in Figure 24.36(b) overcomes this problem as the numbers of gates in each path are equal. However, the S output is generated using an Exclusive OR gate, which is not a simple gate but a collection of gates that have a logical depth of more than 1. The problem can be reduced by using the circuit in Figure 24.36(c) to implement the half adder. The logical depths of the C and S outputs are 2 and 3, respectively, which,

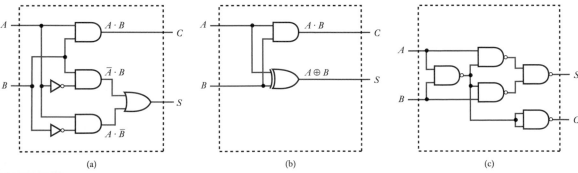

(a)   (b)   (c)

Figure 24.36
Implementation of the half adder.

although not equal, are closer than those of Figure 24.36(a). This implementation also has the advantage of using only one type of gate. You may like to convince yourself that the circuit of Figure 24.36(c) is functionally equivalent to those of Figures 24.36(a) and 24.36(b).

Adding multiple-digit numbers

The design of a circuit to add together two 2-bit binary numbers could be tackled by treating it as a network with four inputs and generating an appropriate truth table and hence a logic circuit. However, this approach becomes unwieldy as we consider circuits to add together longer binary numbers. For example, an arrangement to add two 8-bit numbers would have 16 inputs, giving a truth table with over 65,000 rows! When *we* perform addition we add the digits separately and this seems a sensible approach in the design of a circuit to perform this task.

In order to add together multiple-digit numbers we need a circuit slightly more complicated than the simple half adder described above. This is because, when adding all but the rightmost digit, we need to cater for 'carries' from the previous digit. This can be illustrated by considering the addition of two decimal numbers:

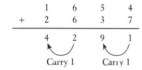

When adding the first (right most) pair of digits, only these two numbers are summed. However, for all the following pairs, any carry from the previous stage must also be added. In binary addition, the process is similar:

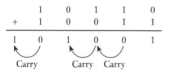

The half adder can be used to sum the right most (least significant) digits of the numbers, but for all the following digits a circuit is required that can add together not only the two digits of the input numbers but also a carry from the previous stage. The carry from one stage to the next can be either 0 or 1. A circuit that can add together the two digits of a number and any carry input is called a **full adder** to distinguish it from the simpler half adder described earlier.

Figure 24.37 shows an arrangement for adding two 4-bit binary numbers. The least significant bits of each word (the right-hand bits) are added together by a half adder circuit, as discussed earlier. The sum output S from this circuit forms the least significant bit of the result, while the carry output C is fed forward to the next 'column' in the addition. This next stage needs to add together the second digit from each input word, plus the carry digit from the previous stage. Therefore, we need a full adder. As the full adder has a carry input from the previous stage and a carry output that feeds to the next stage, these are labelled C_i and C_o to avoid confusion. By adding an appropriate number of full adders, it is possible to add together words of any length. In each case, the S output produces the corresponding element in the output result. The carry output from the final stage then forms the most significant bit of the result.

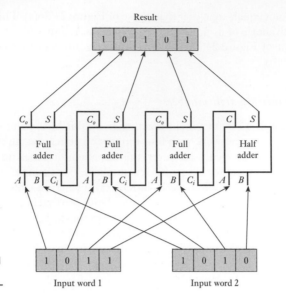

Figure 24.37
An arrangement to add
two 4-bit numbers.

The full adder

It can be seen that each full adder has three inputs (A, B and the carry input, C_i) and two outputs (the sum S and the carry output C_o). A block diagram of a full adder is shown in Figure 24.38(a). The function of the full adder is described by the truth table of Figure 24.38(b).

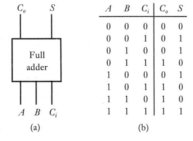

Figure 24.38
The full adder.

A	B	C_i	C_o	S
0	0	0	0	0
0	0	1	0	1
0	1	0	0	1
0	1	1	1	0
1	0	0	0	1
1	0	1	1	0
1	1	0	1	0
1	1	1	1	1

(a)   (b)

Boolean expressions can be obtained directly from this truth table and simplified using algebraic manipulation. Alternatively, the data can be represented using Karnaugh maps, as shown in Figure 24.39.

Figure 24.39
Representation of
a full adder using
Karnaugh maps.

C_o

		AB		
	00	01	11	10
C_i 0	0	0	1	0
1	0	1	1	1

S

		AB		
	00	01	11	10
C_i 0	0	1	0	1
1	1	0	1	0

From either method of simplification, we find that

$$C_o = AB + AC_i + BC_i$$

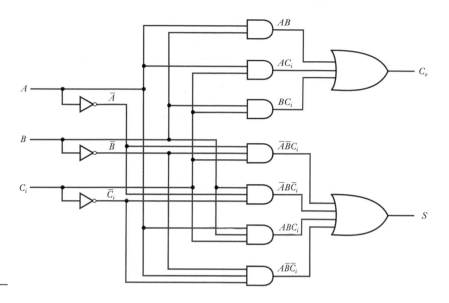

Figure 24.40
Implementation
of a full adder.

and

$$S = \overline{A}\,\overline{B}C_i + \overline{A}B\overline{C}_i + ABC_i + A\overline{B}\,\overline{C}_i$$

These functions can be implemented directly, as shown in Figure 24.40.

An alternative method for producing the function of a full adder is to break down its operation into two parts. Clearly, adding the three components together is equivalent to adding two together and then adding the third. We can therefore produce the same effect using two half adders, as shown in Figure 24.41. The solution using two half adders requires less circuitry than that required to implement the circuit in Figure 24.40, but has a greater *logical depth*, so is slower to respond. The full adder has a logical depth of 3 for both outputs whereas the half adder approach has a logical depth of at least 6 for the sum output and at least 5 for the carry.

When adding together numbers of more than a few bits, it is unusual to construct adder networks directly from basic gates. Integrated circuits are available that provide a number of full adders within a single circuit, simplifying design and construction. A typical arrangement might incorporate circuitry to allow two 4-bit numbers to be added together. Such a circuit would be similar to that in Figure 24.37, but would normally use four full adders rather than three full adders and a half adder. This provides both a *carry in* and a *carry out* for the circuit, as shown in Figure 24.42(a). This allows a number of these circuits to be cascaded to permit binary words with any number of bits to be added. When used for the least significant (right most) bits, the carry input is connected to '0' to indicate the absence of any carry in. This arrangement is shown in Figure 24.42(b). It is common to number the bits of binary numbers from the right and to start from 0. Thus an n-bit number has digits from 0 to $n - 1$.

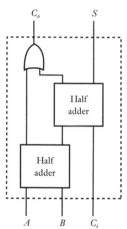

Figure 24.41
Forming a full adder
from two half adders.

Binary subtraction

The process of binary subtraction can be tackled in a similar manner to that of addition. We can construct a **half subtractor**, as shown in Figure 24.43(a), with a truth table, as given in Figure 24.43(b). As we are now concerned with subtraction rather than addition, we have *difference* (D) and *borrow* (B_o) outputs rather than *sum*

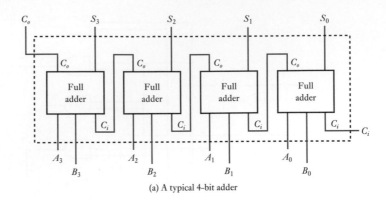

(a) A typical 4–bit adder

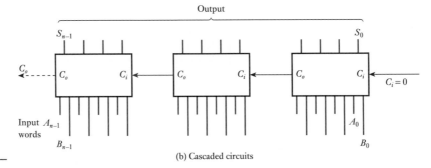

Figure 24.42
A cascadable
4-bit adder.

(b) Cascaded circuits

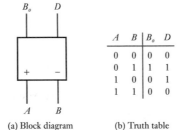

Figure 24.43
A half subtractor.

(a) Block diagram   (b) Truth table

A	B	B_o	D
0	0	0	0
0	1	1	1
1	0	0	1
1	1	0	0

and *carry*. It is also necessary to differentiate between the two inputs A and B to determine which is subtracted from which. In the example shown, the output is equal to $(A - B)$.

From the truth table we can see that

$$B_o = \overline{A} \cdot B$$

and

$$D = \overline{A}B + A\overline{B} = A \oplus B$$

You will notice that D is identical to S for a half adder, but the borrow output is not the same as the carry.

In order to perform multiple-bit subtraction, we again need to consider the effect of one stage on the next. Figure 24.44 shows a 4-bit subtractor using four **full subtractors**. This circuit can be cascaded to allow larger numbers to be used.

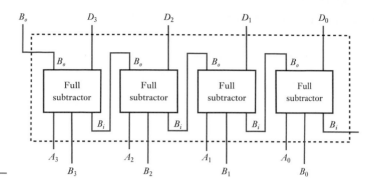

Figure 24.44
A 4-bit subtractor.

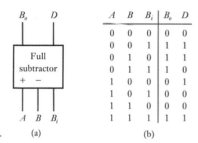

Figure 24.45
A full subtractor.

A	B	B_i	B_o	D
0	0	0	0	0
0	0	1	1	1
0	1	0	1	1
0	1	1	1	0
1	0	0	0	1
1	0	1	0	0
1	1	0	0	0
1	1	1	1	1

(a)   (b)

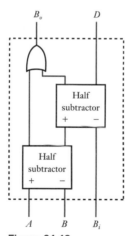

Figure 24.46
Constructing a full
subtractor from two
half subtractors.

Figure 24.45 shows a block diagram and the truth table for a full subtractor. The outputs can be represented by the Boolean expressions

$$B_o = \overline{A}B + \overline{A}B_i + BB_i$$

and

$$D = \overline{A}\overline{B}B_i + \overline{A}B\overline{B}_i + A\overline{B}\overline{B}_i + ABB_i$$

These functions can be implemented directly, as described earlier for the full adder, or constructed from two half subtractors, as shown in Figure 24.46.

The subtractor circuits described above work as we would expect, provided that the result is not negative. We will leave further discussion of the representation of negative numbers until we deal with arithmetic within microprocessors (see Chapter 27).

Binary multiplication and division

Although it is possible to construct circuits to perform multiplication and division using simple logic gates, it is fairly unusual as the complexity of the circuits makes them impractical. It is more usual to perform these functions using dedicated logic circuits containing a large number of gates or a microprocessor. We will again leave discussion of this topic until Chapter 27.

Video 24G

24.12   Numeric and alphabetic codes

24.12.1   Binary code

By far the most frequently used method for representing numeric information within digital systems is to use the simple binary code described earlier. This has the advantages of simplicity of arithmetic and efficiency of storage. However, there are some applications in which other representations are used for specific purposes.

24.12.2 Binary-coded decimal (BCD) code

Binary-coded decimal (BCD) code, as its name implies, is formed by converting each digit of a decimal number individually into its binary form.

Example 24.23

Convert 9450_{10} to BCD.

$$9450_{10} = (1001)(0100)(0101)(0000)_{BCD}$$

Conversion from BCD to decimal is just as simple to do and is achieved by dividing the number into groups of four, starting with the least significant digit and then converting each digit into decimal. Additional leading zeros can be used to complete the last group, if required.

Example 24.24

Convert 11100001110110_{BCD} to decimal.

$$11100001110110_{BCD} = (0011)(1000)(0111)(0110)_{BCD}$$

$$= 3876_{10}$$

BCD requires more digits than the straight binary form, so it is less efficient, but it has the advantage of very simple conversions to, and from, decimal. It is therefore widely used in situations where input and output data are in a decimal form, such as in pocket calculators.

24.12.3 Gray code

We met the Gray code earlier in this chapter when we discussed Karnaugh maps. There, we wished to number the various elements of the map so that the codes for adjacent elements varied in only one bit. Gray code has this property, as is illustrated in Table 24.4.

Table 24.4 Gray code.

Decimal	Gray code
0	0000
1	0001
2	0011
3	0010
4	0110
5	0111
6	0101
7	0100
8	1100
9	1101
10	1111
11	1110
12	1010
13	1011
14	1001
15	1000

As with simple binary numbers, the sequence can be continued to represent arbitrarily large numbers and leading zeros have no effect. At first sight, the order may seem rather strange, but there is a simple and systematic method for producing the sequence, removing the need to memorise it. This is done by first writing down the first two numbers (these are simply 0 and 1 so are not difficult to remember), then writing these numbers down again in reverse order with a '1' in front. This gives the sequence

0

1

11

10

This sequence is then repeated again in reverse order with a '1' in front to give a total of eight numbers

0

1

11

10

110

111

101

100

This process is repeated as often as is necessary, each time repeating the previous sequence, in reverse order, with the addition of a leading '1'. This method for generating the sequence gives rise to its alternative name, which is **reflected binary code**.

In addition to its use within Karnaugh maps, Gray code is found in numerous applications where changing quantities are to be read. The reason for this can be illustrated by considering an imaginary transducer that produces simple binary code as its output. Let us imagine that the output from the device at a particular time changes from 7_{10} to 8_{10}. This represents an output change from 0111_2 to 1000_2. If we now consider some external device connected to the transducer, it is interesting to note the effect of reading the output from the transducer at the exact instant that its value changes. In changing from 0111 to 1000, all four digits change. If the digits are read while they are changing, the value obtained is indeterminate. If, for example, the leading '0' happened to change to a '1' slightly faster than the '1's changed to '0's, the value could be read as 1111 (15_{10}). Alternatively, if the leading '0' turned to '1' slightly slower than the '1's changed to '0's, the number could be read as 0000 (0_{10}). This means that any combination of the four digits could be obtained, giving any number in the range 0 to 15.

If Gray code is used as the output from the transducer this problem cannot occur. In changing from 7_{10} to 8_{10} the output changes from 0100 to 1100. As only one digit has changed, reading the output during this transition can only lead to one digit being uncertain. Thus, the number read will be either 0100 or 1100. Therefore, the number read will always be either the old or the new number – an ideal arrangement. As all adjacent numbers in Gray code differ by only one digit, this property exists for all transitions.

Gray code is widely used in **counters** that must be read asynchronously (counters are discussed in Chapter 25) and absolute position encoders, as discussed in Section 12.6.5. If you look at the pattern of stripes on the encoder of Figure 12.10 you will see that this is in Gray code.

24.12.4 ASCII code

So far we have concentrated on codes that are used to represent numeric quantities. Often it is also necessary to store and transmit alphabetic data in digital form – for example, when storing text within a computer or mobile phone. Various standard codes are used for this purpose, but by far the most widely used is the **American Standard Code for Information Interchange** – normally abbreviated to **ASCII** (and pronounced 'ass-key').

ASCII represents each character using a 7-bit code, allowing 128 possible values. Codes are defined for both upper- and lower-case alphabetic characters, the digits 0 to 9, punctuation marks (such as commas, full stops and question marks) and various non-printable codes that are used as control characters. These control characters include codes to produce a line feed, carriage return and backspace on a printer.

Codes are included for both alphabetic and numeric characters, so this and other codes of this form are often referred to as **alphanumeric** codes. It should be noted, however, that the numeric codes represent *numeric characters*, not the corresponding *quantities*. A partial listing of the ASCII character set is given in Table 24.5.

While ASCII is the most widely known alphanumeric code, the use of 7 bits per character allows it to represent only 128 different characters. This, combined with the standard's pronounced English language bias, has led to the development of a wide range of extensions or variants to the code over the years. More modern codes, such as Unicode and the ISO/IEC 10646 universal character set, offer a much wider range of characters (including the special characters associated with a range of non-English languages) and these are replacing ASCII in a number of situations.

24.12.5 Error detection and correction techniques

All electronic systems suffer from noise (see Chapter 22). One possible effect of noise within digital systems is the corruption of data. This is a particular problem when data must be transmitted from one place to another.

Parity checking

One of the simplest ways in which to tackle the problem of errors is to use **parity** testing. This is done by adding a small amount of redundant information to each word of data to allow it to be checked. The extra information takes the form of a **parity bit**, which is added at the end of each data word. The polarity of the added bit is chosen so that the total number of '1's within the word (including the added parity bit) is either always even (*even parity*) or always odd (*odd parity*). For example, consider an even parity system. The ASCII character for 'S' is

Table 24.5 A partial listing of the ASCII character set.

Character	7-bit ASCII	Hex	Character	7-bit ASCII	Hex
A	100 0001	41	0	011 0000	30
B	100 0010	42	1	011 0001	31
C	100 0011	43	2	011 0010	32
D	100 0100	44	3	011 0011	33
E	100 0101	45	4	011 0100	34
F	100 0110	46	5	011 0101	35
G	100 0111	47	6	011 0110	36
H	100 1000	48	7	011 0111	37
I	100 1001	49	8	011 1000	38
J	100 1010	4A	9	011 1001	39
K	100 1011	4B	blank	010 0000	20
L	100 1100	4C	!	010 0001	21
M	100 1101	4D	"	010 0010	22
N	100 1110	4E	#	010 0011	23
O	100 1111	4F	$	010 0100	24
P	101 0000	50	%	010 0101	25
Q	101 0001	51	&	010 0110	26
R	101 0010	52	'	010 0111	27
S	101 0011	53	(	010 1000	28
T	101 0100	54	)	010 1001	29
U	101 0101	55	*	010 1010	2A
V	101 0110	56	+	010 1011	2B
W	101 0111	57	,	010 1100	2C
X	101 1000	58	−	010 1101	2D
Y	101 1001	59	.	010 1110	2E
Z	101 1010	5A	/	010 1111	2F
a	110 0001	61	:	011 1010	3A
b	110 0010	62	;	011 1011	3B
c	110 0011	63	<	011 1100	3C
d	110 0100	64	=	011 1101	3D
e	110 0101	65	>	011 1110	3E
f	110 0110	66	?	011 1111	3F
g	110 0111	67	[	101 1011	5B
h	110 1000	68	\	101 1100	5C
i	110 1001	69	]	101 1101	5D
j	110 1010	6A	^	101 1110	5E
k	110 1011	6B	_	101 1111	5F
l	110 1100	6C	{	111 1011	7B
m	110 1101	6D	|	111 1100	7C
n	110 1110	6E	}	111 1101	7D
o	110 1111	6F	~	111 1110	7E
p	111 0000	70	delete	111 1111	7F
q	111 0001	71	bell	000 0111	07
r	111 0010	72	backspace	000 1000	08
s	111 0011	73	carriage return	000 1101	0D
t	111 0100	74	escape	001 1011	1B
u	111 0101	75	form feed	000 1100	0C
v	111 0110	76	line feed	000 1010	0A
w	111 0111	77	horizontal tab	000 1001	09
x	111 1000	78	vertical tab	000 1011	0B
y	111 1001	79	start text	000 0010	02
z	111 1010	7A	end text	000 0011	03

which has an even number of '1's. Therefore a '0' parity bit is added to make an 8-bit word, which still has an even number of '1's

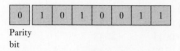

Parity
bit

On reception, the parity of the word is tested by counting the number of '1's. If it is still even, the parity bit is removed and the original 7 bits are passed to their destination. If the parity is incorrect on reception, an error has been detected and the system must take appropriate action.

Although this technique indicates that an error has occurred, it cannot determine which bit or bits are incorrect. It should also be noted that, if two errors are present, the parity of the resultant word will again be even and the errors will not be detected.

This simple error-detecting technique will detect any odd number of errors, but not an even number of errors. Random numbers thus have a 50 per cent chance of passing the parity test. Parity testing is often used on communications channels to give confidence that the line is working correctly. Although the reliability of testing any one word is low, when applied to a large number of words it is sure to detect errors if the line is unreliable.

Checksum

An alternative method that can be used to check the correctness of data is a **checksum**. This provides a test of the integrity of a block of data rather than individual words. When a group of words is to be transmitted, they are summed at the transmitter and the sum is transmitted after the data. At the receiver, the words are again summed and the result is compared with the sum produced by the transmitter. If the results agree, the data is probably correct. If they do not, an error has been detected.

As with the parity check, the test gives no indication as to the location of the error – it simply indicates that one has occurred. The action taken depends on the nature of the system. It might involve sending the data again or sounding an alarm to warn an operator.

Error-detecting and correcting codes

The parity and checksum techniques both send a small amount of redundant information to allow the integrity of the data to be tested. If one is prepared to send additional redundant information, it is possible to construct codes that not only detect the presence of errors but also indicate their location within a word, allowing them to be corrected. An example of this technique is the well-known **Hamming code**.

The performance of these codes in terms of their ability to detect and correct multiple errors depends on the amount of redundant information that can be tolerated. The more redundancy that is incorporated, the greater is the rate at which data must be sent and the more complicated the system. It should also be remembered that it is not possible to construct a code that will allow an unlimited number of errors. This would imply that the system could produce the correct output with a random input – clearly an impossibility.

24.13 Examples of combinational logic design

We have already looked at several design tools for use with combinational logic, including Boolean algebra, truth tables and Karnaugh maps. We now look at a few examples to illustrate their use.

Example 24.25 | Design a circuit to convert 3-bit binary numbers into Gray code.

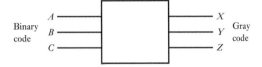

The circuit can be represented by the truth table

A	B	C	X	Y	Z
0	0	0	0	0	0
0	0	1	0	0	1
0	1	0	0	1	1
0	1	1	0	1	0
1	0	0	1	1	0
1	0	1	1	1	1
1	1	0	1	0	1
1	1	1	1	0	0

and from this we can construct three Karnaugh maps

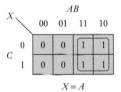

$X = A$

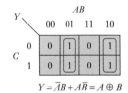

$Y = \bar{A}B + A\bar{B} = A \oplus B$

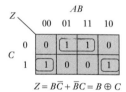

$Z = B\bar{C} + \bar{B}C = B \oplus C$

The circuit may then be implemented using standard gates

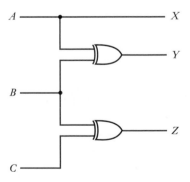

Example 24.26

Design a circuit to take a 4-bit number $ABCD$ and produce a single output Y that is true only if the input represents a prime number.

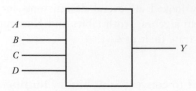

The circuit may be represented by the truth table

Decimal	A	B	C	D	Y
0	0	0	0	0	1
1	0	0	0	1	1
2	0	0	1	0	1
3	0	0	1	1	1
4	0	1	0	0	0
5	0	1	0	1	1
6	0	1	1	0	0
7	0	1	1	1	1
8	1	0	0	0	0
9	1	0	0	1	0
10	1	0	1	0	0
11	1	0	1	1	1
12	1	1	0	0	0
13	1	1	0	1	1
14	1	1	1	0	0
15	1	1	1	1	0

The truth table can be used to form a Karnaugh map

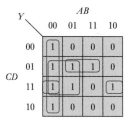

$$Y = \bar{A}\bar{B} + \bar{A}D + B\bar{C}D + \bar{B}CD$$

and the circuit may be implemented using standard gates

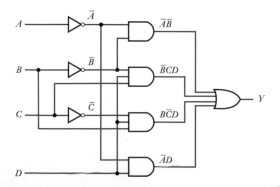

Example 24.27

Design a circuit to take a BCD number *ABCD* and produce a single output *Y* that is true only when the input corresponds to the numbers 1, 2, 5, 6 or 9.

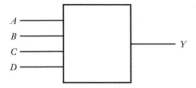

The circuit may be represented by a truth table as follows. Note that, as the input is a BCD number, certain combinations of the input variables cannot occur. This means that the output state is *don't care* for these conditions.

Decimal	A	B	C	D	Y
0	0	0	0	0	0
1	0	0	0	1	1
2	0	0	1	0	1
3	0	0	1	1	0
4	0	1	0	0	0
5	0	1	0	1	1
6	0	1	1	0	1
7	0	1	1	1	0
8	1	0	0	0	0
9	1	0	0	1	1
10	1	0	1	0	X
11	1	0	1	1	X
12	1	1	0	0	X
13	1	1	0	1	X
14	1	1	1	0	X
15	1	1	1	1	X

From this a Karnaugh map can be formed and a simplified expression for *Y* obtained

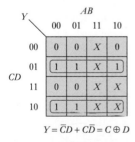

$$Y = \bar{C}D + C\bar{D} = C \oplus D$$

This can then be implemented directly

Example 24.28 | **Design a four-input multiplexer.**

A **multiplexer** is a circuit that can perform the function of a multiway switch by selecting one of a number of input signals and passing this to a single output line. A block diagram of a four-input multiplexer is shown below.

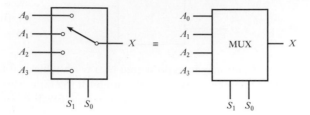

The two control inputs S_1 and S_0 are line select inputs and the signals on these inputs determine which of the four data input lines A_0 to A_3 is selected. The output X then becomes equal to the binary signal on the selected input line. The operation of the multiplexer can therefore be described by the following truth table:

S_0	S_1	X
0	0	A_0
0	1	A_1
1	0	A_2
1	1	A_3

The logic required to implement this function can be designed by treating the multiplexer as a circuit with six inputs (A_0 to A_3, plus S_0 and S_1) and drawing the necessary truth table. This can then be simplified to give the required logic. The required truth table has 64 lines and the construction of this table, and the necessary simplification, is left as an exercise for the reader.

An alternative approach to implementation is to break the design down into two components – the *select logic* and the *gating logic*.

If we look at the truth table of an AND gate we see that any signal ANDed with a '0' gives a '0', while any signal ANDed with a '1' gives the original signal. This allows us to use an AND gate as a gating network. Our multiplexer can therefore be constructed as shown below.

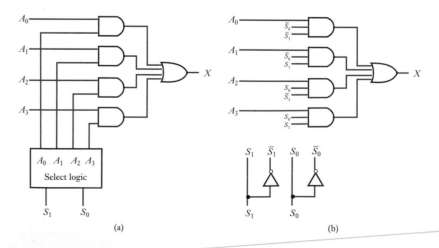

(a)   (b)

In diagram (a) the 'select logic' block produces a '1' on the appropriate select line and a '0' on the remainder. This 'enables' the AND gate connected to the appropriate input line and 'disables' the rest. The outputs from the three disabled AND gates will all be '0' and so will have no effect on the OR gate. If the output from the selected AND gate is also a '0' (because the selected input signal is '0'), then the output from the OR gate will be '0'. However, if the output from the selected AND gate is a '1' (because the selected input signal is '1'), then the output from the OR gate will be a '1'. Therefore, the output X will always be equal to the value of the selected input signal.

If we now turn our attention to the form of the 'select logic', we see that this is very simple. Line A_0 should be selected if both S_1 and S_0 are '0' and so the appropriate select line can be formed by ANDing together the inverted form of these two signals. Similarly, the select signal for A_1 can be formed by ANDing S_1 with the inverse of S_0 and so on. Thus, our four select lines may be formed by individually ANDing together the four combinations of the two select lines and their inverses. In fact, as the select lines are themselves ANDed with the data input lines, a single AND gate may be used to select each line, as shown in diagram (b) above.

This implementation of the multiplexer can be obtained using a truth table and Karnaugh maps, as with earlier circuits, but this example illustrates that it may not always be the simplest approach. If the problem had been to design a multiplexer with eight data lines, this would have required a truth table with 11 inputs (eight data lines plus three select lines). Such a truth table has over 2000 lines. However, the approach described above can easily be extended to a circuit with eight data lines.

The circuit described in this example takes 'logical' signals on its inputs and uses them to determine the state of its output. Such a circuit is often called a **digital multiplexer**. Circuits are also available that can perform the same function with analogue as well as digital signals. Such components are described as **analogue multiplexers**.

A circuit related to the multiplexer is the **demultiplexer**. This takes a single input signal and uses it to determine the state of one of a number of output lines under the control of an appropriate number of line select lines.

Example 24.29

Design a BCD to seven-segment decoder, which takes a 4-bit input representing a number in the range 0 to 9, and generates an appropriate 7-bit output to illuminate the appropriate elements of the display to represent this digit.

In Section 13.3.1 we looked at a seven-segment LED display and here we consider the circuitry required to drive it. We can represent the required arrangement using the following block diagram.

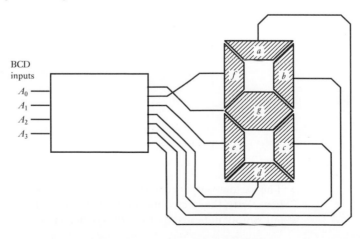

We can form the various digits by illuminating combinations of segments, as shown below:

$$0\ 1\ 2\ 3\ 4\ 5\ 6\ 7\ 8\ 9$$

We can then describe the required system by means of the following truth table

Number	A_3	A_2	A_1	A_0	a	b	c	d	e	f	g
0	0	0	0	0	1	1	1	1	1	1	0
1	0	0	0	1	0	1	1	0	0	0	0
2	0	0	1	0	1	1	0	1	1	0	1
3	0	0	1	1	1	1	1	1	0	0	1
4	0	1	0	0	0	1	1	0	0	1	1
5	0	1	0	1	1	0	1	1	0	1	1
6	0	1	1	0	1	0	1	1	1	1	1
7	0	1	1	1	1	1	1	0	0	0	0
8	1	0	0	0	1	1	1	1	1	1	1
9	1	0	0	1	1	1	1	0	0	1	1
10	1	0	1	0	X	X	X	X	X	X	X
11	1	0	1	1	X	X	X	X	X	X	X
12	1	1	0	0	X	X	X	X	X	X	X
13	1	1	0	1	X	X	X	X	X	X	X
14	1	1	1	0	X	X	X	X	X	X	X
15	1	1	1	1	X	X	X	X	X	X	X

The outputs corresponding to the input numbers 10 to 15 are don't care conditions as these input combinations will not occur. The seven outputs can be represented by the series of Karnaugh maps

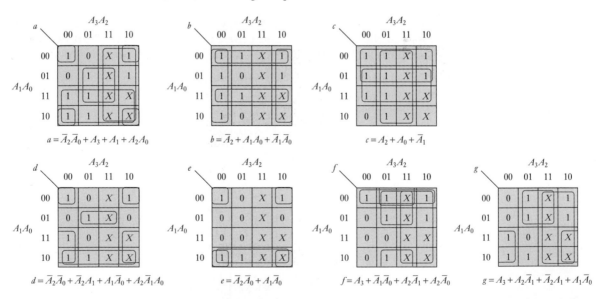

$$a = \bar{A}_2\bar{A}_0 + A_3 + A_1 + A_2A_0$$

$$b = \bar{A}_2 + A_1A_0 + \bar{A}_1\bar{A}_0$$

$$c = A_2 + A_0 + \bar{A}_1$$

$$d = \bar{A}_2\bar{A}_0 + \bar{A}_2A_1 + A_1\bar{A}_0 + A_2\bar{A}_1A_0$$

$$e = \bar{A}_2\bar{A}_0 + A_1\bar{A}_0$$

$$f = A_3 + \bar{A}_1\bar{A}_0 + A_2\bar{A}_1 + A_2\bar{A}_0$$

$$g = A_3 + A_2\bar{A}_1 + \bar{A}_2A_1 + A_1\bar{A}_0$$

The various outputs can then be generated using simple logic gates, as shown below. For clarity, the interconnections between the inputs and the logic gates are not shown but simply indicated by their functional names.

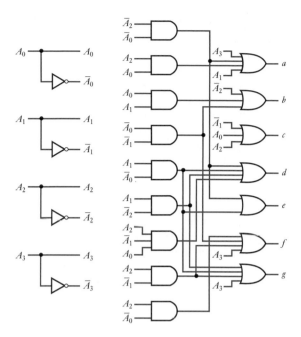

Video 24H

Further study

While it is generally important that all electronic systems work correctly, in some cases it is vital. Consider for example some of the systems used within aircraft, where failure of a system could result in a major accident.

One approach often used in the design of critical systems is to incorporate some form of 'fault tolerance', which permits systems to operate correctly even in the presence of component failure. A simple example of this is shown below.

The diagram shows a module within our critical system. It has a single input and a single output. If the module functions correctly all is well, but what happens if it fails? A more robust arrangement is shown below.

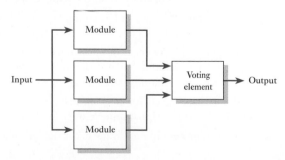

Here three identical modules are employed to provide a degree of fault tolerance. Each module is fed with the same input and thus should produce the same output. The three output signals are fed to a 'voting element' that compares their outputs. If the three outputs agree then the voting element simply passes this signal to its output. However, if one of the outputs disagrees with the others, the voting element takes the majority view and passes this to the output, thus ignoring the faulty device. Thus a single unit can fail without affecting the operation of the system.

The nature and complexity of the voting element is affected by the nature of the signals produced by the modules. If these are simple binary signals then the voting element is relatively simple. Your task here is to design such a voting element that takes three binary inputs and produces a single binary output that represents the 'majority view' of the inputs.

As an extension to this task consider what happens to the safety of the system when one of the modules fails. While this arrangement will tolerate the failure of a single module it cannot tolerate the failure of two modules. If a module fails the system is therefore severely compromised and its safety is reduced. Consider how this arrangement could be modified to reduce this problem.

Key points

- To simplify the description of binary variables, it is common to represent their two states by the symbols '1' and '0'. These might represent on and off, true and false or any other pair of binary conditions.

- In some simple cases, it is possible to implement binary systems using switches. However, generally it is more useful to design such systems using logic gates.

- Our basic building blocks are a small number of simple gates. Three elementary forms – AND, OR and NOT – can be used to form any logic function, although it is often more useful to work with compound gates such as NAND, NOR and Exclusive OR.

- Combinational logic circuits can be described by a truth table that lists all the possible combinations of the inputs and indicates the corresponding values of the outputs.

- It is also possible to define a logic function using Boolean algebra. This notation and set of rules and identities allows binary relationships to be described and simplified.

- Simplification may also be performed graphically, using Karnaugh maps.

- In addition to binary variables, digital systems often use many-valued quantities, which are represented using binary words of an appropriate length.

- In digital electronics, several number systems are used, the most common being decimal, binary, octal and hexadecimal.

- As binary numbers use only two digits, 0 and 1, arithmetic is simpler than it is for decimal numbers.

- Although simple binary code is the most frequently used option for representing numeric information, it is not the only method. In some applications, the use of other representations, such as Gray code, may be more appropriate.

- Codes are also used for non-numeric information, such as the ASCII code, which is used for alphanumeric data.

- Some coding techniques allow error detection and, in some cases, correction.

Exercises

24.1 Show how a power source, a lamp and a number of switches can be used to represent the following logical functions

$$L = A \cdot B \cdot C$$

$$L = A + B + C$$

$$L = (A \cdot B) + (C \cdot D)$$

$$L = A \oplus B$$

24.2 Derive expressions for the following arrangements using AND, OR and NOT operations.

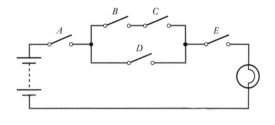

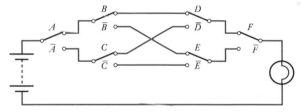

24.3 If the two circuits given in the previous exercise were described by truth tables, how many rows would each table require?

24.4 Sketch the truth table of a three-input NAND gate.

24.5 Sketch the truth table of a three-input NOR gate.

24.6 Show that the two circuits (a) and (b) below are equivalent by drawing truth tables for each circuit.

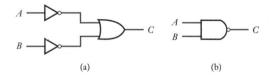

24.7 Repeat the operations in Exercise 24.6 for the following circuits.

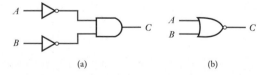

24.8 Simulate the pairs of circuits in Exercises 24.6 and 24.7 and confirm that each pair produces the same output for every possible combination of the inputs.

24.9 List all the possible values of a Boolean constant.

24.10 List all the possible values of a Boolean variable.

24.11 What symbols are used in Boolean algebra to represent the functions AND, OR, NOT and Exclusive OR?

24.12 Write the function of a three-input NOR gate as a Boolean expression.

24.13 Given that A is a Boolean variable, evaluate and then simplify the following expressions: $A \cdot 1; A \cdot \bar{A}; 1 + A; A + \bar{A}; 1 \cdot 0; 1 + 0.$

24.14 Exercises 24.6 and 24.7 illustrate fundamental laws of Boolean algebra. What is the name given to these laws?

24.15 What is the difference between combination and sequential logic?

24.16 Implement the following expressions using standard logic gates.

$$X = (\overline{A + B}) \cdot C$$

$$Y = A\bar{B}C + \bar{A}D + C\bar{D}$$

$$Z = \overline{(A \cdot B) + (\overline{C + D})}$$

24.17 Derive a Boolean expression for the following circuit.

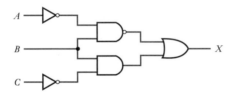

24.18 Design a logic circuit to take three inputs – A, B and C – and produce a single output X, such that X is true if, and only if, precisely two of its inputs are true.

24.19 Use circuit simulation to investigate your solution to Exercise 24.18 and hence demonstrate that it behaves as required.

24.20 Derive Boolean expressions to describe the operation of the following circuit. Minimise these expressions by means of algebraic manipulation and hence simplify the circuit.

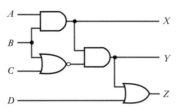

24.21 Use Karnaugh maps to obtain minimised Boolean expressions for the following functions

$$X = \bar{A}\bar{B} + A\bar{B}\bar{C} + A\bar{B}C + ABC$$

$$Y = \bar{A}\bar{B}\bar{C} + \bar{A}B\bar{C}D + A\bar{C}\bar{D} + A\bar{C}D + A\bar{B}C\bar{D}$$

24.22 Use a Karnaugh map to obtain a minimised Boolean expression for the function described by the following truth table.

A	B	C	D	Z
0	0	0	0	1
0	0	0	1	0
0	0	1	0	X
0	0	1	1	0
0	1	0	0	1
0	1	0	1	X
0	1	1	0	1
0	1	1	1	1
1	0	0	0	X
1	0	0	1	0
1	0	1	0	1
1	0	1	1	0
1	1	0	0	0
1	1	0	1	1
1	1	1	0	0
1	1	1	1	X

24.23 Convert the following binary numbers into decimal: 1100, 110001, 10111, 1.011.

24.24 Convert the following decimal numbers into binary: 56, 132, 67, 5.625.

24.25 Convert the following hexadecimal numbers into decimal: A4C3, CB45, 87, 3FF.

24.26 Convert the following decimal numbers into hexadecimal: 52708, 726, 8900.

24.27 Convert $A4C7_{16}$ into binary.

24.28 Convert 10110010100101_2 into hexadecimal.

24.29 Perform the following binary arithmetic

$$
\begin{array}{cccc}
10111 & 110101 & 1011 & 101010 \\
+1001 & -11010 & \times 111 & \div 10 \\
\hline
\end{array}
$$

24.30 Design a circuit to convert 3-bit Gray code numbers into simple binary.

24.31 Design an eight-input digital multiplexer along the lines of the circuit described in Example 24.28. The circuit should have eight data inputs, three line select inputs and a single output.

24.32 Simulate your solution to the previous exercise to confirm that the circuit functions as expected.

24.33 Design a four-output digital demultiplexer. The circuit should have one data input, four data outputs and two select inputs.

24.34 Simulate your solution to the previous exercise to confirm that the circuit functions as expected.

Chapter 25

Sequential Logic

Objectives

When you have studied the material in this chapter, you should be able to:

- describe the characteristics of a wide range of sequential logic circuits, including bistables, monostables and astables
- explain the differences between various forms of bistable, including latches, flip-flops and pulse-triggered versions
- discuss the use of bistables in the construction of memory registers and shift registers
- design simple binary ripple counters and modulo-*N* counters of any length
- appreciate the role of specialised sequential integrated circuits such as monostables, astables and timers.

Video 25A

25.1 Introduction

We have seen that in combination logic, the outputs are determined only by the current state of the inputs. In sequential logic, however, the outputs are determined not only by the current inputs but also by the sequence of inputs that led to the current state. In other words, the circuit has the characteristic of **memory**.

When constructing combinational logic, our basic building blocks are normally the various gates described in Chapter 24. When constructing sequential logic, we generally use slightly larger building blocks, which are often some form of **multivibrator**. This term describes a range of circuits that share certain characteristics. They each have two outputs that are the inverse of each other and these are conventionally given the names Q and $\bar{Q}$. Having only these two outputs means that the circuits have only two possible output states, namely $Q = 1$, $\bar{Q} = 0$ and $Q = 0$, $\bar{Q} = 1$. Different forms of multivibrator are defined by the behaviour of the circuits in these two states:

- **Bistable multivibrators**, in which both output states are stable. When in one state the circuit will remain in that state until an input signal causes it to change state. There are several types of bistable multivibrator but unfortunately there is no general agreement on the names used for these classes of device. Some engineers refer to all bistable devices as **flip-flops**, while others use the term **latch** for level-sensitive devices and **flip-flop** for edge-triggered and pulse-triggered devices (the meanings of these terms will be explained shortly). Here we adopt the latter terminology as it gives additional information about the form of the device.
- **Monostable multivibrators**, in which one state is stable and the other is metastable (or quasistable). The circuit will remain in its stable state until acted on

by an appropriate input signal, whereupon it will change to its metastable state. It will remain in its metastable state for a fixed period of time (determined by circuit parameters) and then will automatically revert to its stable state. The circuit behaves as a single pulse generator. When *triggered* it enters its metastable state causing the outputs to change for a fixed period of time. This circuit is also known as a **one-shot**.

■ **Astable multivibrators**, in which both states are metastable. The circuit stays in each state for a fixed period of time (determined by circuit parameters) before switching to its other state. This produces a circuit that continually oscillates from one state to the other – a digital oscillator.

Of the three forms of multivibrator, bistable circuits are by far the most important and widely used. We will therefore start by looking at a few basic forms of bistable and then go on to look at some of their uses.

25.2 Bistables

Consider the circuit in Figure 25.1. The figure shows two inverters connected in a ring. If the output of the first inverter Q is equal to 1, this signal is fed to the input of the second inverter, making its output P equal to 0. This in turn forms the input to the first inverter, which makes its output 1. Thus the circuit is stable with $Q = 1$ and $P = 0$. Alternatively, if Q is equal to 0, this corresponds to a stable state with $Q = 0$ and $P = 1$. The circuit therefore has two stable states. It also has two outputs – Q and P, where $P = \bar{Q}$. We could therefore consider this circuit to be a form of bistable multivibrator. This arrangement is an example of **regenerative switching** in which the output of one stage is amplified and fed back to reinforce that output signal, forcing the circuit into one state or the other.

Figure 25.1
A regenerative
switching circuit.

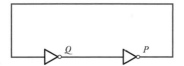

Although the circuit in Figure 25.1 has the characteristics of a bistable, it is of little practical use. Its state is determined when power is applied and it then remains in that state until power is removed.

25.2.1 The S–R latch

The arrangement in Figure 25.1 becomes more interesting if we substitute two input NOR gates for the inverters, as shown in Figure 25.2.

Figure 25.2
A latch formed using
two NOR gates.

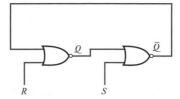

We now have a circuit with two input signals, R and S, and two outputs, which are now labelled Q and $\bar{Q}$. If one input of a two-input NOR gate is held at 0, the relationship between the other input and the output is that of an inverter. Therefore, if R and S are both held at 0, the circuit behaves in the same manner as the previous circuit and will stay in whichever state it finds itself. We could call this condition the **memory mode** of the circuit. If now R is taken to 1 while S remains at 0, Q will be reset to 0 regardless of its previous state. In turn, this will set $\bar{Q}$ to 1. If now R is returned to 0, the circuit will re-enter its memory mode and will stay in this state. Similarly, if S is taken to 1 while R remains at 0, $\bar{Q}$ will be cleared to 0 and Q will be set to 1. Again, if S returns to 0, the circuit will re-enter its memory mode and will stay in this state. Thus, the R input 'resets' Q to 0 and the S input 'sets' Q to 1, while the other input is at 0. When both inputs are at 0 the circuit remembers the last state in which it was placed. This circuit is called a **SET–RESET latch**, or simply an **S–R latch**.

It should be noted that the condition $S = R = 1$ results in both outputs being 0. Under these circumstances the two outputs are no longer the inverse of each other and the circuit is not functioning as a bistable. For this reason, this combination of inputs is normally prohibited.

Generally the circuit diagram for the latch is redrawn as shown in Figure 25.3(a). This form emphasises the basic symmetry of the circuit. An S–R latch can also be produced using two NAND gates, as shown in Figure 25.3(b).

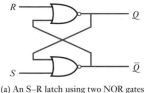

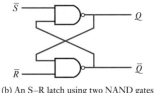

Figure 25.3
S–R latch circuits.

(a) An S–R latch using two NOR gates   (b) An S–R latch using two NAND gates

Comparing the operation of a NAND gate with that of a NOR gate, we note that while the NOR gate resembles an inverter when one of its inputs is connected to 0, the NAND gate resembles an inverter when one of its inputs is connected to 1. Therefore, the memory mode of the circuit in Figure 25.3(b) corresponds to both inputs being at 1. Investigation of the operation of the circuit shows that taking $\bar{S}$ low now *sets* Q to 1, while taking $\bar{R}$ low *resets* Q to 0. For this reason, these inputs are called **active low inputs** and their names are given as $\bar{S}$ and $\bar{R}$ rather than S and R. In logic diagrams, any signal name that has a bar above it is active low and the function described by the name is achieved by taking the appropriate line low. Functions that are achieved by taking signals to logic 1 (such as S and R in the circuit in Figure 25.3(a)) are referred to as **active high inputs**.

The symbols used for S–R latches are shown in Figure 25.4. Circuits with active low inputs can be represented in two ways – either by labelling the inputs as $\bar{S}$ and $\bar{R}$

Figure 25.4
S–R latch logic symbols.

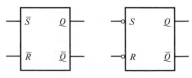

(a) Active high inputs   (b) Active low inputs

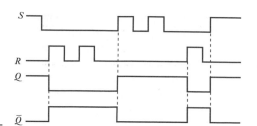

Figure 25.5
Sample input and output waveforms for an S–R latch.

or by showing inverting circles at the inputs. In either case, the signals applied to the input lines correspond to the active low signals $\bar{S}$ and $\bar{R}$.

Figure 25.5 illustrates the operation of an active high input S–R latch by showing its response to a series of input changes. This diagram assumes that the latch responds immediately to changes at its input. In practice, there is a slight delay caused by the **propagation delay** time of the circuit. You will notice that transitions of the S input have no effect, while the Q output is already at 1. Similarly, changes in the R input are ineffective while Q is 0. The only changes that are significant are those that toggle the outputs from one state to the other.

The S–R latch may be thought of as a simple form of electronic **memory** as it remembers which of its two inputs last became active. In fact, latches of one form or another form the basis of a large proportion of the memory circuits used within all computers. However, we will see that slight variants of this basic circuit are more convenient in many applications. Before moving on to look at these alternative forms of bistable, it is perhaps worth looking at some examples of the uses of S–R latches.

Example 25.1 | **Use of an S–R latch for switch debouncing.**

We saw in Section 12.9.2 that switches play a major role in the construction of electronic sensors. We also noted that all mechanical switches suffer from **switch bounce**. This effect is illustrated below in a typical switch arrangement.

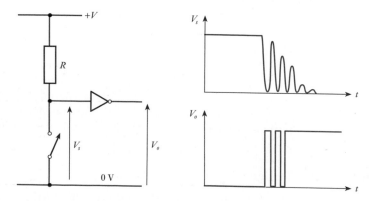

When the switch is open, the voltage V_s is equal to the supply voltage V. When the switch is closed, V_s is pulled to $0\,\text{V}$. If V corresponds to the voltage representing logical 1 and $0\,\text{V}$ corresponds to logical 0, the voltage V_s can be used as the input to a logic gate.

In practice, there is not a clean transition from one voltage level to another. The voltage across the switch oscillates between the two voltage levels as the contacts make and break the circuit as a result of contact bounce. This oscillation generally lasts of the order of a few milliseconds for small switches and somewhat longer for larger devices. The input to the logic gate sees a signal that alternates between a voltage representing a logic 1 and that representing a logic 0 and it therefore produces an output that itself alternates between 0 and 1. Several transitions are thus produced rather than the single step expected. In the example waveform shown, the inverter produces three positive-going edges rather than one. If this output were connected to circuitry designed to count the number of switch closures, it would produce a count of three rather than one.

This problem can be overcome by replacing the 'make-or-break' switch with a 'changeover' switch. This connects an input terminal to one of two output terminals depending on whether or not the switch is pressed. This switch is then connected to an S–R latch, as shown here.

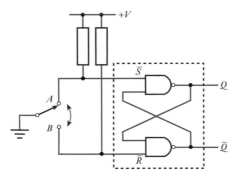

The switch now has three possible positions. The input terminal may be connected to terminal A, terminal B or neither. The circuit is arranged such that, when the switch is connected to neither terminal, the two lines to which they are connected are pulled high by resistors connected to V. These two signals form the $\overline{S}$ and $\overline{R}$ inputs to an active low S–R latch. When both are pulled high, the circuit is in its 'memory mode', in which it simply remains in its present state. When the input terminal is connected to terminal A, $\overline{S}$ is pulled low, setting Q to 1. If the contact bounces at this stage, the input terminal will alternate from being connected to A to being connected to nothing. Thus, the circuit alternates from the set mode to the memory mode. Under these circumstances, the circuit will simply remain with $Q = 1$ and the output will not bounce. Similarly, if the input terminal is connected to terminal B, $\overline{R}$ will be activated, resetting Q to 0. Contact bounce will alternate the circuit between its reset mode and its memory mode and again not affect its output. Thus the circuit removes the effects of switch bounce, provided that the moving contact cannot bounce between the two output terminals. This is normally assured by its mechanical design.

Example 25.2 **Design of a simple burglar alarm.**

A simple burglar alarm can be formed using a series of switches connected to doors and windows throughout a building. Opening any door or window opens the corresponding switch, which should sound the alarm. It is essential that the alarm continues to sound if the door or window concerned is subsequently closed. Some method must be incorporated to silence the alarm when the building has been checked. A suitable arrangement is shown opposite.

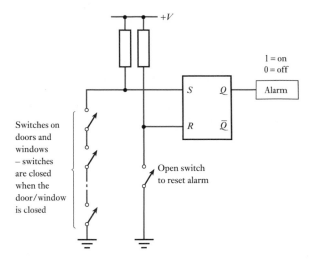

An S–R latch has two pull-up resistors connected to its S and R inputs. The various door and window switches are connected in series and wired so that, when all are closed, they short the S input to ground. The R input is similarly shorted to ground by a reset switch that is normally closed.

Initially, the system is reset by momentarily opening the reset switch with all the sensor switches closed. The latch will be reset with $Q = 0$ and the alarm will be off. Once the reset switch has been closed, the system is armed. If one of the sensor switches is opened, by the opening of a door or window, the S input will go high, setting Q to 1 and sounding the alarm. If, then, the sensor switch is closed, the system will remain in the alarm state until it is reset by opening the reset switch.

25.2.2 The gated S–R latch

It is often useful to be able to control the operation of a latch so that the inputs can be enabled at some times and disabled at others. The circuit in Figure 25.6 provides this facility.

Figure 25.6
A gated S–R latch.

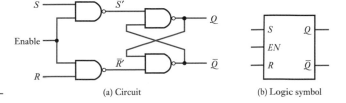

(a) Circuit   (b) Logic symbol

Two NAND gates are used to 'gate' the S and R input signals before they are applied to the latch. A third input, **latch enable** (EN), can be used to allow or inhibit the actions of the other inputs. When the enable signal is low, the signals $\bar{S}'$ and $\bar{R}'$ are both high, regardless of the signals applied to the S and R inputs. This puts the active low input latch into its memory mode, preventing any change to its state. When the enable input is taken high, the S and R signals are inverted by the gating arrangement and then applied to the latch. Thus, when the enable is high the circuit acts as a conventional active high input S–R latch, but when the enable is low the circuit ignores any signals applied to the S and R inputs. Figure 25.7 illustrates the response of the circuit to a series of input changes. As $\bar{Q}$ is simply the inverse of Q, it is not shown in this waveform diagram.

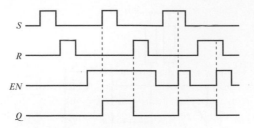

Figure 25.7
Sample input and output waveforms for a gated S–R latch.

25.2.3 The D latch

Another form of latch that is widely used is the **D latch**, which is also known as the **transparent D latch**. This circuit has two inputs, D and EN, as shown in Figure 25.8.

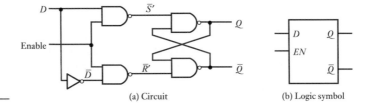

Figure 25.8
A D latch.

(a) Circuit   (b) Logic symbol

Clearly, the circuit bears a striking resemblance to that of the gated S–R latch shown in Figure 25.6, but uses a single signal D and its inverse $\bar{D}$ to act as inputs to the gating network. As before, when the enable input is low the signals fed to the latch are both high and the latch is placed in its memory mode, preventing any change of state. If the enable is taken high, the D input determines the signals applied to the latch inputs $\bar{S}'$ and $\bar{R}'$. If D is high, $\bar{S}'$ will be low and $\bar{R}'$ will be high and the latch will be set with $Q = 1$. If D is low, $\bar{S}'$ will be high and $\bar{R}'$ will be low, which will reset the latch with $Q = 0$. Thus, when the enable is high, the Q output takes the present value of D, but when the enable is low, the Q output will remain in its present state. The D latch may therefore be thought of as a digital equivalent of an analogue sample and hold gate (as described in Section 18.8.3). When the enable is high, the Q output follows the input data D, but when the enable goes low, the output remembers the value of D when the enable went low. The operation of the D latch is illustrated in Figure 25.9.

In addition to storing single bits of information, D latches are often used in groups to store words of information. It is common to combine a number of latches within a single integrated circuit to give, perhaps, 4 bits (a quad latch) or 8 bits (an octal latch) of storage within a single device.

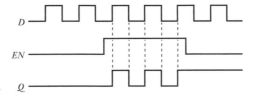

Figure 25.9
Sample input and output waveforms for a D latch.

25.2.4 Edge-triggered devices and the D flip-flop

In many situations, it is necessary to synchronise the operation of a number of different circuits and it is useful to be able to control precisely when a circuit will change state.

Some bistables are constructed so that they only change state on the application of a **trigger** signal. This is defined as the rising or falling edge of an input signal called the **clock**. These devices are termed **edge-triggered bistables** or, more commonly, **flip-flops**. These are divided into those that are triggered by the rising edge of the clock signal (so-called *positive* edge-triggered devices) and those that are triggered on the falling edge of the clock (*negative* edge-triggered devices).

Flip-flops are available in a number of different forms, including the **S–R flip-flop** and **D flip-flop**, which are edge-triggered versions of the latches discussed earlier. The circuit symbols used for these circuits are similar to those of the corresponding latch, except that the enable input is replaced with a clock input. The clock input is conventionally indicated by a triangle, while an inverting circle is used to show a negative edge-triggered device. Figure 25.10 shows circuit symbols for both a positive edge-triggered and a negative edge-triggered D flip-flop, and Figure 25.11 shows a set of sample waveforms for a positive edge-triggered device.

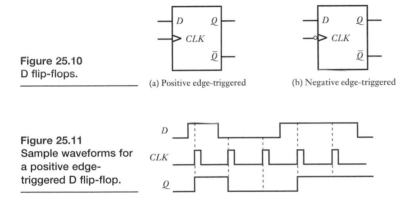

Figure 25.10
D flip-flops.

(a) Positive edge-triggered   (b) Negative edge-triggered

Figure 25.11
Sample waveforms for
a positive edge-
triggered D flip-flop.

25.2.5 J–K flip-flop

The J–K flip-flop is perhaps one of the most widely used forms of bistable. As its name suggests, it has two inputs, J and K, and these are similar in some respects to the S and R inputs of an S–R flip-flop. Taking J to 1 while K is at 0 sets Q to 1, whereas taking K to 1 while J is at 0 resets Q to 0. As in the S–R device, when neither input is active, the circuit is in its memory state, but the operation of the arrangement is different when both inputs are active simultaneously. This is an ambiguous situation in the case of an S–R bistable and so it is avoided. In the case of the J–K device, when both inputs are active, the circuit changes state (or toggles) when a trigger event occurs. Figure 25.12 shows the circuit symbol for a negative edge-triggered J–K flip-flop and a set of sample waveforms. As this is an edge-triggered device, the state of the inputs is only of importance at the instant of the trigger event, which in this case is the falling edge of the clock. Therefore, Figure 25.12(b) marks these events with dashed lines and indicates the state of J and K at these times, together with the corresponding action.

One of the reasons for the J–K flip-flop being so widely used is its great versatility. Several different operating modes are possible, including using it to reproduce the functions of other types of flip-flop. This is illustrated in Figure 25.13, which shows several common configurations. Figure 25.13(a) shows that a J–K flip-flop can be used as a direct replacement for an S–R device as they are identical in operation for all the allowable input combinations of the latter. A J–K bistable can also be used to produce the function of a D flip-flop, as shown in Figure 25.13(b). The arrangement in Figure 25.13(c)

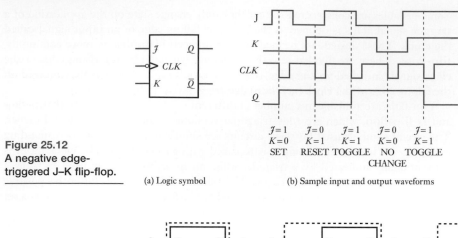

Figure 25.12
A negative edge-
triggered J–K flip-flop.

(a) Logic symbol   (b) Sample input and output waveforms

Figure 25.13
Use of a J–K flip-flop to
reproduce other
flip-flop functions.

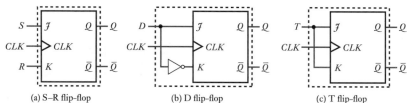

(a) S–R flip-flop   (b) D flip-flop   (c) T flip-flop

shows a J–K flip-flop configured with its J and K inputs joined to form a single input, T. If T is 0, the device is in its memory mode and simply stays in its present state. If T is 1, both J and K are 1 and the device will TOGGLE on every clock pulse. This arrangement is called a **toggle flip-flop**, or simply a **T flip-flop**. Figure 25.14 shows the operation of this form of bistable and we will look at applications for this device later in this chapter.

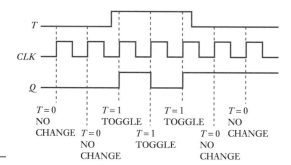

Figure 25.14
Typical waveforms for
a positive edge-
triggered T flip-flop.

25.2.6 Asynchronous inputs

We have seen that in flip-flops the *control inputs* (for example, the J and K inputs in a J–K flip-flop) affect the operation of the circuit only at the moment of an appropriate transition of the clock signal (CLK). We therefore refer to these inputs as **synchronous**, because their operation is synchronised to the clock input.

In many applications it is advantageous to be able to set or clear the output at other times, independently of the clock. Therefore, some devices have additional inputs to perform these functions. These are termed **asynchronous inputs** as they are not bound by the state of the clock. Unfortunately, IC manufacturers are unable to agree on common names for these inputs, so they may be called PRESET and CLEAR, DC SET and DC CLEAR, SET and RESET, or DIRECT SET and DIRECT CLEAR. Here, we use the names PRESET (*PRE*) and CLEAR (*CLR*). As with control input, these lines can be active high or active low, although more often than not they are active low. Figure 25.15(a) illustrates how these inputs are shown in the circuit symbol for a J–K flip-flop, and Figure 25.15(b) gives sample waveforms for them.

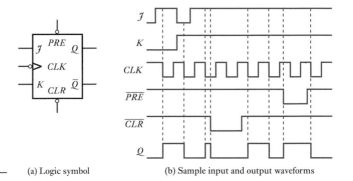

Figure 25.15
A J–K flip-flop with
PRESET and CLEAR.

(a) Logic symbol   (b) Sample input and output waveforms

25.2.7 Propagation delay and races

We noted in the last chapter that logic gates take a finite time to operate. As bistables are constructed from gates, these will also exhibit a delay between their inputs and outputs. As in logic gates, this delay is termed the propagation delay time of the device and, under certain circumstances, this can lead to problems.

Consider the circuit in Figure 25.16. This shows a situation where two edge-triggered devices are connected to the same clock signal, while the output of one device forms an input to the other. On the rising edge of the clock signal, the first flip-flop will respond by changing its output. This process will take a finite time, determined by the propagation delay time of the device. During this time, the second flip-flop will also be responding to the clock signal, which it shares with the first device. If it is fast enough, it may have responded before Q_1 changes, but if not, its input will change before it has had a chance to act. The final output state of the second flip-flop is far from certain and will depend on the relative speeds of the two devices. This uncertainty of operation is referred to as a **race** condition, because the outcome is determined by a race between the two components.

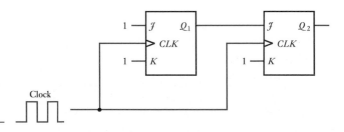

Figure 25.16
A possible race
condition.

Race hazards may be tackled by means of careful design and, in fact, the design of edge-triggered devices aims to prevent such problems by arranging that the input signals must be stable for a certain time (the **hold time**) before the clock event.

25.2.8 Pulse-triggered bistables or master/slave flip-flops

Another way to overcome race problems is to use **master/slave flip-flops** rather than edge-triggered devices. These are also known as **pulse-triggered flip-flops**.

The construction of these devices involves using two bistables in series – a master and then a slave device. However, the resultant circuit behaves like a single bistable in which the outputs are determined while the clock signal is high, but where the outputs change only when the clock falls. Thus, on the falling edge of the clock, the outputs take up the value determined by the state of the inputs a short time before.

Master/slave versions are available for a range of bistables, such as S–R, D and J–K types. The logic symbol usually includes the label 'M/S', and the triangle used on the clock of edge-triggered devices is omitted. Figure 25.17 shows the circuit symbol for a J–K master/slave flip-flop together with some sample waveforms.

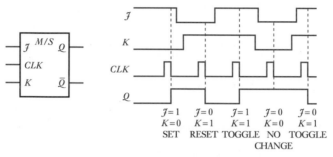

(a) Logic symbol   (b) Sample input and output waveforms

Figure 25.17
A J–K master/slave flip-flop.

The circuit of a basic J–K master/slave flip-flop is shown in Figure 25.18. Essentially, the circuit is two gated S–R bistables connected in series, with some additional feedback to remove the ambiguous input combination. The outputs of the first device form the inputs to the second, and the clock signal being used to form the enable for the first is inverted to form the enable for the second. When the clock input signal goes high, the master is enabled and the slave disabled. The master reacts as an ordinary gated S–R latch, so its outputs are configured accordingly. The slave, being disabled, simply stays in its previous state and the outputs remain unchanged. When the clock goes low, the master is disabled and holds its previous state. The slave is now enabled and responds to its inputs, which are the outputs of the master. As the two inputs of the slave are fed from the two outputs of the master, the slave will always see a 1 on one input and a 0 on the other. The operation of the slave then dictates that, when it is enabled, it will take up the output state of the master. Thus, the device has two operating modes. When the clock is high, the master responds but the slave maintains its previous outputs, but, when the clock is low, the master is disabled and the slave transfers the master's state to the outputs. The device responds to its inputs while the clock is high, but the outputs are not updated until the clock goes low.

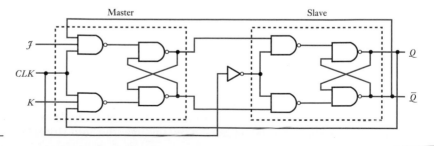

Figure 25.18
The circuit of a basic J–K master/slave flip-flop.

It might seem that this arrangement would suffer from the 'race' problem discussed earlier as, when the clock goes high, the master's outputs might change just as the slave is being disabled. In fact, this is not the case. The design of the circuit ensures that the delay produced by the master latch is greater than that of the inverter used for the clock signal. This ensures that the slave is disabled shortly before the master is enabled, preventing any possible race condition.

25.3 Monostables or one-shots

In Figure 25.2, we considered a bistable formed from two NOR gates connected in a ring. Now consider the circuit of Figure 25.19(a).

Let us assume initially that the input signal T is at 0. The resistor R, as it is connected to a voltage equal to logical 1, will cause the capacitor C to charge or discharge so that V_1 is equal to this voltage. This will cause Q to be 0 and, as both inputs to gate 1 are 0, $\bar{Q}$ will be 1. Both $\bar{Q}$ and V_1 are at a voltage corresponding to logical 1, so there will be no voltage across the capacitor. This condition is **stable** and the circuit will remain in this state indefinitely unless the input changes.

Let us now consider what happens if the input T goes high for a short time. When T goes to 1, $\bar{Q}$ will go low. As the voltage across the capacitor cannot change instantaneously, this will pull V_1 low, which in turn will take Q high. While Q is high, $\bar{Q}$ will be held low even if T reverts to 0. Thus, the application of a positive pulse to T switches the circuit into a state where Q is 1, and it will remain in this state even after T returns to 0. However, this state is not stable, only **metastable**. While in this state, $\bar{Q}$ is at 0, so a voltage exists across the series combination of R and C. This

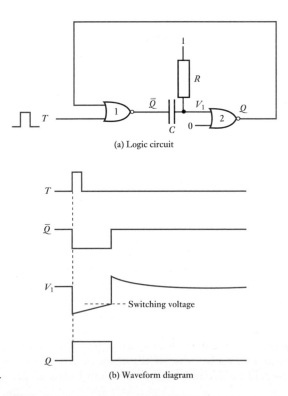

(a) Logic circuit

(b) Waveform diagram

Figure 25.19
A simple monostable.

produces a current through the resistor to charge the capacitor, which results in V_1 increasing exponentially towards the voltage representing logical 1. At some point, it becomes sufficiently large for the input of gate 2 to interpret this voltage as a logical 1. At this point, Q will return to zero, $\bar{Q}$ will go to 1 and V_1 will be pushed above logical 1 by the voltage across the capacitor. This voltage will gradually decay as the capacitor discharges under the influence of R. The circuit is now back in its stable state. The waveforms produced at various points in the circuit are shown in Figure 25.19(b).

The circuit thus has one stable state and one metastable state. It can be forced to enter its metastable state by applying an appropriate input signal. It will then stay in that state for a fixed period of time determined by the values of R, C and the switching voltages within the circuit. The circuit is therefore given the name **monostable** or **one-shot**. The label T given to the input signal stands for **trigger input**.

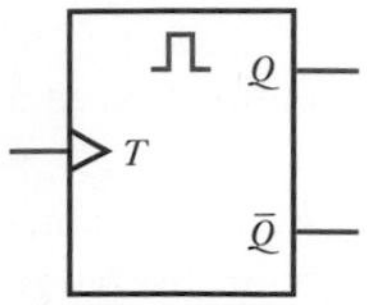

Although it is quite possible to construct monostables from simple logic gates, it is more usual to use dedicated integrated circuits. These use more sophisticated circuit techniques than those described above but have similar characteristics. The logic symbol for a monostable is shown in Figure 25.20.

Monostables can be divided into two types in terms of the way in which they respond to trigger inputs. **Non-retriggerable monostables** ignore any trigger pulses that occur while the circuit is outputting a pulse, while **retriggerable monostables** extend the output pulse if a second trigger occurs. The length of the output pulse τ is normally set by the values of a resistor and a capacitor that form part of the circuit. Figure 25.21 shows typical monostable output waveforms.

Figure 25.20
The logic symbol for a monostable.

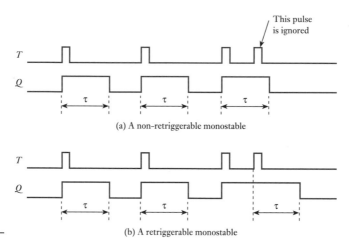

(a) A non-retriggerable monostable

(b) A retriggerable monostable

Figure 25.21
Monostable waveforms.

25.4   Astables

An astable has two metastable states and repeatedly switches between these two states. It therefore produces the function of a **digital oscillator**.

By comparing the circuits in Figures 25.2 and 25.19, it is clear that the addition of a resistor–capacitor combination converts a stable state into a metastable state. It is perhaps not surprising then that if we add a second resistor–capacitor combination, we can generate a circuit with *two* metastable states. This circuit requires no input signals, so we may replace the NOR gates with inverters (as in our original circuit in Figure 25.1). The resultant circuit is shown in Figure 25.22.

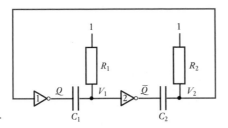

Figure 25.22
A simple astable
arrangement.

Let us assume that initially C_1 and C_2 are discharged and that Q is 0. As there is no voltage across C_1, V_1 will be equal to Q, that is 0. This will cause the output of the second gate to be at 1. As C_2 is discharged, this logic level will be applied to the input of gate 1. This in turn will generate a logic 0 at Q, which is consistent with our original assumption.

This state is metastable as C_1 will now charge up until V_1 is greater than the switching voltage of gate 2. At this time, $\bar{Q}$ will go to 0, pushing V_2 down and making Q go to 1. C_1 will now discharge and C_2 will charge up until V_2 is greater than the switching voltage of gate 1, whereupon the circuit will again change state and the cycle will restart. The circuit will thus oscillate continuously from one state to the other, with both Q and $\bar{Q}$ producing regular pulse waveforms. The length of time spent in each state is determined by the values of the resistors and capacitors, as well as by the switching and logic voltages of the gates. If the gates are identical and the products C_1R_1 and C_2R_2 are equal, the circuit will produce a **square wave**. Figure 25.23 illustrates the waveforms produced within the circuit.

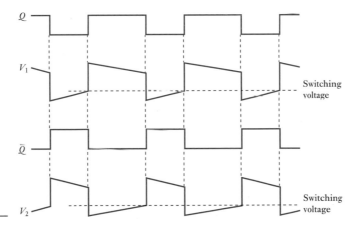

Figure 25.23
Waveforms of the
simple astable circuit.

As with monostables, astables are usually used in IC form. Astables can also be constructed using two monostables connected in a ring, as shown in Figure 25.24. Here, the trailing edge of the pulse generated by the first monostable triggers the second. The trailing edge of the pulse from the second monostable then triggers the first and the cycle repeats.

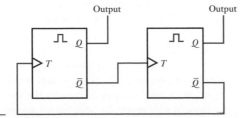

Figure 25.24
An astable formed by
two monostables.

25.5   Timers

While monostables and astables are available in integrated circuit form, a generally more useful component is the integrated circuit timer, which can perform a range of functions. An example of such a device is the **555 timer**, which can be used as a monostable, bistable or astable, as well as for a number of other applications. Such circuits are very versatile and can be configured to perform a range of functions using just a couple of external passive components.

The internal circuitry of the 555 timer employs some 25 transistors, a few diodes and several resistors. While the circuitry is fairly involved, it consists basically of a flip-flop, two comparators, a switching transistor and a resistive potential divider network. A simplified block diagram of the device is shown in Figure 25.25.

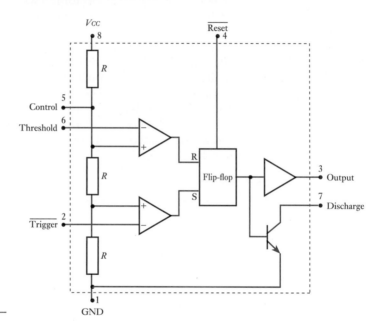

Figure 25.25
Internal structure of a 555 timer.

A network of three equal resistors is connected between the positive supply V_{CC} and ground, producing internal voltage references of $\frac{1}{3}V_{CC}$ and $\frac{2}{3}V_{CC}$. These nodes are connected to two comparators which allow external voltages to be compared with these reference voltages. The outputs of these comparators are then used to control a flip-flop. The output of the flip-flop is buffered to produce the output of the timer and is also used to drive a switching transistor that can be used to drive an external voltage down to close to zero volts. The addition of a small number of external components can then configure the device to perform a range of functions. Note that in the diagram of Figure 25.25 the *reset* and *trigger* inputs are labelled as $\overline{\text{Reset}}$ and $\overline{\text{Trigger}}$. As in our earlier discussions of the S-R bistable, this indicates that these inputs are *active low*. Therefore the reset and trigger functions are achieved by taking the corresponding inputs to 0.

When the device is being used as a monostable, an external resistor and capacitor would typically be connected across the supply with their midpoint connected to the 'threshold' input, as shown in Figure 25.26. The voltage on the threshold input (and therefore across the capacitor) is held at close to zero by the switching transistor which is connected across the capacitor through the 'discharge' output. When an appropriate

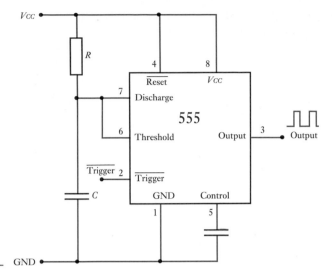

Figure 25.26
A 555 timer configured
as a monostable.

signal is applied to the 'trigger' input this changes the state of the flip-flop, turning off the switching transistor and allowing the external capacitor to start charging up. When the capacitor voltage, and therefore the voltage on the threshold input, reaches $\frac{2}{3}V_{CC}$ the comparator connected to this input will switch states, which in turn will toggle the flip-flop. This will again turn on the switching transistor and the voltage across the capacitor will again be clamped to close to zero. Since the 'output' of the timer corresponds to the state of the flip-flop, it follows that the application of a pulse to the trigger input produces a fixed length pulse at the output. The length of this pulse is determined by the time that it takes for the capacitor to charge up to $\frac{2}{3}V_{CC}$, which is in turn determined by the choice of the external capacitor and resistor.

When used as an astable a slightly different configuration is used, a typical circuit being shown in Figure 25.27. Again a resistor and a capacitor are connected in series but now a second resistor is added to the combination. The junction of these two resistors is connected to the 'discharge output' which allows the timer to control the

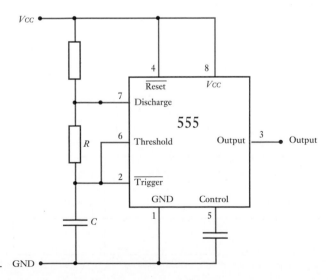

Figure 25.27
A 555 timer configured
as an astable.

voltage applied across the RC network. When the flip-flop is in one state the switching transistor is turned OFF, so most of the supply voltage is connected across the RC network and the capacitor charges up. When the flip-flop is in the other state the switching transistor is turned ON and the voltage across the RC network is forced down to close to zero and the capacitor discharges. In this circuit the 'trigger' and 'threshold' inputs are both connected to the top of the capacitor and both see an increasing voltage as the capacitor charges up. Each of these inputs compares the capacitor voltage with a different reference voltage, and each switches the flip-flop in a different direction. The effect is that if initially the voltage on the capacitor is between $\frac{1}{3}V_{CC}$ and $\frac{2}{3}V_{CC}$ the 'discharge' output will be turned off and voltage on the capacitor will increase. However, when this voltage reaches $\frac{2}{3}V_{CC}$ the comparator connected to the threshold input will change state, switching the flip-flop and turning on the discharge output. This will clamp the voltage across the RC network and the capacitor voltage will start to ramp down. However, when this voltage falls to $\frac{1}{3}V_{CC}$, the comparator connected to the trigger input will change state, again switching the flip-flop and turning off the discharge output. Thus the capacitor voltage will start to increase again and the operation will cycle with the capacitor voltage going between $\frac{1}{3}V_{CC}$ and $\frac{2}{3}V_{CC}$. Since the output of the timer comes from the flip-flop, this will be in the form of a regular pulse waveform at a frequency determined by the component values of the RC network.

While the monostable and astable arrangements shown above are widely used, 555 timers, and a range of similar devices, can be used in a multitude of circuit configurations and applications.

<table>
<tr><td>25.6</td><td></td></tr>
</table>

25.6 Memory registers

Having looked earlier at a few forms of bistable, we are now in a position to look at some circuits that make use of them. We will start by looking at **registers**.

Registers of one kind or another are extensively used in almost all fields of digital electronics. One of the most widely used forms of register is that used to store words of information in computers and calculators. These registers can be used directly in calculations, as in the case of the *accumulator* in a processor or calculator, or for general memory applications, where thousands, or perhaps millions, of registers are used to store programs and data.

A simple memory register is shown in Figure 25.28. Here four D master/slave flip-flops are used to create a 4-bit register. Clearly, additional flip-flops could be used to create a register of any required length.

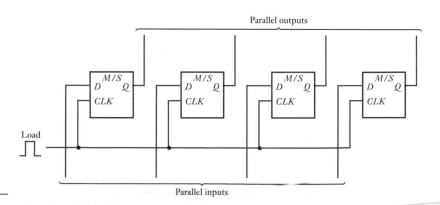

Figure 25.28
A 4-bit memory register.

Memory registers can be created by using a range of different types of bistable, but in practice we are less concerned with their internal circuitry than with their external behaviour. Registers are normally constructed as a single circuit and are used as building blocks. Figure 25.29 shows a typical 8-bit memory register. This could be used independently or together with other devices to produce a longer register. Note that it is normal to number the individual bits of a digital word from 0, starting with the least significant bit (LSB), and draw them with the most significant bit on the left, as in a conventional number.

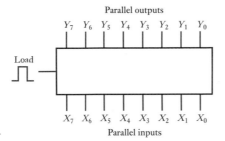

Figure 25.29
An 8-bit memory register.

Video 25B

25.7 Shift registers

Shift registers are used to convert parallel words of information into a stream of bits on a single line. One application of this technique is in long-distance communication where parallel words of information can be converted into serial form to be sent down a single wire rather than needing a number of parallel wires. A shift register can also be used to take a serial datastream and generate from it a parallel data word. This process is illustrated in Figure 25.30.

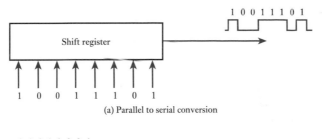

(a) Parallel to serial conversion

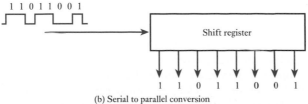

Figure 25.30
The operation of a shift register.

(b) Serial to parallel conversion

To understand the operation of a shift register, consider the circuit of Figure 25.31(a). The circuit consists of four D master/slave flip–flops connected in series. A sequence of regular positive-going shift pulses is applied to the arrangement, which forms the clock signal for each stage.

If we assume that initially the serial data input is 0 and the output of each flip-flop is 0, then repeated shift pulses will have no effect on the circuit. If now the serial data

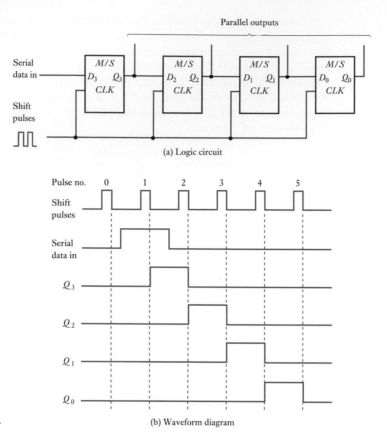

Figure 25.31
A simple shift register.

input goes to 1 during one clock pulse and then returns to 0, the effect of this input will ripple along the register, as shown in Figure 25.31(b). During pulse number 1, the data input is 1, therefore D_3 is high immediately before the falling edge of the clock signal. Thus, when the clock falls to 0, Q_3 goes to 1. None of the other flip-flops is affected by this transition because it occurs as the clock goes low. Immediately before the end of pulse number 2, Q_3 and D_2 are high but D_3, D_1 and D_0 are low. Thus, after the clock pulse, Q_2 goes high, Q_3 returns to 0, while Q_1 and Q_0 are unaffected. On successive clock pulses, each output goes high in turn for one clock cycle and then returns to 0. After pulse number 5, the register is back to its original state.

It should be clear from the above that any pattern of '1's and '0's within the register will move 1 bit to the right at the end of each shift pulse. We therefore have a **shift register**. A pattern of 4 bits of serial data will be shifted into the register after four shift pulses and will appear as a parallel word at the outputs Q_3–Q_0. By adding more flip-flops, registers of any desired length can be formed, allowing longer words to be used. The register therefore performs **serial to parallel conversion**.

As data is shifted into the register, the current contents are also shifted out at the other end. However, in order to perform **parallel to serial conversion**, it is necessary to modify the circuit in Figure 25.31(a) so that parallel data can be loaded into the shift register. We have already seen in Figure 25.28 how a series of D flip-flops can be used as a memory register. What we now require is some circuitry to allow a set of flip-flops to be switched from a memory register, which may be loaded in parallel, to a shift register, which can output data in serial form. Such an arrangement is shown in Figure 25.32.

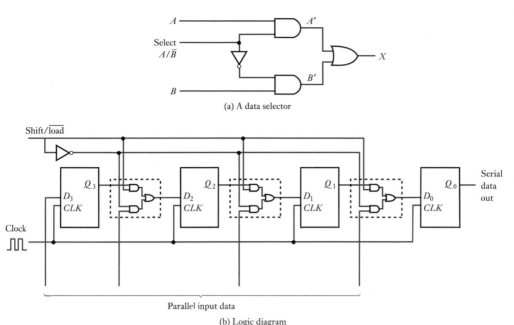

(a) A data selector

Shift/load

Clock

Parallel input data

(b) Logic diagram

Figure 25.32
A 4-bit parallel load shift register.

Figure 25.32(a) shows a subsystem required to implement this function. It is a **data selector**, the function of which is to allow a control signal, SELECT, to determine which of two inputs, A and B, is connected to the output, X. If SELECT is 1, A' will always be equal to A, but B' will always be 0. Thus, X will be equal to A. If SELECT is 0, A' will be equal to 0 and B' will be equal to B, making X equal to B. Thus, when SELECT is high, $X = A$ and, when it is low, $X = B$. The data selector is therefore a two-input multiplexer. We have already met the convention of identifying active low signals by placing a 'bar' over their symbolic name. Here, we describe the SELECT line by the symbol $A/\overline{B}$, which indicates that, when the line is high, A is selected, but when it is low, B is selected.

Figure 25.32(b) shows a 4-bit parallel load shift register. The operation of the circuit is determined by a single control signal that is inverted to drive three data selectors. When the control signal is high, the Q output from each stage is fed to the D input of the next and the arrangement is electrically equivalent to the circuit in Figure 25.31. Each time a clock pulse is applied, the contents of the register will be shifted one place to the right to appear as serial output data at Q_0. When the control signal is low, the D input of each stage is connected to the parallel input lines. The application of a clock pulse will cause the pattern of '1's and '0's on these inputs to be loaded into the register. We therefore have a register that can be used to shift or load data under the control of a single input. This input may be labelled $shift/\overline{load}$.

The circuit in Figure 25.32 can be used for both serial to parallel and parallel to serial conversion by applying and sensing signals appropriately. It is also possible, by slightly modifying this circuit, to construct a register that will shift in either direction. This is achieved by using a data selector arrangement to determine whether the input to a particular stage comes from the output of the stage to the right or that to the left.

Although it is quite feasible to construct such circuits from standard simple gates, it is usual to use specialised integrated circuits that provide all the components of a shift register within a single package. Typical devices provide 4- or 8-bit registers with a range of features. Longer registers can be produced by connecting several devices in series. This is achieved by taking the serial output of one device to the serial input of the next.

Example 25.3

Application of shift registers.

One of the most frequently encountered uses of shift registers is in **serial communications** systems. This involves converting parallel data into a serial form at the *transmitter*, conveying the serial data over some distance and then converting it back into a parallel form at the *receiver*. This process is illustrated below.

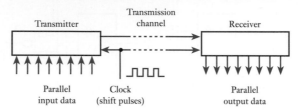

At the heart of the transmitter is a shift register and this loads the input data in parallel, then outputs it in a serial form at a rate determined by a local clock signal. The serial datastream is then transmitted over some form of *transmission channel* to the receiver. This channel may take the form of a piece of wire, radio signal, series of laser pulses or some other information medium.

At the receiver, a second shift register loads the serial data and outputs the information in parallel form. To enable it to load the information, it must receive not only the serial data but also the clock signal to allow it to *synchronise* with the transmitter.

The main advantage of this method of transmission is that it requires fewer lines for the information to be communicated – only two lines (one for data and one for the clock), rather than one line for each bit of the parallel data.

Serial techniques are used extensively for long-distance communication. They are also used for short-range applications, sometimes down to a few centimetres. In some systems, the requirement to transmit the clock signal along with the data is removed by generating (or recovering) the clock signal at the receiver. This reduces the number of signal lines required to one. These techniques will be discussed in Chapter 27 when we look at computer input/output techniques and we will look at communications systems in more detail in Chapter 29.

Video 25C

25.8 Counters

Among the most important classes of sequential circuits are the various forms of counter. These can be used to count events, but are also used to count regular clock 'ticks' and so measure time. Counters form the basis of a wide range of timing and sequencing applications, in everything from quartz watches to digital computers.

25.8.1 Ripple counters

Consider the circuit in Figure 25.33. The circuit consists of four negative edge-triggered J–K flip-flops, with the Q output of each device forming the clock input to the next. The J and K inputs of each device are connected to 1, so each will toggle on the negative-going edge of the signal connected to its clock input (when connected in this way, the devices are acting as **T-type** or **toggle bistables**, which toggle their output in response to each clock trigger).

Figure 25.33
A simple ripple counter.

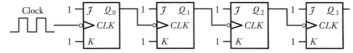

Figure 25.34 shows the resultant waveforms in the circuit when a square-wave clock signal is applied (a square wave is a repetitive pulse waveform with equal-length pulses and spaces). It can be seen that Q_0 toggles on each falling edge of the clock, producing a square waveform at half the clock frequency. Q_1 toggles at each falling edge of Q_0, and so produces a square waveform at half the frequency of Q_0 – that is, one-quarter the frequency of the clock. Similarly, each further stage divides the signal frequency by a factor of 2, producing successively lower frequencies. Such a circuit may be thought of as a **frequency divider** and each stage represents a **frequency halver**.

Figure 25.34
Waveform diagram for
the simple ripple
counter.

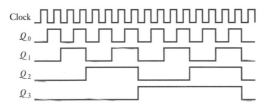

Example 25.4 | **Application of a frequency divider.**

A common example of the use of a frequency divider is found in a digital watch. Most digital watches use a crystal oscillator, which produces a stable timing waveform of 32,768 Hz. This particular frequency is used because it is an exact power of 2 (32,768 $= 2^{15}$), which simplifies the process of frequency division. Following the oscillator, a 15-stage binary divider is used to produce a 1 Hz signal, which is suitable for driving a stepper motor (for watches with an analogue display using hands) or a digital display.

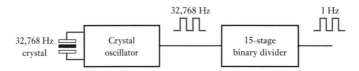

It is interesting to look at the sequence of '1's and '0's that appear on the outputs of the flip-flops. These can be deduced from Figure 25.34, while Figure 25.35 shows the values of the outputs after each clock pulse. It can be seen that the pattern of the outputs represents the binary code for the number of pulses applied to the circuit. The arrangement therefore represents a *counter*. In this particular circuit, the effects of the input are propagated along the series of flip-flops, the outputs changing sequentially along the line. For this reason, this form of counter is called a **ripple counter**. One consequence of the operation of this form of circuit is that the various stages change at slightly different times. For this reason, they are known as **asynchronous counters**.

From Figure 25.35 it can be seen that the circuit counts up to the binary equivalent of 15 and then restarts at zero. The counter therefore takes 16 distinct values. We refer to such a counter as a **modulo-16 counter** or sometimes simply a **mod-16 counter**.

Number of clock pulses	Q_3	Q_2	Q_1	Q_0
0	0	0	0	0
1	0	0	0	1
2	0	0	1	0
3	0	0	1	1
4	0	1	0	0
5	0	1	0	1
6	0	1	1	0
7	0	1	1	1
8	1	0	0	0
9	1	0	0	1
10	1	0	1	0
11	1	0	1	1
12	1	1	0	0
13	1	1	0	1
14	1	1	1	0
15	1	1	1	1
16	0	0	0	0
17	0	0	0	1
18	0	0	1	0
19	0	0	1	1
20	0	1	0	0

Figure 25.35
The output sequence for the ripple counter.

This circuit is also called a **4-bit counter** as its outputs represent a 4-bit digital number. By using more or fewer stages, we can construct a range of circuits that count modulo-2^n, where n is the number of flip-flops used. These counters will count from 0 to 2^{n-1} and then repeat.

File 25A

> ### Computer simulation exercise 25.1
> Simulate the circuit in Figure 25.33 and apply a square-wave input. Confirm that this produces waveforms of the kind shown in Figure 25.34. Now, modify your circuit by adding additional stages and confirm that you can produce frequency dividers of arbitrary length.

25.8.2 Modulo-*N* counters

We have seen that by varying the number of stages we can modify our simple ripple counter to produce circuits that count up to different numbers before restarting at zero. However, varying the number of stages only allows us to choose values for the modulus of the counter that are powers of 2. In many applications, we wish to count up to particular numbers that may not fulfil this requirement. We therefore need a method of constructing a generalised counter that can count up to any number. Such a circuit is usually called a **modulo-*N* counter**.

In order to produce a counter with a modulus of N, we simply need to ensure that, on the clock pulse after the counter reaches $N - 1$, it returns to zero. An example of such a circuit, for a value of $N = 10$, is given in Figure 25.36(a). Counters that count modulo-10 are referred to as **decade counters**. Decade counters that output binary numbers in the range 0000 to 1001 are often called **binary-coded decimal counters**, or simply **BCD counters**.

The circuit of the decade counter is similar to that of the simple ripple counter in Figure 25.33, but it has an extra reset circuit to clear all the flip-flops when they reach a count of 10.

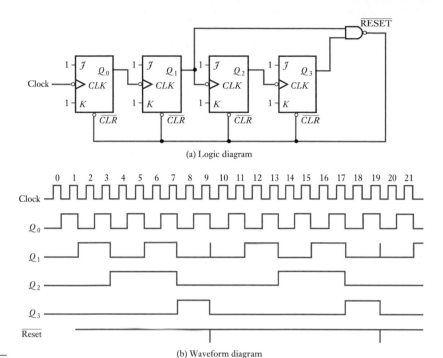

(a) Logic diagram

(b) Waveform diagram

Figure 25.36
A decade counter.

The clearing operation is achieved by using flip-flops that have **clear** inputs. The signal applied to this input comes from circuitry that detects a count of 10. As 10 is binary 1010, a simple two–input NAND gate can be used to detect the first occasion on which bits 1 and 3 are high (remember that the bits start from bit 0). As soon as this is detected, the $\overline{\text{RESET}}$ line goes low, clearing the counter to zero, which also sets $\overline{\text{RESET}}$ back to 1. The counter then continues from zero as before. Figure 25.36(b) illustrates the waveforms produced during the operation of the counter.

A modulo-N counter can be constructed for any value of N by forming a counter with n stages, where $2^n > N$, then adding a reset circuit that detects a count of N.

File 25B

Computer simulation exercise 25.2

Simulate the circuit in Figure 25.36 and apply a square-wave input. Confirm that this functions as a decade counter. Modify the circuit to produce a modulo-12 counter.

25.8.3 Down counters

In some applications it is necessary to have a counter that counts down rather than up. This can be achieved using a similar circuit to that of the earlier ripple counter by taking the clock signal for following stages from $\overline{Q}$ rather than Q. This is illustrated in Figure 25.37, in which you should note that the outputs are taken from $Q_0 - Q_3$ and not from $\overline{Q}_0 - \overline{Q}_3$. Figure 25.38 shows the output sequence for the four-stage down counter. This technique can be applied to counters of any modulus.

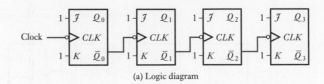

(a) Logic diagram

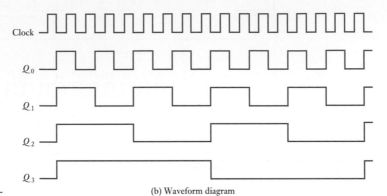

(b) Waveform diagram

Figure 25.37
A ripple down counter.

Number of clock pulses	Q_3	Q_2	Q_1	Q_0	**Count**
0	0	0	0	0	0
1	1	1	1	1	15
2	1	1	1	0	14
3	1	1	0	1	13
4	1	1	0	0	12
5	1	0	1	1	11
6	1	0	1	0	10
7	1	0	0	1	9
8	1	0	0	0	8
9	0	1	1	1	7
10	0	1	1	0	6
11	0	1	0	1	5
12	0	1	0	0	4
13	0	0	1	1	3
14	0	0	1	0	2
15	0	0	0	1	1
16	0	0	0	0	0
17	1	1	1	1	15
18	1	1	1	0	14
19	1	1	0	1	13
20	1	1	0	0	12

Figure 25.38
The output sequence
of the ripple down
counter.

File 25C

Computer simulation exercise 25.3

Design a 4-bit asynchronous down counter using negative edge-triggered J–K
bistables. Take as your starting point the circuit in Computer simulation exercise 25.1. Modify this circuit to produce the required function and simulate the
circuit to confirm its correct operation.

25.8.4 Up/down counters

Combining the circuits in Figures 25.33 and 25.37 using the **data selector** in Figure
25.32, it is possible to construct a counter that can count up or down. Such a circuit
is shown in Figure 25.39.

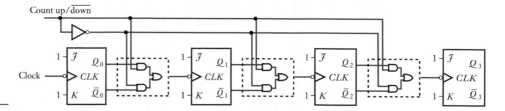

Figure 25.39
An up/down counter.

The direction of counting is controlled by the *up/down* signal. When this line is high, the Q output from each stage is used to provide the clock for the next stage and the circuit counts up. When the control line is low, the $\bar{Q}$ output is fed to the next stage and the circuit counts down.

Example 25.5 | **Application of an up/down counter.**

In Section 12.6.6 we looked at the signals produced by an incremental position encoder and noted that these take the form of two square waves that are phase shifted with respect to each other. In Example 12.1, we looked at the design of a microcomputer mouse and devised an arrangement that produced a similar pair of output signals for each axis of motion. The form of the signals is shown below.

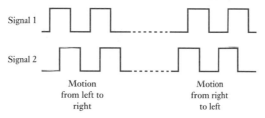

We can see that motion in one direction causes signal 1 to lead signal 2, while motion in the other direction causes signal 2 to lead signal 1. We require a mechanism to count the number of 'steps' in one direction and subtract from it the number of steps in the other direction to give a measure of the absolute position.

To see how this objective may be met, we need to look at the timing of the two signals. In particular, look at the state of signal 2 on the negative-going edge of signal 1. You will see that, for motion from left to right, signal 2 is high on the falling edge of signal 1. However, for motion from right to left, signal 2 is low on the falling edge of signal 1. This allows us to use an up/down counter to determine the absolute position.

Signal 1 is fed to the clock of the counter and signal 2 is connected to the up/down control input. Motion from left to right will now cause the counter to count up and motion from right to left will cause it to count down. Depending on the application, some mechanism might be required to zero the counter to set its absolute value.

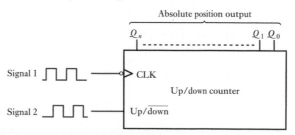

In practice, the pulses produced by a mouse are counted in the microcomputer rather than by an external hardware counter. However, the principles involved are the same.

25.8.5 Propagation delay in ripple counters

Although ripple counters are extremely simple to construct, they do have a major disadvantage, which is particularly apparent when high-speed operation is required. As the output of one flip-flop is triggered by the change of state of the previous stage, delays produced by each flip-flop are summed along the chain. Each flip-flop takes a finite time to respond to changes at its inputs, this being its propagation delay time t_{PD}. In a ripple counter of n stages, it will take $n \times t_{PD}$ for the counter to respond. If the counter is read during this times the value will be garbled, as some stages will have changed while others will not. This produces a fundamental limit to the maximum clock frequency that can be used with the counter. If clock pulses are received by the first stage before the last has responded, at no time will the counter read the correct value.

25.8.6 Synchronous counters

One solution to the problems of propagation delay within ripple counters is to use synchronous techniques. This is achieved by connecting all the flip-flops within a counter to a common clock signal so that they all change state at the same time. This ensures that, a short time after the clock signal changes, the counter can be read with confidence that all stages will have responded.

Clearly, if all the stages of a counter are connected to the same clock, then some method must be used to determine which stages change state and which remain the same. To investigate this, consider the required outputs of a four-stage counter, as shown in Figure 25.40. Looking at this sequence, we can determine rules that define when the various outputs change. We can then produce a counter consisting of a number of edge-triggered J–K flip-flops with circuitry to enforce these rules. As all the flip-flops are triggered simultaneously, they will respond to the values on their control inputs immediately before the clock edge. The slight delay caused by the circuitry used to generate these signals will ensure that they are stable while the flip-flops respond:

Number of clock pulses	Q_3	Q_2	Q_1	Q_0
0	0	0	0	0
1	0	0	0	1
2	0	0	1	0
3	0	0	1	1
4	0	1	0	0
5	0	1	0	1
6	0	1	1	0
7	0	1	1	1
8	1	0	0	0
9	1	0	0	1
10	1	0	1	0
11	1	0	1	1
12	1	1	0	0
13	1	1	0	1
14	1	1	1	0
15	1	1	1	1
16	0	0	0	0
17	0	0	0	1
18	0	0	1	0
19	0	0	1	1
20	0	1	0	0

Figure 25.40
The output sequence of a four-stage up counter.

Q_0 This output is the simplest to define as it changes on every clock pulse. This can be simply produced by configuring flip-flop 0 to toggle on every clock pulse by connecting both J and K to 1.

Q_1 This output changes state after each clock period where Q_0 is high. This can be achieved by connecting both J and K of flip–flop 1 to Q_0. When Q_0 is high, this stage will toggle; when it is low it will remain unchanged.

Q_2 This output changes after each clock period when both Q_0 and Q_1 are high. This can be achieved by ANDing Q_0 and Q_1 together and applying the resultant signal to the J and K inputs of flip–flop 2.

Q_3 This output changes after each clock period when Q_0, Q_1 and Q_2 are all high. This can be achieved by ANDing these three outputs to produce a signal to be applied to the J and K inputs of flip–flop 3.

The resultant circuit is shown in Figure 25.41.

Clearly this technique can be extended to produce synchronous counters of any length. As the number of elements in the counter increases, the number of signals that must be ANDed together at each stage becomes large. This problem can be alleviated by noting that the signals required by the J and K inputs of one stage are simply those required by the previous stage ANDed with the output of the previous stage. This allows a simplification of the circuit, as shown in Figure 25.42. This circuit can be extended by adding as many identical stages as necessary.

Synchronous counters can be configured to produce up, down, up/down and modulo-N counters in a similar manner to ripple counters. They have the advantage that all outputs change at the same time, allowing the counter to be read safely a short time afterwards. They have the disadvantage of increased complexity compared with ripple counters, but can be used at higher clock frequencies.

Figure 25.41
A synchronous four-stage counter.

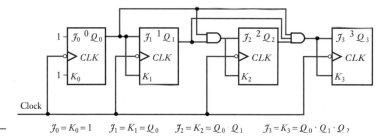

$J_0 = K_0 = 1$ $J_1 = K_1 = Q_0$ $J_2 = K_2 = Q_0 \cdot Q_1$ $J_3 = K_3 = Q_0 \cdot Q_1 \cdot Q_2$

Figure 25.42
A cascadable 4-bit synchronous counter.

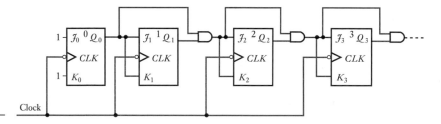

25.8.7 Integrated circuit counters

Although it is feasible to construct counters using individual gates or combinations of flip-flops, it is more usual to use specialised integrated circuits that contain all the functions of a counter in a single package. Both synchronous and asynchronous types are available in a number of sizes and with a range of features. Binary, decade and BCD counters are available, as are up, down and up/down versions. Typical circuits might provide 4- to 14-bit counters or several independent counters in a single package. Some counters allow their contents to be cleared or preset (loaded) with a particular value.

Figure 25.43
Cascading
asynchronous BCD
counters.

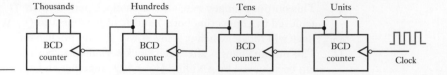

Most integrated counters are designed so that they can be cascaded to form counters of greater length. This is achieved with ripple counters simply by taking the most significant bit of one stage as the clock input for the next. This is illustrated in Figure 25.43, which shows a four-decade BCD counter. The output of each digit of this counter could be used to drive a separate BCD display. With synchronous counters, a common clock is used for all stages and appropriate signals are provided to allow a number of counters to be joined together.

25.9　Design of sequential logic circuits

In this chapter, we have looked at some examples of simple sequential elements in the form of bistables and seen how these can be used to form larger systems, such as counters. In this final section, we will look at the problem of designing sequential circuits to perform particular tasks.

Sequential systems may be either synchronous or asynchronous in nature. Synchronous systems are based on circuits that are controlled by a master clock; the values of the inputs and internal states are only of importance at times determined by transitions of this clock. Such arrangements often use clocked bistables as their building blocks. In asynchronous systems, there is no clock. Internal states, and the outputs, may be affected at any time by changes in the inputs. Such systems are often based on unclocked bistable latches or use time delays within logic elements to represent storage within the feedback path. In general, synchronous design is more straightforward than its asynchronous equivalent, the latter being susceptible to timing problems and instability. We will therefore concentrate on the design of synchronous sequential systems.

25.9.1　Synchronous sequential systems

The approach taken to the design of a sequential system will be greatly affected by the nature of the problem and the way in which it is defined. Here, we will discuss one approach to the design of synchronous systems that uses building blocks already discussed in this chapter.

System states

One of the first tasks is to determine all of the discrete **system states** that exist within the system. In this context, the word state refers to a combination of internal and output variables of the system.

State transition diagram

In simple systems, the system states can be derived directly from the problem definition but, in more complicated cases, it may be necessary to model the system using a **state transition diagram**. By way of a simple example, Figure 25.44 shows a state diagram for a J–K flip-flop.

Figure 25.44
The state transition
diagram of a J–K
flip-flop

The J–K flip-flop has two inputs (J and K), one output (Q) and two states corresponding to $Q = 0$ and $Q = 1$. The states have been named S_0 and S_1 and are represented by the two circles in the diagram. The lines joining the two circles represent possible transitions between these states, while the labels on these lines indicate the input conditions for which these transitions will occur and the resultant output. The notation used for the labels is JK/Q. The diagram also shows lines that leave each state and circle back to terminate on the same state. The labels on these lines indicate the input conditions for the system to stay in that state and the output produced while in that state.

The diagram shows that, when the circuit is in S_0 the output Q is 0 and that it will remain in that state while the inputs, JK, are 01 or 00. A transition from S_0 to S_1 will occur for inputs 10 or 11 and, if this occurs, the output will go to 1. You might like to consider the other aspects of the diagram and confirm that it corresponds to the functions of the J–K flip-flop outlined earlier in this chapter.

The names given to the two states in the above example were arbitrary and it would not affect the diagram if the labels were reversed. Most sequential systems have more than two states and a more typical diagram is shown in Figure 25.45, which shows a system with five states (S_0–S_4). The operation of this circuit is not of paramount importance but is in fact that of a circuit with two inputs, N (next) and R (reset), and one output, Q. While the reset input is low (inactive), the circuit moves from state to state around the loop, moving on one state each time that N is high during a clock pulse. The output Q is high only in state S_0, so the output is high for one in every five counts. Taking the R input high causes the system to jump to S_0, where the output is 1, and remain there as long as R is held high. The arrangement is thus a form of modulo–5, resettable counter.

The state diagram indicates for each state the action (or lack of action) that will result from each of the possible combinations of the inputs. It is thus an unambiguous description of the system, unlike the verbal definition given above, which is open to misinterpretation.

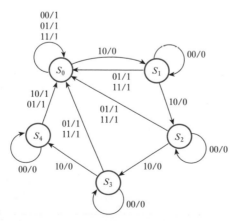

Figure 25.45
The state transition
diagram of a system
with five states.

State transition table

From the state diagram it is possible to construct a **state transition table** (also called simply a **state table** or a **next state table**) that tabulates this information. The table lists all the possible input combinations for each system state and indicates the resultant state (the 'next' state).

Table 25.1 shows the state transition table for the J–K flip-flop shown in Figure 25.44 and Table 25.2 shows the same table simplified by the use of don't care conditions. Table 25.3 shows the more complicated table for the modulo-5 counter shown in Figure 25.45.

Table 25.1 The state transition table for a J–K flip-flop.

Present state	Input conditions JK	Next state	Output Q
S_0	00	S_0	0
	01	S_0	0
	10	S_1	1
	11	S_1	1
S_1	00	S_1	1
	01	S_0	0
	10	S_1	1
	11	S_0	0

Table 25.2 The state transition table for a J–K flip-flop using don't care conditions.

Present state	Input conditions JK	Next state	Output Q
S_0	0X	S_0	0
	1X	S_1	1
S_1	X1	S_0	0
	X0	S_1	1

Table 25.3 The state transition table for the modulo-5 counter in Figure 24.45.

Present state	Input conditions NR	Next state	Output Q
S_0	0X	S_0	1
	10	S_1	0
	11	S_0	1
S_1	00	S_1	0
	X1	S_0	1
	10	S_2	0
S_2	00	S_2	0
	X1	S_0	1
	10	S_3	0
S_3	00	S_3	0
	X1	S_0	1
	10	S_4	0
S_4	00	S_4	0
	X1	S_0	1
	10	S_0	1

State reduction

Inspection of the state table in Table 25.3 shows that all the states included in the table are unique in that they respond differently to the various input combinations. This is not always the case. Consider the state table in Table 25.4, which represents a system with one input and seven states.

Close inspection of Table 25.4 shows that states S_4 and S_6 are identical – in both cases, when the input is 0 the next state is S_0 and the output is 0, and when the input is 1 the next state is S_5 and the output is 1. We can therefore remove S_6 and change all references to it to S_4. Having done this, we now observe that states S_3 and S_5 are identical, so we can remove S_5 and change all references to it to S_3. This process is shown in Table 25.5.

The redundant states can now be removed, leaving a system with five internal states, as shown in the state table in Table 25.6.

Table 25.4 The state table for a system with redundant states.

Present state	Input conditions X	Next state	Output Q
S_0	0	S_0	0
	1	S_1	0
S_1	0	S_2	1
	1	S_3	0
S_2	0	S_0	1
	1	S_3	0
S_3	0	S_4	0
	1	S_5	1
S_4	0	S_0	0
	1	S_5	1
S_5	0	S_6	0
	1	S_5	1
S_6	0	S_0	0
	1	S_5	1

Table 25.5 The state table for a system with redundant states.

Present state	Input conditions X	Next state	Output Q
S_0	0	S_0	0
	1	S_1	0
S_1	0	S_2	1
	1	S_3	0
S_2	0	S_0	1
	1	S_3	0
S_3	0	S_4	0
	1	$S_5 \rightarrow S_3$	1
S_4	0	S_0	0
	1	$S_5 \rightarrow S_3$	1
$S_5 \rightarrow S_3$	0	$S_6 \rightarrow S_4$	0
	1	S_5	1
$S_6 \rightarrow S_4$	0	S_0	0
	1	S_5	1

Table 25.6 The state table for the system in Table 25.4 with redundant states removed.

Present state	Input conditions X	Next state	Output Q
S_0	0	S_0	0
	1	S_1	0
S_1	0	S_2	1
	1	S_3	0
S_2	0	S_0	1
	1	S_3	0
S_3	0	S_4	0
	1	S_3	1
S_4	0	S_0	0
	1	S_3	1

State assignment

Each of the system states must be represented in the final design by a unique combination of internal variables. Clearly, the number of variables required will be determined by the number of independent states. In many cases the internal variables are represented by the outputs of bistables, and in such cases the minimum number of bistables N is related to the number of states S by the relationship:

$$2^N \geq S$$

Thus, a system with two states can be implemented using a single bistable, systems with three or four states require two bistables, systems with five to eight states require three bistables and so on.

Our modulo-5 counter example has five states, so it needs a minimum of three bistables. Having decided on the number of bistables, it is now necessary to assign particular combinations of their outputs to each state. If we represent the Q outputs of our three bistables using the letters A, B and C, we can describe the outputs of the bistables using the 3-digit binary number ABC. We might then choose to make the pattern 000 represent state S_0, pattern 001 represent state S_1 and so on. In synchronous systems any choice of assignments will result in a workable system, although some assignments will result in simpler circuitry than others. We will see later that, in asynchronous systems, this is not the case, so state assignment must take into account the stability of the resulting arrangement. Table 25.7 shows a possible state assignment for our modulo-5 counter, while Table 25.8 shows a state transition table that represents the states by means of their patterns of internal variables.

Table 25.7 The state assignment table for the modulo-5 counter of Figure 25.45.

State	Internal variables ABC
S_0	000
S_1	001
S_2	010
S_3	011
S_4	100

Table 25.8 The state transition table for the modulo-5 counter of Figure 25.45.

Present state ABC	Input conditions NR	Next state ABC	Output Q
000	0X	000	1
	10	001	0
	11	000	1
001	00	001	0
	X1	000	1
	10	010	0
010	00	010	0
	X1	000	1
	10	011	0
011	00	011	0
	X1	000	1
	10	100	0
100	00	100	0
	X1	000	1
	10	000	1

Excitation table

The transition table indicates the action to be taken by each bistable for each combination of the input variables in each state. To design a system from such a table, we need to decide on the nature of the bistables to be used, then determine what the inputs to each bistable must be in order to produce the required actions.

Let us consider the modulo-5 counter and assume that we are to use J–K flip-flops to implement the system. The action of a J–K flip-flop can be deduced from the information given in Table 25.2. This can be reorganised to give the relationship between the desired output transitions and the necessary combinations of the control inputs, as shown in Table 25.9.

Combining the data on the J–K flip-flop given in Table 25.9 with that given in the transition table in Table 25.8 allows us to deduce the control signals required for the three flip-flops.

If we look at the first row of Table 25.8 we see that, when the present state is 000 (ABC) and the inputs are 0X (NR), the next state is also 000. In other words, with this combination of internal variables and inputs, all three bistables go from 0 to 0 (in other words they remain at 0). This 'transition' is represented by the first row in Table 25.9, indicating that the J input for all three bistables must be at 0, whereas the status of the K input is unimportant (don't care).

If we now turn to the second row of Table 25.8 we see that, when the present state is 000 and the inputs are 10, the next state is 001. Thus, bistables A and B go from 0 to 0 as before (and require $J = 0$, $K = X$), but now bistable C goes from 0 to 1 and,

Table 25.9 The control inputs for a J–K flip-flop.

Present state Q_n	Next state Q_{n+1}	Control inputs required	
		J	K
0	0	0	X
	1	1	X
1	0	X	1
	1	X	0

Table 25.10 The excitation table for the modulo-5 counter of Figure 25.45.

Present state ABC	Input conditions NR	Next state ABC	J_A	K_A	J_B	K_B	J_C	K_C	Output Q
000	0X	000	0	X	0	X	0	X	1
	10	001	0	X	0	X	1	X	0
	11	000	0	X	0	X	0	X	1
001	00	001	0	X	0	X	X	0	0
	X1	000	0	X	0	X	X	1	1
	10	010	0	X	1	X	X	1	0
010	00	010	0	X	X	0	0	X	0
	X1	000	0	X	X	1	0	X	1
	10	011	0	X	X	0	1	X	0
011	00	011	0	X	X	0	X	0	0
	X1	000	0	X	X	1	X	1	1
	10	100	1	X	X	1	X	1	0
100	00	100	X	0	0	X	0	X	0
	X1	000	X	1	0	X	0	X	1
	10	000	X	1	0	X	0	X	1
101	XX	XXX	X	X	X	X	X	X	X
110	XX	XXX	X	X	X	X	X	X	X
111	XX	XXX	X	X	X	X	X	X	X

from Table 25.9, requires $J = 1$ and $K = X$. This process can be repeated for each row of the state transition table to identify the inputs required for the J and K inputs of each flip-flop for each state and for each combination of the inputs. This information is used to create an **excitation table**, as shown in Table 25.10, where J_A is the signal on the J input of flip-flop A, K_B is the signal on the K input of flip-flop B and so on. This table also indicates the output of the system for each combination.

You will notice that Table 25.10 has three more rows than Table 25.8. This is because three bistables produce eight possible states, so we must define what the system will do in these **unused states**. Here, we have simply assumed that, as these states are not used, the values of the flip-flop inputs corresponding to these states are unimportant. We will return to this topic a little later.

Circuit design

The excitation table shows which signals must be applied to the inputs of each of the bistables in order to achieve the appropriate transitions between states. The final stage of the design is to define circuitry to produce these signals – and the required outputs – from the inputs and internal variables. This process is shown in Figure 25.46.

The appropriate columns of the excitation table define the signals required for each flip-flop input in terms of the system inputs and internal variables (the flip-flop outputs). In simple systems, Karnaugh maps can be used to produce expressions for these functions. Automated methods would normally be used in more complicated arrangements (as described in Section 24.9).

Figure 25.47 shows Karnaugh maps for the six flip-flop input signals of our modulo-5 counter, and the resultant Boolean expressions.

The logic required to generate the output signals of a system may be defined by an **output logic table**, which lists each combination of the internal states and inputs and indicates the appropriate output. In some cases it is simpler to combine this information with that in the state transition table or excitation table, as in the earlier examples.

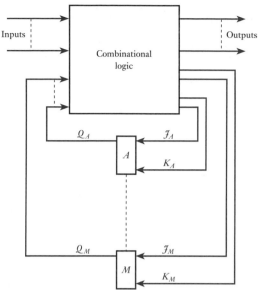

Figure 25.46
Block diagram of a
sequential system.

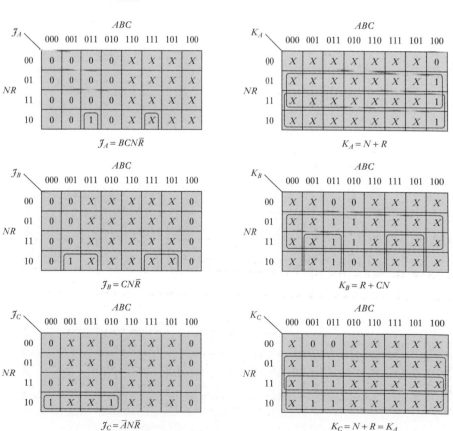

Figure 25.47
Karnaugh maps of the
flip-flop input signals of
the modulo-5 counter.

The excitation table in Table 25.10 indicates the required output of the modulo-5 counter for each combination of the internal states and inputs. From this information, the output logic can be designed as for the control logic for the bistables. Figure 25.48 shows a Karnaugh map that represents this information and indicates a simplified Boolean function for the required logic.

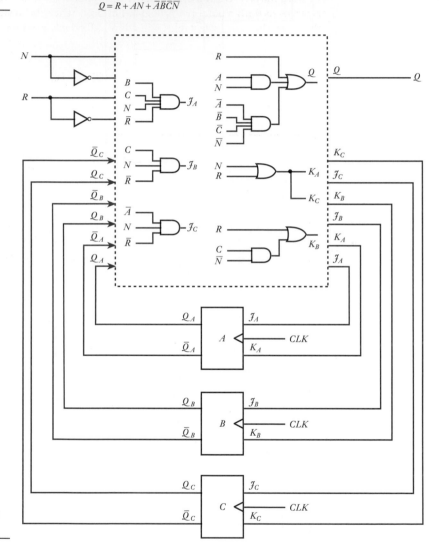

Figure 25.48
A Karnaugh map for
the output logic of the
modulo-5 counter.

$$Q = R + AN + \overline{A}\,\overline{B}\,\overline{C}\,\overline{N}$$

Figure 25.49
Circuit diagram of the
modulo-5 counter.

Figure 25.49 shows the complete modulo–5 counter implemented using a mixture of logic gates. Clearly this design could be adapted to use only NAND gates or only NOR gates if required, as described in Examples 24.7 and 24.8.

The bistables used in this example are synchronous J–K devices. They may be either master/slave or edge-triggered. When using master/slave flip-flops the inputs are sampled before the outputs are allowed to change, thereby preventing any possible hazard problems. If edge-triggered devices are used the delays within the logic gates will prevent the input signals from changing until after each flip-flop has responded to the clock transition, again preventing any hazard condition. Other forms of bistable

– for example R–S flip-flops – could be used in place of the J–K devices by using appropriate control input data in place of that given in Table 25.9.

Unused states

Earlier in the design of the modulo-5 counter we observed that only some of the possible internal states are used. At that time we simply assumed that, as the remaining states were never used, the behaviour of the system in those states was unimportant. This leaves us with a potential problem, however, when the system is first turned on. At that time the various bistables will adopt effectively random output states, resulting in the system state being undefined. In this way, the counter may start in any of the possible system states, including those that are unused. The action taken by the system in these normally unimportant states must therefore be considered.

Figure 25.50 illustrates several possible state diagrams for a simple system with five used states and three unused states. In (a) and (b), the unused states (S_5 to S_7) are linked to the used states and a system starting in any of the unused states would soon move into the area of normal operation. In structures (c) and (d) the unused states do not all lead into the used states, and in these cases it is possible for the system to become 'locked' in an undesirable mode in which it is not operating as intended. In the circuit shown in Figure 25.50(c), for example, the system could start in states S_5, S_6 or S_7 and would then become locked in state S_7 indefinitely.

There are several possible approaches to the problems of unused states. One is to incorporate additional circuitry to force the circuit to start in a particular state at power up. This circuitry can often be very simple, such as a resistor and capacitor arrangement to hold the $\overline{PRESET}$ or $\overline{CLEAR}$ inputs of the bistables low for a short time after the power is applied.

An alternative approach is to specify the action to be taken within the unused states, making them effectively used states. The designer could, for example, specify that the action in the unused states would be as in Figure 25.50(a), ensuring that the circuit

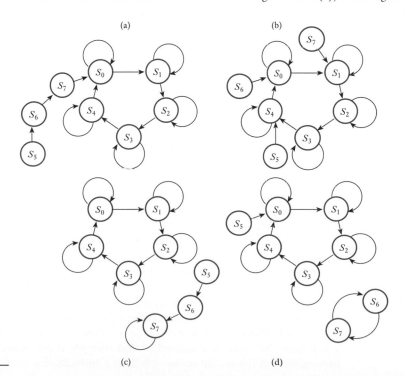

Figure 25.50
Examples of the properties of unused states.

would take up correct operation within a few clock cycles. This approach has the disadvantage that it often leads to a more complicated solution than would otherwise be required as it is not usually apparent which transition routes produce the simplest system.

A third approach is to adopt the technique used in the design of the modulo-5 counter earlier in this section. Here we assumed that the action of the system was unimportant within the unused states. This technique produces simple hardware, but we must then investigate the final design to see if the behaviour in the unused states is acceptable. This is done by working backwards through the design, replacing the don't care conditions in the excitation table with the actual values used. These may be deduced from the logic functions produced or from the Karnaugh maps. From the full excitation table it is then possible to produce a complete transition diagram that includes the unused states.

Figure 25.51 shows the complete state transition diagram for the modulo-5 counter designed earlier. It can be seen that if the system starts in states S_5, S_6 or S_7, it will move into state S_0 as soon as either of the inputs goes to 1. If the system starts in one of the unused states, it will therefore fail to increment on the first clock pulse for which N is 1, but will then work correctly. If the system starts in any of the remaining states except S_0, it will also require a few clock pulses to become established in its correct sequence. If this characteristic is unacceptable, it will be necessary to incorporate circuitry to preset the system to S_0 at power up, as described earlier.

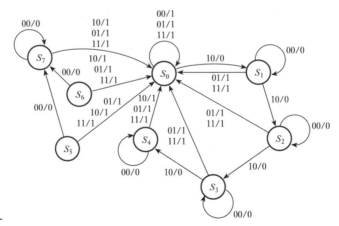

Figure 25.51
The full state transition diagram for the modulo-5 counter.

25.9.2 Asynchronous sequential systems

In the synchronous sequential systems described in the last section, the internal states and outputs are only updated in response to transitions of a clock. In such systems, a single change of state can occur each time the clock pulses – the values of the inputs and existing internal states determine the transitions that occur at that time. Asynchronous systems have no clock and such systems will respond at any time to changes in their inputs. These changes may involve a sequence of transitions between a number of states as the effects ripple through the system. The memory elements in asynchronous systems are usually either unclocked bistables or logic gates providing a time delay. The latter act as memory devices because the logic signals are effectively stored as they propagate through the circuit.

The design of asynchronous sequential systems is more complicated than that of their synchronous counterparts as the timing of the signals plays an important role. In synchronous systems, changes between states can only occur at active transitions of the clock. However, in asynchronous systems, they may occur at any time and the states must therefore be designed so that they are stable. This can be illustrated by looking back at the synchronous system in Figure 25.45.

If we assume that the system is initially in state S_0 and that the input pattern is $NR = 10$, it can be seen that successive clock pulses will cause the circuit to circulate through the states – one state per clock pulse. If we now assume that this diagram represents an asynchronous system, the result is quite different. As soon as the input pattern goes to $NR = 10$, the circuit will leave state S_0 and enter state S_1. However, when in state S_1 this input pattern will immediately cause it to switch to state S_2. This process will continue, with the circuit circulating through the states at a speed determined by the delays within the system. A solution to this problem is to ensure that the pattern of inputs that causes the system to enter a particular state will cause it to remain in that state. A change in the inputs will then be required to move to another state. Such an arrangement is shown in Figure 25.52.

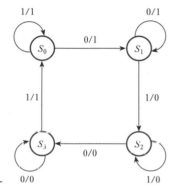

Figure 25.52
A state transition
diagram for a system
with stable states.

The design of asynchronous sequential systems involves many of the same components as synchronous design, although their execution is often more complicated as a result of timing and stability considerations. A detailed description of the design of asynchronous systems is not within the scope of this text and the reader is referred to one of the many books that cover this topic.

Video 25D

Further study

Design a digital stopwatch that displays the time to a resolution of 1 second on four seven-segment displays showing seconds and minutes. The unit should be controlled by three pushbuttons – one to start, one to stop and one to zero the timer.

- Sequential logic circuits have the characteristic of *memory* in that their outputs are affected by the sequence of events that led up to the current set of inputs.
- Among the most important groups of sequential logic elements are the various forms of multivibrator. These circuits can be divided into three types:
  - bistables, which have two stable states
  - monostables, which have one stable and one metastable (quasistable) state
  - astables, which have no stable states but two metastable states.
- The most widely used class of multivibrator is the bistable. These devices may be divided into
  - latches
  - edge-triggered flip-flops
  - pulse-triggered (master/slave) flip-flops.
- Each class of bistable may then be divided into a range of devices with different operating characteristics. These are often described by symbolic names, such as R–S, J–K, D or T-type devices.
- Bistables are frequently used in groups to form registers or counters.
- Registers form the basis of computer memories and are also used for serial to parallel conversion.
- Counters, in their various forms, are used extensively for timing and sequencing functions. They are produced using two basic circuit techniques.
  - asynchronous or ripple counters, in which the clock for one stage is generated from the output of the previous stage. The result is a ripple effect as each stage changes in sequence.
  - synchronous counters, in which all stages are clocked simultaneously so that all the outputs change at the same time.
- Both techniques can be used to produce counters that can count up or down.
- Modulo-N counters can also be produced.
- Standard integrated circuit building blocks are available to simplify the construction of counters and registers. These can normally be cascaded to form units of any desired length.
- Sequential logic circuits can be designed using either synchronous or asynchronous techniques. The former use clock signals to control the operation of the system; the latter have no clock and will respond at any time to changes in their inputs.
- The design of synchronous systems involves a number of stages that will vary depending on the nature of the problem and the circuit techniques used. A fairly straightforward method involves the use of clocked flip-flops and appropriate combination logic. A possible design process might involve:
  - identification of the system states
  - a state transition diagram
  - a state transition table
  - state reduction
  - state assignment
  - generation of an excitation table
  - circuit design
  - investigation of unused states.
- The design of asynchronous systems is a more complicated task because of timing considerations and problems of stability.

Exercises

25.1 Explain the distinction between combinational and sequential logic.

25.2 Define the terms bistable, monostable and astable.

25.3 Explain the origins of the labels S and R given to the inputs of an S–R bistable.

25.4 In an S–R bistable formed using two NOR gates, are the inputs active high or active low?

25.5 Under what circumstances does switch bounce cause problems in digital systems? Design a circuit using two input NOR gates to remove the effects of switch bounce from a changeover switch.

25.6 Deduce the waveform at the Q output of the following circuits.

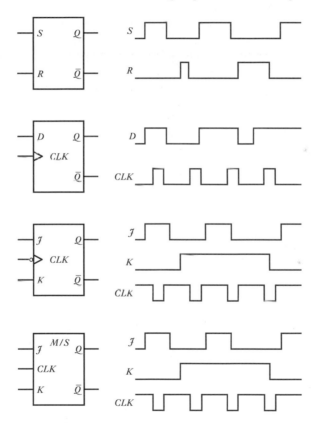

25.7 What is meant by a race in sequential logic?

25.8 Explain how master/slave bistables overcome problems associated with races.

25.9 Explain the difference between a retriggerable and non-retriggerable monostable.

25.10 What form of multivibrator has the characteristics of a digital oscillator?

25.11 Design an 8-bit memory register using D master/slave flip-flops, being careful to number your inputs and outputs appropriately.

25.12 Describe the operation of a simple shift register and explain how this can be used to perform serial to parallel and parallel to serial conversion.

25.13 Explain the meanings of the terms 'synchronous' and 'asynchronous' when applied to counters.

25.14 Design a modulo-5 ripple counter using negative edge-triggered J–K flip-flops.

25.15 Simulate your circuit for Exercise 25.14 and confirm that it functions as expected.

25.16 The decade counter in Figure 25.36 can be used to reduce the frequency of a clock waveform by a factor of ten by taking the output from Q_3. For some applications, this has the disadvantage that the waveform produced is not a square wave. Design a 'divide by ten' counter that does produce a squarewave output. *Hint*: the circuit from the last exercise might be useful.

25.17 Simulate your solution to the previous exercise to confirm that the circuit functions as expected.

25.18 Design a modulo-10 ripple down counter using negative edge-triggered J–K bistables. The circuit should count down to zero and then reset to 9.

25.19 Simulate your solution to the previous exercise to confirm that the circuit functions as expected.

25.20 Design a 4-bit up/down counter that does not overflow or underflow. That is, counting up is disabled when it reaches its maximum value and counting down is disabled when it reaches its minimum value. Can you think of any application for such a counter?

25.21 In Example 25.2 we looked at the design of a simple burglar alarm. Design a more sophisticated alarm that allows the houseowner to leave the house after turning it on and to re-enter the house and turn it off without the alarm sounding. The unit should arm itself 30 seconds after a switch is closed and allow the user 30 seconds to turn it off before sounding the alarm.

25.22 Design a digital clock that displays the time in seconds, minutes and hours on six seven-segment displays. Your circuit should take into account the fact that such clocks display hours in the range 1 to 12, not 0 to 11.

25.23 Modify your design for Exercise 25.22 to allow the time to be set by depressing buttons to increment the seconds, minutes and hours settings.

25.24 Modify your design for Exercise 25.22 to allow the circuit to display time in either a 12-hour or 24-hour format. The display mode should be controlled by the setting of a switch.

25.25 Describe the effects of propagation delay on the maximum operating speed of a ripple counter. How are these problems tackled in a synchronous counter?

25.26 Design a synchronous modulo-5 counter with a single input D (direction), and a single output Q. The counter should count up when the D input is 1 and down when D is 0. The Q output should be 1 in one of the counter's five states and 0 in the remainder.

25.27 Explain the function of the arrangement described by the following state transition diagram.

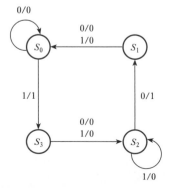

25.28 Design a synchronous circuit to implement the arrangement described by the transition diagram in Exercise 25.27.

25.29 Design a synchronous circuit to implement the following state transition diagram.

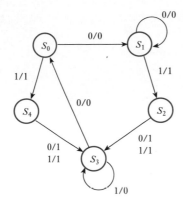

Chapter 26

Digital Devices

Objectives

When you have studied the material in this chapter, you should be able to:

- discuss the use of both field-effect transistors and bipolar transistors in the construction of logic circuits
- list a range of digital electronic 'families' and describe, in broad terms, their methods of operation and characteristics
- outline the form and characteristics of various members of the TTL and CMOS families of logic circuits
- design appropriate arrangements of MOSFET devices for the construction of complex CMOS logic gates
- explain the primary causes and effects of noise in digital systems and techniques that can be used to reduce it.

26.1   Introduction

In the last two chapters we looked at a number of digital applications based on the use of standard logic gates. In this chapter we will consider a range of electronic circuits that can be used to implement these gates.

In the early days of semiconductor manufacture, components were limited to single transistors. As device fabrication techniques improved, it became possible to incorporate a number of active components within a single package and, over the years, the densities of such **integrated circuits** (ICs) have steadily increased. Today, it is possible to place millions of both active and passive components within a single 'chip', enabling complete computers to be constructed on a piece of silicon only a few millimetres square.

Although it is possible to integrate many devices within a single package there are still applications for which single transistors are required and others that require only a handful of components. Electronic devices that contain more than one active component may be classified by their **integration level** and Table 26.1 shows a common way of defining the various levels of integration.

In this chapter we will concentrate on the construction of small- and medium-scale integrated devices. This means that we will be looking at techniques used to construct devices containing a handful of individual gates or simple circuits, such as flip-flops, counters or registers. Figure 26.1 illustrates some typical logic devices and their pin connections (pin-outs). In Chapter 27 we will look at the construction of more sophisticated components (such as microcomputers, memories and programmable devices) and will see that these use many of the same techniques.

Table 26.1 Integration levels for electronic devices.

Integration level	Number of transistors
Zero-scale integration (ZSI)	1
Small-scale integration (SSI)	2–30
Medium-scale integration (MSI)	$30–10^3$
Large-scale integration (LSI)	$10^3–10^5$
Very large-scale integration (VLSI)	$10^5–10^7$
Ultra large-scale integration (ULSI)	$10^7–10^9$
Giga-scale integration (GSI)	$10^9–10^{11}$
Tera-scale integration (TSI)	$10^{11}–10^{13}$

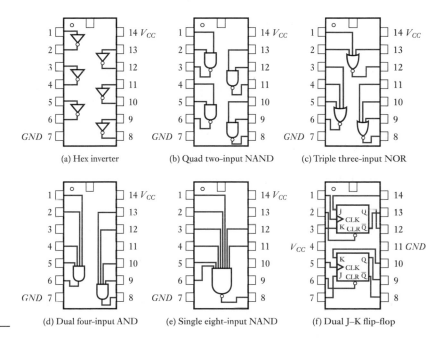

Figure 26.1
Typical logic device pin-outs.

(a) Hex inverter   (b) Quad two-input NAND   (c) Triple three-input NOR

(d) Dual four-input AND   (e) Single eight-input NAND   (f) Dual J–K flip-flop

Although modern digital electronic components are the result of many years of development and evolution, there is no single, ideal set of logic circuits that fulfils all requirements. Over the years, a number of **logic families** have evolved, each offering particular advantages. Some, for example, work at very high speeds, others consume small amounts of power, while others are very tolerant of electrical noise. Part of the designer's function is to select an appropriate logic family for a given application. The information given in this chapter should help in making this choice.

Integrated circuit logic families may be divided into two main groups:

■ those based on field-effect transistors;
■ those that use bipolar transistors (although there are also families that incorporate both types of device).

In this chapter we will start by looking at the characteristics of logic gates and the terminology used to describe them before considering the operation of both FETs and bipolar transistors when used as logical switches. We will then consider a number of logic families and look in more detail at two of the most important families – transistor–transistor logic (TTL) and complementary metal oxide semiconductor (CMOS) logic. The chapter ends with a discussion of the effects of noise on digital systems.

26.2 | Gate characteristics

We have already seen that there are many different forms of logic gate (for example, AND, OR and inverter (NOT) gates). Gates also differ in the circuitry used to implement them and these implementation differences give the gates very different characteristics. In order to understand them, we will start by looking at the simplest form of logic gate – the **inverter** or **NOT gate**.

26.2.1　The inverter

When looking at the characteristics of linear amplifiers in earlier chapters, we saw that all real amplifiers have an output swing that is limited by the supply voltages used. A typical inverting linear amplifier might have a characteristic as shown in Figure 26.2. If we wish to use such a device as a linear amplifier, we must ensure that the input is restricted so that the operation is maintained within the linear range of the device.

It is also possible to use the amplifier of Figure 26.2 as a logical device. If we restrict the input signal so that it is always *outside* the linear region of the amplifier, we are left with two allowable ranges for the input voltage, as shown in Figure 26.3. We may consider these ranges as representing two possible input states – '0' and '1'.

Clearly, when the input voltage corresponds to state '0', the output voltage is at its maximum value, but when the input state is '1', the output is at its minimum value. If we choose component values appropriately, we can arrange that the maximum and minimum output voltages lie within the voltage ranges defined at the input to represent '0' and '1', as shown in Figure 26.4(a).

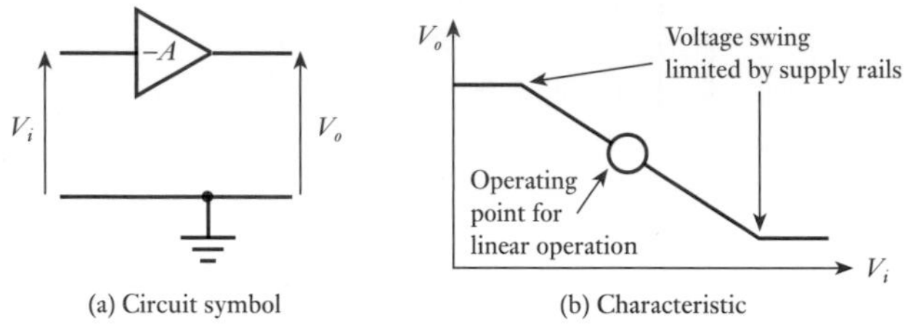

Figure 26.2
An inverting linear amplifier.

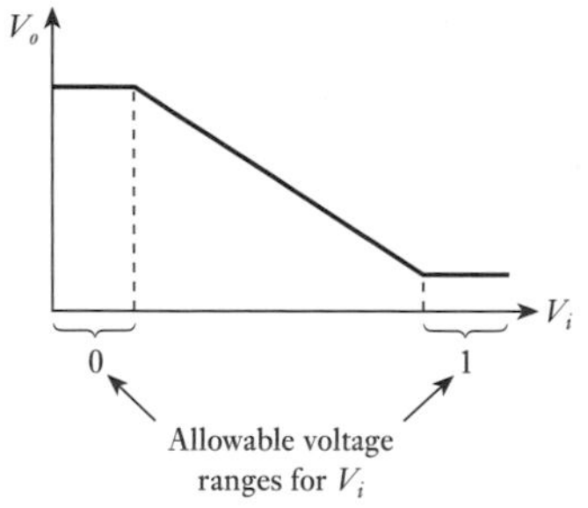

Figure 26.3
Using an inverting amplifier as a logical device.

Figure 26.4
Transfer characteristics for logical inverters.

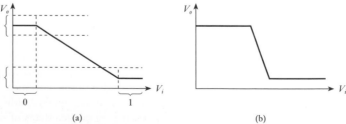

From Figure 26.4(a) it is clear that, when the input is '0', the output is '1' and vice versa. The circuit therefore has the characteristics of a logical inverter. As the input and output voltages are compatible, the output of this arrangement could be fed to the input of a similar gate.

When designing linear amplifiers, we aim to produce an extended linear region to permit a large output swing. When producing a logical inverter we wish the linear portion of the characteristic to be as small as possible to reduce the region of uncertainty. Such circuits therefore have a very high gain and a rapid transition from one state to the other, as shown in Figure 26.4(b).

Logical inverters can be produced using a range of circuit techniques and slight modifications to these basic circuits will produce the functions of the other basic gates. These, in turn, can be combined to form more complicated digital functions. We will look at examples of logic circuits later in this chapter.

26.2.2 Logic levels

The voltage ranges representing '0' and '1' in Figure 26.4(a) represent the **logic levels** of the circuit. In many gates logical 0 is represented by a voltage close to zero, but the range of allowable voltages varies considerably. There is also great variation in the voltage used to represent logical 1, which might be 2–4 V in some components, but 12–15 V in others. In order for one logic gate to work with another, the logic levels used must be compatible.

26.2.3 Noise immunity

Noise is present in any real system. This has the effect of adding random fluctuations to the voltages representing the logic levels. To enable the system to tolerate a certain amount of noise, the voltage ranges defining '0' and '1' at the output of a gate are more tightly constrained than those at the input. This ensures that small perturbations of an output signal caused by noise will not take the signal outside the defined ranges of the input of another gate. Thus, the circuit is effectively immune to small amounts of noise, but may be affected if the magnitude of the noise is large enough to take the logic signal outside the allowable logic bands. The maximum noise voltage that can be tolerated by a circuit is termed the **noise immunity**, V_{NI}, of the circuit. It is also known as the **noise margin**.

The voltage ranges representing the two logic states at the input and output of the gate are defined by four parameters:

- V_{IH} – the minimum voltage required at an input for an input voltage to be interpreted as a '1' (high);
- V_{IL} – the maximum voltage allowed at an input for an input voltage to be interpreted as a '0' (low);
- V_{OH} – the minimum voltage produced at the output of a gate to represent a '1' (high);
- V_{OL} – the maximum voltage produced at the output of a gate to represent a '0' (low).

Clearly, for any real gate there is some variation in these values and it is normal to specify maximum and minimum values (as appropriate) for each quantity. The noise immunity of a logic circuit is determined by the difference between the voltages representing the logic levels at the output of one gate, compared with those specifying the levels at the input of the next. Therefore

noise immunity in logical 1 (high) state, $V_{NIH} = V_{OH}(min) - V_{IH}(min)$ (26.1)

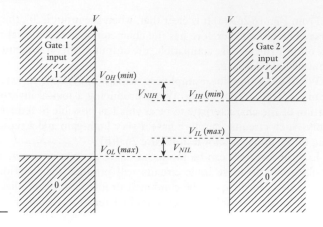

Figure 26.5
Noise immunity.

and

$$\text{noise immunity in logical 0 (low) state, } V_{NIL} = V_{IL}(max) - V_{OL}(max) \qquad (26.2)$$

This relationship is illustrated in Figure 26.5.

Video 26A

26.2.4 **Transistors as switches**

The functions of the various logic gates are invariably implemented using some form of transistor. In Chapters 18 and 19 we looked at the characteristics of FETs and bipolar transistors, and in each case we considered their use as logical switches. Both forms of transistor make good switches, but neither is ideal and their characteristics are somewhat different.

The FET as a logical switch

MOSFETs are the dominant form of FET for digital applications. While in analogue applications it is common to describe such devices as FETs, in digital systems it is more common to talk of **MOS devices**, describing their method of construction rather than their principle of operation.

MOSFETs make excellent switches and the vast majority of modern digital circuitry is based on these devices. The major advantages of MOS technology over circuits based on bipolar transistors are that MOS devices are simpler and less expensive to fabricate. Each MOS gate requires a much smaller area of silicon, allowing a greater number of devices to be produced on a given chip. When used in CMOS gates, MOSFETs can also be used to produce logic circuits with extremely low power consumption. This reduces the amount of waste heat that must be dissipated and allows greater packing densities.

In the early days of digital technology, MOS circuits tended to be slower than those based on bipolar components. However, modern devices can be extremely fast and MOSFETs have largely replaced bipolar transistors in digital applications.

When we use a MOSFET as a logical switch, we ensure that it is driven into one of two states. In the first, the device is effectively turned ON and the channel from the drain to the source has a relatively low resistance and resembles a closed switch. In the other, the device is turned OFF and the effective resistance of the device is so high that it resembles an open switch.

Figure 26.6(a) illustrates the use of a MOSFET as a logical switch. The input to this arrangement is restricted to being either close to zero or close to the supply voltage

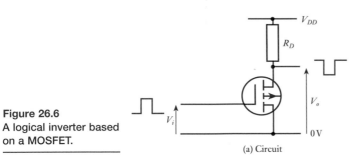

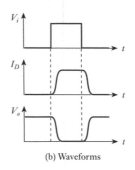

(a) Circuit

(b) Waveforms

Figure 26.6
A logical inverter based on a MOSFET.

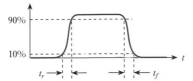

Figure 26.7
Rise and fall times.

V_{DD}. When the input voltage is close to zero the MOSFET is turned off and negligible drain current flows. The output voltage is therefore close to the supply voltage V_{DD}. When the input voltage is close to V_{DD} the MOSFET is turned on and the output voltage is pulled down close to zero. Therefore, the circuit acts as a simple logic inverter, with voltages close to V_{DD} representing logical 1 and voltages close to 0 V representing logical 0.

Figure 26.6(b) shows the relationship between a pulse applied to the input of the inverter and the corresponding drain current and output voltage. The figure shows that there is a delay between a change in the input voltage and the response of the output. Because the waveforms produced are not 'square', we generally quantify the time taken for them to change by defining the **rise time** t_r as the time it takes for the waveform to increase from 10 per cent to 90 per cent of the height of the step, and the **fall time** t_f as the time taken for the waveform to fall from 90 per cent to 10 per cent of the height of the step. These two measures are shown in Figure 26.7.

The circuit in Figure 26.6(a) is not normally used in integrated circuits because the resistor is 'expensive' in terms of circuit area (as noted in Section 18.7.4). It also produces a circuit that is slow to respond as, when the MOSFET turns off, the drain resistor produces a relatively high output resistance, making it slow to charge circuit capacitances. This problem is overcome in CMOS gates (as discussed in Section 18.7.5) by using a push–pull arrangement, which provides a low-resistance output in either state. We will look at CMOS gates later in this chapter.

File 26A

Computer simulation exercise 26.1

Use simulation to investigate the characteristics of the circuit of Figure 26.6(a). A suitable arrangement would use an IRF150 MOSFET with $R_D = 10\ \Omega$ and $V_{CC} = 5$ V. Apply a suitable input waveform and plot V_i, I_D and V_o against time. Repeat this procedure using a 20 Ω drain resistor and compare your results. What is the effect of further increasing the size of the drain resistor?

The bipolar transistor as a logical switch

Bipolar transistors also make good logical switches; a simple switching arrangement is illustrated in Figure 26.8(a). When the input voltage is close to zero, the transistor

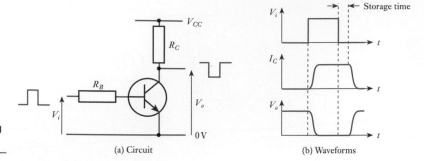

Figure 26.8
A logical inverter using
a bipolar transistor.

(a) Circuit   (b) Waveforms

is turned off and negligible collector current flows. The output voltage is therefore close to the supply voltage V_{CC}. R_B is chosen so that when the input voltage is high the transistor is turned on and the output voltage is equal to the **saturation voltage** of the device, which is generally about 0.1 V. Therefore, the circuit acts as a simple logic inverter, with voltages close to V_{CC} representing logical 1 and voltages close to 0 V representing logical 0.

Figure 26.8(b) shows the relationship between a pulse applied to the input of the inverter and the corresponding collector current and output voltage. As in the MOSFET, there is a delay between a change in the input voltage and the response of the transistor. However, in this case, the time taken for the device to turn off is much greater than the time taken for it to turn on. This increase in switching time results from the **saturation** of the transistor. This causes a large quantity of **minority charge carriers** to be injected into the base of the device, which in turn causes a build-up of charge within the base region. This charge is called the **saturation storage charge**. When the current into the base of the transistor is removed, there is a delay while this stored charge is removed before the collector current falls and the device turns off. This delay is termed the **storage time** of the device and it increases with the magnitude of the base current pumped into the base when it is turned on. A typical value for the storage time of a general-purpose transistor might be about 200 ns, which is several times greater than the time required for the device to turn off after the stored charge has been removed.

The storage time of a device can be reduced by adding impurities to the base region of the transistor to reduce the **minority carrier lifetime**. A common method used to achieve this is **gold doping**. This reduces the storage time, but unfortunately decreases h_{FE} and increases leakage currents.

Some logic circuits overcome the problem of storage time by preventing the transistors from entering their saturation regions. This results in turn-off times that are comparable with the turn-on times of the transistors. We will look at examples of this technique when we consider emitter-coupled logic and the use of Schottky diodes, later in this chapter.

File 26B

Computer simulation exercise 26.2

Use simulation to investigate the characteristics of the circuit in Figure 26.8(a). A suitable arrangement would use a 2N2222 transistor with $R_B = 10$ kΩ, $R_C = 1$ kΩ and $V_{CC} = 5$ V.

Apply a suitable input waveform and plot V_i, I_C and V_o against time. Repeat this procedure using a base resistor R_B of 1 kΩ (to increase the base current) and compare your results.

26.2.5 **Timing considerations**

Propagation delay time

Logic gates invariably consist of a number of transistors (either MOSFET or bipolar), each producing a slight delay as signals pass through them. Inevitably, the resultant delay is different for changes in each direction, and two **propagation delay times** are used to describe the speed of response of the circuit. These are t_{PHL}, which is the time taken for the output to change from high to low, and t_{PLH}, which is the time taken for the output to change from low to high. In some cases a single value is used corresponding to the average time for the two transitions. This average propagation delay time, t_{PD}, is given by

$$t_{PD} = \frac{1}{2}(t_{PHL} + t_{PLH})$$

As, in general, the input waveform will not be a perfect square pulse, t_{PHL} and t_{PLH} are measured between the points at which the input and output signals cross a reference voltage corresponding to 50 per cent of the voltage difference between the logic levels. This is illustrated in Figure 26.9.

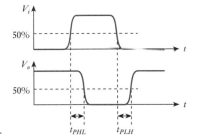

Figure 26.9
Propagation delay
times.

The speed of operation of a logic circuit is greatly affected by the load that it must drive. When an integrated circuit is connected to a printed circuit board, the conductive track joined to each output represents a capacitive load (in addition to the load represented by the components it is linked to). In order to change from one output state to another the circuit must charge or discharge this stray capacitance, and the output stage of the device must be designed to supply sufficient current to do this. Circuits that are only connected to other circuitry *within* the integrated circuit will have a reduced load capacitance, which might be only 1 per cent of that of an external connection. Consequently, gates and other logic circuitry can operate much faster when driving only elements within the same integrated circuit than when driving external components.

Set-up time

In logic circuits that have clock input signals (such as flip–flops), it is often necessary for control inputs to be applied a short while *before* an active transition of the clock to ensure correct operation. The time for which the control input is stable before the clock trigger occurs is termed the **set-up time** t_S. Device manufacturers normally specify a minimum value for this quantity, which might be in a range from less than 1 ns to more than 50 ns.

Hold time

It is also often required that control signals (in circuits such as flip-flops) should not change for a short interval *after* the active transition of a clock input. The time for which a control input is stable after a clock trigger occurs is termed the **hold time** t_H. Typical minimum values might range from 0 to 10 ns.

26.2.6 Fan-out

In many cases, it is necessary to connect the output of one logic gate to the input of a number of gates. As each input draws current from this output, there will be a limit to the number of gates that a single output can supply. This is termed the **fan-out** of the circuit. The fan-out is clearly determined by the output resistance of the gate and by the input resistance of the gates it is driving. Because of the very high input resistance of MOS devices, it is the input capacitance that is of importance. Circuits based on this technology generally have a greater fan-out than those based on bipolar transistors, but in some cases (such as CMOS) the speed of the gate is reduced as the output is more heavily loaded.

Video 26B

26.3 Logic families

Over the years many logic families have evolved, the most successful being:

- resistor–transistor logic (RTL)
- diode logic
- diode–transistor logic (DTL)
- transistor–transistor logic (TTL)
- emitter-coupled logic (ECL)
- metal oxide semiconductor (MOS)
- complementary metal oxide semiconductor (CMOS)
- bipolar complementary metal oxide semiconductor (BiCMOS).

All but the second of these have been used for integrated circuit devices, although RTL and DTL are now of historical interest only. Diode logic is used mainly in discrete (that is, not integrated) circuits.

By far the most commonly used logic family is **complementary metal oxide semiconductor logic (CMOS)**, although **transistor–transistor logic (TTL)** is also widely used. For this reason, later in this chapter we will look in some detail at the operation of these two forms of logic circuit. However, the operation and characteristics of these fairly sophisticated technologies are easier to understand if we look at the evolution of these circuit techniques from more basic circuits. We will therefore look briefly at each of the logic families listed above.

26.3.1 Resistor–transistor logic (RTL)

The simple inverter circuit in Figure 26.8 can be used as the basis for a family of logic devices. The addition of a second transistor forms a two-input NOR gate and extra devices may be added to form gates with any number of inputs. This is illustrated in Figure 26.10. Suitable combinations of NOR gates and inverters can be used to form any required logic function.

The voltages representing logical 0 and 1 in RTL are defined by the output of the gate since these voltages then form the input to the next gate. Logical 0 is given by the saturation voltage of the transistor, which is typically 0.1–0.2 V. The logical 1 voltage

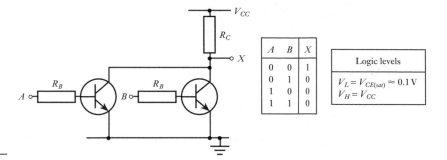

Figure 26.10
An RTL NOR gate.

is simply V_{CC}. The voltages representing logical 1 and 0 are usually referred to as V_H and V_L, respectively (standing for the high and low logic states). This is preferable to the use of V_1 and V_0, which might be mistaken for V_i and V_o, the input and output voltages.

The advantages of RTL are its simplicity and compactness. Its disadvantages are that it has a relatively low **noise immunity** (even small noise spikes added to a logical 0 input voltage will start to turn on the transistor) and slow speed of operation (caused by the presence of the base resistors, which limit the base drive currents).

26.3.2 Diode logic

One of the simplest forms of logic circuit is **diode logic**, examples of which are shown in Figure 26.11. Both AND and OR gates can be constructed using only diodes and resistors. Gates with any number of inputs can be constructed by adding additional diodes. Looking first at the AND gate in Figure 26.11(a), it is clear that, if input logic voltages of 0 V (logical 0) and V_{CC} (logical 1) are used, when either A or B is low, the output X is pulled low. Only if both A and B are high is the resistor R able to pull the output high. It should be noted that the output voltage corresponding to logic 1

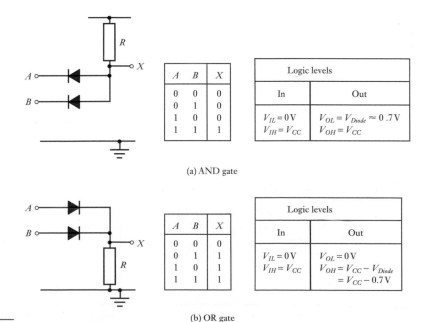

Figure 26.11
Examples of diode logic.

is V_{CC} as at the input. However, the voltage corresponding to logical 0 at the output is not equal to that at the input as it is increased by the voltage across the diode, which is normally about 0.7 V.

In the OR gate circuit shown in Figure 26.11(b), X will be high if either A or B is high. In this case, the output voltage corresponding to logical 0 is equal to that at the input, while that for logical 1 is reduced by about 0.7 V.

Diode logic has the advantage of simplicity, but suffers from being totally passive. As signals pass through a series of gates, the logic levels are gradually eroded, with the voltage representing logical 0 increasing and that representing logical 1 decreasing until the distinction between them disappears. For this reason, diode logic is normally used only for simple logic operations that do not require the complexity of the more sophisticated gate designs.

26.3.3 Diode–transistor logic (DTL)

Many of the problems of diode logic can be overcome by adding an amplifier at the output of the gate to re-establish the logic levels. A simple one-transistor amplifier will suffice, which provides power amplification and reduces the output resistance, increasing the fan-out of the circuit. The circuit produced contains both diodes and transistors, so it is called **diode–transistor logic** (DTL). Figure 26.12 illustrates such a circuit.

The circuit represents the simple AND gate of Figure 26.11(a) with a simple inverting amplifier added to its output. This produces the function of a NAND gate. In fact, the circuit in Figure 26.12 is unsatisfactory as taking either A or B low takes the base of the transistor down to about 0.7 V, which will not, in general, turn it off.

This problem can be overcome by placing a diode in series with the base of the transistor, as shown in Figure 26.13. In order for the transistor in this circuit to be

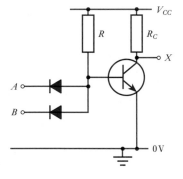

Figure 26.12
A simple DTL
NAND gate.

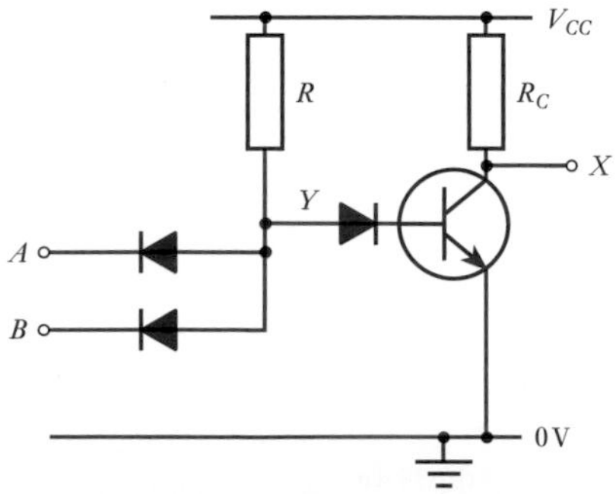

Figure 26.13
A DTL NAND gate.

A	B	X
0	0	1
0	1	1
1	0	1
1	1	0

Logic levels
$V_L = V_{CE(sat)} \approx 0.1\,\text{V}$
$V_H = V_{CC}$

turned on, the voltage at point Y would need to be greater than the sum of the turn-on voltages of the transistor and the diode in series with its base. As each has a turn-on voltage of about 0.5 V, no significant current will flow in R_C unless the voltage at Y exceeds about 1 V. When a 0 logic level is applied to either input, the voltage at point Y is pulled down to about 0.7 V, ensuring that the transistor is turned off and a logical 1 is produced at the output.

26.3.4 Transistor–transistor logic (TTL)

Looking at the circuit of the DTL gate shown in Figure 26.13 it is clear that each input sees a series combination of two back-to-back diodes. This represents an *np–pn* structure, which could be replaced with the *npn* structure of a transistor, as shown in Figure 26.14. Instead of a 'diode–transistor logic' circuit, we now have a 'transistor–transistor logic' (TTL) circuit.

It can be seen that the input transistors in the circuit of Figure 26.14(b) have common connections to their bases and collectors. When using integrated circuit techniques, the circuit can be improved by combining the functions of the input transistors into a single device. This is shown in Figure 26.15(a). The multi-emitter transistor is produced by forming a number of emitter regions within a single base region, as illustrated in Figure 26.15(b). The circuit in Figure 26.15(a) cannot be formed using discrete transistors, so the transistors within the figure (and later diagrams in this chapter) are shown without the customary circle in their symbol to show that they are part of an integrated circuit rather than discrete components.

Commercial TTL gates improve on the basic circuit in Figure 26.15(a) by using additional components to increase speed and drive capabilities. Such circuits will be discussed in more detail in Section 26.4.

Figure 26.14
Replacing the diodes of a DTL gate with transistors.

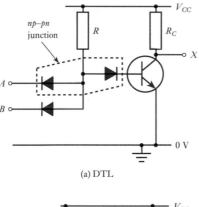

(a) DTL

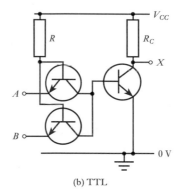

(b) TTL

Figure 26.15
A TTL NAND gate.

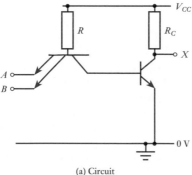

(a) Circuit

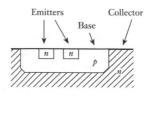

(b) Multi-emitter transistor

26.3.5 Emitter-coupled logic (ECL)

In Section 26.2.4 we discussed the problems of **storage time**, which greatly increases the switching times of bipolar transistors that are driven into saturation. We noted that some circuits overcome this problem by keeping the transistors within their active region, preventing them from becoming saturated. Such a circuit is shown in Figure 26.16.

The circuit is similar to the long-tailed pair amplifier in Figure 19.37, but with the addition of an extra transistor T1. For obvious reasons, circuits of this type are described as **emitter-coupled logic (ECL)**.

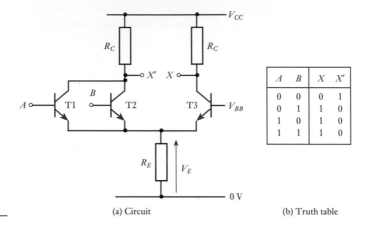

A	B	X	X'
0	0	0	1
0	1	1	0
1	0	1	0
1	1	1	0

(a) Circuit (b) Truth table

Figure 26.16
A non-saturating logic gate.

Further transistors can be added in parallel with T1 and T2 to form a gate with more inputs. Logic inputs are applied to the bases of transistors T1 and T2 and a constant reference voltage V_{BB} is applied to the base of T3. The logic voltages are chosen such that V_L is a little less than V_{BB} and V_H is a little greater than V_{BB}.

If A and B are both at logical 0, the voltages on these inputs are both less than V_{BB}. T3 is turned on and the voltage on the emitters of the transistors V_E is given by

$$V_E = V_{BB} - V_{BE} \approx V_{BB} - 0.7\,\text{V}$$

As the voltages on A and B are less than V_{BB} it follows that the base-to-emitter voltages of T1 and T2 will be less than 0.7 V and they will therefore be turned off. The current I_E flowing through the emitter resistor is given by the expression

$$I_E = \frac{V_E}{R_E} \approx \frac{V_{BB} - 0.7}{R_E}$$

and this flows exclusively through T3 as T1 and T2 are turned off. The voltage at point X is therefore

$$V_X = V_{CC} - R_C I_E = V_{CC} - R_C \frac{(V_{BB} - 0.7)}{R_E} \tag{26.3}$$

and the voltage at X' is approximately V_{CC}. If appropriate values are chosen for V_{BB}, R_C and R_E, it can be arranged that T3 remains within its active region.

If A or B is at logical 1, V_E will be pulled up to $V_H - 0.7\,\text{V}$, reducing the base-to-emitter voltage of T3 to below 0.7 V. This will turn on the corresponding input transistor and turn off T3. The emitter current flowing through the input transistor will now be

$$I_E = \frac{V_E}{R_E} = \frac{V_H - 0.7}{R_E}$$

This current will flow exclusively through the appropriate input transistor and so

$$V_{X'} = V_{CC} - R_C I_E = V_{CC} - R_C \frac{(V_H - 0.7)}{R_E} \tag{26.4}$$

and V_X is approximately V_{CC}. Again, an appropriate choice of component values ensures that the input transistors are not saturated.

If we define the output logic levels such that V_{CC} represents logical 1 and the voltages given by Equations 26.3 and 26.4 fall within the range representing logical 0, the circuit has the truth table given in Figure 26.16(b). The X output therefore represents the OR function and the X' output, being the inverse of X, represents the NOR function. The gate is therefore an OR/NOR gate.

The circuit of Figure 26.16(a) has the advantage that the transistors are never saturated – they are simply switched between cut-off and some fixed current. This enables the circuit to operate at very high speeds because of the absence of the storage time associated with saturated transistors.

The disadvantage of this circuit in its present form is that the input and output logic levels are not the same. This makes it impossible to connect the output of one gate into the input of the next. This problem is overcome by adding a **level-shifting transistor amplifier** to each output. These are simple emitter followers that shift the output voltage by an amount equal to their base-to-emitter voltage (about 0.7 V). A suitable choice of component values enables this to produce equal input and output logic voltages. A typical three-input ECL OR/NOR gate is shown in Figure 26.17.

Compared with the other logic families we have considered, ECL has a relatively small output swing between its two logic levels. One effect of this is that the **noise immunity** of the device is poor at only about 0.2 to 0.25 V. Also, as the transistors are always in their active region, power dissipation is high (typically about 60 mW per gate), resulting in a great deal of waste heat that must be removed to prevent the circuits from overheating. This in turn tends to limit the amount of circuitry that can be integrated into a single chip, increasing system size and cost. However, by keeping the transistors in their active region the switching speed is greatly increased, producing propagation delays of the order of 1 ns. This is considerably faster than saturating logic, such as standard TTL, allowing clock frequencies of up to 500 MHz or more. For many years ECL was the fastest digital logic family and this resulted in it being used in a range of mainframe computers. However, modern high-speed CMOS logic exceeds the speed of ECL while offering many other advantages. For this reason ECL is now rarely used.

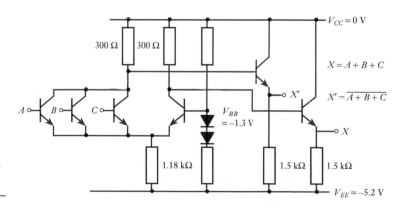

Figure 26.17
A three-input ECL OR/
NOR gate.

26.3.6 Metal oxide semiconductor (MOS) logic

In Figure 18.36(a) we looked at the use of a MOSFET and a resistor in a simple logical inverter. We noted that the use of resistive loads was uneconomical in terms of space in integrated circuits and so, in Figure 18.36(b), we used a second MOSFET as an **active load**. This circuit is reproduced as Figure 26.18(a), which also shows an equivalent circuit of the gate using a switch and a load resistor. The circuit in Figure 26.18(a) uses n-channel MOSFETs and is the basic form of an **NMOS inverter**. Similar circuits can be constructed using p-channel transistors to form PMOS devices.

One of the great attractions of MOS technology is its simplicity. The switching transistors within the circuit (as distinct from those devices that are used as active loads) act as near-ideal switches. Logic levels are equal to the supply rail voltages, giving a large output voltage swing and good noise immunity.

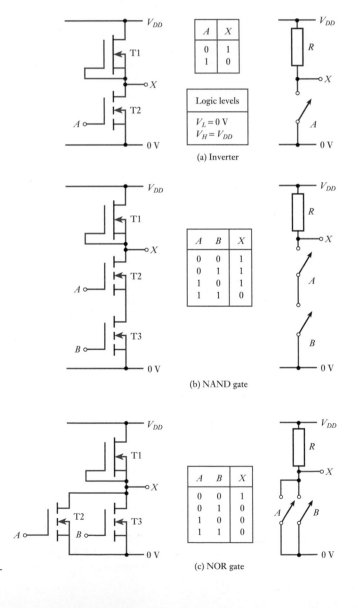

Figure 26.18
NMOS gates.

Operation of the circuit in Figure 26.18(a) is very straightforward. When the input voltage is equal to the positive supply rail (logical 1), transistor T2 is turned on and resembles a closed switch. This pulls the output down to close to 0 V. When the input is taken to 0 V, transistor T2 is turned off and the load transistor T1 pulls the output high, to close to the positive supply V_{DD}.

This circuit technique can be expanded to give other forms of gate. Both **NAND** and **NOR** gates can be formed easily. Other functions can be produced by combining these with the basic inverter circuit. Figures 26.18(b) and 26.18(c) show circuits for two-input NAND and NOR gates. In the NAND gate it is clear that both T2 and T3 must be turned on (*A* AND *B* high) in order for the output to be pulled low. In the NOR gate, the output will be low if either T2 or T3 is turned on (*A* OR *B* high). These circuits can be expanded to produce gates with additional inputs.

When the output of an NMOS gate is at logical 0, the output resistance is very low as the output is shorted to ground by one or more transistors, which are turned on. This enables the circuit to *sink* current efficiently. However, when the output is high, the output resistance is determined by the resistance of the load MOSFET. To enable the circuit to *source* current efficiently in this state, the resistance of this transistor must be low. The power dissipated by the gate is also controlled by the value of the resistance of the load MOSFET. When the output is high, current flows through the load device to the output, but when the output is low, the load transistor is effectively connected directly across the supply rails. In order to minimise power consumption the load MOSFET must have as high a resistance as possible. We therefore have conflicting requirements for the load device. To achieve a low output resistance and thus a good **fan-out**, the load MOSFET must have a low resistance while, to minimise power consumption, this resistance should be as high as possible.

Fortunately, as the input resistance of NMOS gates is so high (generally greater than 10^{12} Ω), it is possible to have a good fan-out (perhaps 50) even with a relatively high output resistance (typically about 100 kΩ), enabling power consumption to be kept to low levels. The power dissipated by the gates is much greater when the output is low than when it is high, with an average value of about 0.1 mW for simple gates. However, the high output resistance, combined with a relatively high input capacitance, does make these devices relatively slow, with a typical **propagation delay time** of about 50 ns.

26.3.7 Complementary metal oxide semiconductor (CMOS) logic

The NMOS logic gates described above suffer from a high output resistance in one of their output states that limits their speed of operation. This problem is common to all amplifiers that use a single-output transistor (remember that the load MOSFET of an NMOS gate is acting as a resistor). One way to eliminate this problem is to use a **push–pull** arrangement (which we looked at earlier in Figure 18.37). For convenience, this circuit is reproduced here as Figure 26.19.

The circuit uses both an *n*-channel and a *p*-channel device and is therefore described as **complementary MOS** (CMOS) logic. As with NMOS circuitry, V_{DD} represents logical 1 and 0 V represents logical 0. Being of different polarities, the two transistors respond in the opposite sense to voltages applied to their gates. While a gate voltage of V_{DD} will turn on the *n*-channel transistor, it will turn off the *p*-channel device. Similarly, a voltage of 0 V will turn off the *n*-channel transistor and turn on the *p*-channel device.

As the gates of the two MOSFETs are joined, input voltages of either logic level will turn one device on and the other off. This arrangement produces a low output

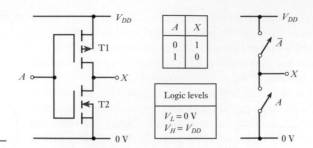

Figure 26.19
A CMOS inverter.

resistance, which can charge load capacitances faster, producing a quicker switching time. The low output resistance also gives a high fan-out of up to about 50 gates.

As one of the two transistors is always turned off, there is no DC path between the supply rails and the only current drawn from the supply is that which is fed to the output. The high input resistance of the gates makes this output current negligibly small, except when the input capacitance of a gate is being charged or discharged after an output has changed. Power is also consumed when the circuit switches from one state to another as, for a short period, both transistors are conducting at the same time. The resultant power consumption is therefore generally negligible when the circuit is static, but increases with the switching rate. Typical values for the power consumption might be a few microwatts per gate when static and about 1 mW when clocked at 1 MHz. It is clear that even when operating at high speeds the gates consume very little power. This makes them ideal for applications in which power consumption is critical – for example, where battery operation is required. Low power dissipation also reduces the amount of waste heat that must be removed, allowing more circuitry to be integrated into a single circuit. We will return to look at power dissipation in more detail in Section 26.7.

The simple inverter in Figure 26.19 can be modified to provide additional logic functions. Examples of two-input NAND and NOR gates are shown in Figure 26.20. Like the inverter, these circuits both provide an active pull-up and active pull-down of the output, giving a low output resistance, and provide no DC path between the supply rails when in either output state.

CMOS circuitry is more difficult to fabricate than NMOS or PMOS as it requires devices of both polarities. However, its increased speed, lower power dissipation and excellent noise immunity make it the dominant technology in modern electronics. We will return to look in more detail at CMOS logic in Section 26.5.

26.3.8 Bipolar CMOS (BiCMOS) logic

While CMOS logic provides a range of excellent characteristics, its load-driving capabilities are not as good as can be achieved with circuitry based on bipolar transistors. For this reason, some gates combine the best features of bipolar and CMOS circuits to produce a hybrid circuit technology called **BiCMOS**. An example of a simple BiCMOS inverter is shown in Figure 26.21.

In the circuit shown in Figure 26.21, transistors T1 and T2 form a CMOS inverter, as in the circuit in Figure 26.19. Transistors T3 and T4 are then used to produce complementary signals to drive the push–pull output stage produced by T5 and T6.

The bipolar transistors produce a much lower output impedance than can be achieved using MOSFETs, making it much better at sourcing and sinking output currents. This results in a circuit that is ideal for high-speed input/output and device driver applications.

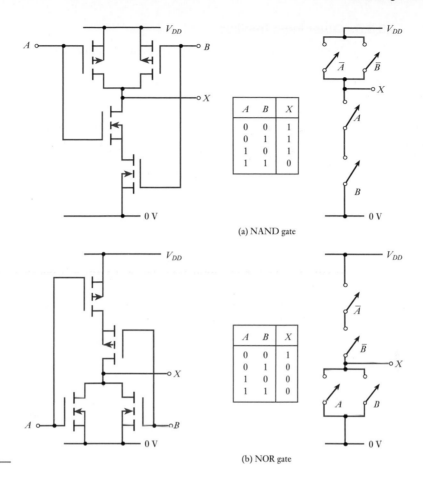

Figure 26.20
Two-input CMOS
gates.

A	B	X
0	0	1
0	1	1
1	0	1
1	1	0

(a) NAND gate

A	B	X
0	0	1
0	1	0
1	0	0
1	1	0

(b) NOR gate

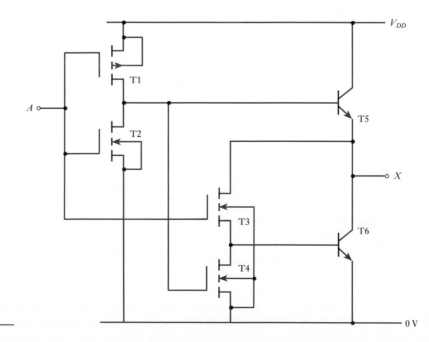

Figure 26.21
A simple BiCMOS
inverter.

26.3.9   Other logic families

While the families listed above represent the most significant technologies, the drive to meet ever more demanding performance requirements forces constant evolution and the production of improved products. In particular, there is a move towards **lower-voltage operation**, which tends to dramatically reduce power consumption.

While most semiconductor devices are constructed using silicon transistors, the introduction of small amounts of germanium into the base region of a silicon bipolar transistor has been found to produce considerable improvements in its performance. The **SiGe transistor** that results is much faster than all-silicon devices and may also have better noise- and power-handling characteristics. Such devices are being used in improved BiCMOS devices and a range of other digital applications.

26.3.10   Logic families – a summary

This section has attempted to outline the characteristics of the most widely used families of logic gates. Several semiconductor logic families have been described and broad comparisons given. It is not within the scope of this text to give detailed descriptions of the operation and characteristics of these technologies, but it is perhaps useful to summarise some of the results in tabular form. Table 26.2 gives a comparison of the three most important logic families in terms of five parameters. Within each of these broad classes a range of variants are available, so the figures attempt to represent the range of the parameters across these series. Consequently, the data should be used with care. Often the values obtained will depend on other factors and the numbers given should be taken simply as a guide for comparison, not as detailed data on a particular device family. The figures given relate to simple logic gates. When used within complex integrated circuits these technologies will often produce circuits that are much faster and consume less power.

Table 26.2 A comparison of logic families.

Parameter	TTL	ECL	CMOS
Basic gate	NAND	OR/NOR	NAND/NOR
Fan-out	10	25	>50
Power per gate (mW)	1–22	4–55	1@1 MHz
Noise immunity	Very good	Good	Excellent
T_{PD} (ns)	1.5–33	1–4	1.5–200

By far the most widely used and important logic family is CMOS and in Section 26.5 we will look in some detail at the characteristics of devices in this family. However, TTL devices are also widely used and, for historical reasons, the characteristics of some modern CMOS gates have been tailored to be compatible with TTL parts. For this reason, we will start by considering the various TTL families of logic gates before turning our attention to CMOS ones.

26.4   TTL

In Section 26.3.4 we looked at the form and operation of a simple TTL gate and traced its development from simpler logic types. Until the mid-1990s, TTL was the most widely used logic family – particularly for applications requiring small- to medium-scale integration (SSI and MSI). More recently, CMOS has taken over this position, although TTL is still used in a range of specialist applications.

A wide range of manufacturers produce TTL circuits and standardisation has been very successful in providing a common specification for such devices. The *standard* TTL family of components contains a broad spectrum of circuits, each specified by a generic serial number starting with the digits 54 or 74. Devices that begin with 54 are specified for operation over a temperature range from −55 to 125 °C, while those starting with 74 are restricted to a range of 0 to 70 °C.

This two–digit prefix is followed by a two– or three–digit code that represents the function of the device. For example, a 7400 contains four two-input NAND gates, while a 7493 is a 4–bit binary counter. These two families are often called the 54XX and 74XX families, the 'XX' implying some combination of two or three digits, or simply the **54/74 families**.

In addition to 'standard' 54XX and 74XX devices, there are other related families with modified characteristics. These are defined by adding alphabetic characters after the 54 or 74 prefix to specify the family. For example, a 74L00 is a low–power version of the 7400 and the 74H00 is a high–speed version. In fact, the 'standard' 74XX parts are effectively obsolete now, having been replaced by other more useful familes, such as the 74LSXX series. We will look at these variants later in this section. Figure 26.22 shows part of a typical TTL data sheet with annotations to indicate important features.

Specification for normal commercial parts

recommended operating conditions

		SN5400			SN7400			UNIT
		MIN	NOM	MAX	MIN	NOM	MAX	
	V_{CC} Supply voltage	4.5	5	5.5	4.75	5	5.25	V
Input voltage limits	V_{IH} High-level input voltage	2			2			V
	V_{IL} Low-level input voltage			0.8			0.8	V
Output current limits	I_{OH} High-level output current			−0.4			0.4	mA
	I_{OL} Low-level output current			16			16	mA
	T_A Operating free-air temperature	−55		125	0		70	°C

electrical characteristics over recommended operating free-air temperature range (unless otherwise noted)

	PARAMETER	TEST CONDITIONS[†]	SN5400			SN7400			UNIT
			MIN	TYP[‡]	MAX	MIN	TYP[‡]	MAX	
	V_{IK}	V_{CC} = MIN, I_I = −12 mA			−1.5			1.5	V
Output voltage limits	V_{OH}	V_{CC} = MIN, V_{IL} = 0.8 V, I_{OH} = −0.4 mA	2.4	3.4		2.4	3.4		V
	V_{OL}	V_{CC} = MIN, V_{IH} = 2 V, I_{OL} = 16 mA		0.2	0.4		0.2	0.4	V
	I_I	V_{CC} = MAX, V_I = 5.5 V			1			1	mA
Input current limits	I_{IH}	V_{CC} = MAX, V_I = 2.4 V			40			40	μA
	I_{IL}	V_{CC} = MAX, V_I = 0.4 V			−1.6			−1.6	mA
	I_{OS}[§]	V_{CC} = MAX	−20		−55	−18		55	mA
	I_{CCH}	V_{CC} = MAX, V_I = 0 V		4	8		4	8	mA
	I_{CCL}	V_{CC} = MAX, V_I = 4.5 V		12	−22		12	22	mA

[†] For conditions shown as MIN or MAX, use the appropriate value specified under recommended operating conditions.
[‡] All typical values are at V_{CC} = 5 V, T_A = 25° C.
[§] Not more than one output should be shorted at a time.

Conditions under which values are measured

switching characteristics V_{CC} = 5 V, T_A = 25°C (see note 2)

	PARAMETER	FROM (INPUT)	TO (OUTPUT)	TEST CONDITIONS		MIN	TYP	MAX	UNIT
Propagation delay times for different transitions	t_{PLH}	A or B	Y	R_L = 400 Ω,	C_L = 15 pF		11	22	n_s
	t_{PHL}						7	15	n_s

NOTE 2: See General Information Section for load circuits and voltage waveforms.

Figure 26.22
Part of a typical TTL data sheet.

The characteristics of the 54XX and 74XX families are similar so, in the following sections, we will refer simply to the 74XX parts as these are the most common. Unless otherwise stated the information given is also relevant for the corresponding 54XX part.

26.4.1 Standard TTL

Integrated circuit TTL gates use slightly more complicated circuitry than that given in Figure 26.15. The circuit of a two-input NAND gate (one of the four gates in a 7400) is shown in Figure 26.23.

The basic operation of this gate is similar to that described in Section 26.3.4 except that a form of **push–pull output** stage has been added to reduce the output resistance, increase the current drive capability and enable the circuit to source as well as sink current. This form of output stage is known as a **totem-pole** output. We will see later that some TTL devices use alternative output circuitry. It must be said that this circuit is primarily designed to *sink* current by providing a path for current to flow into the output terminal to ground when a logical 0 is output. The circuit can also *source* current when a high logic level is output, but this current is very small. This characteristic is acceptable as the gates themselves take very little input current when a high logic signal is applied. Conventional TTL devices operate from a supply voltage of 5.0 V, which must be accurate to within ±0.25 V (±0.5 V for the 54XX family). Typical power dissipation is 10 mW per gate.

If either input A or input B is low, T1 pulls the base of T2 low, turning it off. This causes no current to flow in the 1 kΩ emitter resistor, turning T4 off and allowing the base of T3 to rise, turning it on. The output is thus pulled high, taking a value of V_{CC} less the sum of the voltage drops across the diode D3 and the emitter-to-base junction of T3. Thus, the logical 1 output voltage is given by

$$V_H = V_{CC} - V_{diode} - V_{BE}$$
$$= 5.0 - 0.7 - 0.7$$
$$= 3.6\,\text{V}$$

If both input A and input B are high, T1 pulls the base of T2 high, turning on T2 and T4. This drives the output voltage down to the saturation voltage of T4, thus

$$V_L = V_{CE(sat)} \approx 0.2\,\text{V}$$

The voltage at the collector of T2 falls as a result of the current through its collector resistor, dropping the base of T3 to about 0.9 V (the base-to-emitter voltage of T4 plus

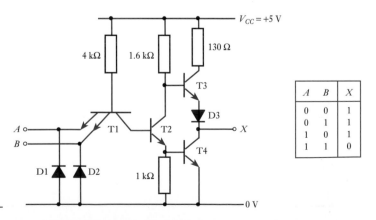

Figure 26.23
A TTL two-input NAND gate.

the saturation voltage of T2). Diode D3 is present to ensure that T3 is turned off under these conditions. If it were absent, the voltage across the base-to-emitter junction of T3 would be equal to the voltage on the base (0.9 V) minus the output voltage (0.2 V), which would be sufficient to turn the device on. The presence of D3 ensures that T3 is held off while the output is low.

Diodes D1 and D2 are **input clamp diodes** that prevent negative-going noise spikes at the input from damaging T1. Negative-going transitions will forward bias the diodes, which therefore prevent the input from going negative by more than the forward voltage of the diode (about 0.7 V).

Transfer characteristic

Figure 26.24 shows the transfer characteristic of the gate – that is, the relationship between the input voltage and the output voltage. The characteristic shows the effect of changing the voltage on one input while the other is held high (if the other input were held low, the output would not change).

The transfer characteristic shows the output logic levels of 3.6 V and 0.2 V derived above and also indicates that the input threshold voltage is 1.4 V. That is, input voltages above 1.4 V will be interpreted as a logical 1, while voltages below 1.4 V will be taken as a logical 0.

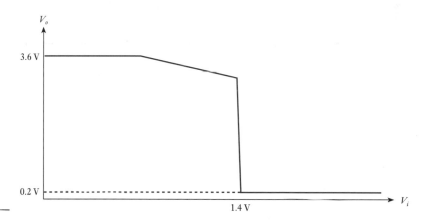

Figure 26.24
TTL transfer
characteristic.

Logic levels and noise immunity

Although the transfer characteristic indicates specific values for the input and output voltages, real devices are subject to variability and the logic levels are specified by bands of voltages. These are shown in Table 26.3.

From Equations 26.1 and 26.2, it is clear that the noise immunities in each of the two logic states are given by

Table 26.3 Input and output voltage levels for 74 family devices.

	Minimum	Typical	Maximum
V_{IL}	–	–	0.8
V_{IH}	2.0	–	–
V_{OL}	–	0.2	0.4
V_{OH}	2.4	3.6	–

$$\text{noise immunity in logical 1 (high) state, } V_{NIH} = V_{OH}(min) - V_{IH}(min)$$

$$= 2.4 - 2.0$$

$$= 0.4 \, \text{V}$$

and

$$\text{noise immunity in logical 0 (low) state, } V_{NIL} = V_{IL}(max) - V_{OL}(max)$$

$$= 0.8 - 0.4$$

$$= 0.4 \, \text{V}$$

Thus, the *minimum* noise immunity of each logic state is 0.4 V. However, taking *typical* values for the logic voltages gives a noise immunity of about 1.0 V for logical 0 and 1.6 V for logical 1.

Input and output currents and fan-out

The input current taken by the gate will clearly be different depending on whether the input is high or low and also varies between the various forms of gate. For the 7400 two-input NAND gate shown in Figure 26.23, the maximum input current for a logical 1 input signal is 40 µA, and for a logical 0 input is −1.6 mA. As currents are conventionally measured *into* a device, the minus sign indicates that this is current flowing *out* of the device.

The specification of the 7400 states that when a logical 1 is output the circuit will source an output current of at least −400 µA, and when a logical 0 is output it will sink at least 16 mA. Again, the negative sign indicates that this current is flowing out of the device. This gives a fan-out of 10.

Switching characteristics

The switching characteristics can be described by the **propagation delay times** for transitions from high to low (t_{PHL}) and from low to high (t_{PLH}). These are shown in Table 26.4 for a typical single gate.

Table 26.4 Propagation delay times for typical 74 family gates.

	Minimum	Typical	Maximum
t_{PHL} (ns)	–	7	15
t_{PLH} (ns)	–	11	22

26.4.2 Open-collector devices

Some 74 family devices use a different output configuration that is referred to as an **open-collector** output stage. The 7401, for example, contains four two-input NAND gates with open-collector outputs. Figure 26.25 shows the circuit of one of these gates.

It is clear that this gate is similar to that of Figure 26.23, with the exception that the output stage has been simplified. The output is taken from the collector of an output transistor that is otherwise unconnected or *open*. In order for the circuit to function, an external **pull-up resistor** must be connected between the output and the positive supply, as shown in Figure 26.26.

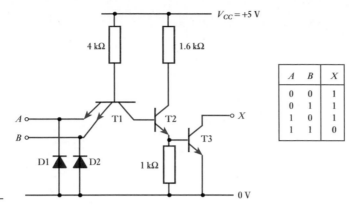

Figure 26.25
The 7401 two-input NAND gate with open-collector output.

A	B	X
0	0	1
0	1	1
1	0	1
1	1	0

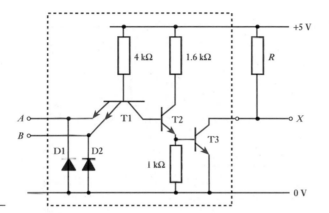

Figure 26.26
Use of an open-collector gate with an external load.

The choice of a value for the pull-up resistor is a compromise between power dissipation and speed of operation. High-value resistors reduce the collector current and, therefore, the power, but also limit the rate at which load capacitances can be charged. Even with relatively low values of resistance, the open-collector circuit is not as fast as the totem-pole arrangement. This is because the latter has an additional output transistor acting as a low-impedance emitter follower, which is able to charge load capacitances quickly.

The output logic levels of an open-collector gate are not the same as those for a totem-pole arrangement. When the output transistor is turned on the output voltage is pulled down to the saturation voltage of the device (about 0.1–0.2 V), but when the transistor is turned off the output voltage rises to the value of the supply attached to the pull-up resistor – normally V_{CC}. Thus, the logic levels are close to the supply rail voltages. These voltages are completely compatible with the inputs of TTL gates, so this difference in logic levels is not usually significant.

Wired-AND operation

One of the advantages of open-collector gates is that their outputs can be connected in parallel to form a **wired-AND** configuration. This is illustrated in Figure 26.27.

Figure 26.27 shows a circuit containing four NAND gates, the outputs of which must be ANDed together to produce the function X. Using conventional totem-pole output devices, the outputs must be combined using a four-input AND gate as shown in Figure 26.27(a). However, if open-collector gates are used, the outputs can simply

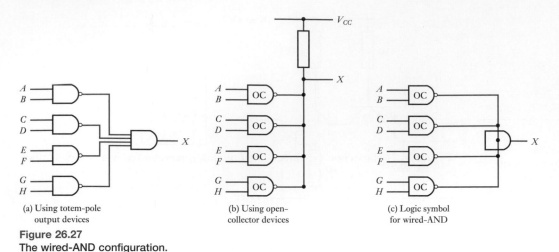

(a) Using totem-pole
output devices

(b) Using open-
collector devices

(c) Logic symbol
for wired-AND

Figure 26.27
The wired-AND configuration.

be connected in parallel to a single resistor to achieve the same result, as shown in Figure 26.27(b). The AND function is obtained as each gate can pull the output low, but all must be high for X to be high. The circuit symbol for the wired-AND operation is shown in Figure 26.27(c).

The wired-AND function is of particular interest when large numbers of signals must be combined as this removes the need for gates with a large number of inputs. It is also of great importance in the production of a **bus** in which a number of devices are connected to a single line. We will consider bus systems in more detail when we look at microprocessor systems (see Chapter 27).

As AND and OR functions may be interchanged (with appropriate signal inversions) by means of **De Morgan's theorem**, it is also possible to use the wired-AND function to produce an OR operation. For this reason, the technique is sometimes referred to as a **wired-OR** configuration.

High-voltage outputs

Some open-collector devices can be used with high output voltages. The device is operated from a standard 5.0 V supply, but the output is connected through a pull-up resistor to a high-voltage supply. The 7406, for example, contains six inverters with high-voltage, open-collector outputs. These can switch up to 30 V at currents of up to 40 mA. The output logic levels are equal to the saturation voltage of the output transistor (about 0.1 to 0.2 V) and the high-level supply voltage.

26.4.3 Three-state devices

Conventional logic gates have two possible output states, namely 0 and 1. In some circumstances, it is convenient to have a third state corresponding to a high-impedance condition, when the output is allowed to float. In such cases the voltage at the output will be determined by whatever external circuitry is connected to it. Circuits with this property are called **three-state logic gates**. The output of the gate is 'enabled' or 'disabled' by a control input, which is usually given the symbol C on simple gates. In more complicated circuits this control signal is often referred to as the **output enable** line.

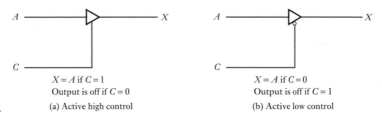

$X = A$ if $C = 1$
Output is off if $C = 0$

(a) Active high control

$X = A$ if $C = 0$
Output is off if $C = 1$

(b) Active low control

Figure 26.28
Symbolic
representation of
three-state logic gates.

Figure 26.28 shows how the three-state function is represented in a circuit symbol. Figure 26.28(a) shows a non-inverting buffer with an active high control input (that is, the output is enabled if $C = 1$). Figure 26.28(b) shows the symbol for a similar gate with an active low control input (the output is enabled if $C = 0$). The first of these could represent one of the gates in a 74126 and the second a gate in a 74125. Both devices contain six such gates.

The output circuit of a three-state gate resembles that of a totem-pole device with the addition of extra components to turn both output transistors off to disable the output. This allows the output to float, independently of the other gate inputs. As, when enabled, the output resembles the conventional totem-pole arrangement, the use of three-state techniques does not incur the speed penalty associated with open-collector circuits.

Three-state devices can be used in the creation of bus systems, in which the outputs of several devices are connected together onto a common line. Each device can then place data on the line provided that the output of only one device is enabled at any time. This arrangement differs from the use of open-collector gates in the wired-AND configuration described earlier. When three-state devices are used, only one gate drives the line at any time and the outputs of disabled gates do not affect the signal on the bus.

Three-state devices are sometimes described as being **tri-state**. This term is a trademark of the National Semiconductor Corporation.

26.4.4 TTL inputs

In certain instances a circuit may not need to use all the inputs of a particular gate. It may be convenient, for example, to OR together three signals using a four-input OR gate because such a device is already available within the circuit. When using TTL gates, if **unused inputs** are left disconnected they will act as if they were connected to a logical 1. Such inputs are said to be **floating**.

Although unused inputs will 'float' to a logical 1, it is inadvisable to leave such inputs disconnected, even if the application requires the input to be at logical 1. Unconnected inputs represent a high impedance to ground, making them very sensitive to electrical noise, which could cause them to switch between states. It is much wiser to 'tie' such inputs high or low, as required.

Inputs that are required to be at logical 1 should not be tied directly to the positive supply rail, but, rather, through a resistor. A typical value for such a resistor might be $1\ \text{k}\Omega$. If appropriate, several inputs can be connected together to the same resistor. Inputs that are required to be low may be tied directly to ground ($0\,\text{V}$). In some circumstances, it may be appropriate to connect an input to ground through a resistance (perhaps to allow it to be pulled high through a switch to logical 1). In this case, the resistor value must be sufficiently small to allow the input current to flow through the resistor without taking the input voltage above the maximum level for a logical 0 input ($V_{IL}(max)$). For standard TTL, a typical maximum value for such a resistor might be $470\ \Omega$.

It is worth noting that when using AND or NAND gates any unused inputs should be tied *high* to prevent them from affecting the output state. When using OR or NOR gates any unused inputs should be tied *low*.

26.4.5 Other TTL families

So far, we have concentrated on the 'standard' 74 TTL family. In fact, these parts are now obsolete, having been replaced by a number of more advanced TTL familes that are optimised for particular operating characteristics.

Low-power TTL (74L)

The 74L family of devices is optimised for low power consumption. Figure 26.29 shows the circuit diagram of a typical gate, the 74L00; a comparison with Figure 26.23 will show that this power reduction is achieved primarily by a change of resistor values.

The average power dissipation of the low-power gate is 1 mW compared with 10 mW for a standard device, but this is achieved at the expense of a reduction in speed. The average propagation delay is increased from 9 ns for a standard gate to 33 ns for the low-power version.

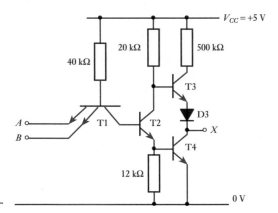

Figure 26.29
A 74L00 two-input
NAND gate.

High-speed TTL (74H)

The 74H family is optimised for speed. Figure 26.30 shows the circuit of the 74H00 which has an average propagation delay time of only 6 ns but an average power dissipation of 22 mW.

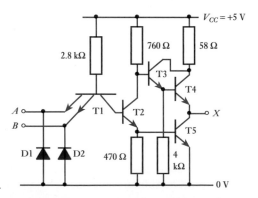

Figure 26.30
A 74H00 two-input
NAND gate.

Schottky TTL (74S)

The 74S family uses circuits similar to those of the 74H devices, except that **Schottky** transistors and diodes are used in place of conventional components.

Schottky diodes are formed by the junction of a metal and a semiconductor, unlike more traditional diodes which consist of a junction between two regions of doped semiconductor (see Section 17.7.2). The circuit symbol for a Schottky diode is shown in Figure 26.31(a).

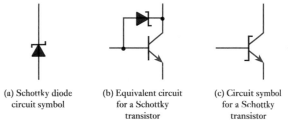

Figure 26.31
Schottky diodes and transistors.

(a) Schottky diode circuit symbol

(b) Equivalent circuit for a Schottky transistor

(c) Circuit symbol for a Schottky transistor

Schottky diodes not only are very fast in operation but also have a forward voltage drop of only about 0.25 V. This allows them to be used to prevent the saturation of a transistor, as shown in Figure 26.31(b). When the transistor is operating well within its active region, the collector is positive in relation to the base and the diode is reverse biased. Under these conditions the diode has no effect on the operation of the transistor. However, as the device nears its saturation region the voltage on the collector drops below that of the base and, in the absence of the diode, would ultimately fall to its saturation value of about 0.1–0.2 V above the emitter voltage. The presence of the Schottky diode across the collector–base junction prevents the transistor from entering saturation as it becomes forward biased, and therefore begins to conduct before the device saturates. Once conducting, the diode robs the transistor of current, inhibiting any further drop in the collector voltage and thus preventing the device from entering its saturation region. The combination of transistor and diode is referred to as a **Schottky transistor**, which has its own circuit symbol, as shown in Figure 26.31(c).

We saw in Section 26.2.4 that saturation of transistors causes a considerable increase in propagation delay because of the presence of **storage time**. The use of Schottky transistors prevents saturation and greatly increases the speed of operation while incurring only a modest increase in power consumption. Consequently, 74S family gates have a propagation delay of about half that of 74H devices with approximately the same power consumption. For this reason, the Schottky devices have largely replaced the older high-speed family.

Figure 26.32 shows an example of a typical device, a 74S00 two-input NAND gate. This has a typical propagation delay of 3 ns and an average power dissipation per gate of about 19 mW. Note that one of the transistors, T4, is not a Schottky type because it does not saturate during normal operation of the circuit.

Advanced Schottky TTL (74AS)

Following on from the original Schottky series of TTL components, several other families also use Schottky transistors. These include the advanced Schottky TTL family, which surpasses the 74S family both in speed and power consumption. These components have a typical propagation delay of about 1.5 ns and a typical power dissipation of 8.5 mW.

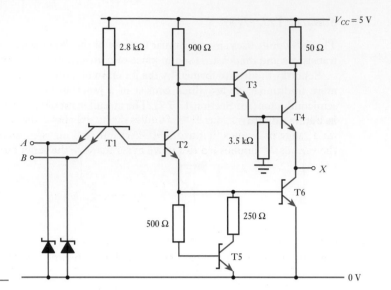

Figure 26.32
A 74S00 two-input
NAND gate.

Low-power Schottky TTL (74LS)

The 74LS low-power Schottky family combines considerations of speed and power consumption. Its typical propagation delay of 9.5 ns is approximately equal to that of standard TTL but, at 2 mW average dissipation per gate, it consumes only one-fifth of the power. Figure 26.33 shows a 74LS00 two-input NAND gate.

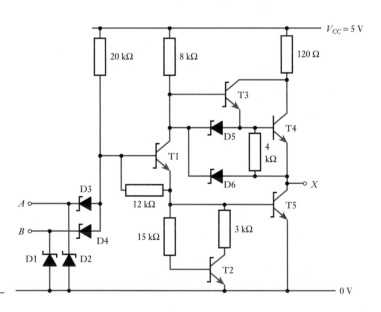

Figure 26.33
A 74LS00 two-input
NAND gate.

Advanced low-power Schottky TTL (74ALS)

A range of high-performance, low-power Schottky devices are available as the 74ALS family. These have a typical propagation delay of 4 ns and a power consumption of 1 mW per gate.

FAST TTL (74F)

The 'FAST' range of TTL logic provides high speed with low power consumption. Typical propagation delay is 2.7 ns and typical power consumption about 4 mW per gate.

The 74 series CMOS families

In addition to the various TTL families discussed above, the '74' notation is also used for a range of familes that use **CMOS** circuitry. This is because the great success of the various 74 families led to a move to standardise the serial numbers given to *all* the various logic devices and led to families of CMOS devices with the same logical functions and similar serial numbers.

Examples of CMOS families include the 74C, 74HC, 74HCT, 74AHC, 74AHCT, 74FCT, 74AC, 74ACT, 74FACT, 74ACQ and 74ACTQ series, where the 'C' in each name stands for CMOS. In addition, there are also several other logic families that are designed for **low-voltage operation**. These also use CMOS circuitry, but often do not have 'C' within their part numbers. These include the 74LV, 74ALV, 74LVCH, 74ALVC, 74LVT, 74LVTZ, 74LCX, 74VCX, 74ALB and 74CBTLV families.

We will leave further discussion of CMOS gates until the next section.

26.4.6 TTL families – a summary

There are myriad TTL families, all with different characteristics, advantages and disadvantages. However, of these only a few families are widely used. The 'LS' family is used for general applications and 'AS' or 'ALS' devices where speed is critical. Increasingly, the varous CMOS ranges are replacing TTL components in all applications, particularly where power consumption is of importance.

Table 26.5 gives a brief comparison between the various TTL families in terms of propagation delay and power consumption. Figures are typical values for standard gates (usually two-input NAND gates) and will vary for other circuits.

Table 26.5 A comparison of TTL logic families.

Family	Descriptor	T_{PD} (ns)	Power per gate (mW)
Standard	74XX	9	10
Low-power	74LXX	33	1
High-speed	74HXX	6	22
Schottky	74SXX	3	19
Advanced Schottky	74ASXX	1.5	8.5
Low-power Schottky	74LSXX	9.5	2
Advanced low-power Schottky	74ALSXX	4	1
FAST	74FXX	2.7	4

Video 26D

26.5 CMOS

In Section 26.3.7 we looked at the basic form of CMOS gates and the circuits of inverter, AND and OR gates. In this section we will look in a little more detail at the characteristics and use of this type of logic.

The first manufacturer to produce CMOS logic was RCA, who described them as the **4000 series**, having numbers 4000, 4001 and so on. These were later replaced by the enhanced 4000B series of components. Some manufacturers adopted the same numbering system, while others devised their own related numbering schemes. Motorola, for example, produced components in the MC14000 and MC14500 series.

Many manufacturers moved away from the original 4000 series parts and produced a range of circuits that follow the circuit functions and pin assignments of the **74XX TTL family** of devices (as discussed in Section 26.4). These are given part numbers, such as 74CXX, 74ACXX and 74HCTXX, where in each case the 'C' stands for CMOS. The 'XX' signifies a two- or three-digit code that represents the function of the device. For example, a 74AC00 contains four two-input NAND gates, while a 74AC163 is a 4–bit synchronous binary counter. Parts with 'A' in their names are 'advanced' devices with improved speed and lower power consumption. 'T' indicates that the devices are unlike those of the other CMOS families in that they are designed to operate with the supply voltages and logic levels of TTL gates. This enables them to be used easily with TTL components, often allowing them to act as direct, low-power replacements for the corresponding TTL parts. Conventional CMOS circuits of the 4000 series or the 74CXX types cannot normally be used directly with TTL parts as their logic levels are different.

We will look in more detail at CMOS logic families in Section 26.5.4.

26.5.1 CMOS characteristics

CMOS gates differ in many respects from the TTL gates described earlier. In this section we will look at the general characteristics of CMOS logic. Figure 26.34 shows part of a typical CMOS data sheet that is annotated to indicate items of importance.

Power supply voltages

Many CMOS gates, such as the 4000B and 74C series devices, are designed to operate using a single supply voltage of 3 to 18 V and, when using such devices, common supply voltages are 5, 10 and 15 V. However, most of the more recent families are designed to operate over a smaller voltage range and many require a supply voltage between 4.5 and 5.5 V (though several will operate over a range of 2 to 6 V). These components are very often used with a supply voltage of 5 V, as for TTL. The speed of operation increases with the supply voltage, as does the power dissipation.

As power dissipation tends to fall with the operating voltage, several CMOS families, such as the 74LV and 74ALVC series, are optimised for low-voltage operation and use a supply voltage below 3.6 V. With such devices, it is common to use a supply voltage of 3.3 V. Other devices, such as the 74VCX series, may be used with a supply voltage of 1.8 V or less.

Logic levels and noise immunity

When considering the logic levels of CMOS gates we need to differentiate between 'conventional' gates and those designed to accept TTL-level input signals. Those in the latter group are designed for easy interfacing to TTL gates and may be recognised by a 'T' within their part numbers – for example, the 74HCTXX and 74ACT families.

The output logic levels of *conventional* CMOS gates are very close to the supply rails and can normally be assumed to be equal to 0 (V_{SS}) and the positive supply voltage (V_{DD}). Therefore, for most purposes, it is reasonable to assume that

$$V_{OL}(max) = 0$$

Maximum ratings before the chip is damaged

absolute maximum ratings over operating free-air temperature range[†]

Supply voltage, V_{CC}	−0.5 V to 7 V
Input clamp current, I_{IK} ($V_I < 0$ or $V_I > V_{CC}$)	±20 mA
Output clamp current, I_{OK} ($V_O < 0$ or $V_O > V_{CC}$)	±20 mA
Continuous output current, I_O ($V_O = 0$, to V_{CC})	±25 mA
Continuous current through V_{CC} or GND pins	±50 mA
Lead temperature 1.6 mm (1/16 in) from case for 60 s: FK or J package	300°C
Lead temperature 1.6 mm (1/16 in) from case for 10 s: D or N package	260°C
Storage temperature range	−65°C to 150°C

Specification for normal commercial parts

[†] Stresses beyond those listed under 'absolute maximum ratings' may cause permanent damage to the device. These are stress ratings only, and functional operation of the device at these or any other conditions beyond those indicated under 'recommended operating conditions' is not implied. Exposure to absolute-maximum-rated conditions for extended periods may affect device reliability.

recommended operating conditions

		SN54HCOO			SN74HCOO			UNIT
		MIN	NOM	MAX	MIN	NOM	MAX	
V_{CC} Supply voltage		2	5	6	2	5	6	V
V_{IH} High-level input voltage	$V_{CC} = 2$ V	1.5			1.5			V
	$V_{CC} = 4.5$ V	3.15			3.15			
	$V_{CC} = 6$ V	4.2			4.2			
V_{IL} Low-level input voltage	$V_{CC} = 2$ V	0		0.3	0		0.3	V
	$V_{CC} = 4.5$ V	0		0.9	0		0.9	
	$V_{CC} = 6$ V	0		1.2	0		1.2	
V_I Input voltage		0		V_{CC}	0		V_{CC}	V
V_O Output voltage		0		V_{CC}	0		V_{CC}	V
t_t Input transition (rise and fall) times	$V_{CC} = 2$ V	0		1000	0		1000	ns
	$V_{CC} = 4.5$ V	0		500	0		500	
	$V_{CC} = 6$ V	0		400	0		400	
T_A Operating free-air temperature		−55		125	−40		85	°C

Limits of input logic levels

Conditions under which values are measured

electrical characteristics over recommended operating free-air temperature range (unless otherwise noted)

PARAMETER	TEST CONDITIONS	V_{CC}	$T_A - 25°C$			SN54HCOO		SN74HCOO		UNIT
			MIN	TYP	MAX	MIN	MAX	MIN	MAX	
V_{OH}	$V_I = V_{IH}$ or V_{IL}. $I_{OH} = -20$ μA	2 V	1.9	1.998		1.9		1.9		V
		4.5 V	4.4	4.499		4.4		4.4		
		6 V	5.9	5.999		5.9		5.9		
	$V_I = V_{IH}$ or V_{IL}. $I_{OH} = -4$ mA	4.5 V	3.98	4.30		3.7		3.84		
	$V_I = V_{IH}$ or V_{IL}. $I_{OH} = -5.2$ mA	6 V	5.48	5.80		5.2		5.34		
V_{OL}	$V_I = V_{IH}$ or V_{IL}. $I_{OL} = -20$ μA	2 V		0.002	0.1		0.1		0.1	V
		4.5 V		0.001	0.1		0.1		0.1	
		6 V		0.001	0.1		0.1		0.1	
	$V_I = V_{IH}$ or V_{IL}. $I_{OL} = 4$ mA	4.5 V		0.17	0.26		0.4		0.33	
	$V_I = V_{IH}$ or V_{IL}. $I_{OL} = 5.2$ mA	6 V		0.15	0.26		0.4		0.33	
I_I	$V_I = V_{CC}$ or O	6 V		±0.1	±100		±1000		±1000	nA
I_{CC}	$V_I = V_{CC}$ or O. $I_O = 0$	6 V			2		40		20	μA
C_i		2 to 6 V		3	10		10		10	PF

Limits of output logic levels

Maximum quiescent supply current

Figure 26.34
Part of a typical CMOS data sheet.

and

$$V_{OH}(min) = V_{DD}$$

The input logic levels also change with the supply voltage and are defined as

$$V_{IL}(max) = 0.3 \times V_{DD}$$

and

$$V_{IH}(min) = 0.7 \times V_{DD}$$

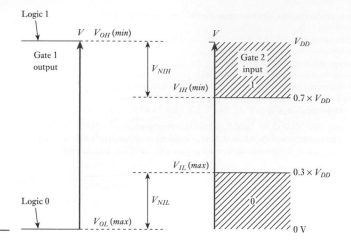

Figure 26.35
CMOS input and
output logic levels.

These definitions are illustrated in Figure 26.35.

From Figure 26.35 and Equations 26.1 and 26.2 it is clear that the noise immunities in each of the two logic states are given by

$$\text{noise immunity in logical 1 (high) state, } V_{NIH} = V_{OH}(min) - V_{IH}(min)$$

$$= V_{DD} - 0.7 \times V_{DD}$$

$$= 0.3 \times V_{DD}$$

and

$$\text{noise immunity in logical 0 (low) state, } V_{NIL} = V_{IL}(max) - V_{OL}(max)$$

$$= 0.3 \times V_{DD} - 0$$

$$= 0.3 \times V_{DD}$$

Thus, the minimum noise immunity in each state is equal to 30 per cent of the supply voltage.

When used with a supply voltage of 5 V (as for TTL), this gives a noise immunity of 1.5 V, which compares very favourably with the figure of 0.4 V for TTL devices. Unfortunately, high noise immunity is not the only criterion for determining the susceptibility of a system to noise. CMOS gates have an output impedance between 3 and 10 times greater than TTL, which increases their sensitivity to capacitively coupled noise. However, CMOS is generally accepted to be one of the most noise-tolerant technologies, provided that appropriate design and layout rules are followed.

It might at first sight seem strange that the input logic levels are not defined such that any voltage greater than 50 per cent of V_{DD} is interpreted as a '1', while any voltage less than this value is interpreted as a '0'. This would produce a system with a noise immunity of 50 per cent of the supply voltage. In practice, this is not possible because of variations in the threshold voltages within the device. The values specified allow for this variability and assure correct operation. In practice, the noise immunity will tend to be greater than the minimum value calculated above, a typical value being 45 per cent of V_{DD}.

Comparing the input and output levels of a conventional CMOS gate with those of the TTL gates discussed earlier shows that they are not directly compatible. When a CMOS gate is used with a supply voltage of 5 V, its output logic levels of approximately

0 V and 5 V will correctly drive the input of a TTL gate (assuming it has the appropriate current driving capability), but the output logic levels of a TTL gate will not correctly drive the input of a CMOS gate.

This can be seen by comparing the output logic levels of a TTL gate shown in Table 26.3 with the input logic levels of a CMOS gate as shown in Figure 26.35. When V_{DD} is equal to 5 V, the CMOS gate requires an input voltage of at least $0.7 \times 5 = 3.6$ V to represent a logical 1, while the TTL gate might produce an output of as little as 2.4 V.

To overcome this difficulty, some CMOS logic gates are designed to have **TTL-compatible inputs**. Devices of this type include the 74HCT and 74ACT families (where the 'T' in each case stands for TTL-compatible), which use modified circuitry to produce a $V_{IH(min)}$ of 2.0 V and a $V_{IL(max)}$ of 0.8 V, as in a TTL device. The resulting components are slightly slower than conventional CMOS gates of a similar form and have a lower noise immunity, but benefit from being able to drive TTL circuits directly.

Power dissipation

One of the most significant characteristics of CMOS logic is its very low power consumption. As was observed in Section 26.3.7, the quiescent power consumption (that is, the consumption when the circuit is static) is extremely low, typically a few microwatts for any supply voltage. However, each time the device changes state, a small amount of power is used to charge capacitances within the circuit and in the load. Some power is also consumed when, for a brief period of time, both halves of the complementary pairs of transistors are on at the same time.

As a small amount of power is dissipated each time the gate changes state, the power consumption of the device increases steadily with the clock rate and also increases with the supply voltage.

For a supply voltage of 5 V, a typical CMOS gate consumes a few microwatts per gate at 1 kHz, but this increases to nearly 1 mW at 1 MHz. At frequencies above 10 MHz, the power consumption is greater than that of 74LSXX TTL gates. With a supply voltage of 15 V, the power consumption is about 10 mW per gate at 1 MHz. We will return to look at power dissipation in more detail in Section 26.7.

Propagation delay

The early 4000 series CMOS logic gates are generally slower in operation than all forms of TTL gate. Because of their relatively high output impedance, the propagation delay time of CMOS devices is greatly affected by the number of gates connected to their output. Their speed is also related to the supply voltage, higher voltages giving a faster response. A typical 4000 series gate operating with a 5 V supply might have a propagation delay of between 50 and 200 ns, depending on the number of gates connected to its output. A similar gate operating from a 15 V supply might have a delay of 20–60 ns.

With technological advancement, the speed of operation of CMOS logic has increased considerably. The 74AC (Advanced CMOS) and 74ACT (Advanced CMOS with TTL pin-outs) families have propagation delay times of only a few nanoseconds, making them comparable with the FAST, AS and S TTL families discussed in the last section. This increase in speed has been achieved without sacrificing the benefit of the very low power consumption of CMOS.

26.5.2 CMOS inputs

A CMOS input looks, to the outside world, like a small capacitor of the order of 1 pF. Because of their very high input impedance, such inputs are very sensitive to **static**

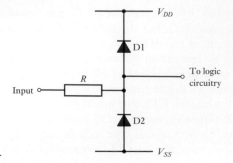

Figure 26.36
CMOS gate protection
circuitry.

electricity, which could easily destroy the device. To reduce these problems, most devices contain gate protection diodes, as shown in Figure 26.36.

The diodes act as **clamp diodes**, preventing the voltage applied to the logic circuitry from going above, or below, safe levels. If the input goes positive with respect to V_{DD}, this will tend to forward-bias D1, which will clamp the voltage applied to the logic circuit to $V_{DD} + V_{diode}$. The resistor R simply limits the current passed through the diode. Similarly, if the input goes below VSS, it will be clamped by diode D2 to $V_{SS} - V_{diode}$, protecting the logic circuitry from damage.

Even with protection circuitry, CMOS gates are very susceptible to static charges – particularly before they are assembled into circuits. Normal precautions include storing these devices in conductive enclosures (for example, electrically conductive plastic tubes) and minimising handling.

Unused inputs

Unused CMOS gates must not be left unconnected. Inputs that are not pulled up or down may cause problems for a number of reasons. First, such inputs are prone to damage caused by static electricity, as described earlier, which could destroy the device. Second, unconnected inputs tend to float midway between the supply rails, which means that they are liable to move above and below the threshold voltage, giving rise to unpredictable behaviour. Third, if inputs are not tied to either logical 1 or logical 0, the corresponding MOSFETs are not switched hard on or off and the current dissipation of the device increases considerably.

All unused inputs must be tied high or low or joined to other inputs. Choosing whether to tie an unused input high or low depends on the function of the gate. As with TTL, unused AND and NAND inputs should be tied *high*, whereas unused OR and NOR inputs should be tied *low*. Unlike in TTL, unused inputs may be connected directly to 0 V or V_{DD} as required.

26.5.3 CMOS outputs

CMOS gates have a typical output resistance of about 250 Ω for a V_{DD} of 5 V. As the input resistance of the gates is so high, a large number of devices can be driven from one output, the main restriction to the **fan-out** being that the propagation delay increases with the number of gates driven (as described earlier). If high-speed operation is not required, at least 50 gates can be driven from a single output.

There is no CMOS equivalent of the TTL open-collector output. However, some CMOS gates have a three-state facility that operates in exactly the same way as in TTL circuits (see Section 26.4.3).

26.5.4 CMOS families

As with TTL, there are many CMOS families, each with different characteristics and strengths. Here, we will look briefly at just a few of the many versions available, but it should be noted that the characteristics of the various families vary considerably from one manufacturer to another.

Standard CMOS (4000B)

This is one of the oldest forms of CMOS gates, but the various devices are still used in certain applications.

These devices will operate with a supply voltage of 3–18 V and are normally used with supplies of 5, 10 or 15 V. When used with a 15 V supply, they produce a very high noise immunity that is typically greater than 5 V, making them suitable for use in very noisy environments. However, with a propagation delay time of typically 45–125 ns (depending on the supply voltage), they are very slow by modern standards.

Standard CMOS with TTL pin-out (74C)

The 74C family of devices adopts a device numbering scheme compatible with corresponding TTL parts. For example, a 74C00 represents a CMOS quad two–input NAND gate (with functionality equivalent to the 7400 TTL gate). As with the 4000B series parts, these devices operate with a supply voltage of 3–18 V and are normally used with supplies of 5, 10 or 15 V. Again, when used with a high supply voltage they produce a high noise immunity, which is typically $0.45 \times V_{DD}$, making them suitable for use in very noisy environments. The 74C parts are somewhat faster than 4000B devices, but, with a typical propagation delay of 30–50 ns (depending on the supply voltage), they are still relatively slow.

High-speed CMOS (74HC)

The 74HC family has an operating voltage range of 2–6 V and the various devices on offer are often used with a 5 V supply. They benefit from a significant increase in speed (with a typical T_{PD} of about 8 ns) and a reduction in power consumption, when compared with standard parts.

High-speed CMOS, TTL-compatible inputs (74HCT)

Another high-speed CMOS family is the 74HCT series, which offers TTL-compatible inputs. These parts operate from a 4.5–5.5 V supply and can be used as direct, low-power replacements for corresponding TTL parts, offering similar speed performance to 74LS parts.

Advanced CMOS (74AC)

This family represents a significant speed improvement over the high–speed CMOS family, with similar power consumption. The supply voltage range is 2–6 V.

Advanced CMOS, TTL-compatible inputs (74ACT)

The 74ACT family offers similar performance to advanced CMOS parts, but has TTL-compatible inputs and a supply voltage range of 4.5–5.5 V.

Low-voltage CMOS (74LV)

This family of devices can be used with supply voltages of 2.0–5.5 V, permitting low-voltage operation where necessary.

Advanced, low-voltage CMOS (74ALVC)

This advanced low-power family can be used with a supply voltage between 1.65 and 3.6 V and provides a considerable speed increase in comparison with the 74LV series. T_{PD} might typically be 3 ns when used with a supply of 3.0 V, increasing to about 4.4 ns when using a supply of 1.8 V.

BiCMOS (74BCT)

The 74BCT family of devices uses BiCMOS technology (as discussed in Section 26.3.8) to provide a range of high-speed input/output and device driver circuits. The devices concerned are often buffers or line drivers rather than simple gates, so direct comparison with the other logic families is difficult. The circuits are normally used with a supply voltage of 5 V, but will accept supply voltages in the range of 4.5–5.5 V. The delay time might typically be 3 to 4 ns.

Low-voltage BiCMOS (74LVT)

This family offers high-speed BiCMOS operation using supply voltages between 2.7 and 3.6 V. Again, these components are normally used for specialist applications where high speed must be combined with the ability to supply quite large currents.

26.5.5 CMOS families – a summary

The operating speed and power consumption of CMOS gates vary tremendously with the supply voltage used and the load capacitance. For this reason, it is very difficult to perform a meaningful comparison of the characteristics of different logic families.

Table 26.6 gives an overview of the speed and power consumption of the various families outlined above and attempts to describe their relative performance. However, the data should be used with care as the characteristics of the various components will be greatly affected by the way in which they are used. Note also that, as mentioned above, the power consumed by a CMOS gate is largely determined by the switching

Table 26.6 A comparison of CMOS logic families.

Family	Descriptor	T_{PD} (ns)	Typical static power per gate (μW)
Standard	4000B	75	50
Standard, TTL pin-out	74CXX	50	50
High-speed	74HCXX	8	25
High-speed, TTL-compatible	74HCTXX	12	25
Advanced	74ACXX	4	25
Advanced, TTL-compatible	74ACTXX	6	25
Low-voltage	74LVXX	9	50
Advanced, low-voltage	74ALVCXX	3	50
BiCMOS	74BCTXX	3.5	600
Low-voltage BiCMOS	74LVTXX	4	400

rate, so will often be much greater than the value given in the table, which relates to static operation.

26.5.6 Implementing complex gates in CMOS

In Section 26.3.7 we looked at a simple CMOS inverter and, for convenience, this circuit is reproduced as Figure 26.37(a). As we noted earlier, the circuit consists of two MOSFET switches – one NMOS and one PMOS. The gate inputs of the two devices are connected together and, because of their different polarities, this means that, for either polarity of input signal, one of the devices will be turned on and the other off.

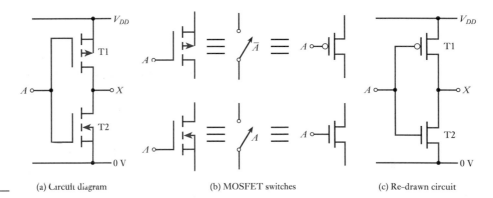

Figure 26.37
A CMOS inverter.

(a) Circuit diagram (b) MOSFET switches (c) Re-drawn circuit

The two switches within the inverter are shown separately in Figure 26.37(b) together with a symbolic representation of the devices as simple switches. In each case the switch is labelled with the logic function for which the switch will be *closed*. For example, the lower of the two switches is labelled A, indicating that when A is true (in other words, when $A = 1$) the switch will be closed. This follows since the lower MOSFET is an *n*-channel device, so when the gate input is *high* (corresponding to the input A being equal to 1) the device will be switched on and will resemble a closed switch. The upper MOSFET is a *p*-channel device. This is turned on when the input to the gate is *low* (corresponding to the input A being 0), so the corresponding switch is labelled $\overline{A}$. When considering complex CMOS gates it is often easier to use slightly simpler symbols for the MOSFET devices and these are also shown in Figure 26.37(b). As in other digital logic symbols, the circle in the symbol for the *p*-channel device indicates a form of *inversion*. Figure 26.37(c) shows the circuit of an inverter using these symbols.

The inverter of Figure 26.37(c) works by 'pulling' the output to the required voltage rail. In Figure 26.38(a) the elements responsible for pulling the output in each direction are highlighted by the two boxes. When the input signal (A) is high, the upper device is turned off and the lower device is turned on, so the output voltage is pulled *down* to 0 V by T2. When the input signal is low, the upper device is turned on and the lower device is turned off, so the output voltage is pulled *up* to V_{DD} by T1. T1 is therefore responsible for pulling *up* the output in response to an appropriate input and T2 is responsible for pulling it *down*.

While the inverter uses a single device to pull the output up and another to pull it down, more complex functions can be implemented by using more sophisticated pull-up and pull-down networks as shown in Figure 26.38(b). In this arrangement the pull-up network would be implemented using *p*-channel MOSFETs and the pull-down network would be implemented using *n*-channel devices, as in the inverter. The

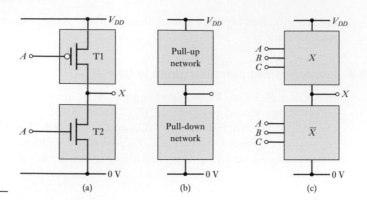

Figure 26.38
CMOS gate structure.

functionality of the two elements must be complementary, such that when one element is turned *on* (that is, it resembles a closed switch) the other will be turned *off* (representing an open switch). Figure 26.38(c) shows an example of an arrangement to implement a Boolean function X, which is dependent on the states of three inputs A, B and C. If the circuitry within the two networks is arranged such that when the combination of inputs corresponds to the condition $X = 1$, the pull-up network will be turned *on* (and the pull-down network will be turned *off*) then the output of the arrangement will correspond to the function X. In this way quite complex Boolean functions can be produced using relatively few components.

Since the elements of our networks are simply switches, combining the elements to produce our required functionality is straightforward. Figure 26.39 shows examples of possible pull-down networks using an appropriate number of *n*-channel devices. The functionality given represents the condition required for conduction through the arrangement. Devices connected in series produce an AND function while those in parallel produce an OR arrangement.

Elements within the pull-up network may be combined in a similar manner, but here *p*-channel devices are used. Figure 26.40 shows some simple examples.

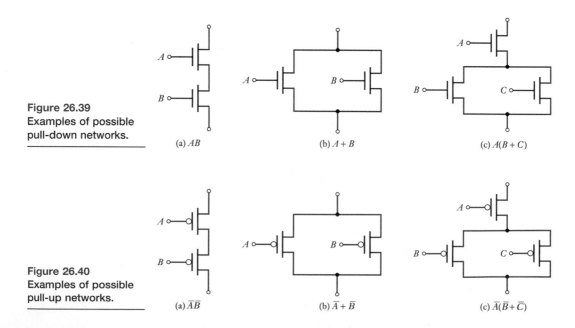

Figure 26.39
Examples of possible pull-down networks.

(a) AB (b) $A + B$ (c) $A(B + C)$

Figure 26.40
Examples of possible pull-up networks.

(a) $\overline{A}\,\overline{B}$ (b) $\overline{A} + \overline{B}$ (c) $\overline{A}(\overline{B} + \overline{C})$

A two-input NAND gate

To illustrate the construction of a simple gate, consider a two-input NAND gate. This can be described by the function

$$X = \overline{AB}$$

From Figure 26.38(c) we know that the form of the *pull-up* network must be such that this conducts when the inputs match the required function. This functionality must be constructed using p-channel devices, and since $\overline{AB} = \overline{A} + \overline{B}$, this can be achieved using the arrangement of Figure 26.40(b). Also from Figure 26.38(c) we know that the form of the *pull-down* network must be such that this conducts when the required function is *not* true. Thus the pull-down network must conduct when A and B are both true, and this can be achieved using the arrangement shown in Figure 26.39(a). Combining these networks gives us the arrangement shown in Figure 26.41(a).

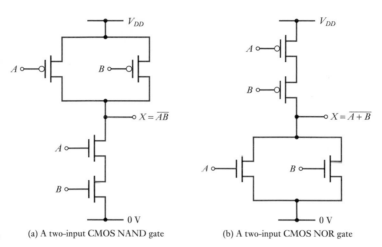

Figure 26.41
Two input CMOS NAND and NOR gates.

(a) A two-input CMOS NAND gate   (b) A two-input CMOS NOR gate

A two-input NOR gate

A similar approach can be used to design a two-input NOR gate. Here the *pull-up* network is designed by noting that $\overline{A + B} = \overline{A}\overline{B}$, which corresponds to the arrangement of Figure 26.40(a), while the pull-down network must conduct when either A or B is true, which can be achieved using the arrangement of Figure 26.39(b). Combining these two networks produces the NOR gate circuit shown in Figure 26.41(b).

You might like to compare the circuits in Figure 26.41 with the circuits discussed in Section 26.3.7 and shown in Figure 26.20. The circuits of Figure 26.41 can be expanded to provide addition inputs simply by increasing the number of devices used in the series/parallel arrangements.

Implementing more complex gates

The process illustrated above can easily be extended to the design of more complex circuits. Consider, for example, the design of a circuit to provide the logic function

$$X = \overline{A + B(C + DE)}$$

From this function it follows that $\overline{X} = A + B(C + DE)$ and so the *pull-down* network can be implemented directly from a series of n-channel devices, following the techniques illustrated in Figure 26.39.

To design the *pull-up* network we need to express X in terms of complemented input variables, which can be done by employing DeMorgan's laws:

$$X = \overline{A + B(C + DE)}$$
$$= \overline{A} \cdot \overline{B(C + DE)}$$
$$= \overline{A} \cdot (\overline{B} + \overline{(C + DE)})$$
$$= \overline{A} \cdot (\overline{B} + (\overline{C} \cdot \overline{(DE)}))$$
$$= \overline{A} \cdot (\overline{B} + (\overline{C} \cdot (\overline{D} + \overline{E})))$$

This function can then be implemented following the techniques illustrated in Figure 26.40.

Combining the pull-up and pull-down networks produces the arrangement shown in Figure 26.42.

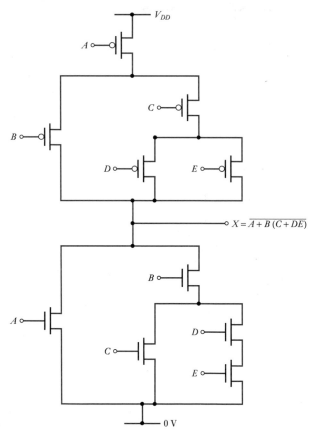

Figure 26.42
A complex CMOS gate.

In the gates we have considered so far, we have had the convenient situation that the function of the pull-up network can be expressed using only *complemented* inputs while the pull-down network can be expressed using only *non-complemented* inputs. This allows the functions to be produced directly from the inputs using the MOSFETs within these networks. However, in many gates this is not the case, and inverters are required to invert the input signals to produce the required control signals. This is illustrated in the following example.

Example 26.1 | **Design a CMOS exclusive-OR (XOR) gate.**

As discussed in Section 24.5.3, the function of an exclusive-OR gate can be described by the expression

$$Y = A\bar{B} + \bar{A}B$$

Since the expression involves both complemented and non-complemented terms, our circuit requires the use of two inverters to produce complemented versions of each input signal. The *pull-up* network can then be implemented directly, as in the earlier circuits.

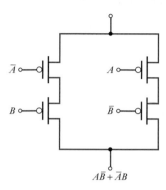

$$A\bar{B} + \bar{A}B$$

Design of the *pull-down* network requires us to determine the form of $\bar{Y}$ which can be found using DeMorgan's laws:

$$\bar{Y} = \overline{A\bar{B} + \bar{A}B}$$
$$= \overline{A\bar{B}} \cdot \overline{\bar{A}B}$$
$$= (\bar{A} + B) \cdot (A + \bar{B})$$

This can then be implemented as shown in (a) below. Note, however, that we could alternatively have continued our manipulation of $\bar{Y}$ as follows:

$$\bar{Y} = (\bar{A} + B) \cdot (A + \bar{B})$$
$$= \bar{A}A + AB + \bar{A}\bar{B} + B\bar{B}$$
$$= AB + \bar{A}\bar{B} \qquad (\text{since } \bar{A}A = B\bar{B} = 0)$$

which can be implemented as shown in (b) below. Thus, as with implementation using discrete gates, the implementation of logic functions using CMOS does not produce a unique structure.

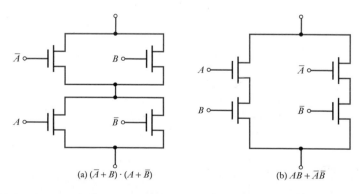

(a) $(\bar{A} + B) \cdot (A + \bar{B})$ (b) $AB + \bar{A}\bar{B}$

Either of these forms can be used and an example of a complete circuit is shown below.

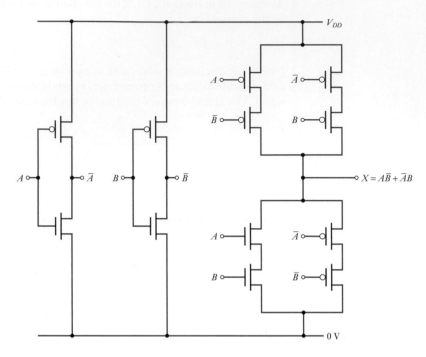

$X = A\bar{B} + \bar{A}B$

Transmission gates

An additional circuit configuration used in the construction of CMOS logic is the *transmission gate*, a technique which is also known as *pass-transistor logic*. This uses MOS transistors to pass or block the flow of signals from one point in a circuit to another.

The basic form of a transmission gate is shown in Figure 26.43. It consists of an *n*-channel transistor and a *p*-channel transistor connected in parallel. Two devices are used since this ensures good transmission for both high and low logic levels. The control signals applied to the gates of the two devices are the complement of each other, and since the gates are turned on by signals of opposite polarities, this means that the devices will both respond in the same way to a given input signal. Thus, in the circuit of Figure 26.43, if *B* is high (logic 1) both devices will be turned *on*, and the output *X* will be driven to equal the input *A*. If *B* is low (logic 0) both devices will be turned *off*, and the output will be disconnected from the input.

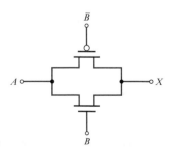

Figure 26.43
A transmission gate.

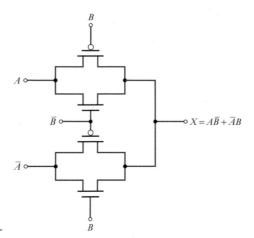

Figure 26.44
An exclusive-OR gate based on transmission gates.

Transmission gates are particularly useful in applications such as multiplexers, where they can be used to select one of a number of input signals. However, they are also used in a range of other applications. Figure 26.44 shows an implementation of an exclusive-OR gate based on the use of transmission gates. This uses fewer transistors than are required in the circuit of Example 26.1, although Figure 26.44 does not show the inverters required to obtain $\overline{A}$ and $\overline{B}$.

26.6   Interfacing TTL and CMOS or logic using different supply voltages

Interfacing is the term used to describe the connecting together of two circuits or systems. As the logic levels and characteristics of TTL and CMOS are different, an output of one cannot normally be connected directly to an input of the other. Similarly, CMOS logic circuits operating with different supply voltages will have correspondingly different logic levels – and, again, may not be directly compatible.

Video 26F

26.6.1   Driving CMOS from TTL

Typical output logic levels for a TTL gate with a totem-pole output are about $3.6\,\text{V}$ for logical 1 and about $0.2\,\text{V}$ for logical 0. When driving other TTL gates, the high logic level will fall as current is taken from the device and $V_{OH(min)}$ may be as low as $2.4\,\text{V}$. The input of a CMOS gate interprets any voltage of less than $0.3 \times V_{DD}$ as a logical 0 and any voltage greater than $0.7 \times V_{DD}$ as a logical 1. For a supply voltage of $5\,\text{V}$, this gives $V_{IL(max)} = 1.5\,\text{V}$ and $V_{IH(min)} = 3.5\,\text{V}$. The logical 0 voltage levels clearly cause no problems, but the TTL logical 1 output is not sufficiently high to guarantee that it will be interpreted as a logical 1 by the input of a CMOS gate.

One approach to this problem is to use a CMOS gate that has **TTL-compatible inputs** (as discussed in Section 26.5.1) such as a member of the 74HCT or 74ACT families. However, this is not the only approach to the problem. We noted in Section 26.4.2 that the logic levels of **open-collector gates** are approximately equal to the supply rail voltages. Therefore, an open-collector gate with a pull-up resistor connected to $5\,\text{V}$ will directly drive CMOS gates operating on a $5\,\text{V}$ supply voltage. Moreover, as CMOS gates have a very high input resistance, they do not load TTL outputs and the addition of a pull-up resistor to a totem-pole output will cause its output voltage to rise to approximately $5\,\text{V}$ when in the high state. This allows both

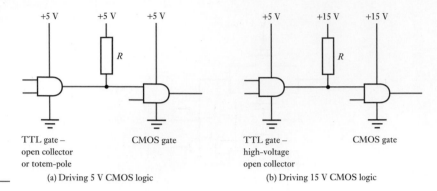

Figure 26.45
Driving CMOS gates from TTL.

TTL gate –
open collector
or totem-pole

CMOS gate

TTL gate –
high-voltage
open collector

CMOS gate

(a) Driving 5 V CMOS logic

(b) Driving 15 V CMOS logic

open-collector and conventional totem-pole devices to be interfaced to 5 V CMOS logic, simply by the addition of a pull-up resistor to 5 V. This is illustrated in Figure 26.45(a).

When driving CMOS gates that are operating with a supply rail higher than 5 V, a high-voltage open-collector TTL gate can be used. The pull-up resistor is taken to the supply rail of the CMOS logic to produce appropriate logic levels. This is shown in Figure 26.45(b) for a system using 15 V CMOS logic. A similar approach can be taken when driving a CMOS gate operating with a supply rail lower than 5 V, although, in this situation, a conventional (rather than high-voltage) open-collector gate can be used.

An alternative way to drive high-voltage CMOS is to use a special-purpose **voltage translator**. These CMOS circuits are specifically designed to allow logic operating at a low supply voltage (such as TTL or low-voltage CMOS) to drive high-voltage CMOS logic. A range of translators are available, with a number of devices normally being provided within a single package. An arrangement using such a translator is shown in Figure 26.46.

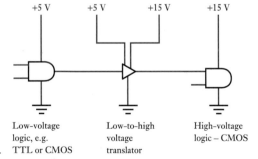

Figure 26.46
Use of a low-to-high voltage translator.

Low-voltage
logic, e.g.
TTL or CMOS

Low-to-high
voltage
translator

High-voltage
logic – CMOS

26.6.2 Driving TTL from CMOS

The logic output levels of CMOS gates operating from a 5 V supply are approximately 0 V and 5 V, so they are compatible with the input logic levels of all forms of TTL. However, the output impedance of CMOS is too high to enable it to provide the input current required to drive *standard* TTL gates.

Fortunately, most modern TTL families, such as the low-power Schottky family of devices (74LS), require a much lower input current than standard TTL. Most CMOS gates will provide sufficient output current to drive correctly a single 'LS' gate, which can then be used to drive other 'LS' gates. Alternatively, the 'LS' gate can then be used to drive standard TTL circuits as 'LS' gates provide sufficient output current to drive at least one standard TTL load (although few applications now use standard TTL).

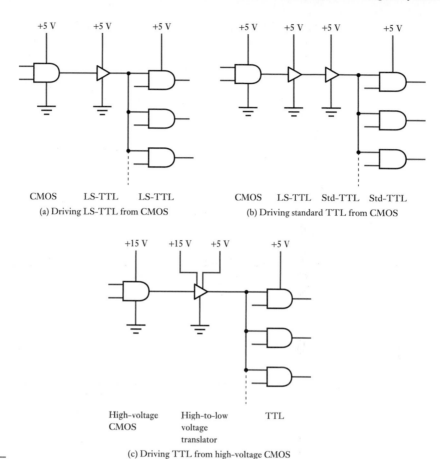

Figure 26.47
Driving TTL from CMOS.

(a) Driving LS-TTL from CMOS

(b) Driving standard TTL from CMOS

(c) Driving TTL from high-voltage CMOS

High-voltage CMOS logic can be interfaced to TTL using appropriate high-to-low **voltage translators**. As with the low-to-high voltage translators mentioned earlier, various devices are available to suit a range of situations.

Figure 26.47 illustrates various methods for driving TTL from CMOS logic.

26.6.3 Interfacing CMOS logic with different supply voltages

Interconnection between CMOS circuits using different supply voltages can be achieved using appropriate voltage translators in a manner similar to that used to interface TTL and CMOS circuits that have different supply voltages. Some CMOS families also simplify interconnection by allowing circuits operating from a 3.3 V supply to accept input signals up to 5 V.

Video 26G

26.7 Power dissipation in digital systems

A factor of great importance in the design of many digital systems is their power consumption. In battery operated applications this determines their battery life and in many cases will limit the functions and features that can be incorporated. However, in *all* applications, power *consumption* is related to power *dissipation*, which determines the amount of heat produced within the system. Dealing with this waste heat is often expensive, and power dissipation imposes constraints on the design of a wide range of electronic systems. In highly integrated devices the ultimate complexity of the component is often limited by considerations of power dissipation.

Since almost all modern electronic systems make use of CMOS technology, in this section we will look at power dissipation in such devices and consider what factors affect the amount of heat produced. Power dissipation within CMOS devices may be divided into two forms, namely *static* dissipation and *dynamic* dissipation.

Static dissipation relates to the power dissipated by a device when in its steady state. Because CMOS circuits use a complementary structure, where one or other of its two switching networks is turned off in the steady state, the static current consists of only a very small leakage current. For this reason the static dissipation is almost always negligible and will not be considered further here.

Although CMOS devices consume negligible power when static, they do consume a small amount of power whenever they switch between one state and another. This is termed their dynamic power dissipation. The input of each gate resembles a small capacitance and energy is dissipated in charging and discharging this each time the device changes state. Capacitance is also associated with any wires connected to the output, and with the source and drain elements of the various devices. To illustrate this process, consider the circuit of Figure 26.48, which shows a simple CMOS inverter connected to a capacitance which represents the input of another gate plus any other load capacitance.

Consider initially the situation where the output of the inverter is low and the capacitor is fully discharged. Imagine now that the input is changed so that the output switches to a high output state. This is achieved by turning T1 *on* and turning T2 *off*. In this state, T1 conducts and passes current from the positive supply line into the capacitor C, charging it up. The instantaneous energy drawn from the supply is equal to the product of the current and the voltage which is iV_{DD}, and the total energy E drawn from the supply to charge the capacitor is therefore

$$E = \int i\, V_{DD}\, \mathrm{d}t$$
$$= V_{DD} \int i\, \mathrm{d}t$$
$$= V_{DD}Q$$

where Q is the charge supplied to the capacitor. Now, from the discussions in Section 4.7, we know that for a capacitor $Q = VC$, and therefore

$$E = V_{DD}Q$$
$$= V_{DD}(V_{DD}C)$$
$$= CV_{DD}^{2}$$

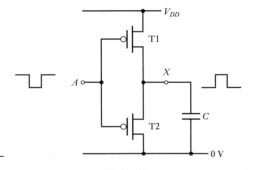

Figure 26.48
A CMOS inverter driving a capacitive load.

From Section 4.9 we know that the energy stored in a charged capacitor is equal to $^1/_2 CV^2$, and therefore the energy dissipated in T1 is equal to the total energy taken from the supply, minus the energy stored in the capacitor, which gives

$$\text{Energy dissipated in T1} = CV_{DD}{}^2 - {}^1/_2 CV_{DD}{}^2 = {}^1/_2 CV_{DD}{}^2$$

Thus the energy dissipated in T1 is equal to that stored in the capacitor.

Now consider what happens when the input changes again, so that the output is driven low. Under these circumstances T1 is turned *off* and T2 is turned *on*. The energy stored within the capacitor is now dissipated within T2 as the capacitor is discharged. Clearly the energy dissipated in this way is that stored within the capacitor, which we have shown to be $^1/_2 CV_{DD}{}^2$.

Therefore, if the input is repeatedly switched, in every cycle $^1/_2 CV_{DD}{}^2$ of energy will be dissipated in T1 and $^1/_2 CV_{DD}{}^2$ of energy will be dissipated in T2, making a total dissipation of $CV_{DD}{}^2$ in each cycle. If the switching waveform has a frequency of f cycles per second, then the dynamic power dissipation P_D is given by

$$P_D = f\, CV_{DD}{}^2 \tag{26.5}$$

While this expression has been determined by considering a simple inverter, a similar analysis could be performed for other CMOS gates, yielding similar results.

In practice, most CMOS gates are *not* switched at a constant frequency and within complex circuits the various gates will be switched at very different speeds. However, what this relationship shows is that in such circuits, the overall power consumption is likely to be directly proportional to the clock speed used. The relationship also shows the great importance of the supply voltage in determining power dissipation and explains why modern logic families are aimed at lower and lower operating voltages.

26.8   Noise and EMC in digital systems

In Chapter 22 we looked at noise and EMC as they apply to analogue circuits and electronic systems in general. In this section we will look at the implications of noise and EMC for *digital* circuitry.

One of the great advantages of digital systems compared with analogue equipment is their greater tolerance of noise. We have already looked at the noise immunity of various forms of logic gate and noted that this is an important factor in determining the ability of a system to work correctly in the presence of noise. Unfortunately, while digital systems tend to have a high *tolerance* of noise, they also tend to *produce* more noise than analogue equipment. Therefore, from an EMC standpoint, digital systems have certain advantages and certain disadvantages when compared with analogue systems.

In the following sections we will look at sources of noise in digital systems and the ways in which noise affects their operation. We will then discuss various design techniques that can be used to improve the EMC performance of such systems.

26.8.1   Digital noise sources

Electronic noise

Electronic logic gates are constructed from electronic components. They will therefore have the same noise sources as analogue circuits, as discussed in Section 22.2. In general, the designer of the gates will have taken these noise sources into account and chosen threshold voltages and logic levels so that the optimum noise performance is achieved.

Interference

Often less predictable is noise generated by interference. Nearby electrical and electronic equipment can produce large amounts of **electromagnetic radiation**, which will induce voltages in any conductor. Long wires act like aerials, picking up the interference and producing noise voltages within the system. Interference can enter a system via radiation pick-up, the power supply or external lines to sensors and actuators.

Internal noise

Often the most important source of noise within an electronic system is the system itself! Signals in one part of a circuit may propagate throughout the system, producing noise elsewhere. The **power supply** is a common source of noise. With simple linear supplies, the fluctuating field from the transformer can induce noise currents at the supply voltage frequency. With **switch-mode power supplies**, the high-frequency switching currents often propagate throughout the system, acting as a powerful noise source.

Power supply noise

Noise carried along the power supply lines is one of the most common forms of noise in digital systems and one of the hardest to remove. When digital devices change state, there is often a step change in the current being taken from the supply. Usually, the operation of many devices within a circuit is synchronised to a master clock or oscillator, so large numbers of devices often change state simultaneously. When this happens, extra current is taken from the supply to charge, or discharge, capacitances within the circuit. The result is that the supply rails in digital systems are usually perpetually ringing with noise spikes that are fed to all parts of the system.

Noise on the power supply rails can also originate from outside the system, having entered the unit via the power supply. Noise from motors or other high-powered actuators propagates along the AC supply lines and is not always removed by the smoothing inside the power supply. This is normally tackled by fitting **mains filters** at the point where these lines enter the unit.

CMOS switching transients

One of the prime offenders in generating noise spikes is CMOS logic. As discussed in Section 26.7, CMOS takes almost no current when static, but passes a surge of current when switching from one state to another. The net result is that in systems with many CMOS gates, all clocked simultaneously, this surge is many times greater than the average current. Any resistance, or inductance, within the supply lines converts this current surge into a voltage spike.

26.8.2 The effects of noise in digital systems

Small amounts of noise (less than the noise immunity of the system) usually have no effect on the operation of the system. This is in strict contrast to analogue techniques in which noise cannot normally be removed from a signal once it has been added. However, large amounts of noise can cause problems for digital systems in two ways:

- excessive noise can cause the operation of a system to be incorrect;
- excessive noise can cause permanent damage to the system.

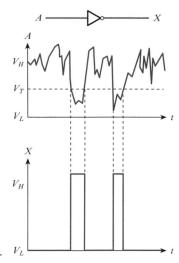

Figure 26.49
The effects of noise on steady logic voltages.

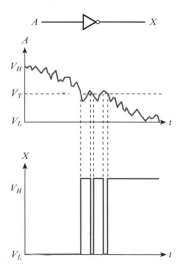

Figure 26.50
The effects of noise on slowly varying logic voltages.

Noise-induced errors

Noise signals in excess of the noise immunity of a circuit can cause problems when added to steady logic voltages, as illustrated in Figure 26.49.

When added to slowly varying logic signals, amounts of noise considerably less than the noise immunity of the system can produce problems, as shown in Figure 26.50. Small fluctuations of the input signal cause it to cross the threshold voltage several times. This generates *several* transitions of the output rather than the *single* transition expected. If this signal were to be used as the input for a counter, an incorrect number of events would be detected.

Maximum ratings

We have seen that many logic families operate from a supply voltage of 5 V. The manufacturers of such gates also specify absolute **maximum ratings** defining the limits of safe operation of the circuit. For TTL, the maximum supply voltage is usually 7 V and the maximum allowable voltage on any input line is about 5.5 V. Maximum currents are also defined. Exceeding the maximum values can destroy or permanently damage the device, sometimes in a fraction of a microsecond!

The maximum supply voltage for a CMOS gate is related to its normal supply voltage range. For a device that permits 3–15 V operation, the allowable supply voltage range might be −0.5 to +18 V. For a device that operates with a 2–6 V supply, the allowable range might be −0.5 to +7.0 V. A low-voltage device that normally operates with a supply of 1.65–3.6 V might have an allowable supply voltage range of −0.5 to +4.6 V. The allowable voltage on any input varies with the supply voltage in use and is generally −0.5 to V_{DD} + 0.5 V. The maximum direct current into any input or output might be of the order of 10 mA.

Because of the very fast response of electronic circuitry, noise spikes of only a few nanoseconds can cause serious damage. Because of the very high input resistances of many logic circuits, **static electricity** is a real problem.

26.8.3 Designing digital systems for EMC

In Section 22.10 we looked at various ways in which we could improve a system's EMC performance. In this section, we will look at issues of particular relevance to the design of digital equipment.

Enclosures and cable shielding

In Chapter 22 we looked at the importance of **screening** in reducing both the susceptibility of a circuit to interference and the amount of noise that it radiates. The use of enclosures and shielded cables is equally important to the design of analogue and digital equipment.

Cables should be kept as short as possible and any long cables shielded. The maximum recommended distance between standard logic chips is only a few centimetres. If longer distances are required, then special-purpose line driver/receiver chips should be used. These are designed to cope with the increased line capacitance and to reject noise.

Clock lines cause particular problems in digital circuits as they carry fast switching signals that have very high-frequency components. A Fourier analysis of a square waveform shows that it contains frequency components at odd harmonics to its fundamental frequency. Therefore, even a relatively low-frequency clock waveform will often contain components at very high frequencies.

Interference between conductors within a system can be a particular problem in computer **backplanes** where a large number of conductors run parallel to each other. **Cross-talk** can be reduced by placing ground conductors between clock and signal lines to provide screening. This is illustrated in Figure 26.51.

Opto-isolation

Noise from input and output lines can be reduced by **opto-isolation** (as described in Section 13.3.1). This is illustrated in Figure 26.52.

The **opto-isolator** consists of an **LED** and a **phototransistor** within a single package. When zero volts (logical 0) is applied to the input of the opto–isolator, the LED is turned off. The phototransistor receives no light and is also turned off. The pull–up resistor R therefore produces an output voltage of approximately V_{CC}. If a positive voltage (logical 1) is applied to the input of the opto–isolator, the LED will be illuminated, causing the phototransistor to turn on. This will pull down the output voltage to close to zero volts. Thus, the opto–isolator acts as a logical inverter. The important aspect of the operation of this arrangement is that there is no electrical

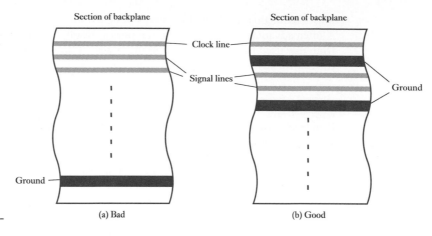

Figure 26.51
Examples of good and bad backplane arrangements.

(a) Bad (b) Good

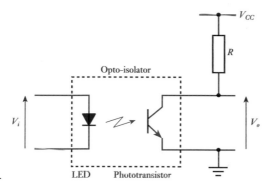

Figure 26.52
The use of opto-isolation.

connection between the input and the output – they are linked only by light. Typical devices produce several kilovolts of electrical isolation. Both the input and the output lines of a system can be protected in this way.

Diode clamps

In extremely noisy environments it may be prudent to provide extra **gate protection** circuitry of the form shown in Figure 26.36. Even if **clamp diodes** are fitted within the logic circuit, they are of limited power-handling ability and very large spikes will simply vaporise them an instant before the rest of the device is destroyed. External diodes with a fast response and capable of taking large current surges will provide improved protection for the circuit at very low cost.

Decoupling capacitors and earthing

Most problems associated with internal noise sources are tackled by the intelligent use of capacitors and careful layout. It is normal to fit **decoupling capacitors** adjacent to *every* digital IC and low-inductance capacitors across the supply lines. Earthing is of particular importance and should be as direct, and of as low a resistance, as possible.

Power supply isolation

In circuits combining both analogue and digital circuits, it is often necessary to use separate power supplies to prevent noise from the digital sections interfering with low-noise analogue stages. In systems designed for use in high-noise environments, it is often advantageous to separate the power supply used for the input/output stages from that used for the remainder of the system. This, combined with opto-isolation of the control signals, prevents noise entering the system through the input/output lines.

Schmitt trigger inputs

Problems associated with slowly varying inputs can be alleviated by using logic gates with **hysteresis**. One of the most frequently used arrangements with this characteristic is the **Schmitt trigger** circuit, which will produce a single change in logic state, even for a noisy, slowly changing input. The transfer characteristic of such a circuit is shown in Figure 26.53.

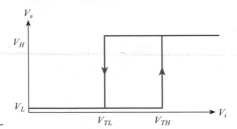

Figure 26.53
The transfer characteristic of a Schmitt trigger.

From the transfer characteristic, it can be seen that the output will only change from low to high if the input goes above the higher threshold voltage, V_{TH}, and will only change from high to low if the input goes below the lower threshold voltage, V_{TL}. This prevents small amounts of noise from switching the output repeatedly between the two output states.

There are many forms of Schmitt trigger circuit. Figure 26.54 shows a simple inverting arrangement based on an operational amplifier. The combination of resistors provides **positive feedback**, which tends to hold the circuit in its existing state. The output, in the absence of any negative feedback, tends to switch between voltages close to the two supply rails (V_{pos} and V_{neg}). The operational amplifier will change state when the input voltage becomes greater than, or less than, the voltage on its non-inverting input. When the output is positive, the voltage on the non-inverting input of the amplifier is given by

$$V_T = V_{pos} \times \frac{R_2}{R_1 + R_2}$$

and when the output is negative, the voltage on the non-inverting input is

$$V_T = V_{neg} \times \frac{R_2}{R_1 + R_2}$$

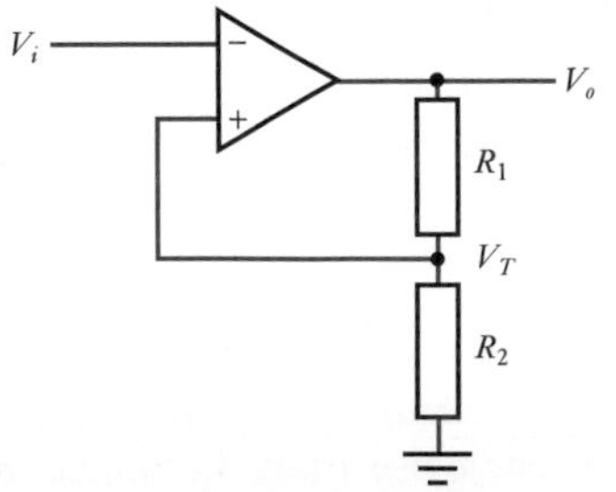

Figure 26.54
A simple Schmitt trigger.

The switching voltage therefore changes with the output state to produce the required hysteresis. Similar circuit techniques, with appropriate resistor values and supply voltages, can be used to generate threshold and output voltages suitable for particular logic devices. The various logic families include a range of gates incorporating Schmitt triggers on their inputs.

Removal of noise by filtering

We have seen that much of the noise associated with digital systems is in the form of spikes or high-frequency fluctuations. If a particular logic signal is known to change slowly, much of this noise can be removed using a low–pass filter. In removing high-frequency components from the signal, the filter will also slow down transitions between the two logic states. It will therefore usually be necessary to follow such filtering with a Schmitt trigger to restore the sharp edges of the waveform.

Video 26H

Further study

An industrial control system takes inputs from two remote mechanical switches, and provides output signals to control an AC heater and an AC motor.

The controller is required to sense the state of the switches and to turn the heater and motor on and off according to some control algorithm. The system is to be used in an environment with a high level of electrical noise. Suggest how the unit could be designed to minimise the effects of noise on its operation.

Key points

- Integrated circuits combine the functions of many gates within a single device. Various levels of integration are available, from 'small-scale integration' (SSI), which combines just a handful of gates, to 'tera-scale integration' (TSI), which may provide more than 10^{12} gates within a single package.

- Logic gates can be produced using either MOS devices (FETs) or bipolar transistors.

- Logic circuits based on MOS transistors have the advantages of low power dissipation and very high circuit density. The input resistance is high, producing a high fan-out, but the output resistance is also high, limiting the rate at which load capacitance can be charged.

- Bipolar transistors generally switch faster than MOS devices, but occupy more space and consume more power.

- Although bipolar transistors switch quickly, their speed is reduced if they are allowed to saturate as this produces a delay, known as storage time. This problem can be tackled in a number of ways, including gold doping to reduce the lifetime of charge carriers, Schottky transistors to prevent saturation, or keeping the transistors within their active region.

- The most popular logic families based on bipolar transistors are the various forms of transistor–transistor logic (TTL). These range from standard TTL circuits of the 54/74 families, through a spectrum of devices optimised for speed, power consumption or other characteristics.

- Conventional TTL devices have a totem-pole output, producing typical logic levels of 3.6 and 0.2 V. Other devices have different output configurations, such as open-collector or three-state outputs.

- CMOS is the dominant MOS logic family. Although more complicated to fabricate than other MOS technologies, CMOS offers very low power consumption coupled with a high operating speed and excellent noise tolerance.

- Complex CMOS circuits can be implemented by constructing pull-up and pull-down networks that implement complementary logic functions. The pull-up network is implemented using only p-channel devices while the pull-down network uses only n-channel devices.

- In some applications it is necessary to use a combination of TTL and CMOS. As the inputs and outputs of these two families are not directly compatible, some form of interfacing is required, although the circuitry needed is often very simple.

- The power dissipation in CMOS circuitry is due almost entirely to the power consumed when gates switch from one state to another. This can be shown to be directly proportional to the switching frequency used, and proportional to the square of the supply voltage.

- All electronic circuits are susceptible to noise of one form or another.

- Within digital systems, noise can cause problems in a number of ways. First, it can disrupt the correct operation of a system by causing errors in its operation. Second, in severe cases, it can cause physical damage to devices within the circuit.

- Various design techniques are available to reduce the effects of noise on the operation of a circuit and provide a high level of protection. Unfortunately, complete immunity from noise is not possible.

Exercises

26.1   What is meant by very large-scale integration (VLSI)?

26.2   Sketch the transfer function of a logical inverter.

26.3   What is meant by the noise immunity of a logic gate?

26.4   What are the logic levels of the simple inverter shown in Figure 26.6?

26.5   Why is the turn-on time of a bipolar transistor less than the turn-off time?

26.6   What is meant by storage time in a bipolar transistor and how may this be reduced?

26.7   Define the terms 'propagation delay time', 'set-up time' and 'hold time'.

26.8   What points in the input and output waveforms are used to measure the propagation delay times?

26.9   What is meant by the fan-out of a logic gate?

26.10 Which two logic families are most widely used for the production of integrated logic circuits?

26.11 Why is ECL faster in operation than TTL?

26.12 What are the main disadvantages of ECL?

26.13 Explain the terms 'PMOS', 'NMOS' and 'CMOS'.

26.14 What are the main disadvantages of NMOS and PMOS logic and how are these overcome in CMOS?

26.15 Why is the power consumption of CMOS gates greatly affected by its clock frequency?

26.16 Explain what is meant by BiCMOS. What are the characteristics and uses of BiCMOS gates?

26.17 What is the difference between a 7400 and a 5400 device?

26.18 What is the normal supply voltage for TTL gates? What are their normal logic levels?

26.19 Explain the function of the input clamp diodes in the circuit in Figure 26.23.

26.20 Sketch the transfer characteristic of a typical TTL gate.

26.21 Simulate a TTL inverter using the circuit in Figure 26.23, but assuming only one input. Connect a controlled DC voltage source to the input of the circuit and a 1 kΩ resistor as a load from the output to ground. Sweep the input voltage from 0 to 5 V and observe the output voltage. Plot the transfer function of the gate and compare this with that given in Figure 26.24. What is the effect of removing the load resistor?

26.22 What is the *minimum* noise immunity of a TTL gate? What would be a *typical* value for the noise immunity of such a gate?

26.23 Why might an open-collector device be used in preference to a totem-pole output device?

26.24 What is meant by a three-state output gate?

26.25 A four-input TTL NAND gate is to be used in an application that requires a three-input gate. What should be done to the unused input in this circuit?

26.26 A four-input TTL NOR gate is to be used in an application that requires a three-input gate. What should be done to the unused input in this circuit?

26.27 What accounts for the increase in speed of Schottky TTL compared with standard devices?

26.28 How does a 74C00 differ from a 7400 gate?

26.29 What supply voltages are normally used with CMOS gates?

26.30 How do the logic levels and noise immunity of a CMOS gate relate to its supply voltage?

26.31 What would be the *minimum* noise immunity of a conventional CMOS gate operating from a 5 V supply? What would be a *typical* value for the noise immunity of such a gate?

26.32 What distinguishes the 74HCT family of devices from the 74HC family?

26.33 Why are special precautions required when handling CMOS components?

26.34 A four-input CMOS NAND gate is to be used in an application that requires a three-input gate. What should be done to the unused input in this circuit?

26.35 A four-input CMOS NOR gate is to be used in an application that requires a three-input gate. What should be done to the unused input in this circuit?

26.36 What is the typical fan-out of a CMOS gate? How is the speed of operation of the gate affected by the number of gates connected to its output?

26.37 What form of CMOS gate is widely used for buffers and line drivers? What makes this form of gate suitable for such applications?

26.38 Design a CMOS pull-down network to implement the function $(A\ B\ C) + (DE)$.

26.39 Design a CMOS pull-up network to implement the function $\overline{(A\ B\ C) + (DE)}$.

26.40 Design a CMOS gate to implement the function $X = \overline{(A\ B\ C) + (DE)}$.

26.41 Sketch the circuit of a three-input CMOS NOR gate.

26.42 Design a CMOS gate to implement the function $X = A\bar{B}C + \bar{A}B\bar{C}$.

26.43 Why is special consideration needed when connecting together TTL and CMOS logic gates?

26.44 Why is power consumption of importance in logic design?

26.45 What is meant by *static* and *dynamic* dissipation in CMOS gates? Which of these is most important?

26.46 What factors determine the power dissipation of CMOS gates?

26.47 Explain why CMOS circuits are a common cause of noise spikes.

26.48 Explain the function of opto-isolators in reducing noise pick-up in digital systems.

26.49 Describe the techniques used within CMOS gates to protect their inputs from static electricity. How can the device be further protected?

26.50 Describe the operation and function of a Schmitt trigger.

26.51 Suggest a method of controlling a relay requiring a 24 V drive signal from TTL logic. What special precautions should be taken because of the inductive nature of the load?

Chapter 27 Implementing Digital Systems

Objectives

When you have studied the material in this chapter, you should be able to:

- describe the evolution of integrated circuit complexity with time
- discuss the use of array logic in allowing widespread use of VLSI technology
- outline the implementation of combinational or sequential logic functions using simple programmable logic devices (PLDs)
- describe the architecture of a simple microcomputer system and explain how such a system can be used to produce the functions of combinational or sequential logic circuits
- explain the nature and uses of the various forms of semiconductor memory
- describe the general form of a programmable logic controller (PLC)
- suggest a range of implementation strategies that might be appropriate for the production of a digital system.

Video 27A

27.1 Introduction

In Chapter 26 we saw how a number of basic gates may be implemented in a single integrated circuit and looked at the various levels of integration available. In that chapter we looked at technologies for implementing SSI and MSI devices and the basic forms of these types of integrated circuit. More highly integrated devices make use of the same underlying technologies but are far more complicated in their structure and operation. The production of these circuits is totally dominated by CMOS technology, which is evolving and growing ever more powerful year by year.

27.1.1 The evolution of integrated circuit complexity

Back in 1965, Gordon Moore, one of the founders of Fairchild's Semiconductor Division, noted an exponential growth in the number of transistors that could be placed in a single integrated circuit and suggested that this trend would continue. **Moore's law**, as it is generally known, predicts that the number of transistors that can be integrated in a single device will double every couple of years. This prediction has proved remarkably accurate, as can be seen from Figure 27.1, which shows the number of transistors in a series of processors produced by a single manufacturer, Intel.

Using modern device production techniques, it is possible to combine millions of gates in a single integrated circuit. However, there are practical limits to the number of separate gates that may usefully be put in a single package. One of the major constraints is simply the number of pins that are required to connect to the inputs and

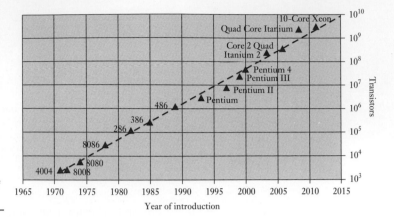

Figure 27.1
Integration densities of
Intel processors.

outputs of the gates. A circuit with 1000 separate gates would require several thousand pins and would inevitably occupy a large amount of space on a circuit board. The external interconnections between these pins would also require a large amount of board area.

To take full advantage of large-scale integration, it is necessary to implement not only the gates required by a circuit but also their interconnections. If this is done, then only the circuit's inputs and outputs need to be brought to the outside world, rather than connections to each node of the circuit. This produces an arrangement in which most of the interconnections are internal, greatly reducing stray capacitance and increasing the speed of operation of the circuit (as discussed in Chapter 26). Internally connecting the gates in a package permits complicated circuits to be implemented in a single device, but results in a device that is dedicated to a particular function.

Complicated integrated circuits are relatively inexpensive to mass-produce, but are very expensive to design. Consequently, it is quite feasible to design a special-purpose device for an application requiring a million such components (where the design costs can be distributed between the million units), but it is unattractive to do so for a project requiring only a handful of systems. One way to tackle this problem is to create a standardised component that can be produced in high volumes and then customised (or programmed) for a particular application.

In this chapter, we will look at two forms of highly integrated general-purpose devices, namely **array logic** and **microprocessors**. Having looked at each of these in turn, we will then consider a packaged form of the microprocessor, the **programmable logic controller** (PLC), before looking at the process of choosing a technology for a particular application.

27.2 Array logic

Modern microprocessors, memories and interface circuits often contain thousands or tens of thousands of gates within a single 'chip'. The development costs associated with such devices are very high and can only be justified for components that are used in very large numbers.

Microprocessors are general-purpose devices that can be used in a multitude of applications as their operation is determined by software. Similarly, memory devices and several other complicated components can be used in a wide range of applications, allowing them to be produced in large quantities.

Unfortunately, not all electronic circuitry is manufactured in great numbers and, in many cases, the circuits used are unique to a particular application. Even systems based on the use of standardised components, such as microprocessors, normally require a certain amount of specialised logic to 'bolt' the major components together. This circuitry is often referred to as **glue logic**, for obvious reasons. While the same microprocessor may be used in thousands of designs, the glue logic varies from one application to another to give the system its own unique hardware characteristics. As this circuitry tends to be specific to individual designs, it is often called **random logic**, this term referring to the selection of functions rather than indicating any non-causal form of operation! This random nature makes it impractical for a manufacturer to produce a single *conventional* integrated circuit combining all the functions within a single chip.

In very small systems it may be possible to implement the required random logic using a handful of small- or medium-scale logic devices (as discussed in the previous chapter). However, as the complexity of the system increases, the number of components required becomes prohibitive. A typical desktop computer, for example, might require only a handful of VLSI chips for the functions of the processor, memory and input/output sections, but would need several hundred additional chips if the glue logic were implemented using simple logic circuits. What is required is a method for providing large numbers of gates within a single, mass-produced device, while allowing them to be interconnected in a manner to suit a particular application. Devices of this type come under the general heading of **programmable logic devices** (PLDs).

A PLD contains a large number of logic gates within a single package, but allows a user to determine how they are interconnected. This technology is also known as **uncommitted logic** as the gates are not committed to any specific function at the time of manufacture.

The various gates within a device, and their interconnections, are arranged within one or more 'arrays'. For this reason, this form of logic is also known as **array logic**. There are many forms of array logic and here we will look at just a few of the more important examples. Unfortunately, a study of this area is complicated by the plethora of names given to different types of PLDs. Here, we will restrict ourselves to some of the more widely accepted terms, which include:

- programmable logic array (PLA)
- programmable array logic (PAL)
- generic array logic (GAL)
- erasable programmable logic device (EPLD)
- programmable electrically erasable logic (PEEL)
- programmable read-only memory (PROM)
- complex programmable logic device (CPLD)
- field programmable gate array (FPGA).

27.2.1 Programmable logic array (PLA)

Video 27B

In Section 24.5.4 we saw that a combinational expression can always be represented by a series of **minterms** that may be derived directly from a truth table. For example, a system with four inputs – A, B, C and D – might have outputs – X, Y and Z – where

$$X = \overline{A}\,\overline{B}\overline{C}D + \overline{A}BCD$$

$$Y = \overline{A}\,\overline{B}CD + ABC\overline{D}$$

$$Z = \overline{A}\,\overline{B}\overline{C}D + \overline{A}BCD + ABC\overline{D}$$

One way to implement such a system is to use a number of inverters to produce the inverted input signals ($\overline{A}, \overline{B}, \overline{C}$ and $\overline{D}$) and then to use a series of AND and OR gates to generate and combine the various minterms. A PLA has a structure that allows such functions to be produced easily.

The structure of a simple PLA is shown in Figure 27.2. This shows an arrangement with four inputs (A, B, C and D) that are inverted to produce four pairs of complementary inputs. These eight signals are then each connected to the inputs of a number of AND gates through an array of fusible links. These **fuses** are initially all intact, but may be blown selectively to determine the pattern of connections between the input signals and the AND gates. In this way, each AND gate is used to detect the input pattern corresponding to an individual minterm. A second array of fuses is used to connect the outputs of the AND gates to a collection of OR gates. These OR gates combine the relevant minterms to produce the various outputs. This process is illustrated in Figure 27.3, which shows the earlier simplified PLA configured to implement the system given in the above example. Here, most of the fuses linking the inputs to the AND gates have been blown, leaving only those connecting the required signals to each gate. Similarly, the fuses connected to the inputs to the OR gates have been selectively blown to produce the required three output signals.

A PLA would normally have more inputs and outputs than the simplified example shown and would also have a greater number of AND gates, allowing more complicated functions to be implemented.

In order to represent symbolically the large numbers of gates and interconnections within a typical device, it is convenient to adopt a more compact notation that reduces the large numbers of interconnecting wires within the various arrays. The symbols used when drawing logic arrays are shown in Figure 27.4. Here, a single line is drawn to represent all the inputs to a gate and a cross is used to indicate those input lines that are connected to that gate. Figure 27.4(a) shows this approach applied to an array of AND gates and Figure 27.4(b) shows how it may be used with OR gates.

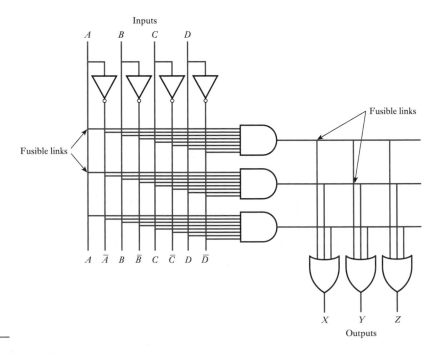

Figure 27.2
The structure of a
simple PLA.

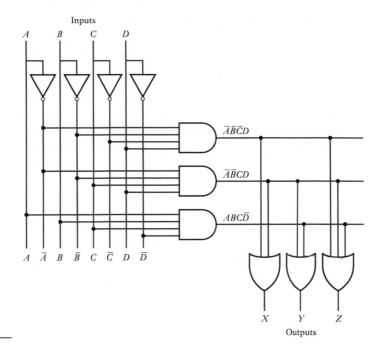

Figure 27.3
A configured PLA.

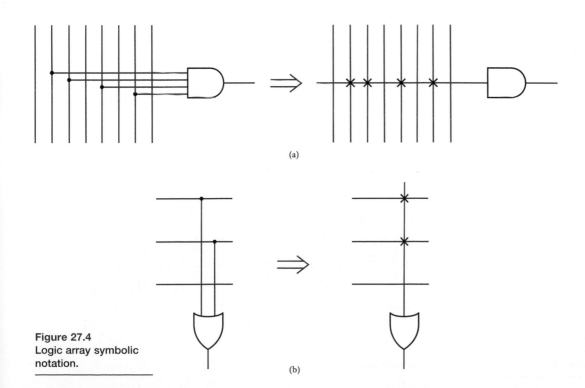

Figure 27.4
Logic array symbolic
notation.

In the PLA shown in Figure 27.3, the input signals are inverted to produce complementary input signals using a conventional inverter, as shown in Figure 27.5(a). A disadvantage of this arrangement is that the propagation delay of the gate will cause the inverted input to change a short time after the non-inverted signal. This is overcome by using a circuit that produces both an inverted and a non-inverted output, with equal propagation times. Such a circuit is given the symbol shown in Figure 27.5(b).

Figure 27.5
Representation of inverters within logic arrays.

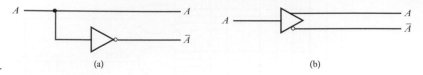

(a)   (b)

Figure 27.6 shows a PLA with 6 inputs, 4 outputs and 16 **product terms**. Devices are manufactured with all their fuses intact and this is indicated by the circle at each location in the two arrays of interconnections. In order to use the device, it must be **programmed** by selectively blowing unwanted fuses. This task is performed by a **PLD programmer** that reads and interprets a **fuse map** supplied by the user. The fuse map may be produced manually, but is more often produced by a dedicated software package that deduces the required fuse pattern from a description of the desired functionality. If all the fuses connected to a particular AND gate are left intact, then

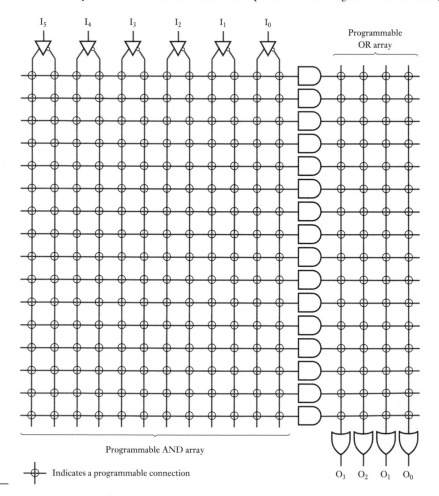

Figure 27.6
A PLA with 6 inputs, 4 outputs and 16 product terms.

its output will remain low as a high output would require both the inverted and non-inverted inputs to be high simultaneously. In this way, any unnecessary AND gates may simply be left unprogrammed and ignored. If all the fuses connected to an AND gate are blown, the output will be permanently high.

It can be seen that the structure of a PLA includes two arrays of interconnections. These are the **AND array**, which is used to select the components of the various minterms to be implemented, and the **OR array**, which combines the various minterms to produce the desired output functions. Within a PLA, both these arrays are programmable, giving great flexibility and the ability to make maximum use of the available product terms. However, the use of two programmable arrays makes these devices complicated and relatively slow. To overcome these problems, other forms of array logic have evolved in which only one of these two arrays is programmable.

Video 27C

27.2.2 Programmable array logic (PAL)

Despite the flexibility of the PLA structure, it became evident that there was some advantage to having a less complicated arrangement. One such arrangement was first developed by Monolithic Memories Inc. (MMI) in the form of **programmable array logic** (PAL). Figure 27.7 shows a PAL structure with 6 inputs, 4 outputs and 16 product terms.

At first sight, this might seem identical to the PLA shown in Figure 27.6, but it differs from the earlier circuit in that only one of its two arrays is programmable. The AND array is equivalent to that of the PLA and can be programmed to select the minterms required for a given function. However, the programmable OR array has been replaced by a fixed pattern of connections to a set of OR gates. The user now constructs the required functions by using the AND array to select the combinations of minterms that are fed into each OR gate. PAL manufacturers compensate for the absence of a programmable OR array by providing a range of devices with different numbers of OR gates and different numbers of inputs on each OR gate.

Because the OR array is fixed in a PAL, it is common to omit this array from its symbolic representation. It is also common to move the inputs to the left of the diagram to produce a circuit with the inputs on the left and the outputs on the right, as shown in Figure 27.8.

To give increased flexibility, many PALs use a technique that permits some of their pins to be used as either inputs or as outputs. Figure 27.9 shows such an arrangement. Here, the output from one of the OR gates of the device is passed through a three-state inverter before being fed to the output pin. The operation of the inverter is controlled by an **output enable** signal that is derived from the AND array in a manner similar to the various minterms. If all the fuses connected to the inputs to this AND gate are blown, its output will remain high, *enabling* the output of the inverter. This will configure this line as an output. If all the fuses connected to the inputs of this AND gate are left intact, its output will remain low and the output of the inverter will be *disabled*, converting the pin into an input. The signal on the pin is used to generate complementary signals that are fed to lines of the input array. Depending on the state of the output enable line, these complementary signals may represent either an input signal or the current state of the output. In the latter case, these lines may be used to allow the output of one OR gate to be fed back to the inputs of other gates. Rather than being set continuously high or continuously low, the output enable signal can be configured to respond to the state of lines in the input array. This can be used, for example, to allow input signals to enable or disable an output.

The circuit example in Figure 27.9 shows the output being controlled by a three-state inverter. It would be equally possible to use a three-state non-inverting buffer in

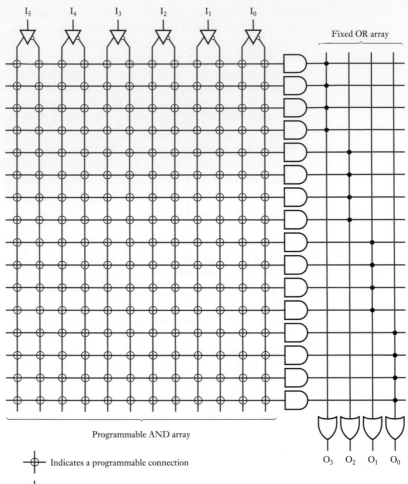

Figure 27.7
A PAL with 6 inputs, 4
outputs and 16 product
terms.

⊕— Indicates a programmable connection

●— Indicates a fixed connection

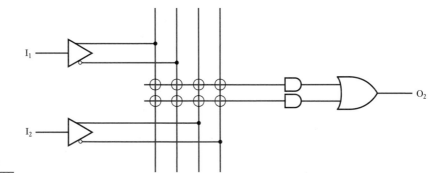

Figure 27.8
A fragment of a PAL.

this arrangement, which would produce a functionality equivalent to that of earlier circuits. The use of an inverting or non-inverting buffer affects the fuse map that must be used – some functions are easier to implement when an inverter is used, while others are easier with a non-inverting buffer. An inverting buffer is shown in the figure as this is the usual configuration.

Inputs

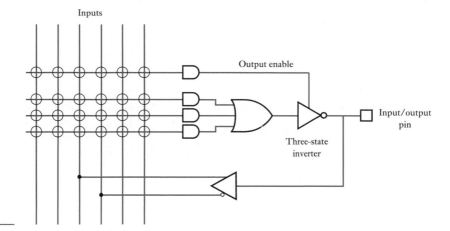

Figure 27.9
A typical PAL input/
output circuit.

Some PALs include an Exclusive OR gate in their output circuit, as shown in Figure 27.10. One input to this gate is connected by a fusible link to ground. If the fuse is intact, this input is pulled low and the relationship between the other input and the output is that of a non-inverting buffer. However, if the fuse is blown, this input will take on a high logic level and the device will then act as an inverter. This allows each output to be individually configured to be inverting or non-inverting.

Figure 27.10
Use of an Exclusive OR
gate as a
programmable inverter.

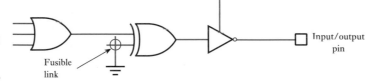

Figure 27.11 shows the functionality of the 16L8 PAL, a 20-pin device that has 10 dedicated inputs, 2 dedicated outputs and 6 lines that can be used as either inputs or outputs. The device can provide up to 16 inputs and up to 8 outputs (though not at the same time). Each output comes from an OR gate with seven input lines, so the device has seven product terms for each output.

More sophisticated PALs replace the 'combinatorial' outputs used in the 16L8 with registered outputs with feedback. An example of such a device is the 16R8 PAL shown in Figure 27.12. Here, each product term is stored into a D–type flip-flop on the rising edge of a clock signal. The output from this flip-flop is used to generate an output signal, but is also fed back to the input array to allow this signal to be used by other parts of the PAL. The ability of the flip-flops to remember the previous state of the device permits the implementation of a range of sequential circuits, such as counters, shift registers and state machines.

More advanced components remove the need to choose between devices with combinatorial and registered outputs by providing a *variable* output structure that can be made to emulate either form. A widely used example of this type of device is the 22V10 PAL, which is shown in Figure 27.13. Here, the output circuit takes the form of a **macrocell** that can be individually configured for each output. In addition to providing a combinatorial or registered output, the macrocell allows outputs to be selectively inverted and provides an output enable function. This device has 10 OR gates which have numbers of inputs ranging from 8 to 16.

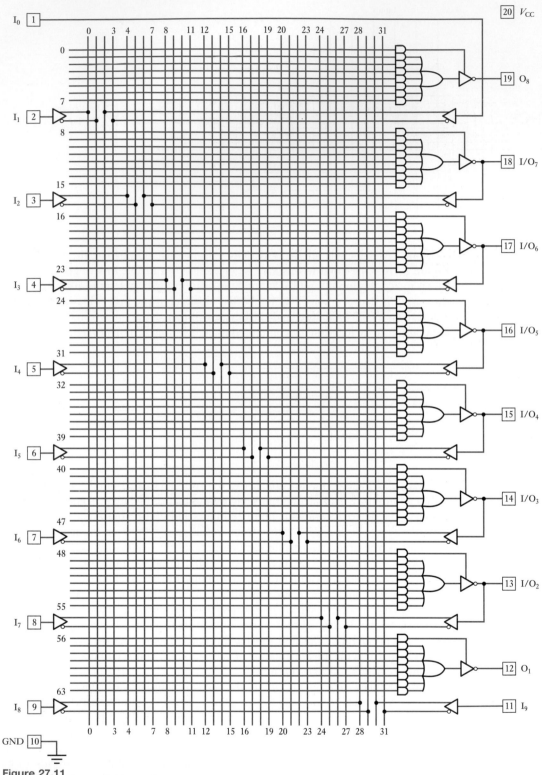

Figure 27.11
A logic diagram of the 16L8 PAL.

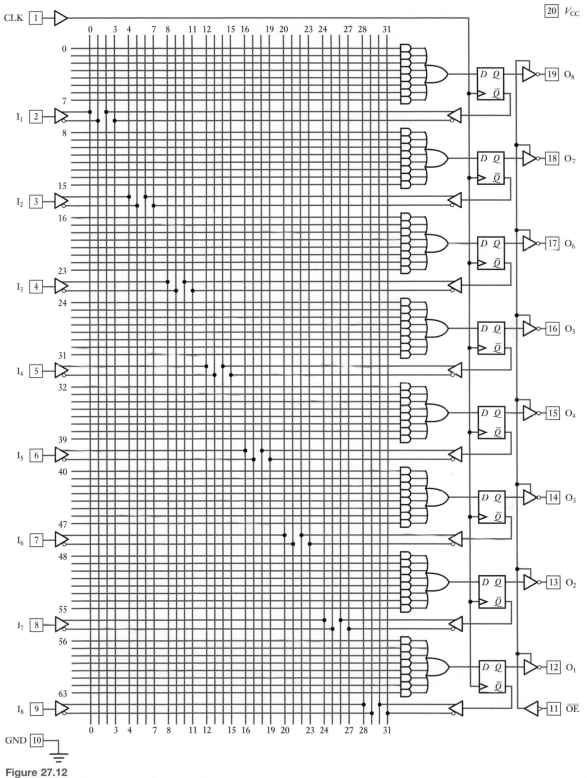

Figure 27.12
A logic diagram of the 16R8 PAL with registered outputs.

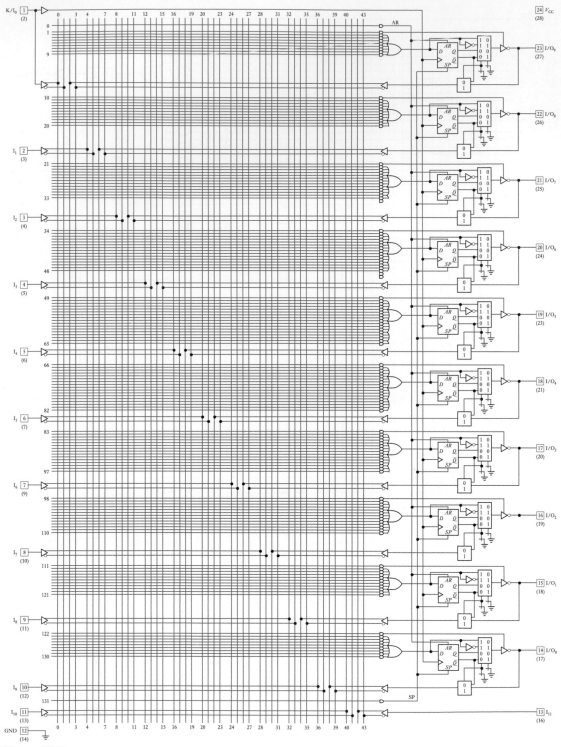

Figure 27.13
A logic diagram of the 22V10 PAL with macrocell outputs.

PALs derive their generic part name (for example, 16L8 or 22V10) from their input/output characteristics:

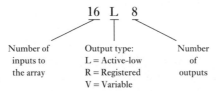

Depending on the device, the number of array inputs varies from 16 to more than 40, while the number of outputs is normally in the range of 4–12. In addition to the three main output types listed above, there are several variants that are given different designations. Generally, the differences relate to variations in the form of the macrocell used at each output.

PALs have evolved over the years since their conception in the early 1980s and now provide high functionality at low cost. One of the attractive characteristics of PALs is that they have a predictable propagation delay time, which in modern high-speed devices is of the order of a few nanoseconds. This allows them to be used at clock speeds of more than 100 MHz.

27.2.3 GALs and EPLDs

Because of their fuse-based construction, PALs can only be programmed once. They are therefore described as **one-time programmable** (OTP) parts. Soon after the development of the PAL, Lattice Semiconductor produced an equivalent device that could be programmed repeatedly. These **generic array logic** (GAL) devices are pin-compatible with conventional PALs, but use **electrically erasable and programmable read-only memory** (EEPROM) technology in place of fuses to achieve reprogrammability (we will look at EEPROM techniques later in this chapter when we look at memory devices). Early GALs were much slower then PALs, but more recent devices have speeds comparable with fuse-based devices.

Other reprogrammable PAL-like devices include **erasable PLDs** (EPLDs). This term is normally applied to parts of a form originally developed by Altera. These are similar to PALs, but use **erasable and programmable read-only memory** (EPROM) techniques in place of fuses (we will look at EPROM in Section 27.3.6). This allows the devices to be erased by exposure to ultraviolet light. Once erased, they can then be reprogrammed. EPLDs generally offer more facilities than PALs and are more flexible. However, they are somewhat slower than PALs, with typical delay times of between 10 and 20 ns.

27.2.4 Programmable electrically erasable logic (PEEL)

Programmable electrical erasible logic (PEEL®) uses CMOS EEPROM technology to provide devices that can be used as direct, low-power replacements for PALs, GALs and EPLDs. The PEEL 18CV8, for example, is a 20-pin device that can be programmed to replicate the functions of a wide range of 20-pin PLDs including the 16L8 and 16R8 PAL devices, while the PEEL 22CV10A is a 22-pin device that can replace devices such as the 22V10 PAL or 22V10 GAL.

The architecture of PEELs is similar to that of the macrocell PALs described earlier, except that they use a more sophisticated macrocell structure. This allows greater flexibility in the configuration of the device and often permits more complicated functions to be incorporated within a single device. The use of CMOS technology permits low power consumption and PEEL devices are available for both 5 V and 3 V operation. Low-power versions might require only a few milliamps when operational and only a few microamps when in their standby mode. Figure 27.14 shows the basic structure of an 18CV8 PEEL device.

27.2.5 Programmable read-only memory (PROM)

We noted earlier that, despite the flexibility of the PLA provided by its two programmable arrays, it is often more efficient to use a less complicated configuration with a single programmable array. This is done in the PAL by replacing the programmable OR array with a fixed series of interconnections. An alternative way in which to simplify the PLA structure is to remove the programmable AND array to form a structure like that shown in Figure 27.15. This forms a **programmable read-only memory (PROM)**.

A PROM may be visualised as a PLA that has one fixed product term (AND gate) for every possible combination of the input variables. Thus, a device with eight inputs would have 2^8 or 256 product terms. As all possible combinations of the inputs are represented by *one* of the product terms, there is no longer any need to program the AND array and this becomes a fixed **decoder**. Each input combination selects a single AND gate and the OR array is used to determine which of the various outputs is activated (taken high) for that input combination. The pattern written into the OR array therefore determines the output pattern that will be produced for each possible set of inputs.

PROMs were one of the earliest forms of array logic and predate both PLAs and PALs. However, the use of a full decoder is inefficient for most logic applications and they are more commonly used for the storage of programs or data than for implementing logic functions. When used for storage of data, the input pattern represents the address and the corresponding pattern in the OR array represents the stored data. When devices of this type are designed for program or data storage, they are more often referred to as **ROMs** and we will look at the characteristics of these devices later in this chapter when we look at computer memory.

Video 27D

27.2.6 Complex programmable logic device (CPLD)

PLAs and PALs – and equivalent reprogrammable devices such as GALs, EPLDs and PEELs – are often collectively referred to as **simple programmable logic devices (SPLDs)**. To make more complicated devices, the architecture of the device has to change. If SPLDs are simply expanded to add extra inputs, the programmable logic array soon becomes too large to be practical. A more attractive approach is to put several SPLDs on a single chip, with programmable interconnectivity. Such devices are termed **complex PLDs** or **CPLDs**.

CPLDs were pioneered by Altera, but are now produced by a range of manufacturers. In addition to providing a large number of array elements, they also offer a powerful method for interconnecting inputs and outputs to allow complicated circuits to be implemented within a single package. A block diagram of a typical CPLD configuration is shown in Figure 27.16.

CPLDs are normally implemented using EPROM or EEPROM techniques rather than fuses and so are reprogrammable. CPLDs are the subject of a great deal of development work and the capabilities of these devices are increasing rapidly. Devices with several thousands of gates are available, with delay times of only a few nanoseconds. As with SPLDs (but unlike FPGAs), the propagation delay time can be predicted when performing the design.

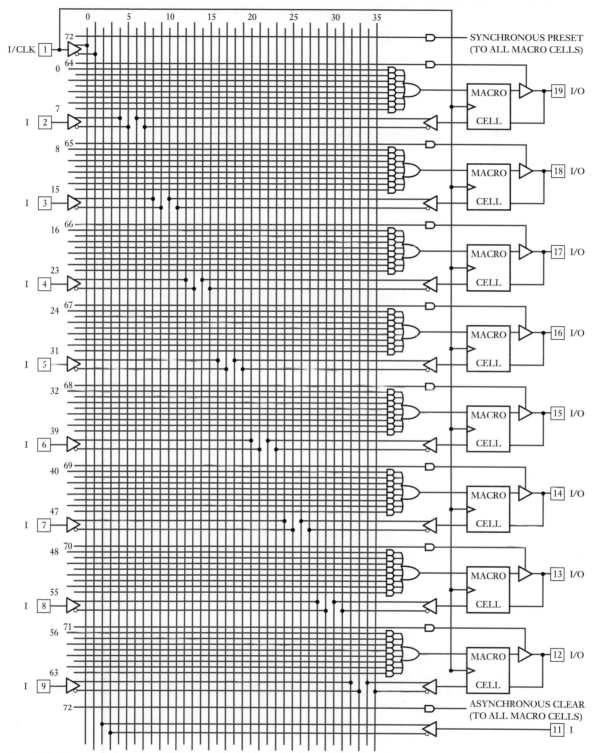

Figure 27.14
A logic diagram of the 18CV8 PEEL.

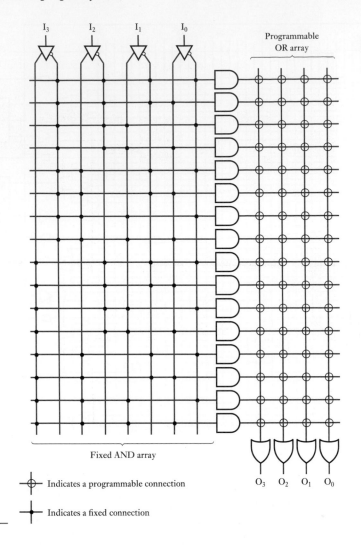

Figure 27.15
A programmable
read-only memory
(PROM).

A single CPLD might typically be used to implement a mixture of registers, decoders, multiplexers and counters that a design using PALs or other SPLDs could only achieve by using several ICs. For example, a complete 32-bit counter can be produced using a single device.

27.2.7 Field programmable gate array (FPGA)

FPGAs take the form of a two-dimensional array of logic cells. These may be arranged in rows or, more commonly, in a rectangular grid, as shown in Figure 27.17. The size of the array varies considerably, with larger parts having tens of thousands of logic elements, and perhaps several megabytes of computer memory. Between the cells of the array run groups of vertical and horizontal channels that can be used to route signals through the circuit. Programmable switches are used to interconnect these conductors and so provide point-to-point connections.

The functionality of each logic cell varies considerably from one manufacturer to another, but a typical cell might contain a register, a few multiplexers and a small look-up table. The look-up table is a small user-programmable memory that takes as its

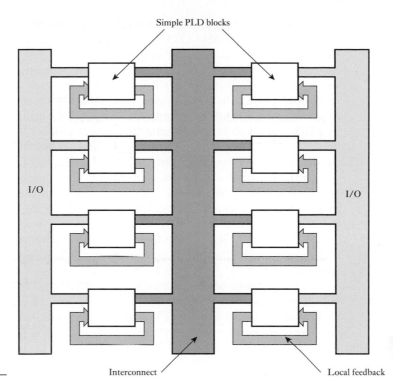

Figure 27.16
Block diagram of a
typical CPLD.

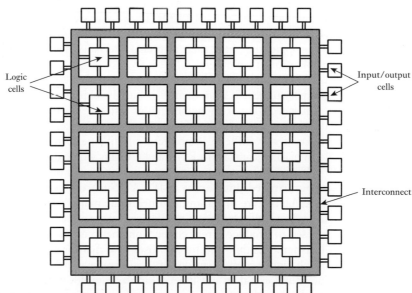

Figure 27.17
A simplified FPGA
arrangement.

address the inputs to the cell. The table can be programmed to implement any logic
function in a manner similar to that of a PROM. More sophisticated devices have
much more complicated logic cells that might contain several registers, look–up tables
and other logic gates. Specialist functions, such as hardware multipliers, memory
blocks, phase-locked loops and input/output buffers, are also provided. The versatility
of this arrangement – and the very flexible interconnection mechanism – allows complicated circuits to be implemented using a single chip.

The programmable switches within FPGAs may be either one-time programmable (OTP) or reprogrammable. OTP parts are based on the use of **antifuses** rather than the conventional fuses used in PALs. When manufactured, antifuses are *open circuit* rather than closed circuit, as in an ordinary fuse. The unprogrammed antifuse resembles two back-to-back diodes that will not conduct in either direction. Programming involves forcing a large current through the diodes by breaking down the appropriate junction. Sufficient current is passed to short-out this junction permanently, removing its effect. The result is a single diode that acts as a closed switch in this circuit configuration. As the breakdown is permanent, the programming cannot be reversed and the device is **one-time programmable**, as in a PAL.

Reprogrammability is achieved in FPGAs by replacing each antifuse with a transistor switch (a MOSFET). The state of the switch is then determined by a memory element that can be set either to open or close the switch. You can visualise the memory as a bistable element. This technique is referred to as a **static random access memory** (SRAM) approach and has the advantage that the contents of the memory can be changed as often as desired. The device is therefore completely reprogrammable.

One disadvantage of this approach is that the content of the memory element is volatile and is lost when power is removed. To overcome this problem, the states of the various interconnections must be loaded from some non-volatile memory (typically a **read-only memory** (ROM) or a computer disk) when power is first applied. In practice, this procedure is not difficult to perform and the advantages of reprogrammability often outweigh this minor drawback.

FPGAs currently represent the most complicated forms of PLDs, with the largest parts being equivalent to tens of millions of gates. This represents a complexity many times that of the largest CPLDs. They are therefore capable of implementing systems that are beyond the capabilities of other forms of array. However, like all forms of array logic, they have their own characteristics that make them more suited to some applications than others.

FPGAs are particularly useful when implementing systems that require on-chip memory or benefit from their distributed architecture. Modern FPGAs operate at very high speeds, but are generally not as fast as PALs or CPLDs. Also, their propagation delay times are greatly affected by the route taken by signals within the chip. This makes it very difficult to predict a circuit's performance before it is completed. This is in marked contrast to PALs and CPLDs where delay times are totally predictable.

For very high-volume applications, it becomes practical to consider using a **mask-programmed gate array** (MPGA). Such devices have an architecture similar to FPGAs, but are programmed during the manufacturing process rather than by the user. The configuration of the device is determined by a photolithographic mask that is used to produce direct connections between the appropriate nodes in the circuit. This removes the need for antifuses or transistor switches and results in a device of much higher component density and greater speed. It can also produce a part with a lower unit cost than an FPGA of similar complexity. However, the disadvantage of this approach is that the mask is extremely expensive to produce. For this reason, use of MPGAs is normally only feasible in situations where tens or hundreds of thousands of similar devices are required.

27.2.8 Programming tools for array logic

The task of configuring a user-programmable logic device for a particular application involves determining the appropriate pattern to be used for the various fuses, antifuses or switches, and then programming this pattern into a target component. For very simple

PLDs it would be possible to derive the necessary **fuse map** manually by studying the functions required. However, this is a complicated and error-prone task and, in practice, automated tools are used for even the simplest parts. More complicated components, such as CPLDs and FPGAs, are far too complicated to be configured manually.

The programming of a device normally makes use of a range of computer-aided design (CAD) tools, although these may be combined within a single package. Figure 27.18

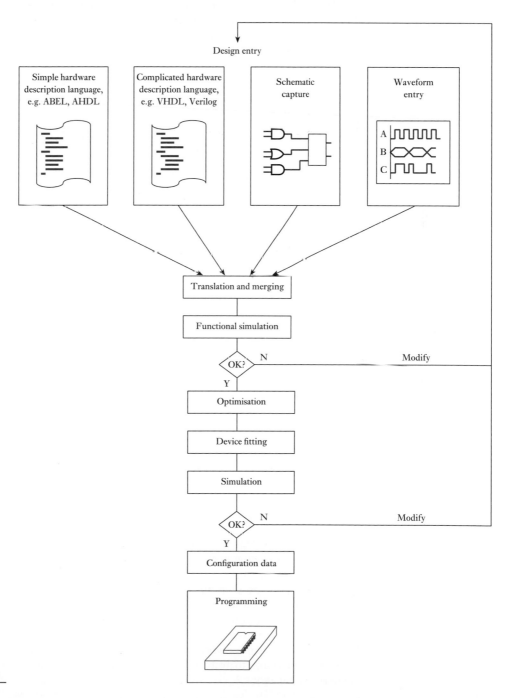

Figure 27.18
A typical design process for a programmable device.

shows the main stages in the process and indicates common methods for design entry. One of the most widely used techniques for specifying the functionality of relatively simple components is the use of a simple **hardware description language** (HDL) such as **ABEL**, **CUPL**, **PALASM** or **AHDL**. These resemble programming languages, but are used to describe the functionality of hardware rather than define a list of instructions for a computer. They describe the required characteristics of the device using elements such as Boolean equations, truth tables and IF–THEN–ELSE constructs. The use of such an HDL is probably the commonest form of design entry used for simple PLDs and also for some more complicated devices.

More sophisticated HDLs, such as **VHDL** and **Verilog**, are also used to specify the functionality of programmable devices, particularly when using complex parts such as CPLDs or FPGAs. We will discuss the use of HDLs when we look at CAD tools in more detail (in Chapter 30).

Design entry can also be achieved by using graphical tools. One approach is to employ a schematics capture package to enter a circuit that represents the functionality of the required device. Alternatively, circuit elements can be synthesised from timing waveforms.

Many CAD packages allow the use of a range of design entry methods and permit elements defined in different ways to be merged to form a complete design. Once the design has been entered and its various components have been merged, the characteristics of the design can be investigated using **functional simulation**. This process does not investigate the exact timing of the final device, but is used to confirm that the circuit is logically correct. This allows design errors to be located quickly, so that time is not wasted in implementing an incorrect design.

When the design has been shown to be functionally correct, automated tools then perform optimisation to simplify its implementation. There then follows a process of **fitting** the optimised design to the selected device. The difficulty of this task depends on the nature of the device concerned. For SPLDs it is a relatively straightforward process of allocating product terms to those available within the chosen part. When using CPLDs the task is somewhat more difficult because of the additional interblock routing. The process is most demanding in the case of FPGAs because of their cellular architecture and the vast number of possible ways to route the interconnections. As a result, the software tools required for FPGA design are perhaps 100 times more complicated than those needed for SPLDs. This difference is reflected in their speed of operation and, of course, in their cost.

When the design has been fitted to its target device, the resultant configuration is subjected to a detailed **timing simulation** to confirm that it will function correctly when programmed into a real device. This simulation takes into account both the functional characteristics of the design and the temporal properties of the target part. If the results of this simulation are satisfactory, a configuration file is produced that can be passed to a **PLD programmer**, which can then be used to configure any number of identical target devices. Alternatively, the file could be used to configure or reconfigure a reprogrammable device by downloading the configuration data directly into the device.

Following the programming process it is normal to read out the configuration pattern placed into a device to verify that it is correct. Once this has been done it is normal to program a **security bit**, which prevents the contents of the component from being read. This makes it impossible for anyone to produce a copy of the part. The contents of a programmable device often represent a great deal of development effort and the ability to protect this investment from piracy is a great advantage of this form of technology.

27.2.9 Custom and semi-custom ICs

When producing systems in very large quantities, it may be practical to design a VLSI circuit from scratch specifically for a given application. Such a **custom design** can produce a very efficient implementation by choosing an architecture to match exactly the functional requirements. Unfortunately, the cost of developing such a specialised device is very high and generally this approach is only attractive when producing components in quantities of hundreds of thousands, or millions. However, for such high-volume applications this may be an attractive option as custom design allows a chip to be optimised for its given function.

For more modest projects, an alternative is to use a **semi-custom IC**. Such devices are also known as **application-specific integrated circuits** (ASICs). This term is used to refer to a range of devices and in the past it was common for FPGAs and MPGAs to be described in this way. More recently, the term ASIC has tended to be used for devices that are manufactured from a range of **standard cells**. Such components are produced by combining standard modules to produce the layout of a dedicated chip. These modules might include registers, counters, input/output circuitry and blocks of memory.

This semi-custom approach is much less costly than developing a complete design from scratch, and so is practical for components that are produced in more modest quantities. However, the design is also less efficient than a full custom design, resulting in a larger chip that is consequently more expensive to produce.

27.2.10 Choosing between the various forms of implementation

An important factor in determining the most suitable form of implementation for a given application is the complexity of the required functions. For systems that require only a few gates, it is likely that basic CMOS logic gates will be most appropriate. However, if more than a handful of gates are needed, it will probably be more attractive to use an SPLD of some form. The choice here is likely to be between PALs, GALs, EPLDs and PEELs, depending on the functionality required and the importance of reprogrammability.

SPLDs are the preferred option for applications requiring up to 100 or so gates, even when devices are required in large quantities. However, for more complicated systems, simple PLDs are insufficient and a designer will normally turn to either a CPLD or an FPGA. Which of these two options is best depends on the functionality required. Some applications will fit in well with the linear structure of a CPLD, while others will be more suited to the cellular form of an FPGA. Very complicated applications may be beyond the scope of CPLDs. The logic capacity of the most sophisticated FPGAs is about 10 times that of the largest CPLDs.

When designing systems that are to be produced in very large numbers, it may be appropriate to consider the use of devices that are programmed by the manufacturer rather than the user. MPGAs, ASICs and full-custom chips come within this category. Such parts usually have a lower unit cost than user-programmable components, but have the disadvantage of high development and tooling costs. These costs can only be justified if very large numbers of devices are required. The development cost of full-custom parts is considerably greater than that of MPGAs or ASICs, restricting their use to very high-volume applications.

Figure 27.19 attempts to summarise the effects of system complexity and volume of production on the device selection process. This figure gives only an overview of the various technologies and the numbers given must not be taken as definitive. The capabilities of the various device families are changing constantly, so the relative merits of the different techniques are subject to change.

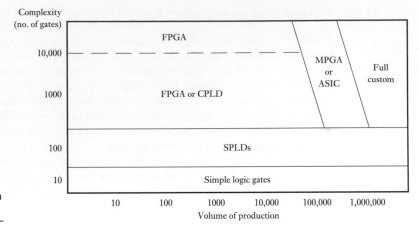

Figure 27.19
A comparison of
various implementation
methods.

While many early PLDs operated from a 5 V supply, more modern components invariably use supply voltages of 3.3 V, 2.5 V, 1.5 V or lower. Low-voltage parts are often slightly slower than those operating from higher voltages, but (as discussed in Section 26.7) their power consumption is considerably lower.

27.3   Microprocessors

Earlier, we noted the problems of creating a single, high-volume VLSI component that can be used in a wide range of applications. In array logic, this problem is tackled by producing a device with a large number of undedicated functions that can be configured for a given application. An alternative strategy is adopted in the case of the microprocessor, which uses a unit that is capable of executing a range of instructions and a program that is used to adapt this to a given situation.

Microcomputers, under the control of a program, can sense input signals, use this information in calculations and then produce relevant output signals. Therefore, given an appropriate program, they can be made to perform the functions of any combinational or sequential logic circuit. This potentially allows large amounts of logic circuitry to be replaced by a single microcomputer chip, with a considerable cost saving. Moreover, as the operation of the microcomputer is controlled by a program, its operation can be modified without changing the physical structure of the system. This flexibility is invaluable in allowing products to be updated easily and cost-effectively. These advantages have led to the widespread use of microcomputers in a range of applications, from laptop computers to aircraft autopilots and from washing machines to the controllers of nuclear power stations.

It should be noted that both array logic and microprocessors are *programmable*, but the meaning of this term is slightly different in these two cases. A PLD or FPGA is programmed by altering the configuration of the device to suit a particular application. However, a microcomputer is programmed by giving it a list of instructions that enable it to perform a given task.

The physical components of a computer are collectively known as the **hardware** of the system, while the programs that control it are referred to as the **software**.

Video 27E

27.3.1   Microcomputer systems

While we often talk of functions being performed by a *microprocessor*, it is more accurate to say that a *microcomputer* is being used. A microprocessor is simply one

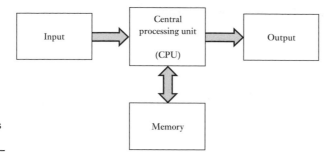

Figure 27.20
The essential elements
of a computer.

element in a microcomputer and other components are needed to make a functioning unit. All computers – from mainframes to microcomputers – consist of a number of primary elements and these are shown in Figure 27.20. The **central processing unit** (CPU) is the heart of the computer. It is responsible for reading the various instructions in a program and performing the operations that these involve. The CPU is often referred to as the **processor** and, in the case of a microcomputer, this is a **microprocessor**. A major element in the CPU is the **arithmetic logic unit (ALU)**, which is responsible for performing the arithmetic and logical operations required by the program.

Memory is used to store both **programs** (a sequence of instructions) and **data** (the information used or produced by a program). A typical small computer might have several thousand memory locations, while large computers might have millions – or sometimes thousands of millions – of locations.

The most powerful computer would be of little practical use without some means of communicating with it. This is achieved by the input and output sections of the computer, which are often collectively called the **input/output (I/O) section**. The form of these elements varies considerably depending on the nature of the information concerned. In a desktop computer, the input/output section would consist largely of circuitry designed to communicate with the keyboard, monitor and printer, and link to peripheral devices. In a small microcomputer in a washing machine, the input and output information would relate to such items as the water temperature and motor speed, so the input/output section would be quite different from that of the computer.

In some cases, the processor, memory and input/output sections of a computer are combined in a single integrated circuit. Such a device is not a microprocessor but a **single-chip microcomputer**. Such components are particularly useful in control and instrumentation applications, where the requirements for memory are normally modest.

Many microcomputers are used in situations in which their presence is far from obvious. These include such application areas as cars, domestic machines and consumer electronics. These arrangements are generally referred to as **embedded systems** because the computer is embedded out of sight, within the equipment. Microcomputers designed specifically for such applications are often called **microcontrollers**, to distinguish them from desktop microcomputers.

Wordlength

Within a microcomputer, information is manipulated and stored in groups of bits. The size of the group that is used within a given machine is called its **wordlength**. The wordlength of a computer is one of the factors that determines its processing 'power' as it controls the amount of data that the machine manipulates at one time.

The earliest microprocessors had a wordlength of 4 bits, but such small machines are rarely used nowadays and then only for simple applications. Most small microcomputers have a wordlength of 8 bits, with more powerful machines using 16, 32 or 64 bits. Machines with wordlengths of 128 or 256 bits are available, but are normally reserved for specialised applications.

It is common to refer to a group of 8 bits as a **byte** and to a group of 4 bits as a **nibble**. The term wordlength has a different meaning for different machines. On an 8-bit machine (a machine with a wordlength of 8), the wordlength is equal to a byte. This is coincidental and should not be allowed to confuse the distinction between the two words.

The wordlength of a machine determines the maximum number of bits of data that can be transferred or processed *at one time*. It does not indicate any fundamental limit to the accuracy of any computation performed by the machine. Equivalent calculations may simply require more operations on a machine with a shorter wordlength.

Communication within the microcomputer

Figure 27.20 shows that information flows between the processor (CPU) and memory, as well as between the processor and the input and output sections. In some circumstances, information may pass directly between the memory and the input or output sections, and this will be discussed in Section 27.3.8.

Communication between the various sections of the computer takes place over a number of **buses**. These are parallel data highways that permit information to flow in one or both direction(s). Figure 27.21 shows the bus structure of a typical microcomputer.

The buses may be considered to be a collection of parallel conductors (wires). Three buses are used to carry data, address information and control signals. For example, if the processor wished to store a data word at a particular memory location, it would place the data on the **data bus**, the address where the information was to be stored on the **address bus**, and various control signals to synchronise the storage operation on the **control bus**.

The number of lines in the data bus is equal to the wordlength of the device and thus determines the number of bits of data that can be moved about the machine at any one time. Therefore, in an 8-bit microprocessor, the data bus would be 8 bits wide, whereas in a 16-bit computer it would be 16 bits wide.

The number of lines in the address bus determines the number of memory locations that can be specified by the processor. This is called the **addressing range** of the device. An 8-bit address bus would be able to specify only 2^8 (256) addresses, which is

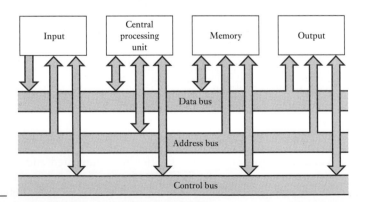

Figure 27.21
A typical microcomputer bus system.

rather limiting for the majority of applications. Most 8–bit computers use a 16–bit address bus, giving an addressing range of 2^{16} or 65,536. Thus, in 8–bit devices, addresses are usually represented by 2 bytes of information. Most 16–bit machines are designed for more demanding applications, which require a greater addressing range than is possible with a 16–bit address bus. Many machines use a 20–bit bus, giving an addressing range of over 1 million, while others use a 24– or even a 32–bit address bus. A 32–bit bus gives an addressing range of over 4,000,000,000 locations, which is sufficient for most applications. More powerful processors will use a wider address bus to achieve a proportionately larger addressing range.

The lines of the control bus are used by the processor to produce actions in external components and to synchronise these operations. The exact nature of these lines varies from one machine to another and we will discuss them in more detail in later sections.

Example 27.1	Determine the addressing range of a microprocessor that uses 24-bit addresses.

A 24–bit word has a range of 2^{24}, which is 16,777,216 locations.

Registers

The memory section of the computer consists of a large number of memory registers that can be used to store both data and programs. The processor and the input/output sections also contain registers that are used for a range of purposes (these will be discussed in the sections that follow). The register is thus a fundamental building block within a computer system, so an understanding of its operation is essential.

We have already looked at the use of D flip-flops in the construction of memory registers (see Section 25.6). Within a microprocessor, communication between registers takes place over a bus system, which imposes some restrictions on their design. We have also seen how gates with **three-state outputs** may be connected together, provided that the output of only one gate is enabled at any time (see Section 26.4.3).

By using D flip-flops with three-state outputs, it is possible to produce memory registers that can be connected directly to a bus system. Figure 27.22 shows such an arrangement. Here, an 8-bit register is connected to a data bus. The register has two control inputs – one to write data from the bus into the register (this corresponds to the clock input to the flip-flops) and the other to enable the outputs of the register to drive the bus (this is connected to the three-state control of the gates). The input and output of each flip-flop are connected together to the corresponding bit of the data bus.

Communication between a number of registers is achieved simply by enabling both the output of one register and the input of another, as illustrated in Figure 27.23. As all the registers within a system are connected by the same data bus, only one piece of information can be communicated at any one time. If many pieces of data are to be

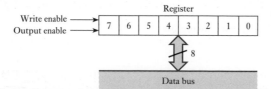

Figure 27.22
A simple 8-bit register.

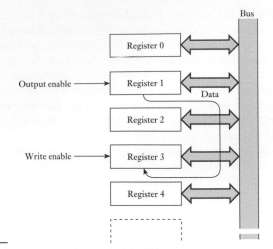

Figure 27.23
Communication
between registers.

transferred, this will require several operations. Thus, this arrangement is a **sequential** communication system.

The process of taking information from a register and placing it onto the bus is referred to as **reading**, while the process of storing information in a register is called **writing**. It is important to note that enabling the output of a register does not change its contents. Therefore, the read operation is *non-destructive* and information can be read from a register many times without changing its contents. The only way in which a data word within a register is changed is if a new data word is written into it (unless, of course, the machine is turned off).

Processor architectures

Most microprocessors are arranged with a single block of memory that is used to store both programs and data. Such processors are normally described as having a **Von Neumann architecture**, after the Hungarian mathematician who did much of the early work in this area.

The ability to use any memory element for either program or data storage gives great flexibility and also produces a simple hardware structure. However, the need to access both instructions and data over the same bus system can lead to bus congestion, which limits the speed of operation of the processor.

An alternative arrangement used in some processors is to have separate memory blocks for programs and data, with separate buses to access each form of information. This **Harvard architecture** can speed up processor operation (as instructions and related data can be obtained at the same time over separate buses), but results in a more complicated architecture.

Here, we will concentrate on processors using a Von Neumann approach as these are the most numerous, although we will look at an example of the use of a Harvard architecture in Section 27.3.9 when we discuss PIC microcontrollers.

27.3.2 Data and program storage

Numeric data

The storage of numeric data within a microcomputer is straightforward. In many cases, the memory within a computer is arranged as a series of single-byte memory elements. Pieces of information that can be represented by a single byte can simply be stored within a single memory location at an address determined by the programmer. A single byte of data can store numbers in the range 0–255. If a piece of information requires more than a single byte, a series of memory locations can be used together. For example, two adjacent memory locations can be used to store a 16-bit number, giving a range of 0–65,535, or four locations could store a 32-bit number, with a range of 0–4,294,967,295. Clearly, numbers of arbitrary size can be stored by using an appropriate number of memory locations.

Example 27.2

One estimate suggests that there are about 10^{75} atoms in the Universe. If this quantity were to be stored as a binary number, how many bits would be required?

Strictly, the smallest word that could be used requires 250 bits as $2^{250} \approx 1.8 \times 10^{75}$. However, we would normally use a whole number of bytes, so we would probably use 32 bytes, which is 256 bits. This gives a range of about 10^{77}. Thus, to store this huge number, we would probably use 32 bytes of memory.

Negative number representation

In some applications it is important to represent not only the magnitude of a quantity but also its polarity. One way to do this is to adopt a technique similar to that used in ordinary decimal arithmetic and use one of the bits of an n-bit number as a sign bit. This is usually the most significant digit (MSD), which is 0 if the number is positive and 1 if it is negative. Thus, the MSD represents the sign of the quantity and the remainder shows its magnitude. This technique is referred to as the **sign–magnitude number representation**. The method, while simple in concept, does give rise to some problems. For example, it produces two ways to represent zero, corresponding to +0 and −0.

Most microcomputer systems use an alternative representation for positive and negative numbers, which is the **two's complement number representation**. This is most simply understood by recalling the characteristics of an up/down counter (as described in Section 25.8.4). Let us imagine that we have an 8-bit counter and that we start initially from some positive value. As we count down through zero, we will get the output sequence shown below:

00000011	3
00000010	2
00000001	1
00000000	0
11111111	−1
11111110	−2
11111101	−3

Clearly, the pattern of all '1's represents −1 and counting down from this number we have −2, −3 and so on. This is the two's complement form.

As with sign–magnitude notation, the MSD again indicates the polarity of the number – 0 for positive and 1 for negative. The largest positive quantity that can be represented by an 8-bit number is clearly 01111111 (127) and the largest negative number is 10000000 (−128). The two's complement representation of positive numbers in this range is the same as its conventional binary form. The two's complement form of a negative quantity can be found by subtracting the binary equivalent of the quantity from 2^n, where n is the number of bits in the two's complement word. For example, the 8-bit two's complement form of −3 may be found as follows:

$$3 \quad = \quad 00000011$$
$$2^n = 256 = 100000000$$

Therefore

$$-3 = \quad 100000000$$
$$\underline{-00000011}$$
$$= \quad 11111101$$

Numbers that represent only magnitude are referred to as **unsigned numbers**, while those that represent polarity and magnitude are called **signed numbers**. An 8-bit unsigned quantity can represent numbers in the range 0–255, whereas a similar signed quantity has the range −128 to +127.

Example 27.3 | **What is the range of a 16-bit signed quantity?**

A 16-bit quantity can represent unsigned numbers in the range 0 to 65,535, so it can represent a signed number in the range −32,768 to +32,767.

Example 27.4 | **Express −5 as a 16-bit signed number.**

In a similar manner to the earlier calculation

$$5 \quad = \quad 00000000\ 00000101$$
$$2^n = 65,536 = 1\ 00000000\ 00000000$$

Therefore

$$-5 \quad = 1\ 00000000\ 00000000$$
$$\underline{-00000000\ 00000101}$$
$$= \quad 11111111\ 11111011$$

Floating point numbers

In addition to the integer number representations outlined above, it is common to use a floating point format to allow both very large and very small quantities to be used. This is done by grouping a number of bytes of information together, then using some of the resultant bits to represent a signed mantissa and the remainder a signed exponent. A typical format uses 4 bytes (32 bits) to represent each number, producing an accuracy of about seven or eight decimal places with a range of 10^{-38}–10^{+38}.

Text

Text information can be stored using a sequence of memory locations, each character being represented by an appropriate code (for example, the ASCII code described in

Section 24.12.4). This usually results in one memory location being used for each character. Word processing systems often use very large character sets that include characters from a range of languages, special characters and graphics elements. In such applications, each character requires more than a single byte of data.

Program storage

A computer program is a list of instructions to the processor. All microprocessors have a set of instructions that they can execute and these make up what is called the **instruction set** of the machine. Each type of processor has its own instruction set and, generally, programs written for one machine will not operate on another.

A typical instruction set will include instructions for: transferring data between registers; transferring data between registers and memory; performing various arithmetic and logical operations; comparisons and tests on register contents; and controlling the sequence of program execution.

Within an 8-bit microprocessor the operation to be performed by a particular instruction is usually defined by a single-byte **operation code**, or **opcode**. The use of a single byte to define the operation limits the number of possible members of the instruction set to 256, though most devices use somewhat fewer than this. In some cases the opcode is all that is required to specify an instruction. For example, an instruction to increment (increase by one) the contents of a specified accumulator requires no further data. However, with some instructions, extra information is required. For example, an instruction to store the contents of an accumulator in memory needs to include the address where the data word is to be stored. This leads to a situation in which some instructions are 1 byte in length (a 1-byte opcode), some are 2 bytes (an opcode plus a single byte of data) and some are 3 bytes long (an opcode plus 2 bytes of data). The information that accompanies the opcode is called the **operand**, which might represent data used by the program or an address. A section of program stored in memory might be of the form shown in Figure 27.24.

It is worth noting that microcomputers differ in the way in which they store the bytes of a 16-bit address within memory. Some machines store the MSB (Most Significant Byte) at the lower address of a pair of adjacent locations, while others store it at the higher address. This variation between machines is called **byte sexing** and there are arguments in favour of both methods. When considering program storage in detail, it is vital to know the orientation used by the processor in question.

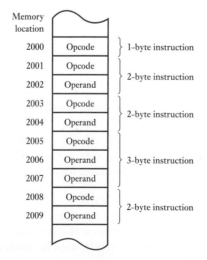

Figure 27.24
Storage of a program in an 8-bit microcomputer.

Some advanced designs of microprocessor increase the number of available opcodes – and, therefore, the size of their instruction set – by defining one opcode that indicates that a second opcode byte should be fetched to select one of a secondary set of instructions. This allows a much larger range of instruction types and improves program efficiency.

16- and 32-bit microcomputers have correspondingly larger opcodes, allowing for a much more extensive range of instructions. Such computers often have very sophisticated and complicated instruction sets.

Reduced instruction set computers (RISC)

The use of sophisticated instruction sets allows complicated sequences of operations to be defined by individual instructions. This reduces the length of programs and so reduces the time that the processor takes to fetch instructions. Processors that adopt this approach are described as **complex instruction set computers** (**CISC**) machines. Unfortunately, while the CISC approach reduces the *length* of programs, the hardware required to execute the complex instructions becomes very complicated, making the processors large and reducing their speed of operation.

An alternative approach is taken by **reduced instruction set computers**, which are generally referred to as **RISC** machines. These have a limited number of relatively simple instructions, resulting in a simple architecture that can run at very high speed. Many high-performance processors adopt a RISC approach.

27.3.3 The processor

The processor of a microcomputer consists of:

- a series of **registers**;
- some electronic circuitry to perform arithmetic and logical operations on the contents of these registers;
- some circuitry to decode and execute a sequence of instructions;
- buffers to interface the signals within the processor to the real world;
- a series of buses that join the various components together.

The structure of a microprocessor is referred to as its **architecture**. As one would expect, 8-bit devices are generally much simpler in form than 16- or 32-bit devices, although their major elements are often similar. More powerful processors tend to include features such as memory caches and memory management units that are not normally found in 8-bit devices.

A detailed discussion of microprocessor architectures is beyond the scope of this text, but it is useful to look at the major components within a typical device. In the sections that follow we will look primarily at 8-bit devices as they are considerably simpler in construction. Much of the discussion is equally relevant to more powerful processors, although factors such as register and memory sizes will clearly be different.

Accumulators

An accumulator is a general-purpose working register. Most operations within the processor are performed on, or are affected by, the contents of such registers. Machines vary considerably in the number of accumulators that they have. Some simple devices provide only a single 8-bit accumulator, while others provide two that can be used individually or together to form a 16-bit register. Other processors have a bank of general-purpose registers that can be used individually or in pairs.

Index registers

Most microprocessors have one or more **index registers**, which are normally used to store address information. These registers can usually be automatically incremented (increased by 1) or decremented (decreased by 1) to allow a sequence of addresses to be used in order. As most 8-bit microprocessors use 16-bit addresses, these registers are normally 16 bits long. However, this is not always the case.

Program counter

A computer program consists of a list of instructions to the processor. These instructions are normally stored in order within a section of the memory and are fetched, one at a time, to be executed. Clearly, the device must keep track of the location of the next byte of the program, and this is done using the **program counter**. This register normally contains the address of the next byte of the program. The value in the program counter is incremented each time a byte is fetched so that it automatically points to the next. The size of the program counter is determined by the length of the addresses used.

Instruction register

As a program runs, the opcode for each instruction is brought, in turn, to the **instruction register** for execution. This register is connected to the instruction decoding and control unit (see later), which produces an appropriate sequence of control signals to implement this instruction. In 8-bit microprocessors, the instruction register is 8 bits long to accommodate an 8-bit opcode.

Arithmetic and logic unit (ALU)

The ALU is responsible for performing all the arithmetic and logical operations required to implement the various opcodes. In general, it is able to add, subtract, increment and decrement 8-bit quantities and perform a range of logical operations on a pair of bytes. Arithmetic operations can be carried out on either **signed** or **unsigned** data and some machines can also work with **binary-coded decimal** (BCD) numbers. Logical functions normally include AND, OR and Exclusive OR and are performed bit by bit on the corresponding bits of the two data words, as illustrated in Figure 27.25.

In addition to arithmetic and logical operations, the ALU can manipulate data words by shifting them to the right or left (as in the **shift register**, described in Section 25.7) or by rotating them in either direction. Rotation is similar to shifting, except that any data bits that fall off one end are re-inserted at the other.

In simple 8-bit processors the ALU cannot perform **multiplication** or **division** directly. Instead, these functions must be carried out by means of a sequence of add/subtract and shift instructions. More sophisticated devices do provide an 8-bit multiply operation, producing an increased speed of execution for this function.

Some 8-bit processors also provide a range of 16-bit operations, which are performed using 16-bit registers or pairs of 8-bit registers. These include 2-byte read and write operations and a range of data manipulation functions.

Figure 27.25
Logical operations of the ALU.

A 1 0 1 1 0 1 1 0	A 1 0 1 1 0 1 1 0	A 1 0 1 1 0 1 1 0
B 0 1 1 1 0 0 0 1	B 0 1 1 1 0 0 0 1	B 0 1 1 1 0 0 0 1
$A{\cdot}B$ 0 0 1 1 0 0 0 0	$A{+}B$ 1 1 1 1 0 1 1 1	$A{\oplus}B$ 1 1 0 0 0 1 1 1
(a) AND	(b) OR	(c) Exclusive OR

Processor status register (flags register)

Associated with the ALU is a **processor status register**. The individual bits of this register are set and cleared to indicate certain aspects of the machine's status and the results of ALU operations. The bits are often referred to as **flags** because they signal a particular event or condition. For this reason, the register is sometimes called the **flags register**. The format of this register differs from one machine to another, but common 'flags' are as shown in Table 27.1.

Although the flags reflect the results produced by the ALU, they are also affected by other operations, such as writing data into an accumulator. Each flag has a set of rules determining when it will change and when it will not.

To make use of the information within the processor status register, the processor has, within its instruction set, a number of **conditional instructions**, the operation of which depends on the status of the flags. The form of these instructions varies considerably between machines, but usually the instructions control the sequence of program execution. This allows the programmer to specify different sections of program that will be executed depending on the result of some calculation or manipulation.

Table 27.1 Common flags.

Symbol	Flag	Meaning
Z	Zero	Set if an ALU operation generates a zero result
N	Negative	Set if the result of an ALU operation is negative
C	Carry	Set if an arithmetic operation generates a carry
V	oVerflow	Set if an arithmetic operation generates an overflow into the sign bit

Instruction decoding and control unit

The instruction decoding and control unit is the heart of the processor. It is responsible for sequentially *fetching* instructions from memory and then *executing* them. Attached to the control unit is a **clock generator**, which normally uses a **crystal oscillator** to produce a very accurate clock signal. This is counted to divide time into a number of **clock cycles**. The operation of the unit can be divided into two parts, which are the **instruction fetch cycle** and **instruction execute cycle**. Both of these operations may take a number of clock cycles.

The control unit performs an instruction fetch cycle by generating a fixed sequence of control signals over a number of clock periods. These control signals correspond to the write enable and output enable signals associated with the various registers, as shown in Figures 27.22 and 27.23. The sequence produced transfers the contents of the program counter to the address buffer registers, sends a *read* command to external memory devices via the control bus, reads in data from the data bus and, if this is the first byte of an instruction, transfers this byte to the **instruction register**.

The instruction is then executed – again by means of the generation of a sequence of control signals over a number of clock periods. However, in this case, the sequence depends on the nature of the instruction. The instruction register is connected to a decoding network, which selects the appropriate sequence of control signals for each known opcode. The selected sequence determines how many further bytes of data are required. These are then loaded by repeated fetch cycles, as above. The number of cycles required to execute the instruction depends on the particular opcode.

When execution has been completed, the machine automatically commences another instruction fetch cycle to obtain the next instruction in the program.

Program execution is therefore a continuous sequence of fetch and execute cycles. Each fetch cycle is the same, but the nature of the execute cycles is determined by the opcode.

Address and data bus buffers

The address and data bus buffers are interfaces between the microprocessor and the outside world. Inside the chip, resistances are high and capacitances low, enabling very low power signals to be used. When signals are required to drive external loads, with their associated capacitances, more power is required. The buffers provide this extra power and improve the noise performance of the circuit.

Stack pointer

The stack pointer is a special-purpose register found in many microprocessors. The register stores an address that *points* to a location in memory (locations that store addresses are often referred to as **pointers**). Within the instruction set of the machine are opcodes that cause information to be stored at the location defined by the stack pointer. When this is done, the value in the stack pointer is automatically incremented so that it points to the next location. Instructions of this type are often given the name **push**. If several pieces of data are 'pushed' onto the stack, they will be stored at sequential locations in memory, forming a *stack* of data. Also provided are instructions for removing data from the stack. These are sometimes given the name **pull**, and sometimes **pop**. These first decrement the address in the stack pointer, then fetch the contents of the address to which it points. The combined effect of these two instructions is that, if several pieces of information are pushed onto the stack, they will emerge again, in reverse order, as a result of repeated pull operations. The use of these complementary instructions is best understood by an example, which is given in Figure 27.26.

The initial value in the stack pointer (SP) can be set using a *load stack pointer* command. Let us assume that this has been set initially to 1000. This condition is illustrated in Figure 27.26(a). If now a number A is pushed onto the stack it will be

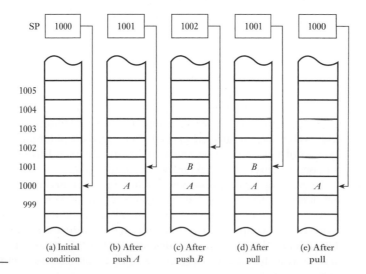

Figure 27.26
Pushing and pulling information with a stack.

stored at the location pointed to by the stack pointer (1000) and the stack pointer will be incremented (to 1001), as shown in Figure 27.26(b). Pushing a second number B onto the stack will repeat the procedure, as shown in Figure 27.26(c). If now a data word is pulled from the stack using a pull command, this will cause the stack pointer to be decremented (to 1001) and data to be read from this location. This will give the number B, as shown in Figure 27.26(d). If this is repeated, the next number to be obtained will be A, as shown in Figure 27.26(e). Thus, data words are pulled from the stack in reverse order.

Although it may not seem particularly apparent yet, this **first-in–last-out** (FILO) mechanism is of great significance. We will see why it is so useful when we come to consider the implementation of *interrupts* later in this chapter. The stack is also used in the handling of subroutines.

The number of items that can be stored and retrieved in this way is limited only by the amount of memory available and the size of the stack pointer register. Most 8-bit processors have a 16-bit stack pointer, allowing the stack to occupy any portion of the address space. However, some devices use shorter registers and these limit stack size. This is a considerable weakness in design and makes such processors very inefficient in the execution of many **high-level languages**.

27.3.4　Communication with external components

The microprocessor communicates with external components using the bus system mentioned above. External devices, whether used for memory or input/output, appear to the processor as registers with specific addresses. To send data to one of these components, the processor places the address of the chosen register on the *address* bus, the data to be transferred on the *data* bus and appropriate synchronisation signals on lines of the *control* bus. To receive information from a component, the processor places its address on the *address* bus, sends appropriate command signals on the *control* bus and the device places the required data on the *data* bus, which the processor reads.

Bus multiplexing

One of the constraints on the features that can be provided within a microprocessor is the number of pins that are available on the device. To reduce the number of pins required, some microprocessors use several pins to perform two functions and many devices use eight lines to represent both the data bus and part of the address bus. This process, which is referred to as **bus multiplexing**, is illustrated in Figure 27.27.

Figure 27.27 is an example of a **timing diagram**, which shows the temporal relationships between various signals within the system. Each signal, or group of signals, is plotted against time and its state is shown using a symbolic representation, as outlined in Figure 27.28.

Returning to Figure 27.27, we see that the high-order byte of the address bus (A_8–A_{15}) is present throughout the period shown. However, the low-order byte of the address bus (A_0–A_7) and the data are each represented for only a fraction of the period by a set of lines that is now called the **multiplexed address/data bus**. The lines of this bus are given the symbolic names AD_0 to AD_7.

In order for external devices to be able to decipher this information, two control signals (or clock signals) are provided that allow external components to extract the information they require. The forms of these control signals differ depending on the machine used. In the example shown, they are called the **address strobe** and the **data strobe**. The falling edges of these signals indicate that the appropriate signals are stable and may be read by external devices. The full 16-bit address is obtained by reading the

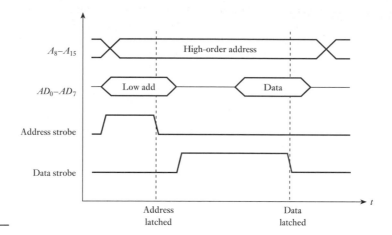

Figure 27.27
Bus multiplexing.

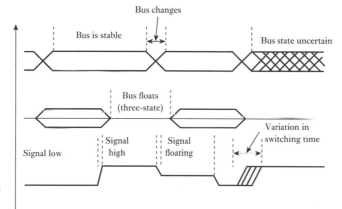

Figure 27.28
Symbolic representations used in timing diagrams.

low-order byte of the address using an 8-bit latch (which is usually called an **address latch** when used in this way). The enable signal for the latch comes directly from the address strobe of the microprocessor. Consequently, this signal is also called the **address latch enable** (ALE) signal. This arrangement is shown in Figure 27.29.

Following the falling edge of the address strobe, the full 16-bit address is available to external components. On the falling edge of the data strobe, such components have both address and data information available.

On devices that do not use bus multiplexing, the address and data buses are separate and an address latch is not required. A single data strobe can now be used to synchronise transfers between the processor and external devices.

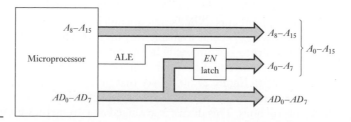

Figure 27.29
The use of an address latch.

27.3.5 Memory

We have seen that the memory section of a microcomputer consists of a large number of registers. Modern IC memory devices may contain thousands or even millions of memory registers. Clearly, in such circumstances it is not possible to use individual write enable and output enable control lines for each register, as shown in Figure 27.22. Instead, all the registers within a device are controlled by a small number of lines, with internal **address decoding** logic being used to select the appropriate location. A typical arrangement is shown in Figure 27.30.

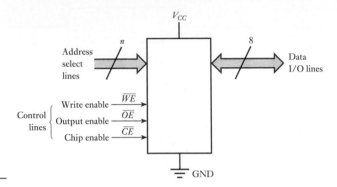

Figure 27.30
A typical memory device.

The number of memory locations within a memory device is invariably a power of 2 as n address lines can specify 2^n memory addresses. Because the numbers concerned are large, it is normal to express memory sizes in **kilobytes** (kbytes), where 1 kbyte is equal to 1024 bytes. This rather strange notation comes about because 1024 is 2^{10}, which is conveniently close to 1000 – normally represented by 'kilo' (as in kilogram and kilometre). Thus, a device with 1024 8-bit memory registers would be called a 1-kbyte memory and would have 10 address lines, while a device with 4096 memory locations would be called a 4-kbyte device and would have 12 address lines. The 16 address lines of a typical 8-bit microprocessor give an addressing range of 64 kbytes. A similar notation is used for very large amounts of memory, where a block of 2^{20} (1,048,576) bytes of memory is referred to as a **megabyte** (Mbyte) and a block of 2^{30} (1,073,741,824) bytes is called a **gigabyte** (Gbyte).

Unfortunately, the indication of memory size is made slightly more complicated by the fact that, although the memory of a system is normally expressed in kilobytes, the capacity of an individual device is often given in kilobits (kbits) or megabits (Mbits). This is because many devices are not organised as a group of 8-bit registers. This alternative notation should not cause problems, but can occasionally lead to misunderstandings, particularly when people abbreviate the unit of memory size to 'k' or 'megs'. If an arrangement has 1 M of memory, it makes a considerable difference whether this is 1 Mbit or 1 Mbyte! Large memory devices are usually arranged as a series of 8-, 16-, 32- or, in some cases, 64-bit wide data words.

In the early days of microcomputing, memory devices had typical capacities of a few kilobytes. This meant that most systems required many chips to provide their memory requirements. Now, single memory devices often contain many megabytes and larger devices become available year by year. The full addressing range of an 8- or 16-bit microprocessor can easily be provided by a single chip, but it is more usual to use more than one chip to combine memory devices with different characteristics.

Returning to Figure 27.30, we see that the device shown has n address lines (where n is determined by the device's capacity), eight bidirectional input/output lines and three control lines. This device is arranged as a group of 8-bit wide registers. Devices

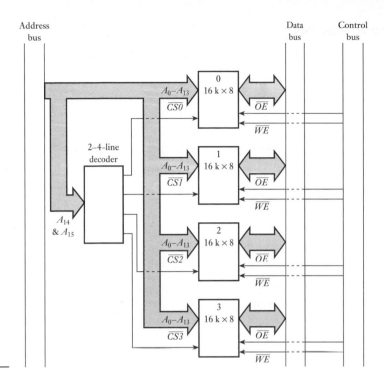

Figure 27.31
A typical memory arrangement.

using 16-, 32- or 64-bit wide structures would have a correspondingly larger number of input/output lines.

Two of the control lines are the **output enable** and **write enable** lines of Figure 27.22. These lines are invariably **active low inputs** and are therefore given the symbolic names $\overline{OE}$ and $\overline{WE}$.

The third control input is the **chip enable** line, $\overline{CE}$. When this line is active (low) the device is 'enabled' and will respond to the two enable signals. When the chip enable line is not active (that is, high) the device will not respond to the other control lines, thus preventing it from being accessed. The chip enable line is used to determine which of a number of devices is used at any one time. When used in this way, this control line takes on the role of a **chip select** line, $\overline{CS}$. The address lines of all the devices will be connected to the low-order bits of the address bus. Signals fed to the chip enable line of each device will determine which is read from or written into. This arrangement is illustrated in Figure 27.31.

The figure shows an arrangement using four 16-kbyte memory devices. Each has 14 address lines ($2^{14} = 16,384 = 16$ kbytes), which are connected to bits 0 to 13 of the address bus (A_0–A_{13}). The most significant address lines (A_{14} and A_{15}) are connected to a '2–4-line decoder'. This selects one of four output lines, depending on the combination of signals applied to its two inputs. If A_{14} and A_{15} are both 0, the first of its outputs, $\overline{CS0}$, will be low and the others will be high. This will select the first of the memory devices and deselect the other three. Similarly, if A_{14} and A_{15} are both 1, memory device 3 will be selected. For each combination of the two address lines, only one of the memory devices will be selected. As A_{14} and A_{15} are the most significant lines of the address bus, addresses placed on the bus will automatically select one of the memory devices depending on the two most significant bits. The location used within that device will be determined by the remaining 14 bits of the address. The address space is therefore partitioned, as shown in Figure 27.32. This graphic representation

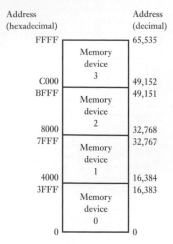

Figure 27.32
The memory map of
the system of Figure
27.31.

of the allocation of memory within a system is called a **memory map** or sometimes
an **address map**.

You will notice that the range of addresses corresponding to each memory device
is shown in both decimal and **hexadecimal**. In fact, hexadecimal is much more commonly used in this application than decimal as hexadecimal is more readily converted
to binary, which gives the pattern of '1's and '0's on the bus.

The arrangement of Figure 27.31 shows a system using a non-multiplexed address
and data bus. In a **multiplexed** system, the circuitry is similar, except that an address
latch is added, as shown in Figure 27.29. When the processor performs a *write* operation,
the write enable signal $\overline{WE}$ is synchronised with the falling edge of the data strobe
signal of Figure 27.27 to ensure that the data word is stable on the data/address bus
when the write is performed. When the processor performs a *read* operation, the
output enable line $\overline{OE}$ is activated, causing the selected memory device to place its
information on the bus. The processor then waits for the data to become stable before
reading the data from the bus.

27.3.6 Memory types

So far, we have considered memory as a simple collection of registers that can be written
and read at will. In fact, there are several different forms of memory that have very
different characteristics. Memory is used for a variety of purposes within a computer
system. In some cases it stores programs that will never be changed while, in other cases,
it is used for data that is constantly being modified by the processor. Some systems
also have **secondary memory**, which is used for bulk storage of programs and data.
Secondary memory is often some form of **disk drive** and so is relatively slow to access.
Here we will concentrate on the primary semiconductor memory of the computer,
which may be categorised as either RAM or ROM.

RAM

Data that must be changed during the operation of a program is normally stored in
random access memory (RAM), this being the name given to memory that can be
both written and read quickly.

The name stems from the fact that, in these devices, any byte of data can be
accessed in an equal amount of time. It is tempting to suppose that this would be true

of all memories. However, it is clearly not true of such storage devices as magnetic tapes, as programs at the end of the tape will take much longer to access than those at the beginning. In fact, the use of the term *random access* is largely historical and is unfortunate as most forms of memory used within microcomputers (other than those used for bulk storage) are random access, even those that are *not* RAMs. A more appropriate name would be read/write memory, but the word RAM is used universally and has therefore become established.

RAM in modern microcomputers is implemented using one of two circuit techniques. **Static RAM** uses a bistable arrangement similar to that described in Section 25.6. Information written into such a device is retained indefinitely provided power is maintained. **Dynamic RAM** stores information by charging or discharging an array of capacitors. Dynamic RAM requires far fewer components for each bit of information stored, permitting more storage elements to be integrated within a single chip. However, it suffers from the disadvantage that the charges on the capacitors tend to decay over time, making it necessary to **refresh** the devices periodically by applying an appropriate sequence of control signals.

One characteristic of RAM is that it is **volatile**. That is, it loses its contents when power is removed. In many applications this is unimportant as, when power is lost, the computer itself stops functioning. However, in some applications the contents of memory must not be lost when power is removed. For example, the storage of the control program for an embedded system must be non-volatile as it must be present within the system when it is first turned on. For this reason, programs are normally stored in ROM, as described in the next section, rather than in RAM.

There are many situations where it is necessary to be able to read and write the contents of memory that must also be non-volatile. In such cases it is normal to use some form of **battery backup** to protect the contents of the RAM from power failure. Fortunately, when CMOS memory is not being clocked, its power consumption is extremely low (as discussed in Section 26.7) allowing even a small battery to maintain its contents for extended periods.

ROM

ROM is **read-only memory** – that is, it can be read by the processor, but the processor cannot write into it. Such devices are non-volatile and so are suitable for storing programs or any non-changing data. There are many forms of ROM. Some must be programmed by the device manufacturer, while others can be programmed by the user, perhaps using special equipment. Table 27.2 lists the acronyms of several forms of ROM and gives their meanings.

The general term *ROM* is applied to all forms of read-only device. Some of these are **mask programmed**, which means that the device is programmed photolithographically by the chip manufacturer as the last stage of production. A designer using this approach must supply the manufacturer with the program and pay a fee for the production of the mask. However, when the mask has been made the unit cost of the device is low. This is the most attractive option for high-volume production, but is unsuitable for development and low-volume applications.

Table 27.2 Acronyms for various types of ROM.

ROM	Read-only memory
PROM	Programmable read-only memory
EPROM	Erasable and programmable read-only memory
EEPROM	Electrically erasable and programmable read-only memory

An alternative for small-scale projects is to use one of the range of **programmable read-only memories** (PROMs). These are available in a number of variants, but all allow users to program the device themselves, saving the high cost of mask production. In fact, the term *PROM* is normally reserved for small, fusible link devices of the type described in Section 27.2.5. These were one of the earliest forms of PLD but have now been overtaken by more modern devices. A characteristic of these devices is that once programmed they cannot be modified.

For flexible system development it is advantageous to have a memory device that can be programmed, and then reprogrammed if necessary. These features are provided by **erasable and programmable read-only memories** (EPROMs). Although the term can be applied to a number of components, usually this description is used for memory that is erased by exposure to ultraviolet (UV) light. The chips are fitted with a transparent window that allows UV light to reach the silicon surface. Programming is normally performed using an **EPROM programmer**, which provides the appropriate voltages and control signals. After use, EPROMs can be erased in 20 to 30 minutes using a UV light source.

EPROMs have the disadvantage that they must normally be removed from their circuit and placed in a special eraser and a programmer to allow them to be modified. Another form of PROM, the **EEPROM**, can be modified electrically without the need for a UV light source. This allows a program to be modified or changed while the chip is in place.

It might seem that the EEPROM should be classified as a RAM because it can be written (programmed) as well as read. However, it should be noted that a RAM can normally be written and read in a fraction of a microsecond. An EEPROM can be read at this speed, but may require several milliseconds to write a single byte. The EEPROM is thus a read quickly but write slowly device and is better described as a ROM than a RAM. The relatively slow programming speed of EEPROM can be rather inconvenient when using large amounts of memory. In such cases it is more common to use **FLASH** memory, which provides the ability to program and reprogram a device electrically at very high speeds. This permits even the largest devices to be programmed in a few seconds.

Memory device standards

Over the years a number of standards have emerged for various forms of memory device, making it possible to purchase these components from a host of different manufacturers with confidence that there will be no compatibility problems. One of the most widely used standards is the **JEDEC** standard for byte-wide memory devices. This defines a standard pin-out for both RAMs and ROMs of a range of sizes. One advantage of this arrangement is that it allows a computer manufacturer to design a board with sockets that will accept RAM, ROM, EPROM, EEPROM or FLASH memory. These devices can be of any size up to the present maximum. This allows designers to build expandability into their products by allowing systems to be compatible with components that are not yet available.

27.3.7 Microcomputer programming

The creation of computer programs (or *software*) can be performed in a number of ways. In the vast majority of cases, software is produced using sophisticated high-level programming languages, which greatly simplifies the task. However, it is useful to understand a little of other techniques, so in this section we will begin by looking at more basic methods.

Machine code

We noted in Section 27.3.2 that computer programs are stored as a sequence of instructions, where each instruction consists of an opcode which may be followed by an operand. In an 8-bit computer the opcode will normally be a single byte and the operand (if present) will normally consist of one or two additional bytes. Each byte consists of eight binary digits, and therefore a short sequence of code might look like

10010110

01110000

10011011

01110001

10010111

10000000

Programs written in this form are referred to as *machine code*, and this is the language that the processor understands. However, one problem with this form is that it is not particularly easy for humans to understand!

One way of making the code slightly more manageable is to convert it into hexadecimal rather than binary format. The code sequence given above then becomes

96

70

9B

71

97

80

This representation is easier to read than straight binary, but still gives no insight into the function of the code. To create programs by writing directly in machine code requires the programmer to memorise (or to continually look up) the various opcodes. This is both time consuming and error prone and in practice this method of programming is almost always avoided.

Assembly code

A much more efficient method of producing computer programs is to write them in a mnemonic form using *assembly code*. Here each instruction within the instruction set of the processor is given an abbreviated name (or *mnemonic*) that makes it much easier to remember. The various instructions (including any operand) are then shown on separate lines. For example, our earlier program segment might be written as

LDAA   70h

ADDA   71h

STAA   80h

These three instructions translate as: 'load accumulator A with the contents of memory location 70 (expressed in hexadecimal)'; 'add the contents of location 71 to accumulator

A'; and 'store the contents of accumulator A in memory location 80'. Thus our program adds the contents of locations 70 and 71 and stores the result in location 80.

Programs written in assembly code must be converted into machine code before they can be executed by the processor, and this translation is performed by a piece of software called an *assembler*. The sophistication of assemblers varies, but most provide a range of features such as: the ability to embed comments within the program; the use of symbolic names for variables and memory locations; and the use of macro instructions. More sophisticated assemblers provide additional advanced features such as the automatic generation of control structures.

One of the characteristics of assemblers and assembly code is that they are *processor dependent*. Since each processor has its own unique instruction set, assembly code (like machine code) is specific to a particular device. This lack of 'portability' means that assembly code programs written for one computer will usually not run on another.

High-level languages

High-level languages include languages such as BASIC, C, C++, C#, Java, Pascal and many others. They overcome problems of portability by providing a level of abstraction, or independence, from the detailed structure and operation of the computer. The task in hand is described in a manner closely related to *what is to be done*, rather than to *how this will be achieved* on a particular processor. For example, our earlier problem of adding two variables might be simply expressed as

$$VariableA = VariableB + VariableC$$

with no specific reference to where these variables will be stored or which particular processor instructions will be used to add them together.

High-level languages make complex programming simpler and produce programs that are considerably smaller and easier to understand. Also, since the programs are not linked directly with the instruction set of the processor, a program written in a high-level language can be run on any computer that supports that language.

As with assembly code, programs written in a high-level language need to be translated into machine code before they can run on a particular processor. However, when using high-level languages this translation can be done in two ways. The first uses a *compiler* to translate the program into machine code, in much the same way that an assembler translates assembly code into machine code. The resultant code is then loaded into the computer and executed. The second approach uses an *interpreter* to continually translate the high-level program into machine code during the execution of the code. This second approach has the advantage of removing the need for a separate compilation stage, but tends to produce slower execution because of the need to do translation in real-time. Most widely-used languages are usually compiled, while others (such as PostScript) are normally interpreted. Other languages, such as BASIC and Pascal, can use either approach.

Choice of programming technique

Having noted earlier that programming directly in machine code is not a sensible approach, we are left with a choice between writing programs in assembly code or in a high-level language.

The use of high-level languages has many advantages since it is easier and faster, and generates source programs that are smaller, easier to understand and easier to maintain. However, while the *source code* of a high-level language program may be smaller, often the resultant *machine code* is much larger and less efficient than that produced by assembly code. For this reason, historically, assembly code has been used in applications that require very compact or very fast code. However, modern optimising compilers now produce code that is arguably as fast as hand-crafted assembly code, and the use of low-level programming is becoming increasingly difficult to justify.

Inevitably, the choice of programming language will be greatly affected by the nature of the application since some languages have features that are particularly useful in specific applications. When programming small embedded systems, using the kind of microprocessors that we have been concentrating on within this chapter, by far the most common programming language is *C* or one of its many variants. In many ways this offers a compromise between high- and low-level programming techniques since it provides many high-level languages features, while allowing direct control of processor elements and allowing programmers to embed assembly code segments directly into the source code.

Video 27F

27.3.8 Input/output

The input/output section of the computer is responsible for communicating with the real world. When a microcomputer is used as the heart of a desktop personal computer, its inputs and outputs generally come from keyboards, displays, printers and communications equipment. When it is used to form an **embedded system**, its input and output devices will be far more varied and may include a wide range of devices including those discussed in Chapters 12 and 13.

In many cases the original input signals will not be directly compatible with the computer. They may, for example, be analogue in nature or too large or small in magnitude. It is therefore usually necessary to provide some form of **interface** to make the signals produced by the sensors compatible with those required by the computer system. Similarly, the signals produced at the output of the computer may not be suitable to drive the necessary actuators directly. Again, an interface may be required to overcome this incompatibility.

The operations performed by the interface circuitry will depend on the nature of the sensors and actuators within the system. This process is often referred to as **signal conditioning**. Typical functions include amplification, filtering (to remove noise), isolation and analogue-to-digital and digital-to-analogue conversion. Several of these topics have already been discussed in earlier chapters (amplification in Chapter 14, filtering in Chapter 8 and isolation in Chapter 26) and the conversion of signals from analogue to digital and from digital to analogue will be discussed in Chapter 28.

The end result of the process of signal conditioning at the input is a clean digital signal that is compatible with the input stage of the computer. Similarly, at the output, the interface is required to take signals from the output stage of the computer and generate from them suitable voltage or current waveforms to drive the output devices. For the moment, we will leave any further discussion of the interface and concentrate on the nature of the input and output stages of the computer.

Input/output organisation

Memory-mapped input/output

The simplest and most commonly used method of computer input/output is to make external devices appear as if they were memory registers to the processor. In this way, part of the **memory map** of the system is given over to input/output devices. This technique is called, for obvious reasons, **memory-mapped input/output**.

As well as its basic simplicity, this approach also has the advantage that all of the instructions within the instruction set of the machine can be used to act on input/output registers in the same way that they can be used on memory locations. Disadvantages of this approach are that the full address bus must be decoded to obtain the chip select signals for the devices and full-width addresses must be used to reference these registers, creating long I/O instructions. The importance of the length of the instructions is that computers spend most of their time fetching, rather than executing, instructions, so a shorter program runs more quickly.

Input/output using a separate address space

Some microprocessors have a separate input/output address space that is accessed by a range of special input/output instructions. Usually 256 input and 256 output addresses are provided, allowing the device address to be expressed in a single byte. External devices are addressed using the low-order byte of the address bus with a special control line being used to distinguish I/O from memory operations. This approach has the advantages that the special I/O instructions are short and only the low-order byte of the address bus has to be decoded.

Its disadvantages are that it is limited to only 256 input and output addresses and only a small number of dedicated instructions can be used for I/O operations. Most microprocessors with a dedicated I/O space also support memory-mapped input/output.

Input/output registers

One of the most frequently encountered methods of input/output is to use input and output registers. These appear to the processor to be ordinary registers, each having its own unique address but, unlike conventional memory registers, they have connections to the outside world.

When the processor writes into an **output register** the pattern of '1's and '0's placed in the register is reproduced on a series of output lines that can be used to drive external circuitry. Similarly, when the processor reads from an **input register** the value returned reflects the pattern of '1's and '0's on a set of input lines.

Although separate input and output registers are not unheard of, it is much more common to have registers that can perform both tasks. These are thus **bidirectional input/output registers**. These devices are often referred to as **input/output ports,** this term reflecting the use of the word *port* to mean a *gateway.* They are also called **parallel I/O ports**.

Serial input/output

When data must be transmitted over a distance it is often advantageous to use serial rather than parallel techniques to reduce the number of wires or channels required. We have discussed the use of shift registers to perform **serial to parallel** and **parallel to serial** conversions and noted their use in serial communications (see Section 25.7). Example 25.3 illustrates this technique and shows the need for a separate connection to carry the clock signal as well as that required for the data. The provision of a separate clock channel is inconvenient and most serial communications systems use alternative methods to synchronise the transmitter and receiver. Two basic techniques are used to tackle this problem – the asynchronous and synchronous approaches.

Asynchronous serial communications

Asynchronous communication removes the need for a separate clock signal and allows data to be transmitted intermittently as well as in a continuous stream. This is achieved by embedding synchronising information within the words of information sent. A simple example of this technique is illustrated in Figure 27.33. Here each byte of information to be transmitted is augmented by the addition of a **start bit**, which is always 0, and one or more **stop bits**, which are always 1. A **parity bit** may also be included to provide error correction (as discussed in Section 24.12.5).

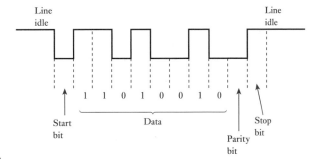

Figure 27.33
The structure of an asynchronous word.

The data word is generated using an accurate clock signal at the transmitter (usually produced by a crystal oscillator) and applied to the communications channel. The receiver also has an oscillator, which produces its own local clock signal. Because of the high level of accuracy and stability of crystal oscillators, the receiver's clock will be very close in frequency to that used at the transmitter.

When the transmission line is not in use, it will idle at logical 1. Thus, the reception of a start bit will always represent a transition from 1 to 0. When the receiver detects this transition it synchronises its own internal clock to that of the incoming signal. It then samples the incoming waveform at appropriate intervals to detect each of the data bits, the parity bit (if present) and the one or more stop bits.

Because of the great stability of the two oscillators used at the transmitter and receiver, the variation in phase during a single word will generally be negligible, ensuring correct reception. However, the presence of the stop bit or bits, which are known to be at logical 1, allows the receiver to check that everything is correct. If the receiver's clock were to run too quickly or too slowly, it would result in the parity bit, or a following start bit, being sampled in place of a stop bit. The receiver continuously monitors the polarity of the detected stop bits and, if an error is detected, flags a **framing error** to the processor. If parity is in use, this will also be calculated at the receiver and any errors reported as **parity errors**.

The asynchronous system can be used with a range of data wordlengths and at a number of signalling speeds. It is, however, essential that both the transmitter and receiver use the same arrangement. Asynchronous communications are widely used within microcomputer systems, particularly for applications producing fairly irregular data, such as **keyboards**.

Synchronous serial communications

Synchronous communications systems are used when a continuous stream of data is available. This makes it possible for the receiver to use **clock recovery** circuitry to reconstruct the clock signal being used at the transmitter by looking at the transitions of the incoming signal. This allows the incoming data to be directly clocked into a shift register for conversion to a parallel form.

One problem remaining is the detection of the start and end of each word of data. This is achieved by periodically sending a unique **synchronising pattern** to indicate the beginning of a block of data. Once this has been detected at the receiver, incoming bits are counted to keep track of the beginning and end of each word.

Unlike asynchronous data transfers, for which error checking is generally performed using parity testing of each word, in a synchronous system it is usual to perform checking over a complete block of data. This can be achieved using a **checksum** (as described in Section 24.12.5), or by using more sophisticated techniques, such as the inclusion of **cyclic redundancy codes**.

Signalling rate

In serial communications systems it is important that both the transmitter and receiver operate at the same speed. The rate at which bits are sent down the line is described in terms of its **baud rate**, which is the number of transitions per second on the line. This in turn determines the rate at which data can be communicated through the channel, which is termed the **data transfer rate** or simply the **data rate**. High speed links may have data rates of many Mbits/s or perhaps several Gbits/s.

Simplex and duplex communications

Communication channels can operate in either one or both directions. The simplest form of link can convey information in only one direction. This is termed a **simplex** channel. Systems that can communicate in both directions are termed **duplex**. These may be subdivided into **half-duplex** systems, which can communicate in only one direction at a time, and **full-duplex** systems, which can communicate in both directions simultaneously.

Serial I/O devices

Serial I/O can be performed by a range of special-purpose integrated circuits, such as the universal asynchronous receiver/transmitter (UART), asynchronous communications interface adapter (ACIA), synchronous serial I/O controller (SIO) and universal synchronous and asynchronous receiver/transmitter (USART). They all provide a complete full-duplex communication channel, usually with the minimum of external components. However, when using modern single-chip microcomputers these functions are invariably built into the device and so no additional hardware is required.

Serial I/O devices usually produce and accept standard logic signals that are not themselves suitable for long-distance transmission. These logic levels are normally buffered using a **line driver** or a **line receiver** to provide an increased range and improved **noise rejection**. There are several standards for serial communication signals to ensure compatibility between diverse systems.

Serial communications standards

While there are several standards for serial communications, perhaps the most important is the **Universal Serial Bus** or **USB**. This was originally devised in the mid-1990s to standardise the connection between computers and peripherals such as keyboards, pointing devices, printers and memory devices. This standard was widely adopted for a range of applications and has since been revised several times with significant increases in performance.

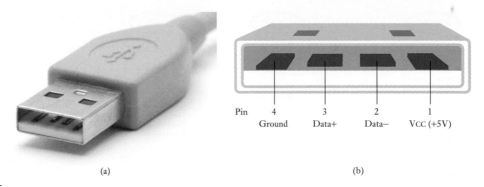

Pin   4   3   2   1
     Ground   Data+   Data−   Vcc (+5V)

(a)   (b)

Figure 27.34
A standard USB type A
plug.

Probably the most widely used version of the standard is USB 2.0, which was released in 2000. This describes an arrangement providing *half-duplex* communication at speeds up to 480 Mbits/s. It uses a complex architecture and a range of techniques including both synchronous and asynchronous methods. The most widely used connector used for USB 2.0 applications is the **USB type A** plug and socket arrangement as shown in Figure 27.34. The cables use a twisted pair to reduce noise and crosstalk as discussed in Section 22.10.7. The cables also have a ground connection and a 5 V line to provide power to external devices.

A newer version of the standard, USB 3.0, uses additional wires to provide *full-duplex* communications at speeds of up to 5 Gbits/s. While this version provides a considerable increase in performance, this has been achieved at the cost of a considerable increase in complexity. For this reason, USB 2.0 remains the most widely used version for less-demanding applications.

Program-controlled input/output

I/O operations can be initiated in a number of ways. Often the programmer decides that at a particular point in a program some data should be read from an input device or written to an output device. Logically enough, this is called **program-controlled input/output**. However, in some circumstances, it is not possible to determine at the time that the program is written when data will become available. An example of such a situation is the reading of data from a keyboard. If we assume that the device has a data register associated with it, there is clearly no problem with reading the port at any time, but data will only become available when a key has been depressed. A solution to this problem, still using program-controlled input/output, is **polling**.

Polling involves the repeated reading of an external device to determine its status. Let us initially consider the above example, involving the reading of data from a keyboard. Simply looking at the contents of the device's data register will not show if any new data has arrived as it might be identical to the previous contents of the register. We therefore require an additional register, called a **status register**, that contains a bit indicating the presence of data. This bit might be called the **data ready flag**. This flag is set when data is placed in the register and cleared by the processor when the data has been taken. The status register can be polled in a number of ways, depending on the application. Figure 27.35 illustrates this process.

Figure 27.35(a) shows a 'straight through' program, which begins, runs through a sequence of instructions and then stops. While such programs are sometimes found on personal and other general-purpose computers, they are rarely found in **embedded**

Figure 27.35
Polling of I/O devices.

(a) A 'straight through' program

(b) A typical control program

(c) Wait for data

(d) Test and continue

systems, which are normally associated with control and instrumentation. Control systems usually have a program structure of the type shown in Figure 27.35(b). This has an initialisation section and then a continuous loop. Once started, the program runs for ever or until the system is turned off. This is a usual requirement of a control system.

Let us now assume that data must be read from an external device at some point during the execution of the main loop of the program. If program execution cannot continue until information is available, the system must sit and wait for it to arrive. This is achieved as shown in Figure 27.35(c). The program reads the status register of the device and looks to see if the data ready bit is set. If it is, it reads the data register, clears the data ready flag and continues with the program. If it is not, it goes back and reads the status register again, repeating this operation until it shows that data has arrived. This technique will result in the program waiting indefinitely for data to arrive, during which time it will not be performing any other task. Such an arrangement might be used when awaiting a start command from a keypad if nothing else can be done until this has been received.

An alternative arrangement is shown in Figure 27.35(d). Here, the status register is tested at an appropriate point in the loop. If the data ready bit is set, the processor reads the data register, performs any necessary actions, and then continues. If it is not, the processor simply continues and repeats the test the next time it reaches this point in the loop. This technique has the advantage that the processor can continue to do useful things until data becomes available. However, it is only practical if the processor can perform useful tasks before the data arrives.

In some applications, there may be not one but a number of external devices, all of which may require action from the processor. If the processor must test them all to see if they have data available, this can waste a great deal of processor time.

Video 27G

Interrupts

The time wasted in polling large numbers of external devices can be reduced by getting these devices to inform the processor when they want attention or 'servicing'. In other words, we get external components to **interrupt** the operation of the processor. This

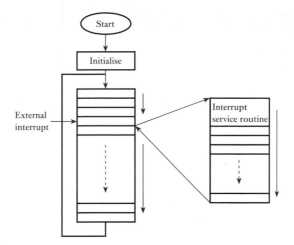

Figure 27.36
The interrupt
mechanism.

is achieved by building into the processor hardware an input line called an **interrupt request line** (IRQ). When an external device activates this line, the processor stops what it is doing and services the device. The interrupt mechanism is illustrated in Figure 27.36.

Suppose that a device generates an interrupt at the time indicated in the figure. This is likely to occur during the execution of an instruction and initially no action is taken. The processor will always complete its current instruction before responding to an interrupt as failure to do so would leave the state of the machine uncertain. When this instruction is complete, the processor leaves the main program and jumps to an **interrupt service routine,** which is simply a sequence of instructions produced by the programmer for execution when an interrupt occurs. On completion of the service routine, the processor jumps back to the main program and continues where it left off.

Although the above description is correct, it leaves many questions unanswered. It would seem, for example, that the operation of the main program would be affected by the execution of the interrupt service routine as this is almost certain to change the contents of the various registers within the processor. It is also not clear how the processor knows where to find the service routine or how it remembers where to re-enter the main program. To answer these questions, we need to look at the interrupt mechanism in a little more detail.

In order that the operation of the main program remains unaffected by the interrupt service routine, the contents of the various registers of the processor are stored before the interrupt service routine is entered. Among the registers saved is the **program counter**, which, you will remember, contains the address of the next instruction in the main program. At the end of the interrupt service routine, the original contents of these registers are restored before the main program is restarted. As the contents of the program counter have been restored, the sequence of instructions in the main program is not disrupted. The processor detects the end of the interrupt service routine when it meets a **return from interrupt instruction** (RTI). This is a member of the instruction set of the machine specifically designed for this purpose. The interrupt mechanism is shown in detail in Figure 27.37.

This leaves one unexplained point. How does the processor know where to find the interrupt service routine? In fact, this begs the question: how does the processor know where to find the main program? To answer these questions, we need to look at the use of *vectors*.

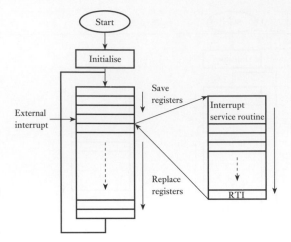

Figure 27.37
A more detailed view of the interrupt mechanism.

Vectors

A vector, as its name implies, *points* to somewhere in memory. In other words, it is an address. Within most 8-bit machines, as addresses are 16 bits long, a vector is a 2-byte quantity requiring two memory locations. Microcomputers use a number of vectors to define the starting addresses of sections of program that should be executed in particular circumstances. For example, the **interrupt vector** defines the start address of the interrupt service routine and the **reset vector** (also called the **restart vector**) defines where the processor will start executing when it is first turned on or after a particular input line (called the **reset** line) has been activated. Some processors store these vectors in fixed locations and Figure 27.38 shows a portion of the memory map of a typical processor, indicating the location of the various vectors.

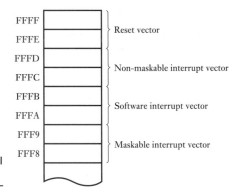

Figure 27.38
The vectors of a typical 8-bit processor.

The processor shown has four vectors, one corresponding to reset and the other three to different forms of interrupt. One of the interrupts is **maskable**, which means that it can be turned on and off under program control using dedicated instructions. There is also a **non-maskable** interrupt, which is always active. A third form of interrupt – termed a **software interrupt** – is unlike the others in that it is not activated by external hardware but as a result of the execution of a dedicated instruction. However, it derives its name from the fact that the action taken by the machine is similar to that taken in the event of a conventional hardware interrupt.

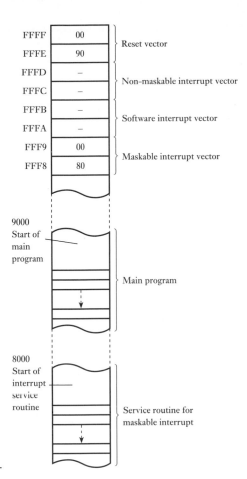

Figure 27.39
Setting the reset and
interrupt vectors.

The programmer, after writing the software necessary to implement the required system functions, places the start address of the main program in the bytes corresponding to the reset vector. When the system is turned on, it will fetch this address and begin executing the program from this location. If the system makes use of interrupt-driven I/O, the programmer will also place the start address of the interrupt service routine(s) in the locations corresponding to the appropriate interrupt vector(s). This is illustrated in Figure 27.39.

In a general-purpose computer, the reset vector points to a program that requests commands from the user and may load further programs. In an **embedded system**, the reset vector points to the start of an application program that will therefore start automatically. This self-starting arrangement is termed a **turn key system** as, conceptually, it starts 'at the turn of a key'.

Some microprocessors use different methods to determine the starting address of the interrupt service routine. They may, for example, get the external device to supply an opcode to the processor that is then used to jump to an appropriate routine. Other processors use a variety of methods to acquire the interrupt vector.

Interrupts and the stack

We have seen that before an interrupt service routine is executed it is necessary to store away the contents of the various registers so that the main program can be restarted successfully. In some machines this process is performed automatically, but in others it is left to the programmer to save those registers that are being used. In either case, it is interesting to consider where this data is stored. Clearly, it would be possible to define certain fixed locations in which the contents of each register would be stored, but this has serious drawbacks. So far we have considered only the situation in which a single interrupt source is used. In many cases there may be several devices capable of generating the same or a variety of interrupts. In this situation it is quite likely that a second interrupt will occur while the processor is busy executing an interrupt service routine. If this happens, when the register information for the second interrupt is saved, it will overwrite the original data and the processor will be unable to return to the main program.

The solution to this problem is to save the register contents using the **stack** described earlier. This is a **first-in–last-out** (FILO) store, which means that information is retrieved in the reverse order to that in which it is saved. Using this arrangement, multiple interrupts will result in the data being stacked in order. At the end of each interrupt service routine the relevant data will be recovered from the stack, allowing the processor to return correctly to the main program. This is illustrated in Figure 27.40.

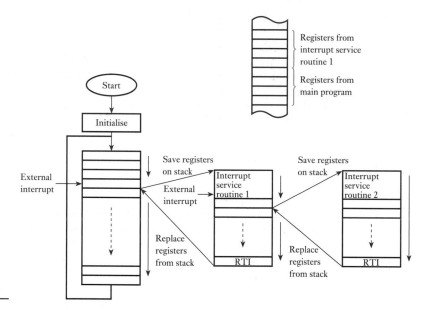

Figure 27.40
Multiple interrupt handling with a stack.

DMA

Although the use of interrupt-driven I/O allows a rapid response to external events, for applications in which large amounts of data need to be transferred it can consume a considerable amount of processor time. Every time an external device requires servicing, the contents of the processor's registers must be stored away and then replaced.

A faster alternative is to allow external devices to transfer data directly into memory without affecting the processor's registers. The processor is made to wait for the short time that a dedicated controller needs to write or read a byte of memory and is then allowed to continue. This is **direct memory access** (DMA) and it is by far the fastest method for computer input/output. Unfortunately, it is also the most expensive as a considerable amount of additional hardware is required to oversee the data transfers. For this reason, DMA is usually used only for applications that can justify its high cost – for example, for disk drives. This method of input/output is sometimes referred to as **cycle stealing** as the DMA unit robs the processor of memory access cycles.

Computer input/output – a summary

We have looked at three methods for computer input/output – program-controlled, interrupt-driven and direct memory access.

The last of these is very fast, but also expensive. It is used primarily for transferring large quantities of data and has the advantage that it consumes very little processor time.

Interrupt-driven I/O can provide a fast response without wasting a great deal of processor time in polling. However, interrupt-driven systems require more complicated hardware than program-controlled ones and are harder to test as their operation is essentially asynchronous.

Program-controlled I/O is the simplest of the three techniques and, as such, should normally be the preferred option. Interrupts should only be used if the application requires them and if the processor time saved can be used meaningfully.

27.3.9 Single-chip microcomputers

We have considered the three main components of a microcomputer – namely, the processor (CPU), memory and I/O sections. For small applications it is possible to obtain all of these functions within a single integrated circuit. Such a device is termed a **single-chip microcomputer**. Combining these components not only saves on physical size but also reduces the number of external connections required, improving reliability and reducing cost. The single-chip microcomputer turns the computer from a system into a component.

Early single-chip microcomputers provided relatively small amounts of ROM and RAM, together with modest I/O facilities. A typical device might have provided a fairly conventional 8-bit CPU with 2 kbytes of mask-programmable ROM, 64 bytes of RAM, 32 bits of parallel I/O, an 8-bit counter and a single interrupt. Even with these limited capabilities, such devices were sufficiently powerful to be used in a large number of applications. Unfortunately, the restricted memory limited their ability to operate using high-level languages, sometimes forcing the programmer to work in assembly code. This was very time consuming and led to high development costs.

In more recent years, the capabilities of single-chip microcomputers have increased enormously. Modern devices normally have a range of memory options with large amounts of ROM, RAM and EEPROM available if required. Both user-programmed and mask-programmed versions are available, simplifying development, while allowing high-volume applications to take advantage of mask-programmed devices, if appropriate. The I/O capabilities of these devices are also extensive, with typical parts providing a large number of I/O lines in addition to multiple counters, interrupts and serial channels.

Single-chip 16- and 32-bit microcomputers are also available for applications that require more power than can be obtained using 8-bit components. These offer the facilities normally associated with more powerful computers, such as on-board memory management and DMA.

Despite the vast capabilities of the most powerful single-chip microcomputers, it is the simplest 8-bit parts that dominate the market for these components. The really high-volume applications of computers are associated with small embedded systems, such as those within cars and domestic appliances. In these systems, relatively modest amounts of computing power are usually required and small 8-bit components are often more than adequate.

PIC microcontrollers

Among the most popular forms of single-chip microcomputer are the various members of the PIC® microcontroller family. These devices, produced by Microchip Technology, use a dual-bus **Harvard architecture** and **RISC**. The range extends from small devices in packages with only 6 pins and having only a few hundred bytes of memory to large devices with perhaps 80 pins and 512 kbytes of memory. The microcontrollers are optimised for embedded applications and offer a wide range of on–chip features, such as control and timing peripherals (including counters, timers and watchdog timers), display drivers (for LED or LCDs), communication support (for standards such as RS232, RS485, SPI, I²C, CAN and USB) and analogue interfaces (including analogue-to-digital and digital-to-analogue converters).

To illustrate the range of devices available, Figure 27.41 shows a simplified block diagram of the PIC10F200 microcontroller. This is a 6-pin device that provides 256 12-bit words of flash program storage, 16 8-bit words of data memory and 3 I/O lines. In contrast, Figure 27.42 shows a block diagram of the PIC18F87J10 microcontroller. This is an 80-pin device that provides 128 kbytes of program storage, 3936 bytes of data storage, 9 parallel data ports (providing up to 67 I/O lines), 5 timers, 2 pulse width modulation modules, 2 serial I/O channels, a parallel communication channel and 15 channels of 10-bit analogue-to–digital conversion.

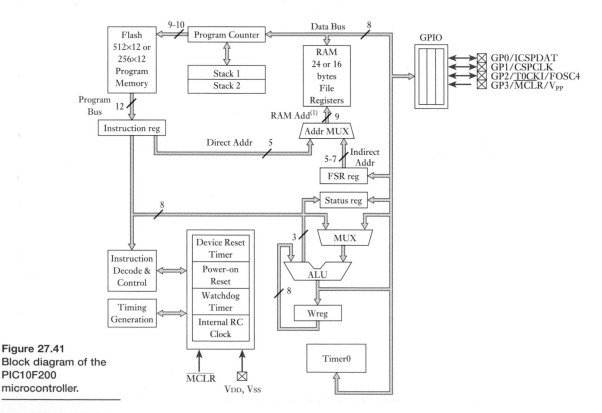

Figure 27.41
Block diagram of the PIC10F200 microcontroller.

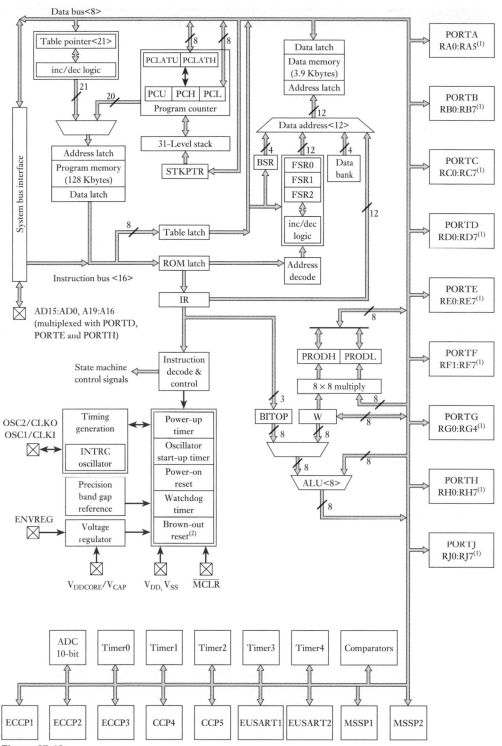

Figure 27.42
Block diagram of the PIC18F87J10 microcontroller.

System-on-a-chip (SOC) devices

A logical extension of the single-chip microcomputer is the *system on a chip*. While the former tends to contain all the main elements of a computer within a single device, the latter aims to provide all the functionality required to implement a complete system. This will often include not only all the elements of a computer, but also additional digital, analogue or mixed-signal components.

In many cases SOCs will incorporate elements of microcomputers *and* of FPGAs, offering the opportunity to select the best features from these two architectural approaches. In addition, devices may contain:

- **Memory** – blocks of RAM, ROM, EEPROM and FLASH memory.
- **Analogue interfaces** – ADCs, DACs and other data acquisition components (as discussed in the next chapter).
- **Communications interfaces** – radio-frequency components, USB, Ethernet or FireWire interfaces.
- **Timing elements** – timers, oscillators and phase-locked loops.
- **Power components** – voltage regulators, power controllers and power amplifiers.
- **Additional components** – application specific interfaces and components.

The design and production of SOC devices represents a significant challenge since they invariably involve a mix of diverse architectures and technologies. However, when successful, they represent a truly innovative and highly integrated approach to embedded system design.

27.5 ## Programmable logic controllers (PLCs)

While array logic, microcomputers and system-on-a-chip devices clearly offer great potential for developing systems to meet almost any requirement, in some situations the development of a system 'from scratch' cannot be justified. In such cases developers may turn to other, ready-made, solutions to their problems.

Programmable logic controllers (PLCs) are self-contained microcomputers that are optimised for industrial control. They consist of one or more processors, together with power supply and interface circuitry in a suitable housing. A range of input and output modules is normally available to allow such units to be used in a wide range of situations without the need for any electronic design or construction. Facilities are also provided for programming and general system development.

PLCs were introduced in the 1970s in order to produce and market computers in large quantities in an industrial area that was characterised by the diversity of its applications. At that time, many simple control systems were based on the use of electromagnetic relays and PLCs were seen initially as a replacement for this form of circuitry. Designers working with relays were used to producing their designs using a graphic notation based on **ladder diagrams** (or **ladder logic**). To simplify the adoption of a new technology, manufacturers of PLCs added a user interface to their products that enabled them to be programmed in a manner that was intuitive to engineers familiar with ladder diagrams. Initially, the controllers had a functionality that was limited to the simple logic functions that could be produced using relays. As time passed, more elaborate features were added to meet the diverse needs of the control engineer. These included sophisticated displays, data logging and communications facilities and the ability to program the devices using a range of graphical or text-based languages.

PLCs are extensively used in machine and process control and have been designed specifically to be very reliable and dependable. The availability of off-the-shelf hardware and interfacing software reduces development time and costs and makes PLCs particularly suited to low-volume applications.

27.6 Selecting an implementation method

In most cases, the method chosen to implement a digital system is likely to be determined by the complexity of the functions to be produced. Where very limited logical operations are required, it may be possible to produce these using simple circuits based on the use of conventional logic gates. Applications that require only a handful of gates will normally use standard CMOS logic packages. Unfortunately, even relatively simple logic arrangements can produce circuits requiring many devices and it soon becomes economical – in terms of both cost and space – to use some form of array logic. The decision as to which of the different forms of array logic to choose is likely to be determined by the number and range of functions required and the programming support available.

For complicated digital applications, simple PLDs will not suffice, so the designer must choose between using a more complicated programmable device, such as a CPLD or an FPGA, or a microcomputer (or, in some cases, an SOC). Further discussion on choosing between programmable and non-programmable approaches will be left until Chapter 30, where we will look in more detail at the selection of design methods.

Video 27H

Further study

In the further study section at the end of Chapter 11 we considered the problem of identifying the inputs and outputs of an electronic controller for a domestic automatic washing machine. At that time we ignored the techniques to be used to implement the controller, but now we are in a position to look more closely at this topic.

Most washing machines use microcomputers within their controllers, and we are now equipped to consider this implementation in more detail. Clearly it is not practical to consider all aspects of such a system, but it is instructive to look at certain aspects of the design.

One such aspect relates to the control of the motor which rotates the drum of the washing machine. At various stages of the washing cycle the drum is required to rotate at different speeds. These include: a low speed of about 30 revolutions per minute (rpm) while clothes are washed; an intermediate speed of about 90 rpm while the water is pumped out; and a high speed of perhaps 500 or 1000 rpm to spin dry the clothes. Consider how the microcomputer should control the speed of the motor.

Key points

- The complexity available within integrated circuits doubles every couple of years.

- In situations where large numbers of gates are required, the use of standard logic gates is impracticable. In such cases, it is common to use a more highly integrated approach.

- Array logic integrates large numbers of gates within a single package. This permits them to be mass produced as a single component that can be programmed for a wide range of applications.

- Many forms of programmable logic device (PLD) are available. These differ in their complexity and in the range of functions that they offer.

- Field programmable gate arrays (FPGAs) offer enhanced functionality due to the use of an array of logic cells.

- PLDs and FPGAs are normally programmed using a sophisticated software package that defines the internal configuration needed to produce a certain functionality. For this reason, the designer does not require a detailed knowledge of the internal structure of the device.

- An alternative approach to the implementation of complicated digital systems is to use microprocessors.

- A microprocessor represents one of the main components of a computer – the processor or CPU.

- Computers consist largely of a set of registers that communicate via a series of buses. There are three main buses – namely, the data bus, address bus and control bus.

- The functionality of a computer is defined by a sequence of instructions (software). This can enable a computer to perform highly complicated functions or replace relatively simple logic circuitry.

- Semiconductor memory can be divided into random access memory (RAM), which is volatile, and read-only memory (ROM), which is not volatile.

- The input/output section is the most application-specific part of the computer.

- In many cases, the processor performs input and output operations directly under the control of the application program. This is termed program-controlled input/output.

- In other instances, it is an external device that initiates such actions by means of interrupts.

- In very high-speed applications the processor may give up control of the buses temporarily to allow external devices to perform direct memory access (DMA).

- A chip that contains the processor, memory and input/output sections of a computer is termed a 'single-chip microcomputer'.

- A chip that contains not only the components of a microcomputer, but also additional elements such as array logic, analogue units, interfaces and radio-frequency circuitry is called a 'system-on-a-chip' (SOC) device.

- A programmable logic controller (PLC) is a self-contained microcomputer that is optimised for industrial control.

- The method selected to implement a digital system is likely to depend on its complexity. Simple systems might be produced using a handful of conventional gates, while more complicated arrangements might suggest the use of a gate array. Where considerable complexity is involved, it will often be necessary to use some form of VLSI device in the form of a logic array or microprocessor.

Exercises

27.1 Explain the principle of Moore's law.

27.2 Why is it unattractive to produce a VLSI device containing 1000 separate gates?

27.3 Explain what is meant by the term 'glue logic'.

27.4 What is meant by random logic?

27.5 Explain the terms 'uncommitted logic' and 'array logic'.

27.6 Explain the basic form of a programmable logic array (PLA).

27.7 Sketch a diagram showing how the PLA of Figure 27.2 could be programmed to implement the following logic functions:

$$X = A\bar{B}CD + \bar{A}B\bar{C}D + ABC\bar{D}$$

$$Y = \bar{A}B\bar{C}D + ABC\bar{D}$$

$$Z = A\bar{B}CD + ABC\bar{D}$$

27.8 How does a PAL differ from a PLA?

27.9 How many inputs and outputs are provided by a 20R6 PAL? What does the 'R' in the part name signify?

27.10 What is meant by an OTP part?

27.11 Describe the characteristics of a PEEL device.

27.12 How does a CPLD differ from an SPLD?

27.13 How does an FPGA differ from a CPLD?

27.14 Describe the process used to program a PLD for a given application.

27.15 What is a microprocessor?

27.16 Sketch a simple block diagram indicating the main elements of a computer and showing the information flow between them.

27.17 Explain what is meant by a single-chip microcomputer.

27.18 What is the addressing range of a computer that uses 22-bit addresses?

27.19 Explain why registers require three-state operation to be used on a computer bus.

27.20 Explain the difference between Von Neumann architecture and Harvard architecture.

27.21 What is the range of a 24-bit unsigned quantity?

27.22 What is the range of a 24-bit signed quantity?

27.23 Explain the terms 'instruction set', 'opcode' and 'operand'.

27.24 What is the difference between CISC and RISC?

27.25 Explain the function of the program counter register.

27.26 During the operation of a program, it is often useful to jump to a subroutine containing a program section that is used several times. At the end of the subroutine, the processor must return to the instruction following the jump to subroutine instruction. Explain how this may be achieved. Your solution should take into account the fact that subroutines can be nested, in that one can call another.

27.27 Explain the difference between multiplexed and non-multiplexed bus systems and describe how the full address bus is derived in the former.

27.28 Explain the meanings of the terms 'memory-mapped I/O', 'polling', 'program-controlled I/O', 'interrupt-driven I/O' and 'DMA'.

27.29 Compare and contrast the use of memory-mapped I/O with the provision of a dedicated I/O space.

27.30 Describe the use of the stack in interrupt handling.

27.31 In what form of memory (ROM or RAM) would it be normal to store the system vectors? Why?

27.32 Why is EEPROM not considered to be a non-volatile RAM?

27.33 In what kind of application is it appropriate to program a microcomputer by the direct production of machine code?

27.34 Under what circumstances might assembly code programming be used in preference to a high-level language?

27.35 What are the advantages of programming in a high-level language rather than in assembly code?

27.36 What is the difference between a compiled language and an interpreted one?

27.37 Describe briefly the characteristics of a USB 2.0 communications channel.

27.38 What methods of computer input/output would seem most appropriate for the following applications?

(a) Reading pushbuttons in a computer-controlled washing machine.

(b) Interfacing a keyboard to a personal computer.

(c) Connecting a high-speed disk drive to a computer.

27.39 Explain the meaning of the term 'FILO' and give examples of the use of such a structure. Within computers, it is often useful to produce a first-in–first-out (FIFO) arrangement. Suggest a possible use of such a structure.

27.40 What are the properties of a turnkey system? How are these properties achieved when using a microcomputer?

27.41 What is the difference between a single-chip microcomputer and a system-on-a–chip device?

27.42 Describe the basic characteristics of a programmable logic controller (PLC).

27.43 What would be a suitable implementation method for a system requiring about six basic gates?

27.44 What would be a suitable implementation method for a system requiring about 20 basic gates?

27.45 What forms of implementation might be appropriate for systems requiring several thousand basic gates?

Chapter 28　Data Acquisition and Conversion

28.1　Introduction

We have seen in earlier chapters that the effects of noise are often less of a problem in digital systems than in those using analogue techniques. Digital data can also be easily processed, transmitted and stored. For these reasons, we often choose to represent analogue quantities in a digital form, which raises the question: how do we translate from one form to the other?

The process of taking analogue information, often from a number of sources, and converting it into a digital form is often termed **data acquisition**. It consists of several stages. This chapter begins by looking at the process of sampling a changing analogue quantity to determine its time-varying nature. It then discusses the hardware required to convert these samples into a digital form and the corresponding reconstruction of this digital information into an analogue signal. Finally, it considers the process of combining information from a number of sources into a single system input and the converse problem of generating a number of analogue output signals from a single information source.

Video 28A

28.2　Sampling

In order to obtain a picture of the changes in a varying quantity, it is necessary to take regular measurements. This process is referred to as **sampling**. Clearly, if a quantity is changing rapidly we will need to take samples more frequently than if it changes slowly, but how can we determine the sampling rate required to give a 'good' representation of a signal? It would seem obvious that the required sampling rate would be determined by the most rapidly changing components in a signal (in other words, by the highest-frequency components), but how do we decide how fast we need to sample to get a 'good picture'?

Fortunately, an answer to this question is available in the form of **Nyquist's sampling theorem**. This states that the sampling rate must be greater than twice the highest frequency present in the signal being sampled. It also states that, under these circumstances, none of the information in the signal is lost by sampling. In other words, it is possible to reconstruct completely the original signal from the samples.

In general, the waveform to be represented will contain components of many frequencies. In order to sample it reliably, we need to know the highest frequency present. Let us assume that we know a certain signal contains no components above a frequency of F Hz. According to Nyquist's theorem, provided that we sample this waveform at a rate greater than $2F$, we will obtain sufficient information to reconstruct the original signal completely. The minimum sampling rate is often called the **Nyquist rate**. This process is illustrated in Figure 28.1.

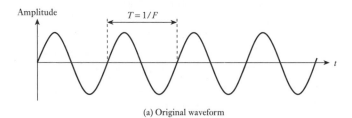

(a) Original waveform

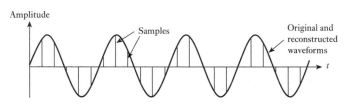

(b) Waveform sampled above the Nyquist rate

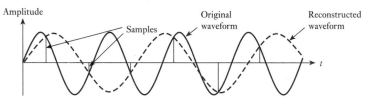

Figure 28.1
The effects of sampling at different rates.

(c) Waveform sampled below the Nyquist rate

While in practice the waveform we wish to sample will consist of many frequency components, for simplicity Figure 28.1(a) shows a sine wave of frequency F. Figure 28.1(b) shows the results of sampling this signal at a rate *greater* than the Nyquist rate. Given these samples, it is possible to reconstruct the original waveform as any other line drawn through the sample points would have frequency components above F. We know that, in this case, the signal has no components above this frequency, so the original waveform is the only possible reconstruction. As this sampling rate allows the reconstruction of a signal of frequency F, it will also allow reconstruction of any signal that contains no components above this frequency.

Figure 28.1(c) shows the results of sampling the waveform at a frequency *below* the Nyquist rate. Here, the samples can be reconstructed in a number of ways, including that shown in the figure. This is clearly not the original waveform. Thus, if a signal is sampled below the Nyquist rate, it will not in general be possible to reconstruct the original signal. The waveform generated appears to have been produced by a signal of a lower frequency than the original. This effect is known as **aliasing** and resembles a *beating* between the signal and sampling waveform.

It should be pointed out that the Nyquist rate is determined by the highest frequencies present in a signal, *not* by the highest frequencies of interest. If a signal contains unwanted high-frequency components, they must be removed before sampling or they will result in spurious signals in the frequency band of interest. It is normal to use filters to remove signals that are above the range of interest to prevent this effect. These filters are referred to as **anti-aliasing filters**. For example, although human speech contains frequencies up to above 10 kHz, it has been found that good intelligibility can be obtained using only those components up to about 3.4 kHz. Therefore, to sample such a signal for transmission over a channel of limited bandwidth, it would be normal to filter the speech signal to remove frequencies above 3.4 kHz, then to sample the waveform at about 8 kHz. This is somewhat above the Nyquist rate (which would be 6.8 kHz) to allow for the fact that the filters are not perfect and some frequency components will be present a little above 3.4 kHz. It is common to sample at about 20 per cent above the Nyquist rate. A typical anti-aliasing arrangement might use a sixth-order Butterworth filter (as discussed in Section 8.13.3).

28.3 Signal reconstruction

In many cases, it is necessary to reconstruct an analogue signal from samples that have been transmitted, processed or stored. Reconstruction requires the removal of the step transitions in the sampled signal and can be seen as removing the high-frequency signal components that these represent. This can be achieved using a low-pass filter to remove these unwanted frequencies. This is called a **reconstruction filter** and would normally have a similar characteristic to the anti-aliasing filter used before sampling. Therefore, a typical reconstruction arrangement would use a sixth-order Butterworth filter.

28.4 Data converters

The process of sampling an analogue signal involves taking an instantaneous reading of its magnitude and converting this into a digital form. Similarly, the process of reconstruction requires us to take digital values and convert these back into their analogue equivalents. These two operations are performed by **data converters**, which can be divided into **analogue-to-digital converters** (ADCs) and **digital-to-analogue converters** (DACs).

A range of converters is available, each providing conversion to a particular **resolution**. This determines the number of steps or **quantisation levels** that are used. An n-bit converter produces or accepts an n-bit parallel word and uses 2^n discrete steps. Thus, an 8-bit converter uses 256 levels and a 10-bit converter uses

1024 levels. It should be noted that the resolution of a converter may be considerably greater than its accuracy. The latter is a measure of the error associated with a particular level, rather than simply the number of levels used. In many applications, a simple 8-bit conversion is sufficient, this giving a resolution of about 0.4 per cent. However, in situations where greater accuracy is required, converters of up to 20-bit resolution or more are available, 20-bit conversion giving a resolution of about 1 part in 1 million, which is sufficient for almost all applications.

Conversions of either form take a finite time, which is referred to as the **conversion time** or **settling time** of the converter. The times taken for conversion differ greatly depending on the form of the converter used, although, in general, digital-to-analogue conversion is faster than the inverse operation.

Although both ADCs and DACs can be constructed fairly simply from basic components, in practice integrated circuit converters are almost always used. These are generally inexpensive, although the cost increases with resolution and speed. When using complex electronic components (such as the single-chip microcomputers discussed in Chapter 27), multiple, high performance data convertors are normally built into the integrated circuits.

28.4.1 Digital-to-analogue converters (DACs)

There are two common forms of DACs – the **binary-weighted resistor** and the **R–2R resistor chain** forms. Let us look briefly at each approach.

Binary-weighted resistor method

This form of DAC is a development of the **current-to-voltage converter** described in Section 16.4.2. A simple form of the converter is shown in Figure 28.2.

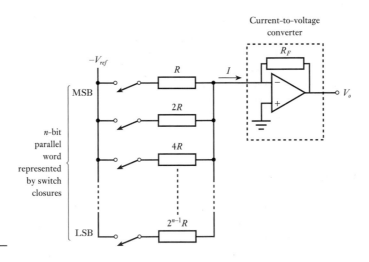

Figure 28.2
A binary-weighted resistor DAC.

Each input controls a switch that connects a resistor to a constant reference voltage, $-V_{ref}$. These switches are closed when the corresponding bit is set to 1. If the switch connected to the most significant bit (MSB) of the input digital word is closed while all the others are open, the reference voltage is connected to one side of the resistor R. The other end of this resistor is connected to the inverting input of the operational amplifier, which is a **virtual earth** point, so is at $0\,V$. The voltage across the resistor is thus equal to the reference voltage and the current flowing into the virtual earth point is given by

$$I = -\frac{V_{ref}}{R}$$

If the next MSB is set to 1 while all the others are at 0, the reference voltage is applied across the resistor $2R$. This will produce a current into the amplifier of

$$I = -\frac{V_{ref}}{2R}$$

which is half that produced by the MSB. The next switch, if closed while all the others are open, will produce a current of one-quarter of that produced by the MSB. The progression continues, each input in turn producing a current of one-half that of its predecessor. The inputs are thus binary weighted.

As the input to the operational amplifier is a virtual earth, its voltage does not change with the current flowing into it. Thus, the fact that one switch is closed will not affect the current injected by another switch. The currents therefore sum to give a value representing the combination of switches closed. The current-to-voltage converter then converts this input current I into an output voltage, V_o (as described in Section 16.4.2), obeying the expression

$$V_o = -IR_F$$

where R_F is the feedback resistor.

When the LSB alone is set to 1, the current I will be given by

$$I = -\frac{V_{ref}}{2^{n-1}R}$$

and the output voltage will therefore be

$$V_o = -IR_F = \frac{V_{ref}R_F}{2^{n-1}R}$$

This represents the output voltage for an input number of 1. For an input number of m the output will therefore be

$$V_o = m \times \frac{V_{ref}R_F}{2^{n-1}R}$$

In practice this form of DAC is implemented using electronic switches (transistors) in place of those shown in Figure 28.2. However, the principles of operation are identical to those just described.

The binary-weighted resistor method of conversion uses a small number of resistors, but requires them to have a broad spread of values (a range of R to $2^{n-1}R$). For a 10-bit converter, for example, this range will have a ratio of over 500 to 1. Unfortunately, resistors of markedly different values tend to have unequal temperature coefficients of resistance, which means that the ratios between them will change with temperature. This limits the temperature stability of this technique.

File 28A

R–2R resistor chain method

The *R–2R* method also makes use of the **current-to-voltage converter** arrangement of Section 16.4.2, but does not require a broad spread of resistor values. The arrangement is illustrated in Figure 28.3.

In many respects this circuit resembles that of the binary-weighted resistor arrangement. Again the binary word controls a series of switches which, in turn, generate currents in a series of resistors. The difference in this case is that all the resistors connected to the switches have the same value. The other end of the resistor in each case is joined to a chain of resistors, which goes from the inverting input of the operational amplifier

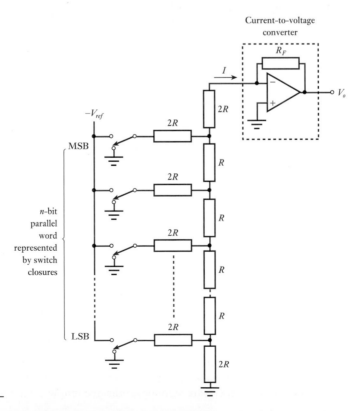

Figure 28.3
An *R–2R* resistor chain DAC.

to earth. The circuit is arranged such that currents flowing through each of the resistors connected to the switches see a resistance of *2R* looking in either direction along the resistor chain. Therefore, half the current will go in each direction. Similarly, currents flowing up the chain see equal resistances in either direction at each node and will again be split. Therefore, each switch contributes half as much current as the switch above, as its current is repeatedly halved at each node on its journey to the op–amp.

Therefore, the currents generated by the switches are binary weighted, as in the previous method, but without the use of a wide range of resistor values. Here, only resistors of *R* and *2R* are required and, if appropriate, these can be formed using only resistors of one value (*R*) by connecting two in series to form the other (*2R*). This allows temperature-matched resistors to be used to provide greatly improved temperature stability.

DAC settling times

The settling times of these two methods of digital-to-analogue conversion are similar and determined by the time taken for the electronic switches to operate and the amplifier to respond. Converters are available with a range of resolutions and, in general, conversion time increases with resolution. A typical general-purpose 8-bit DAC would have a settling time of between 100 ns and 1 μs, while a 16-bit device might have a settling time of a few microseconds. However, for specialist applications, high-speed converters might have settling times down to a few nanoseconds. It is sometimes more convenient to specify the number of samples that can be converted in a second rather than the settling time. Converters used for generating the video signals used in graphics display systems might have a resolution of 10 bits and may have a maximum **sampling rate** of above 100 MHz (100 million samples per second), corresponding to a settling time of less than 10 ns.

It is normal to low-pass-filter the output of a DAC to smooth the resulting waveform and thus remove the effects of sampling. The cut-off frequency of such a **reconstruction filter** would be chosen to remove components at the sampling frequency without attenuating the required signal.

28.4.2 Analogue-to-digital converters (ADCs)

There are a number of techniques available for analogue-to-digital conversion. Of these, the following four types are the most widely used.

Counter or servo

The counter method of conversion gives one of the simplest forms of ADCs. The principle is illustrated in Figure 28.4.

At the heart of the converter is a DAC connected to the parallel outputs of an **up counter**. The output of the DAC is compared with the analogue input signal using a **comparator** (a comparator is a device that produces an output of 0 or 1 depending on which of its two inputs is most positive). The output of the comparator is used to generate a 'stop' control for the counter. Initially, the counter is zeroed and the counter starts to count up. As it does so, the output from the DAC increases. When the DAC voltage becomes equal to the analogue input signal, the output from the comparator will change state, stopping the counter. This signal is also used to generate a 'conversion complete' control signal. At this stage, the digital equivalent of the analogue input signal can be found by reading the parallel output from the counter. When external equipment has read this value, the counter is set to zero and the process begins again.

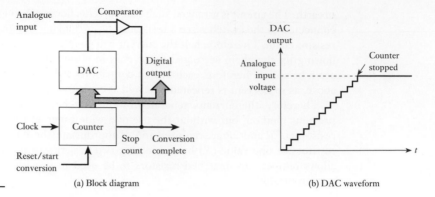

Figure 28.4
A counter ADC.

(a) Block diagram

(b) DAC waveform

The counter ADC is one of the simplest forms of converter, but is relatively slow in operation. For each conversion, the counter must increment from zero, allowing sufficient time after each count for the DAC and the comparator to settle. The conversion time will therefore be at least m times the settling time of the DAC and the comparator, where m is the final digital output value of the converter. For an n-bit conversion, this could take as long as 2^n times this settling time. Conversion times of the order of a few milliseconds are typical.

A modification of the counter ADC is formed by replacing the up counter with an **up/down counter**. The output from the comparator is now used as an up/down control signal, forcing the counter to track the analogue input signal. This arrangement is called a **servo ADC**.

Successive approximation

The counter ADC is slow in operation as it uses a very inefficient method of searching for the correct value. This is perhaps best illustrated by taking an analogy. Let us suppose that we wish to determine which of the 1000 pages of a dictionary contains a particular word. We could perform this task by looking at the first page, seeing if the word was on this page and, if not, moving on to the next. This process would involve us searching progressively through the book until we found the correct page (this technique is similar to that adopted by the counter ADC). A more efficient technique would be to open the book halfway through (at page 500) and look to see whether the appropriate page was before or after this point. This will locate the page within either the first or second half of the book and eliminate 500 pages from our search. Let us suppose that we discover that the page we require is before page 500. We would then open the book at page 250 (halfway through the first half of the book) and again see if the required page was before or after this point. In this way we would home in on the required page, reducing the region of uncertainty by 50 per cent each time we open the book. As 2^{10} is 1024, it will take at most 10 attempts to locate the correct page, which is considerably faster than looking at each page.

The successive approximation ADC is similar in many respects to the counter ADC, except that the simple counter is replaced by logic circuitry, which operates in a manner similar to that described in our dictionary search analogy. The arrangement is shown in Figure 28.5.

The DAC is driven by a digital word produced by the successive approximation logic. Initially, all the bits of this word are set to 0 and then the most significant bit (MSB) is set to 1. This input word is converted by the DAC into an analogue signal corresponding to half of the full range of the DAC. This value is compared with the

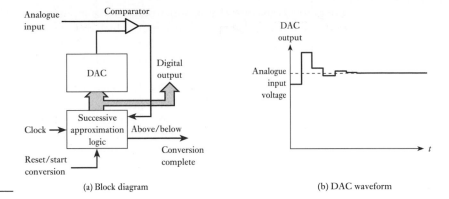

Figure 28.5
A successive
approximation ADC.

(a) Block diagram   (b) DAC waveform

analogue input signal using a comparator and the result is fed back to the control logic. If the comparison shows that the DAC output is less than the analogue input, the MSB will be left at 1; if not, it will be reset to 0. In any event, the logic then sets the next MSB and, again, compares the output of the DAC with the input signal. In this way, each bit of the input to the DAC is set in turn and its correct state determined. The conversion is completed when all the bits of the DAC input have been set correctly. Therefore, for an n-bit conversion, this will take approximately n times the settling time of the DAC and the comparator. This compares favourably with the counter type, which requires up to 2^n times the settling time of the DAC and comparator.

Typical successive approximation converters might have settling times of 1–10 μs for an 8-bit conversion, increasing to perhaps 10–100 μs for a 12-bit device. High-speed variants are available with considerably improved conversion times. The complexity of this form of converter is somewhat greater than that of the counter type. However, its superior speed of operation makes it one of the most commonly used arrangements for integrated circuit converters.

Dual slope

The basic form of this ADC is shown in Figure 28.6. An operational amplifier is used to integrate the input signal for a fixed period of time, producing a charge on the

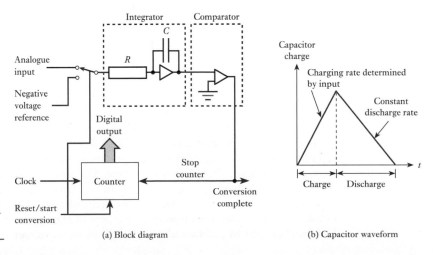

Figure 28.6
A dual-slope integrator
ADC.

(a) Block diagram   (b) Capacitor waveform

integrator's capacitor that is proportional to the input voltage. The integrator is then connected to a constant current source, which discharges the capacitor at a constant rate. The time taken to reduce the charge to zero is measured by counting the cycles of a stable clock oscillator, this time being proportional to the charge on the capacitor and thus the input voltage.

The dual-slope technique has the advantages of high accuracy and low cost and is often used in such applications as digital panel meters. It is also used where a very high resolution is required and can give a resolution of better than 20 bits (a 20-bit conversion represents a resolution of better than 1 part in 1 million). The speed of conversion is relatively slow, with high-resolution devices producing only perhaps 10 to 100 conversions per second.

Parallel or flash

The parallel or flash converter is the fastest of the various forms of ADC. It operates by having a separate comparator to compare the input voltage with every discernible voltage step within the converter's range. The arrangement is illustrated in Figure 28.7.

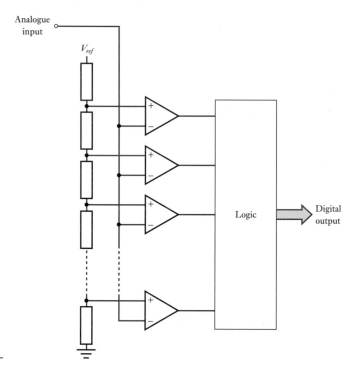

Figure 28.7
A parallel or flash ADC.

The various voltage steps are produced using a precision resistor chain from a reference voltage source. Each voltage increment is connected to a separate comparator that compares it with the input voltage. The result is that all of the comparators connected to points along the resistor chain that have voltages greater than the input voltage will produce an output of one polarity, whereas those connected to voltages below the input voltage will produce voltages in the opposite sense. Combinational logic is then used to determine the value of the input voltage from this pattern.

The great advantage of this method is its high speed of conversion, as all the comparisons are performed simultaneously. This allows sample rates in excess of 150 million conversions per second, with conversion times of only a few nano-seconds. However, as an *n*-bit converter requires 2^n comparators, the hardware is significantly more complicated and therefore more expensive than it is for other techniques.

28.5 Sample and hold gates

With rapidly changing quantities, it is often useful to be able to *sample* a signal and then *hold* its value constant. This may be required when performing analogue-to-digital conversion so that the input signal does not change during the conversion process, upsetting the operation of the converter. It may also be necessary when per-forming digital-to-analogue conversion to maintain a constant output voltage during the conversion period of the DAC.

We have already come across a circuit to perform this function in the form of the **sample and hold gate** described in Section 18.8.3. These gates can be constructed using discrete components or, more commonly, in integrated circuit form. Typical integrated components require a few microseconds to sample the incoming waveform, which then decays (or **droops**) at a rate of a few millivolts per millisecond. Faster devices, such as those used for video applications, can sample an input signal in a few nanoseconds, but are designed to hold the signal for a shorter time. Such high-speed devices may experience a droop rate of a few millivolts per microsecond.

Video 28C

28.6 Multiplexing

Although it is quite possible to have a system with a single analogue input or a single analogue output, it is usual to have multiple inputs and outputs. Clearly, one solution to this problem is to use a separate converter for each input and output signal, but often a more economical solution is to use some form of **multiplexing**. The principle of signal multiplexing is illustrated in Figure 28.8.

A number of analogue input signals can be connected to a single ADC using an **analogue multiplexer**. This is a form of electrically controlled switch based on the use of **analogue switches** (as discussed in Section 18.7.3). Each analogue signal is connected in turn to the ADC for conversion, the sequence and timing being deter-mined by control signals from the system. This is illustrated in Figure 28.8(a).

For certain applications, the arrangement in Figure 28.8(a) is unsuitable as each analogue input signal is sampled at a different time. This may make it impossible to obtain detailed information as to the relationship between the signals, such as their phase difference. The problem can be overcome by sampling all the inputs simultane-ously using a number of sample and hold gates, as shown in Figure 28.8(b). Once the input signals have been sampled, they can be read sequentially without losing the time relationship between the channels.

Figure 28.8(c) shows an arrangement whereby several output channels are pro-duced from a single DAC. Here, the converter is sent data relating to each channel in turn. When the conversion has been completed, a control signal is used to activate the appropriate sample and hold gate. The gate samples the output from the DAC and reproduces this value at its output. The system sets the values of each output channel in turn, updating the values as necessary.

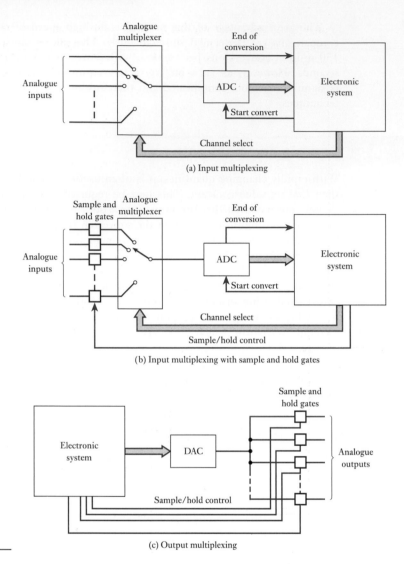

Figure 28.8
Input and output
multiplexing.

(a) Input multiplexing

(b) Input multiplexing with sample and hold gates

(c) Output multiplexing

In practice, anti-aliasing and reconstruction filters would be used on the inputs and output of the arrangements shown in Figure 28.8. These have been omitted from the diagram to aid clarity.

28.6.1 Single-chip data-acquisition systems

Because almost all control and instrumentation systems make use of some form of analogue input/output, a vast array of special purpose devices has been developed to simplify the data-acquisition process. When using single-chip microcomputers these functions will normally be built in the device itself, but where other implementations are being used, or where more specialised functions are required, **single-chip data-acquisition systems** are often employed. Originally this term was used to describe components that contained an ADC and a multiplexer within a single chip. However, modern components provide very sophisticated functionality including sample and hold gates, single-sided or differential inputs, high-resolution analogue to digital conversion, and either parallel or serial communication of the measured data to an associated controller. They therefore remove from the processor the control and computational overhead associated with analogue data acquisition. Though not technically data *acquisition* some single-chip data-acquisition systems also provide a number of DACs.

| **Example 28.1** | **Design a microcomputer-based system to sense eight analogue input signals and produce eight analogue outputs. The inputs come from sensors that produce useful signals with a bandwidth of 1 kHz, but are known to pick up higher-frequency noise. These signals are to be measured to an accuracy of at least 1 per cent. The output signals are to drive actuators with a maximum operating bandwidth of 100 Hz, but which are affected by higher-frequency signals. The actuators require signals to an accuracy of at least 1 per cent.** |

A block diagram of a suitable system is shown below.

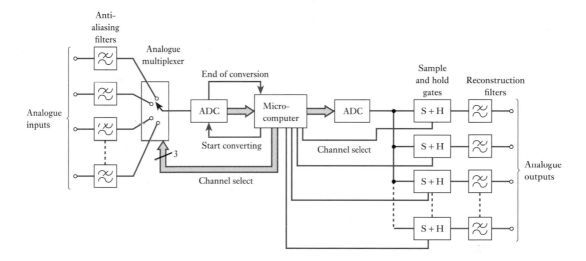

An analogue multiplexer selects one of the eight input signals and applies it to an ADC. The converter needs two control lines from the microcomputer. One of these is an output from the computer, which is used to instruct the ADC to begin a conversion (the **start converting signal**), and the other is an input, which allows the ADC to tell the microcomputer when it has completed the conversion and is ready for more data (the **end of conversion signal**). The multiplexer requires three control lines from the processor to select one of the eight input channels.

As the sensors have a useful signal bandwidth of 1 kHz, the converter will need to sample each channel with a sampling rate of at least twice this frequency. However, the presence of high-frequency noise necessitates the use of **anti-aliasing** filters to remove this noise. Assuming that a slight attenuation near 1 kHz is acceptable, it would be appropriate to use **low-pass filters** with a cut-off frequency of 1 kHz. Sixth-order **Butterworth filters** would be a typical choice. As these are not ideal filters, it is necessary to sample at somewhat above the **Nyquist rate**. An increase of 20 per cent gives a sampling rate of 2.4 kHz. If each channel is to be sampled at this rate, the ADC must be capable of 8×2.4 kHz = 19.2 kHz, which corresponds to a conversion time of 52 μs. This is well within the capability of a general-purpose successive approximation ADC.

In order to achieve an accuracy of 1 per cent, 7-bit resolution is required. In fact, most general-purpose ADCs and DACs have a resolution of at least 8 bits, so it would seem sensible to use such devices for the input and output.

The eight output channels are obtained from a single DAC using a series of **sample and hold gates**. These are individually controlled by lines from the processor, although, if input/output lines were in demand, these could be produced using a '3-line to 8-line decoder'. The outputs from the sample and hold gates would have step changes of voltage as the outputs were updated. These fast transitions represent high-frequency components in the output, which may upset the operation of the actuator. It is therefore necessary to **low-pass-filter** the output signals to remove these unwanted components. As with the input filters, these **reconstruction filters** would typically be sixth-order **Butterworth filters**, in this case with a cut-off frequency of 100 Hz.

Video 28D

Further study

Modern smartphones contain a large number of both sensors and actuators. While some of these produce or accept digital information, others are analogue in nature.

Without concerning yourself with the details of the interfaces required, consider the nature of the analogue sensors in a typical smartphone and estimate the precision and the data sampling rate that would be appropriate for these transducers.

Can you think of any other analogue sensors that might be useful in a smartphone and perhaps suggest some applications that might make use of them? These sensors could be built into the phone or be external devices that would connect to it.

Key points

- The extensive use of digital techniques for information processing, storage and transmission means that conversions between digital and analogue quantities are widely used.

- Converting an analogue signal into a digital form is achieved by sampling the waveform and then performing analogue-to-digital conversion.

- As long as the signal is sampled at a rate above the Nyquist rate, no information is lost as a result of the sampling operation.

- When sampling signals that have a broad frequency spectrum, it is necessary to use anti-aliasing filters to remove components that are at frequencies above half the sampling rate.

- When reconstructing analogue signals from a series of samples, filters are used to remove the high-frequency components associated with the sampling process.

- A wide range of DACs and ADCs are available. These differ in their resolution, accuracy, speed and cost.
- Sample and hold gates may be required to hold an input signal constant while it is sampled or to hold an output signal constant in between the times when it is updated.
- In systems with a number of analogue inputs or outputs, multiplexing can be used to reduce the number of data converters required.

Exercises

28.1 Explain the function of 'sampling' in data acquisition.

28.2 What is meant by the term 'Nyquist rate'? A signal has a frequency spectrum that extends as high as 4 kHz. What is the minimum rate at which the signal may be sampled to obtain a good representation of its form? What would be the effect of sampling below this rate?

28.3 Describe the use of anti-aliasing and reconstruction filters.

28.4 A signal has a frequency range from 20 Hz to 20 kHz. However, for a particular application, only those frequency components up to 10 kHz are of importance. Explain how this signal may be sampled to minimise the amount of data produced while maintaining the amount of useful information. What would be an appropriate sampling rate in your arrangement?

28.5 Explain the terms 'resolution' and 'accuracy' as they apply to data converters.

28.6 How many quantisation levels are used by a 12-bit data converter?

28.7 Discuss the relative advantages and disadvantages of the two forms of digital-to-analogue conversion described in Section 28.4.1.

28.8 Modify the circuit used in Computer simulation exercise 28.1 to investigate the behaviour of the circuit of Figure 28.3. Apply similar input signals and confirm that the circuit functions as expected.

28.9 Explain why successive approximation ADCs are faster than simple counter types.

28.10 Describe the advantages and disadvantages of parallel or flash ADCs.

28.11 Give examples of control signals that might be used with a typical analogue-to-digital converter.

28.12 Explain the use of sample and hold gates in analogue input/output systems. What is meant by the term 'droop' when applied to such gates?

28.13 Explain the function of multiplexing in a data-acquisition system.

28.14 What is meant by a single-chip data-acquisition system?

Communications

When you have studied the material in this chapter, you should be able to:

- outline the basic elements of a communications system and explain terms such as transmitter, receiver and communications channel
- describe the various forms of radio wave propagation and explain how these vary between the different radio frequency bands
- explain the function and characteristics of modulation and identify the most widely used analogue and digital techniques
- describe the complementary process of demodulation
- discuss the need for multiplexing in modern communications systems and describe common techniques for achieving this
- outline the main elements of both tuned radio frequency and superheterodyne radio receivers.

Video 29A

29.1 Introduction

Communications is one of the most important and rapidly growing areas of electronics. In addition to obvious applications such as radio broadcasting, television, mobile phones and the internet, it also plays a major role in the operation of a vast array of control and instrumentation systems. For this reason, an understanding of the basic principles of communications is essential for all those working in this area.

The term *communication* relates to the transmission of information from one place to another. It is used in a wide range of contexts and is often used to describe the exchange of information between people by way of speech or written material. Here we will concentrate on what might be termed **communication systems**, which are responsible for transmitting information of many forms – often over some distance. This process is often referred to as *telecommunication*, the Greek prefix *tele-* meaning 'far'. However, modern communication techniques are used both for long-range and short-range applications.

The basic elements of a simple communication system are shown in Figure 29.1. The first element represents the **source** of the information to be transmitted. Possible sources include: human speech; music; images; video or computer data. The next component in the arrangement is the **transmitter**, which is responsible for converting the information into a form suitable for transmitting over the **channel**. This

Figure 29.1
A simple communications system.

channel can take a number of forms, such as a pair of copper wires or an optical fibre. Alternatively, the channel might simply be the free space used to transmit radio, light or sound waves. Near its destination the **receiver** has the task of converting the signal transmitted over the channel into a form compatible with its destination. In many cases this will be similar to the original form produced by the source. Finally the receiver passes the information on to the **information destination**.

Examples of communication systems are very diverse. A simple example might be the 'system' responsible for the transmission of information from one person to another via speech (assuming that the people concerned are in the same room). Here the information source could be seen as the 'mind' of the speaker, and the transmitter as the biological components (including the vocal chords) responsible for converting thoughts into sounds. The channel takes the form of the air that transmits the sound, and the receiver is the ears and other elements that convert the sound waves into what the listener interprets as speech.

A more electronics-related example of a communication system is a telephone network. Here the information source (the user of the telephones) produces information in the form of sound waves that need to be converted into an appropriate form for communication over the channel. In a simple system this channel might be a pair of copper wires, in which case the transmitter might take the form of a microphone (as discussed in Chapter 12) which converts the sounds into corresponding electrical signals, which can be directly communicated over the copper cables. In such an arrangement the receiver might be a simple loudspeaker (as discussed in Chapter 13) which converts the electrical signal back into sound. However, in a more typical arrangement, the signals from the microphone would be amplified and further modified to make them compatible with more sophisticated communications channels using optical or radio signals. This process normally includes some form of **modulation** (which will be discussed later in this chapter). At the receiver the transmitted signals are converted back into their original form (normally using a process of **demodulation**) before being passed on to their destination.

While the diagram of Figure 29.1 identifies the basic elements of a simple communication system, most real systems are somewhat more complex. Firstly, many systems are bidirectional in nature, meaning that they communicate in both directions. When discussing serial communications systems (in Chapter 27), we noted that bidirectional arrangements are often described as **duplex** systems (as opposed to **simplex** systems), and that these can take two basic forms. Systems that provide communications in both directions, but in only one direction at any one time, are referred to as **half-duplex** arrangements, while those that provide communications in both directions simultaneously are known as **full-duplex** systems. Examples of these two forms of bidirectional communication are shown in Figure 29.2. A possible half-duplex arrangement is shown in Figure 29.2(a). Here the channel is only capable of passing information in one direction at any time and so a switching arrangement is used to control the flow of information. Figure 29.2(b) shows a full-duplex arrangement that uses two independent channels to allow simultaneous communications in both directions.

A further layer of complexity is present in many communication systems since they are used to link several sources to several destinations. This is illustrated in Figure 29.3, which for simplicity shows communication in only one direction – clearly this process could be extended to produce bidirectional communication. Here each *source* could represent an individual telephone user and each *destination* a corresponding recipient of a telephone call. The multiplexer combines the information from the various sources in a manner that allows the demultiplexer to later separate them and pass them to the appropriate destination.

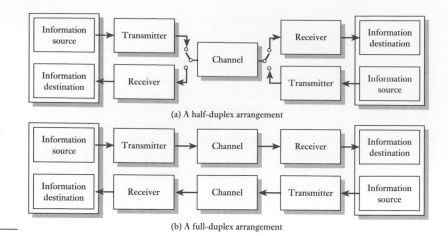

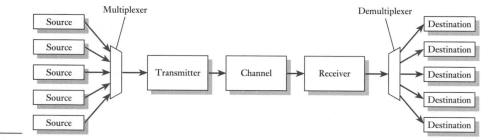

Figure 29.2
Bidirectional communications systems.

(a) A half-duplex arrangement

(b) A full-duplex arrangement

Figure 29.3
A communications system with multiplexing.

| 29.2 | The communications channel |

We have seen that communications channels may take many forms. Some use some kind of physical connection to pass information, such as wires (in the form of twisted-pairs or coaxial cables), waveguides or fibre-optic cables. Others pass information through 'free space' using electromagnetic waves, such as light or radio waves. When used in free space (rather than over optical fibres) light is normally used over relatively short distances (as in a TV remote control). The characteristics and applications of radio waves vary tremendously depending on the frequency range being considered.

29.2.1 Radio wave propagation

Radio signals of different frequencies differ in the way that they propagate from their source, and three basic forms of propagation are illustrated in Figure 29.4.

At relatively low frequencies (up to a few megahertz) radio waves exhibit **ground wave propagation** as shown in Figure 29.4(a). Here waves follow the curvature of the Earth, allowing them to travel beyond the horizon and up to several hundred miles. AM broadcasting takes advantage of these characteristics, as do many other forms of long-distance communication.

Higher-frequency signals are refracted by the ionosphere and 'bounce' back and forth between the ionosphere and the Earth. This permits such signals to propagate over great distances that can (potentially) span the globe. This **sky wave propagation** dominates in the high-frequency (HF) part of the spectrum which extends from 3 to 30 MHz. This is also known as the *short wave band* and is used for amateur radio communication, which takes full advantage of its extended range. However, the

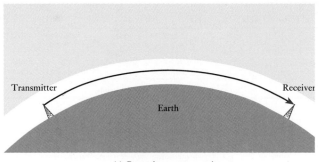

(a) Ground wave propagation

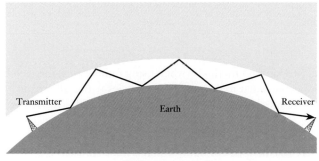

(b) Sky wave propagation

**Figure 29.4
Radio wave
propagation.**

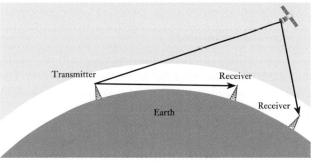

(c) Line-of sight propagation

characteristics of the ionosphere are constantly changing, making quality very variable and making these frequencies less suitable for broadcasting.

At frequencies in the very-high-frequency (VHF) band (30–300 MHz) and above, radio signals are not 'channelled' as at lower frequencies and are also absorbed by any obstructions. For this reason, communication is restricted to situations where there is a **line of sight** between the transmitter and the receiver. However, since these waves will pass through many non-metallic materials, communications are possible within buildings and through some obstacles. Frequencies in this range are used for FM and TV broadcasting, satellite communications and mobile phones.

29.2.2 The radio frequency spectrum

Since the characteristics of radio signals change with frequency, different parts of the spectrum are used for different applications. A number of **frequency bands** are defined to identify different regions of the spectrum and these are shown in Figure 29.5. The figure also indicates typical transmission media (when using physical connections)

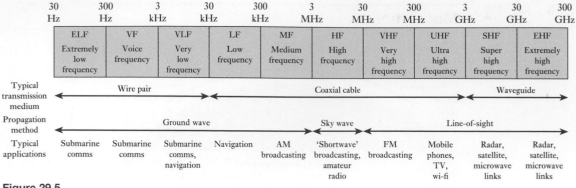

Figure 29.5
The radio frequency spectrum.

and propagation methods (when using radio waves) for each band and suggests typical applications that make use of frequencies within that band.

The wavelength of a radio wave (λ) is related to its frequency (f) by the expression

$$\lambda f = c$$

where c is the speed of light, which is approximately 3×10^8 m/s. Therefore, the range of frequencies shown in Figure 29.5 corresponds to a range of wavelengths from 1 mm (300 GHz) to 10,000 km (30 Hz).

29.2.3 Channel characteristics

The characteristics of a channel may be described in a number of ways, in many cases using parameters that we have met in earlier chapters. For example, key factors in determining the ability of a channel to communicate information are its **bandwidth** and its **signal-to-noise ratio** (*S/N*).

When transmitting *analogue* information, the bandwidth gives an indication of the frequency limitations that will be imposed on any signal. However, when transmitting *digital* information it is more common to give the **channel capacity**, which is a measure of the (theoretical) maximum amount of information that can be transmitted through the channel in a given time. This is normally expressed in bits per second (bps).

The channel capacity (*C*) is related to the bandwidth (*BW*) and the signal-to-noise ratio (*S/N*) of the channel by the **Shannon-Hartley theorem**, which states

$$C = BW \times \log_2(1 + S/N) \tag{29.1}$$

where *BW* is expressed in Hz and *S/N* is expressed as a simple ratio (*not* in dB).

Example 29.1 | A channel has a bandwidth of 10 kHz and a signal-to-noise ratio of 40 dB. What is its channel capacity?

40 dB corresponds to a power ratio of 10,000, so from Equation 29.1

$$C = BW \times \log_2(1 + S/N)$$
$$= 10^4 \times \log_2(1 + 10^4)$$
$$\approx 133 \text{ kbps}$$

Video 29B

29.3	## Modulation

29.3.1 Why do we need modulation?

Modulation performs two basic functions. Firstly it allows a signal to be 'moved' to a different part of the frequency spectrum, and secondly, it allows a number of different signals to be sent simultaneously over the same channel. Appropriate use of modulation can also improve the performance of a communications system in terms of noise and interference.

To understand the importance of the first of these functions, consider the transmission of a simple speech signal. To convey the important aspects of human speech a signal must cover a frequency range from about 300 Hz to about 3 kHz. From Figure 29.5 we see that this corresponds to the VF band of the radio frequency spectrum, and so one possible method of communicating this information would be to simply connect our signal to a suitable antenna and broadcast this signal as radio waves. Unfortunately, this is not an attractive option. The laws of electromagnetic propagation dictate that the size of the antenna must be a significant fraction (perhaps a quarter) of the wavelength of the signal being transmitted. Therefore, in order to transmit a signal with frequency components down to 300 Hz, we would require an antenna approximately 250 km in length! *Modulation* allows us to overcome this problem by translating our signal to a higher frequency. If we combine our signal with one at a much higher frequency, we can effectively *translate* our signal to another part of the frequency spectrum. If we use this technique to move our signal so that its lowest frequency components are at 300 MHz (instead of 300 Hz) our antenna need be only 25 cm long. Another advantage of this translation is that we can also select a frequency range that has properties appropriate to the application (as outlined in Figure 29.5). At the receiver a *demodulator* is used to convert our transmitted signal back into its original form. The basic elements of such an arrangement are shown in Figure 29.6. When describing systems of this form, the low-frequency, unmodulated signal is referred to as the **baseband signal** and the high-frequency signal with which it is combined is termed the **carrier**.

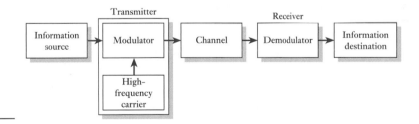

Figure 29.6
A basic communications system.

29.3.2 Basic forms of modulation

Modulation can be divided into a number of basic forms:

- **Analogue modulation** – this includes several approaches, the most widely used being **amplitude modulation** (AM) as used in AM radio broadcasting, and **frequency modulation** (FM) as used in FM radio broadcasting.
- **Digital modulation** – again several types are available, including **amplitude-shift keying** (ASK), **frequency-shift keying** (FSK) and **phase-shift keying** (PSK). These techniques, and more complex variations of them, are used in a wide range of applications such as digital radio, digital TV, mobile phones and data communications.

■ **Pulse modulation** – here modulation is applied to a pulse waveform rather than a sinusoidal carrier wave, and in many ways it combines elements of both analogue and digital modulation. Various aspects of the pulses can be controlled giving rise to several different forms of modulation, including **pulse-amplitude modulation (PAM)**, **pulse-width modulation (PWM)**, **pulse-position modulation (PPM)** and **pulse-code modulation (PCM)**.

In the following sections we will look at each of these forms.

29.3.3 Analogue modulation

Modulation involves using a low-frequency baseband signal to control some aspect of a high-frequency carrier signal. The aspect that is controlled may be the *amplitude*, the *frequency* or the *phase* of the carrier. Of these, the most widely used methods involve control of the amplitude (*amplitude modulation*) or the frequency (*frequency modulation*).

Amplitude modulation

The basic principles of amplitude modulation are illustrated in Figure 29.7. In this simple example the baseband signal is the sine wave shown in Figure 29.7(a), and this is used to modulate the carrier signal of Figure 29.7(b) resulting in the amplitude modulated signal shown in Figure 29.7(c). It can be seen that the 'envelope' representing the amplitude of the modulated waveform corresponds directly to the shape of the baseband signal. In practice the carrier frequency would invariably be at a much

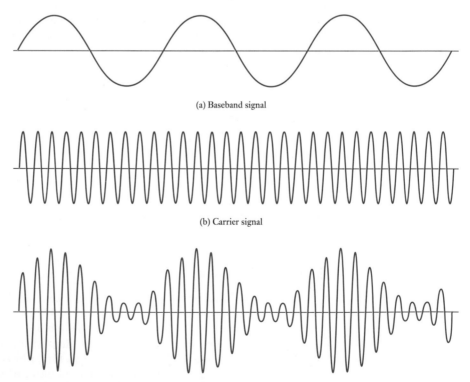

(a) Baseband signal

(b) Carrier signal

(c) Amplitude modulated signal

Figure 29.7
Amplitude modulation.

higher frequency than that shown in the figure and the baseband signal would represent the information to be transmitted (rather than a simple sine wave). However, the figure clearly illustrates the relationship between these three quantities.

The baseband signal of Figure 29.7(a) may be represented by the equation

$$v = V_B \sin(\omega_B t + \theta)$$

where V_B is the peak voltage of the baseband waveform and ω_B is its angular frequency. If we make the simplifying assumption that the phase angle θ is zero, this becomes

$$v = V_B \sin\omega_B t$$

Adopting a similar notation, we can represent the carrier signal of Figure 29.7(b) by the expression

$$v = V_C \sin\omega_C t$$

Combining these two equations we can describe the modulated waveform of Figure 29.7(c) by the expression

$$v = (V_C + V_B \sin\omega_B t) \sin \omega_C t \tag{29.2}$$

This equation is formed by taking the expression for the carrier waveform and replacing the term representing the magnitude of the waveform V_C with the expression $(V_C + V_B \sin\omega_B t)$. Thus the magnitude is equal to a constant value (V_C) plus the instantaneous value of the baseband signal. The relative sizes of V_B and V_C determine the **modulation index** m_a, where $m_a = V_B/V_C$. This is illustrated in Figure 29.8. The modulation index will normally have a value between zero and one. A modulation index of zero corresponds to no modulation while a value of one corresponds to 100 per cent modulation (taking the magnitude of the modulated signal down to zero at the negative extreme of the baseband signal). A modulation index of greater than one corresponds to an *over-modulated* wave and results in severe distortion.

From basic trigonometry we know that

$$\sin x \sin y = \frac{1}{2}\cos(x - y) - \frac{1}{2}\cos(x + y)$$

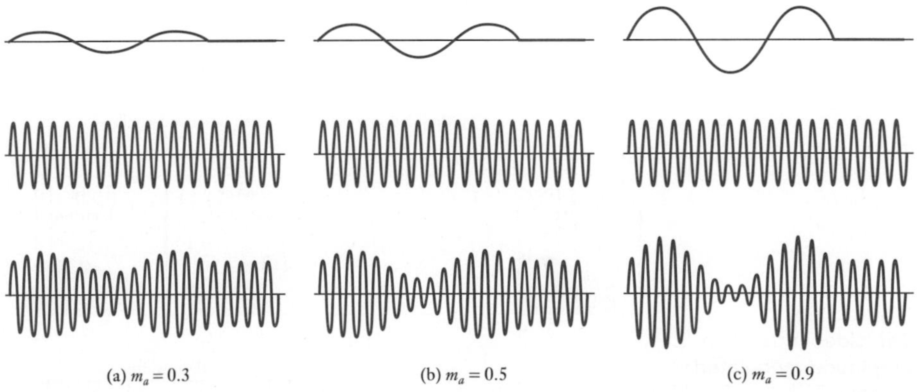

Figure 29.8
The effects of the modulation index.

(a) $m_a = 0.3$ (b) $m_a = 0.5$ (c) $m_a = 0.9$

Applying this to the equation for the modulated waveform given in Equation 29.2 we have

$$v = (V_C + V_B \sin \omega_B t) \sin \omega_C t$$

$$= V_C \sin \omega_C t + V_B \sin \omega_B t \sin \omega_C t$$

$$= V_C \sin \omega_C t + \frac{1}{2} V_B \cos(\omega_C t - \omega_B t) - \frac{1}{2} V_B \cos(\omega_C t + \omega_B t) \qquad (29.3)$$

Component at carrier frequency f_c Component at frequency $f_c - f_b$ Component at frequency $f_c + f_b$

where f_c and f_b are the frequencies of the carrier signal and baseband signal respectively. Since the elements of this expression are all sinusoidal, the minus sign before the third element simply represents a phase inversion. Thus when a sinusoidal carrier wave is modulated by a sinusoidal baseband signal, the resulting signal has three frequency components: one at the carrier frequency; one at the carrier frequency *minus* the baseband frequency; and one at the carrier frequency *plus* the baseband frequency. Note that the modulated signal does *not* have a component at the frequency of the baseband signal. The frequency spectrum of such a signal is shown in Figure 29.9(a). The *frequencies* of the components of the modulated signal are determined by the frequencies of the carrier and baseband signals, while the *magnitudes* of these elements are determined by the magnitudes of these signals.

While it is instructive to look at the nature of a signal modulated by a sinusoidal baseband signal, a more typical situation would use a baseband signal consisting of a range of frequencies. This results in two corresponding **sidebands** as shown in Figure 29.9(b). It can be seen that the bandwidth of the modulated signal is equal to twice the maximum frequency within the baseband signal. Thus a typical speech baseband signal, which might have a frequency range from 300 Hz to 3.4 kHz, would produce an amplitude modulated signal with a bandwidth of about 6.8 kHz. AM broadcasting channels tend to use a wider baseband frequency range and therefore occupy a correspondingly larger bandwidth – typically about 20 kHz.

The modulation method discussed above, and illustrated in Figure 29.9, is described as **full amplitude modulation** (or **full AM**) and this is the form of

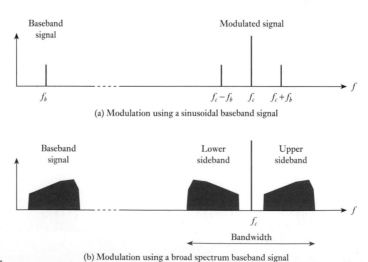

Figure 29.9
The sidebands of an amplitude modulated signal.

(a) Modulation using a sinusoidal baseband signal

(b) Modulation using a broad spectrum baseband signal

modulation most often used for broadcasting and other applications. One of the advantages of this method is that it is very easy to demodulate (as will be discussed later) thus simplifying the design of radio receivers. Unfortunately, this simplicity comes at a cost, since full AM is very inefficient in terms of both bandwidth and power. These inefficiencies can be seen by looking at the form of the modulated signal as described by Equation 29.3. Much of the power associated with this signal is within the carrier component, which conveys *no* useful information. Indeed, even when V_B is equal to V_C, corresponding to a modulation index of one (the most efficient situation), since power is proportional to the square of the voltage, two-thirds of the power of the signal is within the carrier element. Also, while the sidebands *do* convey information about the baseband signal, the use of two sidebands is unnecessary (and wasteful in terms of bandwidth), since each contains all the required information. These limitations result in the use of several other forms of amplitude modulation as shown in Figure 29.10.

Figure 29.10(a) shows a typical full AM signal and Figure 29.10(b) shows a similar signal with the carrier component removed. Signals of this latter type are described as using **double-sideband suppressed carrier (DSBSC)** modulation. The great advantage of this arrangement is that all the power within the signal is now employed within the sidebands to convey useful information. This increases the efficiency of the process considerably, but the bandwidth of the DSBSC signal is the same as that of a full AM signal. The bandwidth of the modulated signal can be reduced by removing not only the carrier (as in a DSBSC signal) but also one of the sidebands, to produce

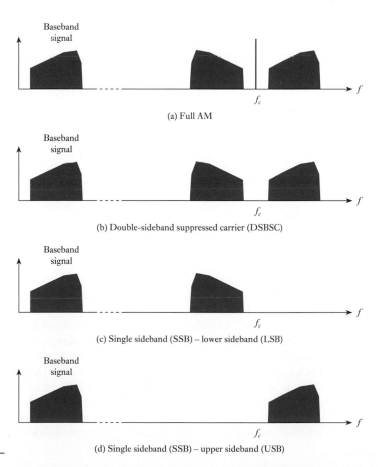

(a) Full AM

(b) Double-sideband suppressed carrier (DSBSC)

(c) Single sideband (SSB) – lower sideband (LSB)

(d) Single sideband (SSB) – upper sideband (USB)

Figure 29.10
Amplitude modulation techniques.

a **single sideband (SSB)** signal. Figures 29.10(c) and (d) show two examples of such signals. The first has been filtered to leave just the lower sideband (LSB) and the second just the upper sideband (USB). Single sideband signals have not only the power efficiency associated with DSBSC techniques, but also occupy a much smaller bandwidth. Unfortunately, the demodulation of DSBSC and SSB signals is much more complicated than that of full AM signals. For this reason these techniques are reserved for more specialist applications.

Example 29.2

A carrier signal with a magnitude of 1 V and a frequency of 1 MHz is amplitude modulated by a sine wave with a magnitude of 0.7 V and a frequency of 10 kHz. Determine the frequencies of the components of the modulated signal, its bandwidth and the modulation index of this arrangement. What percentage of the power in the modulated signal is being used to represent the baseband signal?

From Equation 29.3 we see that the modulated waveform has three frequency components with frequencies of f_c, $f_c - f_b$ and $f_c + f_b$. In this example these correspond to frequencies of 1 MHz, 0.99 MHz and 1.01 MHz.

The bandwidth is the difference between the highest and lowest frequencies, which is 1.01 MHz − 0.99 MHz = 20 kHz.

The modulation index is equal to $V_B/V_C = 0.7/1.0 = 0.7$.

Since power is proportional to the square of the voltage, the power in the sidebands, as a fraction of the power in the complete signal, is given by

$$\text{Percentage of power} = \frac{\left(\frac{1}{2}V_B\right)^2 + \left(\frac{1}{2}V_B\right)^2}{V_C^2 + \left(\frac{1}{2}V_B\right)^2 + \left(\frac{1}{2}V_B\right)^2} \times 100$$

$$= \frac{\left(\frac{1}{2}0.7\right)^2 + \left(\frac{1}{2}0.7\right)^2}{1.0^2 + \left(\frac{1}{2}0.7\right)^2 + \left(\frac{1}{2}0.7\right)^2} \times 100$$

$$\approx 20\%$$

Frequency modulation

In frequency modulation the *amplitude* of the signal remains constant while the instantaneous *frequency* of the waveform is modified to represent the baseband signal. This principle is illustrated in Figure 29.11. Here a sinusoidal baseband signal (shown in Figure 29.11(a)) is used to modulate a sinusoidal carrier signal (shown in Figure 29.11(b)) resulting in the modulated signal shown in Figure 29.11(c).

Frequency modulation is widely used for radio broadcasting of music and speech in the VHF band and for a range of other applications. One reason why FM is used in preference to AM in such applications is that it has a much better noise performance. In general, noise changes the *amplitude* of a signal and so greatly affects the information stored within an AM signal. However, noise has very little effect on the *frequency* of a signal and so has little impact on frequency modulated signals.

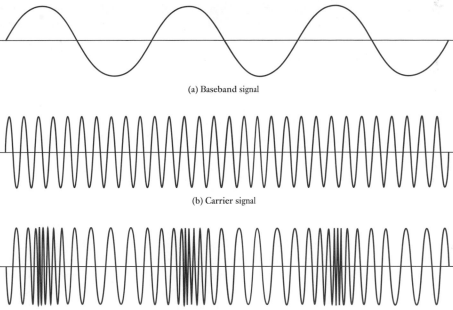

(a) Baseband signal

(b) Carrier signal

Figure 29.11
Frequency modulation.

(c) Frequency modulated signal

Unfortunately, while the general principles of frequency modulation are straight-forward (as outlined above) a detailed mathematical analysis of the process is very involved. For this reason, in this section we will look at the characteristics of this process without looking in too much detail at the related mathematics.

When looking at *amplitude modulation* earlier we considered the situation where a baseband signal described by the equation

$$v = V_B \sin \omega_B t$$

was used to modulate a carrier signal described by the expression

$$v = V_C \sin \omega_C t$$

and in Equation 29.2 we looked at the form of the resultant waveform. We also used a variable called the modulation index m_a to represent the ratio of the peak value of the baseband signal to the peak value of the carrier in the amplitude modulation process.

If we use similar baseband and carrier signals, we can describe the instantaneous value of a *frequency modulated* signal by the expression

$$v = V_C \sin(\omega_C t + m_f \sin \omega_B t) \tag{29.4}$$

Here the amplitude of the waveform is constant (with a peak value of V_C) while the phase (and hence the frequency) of the signal varies with the magnitude of the baseband signal. The degree of modulation is described by the **frequency modulation index**, which is given the symbol m_f. This is defined as the maximum deviation of the instantaneous frequency of the modulated signal (Δf) divided by the frequency of the baseband signal (f_b), and therefore

$$m_f = \Delta f / f_b \qquad \qquad (29.5)$$

At first sight, the form of Equation 29.4 might suggest that the *phase* of the resultant signal would be controlled by the baseband signal and that we therefore have *phase modulation*. In fact, the nature of m_f dictates that the deviation of the *frequency* of the resultant signal is directly proportional to the magnitude of the baseband signal, and we therefore have *frequency modulation* (although *frequency* modulation and *phase* modulation are similar in many ways). You might like to compare the expression of Equation 29.4 with that given for an amplitude modulated signal in Equation 29.2.

When considering amplitude modulated signals earlier in this section we noted that the bandwidth of such a modulated signal is directly related to the bandwidth of the baseband signal and is not affected by the analogue modulation index. Unfortunately, this is not the case in frequency modulated signals. In theory, FM signals contain an infinite number of sidebands and therefore occupy an infinite bandwidth. Fortunately, in practice, many of the sidebands contain negligible power and can therefore be ignored. However, many FM signals will contain a large number of significant sidebands (rather than just two as in a full AM signal). The number and magnitude of these sidebands varies with the modulation index, and the nature of this relationship may be described using **Bessel functions** as shown in Table 29.1. The table shows the number of significant sidebands and their magnitudes for different values of the modulation index m_f. The table indicates the relative magnitudes of each sideband, where J_0 corresponds to the carrier frequency element and a minus sign simply indicates a phase inversion.

Table 29.1 FM sidebands as described by Bessel functions.

		Bessel function order, n													
m_f	Carrier J_0	J_1	J_2	J_3	J_4	J_5	J_6	J_7	J_8	J_9	J_{10}	J_{11}	J_{12}	J_{13}	J_{14}
0.0	1.00														
0.25	0.98	0.12													
0.5	0.94	0.24	0.03												
1.0	0.77	0.44	0.11	0.02											
1.5	0.51	0.56	0.23	0.06	0.01										
2.0	0.22	0.58	0.35	0.13	0.03										
3.0	−0.26	0.34	0.49	0.31	0.13	0.04	0.01								
4.0	−0.40	−0.07	0.36	0.43	0.28	0.13	0.05	0.02							
5.0	−0.18	−0.33	0.05	0.36	0.39	0.26	0.13	0.05	0.02						
6.0	0.15	−0.28	−0.24	0.11	0.36	0.36	0.25	0.13	0.06	0.02					
7.0	0.30	0.00	−0.30	−0.17	0.16	0.35	0.34	0.23	0.13	0.06	0.02				
8.0	0.17	0.23	−0.11	−0.29	−0.10	0.19	0.34	0.32	0.22	0.13	0.06	0.03			
9.0	−0.09	0.25	0.14	−0.18	−0.27	−0.06	0.20	0.33	0.31	0.21	0.12	0.06	0.03	0.01	
10.0	−0.25	0.04	0.25	0.06	−0.22	−0.23	−0.01	0.22	0.32	0.29	0.21	0.12	0.06	0.03	0.01

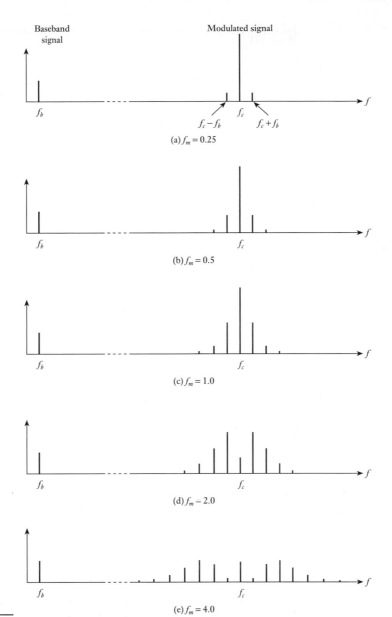

Figure 29.12
The effects of the modulation index on the frequency spectrum of FM signals.

To illustrate the effect of this relationship, Figure 29.12 shows the frequency spectra of a number of waveforms, each formed by modulating a carrier signal of frequency f_c with a baseband signal of frequency f_b, using different values of modulation index.

It can be seen from Figure 29.12(a) that using a modulation index of 0.25 results in a signal with two significant sidebands, producing an overall bandwidth that is similar to that of a full AM signal. However, using higher values of modulation index results in signals with additional sidebands and a significantly greater bandwidth. For a given baseband signal frequency, the modulation index is determined by the maximum frequency deviation of the signal Δf. It is tempting to think that the maximum deviation would directly determine the range of frequencies used and hence the bandwidth – but this is not the case.

Example 29.3 | A frequency modulation arrangement uses a maximum frequency deviation of 20 kHz. Estimate the bandwidth required to transmit a baseband signal of 5 kHz and one of 10 kHz.

From Equation 29.5 we know that

$$m_f = \Delta f / f_b$$

With a baseband frequency of 5 kHz

$$m_f = \Delta f / f_b = 20 \text{ kHz}/5 \text{ kHz} = 4.0$$

From Table 29.1 we see that this will result in seven pairs of significant sidebands (in addition to the carrier frequency component). Therefore the bandwidth required will be $2 \times 7 \times 5$ kHz = 70 kHz.

When the baseband frequency is 10 kHz

$$m_f = \Delta f / f_b = 20 \text{ kHz}/10 \text{ kHz} = 2.0$$

Table 29.1 shows that there are now four pairs of significant sidebands, and the bandwidth is equal to $2 \times 4 \times 10$ kHz = 80 kHz.

Example 29.3 shows that there is not a linear relationship between the baseband frequency and the resultant bandwidth. Note also that while the *analogue* modulation index m_a has a possible range that goes from zero to one, the *frequency* modulation index m_f can take any value from zero to infinity.

While we have concentrated on the use of baseband signals of a single frequency, applications such as broadcasting will inevitably use broadband signals, such as speech or music. As with the AM techniques discussed earlier, this will result in signals with complex spectra, rather than the simple line spectra shown in Figure 29.12. However, the discussions above should make it clear that the bandwidths of such signals will generally be much greater than those used in AM broadcasting. In FM broadcasting it is normal to allocate 200 kHz of bandwidth to each channel, while AM channels might typically use only one tenth of this amount.

Video 29C

29.3.4 Digital modulation

While analogue modulation is used to transmit analogue information over some form of *analogue* channel, digital modulation is used to transmit digital information over a similar *analogue* channel. Most forms of digital modulation are based on some form of **keying**, this term implying that the modulated signal is switched between a limited number of states. Widely used techniques include:

- **amplitude-shift keying (ASK)**
- **frequency-shift keying (FSK)**
- **phase-shift keying (PSK)**.

We will look at each of these techniques in this section.

Amplitude-shift keying (ASK)

Amplitude-shift keying represents information by switching the *amplitude* of a carrier signal between a number of distinct values. A typical arrangement might represent a binary quantity by switching the magnitude of a signal between zero and some fixed value as shown in Figure 29.13.

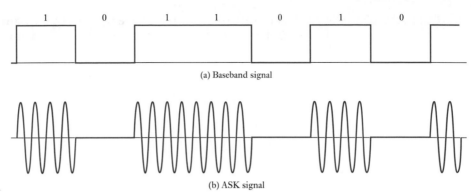

Figure 29.13
An example of
amplitude-shift keying.

(a) Baseband signal

(b) ASK signal

As with linear amplitude modulation, ASK is relatively simple to produce and to demodulate. However, as with AM, ASK is also more susceptible to noise than other techniques, since noise tends to affect the amplitude of a signal rather than its frequency or phase.

Frequency-shift keying (FSK)

Frequency-shift keying represents information by switching the *frequency* of a carrier signal between a number of distinct values. The simplest form is *binary FSK* which uses two frequencies to represent logical 0 and logical 1, as shown in Figure 29.14. The frequency used to represent 1 is generally called the 'mark' frequency and that used to represent 0 is called the 'space' frequency.

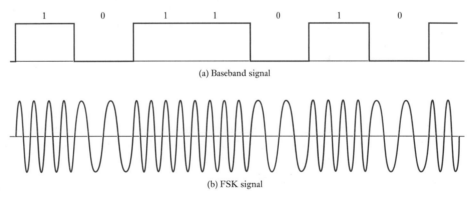

Figure 29.14
An example of
frequency-shift keying.

(a) Baseband signal

(b) FSK signal

In addition to its use within radio frequency applications, FSK is also used to transmit information using audio frequency signals. In the early days of home computing, audio FSK was widely used in telephone modems to send and receive data at speeds of up to about 1200 bits per second. However, while FSK is used in a range of applications, where high data rates are required it is more common to use other approaches, such as the various forms of PSK.

Phase-shift keying (PSK)

Phase-shift keying represents information by switching the *phase* of a carrier signal between a number of distinct values. A simple form is **binary PSK** which uses two

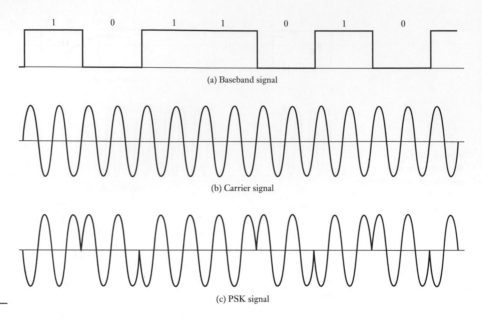

Figure 29.15
An example of binary
phase-shift keying.

phases separated by 180° to represent logical 0 and logical 1. A possible form of this modulation is shown in Figure 29.15. It can be seen that when the baseband signal is 1 the modulated signal is in phase with the carrier signal, and when it is 0 the modulated signal is phase inverted.

One problem with the simple modulation scheme shown in Figure 29.15 is that the receiver requires a knowledge of the phase of the carrier in order to be able to demodulate the signal. This problem can be overcome by the use of **differential phase-shift keying (DPSK)**, which represents data by *changes* to the phase of a signal, rather than its actual value. This process is illustrated in Figure 29.16. Here time is divided into a number of periods, each representing one binary digit. To transmit a 1 the phase of the modulated signal is phase-shifted with respect to the previous period, while to transmit a 0 the phase is left unchanged. At the receiver the demodulator simply needs to compare the phase of successive periods to decide if they are a 1 or a 0 – no knowledge of the phase of the original carrier signal is needed.

In the *binary PSK* described above the two possible values of a single binary digit are represented by two different phase angles in the transmitted signal. An alternative approach combines several binary digits together and uses one of a set of phase values

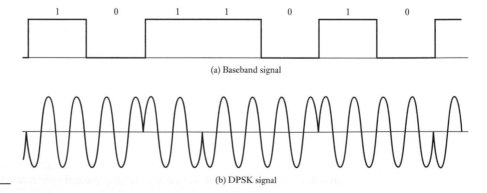

Figure 29.16
A differential phase-
shift keying
arrangement.

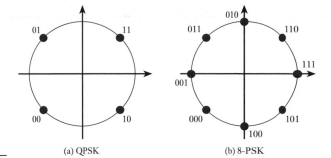

Figure 29.17
Higher-order phase-shift keying arrangements.

to represent each of their possible values. For example, a group of two binary digits has four possible values: 00, 01, 10 and 11. Thus by selecting one of four possible phase angles, a single element of the transmitted waveform can represent two digits rather than one. The phase angles used are normally equally spaced to simplify their recognition and a typical arrangement is shown in Figure 29.17(a). Such an arrangement is referred to as **quadrature phase-shift keying** or **QPSK**. In theory any number of digits can be combined in this way, although in practice it is unusual to use more than eight discrete phase angles. Such an arrangement can code three digits into each element and a typical arrangement is shown in Figure 29.17(b). Arrangements using eight phases are often called **8-PSK**. In the diagrams of Figure 29.17 the position of the element around the circle indicates the phase angle used to represent each binary sequence. You will note that the sequences are arranged in **gray code** order. This is done to minimise bit error rates.

29.3.5 Pulse modulation

The term *pulse modulation* is used to describe several distinct forms of modulation. Some, such as *pulse-amplitude modulation* (PAM), *pulse-width modulation* (PWM) and *pulse-position modulation* (PPM), transfer information by varying the characteristics of a pulse waveform. These are forms of *analogue* modulation and result in a signal that can be transmitted over an analogue channel. Other techniques, such as *pulse-code modulation* (PCM), use *digital* modulation to represent data within a train of pulses that can be transmitted over a digital channel. Figure 29.18 illustrates the basic forms of each of these. Figure 29.18(a) shows a possible baseband signal in the form of a simple sine wave and Figure 29.18(b) shows an unmodulated pulse waveform that replaces the high-frequency carrier signal used in AM and FM techniques.

Pulse-amplitude modulation (PAM)

Figure 29.18(c) shows a PAM waveform. Here the baseband signal is sampled at regular intervals and the magnitude of the samples is used to modulate the *amplitude* of the pulses. The duration and timing of the pulses are unaffected. Both modulation and demodulation are straightforward, the latter using a process of envelope detection (as discussed in the next section). Since the information is carried by the *amplitude* of the pulses, PAM (like AM) is more affected by noise than other techniques.

Pulse-width modulation (PWM)

Figure 29.18(d) shows a PWM waveform. Again the baseband signal is sampled, but now the magnitude of the samples is used to modulate the *width* (or *duration*) of the

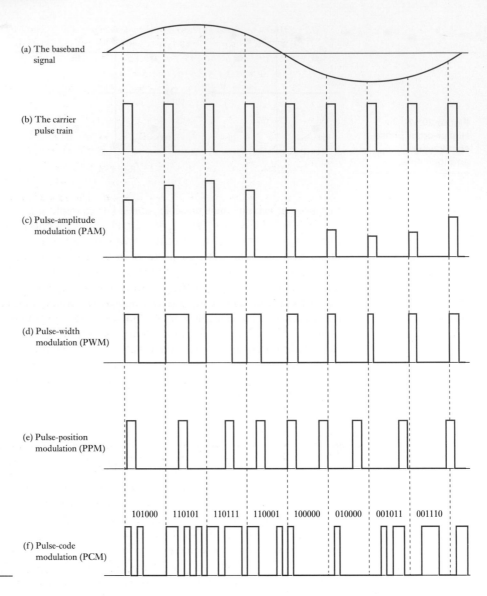

(a) The baseband signal

(b) The carrier pulse train

(c) Pulse-amplitude modulation (PAM)

(d) Pulse-width modulation (PWM)

(e) Pulse-position modulation (PPM)

101000   110101   110111   110001   100000   010000   001011   001110

Figure 29.18
Pulse modulation.

(f) Pulse-code modulation (PCM)

pulses. The magnitude of the pulses is constant. Since the information is carried by the duration of the pulses, PWM is less affected by noise than PAM.

Pulse-position modulation (PPM)

The third of the trio of analogue pulse modulation techniques is PPM, which is shown in Figure 29.18(e). Here the magnitude of the samples is used to modulate the *position* of the pulses, by moving them backwards or forwards in time with respect to the carrier waveform. Since the signal is *not* represented by the amplitude or width of the waveform, such signals are less affected by noise than both PAM and PWM. Its use of position rather than amplitude or duration also makes this approach very power efficient. However, demodulation of the signals requires a detector that is synchronised with the transmitter, making this process more complicated.

Pulse-code modulation (PCM)

PCM differs from the other pulse techniques described above since it utilises *digital* modulation. Samples of the baseband signal are first digitised (or encoded) to form a digital *word* representing its magnitude. The pulse carrier is then modulated to produce a series of pulses representing the binary equivalent of this word. This process is illustrated in Figure 29.18(f), which shows the encoded form of a series of readings representing the baseband signal. PCM is widely used to represent a range of analogue signals and is the standard technique used for digital audio within Blu-ray®, DVD and compact disc equipment.

29.4 Demodulation

The process of *modulation* modifies a carrier signal to allow it to convey information related to its source. The process of *demodulation* reverses this process by extracting the original information from the modulated carrier. This process is also known as **detection**. Since there is a wide range of modulation methods used (as discussed in Section 29.3) there is a similar range of demodulation techniques.

In full AM signals (such as those used for AM broadcasting) the required information is stored within the *envelope* of the waveform. This permits the information to be retrieved relatively easily using an **envelope detector**. The operation of such a circuit was described in Section 17.8.4 and a circuit of a simple detector was shown in Figure 17.22. Other forms of amplitude modulated signals (such as **DSBSC** and **SSB**) require more sophisticated circuitry which often makes use of a local oscillator of the same frequency and phase as the carrier. This results in a considerably more complex detector and is one of the reasons why these modulation techniques are less widely used.

The demodulation of FM and PM signals is much more involved than that of full AM. A common approach uses **quadrature detection**, which multiplies the modulated signal by a similar signal that has been phase-shifted by 90 degrees. The resultant signal is then processed to retrieve the original baseband signal. While FM requires more complex circuitry than AM, its superior fidelity and noise performance make it the dominant technique for quality radio broadcasting.

29.5 Multiplexing

We noted earlier that we often wish to transmit not one, but a number of signals, at the same time, over a single channel. We might, for example, wish to transmit a number of speech signals simultaneously between pairs of telephones. Since each speech signal occupies the same frequency range, we cannot simply combine them and transmit them over a single channel.

This problem may be tackled in a number of ways. One approach is to use the modulation process to transmit the various signals at slightly different frequencies. This is achieved simply by using different carrier frequencies for each signal. This process is known as **frequency division multiplexing** and is illustrated in Figure 29.19. At the receiver, filters are used to separate the various signals so that they can be routed to their appropriate destination.

When using digital signals it is more common to transmit multiple data streams by splitting the data into small blocks. The transmitter then switches rapidly between the streams sending blocks from each in sequence. This process is known as **time**

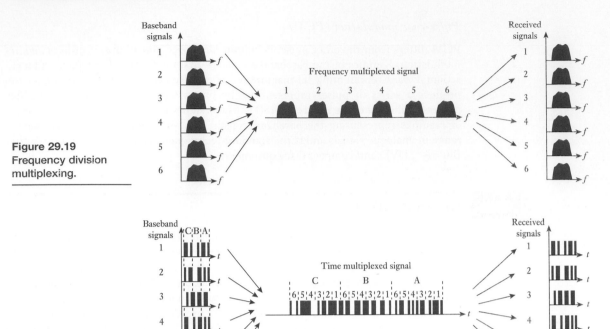

Figure 29.19
Frequency division
multiplexing.

Figure 29.20
Time division
multiplexing.

division multiplexing since the time available on the channel is switched between the information sources. At the receiver a demultiplexer sorts the incoming data and reconstructs the original data streams. This process is illustrated in Figure 29.20.

Video 29D

29.6 Radio receivers

A radio receiver is required to perform several functions. These are: to *select* a single radio station from the range available; to provide radio frequency *amplification*; to *demodulate* the received signal to recover the baseband information; and to *amplify* the resultant audio waveform to drive a speaker. The nature of the demodulation will depend on the form of the transmitted signal (for example, whether this uses AM or FM) but the other aspects of the system are unaffected.

TRF receivers

Early radio broadcasting used full AM modulation and early radio receivers were normally of the form shown in Figure 29.21(a). Here a signal from an **antenna** is first passed to a **tuned radio-frequency amplifier,** which both amplifies the signal and band-pass filters it to select a particular station. The *bandwidth* of this amplifier is set to match that of the transmitted station to eliminate other adjacent transmissions. This also reduces the amount of noise entering the system, since this will be proportional to the bandwidth of the filter. The *centre frequency* of the amplifier is adjusted, using a variable capacitor within a tuned circuit, to select the desired station. The amplified signal is passed to a **detector** which demodulates it to obtain the original baseband waveform. This stage might use a simple *envelope detector* as described earlier.

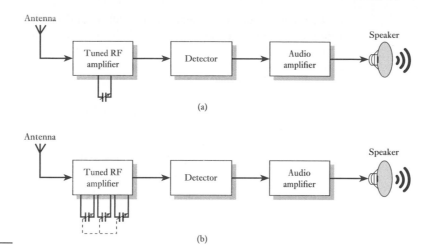

Figure 29.21
A tuned radio
frequency receiver.

The demodulated signal is then passed to an **audio amplifier** to provide the power necessary to drive the speaker. For obvious reasons, receivers of this type are referred to as **tuned radio frequency** or simply **TRF receivers**.

One problem with the simple receiver described above is that a single stage of RF amplification cannot normally produce sufficient gain. The receivers therefore typically have three amplifier stages, with a triple-ganged variable capacitor being used to simultaneously adjust the tuned circuit in each stage. Such an arrangement is shown in Figure 29.21(b).

From our discussion of tuned circuits in Chapter 8 we know that the bandwidth B of such a circuit is related to its quality factor Q and its resonant frequency f_o by the expression $Q = f_o/B$. Therefore, since the Q for such a circuit will be relatively constant, the bandwidth will change as the circuit is tuned from one station to another. Since the frequency at one end of the medium wave band is more than three times the frequency at the other, this means that the bandwidth of the tuned circuit at one end of the band will be three times greater than that at the other end. This is a considerable problem since the bandwidth used by stations is constant throughout the band.

Superheterodyne receivers

The problems of variability of bandwidth in TRF receivers can be overcome by the use of a slightly more complex approach known as a **superheterodyne receiver** which, for simplicity, is often called a **superhet**. A block diagram of such a receiver is shown in Figure 29.22.

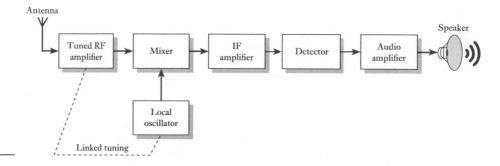

Figure 29.22
A superheterodyne
receiver.

The superhet has at its heart a local oscillator and a mixer which together perform a function somewhat similar to amplitude modulation as described in Section 29.3.3. From our earlier discussions we know that AM combines an incoming signal (the baseband signal) with a locally generated signal (the carrier) in order to shift the incoming signal to a different frequency range. The superhet performs a similar process by combining an incoming signal (in this case the RF signal from the antenna) with a locally generated signal (from the **local oscillator**) to shift the incoming signal to a well-defined frequency range. This range is normally below the frequency of the incoming RF signal but above the frequency of the audio frequency output. For this reason it is known as the **intermediate frequency (IF)**. Because the output of the mixer is at a constant frequency, it can be followed by a high-gain amplifier stage (the **IF amplifier**) which does not need to be tunable. Since this operates at a fixed frequency it can have a well-defined frequency response and it is this stage that is responsible for defining the bandwidth of the receiver. The RF amplifier stage is tuned along with the local oscillator (as in the TRF receiver) but is no longer responsible for defining the selectivity of the device. The IF amplifier is followed by a detector appropriate to the station being received and an audio amplifier.

The superhet is 'tuned' to a particular station by adjusting the frequency of the local oscillator. To see how this works let us consider a numerical example. In medium wave receivers a common choice for the intermediate frequency is 455 kHz. If we wished to receive a station with a centre frequency of 1 MHz, we would set the local oscillator to produce a signal at 1.455 MHz. When combined with the incoming station this would produce components at the sum and difference frequencies of 455 kHz and 2.455 MHz, together with components at the local oscillator frequency. The highly frequency selective IF stage would filter out just the first of these components for amplification and demodulation. If we wished to receive a station with a centre frequency of 1.1 MHz, we would tune the local oscillator to 1.555 MHz, which would again produce a component at 455 kHz as well as one at 2.655 MHz. Again the lower component would be selected and the other rejected. In this way any station can be selected by setting the local oscillator to a frequency equal to the station's centre frequency plus the intermediate frequency.

The superheterodyne offers excellent sensitivity, together with good frequency stability and selectivity. For these reasons it is used in almost all modern radios, televisions, satellite receivers and microwave links. Common intermediate frequencies are: 455 kHz for medium wave AM radio; 10.7 MHz for VHF FM radio; 38.9 or 45 MHz for television; and 70 MHz for satellite receivers.

Video 29E

Further study

Amongst the most frequently used communication systems are those within the various handsets that we use to remotely control our TVs, videos and other consumer gadgets. Most devices employ infrared light as the transmission medium and are limited to line-of-sight operation. However, other technologies are available, such as those based on Bluetooth®.

Consider the requirements of such a communications system and propose appropriate methods for implementing it based on the techniques discussed in this chapter.

Key points

- The term 'communications' relates to the transfer of information from one place to another.
- Information is transferred via a channel using a transmitter and a receiver.
- A commonly used channel involves the use of radio waves. The radio frequency spectrum is divided into a number of bands. Radio wave propagation varies considerably between these bands.
- Modulation is used to transfer a signal to a different part of the frequency spectrum. It can also be used to achieve multiplexing and often improves the performance of a communications system in terms of noise and interference.
- Modulation techniques may take a number of forms, including analogue, digital and pulse modulation.
- Analogue modulation is widely used for radio broadcasting. Amplitude modulation (AM) is used in the medium wave band while frequency modulation (FM) is used in the VHF band.
- Digital modulation, such as ASK, FSK and PSK, is used to transmit digital information over an analogue channel.
- Pulse modulation can take several forms, some using analogue and some using digital channels. Applications include the coding of audio signals on Blu-ray discs and DVDs.
- At the receiver, modulated signals must be demodulated to recover the original baseband signal.
- Multiplexing allows several communications paths to share a common channel. Techniques include frequency division multiplexing and time division multiplexing.
- Early radio receivers were based on tuned radio frequency (TRF) techniques. Almost all modern receivers make use of superheterodyne methods.

Exercises

29.1 Explain what is meant by the term 'communication'.

29.2 Why are some books on communication related to public speaking rather than modulation?

29.3 Explain the functions of the transmitter, receiver and channel in a communications system.

29.4 Give an example of a simplex communication system.

29.5 What is the difference between a full-duplex and a half-duplex arrangement?

29.6 Explain the terms ground wave propagation, sky wave propagation and line-of-sight propagation. Which of these techniques offers the greatest range?

29.7 The various radio frequency bands take as their boundaries values that are of the form 3×10^X Hz. Why are these boundaries chosen rather than frequencies that are a simple power of 10 Hz?

29.8 What is meant by channel capacity and what factors affect this value?

29.9 A channel has a bandwidth of 5 kHz and a signal-to-noise ratio of 30 dB. What is its channel capacity?

29.10 Why is it impractical to transmit voice signals by connecting them directly to an antenna?

29.11 A carrier signal with a frequency of 10 kHz is amplitude modulated by a sine wave with a frequency of 1 kHz. What frequencies are present within the modulated waveform?

29.12 A sinusoidal carrier signal with a frequency of 1 MHz and an amplitude of 2 V is amplitude modulated by a sine wave with a frequency of 10 kHz and an amplitude of 1.5 V. What is the modulation index of this arrangement and what percentage of the power in the modulated signal is used to represent the baseband signal?

29.13 How could the power efficiency of the signal in the previous exercise be improved?

29.14 A 20 kHz sinusoidal baseband signal is used to produce a frequency modulated signal with a maximum frequency deviation of 10 kHz. What is the bandwidth of the resulting signal?

29.15 What are the main advantages of full AM in comparison to FM for radio broadcasting?

29.16 What are the main advantages of FM in comparison to full AM for radio broadcasting?

29.17 Explain briefly the meaning and form of ASK, FSK and PSK.

29.18 Why is ASK more susceptible to noise than other digital modulation methods?

29.19 Explain the advantages of differential phase-shift keying over conventional PSK.

29.20 How does quadrature phase-shift keying differ from binary phase-shift keying?

29.21 Is pulse modulation an analogue or a digital modulation technique?

29.22 Explain the difference between demodulation and detection.

29.23 Describe briefly the operation of the simple envelope detector of Figure 17.22.

29.24 Why is detection of DSBSC or SSB signals more complicated than that of a full AM signal?

29.25 Explain briefly the functions of multiplexing and demultiplexing within a communications system.

29.26 How does frequency division multiplexing differ from time division multiplexing?

29.27 Sketch a block diagram indicating the major components of a TRF radio receiver. What are the advantages and disadvantages of this design?

29.28 Sketch a block diagram indicating the major components of a superheterodyne receiver. What are the advantages and disadvantages of this design?

Chapter 30

System Design

When you have studied the material in this chapter, you should be able to:

- identify the major tasks associated with the design of an electronic system
- suggest a range of alternative system implementation methods, including both analogue and digital techniques
- outline the principal characteristics of the various device technologies and identify appropriate techniques for a variety of applications
- discuss the relative merits of programmable and non-programmable systems and suggest an appropriate strategy for a given task
- list a range of electronic design tools for both analogue and digital systems
- describe the characteristics of system description languages and formal methods for system specification and design.

30.1 Introduction

A good design is one that solves a particular problem in the most appropriate and efficient manner. To achieve this, the designer requires not only a good understanding of the problem but also a wide-ranging knowledge of available techniques and technologies. Inevitably, this means that design ability increases with experience, but this should not be seen as reducing the importance of a systematic and methodical approach. Design is a creative process but must be based on sound engineering principles to achieve a result that is both cost effective and efficient.

A range of automated tools are available to aid the designer at each stage of a project. These include packages that allow circuit diagrams to be drawn easily and quickly, simulate the operation of the circuit for testing, produce layouts for either printed circuit or VLSI implementations and verify that these layouts follow a series of design rules.

Video 30A

30.2 Design methodology

The task of designing an electronic system has many facets and (as discussed in Chapter 11) can be greatly simplified by adopting a methodical, rational approach.

Most experienced engineers agree that a **top-down approach** to system design offers many advantages as it focuses on the major aspects of a system first and fills in the details later. The process of design therefore starts from an idea of 'what we are trying to achieve' and works towards an understanding of 'how we are going to achieve it'.

Customer requirements

The customer requirements represent the problem that the system is to solve. In some circumstances the customer and the designer may be one and the same person, but the principle remains the same. The requirements of the system are usually, and correctly, expressed in terms of the problem rather than the solution. It should be noted, however, that the customer requirements represent the system's *actual* requirements, rather than any verbal or written description of them.

Top-level specification

The top-level specification is an attempt to define a system that will satisfy the customer requirements. The definition is usually in the form of a written description in a natural language (the term 'natural' is used here to distinguish it from a programming or other computer language), but may include appropriate mathematical equations or expressions. We will see later, when we come to look at electronic design tools, that there are other more precise ways in which to define a system. The problem with a written specification in a natural language is that it is extremely difficult to write in a way that is not open to misinterpretation.

It is important to ensure that the specification defines *what* the system is to do, not *how* it is to do it. Such topics as the appropriate device family or the use of analogue or digital techniques do not fall within the realms of the specification.

In large companies it is normal for the specification to be produced by a different team from the one that will ultimately perform the design. This is done to preserve the independence of these two functions. When complete, the specification should be agreed by all concerned, including, if appropriate, the customer. Once the specification has been finalised, work may begin on the next stage of the project. It should be noted that occasionally it may be necessary to make modifications to the specification during the project in the light of new information. There is no reason for not revising the specification, provided that it is done with the agreement of all interested parties. However, it must not be done unilaterally by the designer simply to make the job easier.

In addition to providing a specification for the system, it is common at this stage to define a series of tests that the resultant system must perform to prove its suitability for the task in hand. These will then form the basis of system testing and the ultimate demonstration of the system.

Top-level design

Once the specification of the system has been completed, the design stage may be commenced in a **top-down** manner. For large projects, usually the first task is to divide the system into a number of manageable modules. A specification is then produced for each module, enabling it to be designed and tested independently.

One of the earliest design decisions to be made concerns the choice of technology. Invariably it will be possible to implement a given function in a number of ways, using, for example, analogue, digital or software techniques. Often large systems will include elements of each of these methods. Until these decisions have been made, it is usually impossible to progress to more detailed aspects of the design. This topic is covered in more detail in Section 30.3.

For systems that are to include programmable devices, such as microprocessors or microcontrollers, it is also necessary to perform a **hardware/software trade-off** to decide which functions are to be performed by hardware and which by software. This procedure requires a knowledge of the predicted volume of production of the system (as noted in Chapter 27).

The top-level design results in a block diagram form of description of the system and a specification of each block of hardware and software. Based on this information, work can move to progressively greater levels of detail.

Detailed design

If the top-level design has been performed efficiently, the detailed design stage of the project should be relatively straightforward. Each hardware section will consist of a series of functions that can usually be assembled from standard circuit building blocks, as described in earlier chapters. Each software section will require the provision of a segment of computer program. Again, standard functions and structures are available to simplify this task.

A range of automated tools are available to help the designer at all stages of the project. These are discussed later in this chapter.

Module construction and testing

When the design of a system is complete, the various modules must be constructed and tested to ensure that they perform their required functions correctly.

Unlike design, which is performed in a top-down direction, testing is *usually* performed using a **bottom-up** approach. This involves first verifying the operation of each individual circuit element, then investigating the functioning of progressively larger subsystems. Testing is performed in this way as both error detection and fault location are easier when dealing with small sections of circuitry than with a complete system. Any faults found at this stage must be rectified before continuing to system testing.

System testing

When each module has been tested and any corrective work performed, the complete system may be assembled and tested. Only at this stage is it possible to see whether or not the system meets its top-level specification and to confirm that it does indeed fulfil the customer requirements. It will normally be necessary to demonstrate the system to customers and perform any prescribed system-proving tests.

In systems where incorrect operation could have serious safety or financial implications, the tests required will often be very stringent and it is often necessary to demonstrate that the system is 'safe' or of 'high integrity'.

30.3　Choice of technology

One of the crucial decisions to be made in the design of any system is the choice of technology. In the broadest sense, this could involve implementing a system using perhaps mechanical, hydraulic, pneumatic, electrical or electronic means, but here we are primarily interested in the choice between various forms of electronic circuit. The major options include the use of analogue or digital techniques, programmable or non-programmable methods and bipolar or FET devices. We have already noted that it is normal to subdivide a large project into a number of more manageable modules. Clearly, there is no reason for each module to adopt the same approach and, indeed, it is common for a combination of techniques to be used even within the same section of a system.

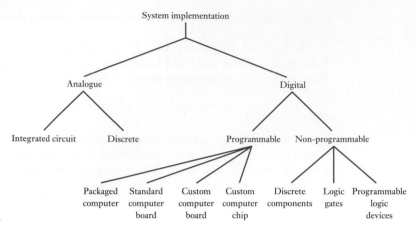

Figure 30.1
Alternative methods of
system implementation.

Figure 30.1 indicates some of the alternative approaches to the design of a particular circuit. One of the most fundamental decisions is whether the solution is to be of an analogue or a digital form. If an analogue solution is chosen, it is then necessary to decide on either a discrete or an integrated approach. If a digital system seems more appropriate, then it is necessary to select either a programmable technique, as in the case of a microcomputer system, or a non-programmable implementation. In either case, several options are available to the designer.

Unfortunately, it is difficult to provide hard and fast rules as to when one solution is more appropriate than another. A more useful approach is perhaps to identify some of the characteristics associated with the various options to allow the designer to assess which method would be appropriate for a given application.

Analogue vs digital

Often the choice between an analogue and a digital solution to a particular problem is indicated by the nature of the input and output signals. Clearly, if a system uses only binary sensors and actuators, then a digital approach is indicated, whereas a system with analogue inputs and outputs could suggest the use of analogue techniques. However, although it is uncommon to use analogue techniques for a system with purely digital inputs and outputs, it is quite common to use digital techniques in applications that use analogue signals. This latter approach requires the use of data convertors (as discussed in Chapter 28) but in many cases the benefits gained more than compensate for this extra complexity.

The appropriateness of a digital approach will vary from application to application, but in general terms one could say that the potential advantages of digital systems are that they offer: improved consistency; better noise performance; easier storage and transmission of signals; and the ability to perform complex signal processing in a straightforward manner. System design is also often easier with digital systems and there is a greater range of automated design tools available.

When we are considering applications of some complexity, and are comparing an analogue implementation of such a system with a complex digital implementation (such as one based on a microcomputer or an FPGA) then there are further aspects to be considered. One might identify the advantages of a such a digital system as being:

- improved consistency
- greater flexibility through programmability

- greater standardisation of hardware
- reduced component count
- lower unit cost (sometimes)
- improved testing and the ability to include self-testing
- the opportunity to easily add additional features
- greater reliability
- the possibility of providing automatic, or simplified, calibration.

In many cases these potential advantages will prove compelling, but one should be aware that the use of complex digital solutions in situations where they are not appropriate can bring disadvantages such as:

- greater development cost
- greater investment required in development equipment
- greater system complexity.

Integrated vs discrete

The choice between the use of integrated circuits and discrete components when producing analogue circuits must be based on a number of factors, including function, noise, power consumption, cost, size and design effort. In general, using ICs is easier as the chip designer has done much of the hard work for you. However, in some very simple or specialised applications, discrete components may be more appropriate. Often, high-power circuitry must be implemented using discrete transistors.

Programmable vs non-programmable

In simple applications, non-programmable solutions are preferable as they require no software development and so have a lower development cost. However, as the complexity of the system increases, the potential advantages of using a programmable (that is, computer-based) approach become considerable.

One of the greatest advantages of computer-based systems is their flexibility. This allows a single standard computer board to be used for a range of applications, reducing the range of subsystems that must be produced. This saves on both design time and inventory costs (the cost of holding stocks of components). It also allows the operation of a system to be updated simply by changing its operating program without having to redesign the hardware. Set against these advantages is the high cost of software development.

In many situations an alternative to the use of a microcomputer is the use of a PLD. In systems that are based on PLDs, much of the design complexity is implemented within the logic device. This enables the functionality of the system to be modified simply by changing the PLD configuration. This allows such systems to be upgraded in much the same way as a computer-based system – often without necessitating modifications to the hardware. When CPLDs are used in this way, the task of configuring the device is actually very similar to that of the production of software. The process uses an array of complicated development tools and can be very time consuming and expensive. One can view microprocessors and PLDs as similar devices that each implement a potentially complicated set of instructions defined by the programmer. The primary difference between the two approaches is that in a computer they are executed in a *serial* manner, while in a PLD they are executed in *parallel*.

Beyond a certain level of complexity, the cost of implementing systems using non-programmable logic becomes prohibitively expensive. In such cases, a computer-based system is the only practical solution, allowing very complicated control algorithms to be constructed in software, rather than by adding more complicated hardware.

Implementing programmable systems

Computer-based systems can be implemented in a number of ways, the strategy adopted depending to a great extent on the volume of production. Low-volume projects will tend to favour the use of ready-made systems, reducing the need for a large amount of expensive design work. Higher-volume applications, on the other hand, will tend to demand a custom approach, with a considerably greater amount of design effort being used to produce specialised circuit boards. For very high-volume projects, it may be appropriate to have **custom integrated circuits** produced. This approach yields a very low unit cost, but is associated with extremely high development costs.

Implementing non-programmable digital systems

With non-programmable digital systems, the method of implementation is likely to be determined by the complexity of the functions to be produced. Where very limited logical operations are required, it may be possible to produce these using simple discrete circuits based on diode logic or the use of a small number of transistors. However, for functions of all but the simplest form, the use of more conventional logic circuits is normal. Applications that require only a handful of gates will normally use standard CMOS logic devices. Unfortunately, even relatively simple logic arrangements can produce circuits requiring many devices and it soon becomes economical, in terms of both cost and space, to use some form of array logic.

30.3.1 Device technologies

Having decided on the method of implementation for a particular system or module, it is then necessary to consider the **device technology** that will be used to produce it. Figure 30.2 outlines some of the major choices to be made.

In both analogue and digital systems, one of the major decisions to be made is whether to use circuits with bipolar transistors, FETs or a combination of the two.

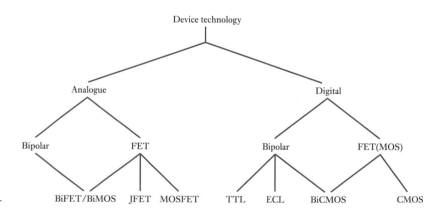

Figure 30.2
Device technologies.

Bipolar vs FET in analogue systems

The characteristics of FETs and bipolar transistors (as discussed in Chapters 18 and 19) are sufficiently different for them to be used in many circuits where they are not interchangeable. In such cases, the choice of component is clear. However, in some applications it is possible to use either device technology and so a decision must be made between them. In general, FETs are used in applications where a high input impedance is required and they can provide a good noise performance in such cases. Bipolar transistors have a lower input resistance, but can often produce a much higher gain and have superior noise performance when used with low impedance sources. In some cases, it is possible to combine the attractive characteristics of the two technologies by using BiFET or BiMOS circuits.

Bipolar vs FET (MOS) in digital systems

In digital systems, the choice between bipolar circuitry and that based on FETs (the latter normally being described as MOS devices when considering digital components) is made largely on considerations of speed, power consumption, noise immunity and component density. These considerations were discussed in some detail in Chapter 26.

In the early days of digital electronics bipolar logic families, such as TTL and ECL, were usually faster in operation than MOS devices, with the non-saturating families, such as ECL and Schottky TTL, being faster than saturating types. However, modern CMOS parts are very fast and have replaced TTL in almost all applications, with bipolar parts being restricted to a small number of specialist applications, such as the construction of high-speed device drivers.

MOS circuits have a better noise immunity than bipolar devices, with CMOS gates being able to tolerate noise of at least 30 per cent of the supply voltage. As these circuits can also be used with a 15 V supply rail, this gives a noise immunity of about 4.5 V compared with about 0.4 V for TTL and somewhat less for ECL. It should be noted, however, that the high input impedance of MOS logic makes it less impressive in terms of its noise performance than these figures might suggest. It should also be remembered that CMOS logic is very often used with much lower supply rail voltages, reducing its noise immunity considerably.

Another important characteristic of the MOS technologies is their very high circuit densities when implemented in IC form. This allows far more circuitry to be combined within a single device than is possible within a bipolar part. For this reason, most microcomputers and their associated memory and support devices are constructed using CMOS techniques. For high-speed applications, bipolar microprocessors are available, but these are generally very expensive and not as sophisticated as MOS types.

30.4   Electronic design tools

A vast range of computer-based electronic aids is available to simplify and speed the process of design. Such aids come under the general title of **computer-aided design (CAD) tools**, but may also be referred to as **electronic computer-aided design (ECAD)**, **computer-aided engineering (CAE)** or **computer-aided software engineering (CASE)** utilities. It is not within the scope of this text to describe the use of these packages or even discuss their functions in any kind of detail. It is, however, relevant to look in general terms at the overall characteristics of the various forms of tools.

Although many of the functions provided by individual packages are distinct, it is often advantageous to use a number of tools in succession. For this reason, it is normal to adopt certain standard forms for the data produced and accepted by the packages, in an attempt to make their data interchangeable.

30.4.1 Schematic capture

A schematic capture package is to the circuit designer what a word processor is to the writer. It allows circuit diagrams to be drawn quickly on a computer screen using both keyboard commands and input from a pointing device such as a mouse or graphics tablet. Components may be added, deleted or moved, and may be joined as appropriate to produce the desired circuit.

However, the tool would be of only limited use if it were simply a graphics package capable of drawing objects on the screen. The power of the technique comes from the use of **component libraries** within the package. These store the technical details of each component, including its **pin-out** and circuit symbol. The designer can select components from the library as required and position them within a circuit without having to draw each one laboriously. Most packages come with an extensive range of standard components that will meet most needs and that are updated as new components become available. It is also normally possible to add new components by drawing the required circuit symbol and adding the appropriate information to the library.

The output from the schematics capture package may be in a number of forms. One of the most obvious is that it can produce a hard copy of the finished diagram via a printer or plotter. It can also provide **components lists** and **net lists** that define their interconnection. This information can be used as input to some of the other packages described below.

Examples of schematic capture packages can be found within the PSpice design suite and the Multisim package, both of which are discussed further in the next section. Both provide huge libraries of standard components and allow new parts to be defined. The programs both produce standard net list files and circuit files that are compatible with other CAD programmes.

30.4.2 Circuit simulation

Having designed a circuit and produced a computer representation of it using a schematic capture package, it is then necessary to find out if it works as required. Obviously one approach to achieving this is simply to build the circuit and test it. This, however, is a time-consuming and often inaccurate method and can result in a waste of time and effort. A more attractive solution is to **simulate** the circuit, so that any required modifications can be made before construction.

A wide range of circuit simulation packages are available, although probably the best-known one is **SPICE**. This is a computer-aided simulation program, the name of which stands for **S**imulation **P**rogram with **I**ntegrated **C**ircuit **E**mphasis. It was originally developed by the Electronic Research Laboratory of the University of California and became available in 1975. The original version of the program produces only numeric and textual output and it is now common to use more modern packages that operate in a similar manner but present the output in a more meaningful graphic format. These packages include **PSpice**, which forms part of a suite of ECAD tools marketed by Cadence, and **Multisim**, which is marketed by National Instruments.

SPICE, and its derivatives, can be used for both analogue and digital circuits, for AC and DC analysis, and for both continuous and transient conditions. As with the schematic capture packages described above, these use a **component library** to store the characteristics of each device. As before, new components can be added by the user. The standard libraries include details of a range of active and passive components, including transistors and integrated circuits.

When the circuit has been defined, the user can then specify a set of initial conditions and input signals and the program will simulate the operation of the circuit and display the results. Simulation is an invaluable tool for the circuit designer and is often used interactively during the development process rather than simply to test a completed design.

30.4.3 PCB layout

When the design has been completed, the next stage is to build it. In the case of circuits that are to be implemented in the form of a **printed circuit board** (PCB) this requires the production of a photographic mask, which is used to define the positions of the copper tracks and holes on the board. Before the advent of electronic tools, the production of these masks was performed manually by placing strips of opaque tape on a transparent film. Today, boards are laid out using one of a number of **PCB layout packages**.

Layout packages take as their input the component list and net list produced by a schematic capture package. The program then allows the user to define the dimensions and shape of the board and position the components. Again, a **component library** is used to store the dimensions and **pin-out** of each component and, as before, the user can specify custom components. Most systems can perform **automatic placement** of components, taking into account their sizes and connections.

Most packages then perform **automatic routing** to join up the required pins on the various components. Until recently, the automatic routing utilities on these packages were very primitive and could not be relied on to perform the complete task unaided. Using such a system, it was normal to position the earth and power supply tracks manually, then allow the program to attempt to route the remaining tracks automatically. The user would then often need to complete the task manually to route any connections that the machine could not manage itself. Newer systems have much more powerful automatic routing algorithms that can often perform the complete layout task without human intervention. This can reduce the time taken to lay out a complicated board from perhaps 50 hours to a few minutes.

For initial checking of the design, the layout package can produce its output on a printer or plotter. For final production, it is normal to use the output from the package to drive a **photoplotter** that produces its output directly onto transparent film to form a printing mask for the board.

30.4.4 PLD design and programming packages

We have looked at the use of **array logic** in the construction of digital electronic systems (see Chapter 27). Many of these devices can be programmed by the user, the components being given the general term of **programmable logic devices** (PLDs). As these arrays may contain several thousands of gates, the task of selecting an appropriate interconnection pattern is not trivial. Consequently, this task is normally delegated to one of a number of automated tools that simplify the process. Such packages take as their input a description of the required functions of the device written in a specification language. This description sets out which of the pins of the PLD will be inputs and which outputs and defines the relationships between them. It also defines a set of **test vectors**, which consist of a set of input combinations and the expected outputs. From this data the package generates a **fuse map** that can be loaded into a **PLD programmer** and this then writes the pattern into a target device. After programming, the PLD is automatically tested using the specified test vectors to ensure that it is functioning as required. The process of programming and verifying a device generally takes a few seconds.

30.4.5 VLSI layout

If a circuit is to be implemented as a VLSI component an alternative form of layout package is required. These have certain similarities to PCB layout programs, but work in dimensions measured in microns (1 micron = 10^{-6} metres) rather than in millimetres. At this level, individual components are assembled from regions of different forms of semiconductor and it is necessary to position these regions and the layers of metallisation that connect them with great accuracy. Several distinct regions are required to form a single transistor but, once this has been designed, it may be stored away as a library component to be used again. A number of transistors might then be connected to form a gate that could also be stored for future use. In this way, a **component library** is assembled. In digital designs there is often a great deal of repetition of circuit components and the use of standard circuit 'cells' greatly simplifies design.

30.4.6 Design verification

In both PCB and VLSI layout, there are several **design rules** that must be obeyed to produce a circuit that will work reliably. These will govern, for example, the minimum separation between conductors, minimum thickness of tracks and relative positions of semiconductor regions. Various packages exist for checking that designs do not breach any of these rules. Sometimes these utilities are provided within a particular layout package. In other cases, they are a separate piece of software.

Another aspect of design verification relates to EMC. Conventional circuit simulation packages allow the functional characteristics of a system to be investigated but do not consider its EMC behaviour. Specialist packages allow the EMC performance of a system to be predicted before construction, to verify that this is acceptable. This can save a great deal of time and effort in fine-tuning the design.

30.4.7 System specification and description

In addition to the CAD packages outlined above, there are various computer languages for use in the specification and design of electronic systems. Examples include such languages as **VDM**, **ELLA**, **HILO**, **HOL** and **Z**. These languages are not used to produce software that will run on a target system to perform a specified task, but are used to define the nature of the system itself. VDM, for example, is essentially a **system specification language** that can be used to describe any system in terms of its inputs, outputs and the relationship between them. This description forms a very precise specification of the system that is independent of the eventual method of implementation. It does not, for example, define whether a particular function will be achieved in hardware or in software.

Once a system has been defined in this way, tools are available to simplify the task of implementing it in hardware, software or a combination of the two. Languages are available to specify circuitry at the logic gate level or to interface directly with VLSI design packages. Techniques also exist for generating conventional programs to implement the required software aspects of the target system. Of paramount importance in this process is the need to confirm that the circuits and software produced correspond directly to the original top-level specification. Various software tools are available to facilitate this process, although they rely heavily on the intellectual abilities of the user.

Because of their rigorous mathematical basis, these forms of specification and design are often described as **formal methods**. The use of these techniques greatly increases the reliability of the design process and gives much greater confidence in its correctness.

Perhaps the most important **hardware description language** is **VHDL**. The name is an acronym standing for **VHSIC Hardware Description Language**, where VHSIC is another acronym standing for very high-speed integrated circuits. VHDL can be used for many purposes, including the documentation, verification, synthesis, simulation and testing of circuits. It can be used to describe the structure, data flow or behaviour of a system and can be used in a number of ways within the development process. For example, VHDL can be used to specify a system by defining the functionality of each section, the interactions between these sections and even acceptance criteria for use in testing. Following specification, VHDL can be used for design capture, as an alternative to a schematic representation. The use of VHDL in this way allows the system to be simulated using some of the very powerful simulation tools that are available. An increasingly important use of VHDL is in the specification of CPLDs (as discussed in Section 27.2.8).

Video 30B

Further study

Robot arms are used in a number of situations and in many ways simulate the operation of a human arm. The unit shown has a number of rotary joints that are each controlled by some form of electric motor, and a clamp that can be opened or closed to hold objects that must be manipulated.

A control system is required to oversee the operation of the arm. This should accept a number of signals representing the required position of the arm (in terms of its end-point and orientation) and the position of the clamp (determining the separation of the 'fingers' of the 'hand'). It should then produce appropriate control signals to drive the arm to the required position.

Without considering the detailed design of the control system, give some thought to the choice of the technology to be used to implement this controller. What factors affect this choice? Would an analogue or a digital approach be more appropriate, and what form should the implementation take?

■ A good design produces an appropriate solution to a problem.

■ The normal method used to achieve such a design is the use of a top-down approach.

■ This begins with the requirements of a system, which are formalised to produce a specification.

■ The specification forms the basis of the top-level design. This may involve subdividing the problem into a number of sections.

■ When this has been done, the detailed design of the hardware and software can begin.

■ Top-down design is normally followed by bottom-up testing. This culminates in the testing of the complete system to ensure that it meets its original specification.

■ At an early stage, it is necessary to decide on the form of implementation to be used. This will involve choosing from a range of technologies.

■ A wide range of automated tools is available to simplify the design of both analogue and digital systems. These include packages to perform schematic capture, circuit simulation, PCB layout, PLD design and programming, VLSI layout and design verification.

■ Several specification languages are also available. Of these, perhaps the most important is VHDL.

Exercises

30.1 List the major tasks associated with the design of an electronic system.

30.2 What is the difference between the customer requirements and the specification of a system?

30.3 Why is it important that the specification describes *what* a system must do, rather than *how* it must do it?

30.4 What factors affect the decisions made in the hardware/software trade-off?

30.5 Define the terms 'top-down' and 'bottom-up'. Which of these methods is appropriate for design and which for testing?

30.6 Why are individual modules tested separately before a complete system is assembled?

30.7 Compare the characteristics of systems constructed using analogue techniques as opposed to those using a digital approach.

30.8 In digital systems, what factors determine whether a microprocessor should be used rather than a circuit based on non-programmable techniques?

30.9 Explain the use of a schematic capture package and describe how the output from this package may be used in association with other software tools.

30.10 Explain the function and characteristics of ECAD tools that perform:

    (a) circuit simulation

    (b) PCB layout

    (c) PLD design and programming

    (d) design verification.

30.11 Describe the function of a system specification language. How does this differ from that of a computer programming language?

Below are the principal symbols used in the text and their meanings.

Symbol	Meaning
α	temperature coefficient of resistance
β	bipolar transistor DC gain (equivalent to h_{FE})
ε	permittivity
ε_0	absolute permittivity, permittivity of free space
ε_r	relative permittivity
ξ	damping factor
μ	permeability
μ_0	permeability of free space
μ_r	relative permeability
ρ	resistivity of a material
σ	conductivity of a material
T	time constant
ϕ	phase difference
Φ	magnetic flux
ω	angular frequency of a sine wave
ω_0	angular centre, corner or resonant frequency of a filter
ω_c	angular cut-off frequency
ω_n	undamped natural frequency
A_i, A_p, A_v	current, power and voltage gains
B	bandwidth, magnetic flux density
C	capacitance
$\overline{CE}$	chip enable
$\overline{CS}$	chip select
D	electric flux density
e	electronic charge
E	electric field strength
E_m	dielectric strength
F	magnetomotive force
f_0	centre, corner or resonant frequency of a filter
f_c	cut-off frequency
f_T	transition frequency
G	overall gain
g_m	transconductance
H	magnetic field strength
h_{FE}	bipolar transistor DC gain in common-emitter configuration
h_{fe}	bipolar transistor small-signal current gain in common-emitter configuration
h_{ie}	bipolar transistor small-signal input resistance in common-emitter configuration
h_{oe}	bipolar transistor small-signal output conductance in common-emitter configuration

Symbol	Meaning
h_{re}	bipolar transistor small-signal reverse voltage gain in common-emitter configuration
I	current
i	small-signal current
I_B, I_C, I_E	DC base, collector and emitter currents
i_b, i_c, i_e	small-signal base, collector and emitter currents
I_{BB}, I_{CC}, I_{EE}	base, collector and emitter supply currents
I_{CBO}	leakage current, collector to base with emitter open circuit
I_{CEO}	leakage current, collector to emitter with base open circuit
I_D, I_G, I_S	DC drain, gate and source currents
i_d, i_g, i_s	small-signal drain, gate and source currents
I_{DD}, I_{GG}, I_{SS}	drain, gate and source supply currents
I_{DSS}	drain-to-source saturation current
I_n	noise current
I_p	peak current of a sine wave
I_{pk-pk}	peak-to-peak current of a sine wave
I_{rms}	root-mean-square current
I_s	reverse saturation current
I_{SC}	short-circuit current
k	Boltzmann's constant
L	inductance
M	mutual inductance
$\overline{OE}$	output enable
P_{av}	average power
P_i, P_o	input and output power
P_n	noise power
P_s	signal power
Q	quality factor, reactive power
q	charge
R	resistance
r_d	drain resistance
R_i, R_o	input and output resistance
r_{gs}	small-signal gate resistance
R_L	load resistance
R_M	meter resistance
R_S	source resistance
R_{SE}	meter series resistance
R_{SH}	meter shunt resistance
S	apparent power, reluctance
T	absolute temperature, period of a repetitive waveform
t_f	fall time
t_H	hold time
t_{PD}	propagation delay time
t_{PHL}	propagation delay for transitions from high to low
t_{PLH}	propagation delay for transitions from low to high
t_r	rise time
t_S	set-up time
V	voltage
v	small-signal voltage
V_+, V_-	non-inverting and inverting op-amp input voltages
V_A	Early voltage

Symbol	Meaning
V_B, V_C, V_E	DC base, collector and emitter voltages
v_b, v_c, v_e	small-signal base, collector and emitter voltages
V_{BB}, V_{CC}, V_{EE}	base, collector and emitter supply voltages
V_{BE}, V_{CE}	DC base-to-emitter and collector-to-emitter voltages
v_{be}, v_{ce}	small-signal base-to-emitter and collector-to-emitter voltages
V_{br}	breakdown voltage
V_D, V_G, V_S	DC drain, gate and source voltages
v_d, v_g, v_s	small-signal drain, gate and source voltages
V_{DD}, V_{GG}, V_{SS}	drain, gate and source supply voltages
V_{DS}, V_{GS}	DC drain-to-source and gate-to-source voltages
v_{ds}, v_{gs}	small-signal drain-to-source and gate-to-source voltages
V_H, V_L	voltage representing logical 1 and logical 0
V_i, V_o	input and output voltages
v_i, v_o	small-signal input and output voltages
V_{IH}, V_{IL}	input voltage representing logical 1 and logical 0
V_{ios}	input offset voltage
V_n	noise voltage
V_{NI}	noise immunity
V_{OC}	open-circuit voltage
V_{OH}, V_{OL}	output voltage representing logical 1 and logical 0
V_P	pinch-off voltage
V_p	peak voltage of a sine wave
V_{pk-pk}	peak-to-peak voltage of a sine wave
V_{pos}, V_{neg}	positive and negative supply voltages for an op-amp
V_{ref}	reference voltage
V_{rms}	root-mean-square voltage
V_S	source voltage
V_T	threshold voltage
V_Z	Zener breakdown voltage
$\overline{WE}$	write enable
X	reactance
X_C	reactance of a capacitor
X_L	reactance of an inductor
Z	impedance

Below is a series of physical quantities and their associated SI units.

Quantity	Quantity symbol	Unit	Unit symbol
Capacitance	C	farad	F
Charge	q	coulomb	C
Conductance	G	siemens	S
Current	I	ampere	A
Electric field strength	E	volts per metre	V/m
Electric flux	ψ	coulomb	C
Electric flux density	D	coulombs per square metre	C/m^2
Electromotive force	E	volt	V
Energy	W	joule	J
Force	F	newton	N
Frequency	f	hertz	Hz
Frequency (angular)	ω	radians per second	rad/s
Impedance	Z	ohm	Ω
Inductance (self)	L	henry	H
Inductance (mutual)	M	henry	H
Magnetic field strength	H	amperes per metre	A/m
Magnetic flux	Φ	weber	Wb
Magnetic flux density	B	tesla	T
Period	T	second	s
Permeability	μ	henries per metre	H/m
Permittivity	ε	farads per metre	F/m
Potential difference	V	volt	V
Power (active)	P	watt	W
Power (apparent)	S	volt ampere	VA
Power (reactive)	Q	volt ampere (reactive)	var
Reactance	X	ohm	Ω
Resistance	R	ohm	Ω
Resistivity	ρ	ohm metre	$\Omega.m$
Temperature	T	kelvin	K
Time	t	second	s
Torque	T	newton metre	N.m
Velocity	V	metres per second	m/s

Below is a list of the most commonly used unit prefixes.

Prefix	Name	Meaning (multiply by)
E	eta	10^{18}
P	peta	10^{15}
T	tera	10^{12}
G	giga	10^{9}
M	mega	10^{6}
k	kilo	10^{3}
h	hecto	10^{2}
da	deca	10^{1}
d	deci	10^{-1}
c	centi	10^{-2}
m	milli	10^{-3}
μ	micro	10^{-6}
n	nano	10^{-9}
p	pico	10^{-12}
f	femto	10^{-15}
a	atto	10^{-18}

Op-amp circuits

The following are examples of basic operational amplifier circuits. These are included for illustrative purposes rather than as a source of definitive circuits. In each case, component values must be chosen with care, taking into account the guidance given in Section 16.6.

Non-inverting amplifier

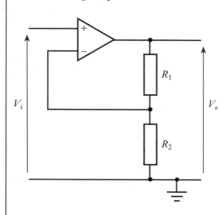

Notes

$$\frac{V_o}{V_i} = \frac{R_1 + R_2}{R_2}$$

High input resistance.
Low output resistance.
Good voltage amplifier.
See Section 16.3.1

Inverting amplifier

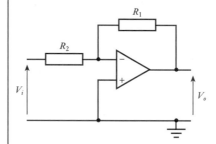

Notes

$$\frac{V_o}{V_i} = -\frac{R_1}{R_2}$$

Input resistance set by R_2.
Low output resistance.
Virtual earth amplifier.
See Section 16.3.2

Unity gain buffer amplifier

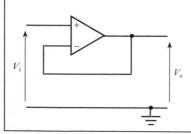

Notes

$$\frac{V_o}{V_i} = 1; \qquad V_o = V_i$$

Very high input resistance.
Very low output resistance.
Excellent buffer amplifier.
See Section 16.4.1

Current-to-voltage converter

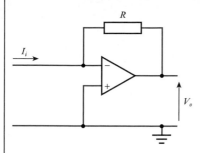

Notes

$$V_o = -I_i R$$

Very low input resistance.
Low output resistance.
Virtual earth circuit.
Also called a trans-resistive or trans-impedance amplifier.
See Section 16.4.2

Differential amplifier (subtractor)

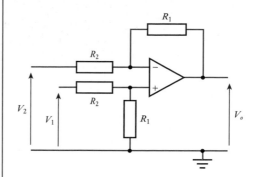

Notes

$$V_o = (V_1 - V_2)\frac{R_1}{R_2}$$

Input resistance generally different for each input.
Low output resistance.
If $R_1 = R_2$ then $V_o = (V_1 - V_2)$.
See Section 16.4.3

Inverting summing amplifier (adder)

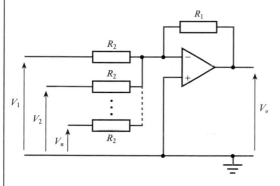

Notes

$$V_o = -(V_1 + V_2 + \cdots + V_n)\frac{R_1}{R_2}$$

Input resistance set by R_2.
Low output resistance.
Virtual earth amplifier.
Any number of inputs.
See Section 16.4.4

Non-inverting summing amplifier (adder)

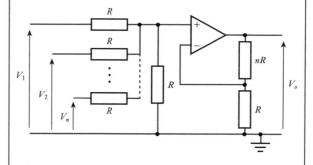

Notes

$$V_o = (V_1 + V_2 + \cdots + V_n)$$

Input resistance determined by resistor values.
Low output resistance.
Any number of inputs.

Differentiator

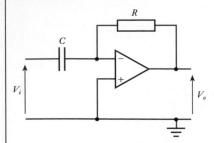

Notes

$$V_o = -RC\frac{dV_i}{dt}$$

Input impedance determined by C.
Low output resistance.
Virtual earth circuit.
Sensitive to noise – resistor in parallel with C
reduces noise.
See Section 16.4.6

Integrator

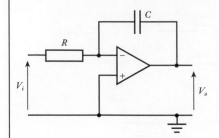

Notes

$$V_o = -\frac{1}{RC}\int_0^t V_i dt$$

Input impedance determined by R.
Low output resistance.
Virtual earth circuit.
DC input will produce a ramp output. Offset
voltages can be a problem.
See Section 16.4.5

Integrator with reset

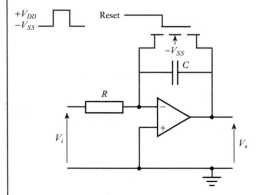

Notes

$$V_o = -\frac{1}{RC}\int_0^t V_i dt$$

Circuit behaves largely like an integrator.
FET acts as a switch, discharging C when closed.
Constant V_i and regular pulses on reset will
produce sawtooth waveform.
See Section 18.8.2

Sample and hold gate

Notes

$$V_o = V_i \text{ (at time of sample)}$$

Very high input resistance.
Very low output resistance.
For best performance, use FET input op-amp
for second amplifier to minimise discharge of C
and maximise hold time.
See Section 18.8.3

A low-pass filter

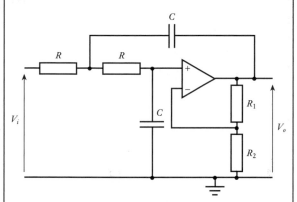

Notes

Two-pole filter.
Values shown give a Butterworth response.

$$f_0 = \frac{1}{2\pi CR}$$

Filter characteristics (and cut-off frequency)
affected by gain set by R_1 and R_2.
See Section 16.4.7

A high-pass filter

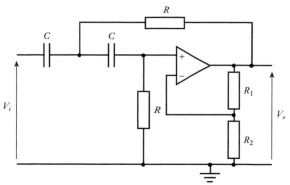

Notes

Two-pole filter.
Values shown give a Butterworth response.

$$f_0 = \frac{1}{2\pi CR}$$

Filter characteristics (and cut-off frequency)
affected by gain set by R_1 and R_2.
See Section 16.4.7

A band-pass filter

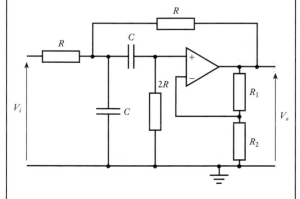

Notes

Two-pole filter.
Values shown give a Butterworth response.

$$f_0 = \frac{1}{2\pi CR}$$

Filter characteristics (and centre frequency)
affected by gain set by R_1 and R_2.
See Section 16.4.7

A band-stop filter

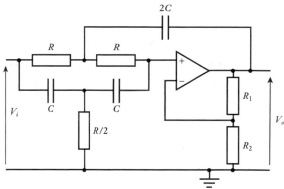

Notes

Two-pole filter.
Values shown give a Butterworth response.

$$f_0 = \frac{1}{2\pi CR}$$

Filter characteristics (and centre frequency)
affected by gain set by R_1 and R_2.
See Section 16.4.7

A Wien-bridge oscillator

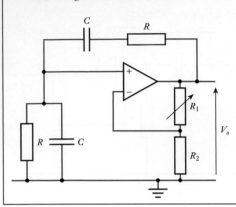

$$f = \frac{1}{2\pi CR}$$

Normally $R_1 \approx 2R_2$.
If gain is too low, oscillation will stop.
If gain is too high, output will saturate and distort.
More sophisticated circuits use automatic gain control.
See Section 23.2.2

Complex numbers

Real, imaginary and complex numbers

Readers will be familiar with the problem of solving quadratic equations of the form

$$ax^2 + bx + c = 0$$

For example, the equation

$$x^2 + x - 6 = 0$$

can be rewritten as

$$(x - 2)(x + 3) = 0$$

which yields the solution that $x = 2$ or $x = -3$.

Unfortunately, some equations, for example

$$x^2 + 1 = 0$$

cannot be solved using *real* numbers. To overcome this problem, mathematicians define an *imaginary* number i, which has the property that

$$i^2 = -1$$

$$i = \sqrt{-1}$$

and this allows all forms of quadratic equation to be solved. While the symbol 'i' is widely used in mathematics, in engineering we generally use the symbol 'j' for this quantity, since 'i' is widely used to represent current.

The existence of imaginary numbers permits us to use several different forms of number, namely **real numbers** (1, 2, 3, etc.); **imaginary numbers**, which are a product of j and a real number (j1, j2, j3, etc.); and **complex numbers**, which are formed by adding a real to an imaginary number (for example, 3 + j4).

A complex number x would therefore be of the form

$$x = a + jb$$

where a and b are real numbers. Here a represents the *real* part of x, and b (not jb) represents the *imaginary* part of x. This is written as

$$\text{Re}(x) = a$$

$$\text{Im}(x) = b$$

Graphical representation of complex numbers

Complex numbers are two-dimensional quantities that can be represented as a point on a rectangular co-ordinate plane called a **complex plane**. This has a real horizontal axis and an imaginary vertical axis, and a complex number can be represented by a line as shown in Figure D.1. This form of representation is known as an **Argand diagram**, and this figure shows the **rectangular form** of $x = a + jb$.

An alternative method of representing x is shown in Figure D.2, where the number is defined by the length r and the angle of rotation θ of a line. This is termed the **polar form** of the complex number. Comparing Figures D.1 and D.2, it is clear that the conversion from the rectangular form to the polar form is straightforward, since by Pythagoras

$$r = \sqrt{a^2 + b^2}$$

and

$$\theta = \tan^{-1}\frac{b}{a}$$

r is called the *magnitude* of the complex number x and may be written as $|x|$. θ is the *angle* of the complex number and may be written as $\angle\theta$. Therefore, the polar form of a complex number can be expressed as

$$x = r\angle\theta \qquad \text{or} \qquad x = |x|\,\angle\theta$$

Conversion from the polar form to the rectangular form is also straightforward, as illustrated in Figure D.3. Clearly, the real part of a complex number x with magnitude r and phase angle θ is equal to $r\cos\theta$, and the imaginary part is equal to $r\sin\theta$. Consequently, x may be written as

$$x = r\cos\theta + jr\sin\theta$$

A further form can be obtained by using Euler's formula, which says that

$$e^{j\theta} = \cos\theta + j\sin\theta$$

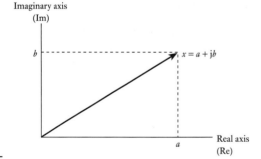

Figure D.1
Rectangular representation of a complex number.

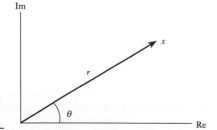

Figure D.2
Polar representation of a complex number.

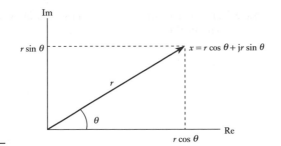

Figure D.3
Conversion from polar to rectangular form.

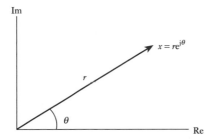

Figure D.4
Exponential representation of a complex number.

Therefore, an alternative form of x is given by

$$x = r \cos \theta + jr \sin \theta$$
$$= re^{j\theta}$$

This is called the **exponential form** of the complex number and is shown in Figure D.4.

The complex conjugate

The **conjugate** of a complex number x is formed by negating the imaginary part of the number and is given the symbol x^*. Therefore, if $x = a + jb$, then

$$x^* = a - jb$$

The relationship between x and x^* is shown in Figure D.5. From the figure it is clear that when using polar notation the magnitude of x^* is equal to the magnitude of x, but that the angle is reversed. Therefore, if $x = r\angle\theta$, then $x^* = r\angle-\theta$. Similarly, when using the exponential form, if $x = re^{j\theta}$, then $x^* = re^{-j\theta}$.

Complex arithmetic

To add (or subtract) complex numbers, we simply add (or subtract) their real parts and their imaginary parts. For example, if $x = a + jb$ and $y = c + jd$, then

$$x + y = (a + jb) + (c + jd)$$
$$= (a + c) + j(b + d)$$

The multiplication of complex numbers is also straightforward, provided that we remember that $j^2 = -1$. If x and y are as before, then

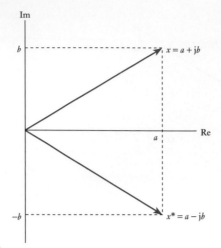

Figure D.5
The relationship between a complex number x and its conjugate x*.

$$xy = (a + jb)(c + jd)$$
$$= ac + jad + jbc + j^2bd$$
$$= ac + jad + jbc - bd$$
$$= (ac - bd) + j(ad + bc)$$

It is interesting to note that the multiplication of a complex number with its conjugate produces a real number. For example, if $x = a + jb$, then

$$xx^* = (a + jb)(a - jb)$$
$$= a^2 - jab + jab - j^2b^2$$
$$= a^2 + b^2$$

The division of complex numbers is simplified by the use of the conjugate. If, as before, $x = a + jb$ and $y = c + jd$, then

$$\frac{x}{y} = \frac{(a + jb)}{(c + jd)}$$

The presence of an imaginary element in the denominator is inconvenient, but it can be removed by multiplying top and bottom by y^*

$$\frac{x}{y} = \frac{(a + jb)}{(c + jd)}$$
$$= \frac{(a + jb)(c - jd)}{(c + jd)(c - jd)}$$
$$= \frac{(ac + bd) + j(bc - ad)}{c^2 + d^2}$$
$$= \frac{ac + bd}{c^2 + d^2} + j\frac{bc - ad}{c^2 + d^2}$$

While multiplication and division of complex numbers are straightforward (as described above), it is often simpler to perform these tasks using the polar form of the number, since

$$\frac{A\angle\alpha}{B\angle\beta} = \frac{A}{B}\angle(\alpha - \beta)$$

Multiplication and division are also easy using the exponential form, since

$$\frac{Ae^{j\alpha}}{Be^{j\beta}} = \frac{A}{B}e^{j(\alpha-\beta)}$$

For this reason, it is common to perform addition or subtraction of complex numbers using the rectangular form but to use the polar or exponential form when performing multiplication or division. Fortunately, converting between these different forms is straightforward (as described above).

Chapter 1

1.4	5 mA
1.5	6 kΩ
1.6	25 W
1.7	10 nW
1.8	50 Ω
1.9	12 Ω
1.10	7.9 kΩ
1.11	600 Ω
1.12	20 Ω, 50 Ω
1.13	1.51 kΩ, 208 Ω
1.14	6 V, 10 V, 8 V
1.15	16 V, 4 V, −10 V
1.16	1 ms
1.17	50 kHz

Chapter 2

2.2	0.1 Hz
2.3	40 ms
2.4	5 V
2.5	20 A
2.6	62.8 rad/s
2.7	25 Hz
2.8	5 V, 10 V, 250 Hz, 1570 rad/s
2.11	75 Hz, 25 V
2.13	6.37 V
2.14	7.85 A
2.17	2 W
2.18	4 W
2.19	6.66 V
2.20	5 V
2.21	1 W
2.22	2 mΩ
2.23	200 kΩ
2.24	11 per cent (high)
2.25	11.1 V
2.32	5.3 V
2.34	60°, B leads A

Chapter 3

3.2	50 C
3.6	100 V, 50 V, 500 μV, −2.35 V
3.7	200 W, 1.25 W, 2.5 mW, 117 μW
3.8	16 mΩ
3.9	150 Ω, 33.3 Ω, 42 Ω
3.10	5 kΩ
3.15	6 V, 375 V, 60 V
3.16	1.8 V
3.18	−446 mA
3.20	2.5 mA
3.22	471 mV
3.24	1.07 V
3.26	−62 mA
3.38	5 V

Chapter 4

4.6	45.5 V
4.7	20 μF
4.10	66 pF
4.11	13 nF
4.14	16.7 MV/m
4.17	67 mC/m^2
4.18	150 μF, 3.75 mF, 5.9 nF, 39.3 μF
4.21	100 ms
4.22	10 μF
4.26	562 mJ
4.27	15.6 mJ

Chapter 5

5.3	0.48 A/m
5.7	3000 ampere-turns, 3333 A/m, 4.19 mT, 1.68 μWb
5.9	3000 A/Wb
5.15	3 mH
5.17	36.3 μH
5.18	20.9 mH

5.19 1.43 H, 35 mH, 10.9 μH, 250 mH
5.23 2 s
5.24 20 H
5.29 49 mJ
5.34 50 V

Chapter 6

6.1 100 rad/s, 15 V
6.2 39.8 Hz, 17.7 V
6.14 12.6 Ω
6.15 200 kΩ
6.16 1.18 A r.m.s
6.17 5 mV peak
6.21 28.0 V, ∠14.6°
6.26 58.6∠−65°
6.28 1000 + j0
6.29 0 − j159
6.30 0 + j6.28
6.31 80 + j124, 40 − j40
6.32 36∠56°, 36e$^{j56°}$
6.33 19.1 + j16.1, 25e$^{-j40°}$

Chapter 7

7.1 1 W
7.4 700 VA, 0.5, 350 W
7.6 1.97 A, 197 VA, 0.786, 155 W, 121 var
7.7 500 VA, 400 W, 300 var, 2 A
7.9 12.7 μF
7.10 3.4 μF

Chapter 8

8.1 15.9 Ω, 2 Ω
8.2 39.8 Hz
8.3 1571 rad/s
8.5 495 μs
8.8 15 Hz, 100 kHz, 8 kHz, 10 MHz, 3 Hz, 50 kHz
8.11 200 μs

Chapter 9

9.8 40 ms
9.9 576 μs

Chapter 10

10.2 2.9 V
10.12 1000 rpm
10.16 18,000 rpm

Chapter 12

12.7 138.5 Ω, 1.385 V
12.18 1 V, 3.85 mV/°C

Chapter 14

14.6 25 V
14.7 0.1
14.8 9.12 V
14.9 18.6
14.10 24 μW, 83 mW, 3.5×10^3
14.12 10.8 V
14.13 439
14.14 2.42 nW, 667 mW, 2.8×10^8
14.16 13.2 V
14.18 30 dB
14.19 22 dB
14.20 7.07
14.21 24 kHz
14.22 5 MHz
14.23 10 V

Chapter 15

15.14 0.04
15.16 6

Chapter 16

16.6 16
16.9 −25
16.14 0.5 V
16.15 −5 V
16.23 40 kHz
16.26 (a) 31.3, 32 GΩ, 3.1 mΩ
       (b) −6.83, 12 kΩ, 680 μΩ
       (c) 46.3, 22 GΩ, 4.6 mΩ
       (d) 1, 1 TΩ, 100 μΩ

Chapter 17

17.21 9.1 V

Chapter 18

18.24 1.22 mS, 1.73 mS, 2.45 mS
18.27 2 MΩ, 4 kΩ, −12, 2.9 Hz
18.28 1 MΩ, 4.7 kΩ, −4.7, 1.6 Hz
18.30 1.33 MΩ, 4.7 kΩ, −113

Chapter 19

19.10	930 μA, 4.8 V
19.12	≈ −240, 730 Ω, 985 Ω
19.13	≈ −180, 730 Ω, 750 Ω
19.14	1.5 mA, 6.15 V, −3.9
19.15	1.8 kΩ, 3.9 kΩ
19.16	f_c = 88 Hz
19.18	6.15 V, −234, ≈ 940 Hz, ≈ 1000 μF
19.23	5.2 mA, 5.2 V, 1
19.24	2.3 kΩ, 1.4 μF
19.26	6.8 V, 90, 11 Ω, 1 kΩ

Chapter 20

20.28	15 W

Chapter 22

22.7	3.9 μV
22.8	28 μV
22.10	4 per cent
22.16	54 dB
22.17	58 dB

Chapter 23

23.3	65 Hz
23.5	159 Hz
23.11	0.00004 per cent

Chapter 24

24.3	32, 64
24.23	12, 49, 23, 1.375
24.24	111000, 10000100, 1000011, 101.101
24.25	42,179, 52,037, 135, 1023
24.26	CDE4, 2D6, 22C4
24.27	1010010011000111
24.28	2CA5
24.29	100000, 11011, 1001101, 111

Chapter 27

27.18	4,194,304
27.21	0 to 16,777,215
27.22	−8,388,608 to +8,388,607

Chapter 28

28.2	8 kHz
28.6	4096

Chapter 29

29.9	≈ 50 kbps
29.11	9 kHz, 10 kHz and 11 kHz
29.12	m_a = 0.75, ≈ 22%
29.14	80 kHz

Index